Handbook of Research on the Adverse Effects of Pesticide Pollution in Aquatic Ecosystems

Khursheed Ahmad Wani
ITM University Gwalior, India

Mamta
Jiwaji University, India

A volume in the Advances in Environmental Engineering and Green Technologies (AEEGT) Book Series

IGI Global
DISSEMINATOR OF KNOWLEDGE

Published in the United States of America by
IGI Global
Engineering Science Reference (an imprint of IGI Global)
701 E. Chocolate Avenue
Hershey PA, USA 17033
Tel: 717-533-8845
Fax: 717-533-8661
E-mail: cust@igi-global.com
Web site: http://www.igi-global.com

Library of Congress Cataloging-in-Publication Data

Names: Wani, Khursheed Ahmad, editor. | Mamta, 1987- editor.
Title: Handbook of research on the adverse effects of pesticide pollution in aquatic ecosystems / Khursheed Ahmad Wani and Mamta, editors.
Other titles: Adverse effects of pesticide pollution in aquatic ecosystems
Description: Hershey, PA : Engineering Science Reference, [2018] | Includes bibliographical references.
Identifiers: LCCN 2018001724| ISBN 9781522561118 (h/c) | ISBN 9781522561125 (eISBN)
Subjects: LCSH: Agricultural pollution. | Agricultural chemicals--Toxicology. | Water--Pollution. | Aquatic ecology.
Classification: LCC TD428.A37 A33 2018 | DDC 577.6/279--dc23 LC record available at https://lccn.loc.gov/2018001724

This book is published in the IGI Global book series Advances in Environmental Engineering and Green Technologies (AEEGT) (ISSN: 2326-9162; eISSN: 2326-9170)

British Cataloguing in Publication Data
A Cataloguing in Publication record for this book is available from the British Library.

For electronic access to this publication, please contact: eresources@igi-global.com.

Advances in Environmental Engineering and Green Technologies (AEEGT) Book Series

Sang-Bing Tsai
University of Electronic Science and Technology of China Zhongshan Institute, China
Ming-Lang Tseng
Lunghwa University of Science and Technology, Taiwan
Yuchi Wang
University of Electronic Science and Technology of China Zhongshan Institute, China

ISSN:2326-9162
EISSN:2326-9170

Mission

Growing awareness and an increased focus on environmental issues such as climate change, energy use, and loss of non-renewable resources have brought about a greater need for research that provides potential solutions to these problems. Research in environmental science and engineering continues to play a vital role in uncovering new opportunities for a "green" future.

The **Advances in Environmental Engineering and Green Technologies (AEEGT)** book series is a mouthpiece for research in all aspects of environmental science, earth science, and green initiatives. This series supports the ongoing research in this field through publishing books that discuss topics within environmental engineering or that deal with the interdisciplinary field of green technologies.

Coverage

- Cleantech
- Renewable energy
- Contaminated Site Remediation
- Green Transportation
- Waste Management
- Alternative Power Sources
- Sustainable Communities
- Biofilters and Biofiltration
- Radioactive Waste Treatment
- Pollution Management

The Advances in Environmental Engineering and Green Technologies (AEEGT) Book Series (ISSN 2326-9162) is published by IGI Global, 701 E. Chocolate Avenue, Hershey, PA 17033-1240, USA, www.igi-global.com. This series is composed of titles available for purchase individually; each title is edited to be contextually exclusive from any other title within the series. For pricing and ordering information please visit http://www.igi-global.com/book-series/advances-environmental-engineering-green-technologies/73679. Postmaster: Send all address changes to above address.

Titles in this Series

For a list of additional titles in this series, please visit: www.igi-global.com/book-series

Biostimulation Remediation Technologies for Groundwater Contaminants
Ashok K. Rathoure (Independent Researcher, India)
Engineering Science Reference • ©2018 • 355pp • H/C (ISBN: 9781522541622) • US $215.00

Innovations and Trends in Environmental and Agricultural Informatics
Petraq Papajorgji (University of New York Tirana, Albania) and Francois Pinet (Irstea Centre de Clermont-Ferrand, France)
Engineering Science Reference • ©2018 • 317pp • H/C (ISBN: 9781522559788) • US $205.00

Effective Solutions to Pollution Mitigation for Public Welfare
Ahmet Ozan Gezerman (Yildiz Technical University, Turkey) Burcu Didem Corbacioglu (Yildiz Technical University, Turkey) and Bhola R. Gurjar (Indian Institute of Technology Roorkee, India)
Engineering Science Reference • ©2018 • 274pp • H/C (ISBN: 9781522533795) • US $180.00

Handbook of Research on Microbial Tools for Environmental Waste Management
Vinay Mohan Pathak (Gurukul Kangri University, India) and Navneet (Gurukul Kangri University, India)
Engineering Science Reference • ©2018 • 496pp • H/C (ISBN: 9781522535409) • US $285.00

Food Systems Sustainability and Environmental Policies in Modern Economies
Abiodun Elijah Obayelu (Federal University of Agriculture – Abeokuta (FUNAAB), Nigeria)
Engineering Science Reference • ©2018 • 371pp • H/C (ISBN: 9781522536314) • US $195.00

Microbial Biotechnology in Environmental Monitoring and Cleanup
Pankaj (G. B. Pant University of Agriculture and Technology, India) and Anita Sharma (G. B. Pant University of Agriculture and Technology, India)
Engineering Science Reference • ©2018 • 427pp • H/C (ISBN: 9781522531265) • US $235.00

Handbook of Research on Renewable Energy and Electric Resources for Sustainable Rural Development
Valeriy Kharchenko (Federal Scientific Agroengineering Center VIM, Russia) and Pandian Vasant (Universiti Teknologi PETRONAS, Malaysia)
Engineering Science Reference • ©2018 • 672pp • H/C (ISBN: 9781522538677) • US $345.00

IGI Global
DISSEMINATOR OF KNOWLEDGE

701 East Chocolate Avenue, Hershey, PA 17033, USA
Tel: 717-533-8845 x100 • Fax: 717-533-8661
E-Mail: cust@igi-global.com • www.igi-global.com

With all our love and affection, we dedicate this book to our beloved parents, Artaza Khursheed & Shambhavi Mishra

Editorial Advisory Board

List of Contributors

Table of Contents

Section 4
Pesticide Hazards, Analysis, and Purification

Detailed Table of Contents

A chemical that is toxic to one animal may also be toxic to other forms of animal life. Although it might take a larger dose of pesticides to harm humans than pests, such as insects, many pesticides are still toxic to humans. The doses needed to kill a pest might not kill us, but may still harm us. Many pesticides classified as herbicides are designed to target plant pests. The exceptions to this are broad spectrum herbicides that are designed to kill a wide variety of plants. An herbicide that is specific to one or more species of plants does not ensure that it is safe to enter the water system. Some of the dangers from these chemicals are yet to be fully understood. Caution should be used to ensure that these products do not unnecessarily enter the water system. Using safe, well-planned applications of materials, such as pesticides, decreases the risk to humans and other animals. The overall picture is not as bleak as one might be imagining as the optimization of these important water resources are available at present and will also be available in the near future. Owing to the threat to water systems and mechanisms, those that may cause water to become polluted are now well-understood and precautions have been taken to protect the quality of water.

Since the industrial revolution, several new chemicals were discovered and introduced in society, and soon after the green revolution, pesticides were also introduced to strengthen food security. However, limited

Section 2
Pesticides and Human Health

Pesticides are known to be one of the extremely useful and incredibly beneficial agents for preventing losses of crops as well as diseases in humans. They are used in a large number of conditions as in farms, orchards, gardens, parks, sports lawn, residences, industrial areas, shops, schools, hospitals, airports, railway lines, drains, on animals, and on people for control of diseases such as scabies and head lice. People are exposed to pesticides in their daily lives through multiple routes of exposure such as occupational or food, water, and air. Many pesticides can be used safely and effectively, but care must be taken while using them. Several pesticides are beneficial in agriculture for killing pests. Yet many times their injurious effects offset the positive ones. Uses of pesticides are apprehension for sustainability of environment and global stability. This chapter aims to discuss pesticides, their types, routes of their exposure, human health concerns related to them, methods to stop using them, and a future scenario of the world after eradicating pesticides.

Pesticides, along with hybrid seeds and fertilizers, are an integral part of the green revolution and are used to control and eradicate disease vectors for the improvement of agricultural production. Pesticides is an umbrella term for insecticides, nematocides, fungicides, herbicides, fumigants, repellents, and attractants. Pesticides are used against unwanted plants and animals to control diseases and losses. Efforts at different levels may help to reduce the impact of pesticides on newborn babies and on pregnant women. Different efforts can be considered at clinical, educational, and policymaking institutes. Use of risk assessment tools, encouragement of organic diets, educating parents working in agricultural fields from hazards of pesticides particularly in pregnancy and breast feeding, implementation of integrated pest management (IPM) programs, and encouraging policies supporting IPM can help in tackling the menace of pesticide hazards.

In the recent years, pesticide research and regulatory efforts have focused on the prevention of acute health effects from pesticide poisonings and pesticide residues on foods, but more attention is being given to the deleterious chronic health effects. Children and pregnant women's exposure to contaminated

Humans made use of pesticides to kill pests infesting crops. This was done to increase agricultural yields and improve public health. Pesticides however turn out to be damaging for the environment, causing many harmful impacts. Certain pesticides after being applied to the environment show long-term residual effects while others show acute fatal effects particularly to aquatic life. For example, organochlorine pesticides are persistent in the environment; as a result of this, these pesticides find their way to contaminate ground water, surface water, food products, air, soil, and may also affect human beings through direct contact. Pesticide exposure to humans has been found to be an important cause of some diseases such as cancer, respiratory diseases, skin diseases, endocrine disruption, and reproduction disorders. It is this aspect of pesticides in the environment that has raised concern among environmental scientists to study their behavior in the environment and then come out with a sound alternative so as to rescue the human population from their adverse effects. Fifty years (half a century) after Rachel Carson's warning to the world about the devastating effect pesticides have on birds and beneficial insects, pesticides continue to be in use. Continued usage of pesticides can be described as a massive chemical assault on our environment which threatens the survival of many birds, fish, insects, and small aquatic organisms that form the basis of the food web. More generally, pesticides reduce species diversity in the animal kingdom and contribute to population decline in animals and plants by destroying habitats, reducing food supplies, and impairing reproduction. Organisms in ecosystems exist in complex interdependent associations such that losses of one keystone species as a result of pesticides (or other causes) can have far reaching and unpredictable effects. A keystone species is a species that is disproportionately connected to more species in the food-web. The many connections that a keystone species holds mean that it maintains the organization and structure of entire communities. The loss of a keystone species results in a range of dramatic effects that alters trophic structure, other food-web connections, and can cause the extinction of other species in the community. A pesticide may eliminate a species essential to the functioning of the entire community, or it may promote the dominance of undesired species or it may simply decrease the number and variety of species present in the community. This may disrupt the dynamics of the food webs in the community by breaking the existing dietary linkages between species.

Chapter 11
Impact of Pesticides on Invertebrates in Aquatic Ecosystem.. 182
Azad Gull, Jiwaji University Gwalior, India
Ashaq Ahmad Dar, Jiwaji University Gwalior, India
Jaya Chaturvedi, Jiwaji University Gwalior, India

Aquatic ecosystems do not contain more than a fragment of the global water resources, but they are exclusive and complex habitats due to the extremely close association between terrestrial and aquatic habitats. The important fish stocks and a unique set of organisms that provides priceless consumer services, such as chemical water purification and organic matter processing, are affected. The pollution of aquatic ecosystems with pesticides applied in agricultural production is widely acknowledged as one of the greatest anthropogenic stressors to stream ecosystems, and agricultural pesticides are known to cause a threat to all living organisms in stream ecosystems. The general objective of this chapter is to study the effects of agricultural pesticides on invertebrates. There are only a few evaluating effects of pesticide contamination resulting from normal agricultural practice on invertebrates, and there is a lack of studies focusing on the indirect effects of pesticides. The importance of physical habitat degradation in the assessment and mitigation of pesticide risk in agricultural streams will be discussed.

Section 4
Pesticide Hazards, Analysis, and Purification

The chapter gives insight into the harmful use of pesticides in different professional environments. It portrays the use of pesticides as the potential risks to the health of users and third parties and a danger to the environment. The use of pesticides has increased at a phenomenal rate. Pesticides and their threat to the biological world have reached almost hysterical proportions. Their residues are found everywhere, particularly those of the so-called "hard pesticides" or organochlorine compounds, DDT. Herein, an attempt has been made to reflect pesticide exposure in different occupational settings and their harmful effects on humans. Excess use of pesticide in agriculture has placed workers in this industry at risk of lethal exposure. Personnel working in domestic pest control service is also from continuous exposure to the pesticide. Further, the chapter highlights various corrective measures to be taken by the people working in different occupational settings to combat the dangerous effects of pesticides in everyday life.

Pesticides that are meant for the demolition of pests have become a nuisance for public health and environmental safety. Hence, the detection of pesticides in various types of samples such as food, environment, and even bodily fluids from various living entities is essential. Even though spectroscopic techniques are widely used for routine detection and quantification of pesticides, these techniques cannot determine pesticides with accuracy assured by chromatographic techniques. This chapter deals with various pesticide analysis techniques and the limitations associated with each technique.

Environmental protection efforts require numerous advanced technologies to prevent and monitor the health and ecological effects associated with abiotic and biotic systems. Development of innovative tools and methodologies with the help of multidisciplinary approach to assess the transport, accumulation, and impact of pesticides will avoid the long-term effects in the environment. The lack of information about the pesticides hampers the labeling requirements that lead to misuse and discharge of pesticide-contaminated effluents into the water resources. This chapter covers the information on major sources of pesticides, chronic impacts, labeling of pesticides, multidisciplinary approach for monitoring, current

Natural or synthetic chemical compounds in pesticides are commonly used to kill pests or weeds. In general, pesticides are potentially toxic to not only organisms but also the environment and should be used safely disposed of expediently. Pesticide residues in foods may cause various crucial diseases in the body. The damage of pesticides can be changed depending on the application dose or type of active compounds. For this reason, identification and quantification of pesticides via highly selective, sensitive, accurate, and renewable techniques are of vital importance due to the large amounts of possible interfering substances during the extraction stages. Analysis of pesticide residues by analytical methods can be fluctuate based on the pesticide types. For food and health safety, maximum residue limits (MRL) of pesticides in foods were determined by the European Community. There are many analytical methods developed for identification and quantification of pesticides. Although there are some limitations, the multi-residue methods sensible for analyzing a great number of pesticides in one single run is the fastest, the most favorite, and efficient choice. However, some of the pesticides need specific methodologies and single-residue methods apply as compulsory for them. In this chapter, recent advances in the various analysis of pesticide residues in crops and their applications and limitations are discussed.

Halogenated organic compounds are usually xenobiotic in nature and used as ingredients for the synthesis of pesticides, solvents, surfactants, and plastics. However, their introduction to the aquatic ecosystems resulted in ecological danger due to their toxic effects. The usual method of toxicity assessment is by performing the experimental approach by considering some model organism. In this aspect the computational techniques such as QSAR (quantitative structure activity relationship) is considered an effective method. By computing several molecular features and the experimental activity, the toxic effect of a compound can be correlated. This chapter describes the aquatic toxicity of the compounds. The information about different computational resources (databases, tools, and modeling tools) have been given. Also, the application of QSAR to predict aquatic toxicity of different halogenated compounds available in the literature has been reviewed.

Water from surface sources is often contaminated by microbes, whereas groundwater is normally safer, but even groundwater can be contaminated by harmful chemicals from human activities or from the natural environment. The purification process of water may reduce the concentration of particulate matter including suspended particles, parasites, bacteria, algae, viruses, fungi, and a range of dissolved and particulate material derived from the surfaces. Water purification is the process of removing undesirable chemicals, materials, and biological contaminants from contaminated water. Most water is purified for human consumption (drinking water), but water purification may also be designed for a variety of other purposes, such as medical, pharmacology, chemical, and industrial applications. In general, the methods used include physical processes such as filtration and sedimentation, biological processes such as slow sand filters or activated sludge, chemical processes such as flocculation and chlorination, and the use of electromagnetic radiation such as ultraviolet light.

Foreword

There have been huge advances in many areas of research directed towards enhancing the amount and nature of sustenance and fiber chemical and other means. The concerns for the use of pesticides and their sustenance in nature, and our comprehension of the ecological effect of agrochemicals has likewise expanded and turned out to be more complex on account of multidisciplinary approaches.

Monitoring data for pesticides are generally poor in much of the world and especially in developing countries. The checking of pesticide utilize is a major part of any strategy on minimisation, from gauging the success of government initiatives, policy changes or fluctuations in support, through to environmental impact assessment at the field, farm, catchment, regional or national level. The proficient utilization of pesticides and different xenobiotics to crops depends on targets. Pesticide monitoring requires highly flexible field and laboratory programs that can respond to periods of pesticide application, which can sample the most appropriate medium (water, sediment, biota), are able to apply detection levels that have meaning for human health and ecosystem protection, and which can discriminate between those pesticides which appear as artifacts of historical use versus those that are in current use.

Pesticides contain poisonous materials that cause both environmental and human health risks. All living organisms can be severely threatened by these chemicals. However, with an aggressive march toward the protection of source waters from pesticide and chemical mixtures, as well as improving technology to better treat suspect waters, there is anticipation that the flow of pesticides into humans via drinking water can be brought to a tiny trickle for future generations.

The amalgam of 24 chapters of this book is complete inventory of the pesticides as pollutants, its sources, pathways, analysis, health effects, status of water bodies, purification technologies, etc. that will help researchers all over the world to discover new ways that can reduce the burden of pesticides on different aquatic ecosystems. The book is written in lucid language with illustrative diagrams and charts that may prove fruitful in long way both in academics and industrial sector.

I wish all the success to this book in its endeavors.

Tahir Ali
Sher-e-Kashmir University of Agricultural Sciences and Technology, India

Preface

Aquatic ecosystem degradation is considered as devastating for the environment since degraded as they could not only lose their ability to perform their valuable functions but may also damage other interconnected ecosystems. Loss of small aquatic ecosystems would have great environmental impacts and it is critical that the few remaining aquatic ecosystems that are healthy need to be protected and conserved as these ecosystems have stable communities that tolerate chronic pollutants.

The use of agrochemicals has become important in agricultural practices because of the increasing demand for quality and quantity of food products that has resulted growing dependency on agrochemicals and at the same time they are produced, purchased and used in large quantities in developing countries. The accumulation of the pesticides in different food chains is considered extremely toxic. Organochlorines pesticides are non biodegradable or degraded very slowly and accumulate in the environment. These pollutants contaminate not only the surrounding soils and water but in most cases may enter into the drinking water. These types of pesticides are widely used in agriculture in all parts of the world due to their relatively low cost, broad spectrum of activity, and high efficiency on insects. The potential risk to the human health, extensive pollution to the environment, their accumulation potential in different trophic levels is a matter of great concern for the researchers.

The use of quantitative structure-activity relationships (QSARs) is used as an alternative to the difficult experimental methods to accesses the potential toxic effects of organic chemicals on aquatic organisms. Therefore, the regulatory agencies have used the QSAR based approach as an essential tools to help prioritize the risk assessments when empirical data are not available to evaluate toxicological effects. Pesticide classes mostly detected involve herbicides used extensively in corn, cotton and rice production, organophosphorus insecticides as well as the banned organochlorines insecticides due to their persistence in the aquatic environment. The compounds most frequently detected were atrazine, simazine, alachlor, metolachlor and trifluralin of the herbicides, diazinon, parathion methyl of the insecticides and lindane, endosulfan and aldrin of the organochlorine pesticides. River and ground water were found to be more polluted than lakes because the detected concentrations of most pesticides follow a seasonal variation, with maximum values occurring during the post spring and summer period followed by a decrease during winter. In many cases it is reported by many scientist that concentrations ranged in low ppb levels.

The toxicity of a pesticide in an aquatic organism is decided by specific chemical and physical property of a pesticide such as persistence in water, stability to degradation and bioaccumulation potential. Pesticides can either bioaccumulate in organisms or excrete off depending upon its metabolism. They can also cause neural issues, sex alterations, genetic mutations and carcinogenic problems in aquatic organisms. Other physiological disorders include histo-architectural changes in liver, kidney, heart and other body parts. Pesticides can act as endocrine disruptors that can act through the receptor medicated response.

Application of herbicides to water bodies can cause fish kills when the dead plants decay and consume the water's oxygen, resulting in suffocation of the fish. Pesticides such as copper sulfite that are applied to kill unwanted plants are toxic to fish and other water animals at concentrations similar to those used to kill the plants. Continued exposure to sublethal concentrations of some pesticides can cause can reduce fish populations by causing physiological and behavioral changes such as abandonment of nests and broods, decrease in immunity to disease and lesser predator avoidance. Application of herbicides to bodies of water can also result in killing of plants on which fish depend for their habitat. Pesticides can accumulate in aquatic ecosystems to levels that kill off zooplankton, the main source of food for young fish. Exposure to organophosphates and carbamates causes, symptoms similar to those of increased neurotransmitter-acetylcholine. These pesticides interfere with the normal nerve signal transduction, and exposure to them causes headaches, dizziness, confusion, nausea and vomiting, muscle and chest pain. Long-term pesticide exposure damages the immune system. Generally the methods included for purification of water are physical processes such as filtration and sedimentation, biological processes such as slow sand filters or activated sludge, chemical processes such as flocculation and chlorination and the use of electromagnetic radiation such as ultraviolet light but these techniques hardly remove the pesticides from water. Therefore, adsorption is considered as ecofriendly method. The adsorbent derived from biomaterial is considered as encouraging adsorbent due to its cost effective and high adsorption capacity.

ORGANIZATION OF THE BOOK

Chapter 1 presents an overview of pesticide use, classification their impact to the different ecosystems. This chapter argues that the pesticides which are specific to any particular species of the plants also does not assured that it is safe when enter into water system, consequently caution needed to insured that that the product does not enter into the water system without any necessity. Approximately 30% of Indian crop yield potential is being lost due to insects, disease and weeds which in terms of quantity would mean 30 million tons of food grain. The value of total loss has been placed at Rs 50,000 million, represents about 18% of the gross national agriculture production. The population of the country is rising at faster rate, in order to feed the population, pesticide use has become a necessary evil.

Chapter 2 defines the different sources of pesticides, based on their point of origin and way it transports pesticides to the aquatic systems. After which the pesticide interaction in an aquatic environment with various organic and inorganic substances has been described. Each interaction is supported with the recent researches and examples. The fate of a pesticide in aquatic ecosystem depends upon the variety of climate factors (temperature and precipitation); soil characteristics, topography and agricultural particles are also discussed. This chapter is concluded with recommended management practices and future research directions.

Chapter 3 advocates that presence of OCPs in aquatic ecosystems is a threat as it not only changes the physico-chemical characteristics of water but also impose drastic impact on aquatic organisms. OCPs get sorbed to soil and owing to its hydrophobic nature become persistent in soil and sediments. Degradation of pesticide through biological means using bacteria and fungi is receiving serious attention as compared to existing conventional methods and such alternative methods should be adopted to minimize environmental deterioration till more realistic policies are developed. It is suggested that stringent government regulations and monitoring of water bodies should be observed. The chemical stability of these compounds, their high lipid solubility and toxicity to human beings and animals has led government and researchers to feel concerned about their presence in the environment. Conclusively, the exposure to OCPs should be reduced so as to minimize the associated environmental and human health hazard.

Chapter 4 provides an overview of the status of wetlands in India. The chapter accounts that the wetlands in India are unique ecosystems that harbour diverse types of species. These wetlands obviously provide livelihood to good number of people in India and have various resources that have increased the GDP of different states of the country. However, the intrusion of pesticides in these ecosystems have caused adverse effects not only to the species in these ecosystems, but have caused great damage to the wetland itself, as a result of that most of these ecosystems are under eutrophication. Hence, there is an urgent need to address the issues of pesticide contamination by making strict rules and regulations.

Chapter 5 aims to discuss about pesticides, their types, routes of their exposure, human health concerns related to them, methods to exterminate the use of pesticides and future scenario of world after eradicating pesticides. If the pesticides are used in appropriate quantities and used only when required or necessary, then pesticide risks can be minimised. In future chemical pesticides can be used in combination with natural treatments and remedies which result in more sustainable elimination of pests and insects. This combination not only promises environmental sustainability, but also has diverse applications in controlling of urban pests and invasive species.

Chapter 6 highlights the concerns of pesticides in drinking water and its repercussions. Pesticides occupy a unique position in drinking water contamination among the number of pollutants, since their low levels have the deleterious effect on organisms. The pesticides occur not only in surface water but also have been detected in groundwater. Pesticides are persistent for biodegradation which affect non-target organisms and have endocrine disrupting and neurotoxic properties. Developing fetus and children are acknowledged objects especially prone to the ill effects of pesticide hazards as most of the organ systems are in developing stage, for example, their immune system may not be able to safeguard them against different diseases and environmental threats. Same is the case with pregnant women; early-life exposure to pesticides during pregnancy and through breastfeeding can have detrimental on child development.

Chapter 7 discussed that small effects of drinking water contamination on all children, but large and statistically significant effects on birth weight and gestation of infants born to less educated mothers. Those mothers who were most affected by contamination were the least likely to move between births in response to contamination. In addition, alteration in sex ratio of normal live births to fathers employed in pesticide application, deep sea divers, and carbon setters were also recorded. In general pesticide illness reports that on exposure, the victims face very serious chronic health outcomes impacting on cancer, diabetes, lung function, neurodegenerative disorders, intelligence quotient, birth defects, and reproductive disorders.

Chapter 8 highlighted the adverse effect of pesticides on environment and on human health. This review helps to seek an attention of researchers, government and non-government organization on health issues that have been associated with the exposure of harmful chemical pesticides and encourage research on finding the new concept in modern agriculture involving the reduction in the use of chemical pesticides.

Chapter 9 is argues that microorganisms are the most dominant natives of aquatic ecosystems, where they fulfill very specific roles in primary productivity, decomposition and nutrient cycling. Thus, the purpose of this chapter is to review the significance of pesticides and their effects on aquatic microorganism and to study their ecological significance. The microorganism can make use of components entering the environment as feeding substrate for building material or a source of energy thus affecting balance in the ecosystem. Natural populations possess a number of responses to these contaminants and quickest reaction has been reported from the microorganisms at the base of trophic cascade, fungi and bacteria. Especially populations in the growth phase are more vulnerable to pesticides but have the capability to recover rapidly from the injury. The following patterns are recognized as effects of pesticides at the ecosystem and community levels like an increase of species richness and diversity and elongation of the food chain, induction of dominance by small species and reduction of energy transfer efficiency from primary producers to top predators.

Chapter 10 discusses the Impact of pesticides on aquatic life. The loss of a keystone species results in a range of dramatic cascading effects that alters trophic dynamics, other food-web connections and can cause the extinction of other species in the community. A pesticide may eliminate a species essential to the functioning of the entire community, or it may promote the dominance of undesired species or it may simply decrease the number and variety of species present in the community. This may disrupt the dynamics of the food webs in the community by breaking the existing dietary linkages between species.

Chapter 11 studies the effects of agricultural pesticides on the structure and function of macroinvertebrates and microinvertebrates. These organisms respond strongly to pesticide contamination, they are thoroughly studied in the field of ecology which provides more available relevant information for the studies and they are very necessary for a series of ecosystem processes. Further, there is only few evaluating effects of pesticide contamination resulting from normal agricultural practice on macro-invertebrates and micro-invertebrates and there is a lack of studies focusing on the indirect effects of pesticides – e.g. pesticide effects on food resources of macro-invertebrates. Agricultural pesticides continue to impair surface water ecosystems, although there are few assessments of interactions with other modifications such as fine sediment and physical alteration for flood drainage. The importance of physical habitat degradation in the assessment and mitigation of pesticide risk in agricultural streams is also discussed.

Chapter 12 assess the health hazards during pesticides handling, storage and use to create awareness to control its exposure. Biological monitoring techniques are discussed to reduce the intensity of absorption of selected pesticides. The different metabolites that are detected in the general population are organophosphates, organochlorines, carbamates, herbicides, pest repellents, and disinfectants are also argued. Indiscriminate use of pesticidal chemicals is responsible for many complications in humans.

Chapter 13 deals with various pesticide analysis techniques and the limitations associated with each techniques. Different techniques are employed for detection of pesticides in variety of samples. Although a best method of detection is one that can detect nanogram quantities of pesticides in short period of time, numerous techniques lack either one of these advantages. From very routinely employed spectrometric techniques to complex chromatographic techniques efforts are taken towards the improvement of accuracy with broader detection range.

Chapter 14 explains major sources of pesticides, chronic impacts, labeling of pesticides, multidisciplinary approach for monitoring, current cost effective technologies, pros and cons of current technologies and future perspectives of the pesticide monitoring technologies. This chapter further argues that upcoming technologies namely, biosensing, electrochemical sensing, FRET, PET, SERS and paper based bioactive strips already have proved its own efficiency in terms of field deployable detection methodologies.

Chapter 15 presents a comprehensive overview on the recent advancement in the utilization of different adsorbents for the removal of pesticides. Overall, this study assists researchers to move forward in exploring a simple and economically-viable technique to produce adsorbents with outstanding physiochemical properties and excellent adsorption capacity, so that the pesticides can be removed from aquatic ecosystem.

Chapter 16 investigated the residual analysis of pesticides in surface water of Nagpur region, India. The pesticides, particularly Lindane, Endosulfan and DDT which are present in significant concentration have come down to a very low level, although, pesticides DDT and Lindane have been banned in India since 1989. Results of monitoring conducted on various specified water bodies from the different zones during pre-monsoon, monsoon and post-monsoon seasons allowed the different samples to show their susceptibility for the above mentioned pesticide residues.

Chapter 17 studied the recent advances in various analytic methods of pesticide residues in crops. Analysis of pesticide residues by analytical methods can be fluctuating based on the pesticide types. For food and health safety, maximum residue limits (MRL) of pesticides in foods were detected by The European Community. There are many analytical methods developed for identification and quantification of pesticides. Although, there are some limitations the multi-residue methods sensible for analysing of a great number of pesticides in one single run is the most fast and the favorite and efficient choice. However, some of the pesticides needed specific methodologies and a single- residue methods apply as compulsorily for them.

Chapter 18 gives an overview of the aquatic toxicity by halogenated organic compounds are usually xenobiotic in natures, and used as ingredients for the synthesis of pesticides, solvents, surfactants, and plastics. The information about different computational resources (databases, tools and modeling tools) have been given. Also, the application of QSAR to predict aquatic toxicity of different halogenated compounds available in the literature has been reviewed.

Chapter 19 advocates that the purification process of water may reduce the concentration of particulate matter including suspended particles, parasites, bacteria, algae, viruses, fungi; and a range of dissolved and particulate material derived from the surfaces that water may have made contact with after falling as rain. Generally the methods included for purification of water are physical processes such as filtration and sedimentation, biological processes such as slow sand filters or activated sludge, chemical processes such as flocculation and chlorination and the use of electromagnetic radiation such as ultraviolet light. The different treatment processes are defined and explained in this chapter.

Chapter 20 presents an up-to-date knowledge on extenuation strategies to diminish pesticide accumulation in aquatic systems, which has remained a major concern for ground water as well as surface water, adversely affecting aquatic ecosystems and humans through bio-magnification are included. Several factors affect the toxicity of pesticides like dosage of concentration, relative toxicity and chemical interactions, etc. The best approach to decrease pesticide pollution in environment is to use safer, non-chemical control methods and industrial or sewage superfluous should not be dumped into water reservoirs without proper pretreatment. Biological and chemical methods used for the control measures of pesticides pollution in aquatic systems. Thus greener approach for remedy of pesticide contaminated aquatic system could be more cost effective and sustainable.

Chapter 21 advocates that biopesticides provides a cost-effective and innovative approach employing bioremediation techniques for the removal of pesticides in waters because of its advantage linked with environmental safety, biodegradability, effectiveness, and target-specificity. Furthermore, biopesticides provide an efficient method for detoxification of pesticides and appropriateness in the integrated pest management (IPM) programs.

Chapter 22 focuses on agricultural chemicals that impact the aquatic environment. The aim of the chapter is to improve ecological sustainability and to minimise the effects of pesticides on aquatic ecosystem. In addition, authors attempted to reveal almost all positive aspects of organic farming in special reference to aquatic pollution. Good organic practices work to build the soil and maintain an ecological balance so that chemical fertilizers and synthetic pesticides are proven unnecessary.

Chapter 23 presents overview of how pesticides use have increased in the wake of climate change. The largest impact of climate change on future water demand, however, will potentially be felt by agriculture and horticulture. The indirect effects of climate-induced changes in demand for water and other natural and agricultural resources and changes in land use may have a greater effect on fate and transport of pesticides in the environment than direct effects. There is a growing body of evidence that climate change will have broad negative impacts on the distribution and toxicity of environmental contaminants. Temperature and precipitation, as altered by climate change, are expected to have the largest influence on the partitioning of chemical toxicants.

The book is intended for researchers, academicians, scientists, industrialists and NGO's. this book has some graphical and illustrative figures that will help the readers to comprehend different chapters in a lucid way. This book will be highly useful to researchers as a reference book as well. The different contributors from different regions have discussed the topics in detail that will help us to understand different aspects of pesticide pollution in a better way.

Acknowledgment

We would like to express our gratitude to the many people who saw us through this book; to all those who provided support, talked things over, read, wrote, offered comments, allowed me to quote their remarks, and assisted in the review, editing, proofreading, and design processes.

We would like to thank our parents and other family members who provided every support during the course of editing this book. Dr Javid Manzoor is warmly acknowledged who supported and encouraged us during this period.

We would like to thank ITM University, Jiwaji University Gwalior, and advanced learning and career counseling for providing all the necessary facilities. Thanks to IGI Global our publisher who encouraged us in many ways.

Last and not least: I beg forgiveness of all those who have been with me over the course of the years and whose names I have failed to mention.

Section 1
Pesticides in Water

Chapter 1
Pesticides as Water Pollutants

Anamika Srivastava
Banasthali Vidyapith, India

Nirmala Kumari Jangid
Banasthali Vidyapith, India

Manish Srivastava
Banasthali Vidyapith, India

Varun Rawat
Amity University, India

ABSTRACT

A chemical that is toxic to one animal may also be toxic to other forms of animal life. Although it might take a larger dose of pesticides to harm humans than pests, such as insects, many pesticides are still toxic to humans. The doses needed to kill a pest might not kill us, but may still harm us. Many pesticides classified as herbicides are designed to target plant pests. The exceptions to this are broad spectrum herbicides that are designed to kill a wide variety of plants. An herbicide that is specific to one or more species of plants does not ensure that it is safe to enter the water system. Some of the dangers from these chemicals are yet to be fully understood. Caution should be used to ensure that these products do not unnecessarily enter the water system. Using safe, well-planned applications of materials, such as pesticides, decreases the risk to humans and other animals. The overall picture is not as bleak as one might be imagining as the optimization of these important water resources are available at present and will also be available in the near future. Owing to the threat to water systems and mechanisms, those that may cause water to become polluted are now well-understood and precautions have been taken to protect the quality of water.

DOI: 10.4018/978-1-5225-6111-8.ch001

INTRODUCTION

Water is an essential source for our life. With the growth of the modern civilization, our life is in danger due to pollution of water both from surface and underground. In India, the scarcity of pure drinking water is so much felt that about 50% of urban people and 80% of water for rural people are affected by water pollution (Hegde, 2012). Historically, people looked at water pollution as a problem affecting people far away. Most people thought that they had clean, unpolluted water but it is not true. However, since 1970s the general public found that some water sources were in fact polluted. In some areas of our country, the water was not safe to drink because they contain high level of heavy metal, pesticide, herbicide and bacterials. These contaminants in water supplies came not only cities and industry's waste but from livestock and field runoff. Not only there were problems with the surface water but problems were beginning to show up in water deep below the surface. Water pollution has emerged as one of the most burning tribulations of this century. The aquatic ecosystems pollution occurs globally which includes an assortment of sources, impacts and is escalating. No other natural resource is more contested than water. It is imperative to differentiate the effects of human actions from natural phenomena, e.g. mud slides, and volcanic eruptions etc. Water pollution linked with anthropogenic activities is characteristically brisk and outstrips the evolutionary potential of ecosystems, leaving them in a depurate state (Kamble & Rao, 2016).

Water pollution threatens the survival of life on this planet and efforts to eradicate sources of pollution and reinstate impacted systems become a main concern with worldwide. Although over 70% of the planet is covered with water less than 3% is available as freshwater of that only 1% is utilized to maintain life. Exponential population growth, urbanization, industrialization and getting higher food production amplify the stipulate for water and additional decrease the limited amount obtainable (Cassardo & Jones, 2011).

With the advent of Green Revolution in the second half of the 20th century farmers began to use advance technology to enhance yields by using synthetic fertilizers, pesticides and herbicides. Now a day the use of pesticides has become common around the world not only on farms but in backyard gardens as well. These chemicals were developed in the lab and are petroleum-based have allowed farmers and gardeners of every stripe to exercise greater control over the plants they want to grow by enriching the immediate environment and warding off pests. But such benefits have not come without environmental costs like pollution of streams, rivers, ponds, lakes and even coastal areas, as these harmful synthetic chemicals run-off into the nearby water ways (FAO, 2013).

When the excess nutrients from all the fertilizers we use, run off into our waterways, they cause algae blooms sometimes big enough to make waterways infertile. When the algae die, sink to the bottom and decompose in water by the process that removes oxygen from the water. Fish and other aquatic species cannot flourish in this environment called as "dead zones" they either die or move on to other greener water pastures. The related issue is the poisoning of aquatic life. According to the United State Centers for Disease Control (CDC), citizens of America alone agitate through 75 million pounds of pesticides each year to keep the bugs off their peapods and petunias. When these harmful chemicals enter into waterways, fish ingest them and become diseased. Humans who eat these diseased fish can themselves become ill (EIFAC, 1968). Industrial processes produce toxic waste containing heavy metals and harmful by products which are lethal to marine life. The constructions of industries are responsible for contaminating our water resources with cement, lubricants, plastics, its byproducts and heavy metals. Rivers and lakes are also polluted from heavy silt or sediment run-off from industries construction sites. Natural misfortunes such as storm, earthquakes, volcano, floods and acid rain are known to gravely affect the ecological system and contaminate in water source. Farmers use chemicals to prevent diseases from crops, these chemicals can run off into lakes, creeks and rivers causes water pollution. Agricultural water pollution is caused by fertilizers, pesticides, farm animal wastes and sediments. Research findings indicate that insignificant application of fertilizers pollute the water through leaching of nitrate from nitrogenous fertilizers and pesticides. The use of various types of pesticides and insecticides in agriculture cause water pollution. Death of aquatic animals has been reported due to application of pesticides (Helfrich *et. al.,* 1996) is known to be hazardous.

Although there are benefits in the use of pesticides, there are also drawbacks, such as potential toxicity to humans and other animals. According to the Stockholm Convention on persistent Organic Pollutants, 10 of 12 most dangerous and persistent organic chemical are pesticides (Jacob & Cherian, 2013).

Pesticides

Pesticides may be a chemical substance, biological agent, antimicrobial, disinfectant or device which can use against any pests. Pests include insects, plant pathogens, weeds, molluscs, birds, mammals, fish, nematodes and microbes that decrease the property, enhance disease or a vector for disease or cause nuisance (Randall, 2013).

Since before 2000 BC, humans have utilized pesticides to protect their crops. About 4,500 years ago in ancient Mesopotamia's citizens have utilized pesticides to protect their crops. The first known pesticide was elemental sulfur dusting used in ancient summer. In 15th century, toxic chemical containing heavy metals like arsenic, mercury and lead were being used on crops to kill and vermin. In the 17th century toxic chemical like nicotine sulfate was extracted from tobacco leaves were being used on insecticide. In 19th century has seen the introduction of two more natural non-toxic pesticides which are derived from chrysanthemums and rotenone extracted from the roots of tropical vegetables (Miller, 2002).

Until the 1900s, there were toxic heavy metals like arsenic based pesticides were dominant. In 1960s herbicides and pesticides are became common which contain triazine and other nitrogen-containing compounds and carboxylic acids such as 2, 4-dichlorophenoxyacetic acid, and glyphosate (Ritter, 2009). The first legislation given that federal authority for regulating pesticides was enacted in 1910 (Goldman, 2007). However, decades later during the 1940s, manufacturers began to produce large amounts of synthetic pesticides and their use became extensive (Daly *et al*., 1998).

Some sources consider that mid 19th century has been called as the "pesticide era" (Graeme, 2005). Although the U.S. Environmental Protection Agency was recognized in 1970 and amendments to the pesticide law in 1972 (Miller, 2004), pesticide use has increased 50-fold since 1950 and 2.3 million tons (2.5 million short tons) of industrial pesticides are now used each year (Miller, 2002). Seventy-five percent of all pesticides in the world are used in developed countries (Miller, 2004). A study of USA pesticide use trends through 1997 was published in 2003 by the National Science Foundation's Center for Integrated Pest Management (Ritter, 2009).

In the 1960s, it was discovered that DDT prevented many fish-eating birds from reproducing, that was a serious threat to biodiversity. The agricultural use of DDT is now banned under the Stockholm Convention on Persistent Organic Pollutants but it is still used in some developing nations to prevent malaria and other tropical diseases by spraying on interior walls to kill or repel mosquitoes (Lobe, 2006). About 10,000 species of insects from 7, 50,000 species are identified as important pests.

Over 50,000 species of fungi are to blame for some 1,500 plant diseases. Over 1,800 species of weeds out of 30,000 cause serious economic loss. Approximately 15,000 species of nematodes produce more than 1,500 serious toxic effects on plants. Over 1, 00,000 species of pests destroy food which could be food for 135 million people. The word pest has no biological meaning. Pests are organisms that diminish the value of resources in which we are interested. In India, crops are affected by over 200 major pests, 100 plant diseases, hundreds of weeds and other pests like nematodes, harmful birds and rodents. About 4,800 million rats cause havoc. Approximately 30% of Indian crop yield potential is being lost due to insects, disease and weeds which in terms of quantity would mean 30 million tons of food grain. The value of total loss has been placed at Rs 50,000 million, represents about 18% of the gross national agriculture production. Pesticides are used in grocery stores and food storage services to manage rodents and insects that eat and distroy food such as grain. Each use of a pesticide are associated to some risk to a level deemed up to standard by pesticide regulatory agencies for instance the United States Environmental Protection Agency (EPA) and the Pest Management Regulatory Agency (PMRA) of Canada. Pesticides can save farmers' money by preventing crop losses to insects and other pests. In the U.S., farmers get an estimated fourfold return on money they spend on pesticides (Kelloge *et al.*, 2000).

DDT a harmful insecticides, sprayed on the walls of houses is containing an organochloride that has been used to fight malaria since the 1900s. Recent policy statements by the WHO have given stronger support to this approach. Many pesticides are considered too hazardous for sale to the general public and are designated restricted use pesticides. Only certified applicators may purchase or supervise the application of restricted pesticides (Willson, 1996).

The EPA regulates pesticides under two main acts, both of which were amended by the Act 1996 of Food Quality Protection. In addition to the EPA, the United States Department of Agriculture (USDA) in 1995 and the United States Food and Drug Administration (FDA,) set many standards for the leveling pesticide residue that is allowed on or in crops (Stephen, 2011). The EPA looks at what the potential human health and environmental effects might be associated with the use of the pesticide (EPA, 2011). The consumption of pesticide in India is about 600 gm / hectare, where as that of developed countries is touching 3000 gm / hectare. There is a wide range of pesticides found used in non-agriculture situations such as industries, public health and for a number of purposes in the home. Domestic use of pesticides is mainly as fly killer, ant killer, moth killer, repellants, rodenticides fungicides etc.

Use of pesticide is of vital importance in the industries such as carpet industry, wood preservation, paint industry, paper industry, board industry, leather industry, building industry and miscellaneous industrial application (Manncsa, 2009).

For each dollar that is spent on pesticides and herbicides for crops yields four dollars in crops were saved (Pimentel, 1991). For this one can conclude that the amount of money spent per year on pesticides about $10 billion. There is an additional $40 billion savings in crop that would be vanished due to damage by insects and weeds. Farmers benefit from having an increase in crop yield and able to grow a variety of crops throughout the year.

Consumers of agricultural products are benefitted from being able to afford the vast quantities of produce available year round (Pimentel, 2005). The general public also gets benefits from the use of pesticides for the control of insect-borne diseases and illnesses such as malaria. The use of pesticides creates a large job market in the industry. Control of pesticides using seven methods evolve over time as knowledge and techniques improved. This includes the development of chemical means of control which become very important because of a number of advantages. Farm chemicals are an economical way of controlling pests. They require low labor input and allow large areas to be related quickly and efficiently (Howell, 1998). A suitable farm chemical is available for most pest problems with variations in activity, selectivity and persistence. The best product can be chosen for the situation. This allows more flexibility in management options and better timeliness of pest control. Farm chemicals ensure a plentiful supply and variety of high quality, wholesome food at a reasonable price. Pesticides are classified in number of ways.

Classification of Pesticide

Pesticides include herbicides, insecticides, fungicides, rodenticides, pediculicides, and biocides (Gilden, *et al.*, 2010). There can be classified by target organism, chemical structure, and physical state. Pesticides can also be classed as inorganic, synthetic, or biological (Biopesticides). Biopesticides include microbial pesticides and biochemical pesticides (EPA, 2009). Plant-derived pesticides, or "botanicals", have been developing quickly. These include the pyrethroids, rotenoids, nicotinoids and a fourth group including strychnine and scilliroside. Many pesticides can be grouped into chemical families. Prominent insecticide families include organochlorines, organophosphates, and carbamates.

Organochlorine hydrocarbons could be separated into dichloro diphenylethanes, cyclodiene compounds and other related compounds. They operate by disrupting the sodium/potassium balance of the nerve fiber, forcing the nerve to transmit continuously. Their toxicities vary greatly but they have been phased out because of their persistence and potential to bioaccumulate. Thiocarbamate and dithiocarbamates continuously and their toxicities vary greatly but they have been phased out because of their persistence and potential to bioaccumulate. Thiocarbamate and dithiocarbamates are subclasses of carbamates. Prominent families of herbicides include pheoxy and benzoic acid herbicides (e.g. 2, 4-D), triazines (atrazine), ureas (diuron) and chloroacetanilides (alachlor). Phenoxy compounds tend to selectively kill broadleaved weeds rather than grasses. The Phenoxy and benzoic acid herbicides function similar to plant growth hormones, and grow cells without normal cell division, crushing the plants nutrient transport system where triazines interfere with photosynthesis (Kamrin, 1997). Many commonly used pesticides are not considered these families including glyphosate. Pesticides can be classified based upon their biological mechanism function or application method. Most pesticides work by poisoning pests. A systemic pesticide moves inside a plant following absorption by the plant. Insecticides and most fungicides move usually upward (through the xylem) and outward. Systemic insecticides which poison pollen and nectar in the flowers, may kill bees and other needed pollinators. In 2009, the development of a new class of fungicides called paldoxins was announced. The paldoxins inhibit the fungi's detoxification enzymes. They are believed to be safer and greener (Eurek, 2009). Though these pesticides are beneficial but always have their impact in environment causing pollution. Classification of pesticides also shown in Table 1.

PESTICIDE POLLUTION

Pesticides are carried in rainwater runoff from farm fields, sub urban lawns or roadside embankments into the nearest creeks streams and rivers. Occasionally they are even calculatedly sprayed into waterways as part of a pest-control effort. Commonly used pesticides can be harmful to environment and living organisms as they enter into creeks streams and river water, air and soil.

Table 1. Classifications of pesticides

By Target		By Mode or Time of Action		By Chemical Structure
Type	Target	Type	Action	
Bactericide (sanitizers or disinfectants)	Bacteria	Contact	Kills by contact with pest	Pesticides can be either organic or inorganic chemicals, most of today's pesticides are organic.
Defoliant	Crop foliage	Eradicant	Effective after infection by pathogens	Today's pesticides are organic.
Fungicides	Fungi	Non selective	Toxic to both crop and weed	
Herbicides	Weeds	Post emergence	Effective when applied after crop and weed emergence	Commonly used inorganic pesticides include copper based fungicides. lime-sulfur used to control fungi and mites, boric acid used for cockroach control and sulfamate herbicides.
Insecticides	Insects	Post emergence	Effective when applied after planting and before crop or weed emergence	
Maticides (acaricides)	Mites and ticks	Preplant	Effective when applied prior to planting	
Molluscicide	Slugs and snails	Protectants	Effective when applied before pathogen infects plants	
Nematicides	Nematodes	Selective	Toxic only to weed	Organic insecticides can either be natural(usually extracted from plants and bacteria) or synthetic. Most pesticides used today arte synthetic organic chemicals. They can be grouped into chemical families based on their structure.
Plant to growth regulation [a]	Crop growth processes		Toxic to all vegetation	
Rodenticides	Rodents	Stomach poison	Kills animal pests after ingestion	
Wood preservative	Wood destroying organisms	Systemic	Transported through crop or pest following absorption	

Pesticides as Water Pollutants

Herbicides and pesticides are used to control weeds and pests. Both of them also contribute to water pollution (Werner, 2002). Their leaching also pollutes ground water. Leaching is influenced by soil texture, pesticide properties, irrigation and rain fall. If soil is sandy and pesticide is water soluble more will be the leaching. Similarly pesticides and herbicides also reach natural water bodies through runoff. These pesticides and herbicides residues when reach to natural water bodies they disturb flora and fauna. Pesticides which do not mortify easily or take time to mortify are more harmful (Pope *et al.,* 2016).

Pesticide classes mostly detected involve herbicides used extensively in corn, cotton and rice production, organophosphorus insecticides as well as the banned organochlorines insecticides due to their persistence in the aquatic environment. The compounds most frequently detected were atrazine, simazine, alachlor, metolachlor and trifluralin of the herbicides, diazinon, parathion methyl of the insecticides and lindane, endosulfan and aldrin of the organochlorine pesticides. River and ground water were found to be more polluted than lakes because the detected concentrations of most pesticides follow a seasonal variation, with maximum values occurring during the post spring and summer period followed by a decrease during winter. In many cases it is reported by many scientist that concentrations ranged in low ppb levels. However, elevated concentrations were recorded in areas of high pesticide use and intense agricultural practices. In generally same trends and levels of pesticides were found in Greek rivers compared to pesticide contamination in other European rivers. Monitoring of the Greek water resources for pesticide residues must carry on, especially in agricultural regions, because the nationwide patterns of pesticide use are constantly changing. Moreover, emphasis should be placed on degradation products not sufficiently studied so far (Ioannis *et al.,* 2006).

Origins of Pesticide Entry Into Water

Research over the last few years has proved that origin of pesticides into the water has been depend on the nature and properties of the pesicides and the prevailing agro climatic conditions (Carter, 1999 & Flury *et al.,* 1994). Investigations have focused on monitoring and understanding the processes which determine fate following application to agricultural land. Entry of pesticides into water has been shown to occur in some circumstances and it is now increasingly recognized that there are also a number of other entry routes shown in Figure 1.

There are mainly two routes for entry of pesticides into water which are shown in Table 2.

Diffuse Sources

Water pollution can come from either diffuse or point sources. Diffuse source water pollution is one of the key impacts on water quality in our waterways. In some waterways it is the largest source of pollutants. It is controlled by regulation. Diffuse source pollution typically comes from unlicensed sources and dispersed land-use activities. Central government has a critical role to play in the transition to more effective management of the risks from diffuse water pollution. This includes strong over-arching regula-

Figure 1. Pathways of a pesticide applied to a crop; ideally, at least one includes its contact with the targeted pest

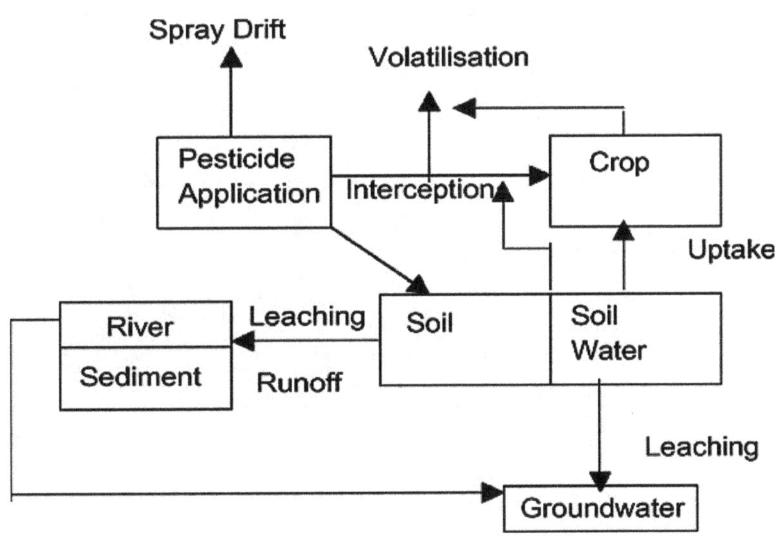

Table 2. Routes for entry of pesticides into water

1.	Diffuse sources	a. Spray Drift
		b. Surface runoff or overland flow
		c. Leaching:
		d. Drain flow and through flow
2.	Point sources	

tory frameworks, stakeholder engagement, and money allocated to initiate experimental projects. In doing so, government sends the right signals to local authorities, stakeholders and investors, and minimizes the cost of water quality management for society as a whole. Pollutant in diffuse sources are overuse of fertilizers, sediments, soil erosion by pathogens like bacteria from leaking septic tanks, toxicants, pesticides, salts, acid sulfate soils in drained wetlands, gross pollutants etc.

Spray Drift

A key potential route to surface water is from spray drift. For plant protection many pesticides are sprayed very close to surface water this causes spray drift shown in Figure 2. This is proved from some field monitoring data concerning the amount of product which does drift to water. For example a ground application of a pesticide to arable crops resulted in drift deposition ranging from 0.3% to 3.5% of the normal field (Clyde, 2013).

Figure 2. Spray drift

Surface Runoff or Overland Flow

When the infiltration capacity of an application surface is exceeded water, soluble residues and residues sorbed to fine particles can move across an application surface. Losses from agricultural fields are typically less than 0.05% unless extreme rainfall falls within 1–2 weeks of application of the applied active substance. Losses can be environmentally significant at the local level due to specific agro climatic conditions and those soils which have the potential to slake or cap, or are situated on steeper slopes, can be particularly susceptible to surface runoff (Carter, 1999) Agricultural runoff cases fresh water body's eutrophication. Half of lakes in US are eutrophic. Phosphate is the main contributor to eutrophication its high concentration promotes cyanobacteria and algae growth which ultimately reduces dissolved oxygen in water. Harmful toxins which accumulate in food chain are produced by cyanobacterial blooms (Schmidt *et al.,* 2013). Nitrogen rich fertilizer compounds causes dissolved oxygen deficiency in rivers, lakes and coastal zones which have devastating effects on oceanic fauna. In America and Northwest Europe nitrogen fertilizer use is controlled from 2006. Nitrogen fertilizers have high water solubility and increased runoff and leaching rate which results in ground water pollution (Van *et al.,* 2012).

Leaching

Leaching is a major process for the transport of soluble to ground and surface water. Leaching residues may move directly to underlying groundwater or may be transported laterally to surface waters (Figure 3). Losses of applied active substance by leaching can, in exceptional cases be as high as 5%, but are typically less than 1%. Losses can be greater when soil water moves rapidly as preferential flow, through the soil via cracks, fissures or macropores since residues have less opportunity to dissipate. In some soil types, residues absorbed to soil or organic particles can be transported from the soil surface suspended in leachate water. Strongly absorbed active substances which do not normally leach can be transported in this way and the transportation of diflufenican from non calcareous topsoil to surface water. The extent and ecotoxicological significance of this transport mechanism in Europe has not been determined but it will be more important for compounds with high sorption values.

Figure 3. Leaching Process

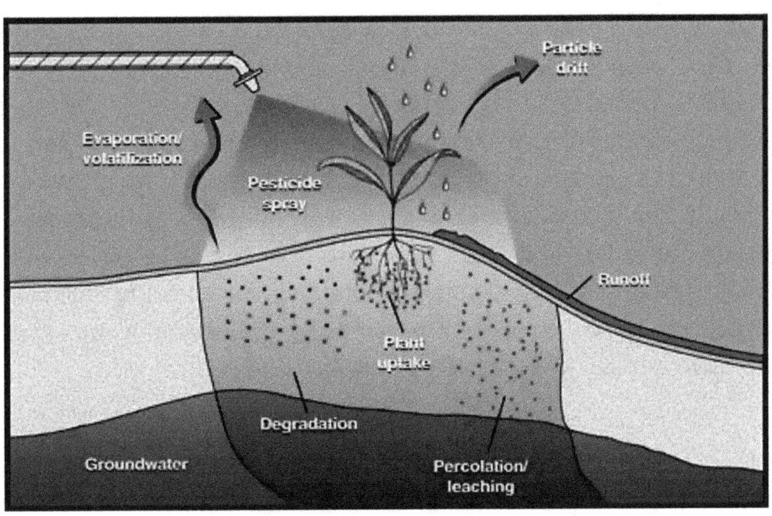

Drain Flow and Through Flow

Land drainage design has as its objective the removal of excess water from soil or the land surface. Large scale attempts have been made for at least two centuries to improve sub-surface water control by the installation of various under drainage systems e.g. stone drains, baked clay horseshoe tiles clay pipes, and now slotted plastic pipe with permeable fill. These drains are effective and responsible for removing water from many slowly permeable soils or those with shallow water tables which would otherwise not have been cultivated. Artificial drainage has been shown to be responsible for the transport of significant quantities of dissolved pesticide or that carried on sediment particularly when rainfall and subsequent drainage occur shortly after application. Losses might represent up to 1% of the applied active substance but are typically less than this. Through flow is the lateral movement of water below the soil surface which occurs naturally in the absence of artificial drainage. Through flow can enter surface water via bankside seepage but will generally contain less residues than drain flow as it has passed through the soil matrix having the opportunity for sorption and degradation of the active substance (Sneyd & Hosking, 1976).

Point Sources

The sources are called point sources because in mathematical modeling, they can be approximated as a mathematical point to simplify analysis (Van, 2010). Pollution point sources are identical to other physics, engineering, optics, and chemistry point sources. A point source pollution includes air pollution from an industrial source, rather than an airport or a road. It is considered a line source or a forest fire which is measured an area source, or volume source. Water pollution from an oil refinery wastewater discharge outlet, noise pollution from a jet engine, disruptive seismic vibration from a localized seismic study, light pollution from an intrusive street light, thermal pollution from an industrial process outfall and radio emissions from an interference-producing electrical device. A point source of pollution is a single restricted source of air, water, thermal, noise or light pollution. A point source has negligible extent, distinguishing it from other pollution source geometries (Hogan, 1973).

Water Pollution Due to Pesticides

Pesticides enter surface and ground water primarily as runoff from crops and are most prevalent in agricultural areas. Pesticides are also used on golf courses, forested areas, along roadsides, and in suburban and urban landscape areas. Since World War II herbicide and insecticide application to crops has grown to an estimated 660 million pounds of active ingredient in 1993 (Aspelin, 1994) without proper safeguards pesticides have the potential to seriously threaten many groundwater supplies in the United States. Approximately 50% of the U.S. population obtains its drinking water from groundwater sources and as much as 95% of the population in agricultural areas uses groundwater as its source of drinking water.

'Pesticide' is a general term for substances which are used to poison pests (weeds, insects, molds, rodents, etc.). The pesticides most acutely dangerous to man are insecticides and rodenticides, although pound for pound, herbicides are the most widely used type of pesticide (Pimentel *et al.,* 1991). Not every pesticide is acutely toxic to humans or other non-target species.

On a national scale less than 2% of wells sampled in multi-state studies were found with pesticide concentrations above the established Maximum Contaminant Level (MCL). Due to repeated detection of various pesticides in U.S. wells, the U.S. Environmental Protection Agency (EPA) recently proposed a State Management Program (SMP), which would control or ban pesticides with the greatest potential to contaminate groundwater (Barbash & Resek, 1996). Five pesticides were initially selected due to the frequency of their occurrence: alachlor, atrazine, cyanazine, metolachlor, and simazine. According to the EPA they all have been detected in many states, and have the potential to reach levels which exceed health based standards. They are all associated with serious health effects including cancer.

The five selected pesticides are herbicides which are used to control broadleaf weeds and grasses. The EPA estimates between 200 and 250 million pounds of these herbicides are applied annually in the U.S. Atrazine, simazine, and cyanazine are applied to agricultural land before and after planting. Alachlor and metolachlor are applied to soil prior to plant growth (pre-emergent).

There are several factors which influence a pesticides' potential to contaminate water:

- The ability of the pesticide to dissolve in water (solubility).
- Environmental factors, such as, soil, weather, season, and distance to water sources
- Application methods and other practices associated with the pesticide use.

Groundwater contamination is higher when there is no crop or a young crop. A large actively growing crop has the ability to reduce pesticide concentration through a variety of mechanisms:

- Larger plants consume more water from the soil and therefore reduce the ability of a pesticide to migrate through the soil and enter streams or groundwater.
- Larger plants can collect precipitation which prevents pooling of water and run-off from the area
- Root zones enrich the microbial community of the soil which then enhances the biodegradation of the pesticide by bacteria.

Pesticides are those chemicals (such as insecticides, fungicides, herbicides, rodenticides, molluscicides, nematocides, plant growth etc.), which have been widely used throughout the world to increase crop yield and to kill the insect-pests responsible for transmitting various diseases to humans and animals. However, according to several reports, these chemicals have been proved to inflict adverse impacts on the health of living beings and their environment. (Igbedioh, 1991 & Jeyaratnam, 1985).

Pesticide impacts on aquatic systems are often studied using ahydrology transport model to study movement and fate of chemicals in rivers and streams. As early as the 1970s quantitative analysis of pesticide runoff was conducted in order to predict amounts of pesticide that would reach surface water (Hogan, 1973)

There are four major routes through which pesticides reach the water: it may drift outside of the intended area when it is sprayed, it may percolate, or leach, through the soil, it may be carried to the water as runoff, or it may be spilled, for example accidentally or through neglect. They may also be carried to water by eroding soil (Papendick *et al.,* 1986). Factors that affect a pesticide's ability to contaminate water include its water solubility, the distance from an application site to a body of water, weather, soil type, presence of a growing crop, and the method used to apply the chemical (Pedersen, 1997).

EFFECTS OF PESTICIDES

Human Health Effects of Pesticides

One of the most important regional examples of pesticide contamination and adverse effect on human health is that of the Aral Sea region (Figure 2). In 1993 UNEP linked the harmful effects of pesticides which leads to cancer, pulmonary and haematological morbidity, as well as inborn deformities and immune system deficiencies".

Inhalation and skin contact during preparation and application of pesticides to crops can be especially harmful to form workers. However, for the majority of the population, adverse effect is through ingestion of food that is contaminated by pesticides. Degradation of water quality by pesticide runoff has two principal human health impacts. The first is the consumption of fish and shellfish that are contaminated by pesticides; this can be a particular problem for subsistence fish economies that lie downstream of major agricultural areas. The second is the direct consumption of pesticide-contaminated water. WHO (1993) has established drinking water guidelines for 33 pesticides. Many health and environmental protection agencies have established "acceptable daily intake" (ADI) values which indicate the maximum allowable daily ingestion over a person's lifetime without appreciable risk to the individual. For example, in a recent paper by Wang and Lin (1995) studying substituted phenols, tetrachlorohydroquinone, a toxic metabolite of the biocide pentachlorophenol, was found to produce "significant and dose-dependent DNA damage".

Ecological Effects of Pesticides

Pesticides are included in a broad range of organic micro pollutants that have ecological impacts. Different categories of pesticides have different types of effects on living organisms, therefore generalization is difficult. Although terrestrial impacts by pesticides do occur, the principal pathway that causes ecological impacts is that of water contaminated by pesticide runoff. The two principal mechanisms are bioconcentration and biomagnification.

- **Bioconcentration:** This is the movement of a chemical from the surrounding medium into an organism. The primary "sink" for some pesticides is fatty tissue ("lipids"). Some pesticides, such as DDT, are "lipophilic", meaning that they are soluble in, and accumulate in, fatty tissue such as edible fish tissue and human fatty tissue. Other pesticides such as glyphosate are metabolized and excreted.
- **Biomagnification:** This term describes the increasing concentration of a chemical as food energy is transformed within the food chain. As smaller organisms are eaten by larger organisms, the concentration of pesticides and other chemicals are increasingly magnified in tissue and other organs. Very high concentrations can be observed in top predators, including man.

The ecological effects of pesticides (and other organic contaminants) are varied and are often interrelated. Effects at the organism or ecological level are usually considered to be an early warning indicator of potential human health impacts. The major types of effects are listed below and will vary depending on the organism under investigation and the type of pesticide. Different pesticides have markedly different effects on aquatic life which makes generalization very difficult. The important point is that many of these effects are chronic (not lethal), are often not noticed by casual observers, yet have consequences for the entire food chain.

- Death of the organism.
- Cancers, tumours and lesions on fish and animals.
- Reproductive inhibition or failure.
- Suppression of immune system.
- Disruption of endocrine (hormonal) system.
- Cellular and DNA damage.
- Teratogenic effects (physical deformities such as hooked beaks on birds).
- Poor fish health marked by low red to white blood cell ratio, excessive slime on fish scales and gills, etc.
- Intergenerational effects (effects are not apparent until subsequent generations of the organism).
- Other physiological effects such as egg shell thinning.

These effects are not necessarily caused solely by exposure to pesticides or other organic contaminants, but may be associated with a combination of environmental stresses such as eutrophication and pathogens. These associated stresses need not be large to have a synergistic effect with organic micro pollutants.

Ecological effects of pesticides extend beyond individual organisms and can extend to ecosystems. Swedish work indicates that application of pesticides is thought to be one of the most significant factors affecting biodiversity. Jonsson *et al.* (1990) report that the continued decline of the Swedish partridge population is linked to changes in land use and the use of chemical weed control. Chemical weed control has the effect of reducing habitat, decreasing the number of weed species, and of shifting the balance of species in the plant community. Swedish studies also show the impact of pesticides on soil fertility, including inhibition of nitrification with concomitant reduced uptake of nitrogen by plants (Torstensson, 1990). These studies also suggest that pesticides adversely affect soil micro-organisms which are responsible for microbial degradation of plant matter (and of some pesticides), and for soil structure. Box 6 presents some regional examples of ecological effects of pesticides.

PESTICIDES MANAGEMENT AND CONTROL METHOD

Predicting the impact of pesticides on water quality is an essential element for developing generic as well as site specified pesticides control methods. The key factors affecting the movement of pesticides are runoff, erosion and sediment transport

The major characteristic of non-point source pollution is that the primary transfer mechanisms from land to water are driven by those hydrological processes that lead to runoff of nutrients, sediment, and pesticides. This is important, not only to understand the nature of agricultural pollution, but also because modelling of hydrological processes is the primary mechanism by which agriculturalists estimate and predict agricultural runoff and aquatic impacts. Except where agricultural chemicals are dumped directly into watercourses, almost all other non-point source control techniques in agriculture involve control or modification of runoff processes through various land and animal (manure) management techniques.

Figure 4. Schematic diagrams showing the major processes that link rainfall and runoff

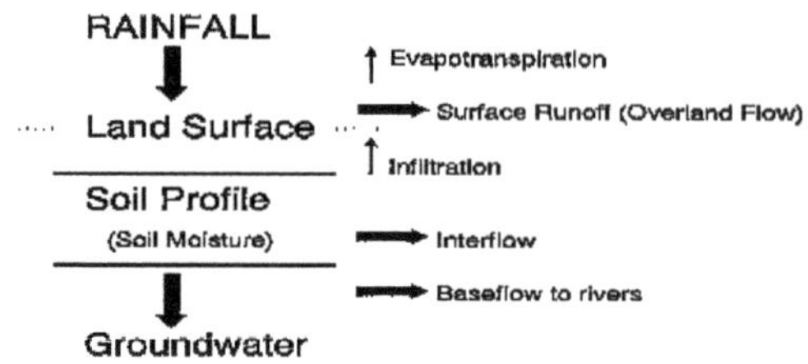

In large parts of the world, precipitation is in the form of rain. However, in those areas where precipitation is in the form of snow, the science becomes more complex. Nevertheless, control measures, whether for areas subject to rain or snow can be easily summarized. Therefore, for the purpose of this publication, focus will be on the relationships between rainfall and runoff.

While the practice of hydrology can be quite theoretical, the principal concepts are easily understood (Figure 4).

- **Rainfall***:* The primary controlling factor is the rate (intensity) of rainfall. This controls the amount of water available at the ground surface, and is closely related to measures of energy that are used in many mathematical formulations to calculate soil detachment by rain drops. Soil detachment makes soil particles available for sediment runoff.
- **Soil Permeability:** Permeability is a physical characteristic of a soil and is a measure of the ability of the soil to pass water, under saturated conditions, through the natural voids that exist in the soil. Permeability is a function of soil texture, mineral and organic composition, etc. In contrast, "porosity" is the measure of the amount of void space in a soil; however, permeability refers to the extent to which the porosity is made up of interconnecting voids that allow water to pass through the soil. As an example, styrofoam is highly porous but impermeable, whereas a sponge is both porous and permeable.
- **Infiltration:** Infiltration rate, the rate at which surface water passes into the soil (cm/hr), is one of the most common terms in hydrologic equations for calculating surface runoff. Infiltration is not identical to permeability; it is mainly controlled by capillary forces in the soil which, in turn, reflect the prevailing conditions of soil moisture, soil texture, degree of surface compaction, etc. Infiltration will vary between and within rainfall events, depending upon factors such as antecedent soil moisture, nature of vegetation, etc. In general, infiltration rate begins at a high value during a precipitation event, and decreases to a small value when the soil has become saturated.
- **Surface Runoff:** This is the amount of water available at the surface after all losses have been accounted for. Losses include evapo-transpiration by plants, water that is stored in surface depressions caused by irregularity in the soil surface, and water that infiltrates into the soil. The interaction between infiltration rate and precipitation rate mainly governs the amount of surface runoff. Intense rainstorms tend to produce much surface runoff because the rate of precipitation greatly exceeds the infiltration rate. Similarly, in areas of monsoonal rain and tropical storms, the length and intensity of precipitation frequently exceeds infiltration capacity. Destruction of protective surface vegetation and compaction of the soil, especially in tropical environments, leads to major erosional phenomena due to the amount of surface runoff. Except for nitrogen which is usually

found in groundwater in agricultural areas, surface runoff is the primary contributor of agricultural chemicals, animal wastes, and sediment to river channels.

- **Interflow:** Because soil horizons have different levels of permeability not all water in the soil will move downward into the groundwater. The residual water in the soil will move along the soil horizons, parallel to the ground surface. Interflow usually emerges near the bottom of slopes and in valley bottoms. Therefore, identification of these hydrologically active zones is an important part of agricultural non-point source control measures. Interflow is the mechanism which has also been linked to soil piping, a potentially destructive characteristic in some soils by which shallow "pipes" form naturally in the soil and are enlarged by interflow to the point where they collapse causing gullies in the agricultural surface.
- **Groundwater:** Groundwater is supplied by water which passes through the soil horizons into the parent material and/or bedrock underlying the soil. Groundwater tends to flow towards rivers channels where it emerges and supports stream flow during periods of little or no rain.
    - This component of stream flow is called "base flow". The chemistry of base flow reflects the soil and bedrock geochemistry, plus any agrochemicals that have been leached into the groundwater.
- **Snowmelt:** The phenomenon of snowmelt greatly complicates prediction of agricultural pollution using conventional hydrologic models. Snowmelt, by itself, is not normally a major producer of surface runoff. However, the combination of spring rain and snowmelt on frozen or thawing soils can produce serious erosional problems. Snowmelt tends to contribute greatly to agricultural non-point source pollution by carrying to adjacent streams the animal wastes, sludges, and other wastes that were spread on frozen agricultural soils during the winter period. Correct management of animal wastes in regions of frozen ground has major beneficial effects on water quality.

REFERENCES

Aspelin, A.L. (1994). Pesticides industry sales and usage, 1992 and 1993 market estimates: U.S. EPA, Office of Pesticides Programs, Biological and Economic Analysis Div. *Economic Analysis Branch Report, 33*, 733-K-94-001.

Barbash, J. E., & Resek, E. A. (1996). Pesticides in Groundwater, Volume 2 of the series Pesticides in the Hydrologic Environment. Ann Arbor Press, Inc.

Carter, A. D. (1999). Leaching Mechanisms. In G. T. Brooks & T. R. Roberts (Eds.), *Pesticide Chemistry and Bioscience, the Food-Environment Challenge* (pp. 291–301). Cambridge, UK: The Royal Society of Chemistry. doi:10.1533/9781845698416.7.291

Cassardo, C., & Jones, J. A. A. (2011). Managing water in a changing world. *Water (Basel), 3*(2), 618–628. doi:10.3390/w3020618

Clyde, L. O. (2013). *Spray drift of pesticide*. Neb guide. Retrieved from https://en.wikipedia.org/wiki/Pesticide_drift

Daly, H., Doyen, J. T., & Purcell, A. H. (1998). *Introduction to insect biology and diversity* (2nd ed.). New York: Oxford University Press.

EIFAC. (1968). *Water Quality Criteria for European Freshwater Fish: Report on Extreme pH Values and Inland Fisheries*. Rome: European Inland Fisheries Advisory Commission.

EPA. (2009). *Types of Pesticides*. Retrieved from https://en.wikipedia.org/wiki/Pesticide

EPA. (2011). *Pesticides industry sales and usage*. Retrieved from https://en.wikipedia.org/wiki/Environmental_impact_of_pesticides

FAO. (2013). *Pesticides as water pollutants in Control of Water Pollution from Agriculture*. Food and Agriculture Organization of the United Nations.

Flury, M., Leuenberger, J., Studer, B., Fluhler, H., Jury, W. A., & Roth, K. (1994). *Pesticide transport through unsaturated field soils:Preferential flow*. Zurich: ETH.

Gilden, R. C., Huffling, K., & Sattler, B. (2010). Pesticides and health risks. *Journal of Obstetric, Gynecologic, and Neonatal Nursing*, *39*(1), 103–110. doi:10.1111/j.1552-6909.2009.01092.x PMID:20409108

Goldman, L. R. (2007). Managing pesticide chronic health risks: U.S. policies. *Journal of Agromedicine*, *12*(1), 67–75. doi:10.1300/J096v12n01_07 PMID:18032337

Graeme, M. (2005). *Pesticide Rotation. Ontario Ministry of Agriculture*. Food and Rural Affairs.

Hegde, N. (2012). Water scarcity and security in India. *BAIF at the Indian Science Congress.*

Helfrich, L. A., Weigmann, D. L., Hipkins, P., & Stinson, E. R. (1996). *Pesticides and aquatic animals: A guide to reducing impacts on aquatic systems*. Virginia Cooperative Extension.

Hogan, C. M., Patmore, L. L. G., & Seidman, H. (1973). *Computer modeling of pesticide transport in soil for five instrumented watersheds, U.S. Environmental Protection Agency Southeast Water laboratory*. Sunnyvale, CA: Athens, Ga. by ESL Inc.

Hooda, P. S., Edwards, H. C., Anderson, H. A., & Miller, A. (2000). A review of water quality concerns in livestock farming areas. *The Science of the Total Environment*, *24*(1-3), 143–167. doi:10.1016/S0048-9697(00)00373-9 PMID:10811258

Howell, V. D., Doyen, J. T., & Purcell, A. H. (1998). *Introduction to insect biology and diversity*. Oxford University Press.

Igbedioh, S. O. (1991). Effects of Agricultural Pesticides on Humans, Animals and Higher Plants in Developing Countries. *Archives of Environmental Health*, *46*(4), 218–223. doi:10.1080/00039896.1991.9937452 PMID:2069430

Ioannis, K., Konstantinou, D. G. H., & Triantafyllos, A. A. (2006). The status of pesticide pollution in surface waters (rivers and lakes) of Greece. Part I. Review on occurrence and levels. *Environmental Pollution*, *141*(3), 555–570. doi:10.1016/j.envpol.2005.07.024 PMID:16226830

Jacob, J., & Cherian, J. (2013). Review of Environmental and Human Exposure to Persistent Organic Pollutants. *Asian Social Science*, *9*(11), 107–120. doi:10.5539/ass.v9n11P107

Jeyaratnam, J. (1985). Health Problems of Pesticide Usage in the Third World. *British Medical Journal, 42*, 505–506. PMID:4016001

Kamble, V. S., & Rao, K. R. (2016). Intentional Study of Bioaccumulative Behaviour of Isomers of Organochlorine Pesticide Thiodan (Endosulfan 35% Ec) In Fresh Water Mollusc, Lamellidens Corrianus During Monsoon Season. *Indian Journal of Applied Research, 6*(4), 264–266.

Kamrin, M. A. (1997). *Pesticide Profiles: toxicity, environmental impact, and fate*. CRC Press. doi:10.1201/9781420049220

Kelloge, R. L., Nehring, R., Grube, A., Goss, D. W., & Plotkin, S. (2000). *Environmental indicators of pesticide leaching and runoff from farm fields*. United States Department of Agriculture Natural Resources Conservation Service.

Lobe, J. (2006). *WHO urges DDT for malaria control Strategies*. Inter Press Service.

Manncsa. (2009). *Pesticides in Agriculture*. Retrieved from http://agropedia.iitk.ac.in/content/pesticides-agriculture

Manuel, A. E., Eugenio, L. P., Elena, M. C., Jesus, S. G., Juan-Carlos, M., & Luis, G. R. (2008). The mobility and degradation of pesticides in soils and the pollution of groundwater resources. *Agriculture, Ecosystems & Environment, 123*(4), 247–260. doi:10.1016/j.agee.2007.07.011

Michael, H. C. (2010). Water pollution. In Encyclopedia of Earth. Cleveland, OH: National Council on Science and the Environment.

Miller, G. T. (2002). Living in the Environment (12th ed.). Belmont: Wadsworth/Thomson Learning.

Murphy, G. (2005). *Resistance Management- Pesticide Rotation*. Ontario Ministry of Agriculture, Food and Rural Affairs.

Ongley, E. D. (1996). *Control of water pollution from Agriculture*. Food and Agriculture Organisation of United Nation Food and Agriculture Org.

Papendick, R. I., Elliott, L. F., & Dahlgren, R. B. (1986). Environmental consequences of modern production agriculture: How can alternative agriculture address these issues and concerns. *American Journal of Alternative Agriculture, 1*(1), 3–10. doi:10.1017/S0889189300000722

Paul, M. B. (2003). *Ground Water in Freshwater-Saltwater Environments of the Atlantic*. Reston, VA: U.S. Geological Survey.

Pedersen, T.L. (1997). *Pesticide residues in drinking water.* Retrieved from extoxnet.orst.edu

Pimentel, D. (2005). Environmental and economic costs of the application of pesticides primarily in the United States. *Environment, Development and Sustainability, 7*(2), 229–252. doi:10.100710668-005-7314-2

Pimentel, D., & McLaughlen, L. (1991). Environmental and economic impacts of reducing U.S. agricultural pesticide use. In D. Pimentel & A. A. Hanson (Eds.), *CRC Handbook of Pest Management in Agriculture* (pp. 679–718). Boca Raton, FL: CRC Press.

Pope, C. A. III, Bhatnagar, A., McCracken, J. P., Abplanalp, W. T., Conklin, D. J., & O'Toole, T. (2016). Exposure to fine particulate air pollution is associated with endothelial injury and systemic inflammation. *Circulation Research, 119*(11), 1204–1214. doi:10.1161/CIRCRESAHA.116.309279 PMID:27780829

Randall, C. (2013). *National Pesticide Applicator Certification Core Manual*. Washington, DC: National Association of State Departments of Agriculture Research Foundation.

Ritter, S. R. (2009). *Pinpointing Trends in Pesticide Use in* E-News. Academic Press.

Schmidt, J. R., Shaskus, M., Estenik, J. F., Oesch, C., Khidekel, R., & Boyer, G. (2013). Variations in the microcystin content of different fish species collected from a eutrophic lake. *Toxin, 5*(5), 992–1009. doi:10.3390/toxins5050992 PMID:23676698

Singh, A., Imtiyaz, M., Isaac, R. K., & Denis, D. M. (2012). Comparison of soil and water assessment tool (SWAT) and multilayer perceptron (MLP) artificial neural network for predicting sediment yield in the Nagwa agricultural watershed in Jharkhand, India. *Agricultural Water Management, 104*, 113–120. doi:10.1016/j.agwat.2011.12.005

Sneyd, A. D., & Hosking, R. J. (1976). Seepage flow through homogeneous soil into a row of drain pipes. *Journal of Hydrology (Amsterdam), 30*(1-2), 127–146. doi:10.1016/0022-1694(76)90094-9

Stephen, W. C. (2011). Determination of organochlorine pesticide residues in fatty foods: A critical review on the analytical methods and their testing capabilities. *Journal of Chromatography. A, 1218*(33), 5555–5567. doi:10.1016/j.chroma.2011.06.066 PMID:21742333

Sustain, D. (2013). Soil compaction impact and modelling A review. *Muhammad Farrakh Nawaz & Guilhem Bourrié & Fabienne Trolard Agron, 33*, 291–309.

Torstensson, L. (1990). *Natruvaardsverket rapport*, 3536. Solan, Sverige.

Van, L. C. J. (2010). Risk Assessment of Chemicals: An Introduction (2nd ed.). Dordrecht, The Netherlands: Springer.

Van Grinsven, H. J. M., Ten Berge, H. F. M., Dalgaard, T., Fraters, B., Durand, P., Hart, A., ... Willems, W. J. (2012). Management, regulation and environmental impacts of nitrogen fertilization in northwestern Europe under the Nitrates Directive; a benchmark study. *Biogeosciences, 9*(12), 5143–5160. doi:10.5194/bg-9-5143-2012

Wang, Y. J., & Lin, J. K. (1995). Estimation of Selected Phenols in Drinking Water with In Situ Acetylation and Study on the DNA Damaging Properties of Polychlorinated Phenols. *Archives of Environmental Contamination and Toxicology, 28*(4), 537–542. doi:10.1007/BF00211639 PMID:7755407

Werner, W. (2002). *Fertilizers, Environmental Aspects. In Ullmann's Encyclopedia of Industrial Chemistry*. Weinheim: Wiley-VCH.

Willson, H. R. (1996). *Pesticides Regulations*. University of Minnesota. Retrieved from https://en.wikipedia.org/wiki/Pesticide

KEY TERMS AND DEFINITIONS

Classification of Pesticide: Pesticides are classified into different type such as insecticides, herbicides, defoliants, desiccants, fungicides, nematicides, avicides, rodenticides, etc.

Pesticide: Pesticide is a chemical or biological agent (virus, bacterium, fungus) that deters, incapacitates, kills, or otherwise discourages pests. These are the chemical substances which are used to control pests, weeds, insects etc. These chemical substances are designed to kill or retard the growth of pests that damage or interfere with the growth of agricultural crops and vegetations.

Pollution: Pollution is the introduction of contaminants into the natural environment that cause adverse change. Pollution can take the form of chemical substances or energy, such as noise, heat, or light. Pollutants, the components of pollution, can be either foreign substances/energies or naturally occurring contaminants.

Chapter 2
Pesticide Sources, Their Fate, and Different Ways to Impact Aquatic Organisms

Samreen Siddiqui
Texas A&M University – Corpus Christi, USA

ABSTRACT

Since the industrial revolution, several new chemicals were discovered and introduced in society, and soon after the green revolution, pesticides were also introduced to strengthen food security. However, limited education on their application, handling, and usage resulted in them making their way into the aquatic ecosystem. This chapter defines the different sources of pesticides, based on their point of origin and the way it transports pesticides to the aquatic systems. After this, the pesticide interaction in an aquatic environment with various organic and inorganic substances is described. Each interaction is supported with the recent researches and examples. Following pesticides sources and interactions, its fate in the aquatic system has been defined through various physical and chemical processes. Ultimately, its impact on aquatic organisms is discussed. This chapter is concluded with recommended management practices and future research directions. Some terms are also defined at the end of this chapter.

INTRODUCTION

Since 1960s pesticides helped farmers to expand agricultural production and support the growing population globally. Food security with an increasing global population intensifies pesticides application in the agricultural fields. USEPA reported (USEPA, 2016) world pesticide expenditure at the producer level nearly $56 billion in 2012, out of which $9 billion was spent by the USA alone. Due to the high application, pesticides can be easily detected in surface waters. They may enter into aquatic systems from the point or non-point sources of pollution. After entry, they may affect biota individually or in a mixture form defined further in this chapter. The fate is decided by the type of aquatic system they enter and interaction with different organic and inorganic substances in the ecosystem.

DOI: 10.4018/978-1-5225-6111-8.ch002

Figure 1. Schematic classification of pesticides on the basis of mode of action and chemical structure

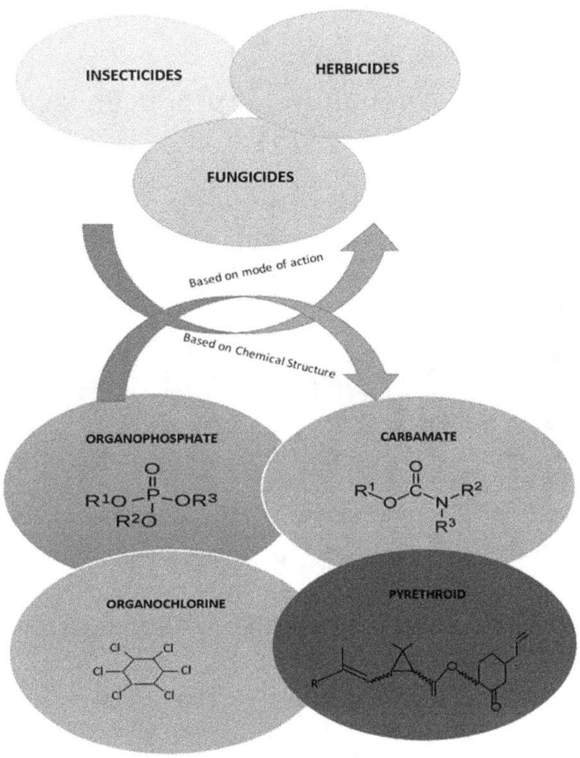

Pesticides are natural and artificially synthesized chemical products used to limit, inhibit, and prevent the growth of harmful organisms, insects, invasive plants, weeds, and fungi. Pesticides can be classified as herbicide, insecticide, and fungicide (based on targeted pests) and organophosphate (OP), carbamate, organochloride (OC) and pyrethroid (based on their chemical structure) (Figure 1). Out of the three major subclasses of pesticides, herbicides are the most frequently detected compounds in surface waters of Europe and USA (Gustavsson, Kreuger, Bundschuh, & Backhaus, 2017) due to generally higher hydrophobicity, shorter application duration, and lower quantity mass application. Herbicides were accounted for nearly $25,000 million expenditure worldwide in 2012 (Figure 2) and mostly used by agriculture with some percentage by home and garden (Figure 3). In contrast, the largest share of insecticides was used in households as compared to other areas (Figure 3).

Pesticides are found at varying concentrations in the environment depending upon their uses. They enter into the environment through their application in public health (e.g. control of Mosquitoes and flies), large structure preservation and maintenance (e.g. Monuments and historic buildings), green area maintenance (e.g. public parks and community gardens), maintenance of water reserves especially use for recreational activities (e.g. fountains and lakes), livestock farming, aquaculture, industrial application (e.g. food preservation) and homes (e.g. insect repellent) (Figure 4). Once a pesticide applied in the field, it enters into the aquatic environment and undergoes one or more transformation product depending upon their physicochemical property. From there pesticides are known to affect aquatic organisms even at trace level concentrations. The severity and duration of toxicity decided by several chemicals and physical factors, including the insecticide chemical structure. The toxicity of a particular insecticide increases with higher concentrations in tissue and its affinity for acetylcholinesterase.

Figure 2. Graphical representation of world pesticide expenditure at producer level (A) from 2008-2012; (B) in 2012
Data source: USEPA, 2016.

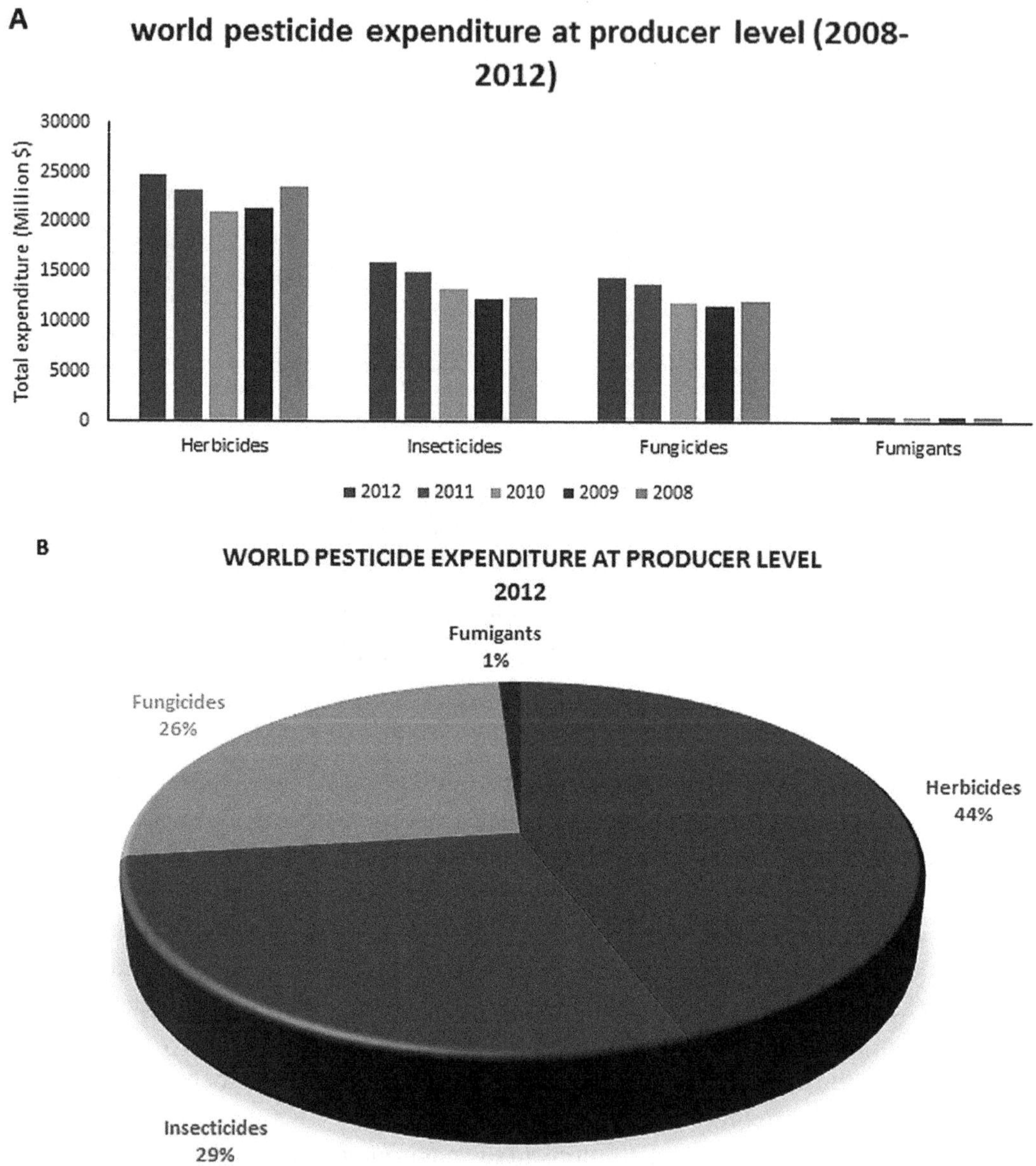

Pesticides are typically observed as mixtures in the environment and found to be associated with a range of unintended organic and inorganic substances. This causes them to affect non-target stream communities by direct toxicity and indirect toxicity. Pesticides also interfere with prey, predators, and competitors and cause broad scale effects in community structure and function. The objective of this chapter is to introduce various sources and practices through which pesticides enter into the aquatic system, its fate, interaction with different substances and ways it may impact aquatic organisms, supported by some management practices.

Figure 3. Individual user expenditure on conventional pesticides in the USA by pesticide type and market sector - 2012 estimates
Data Source, USEPA, 2016.

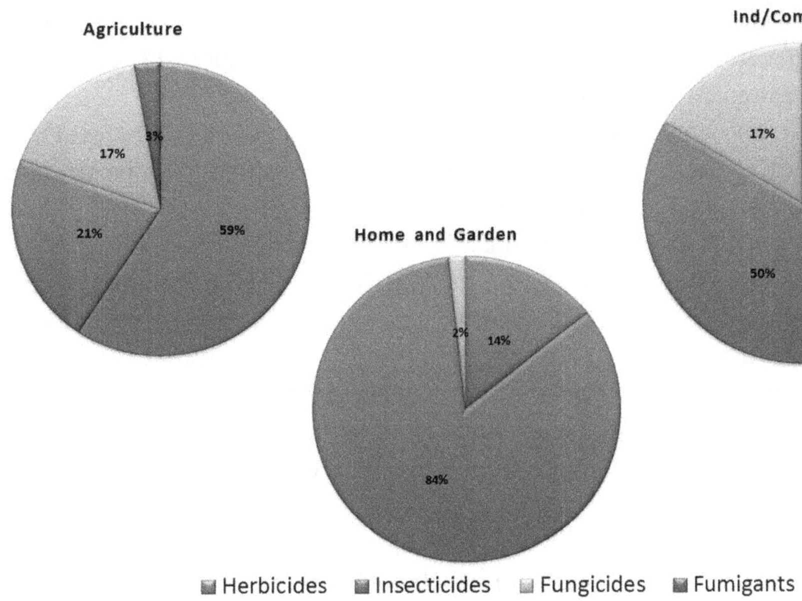

BACKGROUND

Pesticides are the second largest group of applied chemicals after fertilizers (Stokstad & Grullon, 2013). The total consumption of pesticide is about two million per year, out of which theUSA contributes about 25% and Europe about 45% (De et al., 2014). Growing urbanization and modern living style require the high use of pesticides, which in turn shares its non-targeted release in the environment through road runoff, sewer outflows, and roof runoff during high precipitation events. Similarly, overuse of pesticides in agriculture and mishandling processes are an increasing point of concern towards the non-targeted release in aquatic ecosystems (Jallow, Awadh, Albaho, Devi, & Thomas, 2017). Most of the pesticides released during agriculture and fruit packaging processes end up in surface waters (Dahlawi & Siddiqui, 2017). According to some studies in Mediterranean region (Spain and Italy), pesticides are one of the most frequently detected class of micropollutant in water (Rousis, Zuccato, & Castiglioni, 2017). Serious concerns regarding pesticide pollution can be estimated from the EU Water Framework Directive (WFD) in which out of the 41 priority substances listed, 14 are pesticides (Directive 2008/105/EC).

Major sources of pesticide into surface waters are categorized into Point and Non-Point sources. From there, they end up in rivers and nearby coastal waters due to which, recently pesticide pollution is receiving attention globally. Nowell et al (2018) found 183 pesticide compounds in 100 Midwestern US streams sampled during 2013. Similarly, a study in stormwater was conducted in 2012 and 2013 at different locations in Australia and found 19 pesticides out of 27 total monitored pesticides (Rippy et al., 2017). Out of 19 pesticides, 5 were found > 50% of samples. Another study performed in Tyrrhenian Sea (Central Mediterranean Sea), Italy in 2014-2015 on eight organophosphate pesticides at 21 sites calculated that 545.36 Kg pesticides were introduced in the Tyrrhenian Sea every year from Tibet River (Montuori et al., 2016). In China, 102 pesticides were analyzed in Jiulong River and estuary (2009) at 35 sited in different seasons. A total number of 82 pesticides were detected, out of which 14 individual pesticides were detected with above 100 ng/L concentration (Zheng et al., 2016). Such studies raise the need to introduce management practices and outreach towards the responsible application of these chemicals before they contaminate the whole aquatic system.

Being one of the major environmental reservoirs, aquatic ecosystem received recent consideration on the loss of aquatic biodiversity (freshwater) and ecosystem functioning from pesticide contamination (Beketov, Kefford, Schafer, & Liess, 2013; Malaj et al., 2014). Invertebrate community structure and biodiversity are known to get affected by pesticides in surface waters (Beketov et al., 2013; Liess & Von Der Ohe, 2005). Recently Russo, Becker, & Liess (2018), studied the low level of pesticides and temperature stress in crustacean (*Gammarus pulex*) in a macrocosm study in streams located in Germany. He found a 2.7 fold increase in sensitivity in crustacean exposed to two pyrethroid insecticide compared to clean stream water. Algae and other primary producers also getting attention due to indirect effects of pesticides on their photosystem. From primary producers like algae, pesticides enter into the food chain and potentially bioaccumulate in higher trophic organisms causing adverse health effects (Qiu, Zeng, Qiu, Yu, & Cai, 2017). This, in turn, affects organisms at the community level by altering their structure and population size.

The fate of a pesticide in an aquatic ecosystem depends upon the variety of climate factors (temperature and precipitation); soil characteristics, topography and agricultural particles (Leonard, 1990; Wauchope, 1978). The Octanal water partitioning coefficient (Log K_{ow}), hydrophilic nature and ionic properties of specific pesticide are some chemical factors to decide its persistence in an aquatic system. It has been estimated that less than 0.1% of pesticides applied to crops worldwide reach their specific target, remaining large amount of residue moves freely into different environments reservoirs (Nguyen, Hwang, Bui, & Park, 2016) and causes adverse impacts in non-targeted organisms. This effect causes a disturbance at community structure level and its functioning (Nowell et al., 2018; Ralf B. Schäfer et al., 2012; Ralf Bernhard Schäfer et al., 2007). In a mesocosm study, conducted in Germany, a pyrethroid insecticide was exposed to an invertebrate (isopod, *Asellus aquaticus*) (Wieczorek et al., 2018). At 5.3 µg/L concentration, 100% isopods were effected at the individual and population level. This reflects the adverse effect of a pesticide on local community structure and requires further studies to understand this phenomenon in the longer run.

SOURCES OF PESTICIDES

Broad spectrum nature of pesticides delineate their ways into the aquatic ecosystem intentionally and unintentionally (Figure 4). In this section, the major sources of pesticides have been classified on the basis of their initial point of entry. Point sources for pesticide contamination (PSPC) directly introduce pesticides into the aquatic ecosystem through an outfall or source such as Municipal WWTP/ sewer system, industries, livestock and aquaculture (cage, pen, and floating). In contrast, non-point sources of pesticide contamination (NSPC) introduce pesticides into aquatic systems through indirect means (e.g. wind, runoff, and leaching) such as croplands, community parks and gardens, leisure facilities, forestry, and horticulture.

Figure 4. Different point and non-point sources of pesticide contamination in the aquatic system

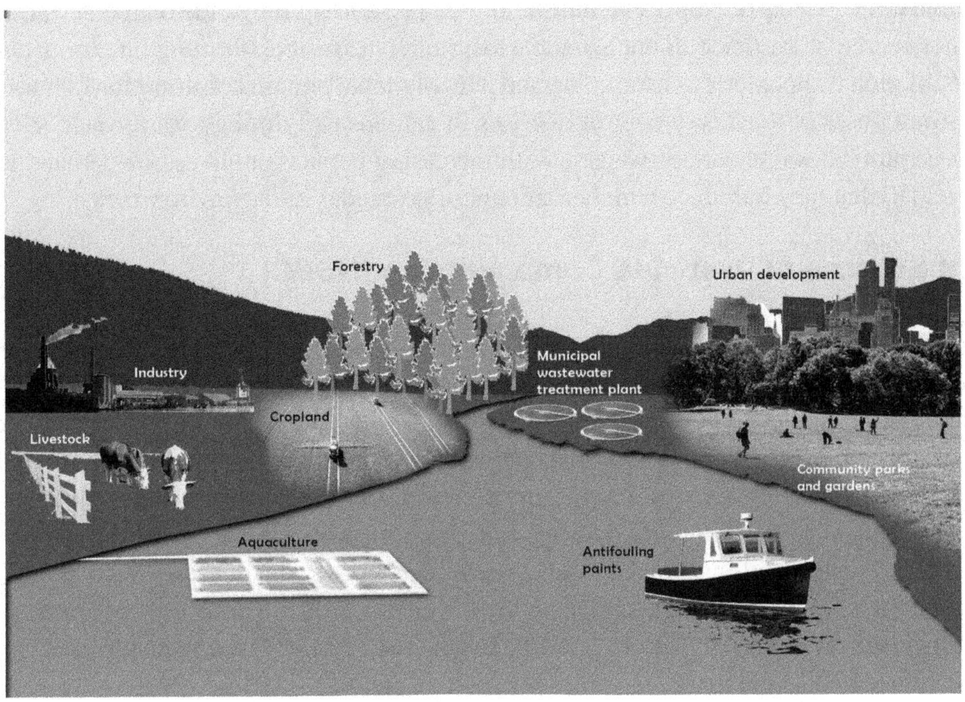

Point Sources of Pesticide Contamination (PSPC)

WWTP and sewer systems are potential sources of PSPC in aquatic systems. Increasing use of pesticides causing higher residues in fruits and vegetables, which often release into the wastewater system during household cleaning and washing. Similarly, use of insecticides inside houses and their storage can make its way into the aquatic ecosystem through sewage systems. Additionally, sources like urban gardening and household application, pesticide runoff in a sewer system and WWTP effluent (Münze et al., 2015). WWTP are specifically designed to remove particulate matters, nutrients, and pathogens rather than pesticides. This makes WWTP less efficient to break or fully degrade pesticide before leaving the WWTP facility.

Some of the food packaging industries also contribute towards pesticide concentration in aquatic systems. After harvesting, fruits and vegetables are sent to packaging plants, where they are treated with fungicides to minimize the losses due to fungal infestation. Pesticides can also use to control physiological disorders during handling and storage. During the postharvest process, these solutions make a large volume of wastewater and released into the surface water with minimum treatment (Karas et al., 2015). The global fruit production in 2013 was estimated at 676.9 million tonnes and global vegetation production was 879.2 million tonnes (Smith, 2014). This can contribute a large amount of pesticides to wastewater and eventually to the aquatic environment.

Aquaculture is one of the fastest growing fields and expected to double its production by 2030 due to marine and freshwater capture fisheries overexploitation. To meet public demand aquaculture uses pesticides for increased production. Aquaculture in ponds and ditches can collect some wastewater and release with minimal treatment in an aquatic system. However, when aquaculture practices in coastal waters, it releases contaminants directly into the aquatic system without any prior treatment, this may increase pesticide contamination. Similarly, ships with antifouling paints, add up fungicides directly into the system. Another major source of Pesticide in the aquatic ecosystem is Livestock farming due to its demand as it provides 17% of global kilocalorie consumption and 33% of global protein consumption (Thornton, 2010). Livestock farming uses pesticides to control diseases in animals and through wastewater release, it may introduce these pesticides into surface waters. With increasing population, livestock demand is projected to double by 2050 that may introduce a higher amount of pesticides in the environment.

Non-Point Sources of Pesticide Contamination (NSPC)

Agriculture is a major source of NSPC introduced in the aquatic system. Pesticides enter into surface waters through drainage, spray drift and/or surface runoff during agriculture application (Schulz, 2004). When the applied pesticide is not used by the plant due to excessive quantity, losses can occur within the field, it can be considered as *endodrift*, whereas, the losses occur even before reaching in the field or plant is considered as *exodrift*. Most of the environmental impact of agriculture use comes from applying higher than recommended procedures and dosages, improper usages, illegal use of banned pesticides, as well as improper storage, handling and disposal of pesticides. For several decades, agricultural soils were detected with a major source of persistent pesticides, which makes it a source of continuous pesticide flux to the freshwater system (Aliyeva et al., 2013; Barth et al., 2007). The upper soil layer in catchment areas can be considered as the main reservoirs for such pesticides.

Forestry is another key source of NSPC in the aquatic ecosystem. Pesticides are used in the forest not only to control insects and diseases but also to help them in regeneration through the use of herbicides. Herbicides like glyphosate are also used to remove non-commercial trees from forest areas, also known as thinning (Sullivan, Sullivan, Lindgren, & Boateng, 2002; Swinfield, Afriandi, Antoni, & Harrison, 2016). Pesticides can be applied through manual and air spray process depending upon the forest management practices. During manual applications, herbicides can leach out into the groundwater or can evaporate to the nearest surface water. Similarly, during air spray, through the drifting process, pesticides can travel to the nearest aquatic ecosystem.

National/State parks and community gardens can also be a potential NSPC in aquatic systems. Most of the time while visiting national parks, tourists and workers use various kinds of insecticides (e.g. mosquitoes, fly, and bees) during camping and leisure. According to National park service press release (2016), more than 305 million people visited USA national parks in 2015. There are some indirect ways like wind and volatilization through which national parks and gardens can get pesticides deposited over time and ultimately release them to the surface waters (Elliott & VanderMeulen, 2017). NPSPC includes a wide range of sources and it needs regulations at the initial point of release to control over time.

DIFFERENT WAYS OF PESTICIDES IN THE AQUATIC ECOSYSTEM

Pesticides may behave differently when they come in contact with other materials/substances in the aquatic environment. There is awide range of materials available in the aquatic environment from inorganic to organic substances. Some of these substances are available naturally into the system, whereas others are being introduced by humans in the form of pollutants (e.g. other pesticides). A brief interaction of pesticides with different substances is described in this section, with relevant examples.

Pesticide Toxicity in Combination With Metals

Pesticides, along with other pollutants like metals can be more toxic than the individual compounds. Chemical nature of pesticides like ionic form and chelating behavior are some important factors in determining its interaction with available metals and resulting action in the aquatic environment. The major effect observed by various studies so far is the adsorption capacity of herbicide in sediment as a result of the metal interaction. Ionic pesticides like Glyphosate (GPS), when interacting with Cu^{2+}, it forms GPS-Cu complexes. These complexes when found a suitable site of adsorption with decreasing pH, results in enhanced GPS adsorption in soil (Morillo et al., 1997, 2000). Some studies also reported Cu^{2+} could replace herbicides from sorption sites within the soil (Pateiro-Moure et al., 2007). Similarly, on interaction with Cd^{2+} GPS demonstrated improved adsorption in soil (Zhou et al., 2004). In contrast, when anionic herbicides like Mefenacet interact with Cu^{2+} demonstrated significantly decreased sorption in the soil, whereas, increased sorption when interacted with Ag^{+}(Liu, Guo, He, & Sun, 2012).

One such investigation was performed by Ghazy, Abdel-Razek, El Nahas, & Mahmoud, (2017) at the Egyptian Northern lake (Lake Bursulus) to assess the impact of heavy metals (Fe, Mn, Zn, Cu, Ni) and pesticide (pyrethroid) on tissue accumulation, mRNA expression levels at Metallothionein, some immune response related genes (IgM) and inflammatory cytokines (*TNF, IL.8 and IL.1*) in fish, Nile tilapia (*Oreochromisniloticus*). The fish tissue concentration for both the metals and pesticide were higher in concentration (3-42 folds) compared to the lake water sample concentration. Additionally, gene expression was also affected due to the pollutants stress in fish.

Pesticide Toxicity in Combination With Humic Substances

Once a pesticide enters into the aquatic environment, it undergoes different behaviors depending on the interaction medium. Some of these interactions within sediments result in absorption/adsorption of chemical substances. Sediments are made up of organic (humic substances) and inorganic (sand/silt) materials. Humic acids are a principal component of humic substances, produced by biodegradation of dead organic matter. Humic substances are the major organic constituents of soil (humus), peat and coal. Humic acids can be found in upland stream sediments, dystrophic lakes, and ocean waters. Pesticides have strong affinity to bind with humic substances and organic matter. Both the pesticides and humic acid chemical nature are important in determining their strength of interaction which decides pesticide transport, bioavailability, and persistence in the aquatic environment. Some environmental factors like pH are important in supporting their binding strength like the sorption capacity of anionic herbicide generally decreases with increasing pH (Liu et al., 2012).

Humic acid content is reported to affect the adsorption capacity of pesticides in soil (Zhou, Wang, Cang, Hao, & Luo, 2004). Depending upon their strength, the overall toxicity of a pesticide is determined. When pesticides bind to organic matter in aqueous solution, it was observed that the overall toxicity of a particular pesticide generally decreases with higher humic substance. As observed in abacterium, *Vibrio fischeria*, the toxicity of pesticide (chlorpyrifos) was reduced by 4.4-100% after addition of compost humic substance at a concentration of 2.1 - 42 mg C/l (Jones & Huang, 2003). This suggests that pesticides decreased their toxicity in the presence of humic acid. This requires further research with organic substances and other aquatic organisms, which would be helpful in defining policies and management practices in the future.

Pesticides Toxicity as Mixture Effects

Other than individual pesticide occurrence, a mixture or cocktail effects have been found in surface waters of the USA, Europe, Australia and China (Allinson et al., 2015; Ccanccapa, Masiá, Andreu, & Picó, 2016; Gilliom, 2001; Hunt et al., 2016; Zhang, Jiang, & Ou, 2011). Their combined impacts on aquatic organisms are far more toxic than the individual organism as investigated by various scientists (Knauert, Dawo, Hollender, Hommen, & Knauer, 2009). This joint toxicity of pesticide mixtures demonstrates the concentration addition (CA), that's defined as the "all components contribute to the overall mixture toxicity, independently of whether they are present at concentration above or below their individual No Observation Effect Concentration (NOEC)or Environmental Quality Standard (EQS)" (Rodney, Teed, & Moore, 2013). CA is a recommended approach for setting environmental quality assessment within the EU water framework directive to assess the risk assessment of pesticide mixture (European Commission, 2016).

In terms of mixture toxicity, only a few compounds contribute substantially to the overall mixture risks. It is observed in the river Meolo, Italy, where just only one or few compounds were responsible for more than 80% of CA estimated mixture toxicity (Verro, Finizio, Otto, & Vighi, 2009). Also in the USA during water quality monitoring of National water-quality assessment (NAWQA) program of US geological survey (USGS)concluded the environmental risk was driven by just one compound after analyzing more than 90% of samples (Vallotton & Price, 2016). Similarly, during ecotoxicology of pesticide mixtures detected in a Swedish monitoring program only a few compounds were often found to contribute towards higher toxicity (Gustavsson et al., 2017). Furthermore, Mansano et al. (2017), discussed the increased toxicity caused by a mixture of carbofuran and diuron compared to an individual compound in microalgae *Raphidoceli subcapitata*, after fitting CA and Independent Action (IA) model. Toxicity was even higher in algae with adominant chemical in the mixture was diuron.

Pesticide Toxicity and Global Warming

Global warming as an outcome of climate change and chemical input in the natural ecosystem are two major threats faced by environment today, especially in the aquatic ecosystem (Bronmark & Hansson, 2002; MEA, 2005). Both these threats have a potential to interact and alter their behavior at the ecosystem level. As studied, warming can alter the toxicity of a pollutant (Noyes & Lema, 2015). Climate-induced toxicant sensitivity is a phenomenon, where pollutants can impair the organism's ability to cope with increased temperature (Hooper et al., 2013). Such complex interaction can limit in understanding and predicting the effects of global warming and toxicant exposure with changing climate scenarios. Furthermore, the response of such interaction can have a strong influence from a geographical location due to which population of same species living in different thermal conditions can behave differently (Dinh Van et al., 2013).

It is not necessary that the increase in temperature always increases the negative impact of a chemical. In case of some chemicals like pesticides, increase in temperature decreases the exposure of pollutants to the organisms because of a higher state of degradation in the medium (Hooper et al., 2013). However, with global climate change, there are different impacts in terms of increasing ocean acidification, increasing precipitation in some places and various other extreme climate conditions. If precipitation increases that means runoff through croplands and forestry may also increase, that may bring more pesticide into the aquatic ecosystem. Also, at some places in the event of high precipitation sewer outflows directly falls into the aquatic system without any treatment, which is another threat to the ecosystem. There is further research required to understand the ongoing changes and impacts of climate change on pesticide inflow and effects on aquatic organisms.

THE FATE OF PESTICIDES IN THE AQUATIC ECOSYSTEM

The fate of pesticides in aquatic ecosystems is defined through three major processes; Transportation of pesticides, Transfer of pesticides into aquatic systems from source and Transformation (Figure 5). The process of transportation involves pesticide movement from the source through physical (e.g. wind, precipitation) and biological (e.g. insects) ways. Depending upon the quantity and chemical properties of pesticides, their travel distance and destination is determined. Pesticide loads in running water depend upon the amount of pesticides applied to the field; the timing and intensity of rainfall; the soil hydrology at the catchment level (Münze et al., 2015). During the transfer, pesticides can dissolve, degrade (photolysis and hydrolysis), adsorbed and absorbed into the aquatic ecosystem. The process of transformation includes the interaction of pesticides with available dissolved organic matter (DOM), humic substances and other pollutants present in the water. The ultimate fate of pesticides in an aquatic ecosystem is decided by the physical interaction of climate and environmental location with physical, chemical and biological properties of a pesticide.

Pesticides can travel from various sources and deposit into the aquatic systems. Transportation can be carried out either from apoint source of contamination as described earlier in this chapter. Pesticides can travel through various physical processes like diffusion, drifting, volatilization, and wind/water runoff. During an event of precipitation, pesticides can travel with rainfall as well as through excessive

Figure 5. Schematic diagram of pesticide fate in an aquatic environment defined by WWTP as a major of pollution and different ways pesticide may travel, transfer and transform in the aquatic environment

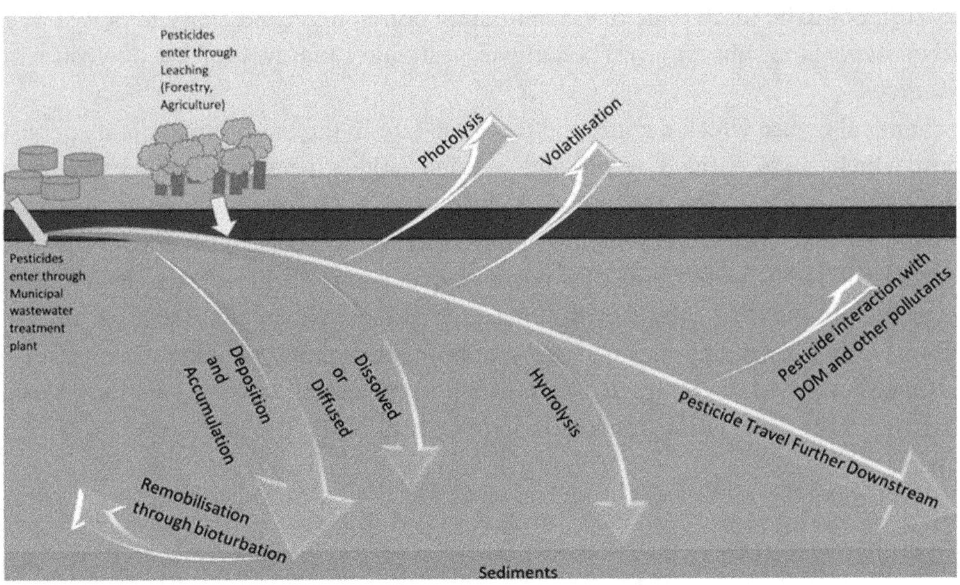

water runoff in the form of stormwater. Streams receiving waters from stormflow observed to have high pesticide concentrations in sediments. Resuspension of contaminated sediments present in agriculture soil during storm flow can be another way to reintroduce suspended pesticides in the aquatic ecosystem. During resuspension, contaminated sediments may alter the partitioning between particle-bound and dissolved phase of pesticides, which becomes readily available for aquatic biota during storm flow conditions. Pesticides can also deposited through a carrier like insects and flies. Some biological carriers can carry pesticides in their body parts like wings, leg, and antennae and can deposit them into the aquatic ecosystem during predation and pollination process. As observed recently in Alaska, some of the pesticides never used in those regions are found in different places and animal tissues (Myers et al., 2008; Weber et al., 2010).

Once Pesticides enter into the aquatic system, they can transfer into the sediments, remain suspended in the water column or bioconcentrate in the organism. Pesticides have a volatile nature and transferred into the atmosphere soon after entering into the aquatic system. Pesticides can adsorb or absorb in sediments and other submerged surfaces like macrophytes (Stehle et al., 2011). The particles absorbed in sediments can remobilize during high discharge events and feeding activity of sediment-dwelling organisms (bioturbation) (Roberts, 2012). Polychaetes and amphipods have been observed to bioturbate and remobilize polychlorinated biphenyls and flame-retardants in aquatic systems (Hedman, Bradshaw, Thorsson, Gilek, & Gunnarsson, 2008; Hedman, Tocca & Gunnarsson, 2009; Josefsson, Leonardsson, Gunnarsson, & Wiberg, 2010).

In the process of transformation, pesticides can react with different substances available in the aquatic system. During this process, the transformed pesticide may behave differently than original pesticides, which is discussed in detail earlier in this chapter. Pesticides can also undergo degradation through hydrolysis and photolysis process. Hydrolysis is a hydrogen molecule-mediated process, which can break down pesticide molecules, whereas, in Photolysis, UV radiations can mediate in their degradation process. The chemical structure and nature of pesticide decides the transformation process and ultimately products in the aquatic environment. Some pesticides are very reactive in nature and undergo a chemical reaction with different organic matter and pollutants present in the water. Pesticides can react with DOM, metals, other pesticides and organic debris to transform into new products. During transformation, products may turn out to be more toxic than their parent compounds and cause toxicity issue in aquatic biota (described in next the subsection). The leftover pesticide can travel further downstream to deposit far from its sources.

Another chemical nature which decides the pesticide fate in the environment is its persistence in an aquatic system, which can be defined by its Log K_{ow} value and persistence nature. Persistent organic pollutants (POP), have an ability to bioaccumulate in biota with a strong resistance to biodegrade through environmental factors like chemical, biological and photolysis. Some OC pesticides come under the POP category are DDT and DDE. Some of the POP pesticides such as hexachlorobenzene (a type of OC) were banned 50 years ago in the USA, however, still detected in the environment (Jones & de Voogt, 1999). The problem of POPs is receiving international attention since the 1960s due to its effects on aquatic organisms at trace levels (Malaj et al., 2014). Various international conventions like Aarhus protocol (UNECE, 2010) and the Stockholm Convention on POPs (UNEP, 2009) are working on its control in the environment.

IMPACTS ON AQUATIC ORGANISMS

Pesticides have a broad range of applications at various environmental levels from agricultural to urban areas. These pesticides can miss their target resulting in high toxicity to aquatic species, out of which crustaceans, insects, and fishes are most sensitive taxa (Wijngaarden, Brock, & Brink, 2005). There are various different ways pesticides can cause toxicity in aquatic organisms depending upon its chemical nature and different physical environmental conditions. Aquatic organisms can come in contact with pesticide directly or indirectly, depending upon the fate of pesticide in the aquatic environment. Pesticide can impact organism interaction from individual to the ecosystem level, including its ability to interact with other organisms in the system like sensitivity to avoid predators and symbiosis behavior. It may enter into the organism's body through ingestion, respiration, and absorption. This, in turn, can cause issues ranges from physiological to high chronic level in aquatic biota.

The toxicity of a pesticide in an aquatic organism is decided by specific chemical and physical property of a pesticide such as persistence in water, stability to degradation and bioaccumulation potential. Pesticides can either bioaccumulate in organisms or excrete off depending upon its metabolism. They can also cause neural issues, sex alterations, genetic mutations and carcinogenic problems in aquatic organisms. Other physiological disorders include histo-architectural changes in liver, kidney, heart and other body parts. Pesticides can act as endocrine disruptors that can act through the receptor medicated response. In some extreme cases, pesticides have been observed to cause increased mortality in microorganisms (Song, Chen, & Zhou, 2017). This is a serious concern and requires further studies to address such issue effectively.

Pesticides may affect an organism through alteration in social responses/ interactions or physiological disorders. During social imbalances, organisms may delay in response to prey due to slow stimulus responses. This makes organisms prone to the prey and ultimately limit their population abundance. Other social issues observed in aquatic organisms include, change in population pattern through food chain alterations, high susceptibility to predators due to lack of sensitivity and habitat loss. Once pesticide contamination initiates, as a first response, organisms move away from the sites. It may result in higher exposure to predators, habitat loss, and reproductive problems. Alternatively, sessile organisms like mussels and oysters, which can't move easily from the site of contamination, can start losing their habitat and becomes extinct over time. These organisms are filter feeders and some species are highly commercial. Due to their feeding habit, they accumulate high amounts of pesticides in their tissues and can pass it on to higher level organisms feed on them including humans. As studied in mussels (*Mytilus galloprovincialis*) at Kastela Bay, Croatia from 2000-2011, for chlorinated pesticides. The mussels from the clean sea were transplanted into the Kastela Bay and sampled bimonthly for the targeted contaminant. The authors reported two times increased DDT concentration after 2 months of exposure (Milun, Grgas, & Dragičević, 2016).

Another effect often observed with herbicides are interference with a primary producer (e.g. algae and phytoplankton) growth by affecting their photosystem. This can alter food chain as well contribute global climate change, due to the fact that, half of the world oxygen is produced by phytoplankton and algae. Additionally, these algae live in a symbiotic relationship with corals to provide them food and receive shelter, for example, algae living inside sea anemone, *Exaiptasia pallida* (Siddiqui, Goddard, & Bielmyer-Fraser, 2015). In case of pesticide contamination, these organisms die off or leave host organisms (bleaching), which may cause stress and other health issues in the host organism. This is increasing issue observed in Great Barrier Reef, Australia due to sugar cane pesticide contamination on the nearby coast (Lewis et al., 2009).

RECOMMENDED MANAGEMENT PRACTICES AND FUTURE RESEARCH DIRECTION

Pesticide monitoring in an aquatic ecosystem is very complex due to their physicochemical properties, chemical structure and persistence in the environment. Some of the management practices like early monitoring help identifying the pesticide before higher levels reach to aquatic biota. Biomarkers (oxidative stress) can be used as "Early warning" signs by providing data on the potential adverse impact on other aquatic organisms (Booth, Bithell, Wratten, & Heppelthwaite, 2003). Technology advances introduce new detection techniques to map chemical pollutants in the environment. Different spatiotemporal software like GIS can also be used to manage and mitigate such issues.

Pesticides can cause health issues in farmers and increased application can introduce resistance issues in pests. As observed in the USA, there was a 10-fold increase in pesticide use after 1945, even after that, the crop damage due to insect-pest resistance from 1945-2000, doubled from 7-13% (Pimentel, 2005). Improper and excessive application of pesticides is an emerging issue and has a link with lack of education and training in pesticide use; lack of pesticide alternatives; inadequate information on related hazards; aesthetic requirement for crop appearance and unwillingness in farmers towards the risk of crop loss. Education to farmers about environmental and human health is important to mitigate this issue. Strong law enforcement is another highly recommended step to prevent overuse of pesticides. There is an emerging requirement for access to extensions support for farmers to assist them in pesticide dosage and application.

Similarly, application dosage and proper application in the household needs attention. People store pesticides in improper ways and dispose of it openly, which may cause high input in storm waters and WWTPs. A pesticide may also leak out from paints and garden application. This requires proper attention and education to the common public. To prevent pesticide discharge through WWTP, higher wastewater treatment processes like ozone and advanced oxidation process should be implemented. There is a lack of high-quality inventory data on pesticide application and emission. Large level studies are required to identify the overall ecological impact of these pesticides, as many invertebrates undergo masking effect and it is difficult to study them to identify the health status of an aquatic ecosystem. The environmental monitoring due to pesticide contamination, also required to understand its pathways and fate in the aquatic environment.

CONCLUSION

Since after Industrial revolution, scientists identified nitrogen as a fertilizer for growing population food security, various other chemicals also introduced into the environment to save crops from pests and insects. The ease of use and availability of common peoples' usage, ultimately bring them into the houses for rodents and other household insects. This makes agriculture and WWTP, major sources of pesticide contamination in the environment. Although with increasing use of pesticide application, scientists believe that safe operating practices of chemical pollution are now transgressed (Diamond et al., 2015), which needs attention to at least stabilize this process. Most of the time pesticides entering into the aquatic ecosystem are non-targeted and cause adverse impacts on biodiversity. The chemical nature of pesticides decides its fate into the aquatic environment, various other organic and inorganic materials may increase or decrease pesticide toxicity in aquatic organisms. Pesticides may cause a wide range of issues in aquatic biota and require further studies and attention towards the responsible handling and application. Some of the management practices and future research also discussed in this chapter that can help in mitigating this problem to a certain extent.

REFERENCES

Aliyeva, G., Halsall, C., Alasgarova, K., Avazova, M., Ibrahimov, Y., & Aghayeva, R. (2013). The legacy of persistent organic pollutants in Azerbaijan: An assessment of past use and current contamination. *Environmental Science and Pollution Research International*, *20*(4), 1993–2008. doi:10.100711356-012-1076-9 PMID:22825638

Allinson, G., Zhang, P., Bui, A. D., Allinson, M., Rose, G., Marshall, S., & Pettigrove, V. (2015). Pesticide and trace metal occurrence and aquatic benchmark exceedances in surface waters and sediments of urban wetlands and retention ponds in Melbourne, Australia. *Environmental Science and Pollution Research International*, *22*(13), 10214–10226. doi:10.100711356-015-4206-3 PMID:25697552

Barth, J. A. C., Steidle, D., Kuntz, D., Gocht, T., Mouvet, C., von Tümpling, W., & Grathwohl, P. (2007). Deposition, persistence and turnover of pollutants: First results from the EU project AquaTerra for selected river basins and aquifers. *The Science of the Total Environment*, *376*(1–3), 40–50. doi:10.1016/j.scitotenv.2007.01.065 PMID:17307233

Beketov, M. A., Kefford, B. J., Schafer, R. B., & Liess, M. (2013). Pesticides reduce regional biodiversity of stream invertebrates. *Proceedings of the National Academy of Sciences of the United States of America*, *110*(27), 11039–11043. doi:10.1073/pnas.1305618110 PMID:23776226

Booth, L. H., Bithell, S. L., Wratten, S. D., & Heppelthwaite, V. J. (2003). Vineyard Pesticides and Their Effects on Invertebrate Biomarkers and Bioindicator Species in New Zealand. *Bulletin of Environmental Contamination and Toxicology*, *71*(6), 1131–1138. doi:10.100700128-003-8879-9 PMID:14756280

Brönmark, C., & Hansson, L.-A. (2002). Environmental issues in lakes and ponds: Current state and perspectives. *Environmental Conservation*, *29*(3), 290–306. doi:10.1017/S0376892902000218

Ccanccapa, A., Masiá, A., Andreu, V., & Picó, Y. (2016). Spatio-temporal patterns of pesticide residues in the Turia and Júcar Rivers (Spain). *The Science of the Total Environment*, *540*, 200–210. doi:10.1016/j.scitotenv.2015.06.063 PMID:26118860

Commission, E. (2016). *Impact Assessment Report on Criteria to identify EDs*. Academic Press.

Dahlawi, S. M., & Siddiqui, S. (2017). Calcium polysulphide, its applications and emerging risk of environmental pollution—a review article. *Environmental Science and Pollution Research International*, *24*(1), 92–102. doi:10.100711356-016-7842-3 PMID:27761854

De, A., Bose, R., Kumar, A., & Mozumdar, S. (2014). Worldwide Pesticide Use. In Targeted Delivery of Pesticides Using Biodegradable Polymeric Nanoparticles, Springer Briefs in Molecular Science. Springer. doi:10.1007/978-81-322-1689-6_2

Diamond, M. L., de Wit, C. A., Molander, S., Scheringer, M., Backhaus, T., Lohmann, R., ... Zetzsch, C. (2015). Exploring the planetary boundary for chemical pollution. *Environment International*, *78*, 8–15. doi:10.1016/j.envint.2015.02.001 PMID:25679962

Dinh Van, K., Janssens, L., Debecker, S., De Jonge, M., Lambret, P., Nilsson-Örtman, V., ... Stoks, R. (2013). Susceptibility to a metal under global warming is shaped by thermal adaptation along a latitudinal gradient. *Global Change Biology*, *19*(9), 2625–2633. doi:10.1111/gcb.12243 PMID:23640735

Elliott, S. M., & VanderMeulen, D. D. (2017). A regional assessment of chemicals of concern in surface waters of four Midwestern United States national parks. *The Science of the Total Environment*, *579*, 1726–1735. doi:10.1016/j.scitotenv.2016.11.114 PMID:27932214

Ghazy, H. A., Abdel-Razek, M. A. S., El Nahas, A. F., & Mahmoud, S. (2017). Assessment of complex water pollution with heavy metals and Pyrethroid pesticides on transcript levels of metallothionein and immune related genes. *Fish & Shellfish Immunology*, *68*, 318–326. doi:10.1016/j.fsi.2017.07.034 PMID:28734967

Gilliom, R. J. (2001). Pesticides in the hydrologic system - What do we know and what's next? *Hydrological Processes*, *15*(16), 3197–3201. doi:10.1002/hyp.501

Gustavsson, M., Kreuger, J., Bundschuh, M., & Backhaus, T. (2017). Pesticide mixtures in the Swedish streams: Environmental risks, contributions of individual compounds and consequences of single-substance oriented risk mitigation. *The Science of the Total Environment*, *598*, 973–983. doi:10.1016/j.scitotenv.2017.04.122 PMID:28468125

Hedman, J. E., Bradshaw, C., Thorsson, M. H., Gilek, M., & Gunnarsson, J. S. (2008). Fate of contaminants in Baltic Sea sediments: Role of bioturbation and settling organic matter. *Marine Ecology Progress Series*, *356*, 25–38. doi:10.3354/meps07218

Hooper, M. J., Ankley, G. T., Cristol, D. A., Maryoung, L. A., Noyes, P. D., & Pinkerton, K. E. (2013). Interactions between chemical and climate stressors: A role for mechanistic toxicology in assessing climate change risks. *Environmental Toxicology and Chemistry*, *32*(1), 32–48. doi:10.1002/etc.2043 PMID:23136056

Hunt, L., Bonetto, C., Resh, V. H., Buss, D. F., Fanelli, S., Marrochi, N., & Lydy, M. J. (2016). Insecticide concentrations in stream sediments of soy production regions of South America. *The Science of the Total Environment*, *547*, 114–124. doi:10.1016/j.scitotenv.2015.12.140 PMID:26780136

Jallow, M. F. A., Awadh, D. G., Albaho, M. S., Devi, V. Y., & Thomas, B. M. (2017). Pesticide risk behaviors and factors influencing pesticide use among farmers in Kuwait. *The Science of the Total Environment*, *574*, 490–498. doi:10.1016/j.scitotenv.2016.09.085 PMID:27644027

Jones, K. C., & de Voogt, P. (1999). Persistent organic pollutants (POPs): State of the science. *Environmental Pollution*, *100*(1), 209–221. doi:10.1016/S0269-7491(99)00098-6 PMID:15093119

Josefsson, S., Leonardsson, K., Gunnarsson, J. S., & Wiberg, K. (2010). Bioturbation-Driven Release of Buried PCBs and PBDEs from Different Depths in Contaminated Sediments. *Environmental Science & Technology*, *44*(19), 7456–7464. doi:10.1021/es100615g PMID:20831254

Karas, P., Metsoviti, A., Zisis, V., Ehaliotis, C., Omirou, M., Papadopoulou, E. S., ... Karpouzas, D. G. (2015). Dissipation, metabolism and sorption of pesticides used in fruit-packaging plants: Towards an optimized depuration of their pesticide-contaminated agro-industrial effluents. *The Science of the Total Environment, 530–531*, 129–139. doi:10.1016/j.scitotenv.2015.05.086 PMID:26042894

Knauert, S., Dawo, U., Hollender, J., Hommen, U., & Knauer, K. (2009). Effects of photosystem ii inhibitors and their mixture on freshwater phytoplankton succession in outdoor mesocosms. *Environmental Toxicology and Chemistry, 28*(4), 836–845. doi:10.1897/08-135R.1 PMID:19391680

Leonard, R. A. (1990). Movement of pesticides into surface waters. In H. H. Cheng (Ed.), *Pesticides in the Soil Environment: Processes, Impacts and Modelling* (pp. 303–349). Maddison, WI: Soil Science Society of America.

Lewis, S. E., Brodie, J. E., Bainbridge, Z. T., Rohde, K. W., Davis, A. M., Masters, B. L., ... Schaffelke, B. (2009). Herbicides: A new threat to the Great Barrier Reef. *Environmental Pollution, 157*(8), 2470–2484. doi:10.1016/j.envpol.2009.03.006 PMID:19349104

Liess, M., & Von Der Ohe, P. C. (2005). Analyzing effects of pesticides on invertebrate communities in streams. *Environmental Toxicology and Chemistry, 24*(4), 954–965. doi:10.1897/03-652.1 PMID:15839571

Liu, Z., Guo, H., He, H., & Sun, C. (2012). Sorption and cosorption of the nonionic herbicide mefenacet and heavy metals on soil and its components. *Journal of Environmental Sciences (China), 24*(3), 427–434. doi:10.1016/S1001-0742(11)60778-X PMID:22655355

Malaj, E., von der Ohe, P. C., Grote, M., Kühne, R., Mondy, C. P., Usseglio-Polatera, P. ... Schäfer, R. B. (2014). Organic chemicals jeopardize the health of freshwater ecosystems on the continental scale. *Proceedings of the National Academy of Sciences, 111*(26), 9549-9554. Retrieved from http://www.pnas.org/content/111/26/9549.abstract

Mansano, A. S., Moreira, R. A., Dornfeld, H. C., Freitas, E. C., Vieira, E. M., Sarmento, H., ... Seleghim, M. H. R. (2017). Effects of diuron and carbofuran and their mixtures on the microalgae Raphidocelis subcapitata. *Ecotoxicology and Environmental Safety, 142*, 312–321. doi:10.1016/j.ecoenv.2017.04.024 PMID:28433596

Millenium Ecosystem Assessment (MEA). (2005). *Ecosystems and Human Well-being: Biodiversity Synthesis*. Washington, DC: World Resources Institute.

Milun, V., Grgas, D., & Dragičević, T. L. (2016). Assessment of PCB and chlorinated pesticide accumulation in mussels at Kaštela Bay (Eastern Adriatic). *The Science of the Total Environment, 562*, 115–127. doi:10.1016/j.scitotenv.2016.03.133 PMID:27096632

Montuori, P., Aurino, S., Garzonio, F., Sarnacchiaro, P., Polichetti, S., Nardone, A., & Triassi, M. (2016). Estimates of Tiber River organophosphate pesticide loads to the Tyrrhenian Sea and ecological risk. *The Science of the Total Environment, 559*, 218–231. doi:10.1016/j.scitotenv.2016.03.156 PMID:27065443

Morillo, E., Undabeytia, T., & Maqueda, C. (1997). Adsorption of glyphosate on the clay mineral montmorillonite: Effect of Cu^{2+} in solution and adsorbed on the mineral. *Environmental Science & Technology, 31*(12), 3588–3592. doi:10.1021/es9703411

Morillo, E., Undabeytia, T., Maqueda, C., & Ramos, A. (2000). Glyphosate adsorption on soils of different characteristics. Influence of copper addition. *Chemosphere, 40*(1), 103–107. doi:10.1016/S0045-6535(99)00255-6 PMID:10665451

Münze, R., Orlinskiy, P., Gunold, R., Paschke, A., Kaske, O., Beketov, M. A., ... Liess, M. (2015). Pesticide impact on aquatic invertebrates identified with Chemcatcher® passive samplers and the SPEARpesticides index. *The Science of the Total Environment, 537*, 69–80. doi:10.1016/j.scitotenv.2015.07.012 PMID:26282741

Myers, M. J., Ylitalo, G. M., Krahn, M. M., Boyd, D., Calkins, D., Burkanov, V., & Atkinson, S. (2008). Organochlorine contaminants in endangered Steller sea lion pups (Eumetopias jubatus) from western Alaska and the Russian Far East. *The Science of the Total Environment, 396*(1), 60–69. doi:10.1016/j.scitotenv.2008.02.008 PMID:18384839

Nguyen, M.-H., Nguyen, T.-H.-N., Hwang, I.-C., Bui, C.-B., & Park, H.-J. (2016). Effects of the physical state of nanocarriers on their penetration into the root and upward transportation to the stem of soybean plants using confocal laser scanning microscopy. *Crop Protection (Guildford, Surrey), 87*, 25–30. doi:10.1016/j.cropro.2016.04.012

Nowell, L. H., Moran, P. W., Schmidt, T. S., Norman, J. E., Nakagaki, N., Shoda, M. E., ... Hladik, M. L. (2018). Complex mixtures of dissolved pesticides show potential aquatic toxicity in a synoptic study of Midwestern U.S. streams. *The Science of the Total Environment, 613–614*, 1469–1488. doi:10.1016/j.scitotenv.2017.06.156 PMID:28802893

Noyes, P. D., & Lema, S. C. (2015). Forecasting the impacts of chemical pollution and climate change interactions on the health of wildlife. *Current Zoology, 61*(4), 669–689.

Pateiro-Moure, M., Perez-Novo, C., Arias-Estevez, M., Lopez-Periago, E., Martinez-Carballo, E., & Simal-Gandara, J. (2007). Influence of copper on the adsorption and desorption of paraquat, diquat, and difenzoquat in vineyard acid soils. *Journal of Agricultural and Food Chemistry, 55*(15), 6219–6226.

Pimentel, D. (2005). Environmental and Economic Costs of the Application of Pesticides Primarily in the United States. *Environment, Development and Sustainability, 7*(2), 229–252. doi:10.100710668-005-7314-2

Rippy, M. A., Deletic, A., Black, J., Aryal, R., Lampard, J. L., Tang, J. Y. M., ... Gernjak, W. (2017). Pesticide occurrence and spatio-temporal variability in urban run-off across Australia. *Water Research, 115*, 245–255. doi:10.1016/j.watres.2017.03.010 PMID:28284091

Roberts, D. A. (2012). Causes and ecological effects of resuspended contaminated sediments (RCS) in marine environments. *Environment International, 40*, 230–243. doi:10.1016/j.envint.2011.11.013 PMID:22244126

Rodney, S. I., Teed, R. S., & Moore, D. R. J. (2013). Estimating the Toxicity of Pesticide Mixtures to Aquatic Organisms: A Review. *Human and Ecological Risk Assessment, 19*(6), 1557–1575. doi:10.1080/10807039.2012.723180

Rousis, N. I., Zuccato, E., & Castiglioni, S. (2017). Wastewater-based epidemiology to assess human exposure to pyrethroid pesticides. *Environment International, 99*, 213–220. doi:10.1016/j.envint.2016.11.020 PMID:27908456

Russo, R., Becker, J. M., & Liess, M. (2018). Sequential exposure to low levels of pesticides and temperature stress increase toxicological sensitivity of crustaceans. *The Science of the Total Environment, 610–611*, 563–569. doi:10.1016/j.scitotenv.2017.08.073 PMID:28822923

Schäfer, R. B., Caquet, T., Siimes, K., Mueller, R., Lagadic, L., & Liess, M. (2007). Effects of pesticides on community structure and ecosystem functions in agricultural streams of three biogeographical regions in Europe. *The Science of the Total Environment, 382*(2–3), 272–285. doi:10.1016/j.scitotenv.2007.04.040 PMID:17555800

Schäfer, R. B., von der Ohe, P. C., Rasmussen, J., Kefford, B. J., Beketov, M. A., Schulz, R., & Liess, M. (2007). Thresholds for the Effects of Pesticides on Invertebrate Communities and Leaf Breakdown in Stream Ecosystems. *Environmental Science & Technology, 46*(9), 5134–5142. doi:10.1021/es2039882 PMID:22455566

Schulz, R. (2004). Field studies on exposure, effects, and risk mitigation of aquatic nonpoint-source insecticide pollution: A review. *Journal of Environmental Quality, 33*(2), 419–448. doi:10.2134/jeq2004.4190 PMID:15074794

Siddiqui, S., Goddard, R. H., & Bielmyer-Fraser, G. K. (2015). Comparative effects of dissolved copper and copper oxide nanoparticle exposure to the sea anemone, Exaiptasia pallida. *Aquatic Toxicology (Amsterdam, Netherlands), 160*, 205–213. doi:10.1016/j.aquatox.2015.01.007 PMID:25661886

Smith, S. (2014). Global Horticulture (2010 - 2014). *Pink and Healthy, 24.*

Song, Y., Chen, M., & Zhou, J. (2017). Effects of three pesticides on superoxide dismutase and glutathione-S-transferase activities and reproduction of Daphnia magna. *Archives of Environmental Protection, 43*(1), 80–86. doi:10.1515/aep-2017-0010

Stehle, S., Elsaesser, D., Gregoire, C., Imfeld, G., Niehaus, E., Passeport, E., ... Schulz, R. (2011). Pesticide risk mitigation by vegetated treatment systems: A meta-analysis. *Journal of Environmental Quality, 40*(4), 1068–1080. doi:10.2134/jeq2010.0510 PMID:21712575

Stokstad, E., & Grullon, G. (2013). Infographic: Pesticide planet. A global look at the uses, benefits, and drawbacks of pesticides. *Science, 341*, 730–731. PMID:23950524

Sullivan, T. P., Sullivan, D. S., Lindgren, P. M. F., & Boateng, J. O. (2002). Influence of conventional and chemical thinning on stand structure and diversity of plant and mammal communities in young lodgepole pine forest. *Forest Ecology and Management, 170*(1), 173–187. doi:10.1016/S0378-1127(01)00775-7

Swinfield, T., Afriandi, R., Antoni, F., & Harrison, R. D. (2016). Accelerating tropical forest restoration through the selective removal of pioneer species. *Forest Ecology and Management, 381*, 209–216. doi:10.1016/j.foreco.2016.09.020

Thornton, P. K. (2010). Livestock production: Recent trends, future prospects. *Philosophical Transactions of the Royal Society of London. Series B, Biological Sciences*, *365*(1554), 2853–2867. doi:10.1098/rstb.2010.0134 PMID:20713389

UNECE. (2010). Hemispheric Transport of Air Pollution. Persistent Organic Pollutants, Economic Commission for Europe. UNECE.

UNEP. (2009). Stockholm Convention on Persistent Organic Pollutants (POPs). *Secr. Stock. Conv. Persistent Org. Pollut., 1.*

USEPA. (2016). *Pesticides Industry Sales and Usage. 2008-2012 Market estimates*. USEPA.

Vallotton, N., & Price, P. S. (2016). Use of the Maximum Cumulative Ratio As an Approach for Prioritizing Aquatic Coexposure to Plant Protection Products: A Case Study of a Large Surface Water Monitoring Database. *Environmental Science & Technology*, *50*(10), 5286–5293. doi:10.1021/acs.est.5b06267 PMID:27057923

Verro, R., Finizio, A., Otto, S., & Vighi, M. (2009). Predicting pesticide environmental risk in intensive agricultural areas. II: Screening level risk assessment of complex mixtures in surface waters. *Environmental Science & Technology*, *43*(2), 530–537. doi:10.1021/es801858h PMID:19238990

Wauchope, R. D. (1978). The Pesticide Content of Surface Water Draining From Agricultural Fields-A Review. *Journal of Environmental Quality*, *7*(4), 459. doi:10.2134/jeq1978.00472425000700040001x

Weber, J., Halsall, C. J., Muir, D., Teixeira, C., Small, J., Solomon, K., & Bidleman, T. (2010). Endosulfan, a global pesticide: A review of its fate in the environment and occurrence in the Arctic. *The Science of the Total Environment*, *408*(15), 2966–2984. doi:10.1016/j.scitotenv.2009.10.077 PMID:19939436

Wieczorek, M. V., Bakanov, N., Bilancia, D., Szöcs, E., Stehle, S., Bundschuh, M., & Schulz, R. (2018). Structural and functional effects of a short-term pyrethroid pulse exposure on invertebrates in outdoor stream mesocosms. *The Science of the Total Environment*, *610–611*, 810–819. doi:10.1016/j.scitotenv.2017.08.048 PMID:28826119

Wijngaarden, R. P. A. V., Brock, T. C. M., & Van Den Brink, P. J. (2005). Threshold Levels for Effects of Insecticides in Freshwater Ecosystems: A Review. *Ecotoxicology (London, England)*, *14*(3), 355–380. doi:10.100710646-004-6371-x PMID:15943110

Zhang, W., Jiang, F., & Ou, J. (2011). Global pesticide consumption and pollution: With China as a focus. *Proceedings of the International Academy of Ecology and Environmental Sciences*, 1.

Zheng, S., Chen, B., Qiu, X., Chen, M., Ma, Z., & Yu, X. (2016). Distribution and risk assessment of 82 pesticides in Jiulong River and estuary in South China. *Chemosphere*, *144*, 1177–1192. doi:10.1016/j.chemosphere.2015.09.050 PMID:26461443

Zhou, D. M., Wang, Y. J., Cang, L., Hao, X. Z., & Luo, X. S. (2004). Adsorption and cosorption of cadmium and glyphosate on two soils with different characteristics. *Chemosphere*, *57*(10), 1237–1244. doi:10.1016/j.chemosphere.2004.08.043 PMID:15519368

ADDITIONAL READING

Baranda, A. B., Fundazuri, O., & Martínez De Marañón, I. (2014). Photodegradation of several triazidic and organophosphorus pesticides in water by pulsed light technology. *Journal of Photochemistry and Photobiology A Chemistry*, *286*, 29–39. doi:10.1016/j.jphotochem.2014.03.015

Belfroid, A. C., Van Drunen, M., Beek, M. A., Schrap, S. M., Van Gestel, C. A. M., & Van Hattum, B. (1998). Relative risks of transformation products of pesticides for aquatic ecosystems. *The Science of the Total Environment*, *222*(3), 167–183. doi:10.1016/S0048-9697(98)00298-8 PMID:9851068

Bertrand, L., Marino, D. J., Monferrán, M. V., & Amé, M. V. (2017). Can a low concentration of an organophosphate insecticide cause negative effects on an aquatic macrophyte? Exposure of Potamogeton pusillus at environmentally relevant chlorpyrifos concentrations. *Environmental and Experimental Botany*, *138*, 139–147. doi:10.1016/j.envexpbot.2017.03.006

Bischof, I., Köster, J., Segner, H., & Schlechtriem, C. (2016). Hepatocytes as in vitro test system to investigate metabolite patterns of pesticides in farmed rainbow trout and common carp: Comparison between in vivo and in vitro and across species. *Comparative Biochemistry and Physiology Part - C. Toxicology and Pharmacology*, *187*, 62–73. PMID:27185525

Boesten, J. J. T. I., Köpp, H., Adriaanse, P. I., Brock, T. C. M., & Forbes, V. E. (2007). Conceptual model for improving the link between exposure and effects in the aquatic risk assessment of pesticides. *Ecotoxicology and Environmental Safety*, *66*(3), 291–308. doi:10.1016/j.ecoenv.2006.10.002 PMID:17141869

Calamari, D., & Zhang, L. (2002). Environmental risk assessment of pesticides on aquatic life in Xiamen, China. *Toxicology Letters*, *128*(1–3), 45–53. doi:10.1016/S0378-4274(01)00532-X PMID:11869816

Cunha, J. P., Chueca, P., Garcerá, C., & Moltó, E. (2012). Risk assessment of pesticide spray drift from citrus applications with air-blast sprayers in Spain. *Crop Protection (Guildford, Surrey)*, *42*, 116–123. doi:10.1016/j.cropro.2012.06.001

Halstead, N. T., Civitello, D. J., & Rohr, J. R. (2015). Comparative toxicities of organophosphate and pyrethroid insecticides to aquatic macroarthropods. *Chemosphere*, *135*, 265–271. doi:10.1016/j.chemosphere.2015.03.091 PMID:25966044

Houbraken, M., Habimana, V., Senaeve, D., López-Dávila, E., & Spanoghe, P. (2017). Multi-residue determination and ecological risk assessment of pesticides in the lakes of Rwanda. *The Science of the Total Environment*, *576*, 888–894. doi:10.1016/j.scitotenv.2016.10.127 PMID:27838579

Okoumassoun, L. E., Brochu, C., Deblois, C., Akponan, S., Marion, M., Averill-Bates, D., & Denizeau, F. (2002). Vitellogenin in tilapia male fishes exposed to organochlorine pesticides in Ouémé River in Republic of Benin. *The Science of the Total Environment*, *299*(1–3), 163–172. doi:10.1016/S0048-9697(01)01053-1 PMID:12462582

KEY TERMS AND DEFINITIONS

Absorption: The process by which one thing absorbs or is absorbed by another.

Adsorption: Adsorption is the adhesion of molecules of gas, liquid, or dissolved solids to a surface.

Aquaculture: The culture of fishes and other aquatic organisms in man-made ponds, ditches, and floating cages.

Drift: A process of slow movement of chemical particles by a current of air and water.

Environmental Quality Standard (EQS): An environmental quality standard is a value, generally defined by regulation, which specifies the maximum permissible concentration of a potentially hazardous chemical in an environmental sample, generally of air or water.

Leaching: A process in which certain material extract or move from a carrier material extract or move from a carrier to liquid. For example, pesticides get washed off from soil to groundwater through percolation.

No Observation Effect Concentration (NOEC): In toxicology, it is specifically the highest tested dose or concentration of a chemical, at which no such adverse effect is found in exposed test organisms where higher doses or concentrations resulted in an adverse effect.

Oxidative Stress: A biological phenomenon includes various stress enzymes that can cause health issues in organisms.

Runoff: A process of draining water due to rainfall, snow, or any other precipitation event from land to surface water.

Storm Flow: The runoff of land surface water from rainfall.

Vitalization: Refers to the vaporization or sublimation of a volatile herbicide.

Chapter 3
Organochlorine Pesticides:
A Threat to Aquatic Ecosystems

Garima Harit
IIS University, India

ABSTRACT

Indiscriminate use of different pesticides in agriculture has increased over the years, especially in the developing countries. This influences the aquatic environment to a great extent. This also poses a great danger to freshwater organisms, including fish, which constitute a major share in the aquatic environment and contribute to the economy of the nation. Water pollution is posing intricate problems that need immediate redress. Organo-chlorine pesticides (OCPs) are a major contributor to aquatic pollution and are amongst the most serious global contaminants. In addition, organochlorine pesticides have a tendency to accumulate in aquatic biota; they also undergo food chain amplification. Lipophilic pollutants are chemically very stable and resistant to microbial, photochemical, chemical, and thermal degradation. The chemical stability of these compounds, their high lipid solubility, and their toxicity to human beings and animals has led government and researchers to feel concerned about their presence in the environment.

INTRODUCTION

Aquatic Ecosystem

An ecosystem in a water body where communities of various organisms depend on each other as well as on their environment is referred as aquatic ecosystem. Organisms which depend on water for their life activities such as feeding, breeding, shelter, etc. in order to survive are called as aquatic organisms. It is of three types namely freshwater, marine and estuarine. Freshwater covers only 8% of the earth and involves lakes, ponds and pools (lentic) and rivers and streams (lotic) habitats. Marine water bodies cover largest surface area and constitute oceans, seas, reefs, sea beds, intertidal zones, etc. Estuarine areas are those which experience the flux of both freshwater and marine water depending on tide and water currents. Animals living in these water bodies have adapted themselves according to their habitats.

DOI: 10.4018/978-1-5225-6111-8.ch003

Estuaries are semi enclosed bodies of brackish water which is less salty than marine waters. Freshwater animals cannot survive in the saline environment of marine water bodies. Freshwater used for irrigation purposes in agricultural land often absorbs levels of salt that might harm freshwater organisms.

Aquatic ecosystem links human populations, land and wildlife through water. Man has shown keen interest in aquatic resources such as sea food, fisheries, fishing sport, swimming, observing natural beauty, etc. The relationship of man with his environment is essentially symbiotic and equilibrium should be maintained between the two. Rapid growth of human population, over-exploitation of natural resources and developments in agriculture and industry has contributed extensively to the presence of various pesticides in the aquatic environment. Time is perhaps not far when pure and clean water, particularly in densely populated and industrialized areas, may be inadequate for maintaining normal living conditions.

Pesticides

Pesticides are the chemicals that were designed for the human beneficial aspect to control the pests. According to FAO pesticide has been defined as, "Any substance or mixture of substances intended for preventing, destroying or controlling any pest, including vectors of human or animal disease, unwanted species of plants or animals causing harm during or otherwise interfering with the production, processing, storage, transport or marketing of food, agricultural commodities, wood and wood products or animal feedstuffs, or substances which may be administered to animals for the control of insects, arachnids or other pests in or on their bodies." Pesticides can protect forests and farms from crop losses and can lead to more production of yield. The application of pesticides in forests is done to reduce the spread of insects and control many insect borne human diseases like malaria, plague, etc. They are generally cost effective and their mode of application is also relatively easy. In fact in certain situations they are the only options left to deal with the pests. Indiscriminate use of different pesticides in agriculture has increased over the years, especially in the developing countries (Prashanth and Neelagund, 2008) this influences the aquatic environment to a great extent leading to a great danger to amphibians, fish, reptiles, aquatic birds and other wildlife which constitute a major share in the aquatic environment and contribute to the economy of the nation.

Protection of wildlife and water quality becomes convenient with the use of pesticides; but these have to be wisely selected and safely applied so as to avoid the surface water pollution and any sort of damage to aquatic life. Excessive use of pesticides can lead to the damage of ecosystems. They may harm untargeted animals, decrease biodiversity and might lead to extinction of species. By disturbing food chains/webs, the balance in ecosystem is disrupted, thereby affecting many aquatic and terrestrial species. Life of microorganisms, plants and fish in aquatic ecosystem is reported to be adversely affected by pesticides (Grande *et al.*, 1994; De Lorenzo *et al.*, 2001; Frankart *et al.*, 2003; Liess *et al.*, 2005 and Castillo *et al.*, 2006). In the India, excessive use of pesticides started since the 1960s with the initiation of the "Green Revolution" and in order to fetch high agricultural, yield maximum agrochemicals were used.

Pesticides have influenced the following at the community and ecosystem levels:

1. Induction of dominance by small species.
2. An increase of species richness and diversity.
3. Elongation of the food chain and reduction of energy transfer efficiency from primary producers to top predators. Pesticides may affect the population dynamics by controlling individual survival and reproduction, and by altering the sex ratio (Hanazato, 2001).

Figure 1. Pesticide circulation in environment

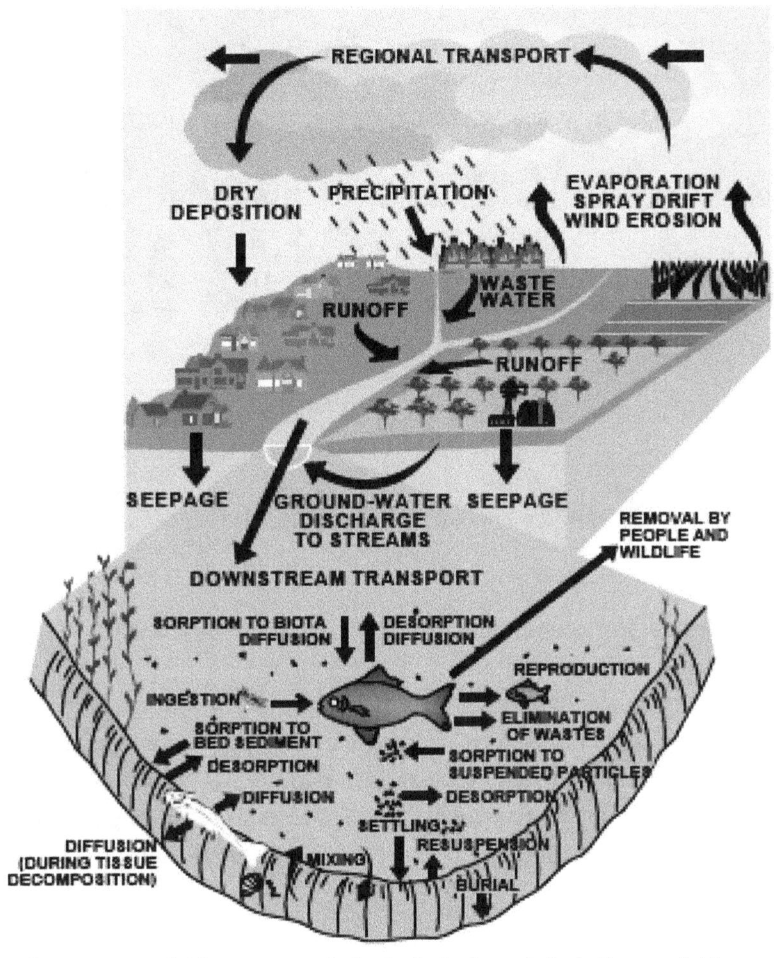

Direct application of pesticides into the water means that these chemicals may drift away from the original exposed site, attacking a much larger area of the lake or pond. This might also affect swimmers or wildlife in areas which may have not posted pesticide warnings. Moreover, persistence of few chemicals in the environment is yet to be known along with their long-term exposure effects on aquatic ecosystem.

Impact on Aquatic Organisms

Aquatic animals as well as aquatic resources are precious natural gifts. Aquatic animals not only enhance national economy but also increase employment opportunities, profits and financial savings. Rapid utilisation of pesticides in agricultural industry depicts unavoidable danger to the human health along with the entire aquatic ecosystem. Pesticide causes undesirable loss that can even cause death of aquatic animals once they enter in aquatic ecosystem. This causes reduction in aquatic organisms.

Impact of pesticides in aquatic invertebrates exposed to experimental concentrations of pesticides have often been observed to be higher than the rates recommended for field application. Knowledge of

Figure 2. Schematic representation of circulation of pesticides in aquatic ecosystem

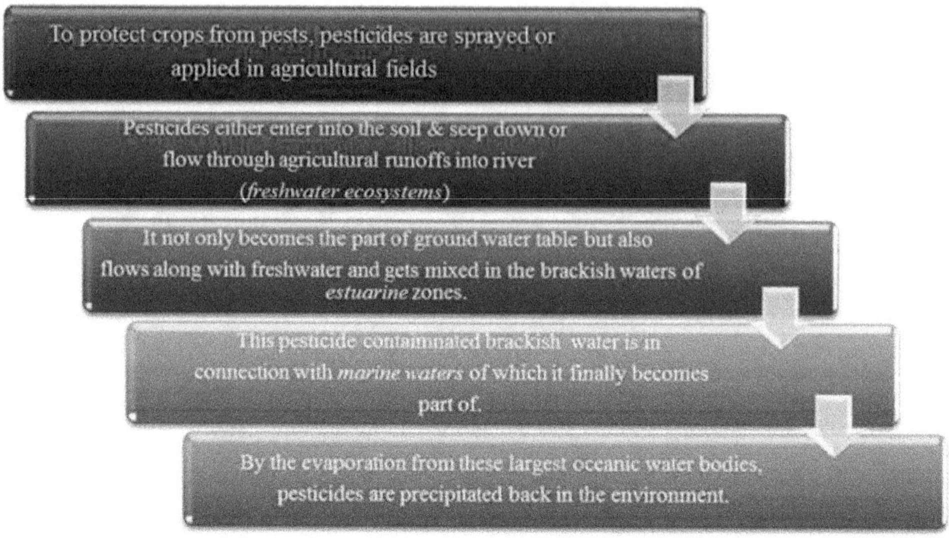

pesticide misused/mishandled and accidental spillage is essential as where these pesticides are applied genuinely, such observations are very limited.

But perspective of pesticide application differs from farmers' point of view, leading to difference in application rates from recommended rates. Farmers often apply more pesticides than the quantity recommended as they believe that if the use of pesticides is more, its effect on pests will be long lasting. Whereas, on the other side, few farmers apply less pesticide than recommended rates so as to make it cost effective. Care must be taken with the application of these pesticides as they also affect the entire food chain of aquatic ecosystem.

BACKGROUND

Zooplanktons play a major role in distributing organochlorines from atmospheric 'fallout' throughout the ocean depth (Harding, 1986). In comparison to phytoplankton, the work done to study the impact of pesticides on zooplankton is more followed by fish and amphibians. The reason behind the selection of zooplanktons and macro-invertebrates for bio-monitoring might be due to the ease of studying the effects of pesticides which could be observed in these aquatic organisms.

Fish plays a vital role in food and nutritional aspects of the country due to the presence of high quality animal proteins, vitamins (A, B, D and E) and minerals essential for vitality and growth (Chitranshi, 2004). Fishes are also amongst the best biological indicators of pollution (Srivastava and Kaushik, 2001) as they can take up contaminants directly from water and diet (Harit and Srivastava, 2017a). The ability of fish to metabolize organochlorines is moderate; therefore, contaminant loading in fish is reflective of the state of pollution in the surrounding environment (Guo *et al.*, 2008). Pesticides are very potent toxicants and even very short exposures for a few hours are sufficient to bring about changes in ethological response of fish (Harit and Srivastava, 2017b). The concentrations of OCP is observed to be higher in fish than in invertebrates like molluscs, and also accumulation is more in freshwater fish than

Table 1. Impact of organochlorines on various aquatic animals

S. No.	Name of Organism	Name of OCP	Toxicity	Reference
1	Phytoplankton/Zooplanktons	DDT	Accumulation	Siriwong et al., 2009
2	Crassostrea gigas (oyster)	DDT and HCH	Accumulation	Liu et al., 2010
3	Mytilus edulis (mussel)	DDT and HCH	Accumulation	Liu et al., 2010
4	Labeo rohita	Endosulfan	Behaviour alterations and changes in protein content of liver, gills, brain and muscle	Ullah et al., 2016
5	Channa punctatus	Endosulfan	Accumulation & Histopathological changes in gills	Harit and Srivastava, 2017a
6	Heteropneustus fossilis	Aldrin	Hyperglycemia & increase in acetylecholine content	Singh and Srivastava, 1995

in marine fish; which is indicative of two things. Firstly, that freshwater fishes are directly influenced by OCP residues from agricultural runoffs as compared to marine fishes and molluscs and secondly, there is difference in accumulation factors for OCPs in fish and molluscs as suggested by Liu *et al.* (2010).

Amphibians and insectivorous reptiles connect invertebrates with vertebrates in the food chain. They act as a food source for other animals and serve to be the channel through which lipophilic pesticides (OCPs) enter into the food chain. These OCPs can be consumed by amphibians through air, skin and food. As pesticides are passed in water bodies via agricultural fields nearby, probability of their exposure to amphibians of those water bodies also enhances. Long term exposure of pesticides to economically important insect like honey bee and other pollinating insects might also affect the quality of crops and by-products obtained from beneficial insects such as honey. Humans get exposed to pesticides through inhalation and ingestion of plant products like vegetables, fruits, and animal products such as meat, milk etc. and accidental exposure due to occupational hazard.

TYPES OF PESTICIDES

Pesticide is a general classification that includes insecticides, rodenticides, fungicides and herbicides; pesticides can be toxic to man if contaminated food and water is ingested. They are classified on the basis of their chemical structure, as their environmental properties (persistent nature, toxicity level, solubility, etc.) are almost similar.

Pest control can also be done by killing pests by burning crop residues or draining wetlands to eliminate breeding sites. A narrow spectrum pesticide that kills all living organisms is called biocide. There are narrow spectrum pesticides that are meant to destroy specific type of pests:

- **Herbicides:** To destroy weeds.
- **Insecticides:** To kill insects.
- **Fungicides:** To kill fungi.
- **Acaricides:** To kill mites, ticks and spiders.
- **Nematicides:** To kill nematodes.
- **Rodenticides:** To kill rodents.

Figure 3. Benefits and problems of pesticides

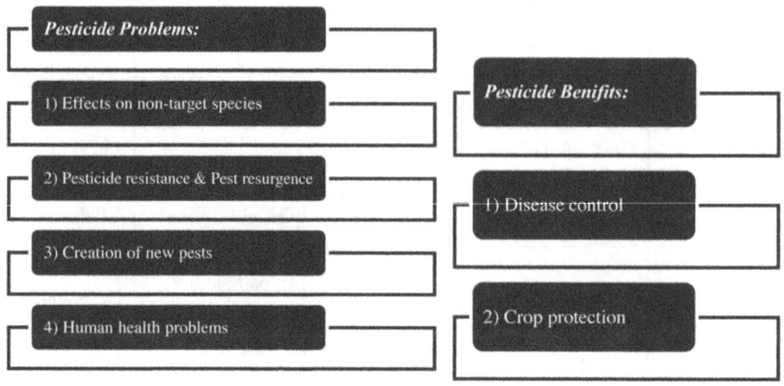

Figure 4. Types of Pesticides

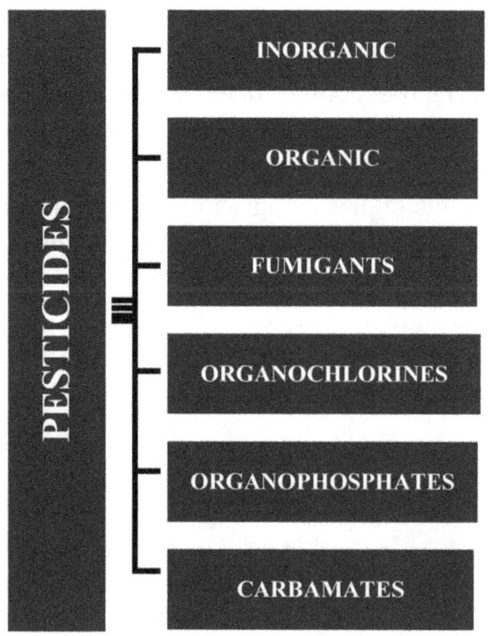

- **Avicides:** To kill birds.
- **Ovicides:** To kill eggs of pests

ORGANOCHLORINE PESTICIDES

Organochlorines, initially developed in the 1930s to be used as insecticides and pesticides, are majorly utilized for controlling pests in soil and crops of agricultural fields from where they make their way to aquatic systems via domestic sewage and industrial wastewater disposal. DDT had become famous worldwide after its use in overcoming a typhus infestation in Naples in 1939 and for the control of vector borne diseases like malaria. The application of organochlorines increased during the 1950s and reached its height in the 1970s. Pesticides are subsequently distributed in the freshwater ecosystem and accumulate in aquatic biota, especially fish; in turn this toxicity is passed on to man through the food chain. Chlorinated pesticides have an affinity for lipids and hence tend to accumulate in the adipose tissues, depending on their structural characteristics and lipid solubility. Lipophilic pollutants are chemically very stable and resistant to chemical degradation, microbial degradation (by biological actions), photo degradation (by sunlight), thermal degradation (by high air or water temperatures) and soil conditions (pH). Chlorinated hydrocarbons such as DDT, Chlordane, aldrin, dieldrin, toxaphene, paradichlorobenzene and lindane are synthetic organic insecticides that inhibit nerve membrane ion transport and block nerve signal transmission and are highly toxic to sensitive organisms (Cunnighum and Saigo, 1998). Overuse and misuse of these OCPs has placed them under the category of pollutants and toxicants from environmental point of view. They find their final destinations in aquatic ecosystems through-

- Surface runoff of pesticides from agriculture and forestry.
- Pesticides from rainfall.
- Accidental spraying and accidental spills.
- Continuous release from industrial waste water as water being one of the primary means by which pesticides are transported from site of application to environment.

Due to volatile property of OCPs, its concentrations are even detected in the Polar Regions, inspite of the fact that they were never applied there (Ritter *et al.*, 1995). Thus, once released into the environment, OCPs make entry in aquatic ecosystems through air or water channels. Once these OCPs are sprayed in the environment, through wind drift, volatilisation and atmospheric transport they tend to accumulate in polar ecosystems far away from the place they were initially applied.

OCPs are amongst the first set of pesticides in use and are still in use, despite their ban due to associated problems of potency and persistency (Iyamu *et al.*, 2007 and Gao *et al.*, 2008). The chemical stability of these compounds, their high lipid solubility and toxicity to human beings and animals (Rathore *et al.*, 2002 and Wolkers *et al.*, 2009), has led government and researchers to feel concerned about their presence in the environment.

Organochlorine pesticides are hydrophobic and are easily adsorbed to the organic matter in the soil. These pesticides not only reach deep into the groundwater table by the processes of desorption, solubilisation and infiltration but also join many small streams and ultimately find their final destination into the sea (Coat *et al.*, 2006). The contamination of the aquatic ecosystem not only threatens the organisms living in it but also affect the non-target orgranisms.

- **DDT:** The most commonly used pesticide in agricultural practice is dichlorodiphenyltrichloroethane (DDT), which is moderately hazardous, with high persistence and a half-life of 2–15 years (Augustijn-Beckers *et al.*, 1994). Impact of DDT on aquatic, wildlife and natural ecosystems is threatening. Being an organochlorine pesticide, DDT bioconcentrates, bioaccumulates, and gets biomagnified up to the trophic level through the food chain, based on the habitat and foraging behavior of selected freshwater species as suggested by Siriwong *et al.* (2009). Aldrin, toxaphene, dieldrin, mirex, endosulfan and heptachlor are other OCPs which are very toxic to fish and wildlife and have been banned from use. The use of DDT and other OCPs is also now banned in many countries but it is illegally used in most of the developing countries.

Figure 5. Adapted from Chemie über den Wolken: ... und darunter
Edited by Reinhard Zellner.

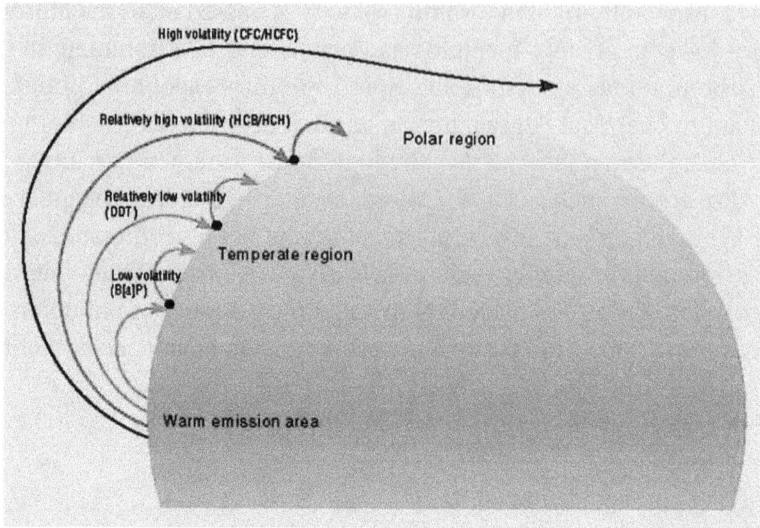

- **Endosulfan:** Besides DDT, endosulfan is also being banned in many developing countries, including India. Endosulfan, an insecticide is used in the production of cashew but has moderate persistence with a half-life of fifty days (Quijano, 2002). Endosulfan, (6, 7, 8, 9, 10, 10-hexachloro-1, 5, 5a, 6, 9, 9a-hexahydro-6, 9-methano-2, 4, 3-benzadioxathiepin-3-oxide), a broad-spectrum cyclodiene pesticide of the organochlorine group, is widely used for controlling a variety of insects. It is water-soluble but also adheres readily to clay particles and persists in soil and water for several years. Endosulfan is moderately hazardous and it act as a central nervous system (CNS) poison (WHO, 1986). Cyclodines, like endosulfan and lindane, are the GABA antagonists and they block neurotransmittors by inhibiting the calcium ion influx and Ca- and Mg-ATPase (Mathew, 2012). Fishes are adversely affected by endosulfan; significant mortality has been reported as a result of endosulfan leakage into rivers (Naqvi and Vaishnavi, 1993). Endosulfan has been banned in most of the developed countries, although it is still produced, sold and used in developing countries (Salvo *et al.*, 2008). In a study conducted by Central Pollution Board of India (CPCB, 2000) it has been identified as one of the main pesticides found in the waters of major rivers; although human intake of endosulfan exceeding the FAO/WHO temporary ADI of 0.006 mg/kg, body weight/day has not been reported (FAO/WHO, 2010). Endosulfan is also included in the EPA's (Environmental Protection Agency) list of priority pollutants. Due to its persistent nature and slow degradation it accumulates easily in fish tissues (Srivastava *et al.*, 2002 and Harit and Srivastava, 2017a) and interferes with their metabolic and physiological functions (Rawat *et al.*, 2002 and Harit and Srivastava, 2007 and 2009).

- **Chlordane:** It is a non-systemic contact and stomach poison. It is comprised of more than 140 different chlorinated hydrocarbons (Dearth and Hites, 1991) and is used in the control of a diverse range of organisms, such as ants, beetles, grasshoppers, termites and earthworms (Worthing and

Figure 6. Normal structure of gills of fish

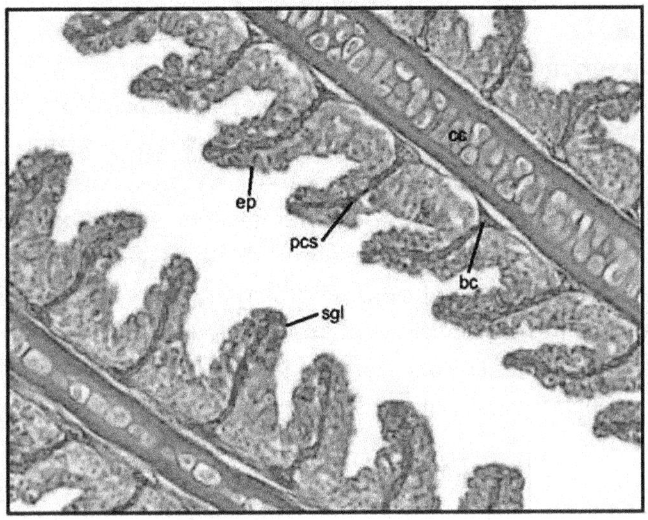

Figure 7. Structural alterations in gills after exposure of fish to endosulfan
Harit and Srivastava, 2017a.
Photomicrograph of the gills of Channa punctatus showing shortened secondary gill lamellae (sgl), fusion (fu) of sgl, extensive hyperplasia (hp), vacuolization (vc) and enlarged blood capillaries (E). Disorganized pillar cell system (D) can be clearly noticed and detachment of basement membrane (dt), blood cell accumulation (bc), proliferation (P) from the base of secondary gill lamellae and telangiectasis (te) in some secondary gill lamellae is also visible, narrow cartilagenous core (th) is observed in affected gills in comparison to normal healthy fish.

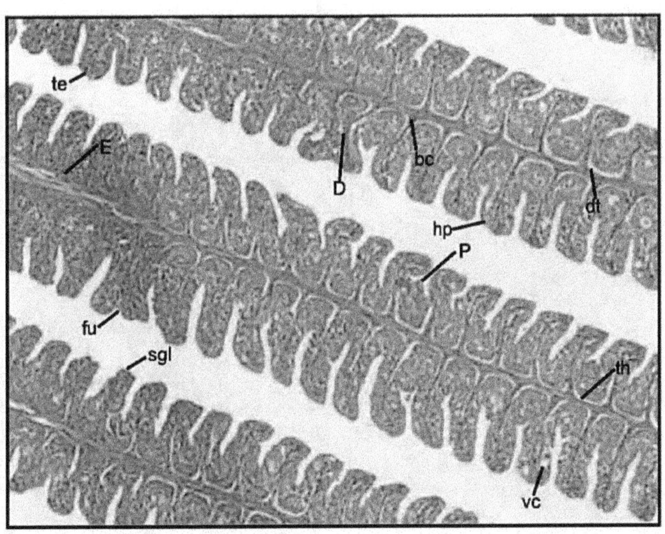

Walker, 1983). Due to its worldwide distribution (Kawano *et al.*, 1988), it is considered a major environmental contaminant (Mowbray, 1988). It is persistent in both terrestrial (Parker *et al.*, 1985) and aquatic ecosystems (Wood *et al.*, 1986). It has been proved toxic to various non-target organisms (Srivastava and Srivastava, 1988). Kawano *et al.* (1988) and Norstrom *et al.* (1988) have reported bioaccumulation and biomagnification of chlordane via food chain.

- **Dieldrin and Aldrin:** Other OCPs that broadly used in agricultural fields are dieldrin and aldrin. Both are chemically made by the process known as the Diels-Alder reaction and were first formulated from a waste product of synthetic rubber, cyclopentadiene. These are toxic as well as bioaccumulative. Marine mammals show a high bioaccumulation factor for dieldrin. The highest bioaccumulation rates for dieldrin are observed in fish. Amphibians show high levels of biomagnification for dieldrin (Jorgenson, 2001) as it exhibits resistance to bacterial and chemical breakdown processes and therefore tends to biomagnify. Like DDT, metabolization of dieldrin is not easy in water and has least chances of being digested and excreted from the body. It tends to be absorbed and transported throughout the body in the blood or body hemolymph of invertebrates. In blood, dieldrin resides in the erythocytes and plasma. Aldrin was used to protect soil in corn and potato fields from pests (namely termites). It tends to break to dieldrin in living organisms. Dieldrin was used on fruit, soil, and seed to control insects. Dieldrin shows persistence in the soil with a half-life of five years at temperate latitudes. Both of these show property of volatilisation from the sediment and hence, get redistributed by air currents, contaminating areas far from their sources. Aldrin and dieldrin have already been banned in most developed countries, but aldrin is still used as a termiticide in various countries.

Box 1.

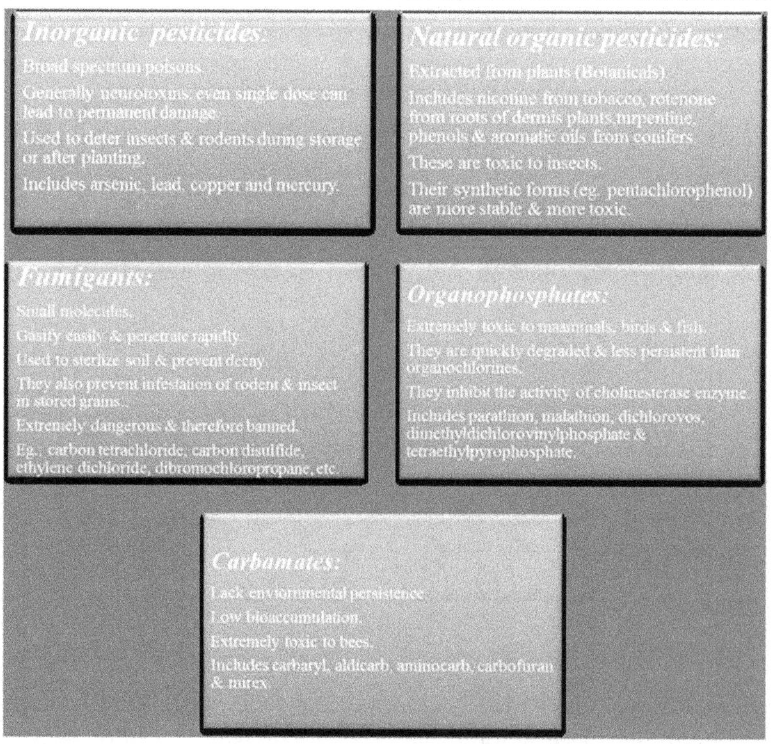

- **Toxaphene:** It is can be called as 'Dirty Dozen', as it comprises of group of 12 highly toxic chemicals that are associated with various diseases in livestock and humans. Toxaphene was an insecticide applied on cereals, vegetables and cash crops. Livestock were protected from such pests as lice, fleas, ticks and scab mites. Being persistent but relatively nontoxic to bees, it was used on flowers. The use of this as a pesticide was cancelled in 1982 and banned in 1990. Moreover, existing stocks were not allowed for sale in the United States after March 1, 1990. But production and utilization of toxaphene-like pesticides is still observed in other countries including India, America, Africa and a few parts of Eastern Europe.
- **Mirex:** It was widely used for killing fire ants and was also commonly used as a fire retardant in plastics, rubber, paint, paper, and electrical goods. It shows slow degradation in the environment and does not evaporate much from surface water or surface soil; may persist on soil and water for decades. It is highly lipophilic in nature, and so is strongly adsorbed on sediments and easily adheres to soil and sediment particles. Accumulation of mirex in aquatic and terrestrial food chains can prove to be harmful. Although its sale, distribution and use are prohibited, it still can be detected in environment.
- **Other Pesticides:** The four major types of agricultural insecticides used today are pyrethroids (PYs), organophosphates (OPs), carbamates (CBs), and biological insecticides (BIs). Though they are more toxic than OCPs but are comparatively less persistent in environment.

AQUATIC TOXICOLOGY

Aquatic toxicology is the study of the effects of manufactured chemicals, other anthropogenic and natural materials (environmental contaminants) on aquatic organisms at various levels of organisations beginning from subcellular to entire ecosystem, such as the impact of pesticides on zooplanktons to the health of fish or other aquatic organisms. A capacity of pesticide to harm aquatic animals is dependent on:

Toxicity

"The dose makes the poison" as said by German scientist Paracelsus, thereby meaning that almost everything is toxic at certain level. Toxicity of the pesticide refers to the extent to which it is poisonous. Few are extremely toxic, while others are comparatively nontoxic. Measuring toxicity depends on the rate, medium and route of entry of the material delivered to the organism. But it is not so easy to measure and compare toxicity as different species exhibit different sensitivity and even individuals within the species respond differently to a given exposure.

Exposure

It refers to the duration the animal is in contact with the pesticide. Single exposure to the toxin may result in immediate health crisis. In long lasting exposures described as chronic, the effects may or may not persist after the removal of toxin. It is difficult to assess the specific health risks of chronic exposures.

Figure 8.

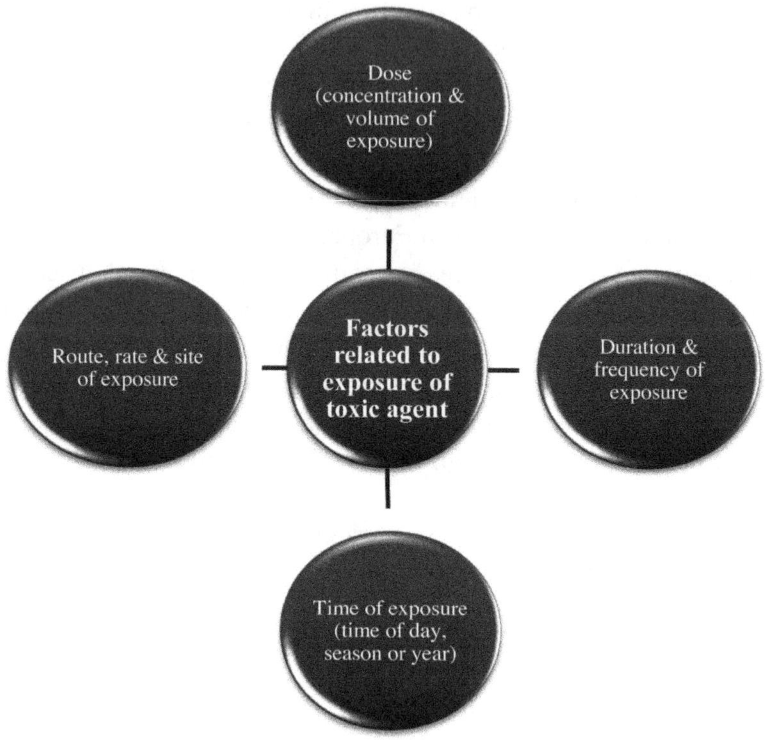

The Dose Rate

It refers to the quantity or amount of pesticide to which an animal is subjected through oral, dermal, or through inhalation route. A small dose of a more toxic chemical may be more deleterious than a large dose of a less toxic chemical. A moderate toxin is considered of about one gram per kilogram of body weight to make a lethal dose. A lethal dose is the amount of pesticide necessary to cause death; the dose to which 50% of the population is sensitive. All the animals of a species do not die at the same dose (as some are more tolerant than others), dose measurement of a standard toxicity, called as lethal concentration 50 (LC_{50}), is used. At this concentration of pesticide, 50% of a test population of animals shows mortality within a set period of time, usually 24 to 96 hours (Helfrich, 2009).

Persistence

Some chemical compounds are very unstable and degrade rapidly under most environmental conditions so that their concentrations decline quickly after release. Other substances are more persistent and last for much longer time such as chlorinated hydrocarbons which are resistant to degradation.

Exposure of aquatic animals to a pesticide depends on the bioavailablility, bioconcentration, biomagnification, and persistence of the pesticide in the environment. The content of pesticide that is available to the fish and other organisms in the environment is termed as bioavailability. Few pesticides degrade rapidly after their application while some adhere tightly to soil particles suspended in the water, thereby reducing their availability. While some others dilute in water or volatize rapidly into the air and are less available to aquatic life.

Bioconcentration is the content of pesticides that gets accumulated in animal tissue at levels greater than those present in the water or soil to which they were applied. Organochlorine pesticides are lipophilic and tend to concentrate in fat tissues of the aquatic organisms that are exposed to them at levels 10 million times greater than in the water.

Biomagnification occurs if the toxicant is persistent, mobile, fat soluble and biologically active. If it is water soluble, it will be excreted out of the organism whereas fat soluble pollutants get retained in the organism for a long time and move up in the food chain when consumed by predators. Biomagnification is the accumulation of pesticides at each successive level of the food chain. The effects of OCPs are magnified in the environment through food chains and food webs, thereby getting an easy entry in non-target organisms and these also become contaminated. If the occurrence of the toxic content of a large number of organisms at lower trophic level is accumulated and concentrated by a predator at a higher trophic level, then it is termed as biomagnification (Cannighum and Saigo, 1998). At each step in the food chain the concentration of pesticide increases. Fish are continuously exposed to these toxins and repeatedly consume contaminated animals, they bio-concentrate high levels of toxins in their body fat. Fish can transfer these poisons on to humans on consumption.

Uptake of a given substance directly from water or through the consumption of food containing the substance is termed as bioaccumulation. Cells have mechanism for bioaccumulation, the selective absorption and storage of great variety of molecules. Toxins that are dilute in the environment can reach

Figure 9.

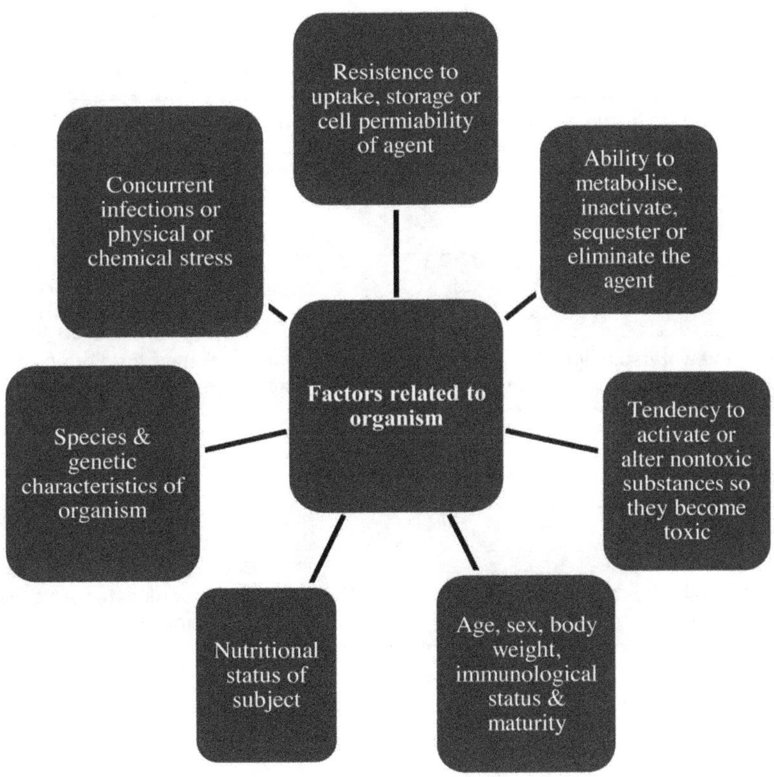

dangerous levels inside cells and tissues through this process of bioaccumulation. Uptake from water is referred as bioconcentration factor (BCF) and how the absorption of contaminant occurs in small organisms and increases as it progresses up the food chain is referred as biomagnification factor (BMF) (Jorgenson, 2001).

Twenty-eight of the bacterial genera that can breakdown aliphatic hydrocarbons have been isolated (David, 2002), and most common among them are the species of *Pseudomonas, Alcaligenes, Bacillus, Arthrobacter, Brevibacterium, Flavobacterium, Klebsiella, Methylococcus*, etc. Several fungi having pesticide degrading potential have also been identified, such as the species of *Aspergillus, Candida, Fusarium, Penicillium, Trichoderma, Rhodotorula, Pleurotus, Phaenerochaete*, etc. Degradation of pesticides is the process divided in three phases (Hatzios, 1991 and Shimabukuro, 1985).

It is essential to know the formulation, method of application, and solubility of a pesticide. Solubility helps in the determination of how, where and when the toxin will move through the environment or the body to its site of action. Chemicals can more readily dissolve in water or in oil. Once the oil\fat soluble toxins get inside the body, they readily enter into tissues and cells as the membranes that enclose cells are themselves made of similar oil soluble chemicals. These compounds get accumulated in the cells and stored in lipid deposits and are protected from metabolic breakdown and persist for many years (Cannighum and Saigo, 1998). Decomposition products formed as a result of pesticide degradation are often toxic than the parent compound. Impact of these chemicals on non-target organisms can be divided into lethal, sublethal, or indirect effects. Lethal effects are observed when concentration of pesticide is enough to cause mortality directly whereas sub lethal effects lead to alteration in behavior and physiol-

Figure 10. Phases of degradation of Ogranochlorine Pesticide

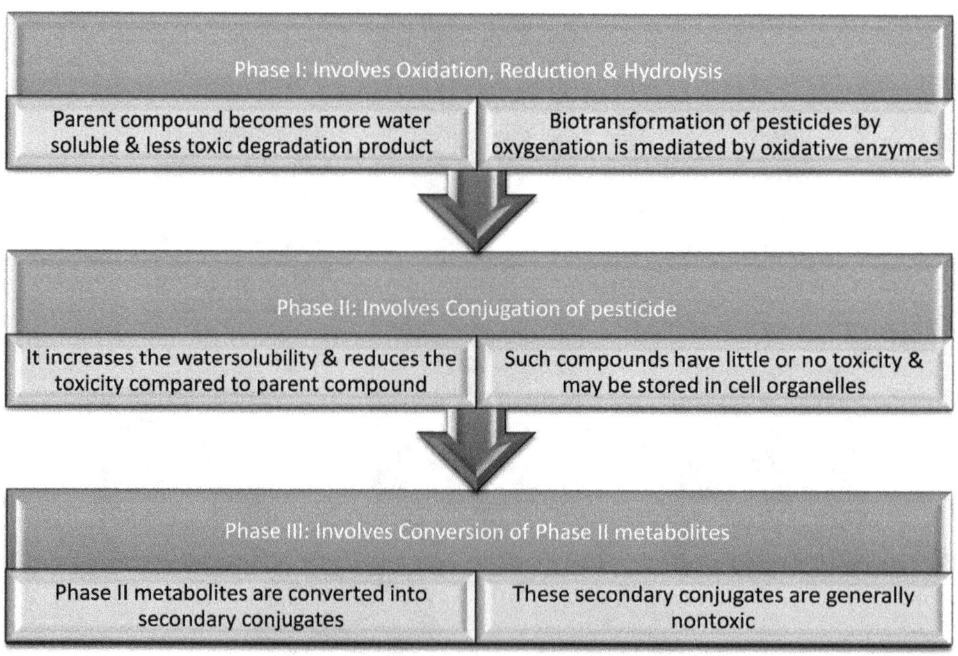

Figure 11. Bioaccumulation and Biomagnification in different Trophic levels

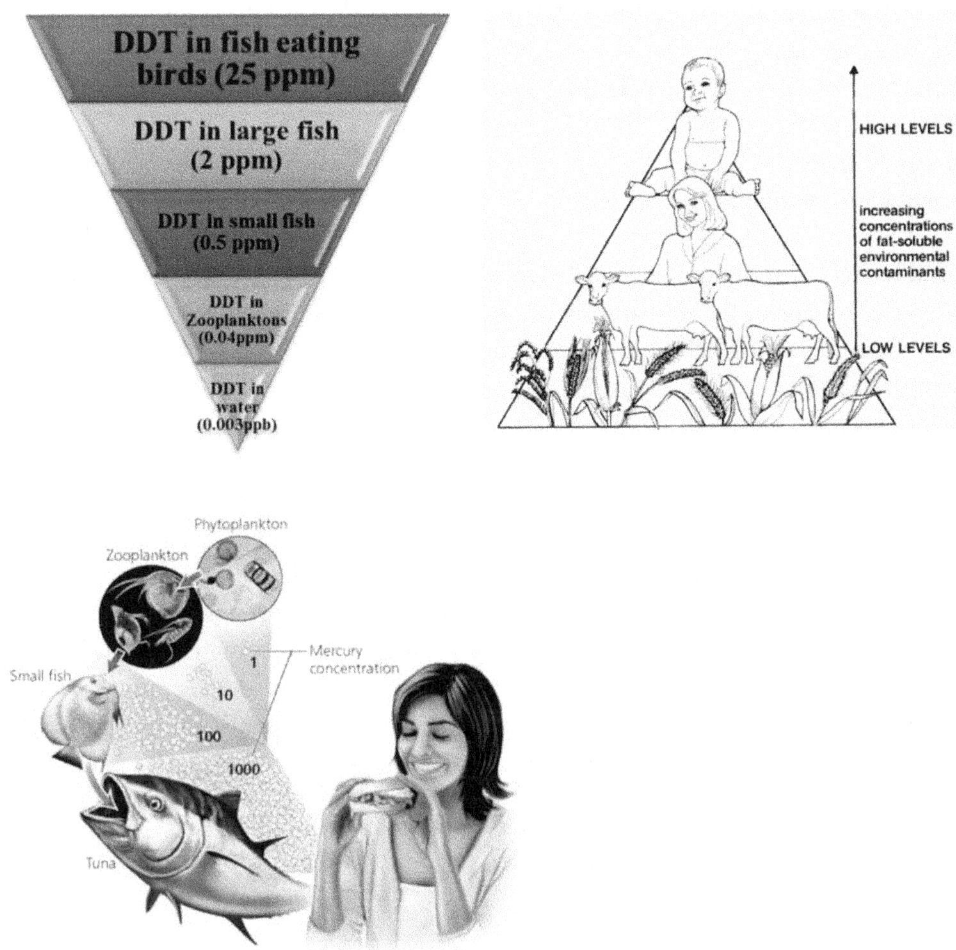

ogy, interference with reproduction and maturation, and changes in morphology of an organism and can even reduce the chances of survival for the individual and the population.

REGISTRATION OF PESTICIDES

Chlorinated pesticides have been applied to protect the crops and prevent vector borne diseases. Due to the persistence of OCPs for prolonged periods in the environment, these undergo bioaccumulation and biomagnifications and impart toxicity to non-target organisms and human beings; it therefore, became need of the hour to either ban or restrict the manufacture and use of these chlorinated pesticides in most of the developed and developing countries (Vivekanandhan and Duraisamy, 2012).

The Environmental Protection Agency (EPA) is responsible to administer the laws and also has the authority to register, restrict, or ban the use of pesticides. Pesticide registration decisions balance the risks involved with the benefits. Registration of pesticide by EPA is based on:

- The ingredients
- Manufacturing process
- Physical and chemical properties
- Environmental state (mobility, volatility, breakdown rates, accumulation potential in plants and fish)
- Toxicity to animals
- Carcinogenic or mutagenic properties.

The EPA under its authority can also approve or disapprove the registration of new pesticides, and may further restrict or cancel the registration of pesticides already in use. For example, DDT in 1972, aldrin and dieldrin in 1974, heptachlor in 1983, mirex in 1977, and toxaphene in 1982 were banned from use i.e. their registration was cancelled in the United States. The use of endrin was highly restricted in 1979. State agencies also require the Registration of pesticides is also needed by state agencies so that they can be used within their boundaries.

The 'Pesticide Label' containing guidelines on its use and safety measures must be attached to all pesticide containers. The label should have the name of product, details of active ingredients, EPA registration number, name and address of the manufacturer, and net contents. There has to be the use classification (general use or restricted use) and hazard classification (danger, warning, or caution) on the label (Helfrich, 2009). Safety statements provide its users the information regarding the handling requirements, procedures and special concerns. Directions for use specify legal application sites, amounts and mixing and handling instructions. The EPA is also planning to identify those pesticides that have the potential to jeopardize endangered or threatened species by a statement on the label. This statement will instruct its users to determine if there are any limitations on the pesticide's use in the county where it is to be applied.

Pesticides Restricted for Use in India

Special care must be taken while handling the "Restricted-Use Pesticides". Use of such pesticides is restricted because they are particularly toxic to fish, birds or mammals. These can be used only by a trained, certified pesticide applicator.

In India, the Insecticides Act, 1968 and Insecticides Rules, 1971 were made to regulate the import, registration, process, manufacture, sale, transport, distribution and use of insecticides with a view to prevent risk to human beings or animals throughout India.

Box 2. List of few pesticides restricted for use in India

□ Aluminium Phosphide	□DDT	□Lindane
□Methyl Bromide	□Methyl Parathion	□Sodium Cyanide
□ Monocrotophos	□Dazomet	□Fenthion
□Endosulfan	□Fenitrothion	□Diazinon
□ Methoxy Ethyl Mercuric Chloride (MEMC)		

Adapted from Web portal of 'Directorate of Plant Protection Quarantine and Storage, Faridabad' as on 15.03.2011.

Pesticide Management Bill of 2008 which is meant to replace the Insecticide Act, 1968, is still waiting for Parliamentary approval.

Global ban on the manufacture and use of endosulfan was negotiated under the Stockholm Convention in April 2011. India opposed this move due to pressure from the endosulfan manufacturing companies.

In 2011: The Supreme Court of India banned manufacture, sale, and use of toxic organochlorine pesticide endosulfan in India.

The production, storage, sale and use of the pesticide were temporarily banned on 13 May 2011, and later permanently by the end of 2011.

PUBLIC AWARENESS

Use of variety of pesticides in tropical countries (where crops are adversely affected by high temperatures and humidity and are conductive to rapid multiplication of pests) has been advised in order to enhance the crop productivity (Kannan *et al.*, 1993; Lakshmi, 1993). As per the World Health Organisation study, 80% of all pesticides are used by developing countries (Veil, 1990). Due to lack of proper government rules, improper market regulations and public ignorance, agricultural workers from developing countries experience a high level of exposure to agricultural chemicals, including pesticides.

Pesticides are still in use even after being banned due to:

- Lack of proper monitoring bodies.
- Stringent actions not taken against the defaulters due to inadequate laws.
- Lack of implementation of awareness programmes in remote areas.
- Alternatives (biological means) to the use of pesticides are expensive. Hence pesticides are preferred more.
- Study of Environmental Sciences although is included in curriculum; could not sensitize the learner and his or her acquainted.

DIRECTIONS FOR FUTURE: NEED OF THE HOUR

To Reduce the Risk

- *To be used as per requirement and only when necessary*: Use of Integrated Pest Management (IPM) practices to control the use of chemical and apply only when necessary. It has to be made sure that the application of pesticide is done only when needed and is accomplished safely and effectively.
- *To use relatively less toxic pesticides*: Selection of pesticides should be based on lowest toxicity of humans, mammals, birds, fish and small invertebrates. Acute toxicity of pesticides should be low. Along with higher soil absorption, lower water solubility (low run off potential) and shorter half-life.
- *To use safe and sensible application method following points can be considered*:
  - Evaluation of chemical control options i.e. selecting that option which has least negative impact on water quality.

- Selection of those products that can minimize waste and applicator exposure.
- Application of pesticides can be avoided just before anticipated rainfall events or when soil moisture conditions are high, or when very windy.
- Prevention of runoff by managing irrigation in such a manner that application rates do not exceed infiltration capacity of the soil.
- Directions given on the label should be read and followed carefully.
- Application of pesticides should be as per the directions as they are legally formulated and not just the advice. Pay careful attention to application site requirements, methods, and rates.
- Mixing and loading of pesticides should be done carefully.
- Functioning of the equipment is to be checked regularly if it is working correctly and is properly calibrated.
- Preparation of only that amount of pesticide to be mixed which is needed for the immediate application.
- Application of pesticides should be done at the proper time.
- Weather conditions and pest life cycle should also be kept in mind while planning applications.
- Store pesticides safely in a ventilated, well lighted, and secure area free from flooding.
- Disposal of empty containers and rinsing water should be done properly.
- Records of all pesticide use should be kept. Records will allow evaluation of pest control efforts and help plan future treatments.

Best Management Practices (BMPs)

These practices are used to protect water quality.

In order to reduce the chances of pesticides and sediments of moving off treated areas into receiving water bodies, science based holistic environmental management should be approached.

The rate, method, timing and type of chemicals to be applied should be controlled.

Appropriate pesticides rates should be used.

Use of pesticide should be as needed only to eliminate routine maintenance programmes.

An integrated pest management (IPM) approach should be adopted for pesticide application to use the least amount and least toxic of pesticides possible to achieve acceptable pest control.

CONCLUSION

The application of pesticides with the perspective of agricultural betterment, has not only affected the agricultural practice, but it has also altered the food chain and the ecosystem. Therefore, the use of pesticides should be under controlled and more of bio-pesticides should be administered. To reduce the pesticides, certain alternative methods can be used such as manual removal, application of heat, plastic covering of weeds, placement of traps and lures, removal of pest breeding sites, maintenance of healthy soils to have healthy breeds of more pest-resistant plants, to crop native species more as they are naturally more resistant to native pests and to support bio-control agents such as birds and other pest predators.

Presence of OCPs in water not only changes the physico-chemical characteristics of water but also impose drastic impact on aquatic organisms. OCPs get sorbed to soil and owing to its hydrophobic nature become persistent in soil and sediments (Tejomyee and Pravin, 2007). Degradation of pesticide through biological means using bacteria and fungi is receiving serious attention as compared to existing conventional methods and such alternative methods should be adopted to minimize environmental deterioration till more realistic policies are developed. It is suggested that stringent government regulations and monitoring of water bodies should be observed. Conclusively, the exposure to OCPs should be reduced so as to minimize the associated environmental and human health hazard.

REFERENCES

Augustijn-Beckers, P.W., Hornsby, A.G., & Wauchope, R.D. (1994). Additional Properties. In *Reviews of Environmental Contamination and Toxicology*. SCS/ARS/CES Pesticide Properties Database for Environmental Decision-making.

Castillo, L. E., Martinez, E., Ruepert, C., Savage, C., Gilek, M., Pinnock, M., & Solis, E. (2006). Water quality and macroinvertebrate community response following pesticide applications in a banana plantation, Limon, Costa Rica. *The Science of the Total Environment*, *367*(1), 418–432. doi:10.1016/j.scitotenv.2006.02.052 PMID:16643988

Chitranshi, U. R. (2004). R & D priorities for sustainable development of inland capture fishery sector in India. *Symposium on advances in Fish Physiology: Ecological Considerations, Varanasi*, 12-14.

Coat, S., Bocquené, G., & Godard, E. (2006). Contamination of some aquatic species with the organochlorine pesticide chlordecone in Martinique. *Aquatic Living Resources*, *19*(2), 181–187. doi:10.1051/alr:2006016

CPCB. (2000). *Water quality status of Yamuna river. Assessment & development of river basin series: ADSORBS 32, Central Pollution Control Board*. Delhi, India: CPCB.

Cunningham, W. P., & Saigo, B. W. (1998). Environmental Science: A Global Concern (5th ed.). WCB/McGraw-Hill.

David, P. (2002). Encyclopedia of Pest Management (vols. 1-2). CRC Press, Taylor & Francis Group.

De Lorenzo, M. E., Scott, G. I., & Ross, P. E. (2001). Toxicity of pesticides to aquatic microorganisms: A review. *Environmental Toxicology and Chemistry*, *20*(1), 84–98. doi:10.1002/etc.5620200108 PMID:11351418

Dearth, M. A., & Hites, R. A. (1991). Conplete analysis of technical chlordane using negative ionization mass spectrometry. *Environmental Science & Technology*, *25*(2), 245–254. doi:10.1021/es00014a005

FAO/WHO. (2010). *FAO Specifications and Evaluations for Agricultural Pesticides*. Endosulfan FAO/WHO evaluation report 89/2010.

Frankart, C., Eullaffroy, P., & Vernet, G. (2003). Comparative effects of four herbicides on non-photochemical fluorescence quenching in Lemna minor. *Environmental and Experimental Botany*, *49*(2), 159–168. doi:10.1016/S0098-8472(02)00067-9

Gao, F., Jia, J., & Wang, X. (2008). Occurrence & ordination of DDT & HCH in agricultural soils from Guangzhou, China. *Archives of Environmental Contamination and Toxicology*, *54*, 155–166. doi:10.100700244-007-9023-3 PMID:17763882

Grande, M., Andersen, S., & Berge, D. (1994). Effects of pesticides on fish. *Norwegian Journal of Agricultural Sciences*, *13*, 195–209.

Guo, Y., Meng, X. Z., Tang, H. L., & Zeng, E. Y. (2008). Tissue distribution of organochlorine pesticides in fish collected from Pearl River Delta, China: Implications for fishery input source & bioaccumulation. *Environmental Pollution*, *155*(1), 150–156. doi:10.1016/j.envpol.2007.10.025 PMID:18063460

Hanazato, T. (2001). Pesticide effects on freshwater zooplankton: An ecological perspective. *Environmental Pollution*, *112*(1), 1–10. doi:10.1016/S0269-7491(00)00110-X PMID:11202648

Harding, G. C. (1986). Organochlorine dynamics between zooplankton and their environment, a reassessment. *Marine Ecology Progress Series*, *33*, 167–191. doi:10.3354/meps033167

Harit, G., & Srivastava, N. (2007). Endosulfan uptake and biochemical profile of the muscle in a freshwater murrel, *Channa punctatus. Bulletin of Pure & Applied Sciences*, *26A*, 23–27.

Harit, G., & Srivastava, N. (2009). Bioaccumulation & elimination of endosulfan in muscle, gills & blood of *Channa punctatus. Journal of Environment & Sociobiology.*, *6*, 13–20.

Harit, G., & Srivastava, N. (2017a). Changes in gill structure induced by uptake of endosulfan in *Channa punctatus* followed by subsequent recovery. *Asian Journal of Applied Science and Technology*, *1*(7), 60–64.

Harit, G., & Srivastava, N. (2017b). Toxicity of endosulfan in *Channa punctatus* (Bloch). *Current Topics in Toxicology*, *13*, 23–30.

Hatzios, K. K. (1991). Biotransformation of herbicides in higher plants. Environmental Chemistry of Herbicides, 2, 141-185.

Helfrich, L. A. (2009). *Pesticides and Aquatic Animals: A Guide to Reducing Impacts on Aquatic Systems*. Blacksburg, VA: Virginia Department of Game and Inland Fisheries.

Iyamu, I., Asia, I. O., & Egwakhide, P. A. (2007). Concentrations of residues from organo-chlorine pesticide in water & fish from some rivers in Edo State Nigeria. *International Journal of Physical Sciences*, *2*, 237–241.

Jorgenson, J. L. (2001). Aldrin and Dieldrin: A Review of Research on Their Production, Environmental Deposition and Fate, Bioaccumulation, Toxicology, and Epidemiology in the United States. *Environmental Health Perspectives*, *109*(1), 113–139. doi:10.1289/ehp.01109s1113 PMID:11250811

Kannan, K., Tanabe, S., Borrell, A., Aguilar, A., Focardi, S., & Tatsukawa, R. (1993). Isomer specifi c analysis and toxic evaluation of polychlorinated biphenyls in food stuff s from India and their implications on human dietary exposure. *Journal of Agricultural and Food Chemistry*, *40*, 158.

Kawano, M., Tsuyoshi, I., Wada, T., Hidaka, H., & Tatsukacva, R. (1988). Bioconcentration and residue patterns of chlordane compounds in marine animals: Invertebrates, fish, mammals, and seabirds. *Environmental Science & Technology*, *22*(7), 792–797. doi:10.1021/es00172a008 PMID:22195663

Lakshmi, A. (1993). Pesticides in India: Risk assessment to aquatic ecosystem. *The Science of the Total Environment*, *134*(1), 243–253. doi:10.1016/S0048-9697(05)80025-7

Liess, M., Brown, C., Dohmen, P., Duquesne, S., Heimbach, F., & Kreuger, J. (2005). *Effects of Pesticides in the Field—EPIF*. Brussels, Belgium: SETAC Press.

Liu, Z., Zhang, H., Tao, M., Yang, S., Wang, L., Liu, Y., ... He, Z. (2010). Organochlorine Pesticides in Consumer Fish and Mollusks of Liaoning Province, China: Distribution and Human Exposure Implications. *Archives of Environmental Contamination and Toxicology*, *59*(3), 444–453. doi:10.100700244-010-9504-7 PMID:20352204

Mathew, L. L. (2012). *Organochloride Pesticide Toxicity, Drugs, Diseases and Procedures*. Medscape References.

Mowbray, D. L. (1988). *Pesticide use In the South Pacific. UNEP Regional Seas Reports and Studies No. 89*. Nairobi: United Nations Environment Programme.

Naqvi, S. M., & Vaishnavi, C. (1993). Bioaccumulative potential & toxicity of endosulfan insecticide to non-target animals. *Comparative Biochemistry & Physiology C*, *3*(3), 347–361. doi:10.1016/0742-8413(93)90071-R PMID:7900959

Norstrom, R. J., Simon, M., Muir, C. G., & Schweinsburq, R. E. (1988). Organochlorine contaminants in arctic marine food chains: Identification, geographical distribution, and temporal trends in polar bears. *Environmental Science & Technology*, *22*(9), 1063–1107. doi:10.1021/es00174a011 PMID:22148661

Parker, L. W., Rydenvhi, J., Thomas, S., & Whitford, W. G. (1985). Effects of oxamyl and chlordane on the activities of nontarget soil organisms. *Biology and Fertility of Soils*, *1*(3), 141–148. doi:10.1007/BF00301781

Prashanth, M. S., & Neelagund, S. E. (2008). Impact of cypermethrin on enzyme activities in the freshwater fish *Cirrhinus mrigala* (Ham.). *Caspian Journal of Environmental Sciences*, *6*, 91–95.

Quijano, R. (2002). *Endosulfan poisoning in Kasaragod, Kerala, India*. Report on a fact finding mission, PAN.

Rathore, M., Bhatnagar, P., Mathur, D., & Saxena, G. N. (2002). Burden of organochlorine pesticide in blood and its effect on thyroid hormones in women. *The Science of the Total Environment*, *295*(1-3), 207–215. doi:10.1016/S0048-9697(02)00094-3 PMID:12186288

Rawat, D. K., Bais, V. S., & Agrawal, N. C. (2002). A correlative study on liver glycogen and endosulfan toxicity in *Heteropneustes fossilis* (Bloch). *Journal of Environmental Biology*, *23*, 205–207. PMID:12602859

Ritter, L., Solomon, K. R., Forget, J., Stemeroff, M., & Leary, C. O. (1995). *An Assessment Report on: DDT, Aldrin, Dieldrin, Endrin, Chlordane, Heptachlor, Hexachlorobenzene, Mirex, Toxaphene, Polychlorinated Biphenyls, Dioxins and Furans. Ritter-Report. International Programme on Chemical Safety*. Genf: IPCS.

Salvo, L. M., Sinhorini, I. L., Malucelli, B. E., Klemz, C., Sanchez, D. C. O., Nicaretta, L., ... Assis, H. C. S. (2008). Effects of endosulfan sub-lethal concentrations on carp (*Cyprinus carpio, Linnaeus*, 1758): Morphometric, hystologic, ultrastructural analyses & cholinesterase activity evaluation. *Brazilian Journal of Veterinary Research & Animal Science, Sao Paulo*, *45*, 87–94. doi:10.11606/issn.1678-4456.bjvras.2008.26704

Shimabukuro, R. H. (1985). Detoxification of herbicides. We Physiology, 2, 215-240.

Singh, N.N., & Srivastava, A.K. (1995). Sublethal physiological effects on catfish, *Heteropneustes fossilis* (Bloch) during longterm exposure to aldrin and subsequent recovery. *Proceedings of National Academy of Science, India, 65*, 33-38.

Siriwong, W., Thirakhupt, K., Sitticharoenchai, D., Rohitrattana, J., Thongkongowm, P., Borjan, M., & Robson, M. (2009). DDT and Derivatives in Indicator Species of the Aquatic Food Web of Rangsit Agricultural Area, Central Thailand. *Ecological Indicators*, *9*(5), 878–882. doi:10.1016/j.ecolind.2008.10.004 PMID:20161116

Srivastava, A. K., & Srivastava, A. K. (1988). Chlordane induced changes in carbohydrate metabolism of the Indian catfish *Heteropneustes fossilis* (Bloch). *Bulletin of the Institute of Zoology, Academia Sinica*, *27*, 211–215.

Srivastava, N., Agrawal, M., Tyagi, A., & Kaushik, N. (2002). Endosulfan induced changes in the fish *Channa punctatus*. *Proceeding of XI National Symposium of Environment*, 356-358.

Srivastava, N. & Kaushik, N. (2001). Use of fish as bioindicator of aquatic pollution. *Proceeding of ICCE*, 227-229.

Tejomyee, S. B., & Pravin, R. P. (2007). Biodegradation of organochlorine pesticide, endosulfan, by a fungal soil isolate, *Aspergillus niger*. *International Biodeterioration & Biodegradation*, *59*(4), 315–321. doi:10.1016/j.ibiod.2006.09.002

Ullah, S., Hasan, Z., & Dhama, K. (2016). Toxic effects of endosulfan on behaviour, protein contents and antioxidant enzyme system in gills, brain, liver and muscle tissues of rohu, *Labeo rohita*. *International Journal of Pharmacology*, *12*(1), 1–10. doi:10.3923/ijp.2016.1.10

Veil, J. F. (1990). Public health impact of pesticides used in agriculture. Geneva, Switzerland: WHO (World Health Organization).

Vivekanandhan, N., & Duraisamy, A. (2012). Ecological Impact of Pesticides Principally Organochlorine Insecticide Endosulfan: A Review. *Universal Journal of Environmental Research and Technology*, *2*(5), 369–376.

Wolkers, H., Boily, F., Gremmels, J. F., Bavel, B. V., Hammill, M. O., & Primireerio, R. (2009). Tissue-Specific contaminant accumulation & associated effects on hepatic serum analytes & cytochrome P450 enzyme activities in hooded seals (*Cystophora cristata*) from the gulf of St. Lawrence. *Archives of Environmental Contamination and Toxicology*, *56*(2), 360–370. doi:10.100700244-008-9186-6 PMID:18626617

Wood, L. W., Jones, P. A., & Richards, A. (1986). Possible sediment scavenging of chlordane and contamination of the aquatic biota in Belmont Lake, New York. *Bulletin of Environmental Contamination and Toxicology*, *36*(1), 159–167. doi:10.1007/BF01623489 PMID:3947750

World Health Organization. (1986). *The WHO recommended classification of pesticide biohazard. Guideline to classification 1986-1987*. Geneva: World Health Organization.

Worthing, C. R., & Walker, S. B. (Eds.). (1983). The pesticide manual, a world compendium (7th ed.). The British Crop Protection Council.

ADDITIONAL READING

Jayaraj, R., Megha, P., & Sreedev, P. (2016). Organochlorine pesticides, their toxic effects on living organisms and their fate in the environment. *Interdisciplinary Toxicology*, *9*(3–4), 90–100. doi:10.1515/intox-2016-0012 PMID:28652852

Pridmore, R. D., Thrush, S. F., Wilcock, R. J., Smith, T. J., Hewitt, J. E., & Cummings, V. J. (1991). Effect of the organochlorine pesticide technical chlordane on the population structure of suspension and deposit feeding bivalves. *Marine Ecology Progress Series*, *76*, 261–271. doi:10.3354/meps076261

KEY TERMS AND DEFINITIONS

Aquatic Biota: Organisms living in or depending on aquatic environment.

Carcinogenic: Having the potential to cause cancer.

Lipophilic Compounds: Compounds that have tendency to get dissolved in lipids or fats.

Mutagenic: Having the potential to cause changes at genetic level.

Organochlorine Pesticides: Chlorinated hydrocarbons that are used extensively in agriculture and vector control.

Pesticide Label: The guide that gives directions for safe and effective use of pesticide.

Chapter 4
Status of Indian Wetlands With Special Reference to Pesticides and Their Impact

Javid Manzoor
Jiwaji University Gwalior, India

Manoj Sharma
Jiwaji University Gwalior, India

Irfan Rashid Sofi
Jiwaji University Gwalior, India

Mufida Fayaz
Jiwaji University Gwalior, India

Musadiq Hussain Bhat
Jiwaji University Gwalior, India

ABSTRACT

Wetlands are home to numerous species of fish, birds, and reptiles. The enormous roots of the mangrove trees act as shelter to small fish, reptiles, and amphibians. Pesticides and agro-chemical fertilizers have been playing a very pivotal role in the degradation of the land and the water bodies. The different herbicides that are present in wetlands are Dicamba, Endothall, MCPA, Triallate, Trifluralin, 2, 4-D, and insecticides Carbaryl, Carbofuran, Fenvalerate, Malathion, Parathion, and Terbufos. These pesticides have been provided with the aim of catering to the security of the crops which are highly vulnerable to the pests. However, harmful effects of pesticides on wetland species have been a concern for long time. Wetlands constitute one such habitat threatened by the pesticides. But there has been a lack of comprehensive research in this direction. The chapter will identify the gaps in the current research and will review the status of Indian wetlands with special reference to pesticides and their impact.

DOI: 10.4018/978-1-5225-6111-8.ch004

INTRODUCTION

Wetlands are the precious life-sustaining water resources in India. Besides playing a vital role in the hydrological cycle, wetlands are the most productive ecosystems of the world and a potential source of carbon sequestration, although they covers about 4% of the earth's ice-free land surface (Prigent *et al.*, 2001). Ramsar Convention on Wetlands, which is an international treaty signed in 1971 for national action and international cooperation for the conservation and wise use of wetlands and their resources, defines wetlands as "areas of marsh, fen, peatland or water, whether natural or artificial, permanent or temporary, with water that is static or flowing, fresh, brackish or salt, including areas of marine water the depth of which at low tide does not exceed six metres".

Globally, the areal extent of wetland ecosystems ranges from 917 million hectares (Lehner and Doll, 2004) to more than 1275 m ha (Finlayson and Spiers, 1999). One of the first widely used wetland classifications systems (devised by Cowardin et al., 1979) categorized wetlands into marine (coastal wetlands), estuarine (including deltas, tidal marshes, and mangrove swamps), lacustarine (lakes), riverine (along rivers and streams), and palustarine ('marshy'– marshes, swamps and bogs) based on their hydrological, ecological and geological characteristics.

The Ramsar Classification is exhaustive, given under three broad groups as Marine/Coastal wetlands, Inland wetlands and Human-made wetlands. There are twelve categories under Marine/Coastal wetlands, twenty categories under Inland wetlands and ten categories under Human-made wetlands, as given below:

- Marine/Coastal wetlands
- Inland wetlands
- Human-made wetlands

Wetlands are ecologically sensitive and adaptive systems (Turner *et al.*, 2000). Wetlands exhibit enormous diversity according to their genesis, geographical location, water regime and chemistry, dominant species, and soil and sediment characteristics (Space Applications Centre, 2011). However, many wetlands which perform potentially valuable functions are continued to be ignored in the policy process. As a result many freshwater wetland ecosystems are threatened and many are already degraded and lost due to urbanization, population growth, and increased economic activities (Central Pollution Control Board, 2008) (Figure 1). Wetlands being dynamic and influenced by both natural and man-made activities, need frequent monitoring. Regular updation of the status of the wetlands is all the more significant in view of the accelerating pressure on the very existence of these resources due to developmental activities and population pressure being witnessed currently.

STATUS OF WETLANDS IN INDIA

India with its annual rainfall of over 1,300 mm, varying topography and climatic regimes, supports diverse and unique wetland habitats (Prasad *et al.*, 2002). The available estimates about the area extent of wetlands in India vary widely from a lowest of 1% to a highest of 5% of geographical area (Space Applications Centre, 2011).

Figure 1. Wetlands under threat

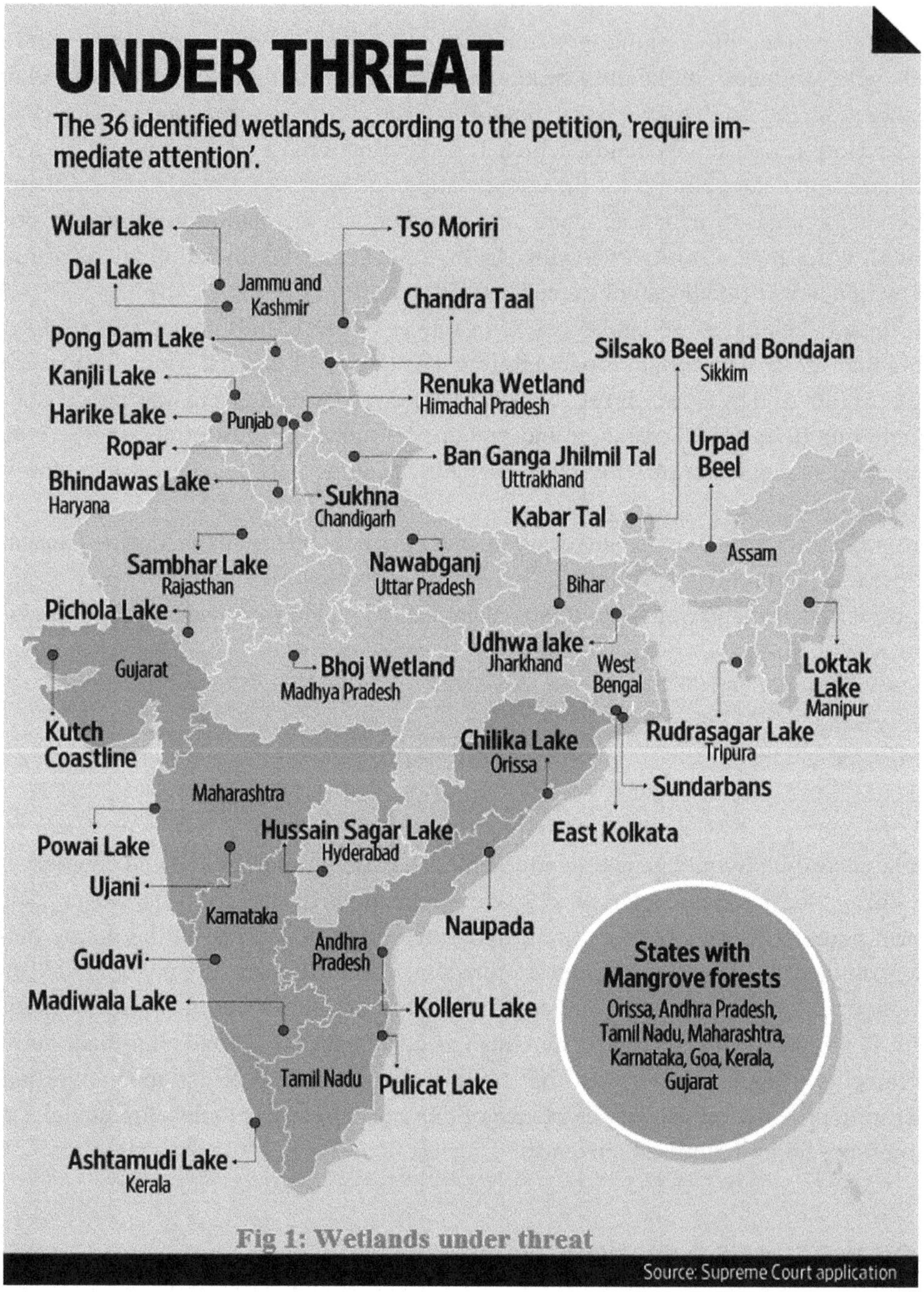

Natural wetlands in India consist of the high-altitude Himalayan lakes, followed by wetlands situated in the flood plains of the major river systems, saline and temporary wetlands of the arid and semi-arid regions, coastal wetlands such as lagoons, backwaters and estuaries, mangrove swamps, coral reefs and marine wetlands. With the exception of bogs, fens and typical salt marshes, Indian wetlands cover the whole range of wetland ecosystem types elsewhere. In addition to the various types of natural wetlands found, a large number of man-made wetlands also contribute in sustaining the faunal and floral diversity.

These wetlands include seasonally flooded as well as permanent marshes and swamps in shallow lakes, large river flood plains and littoral zones of large lakes and reservoirs. Along the entire coast there are several lagoons, estuarine backwaters and extensive mangroves in the deltas and estuaries of numerous rivers. There are also marine wetlands which include coastal beds of marine algae and coral reefs (Gopal and Sah, 1995). India has a wealth of wetland eco-systems distributed in different geographical regions from the cold arid zone of Ladakh in the North to the wet humid climate of Imphal in the East, the warm arid zone of Rajasthan in the west to the tropical monsoonal Central India and the wet and humid zone of Southern Peninsula. Most of the wetlands of India are part of the major river systems such as Ganga, Brahmaputra, Narmada, Tapti, Godavari, Krishna, Kaveri etc (Garg *et al., 1998*). The Ministry of Environment and Forests, Government of India (1992) estimated that 4.7 million ha area are wetlands, of which 1.5 million ha is natural, 2.6 million man made and 0.6 million ha mangrove vegetation. According to the nationwide wetland inventory there are 27,403 wetland units in the country occupying 7.6 million ha of which coastal wetlands occupies 3959 units and 4 million ha whereas inland wetlands are of 23444 units with 3.6 million ha (Ramsar, 2002). During recent decades, the rapid increase in human population and demand for natural resources for food, fuel and fodder have resulted in rapid deterioration and decline of all kinds of wetlands throughout the south Asian region. Diverse human activities on adjacent land and in water further aggravate the problem of wetland decline. It is interesting to also point out the impacts of introduced plants and animals on natural biota and processes in both mangroves and freshwater wetlands. Realizing the importance of wetlands in India, Ministry of Environment and Forests (MoEF), Government of India, has published a directory of wetlands (1990), based on the survey carried out during 1972. The total wetland area in the country excluding the area under paddy, rivers and canals has been estimated to be about 7.6 million hectors out of which 3.6 Million hectors are inland and 4 million hectors are coastal wetlands. (Kaul, 2007).

Out of the total area under wetlands, area under inland wetlands accounts for 69%, coastal wetlands 27%, and other wetlands (smaller than 2.25 ha) 4% (Figure 2). In terms of average area under each type of wetland, natural coastal wetlands have the largest area. The water spread area of wetlands varies greatly. Overall, inland wetlands have a water spread area of 7.4 m ha in post monsoon and 4.8 m ha in pre-monsoon; and coastal wetlands have 1.2 m ha and1 m ha in post monsoon and pre monsoon, respectively. Overall, reduction in water spread area of inland wetlands is highest (35%) followed by that of coastal wetlands (16%).Within inland wetlands, reduction is significantly higher in man-made types (49.5%), such as surface reservoirs and tanks, in comparison to natural types (24%), such as lakes and ponds, as they are under pressure to meet various irrigational and non-irrigational needs and are also subjected to higher evaporation losses (SAC, 2011).

Figure 2. Total area under Wetlands
As per the National Wetland Atlas 2011.

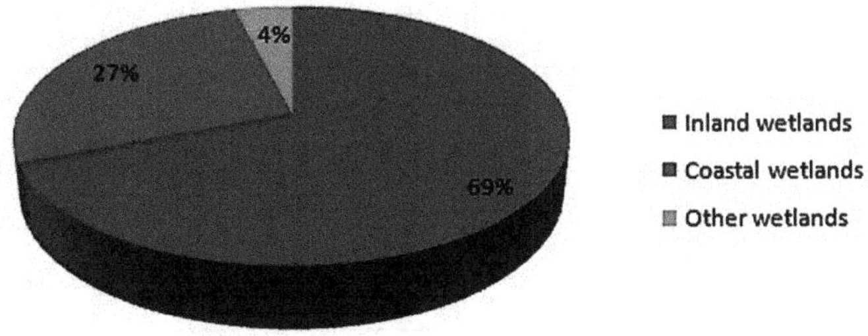

REGIONAL DISTRIBUTION OF WETLANDS IN INDIA

In terms of the proportion of the geographical area, Gujarat has the highest proportion (17.5%) and Mizoram has the lowest proportion (0.66%) of the area under wetlands. Among Union Territories in India, Lakshadweep has the highest proportion (around 96%) and Chandigarh has the least proportion (3%) of geographical area under wetlands. Gujarat has the highest proportion (22.8%) and UT of Chandigarh has nearly negligible part of the total wetland area in the country. Water-spread area of wetlands changes over seasons. The States of Sikkim, Nagaland, Mizoram, Meghalaya, and Jharkhand have more than 90% of the total wetland area as water spread area during post monsoon. Significant reduction in water spread area of wetlands from post monsoon to pre monsoon was found in the States of Uttar Pradesh (28%), Chhattisgarh (29%), Himachal Pradesh (29%), Tripura (29%), Sikkim (30%), Andhra Pradesh (31%), Jharkhand (32.5%), Punjab (33%), Bihar (34%), Gujarat (36%), Karnataka (38.5%), Maharashtra (53.5%), Tamil Nadu (55%),Madhya Pradesh (57%), and Rajasthan (57%).

In terms of contribution of the total water spread area in the country, highest during post monsoon was observed in the State of Gujarat (13.5%) and lowest in Sikkim and Tripura (0.1% each). During pre-monsoon, highest was again in Gujarat (12.6%) and lowest was in Sikkim and Tripura (0.1% each).As regards percentage area under aquatic vegetation, Andhra Pradesh, Delhi, Karnataka, Manipur, Orissa, Punjab, Tamil Nadu, Tripura, and West Bengal have 15–59% of the wetland area under aquatic vegetation. Further, Andhra Pradesh, Gujarat, Karnataka, Orissa, Tamil Nadu, Uttar Pradesh, and West Bengal account for nearly 3/4th of the total area under aquatic vegetation.

Table 1. Significant reduction in water spread area of wetlands from post monsoon to pre-monsoon was found in the States given in the table

S. No	State	Percentage
01	Uttar Pradesh	28
02	Chhattisgarh	29
03	Himachal Pradesh	29
04	Tripura	29
05	Sikkim	30
06	Andhra Pradesh	31
07	Jharkhand	32.5
08	Punjab	33
09	Bihar	34
10	Gujarat	36
11	Karnataka	38.5
12	Maharashtra	53.5
13	Tamil Nadu	55
14	Madhya Pradesh	57
15	Rajasthan	57

The negative economic, social, and environmental consequences of declining water quality in wetlands are also an issue of concern for India. The problem of deteriorating water quality is particularly more alarming in the case of small water bodies such as lakes, tanks and ponds.

Despite all the benefits, many decision-makers and even many of the 'primary stakeholders' think of them as 'wastelands'. Everyone claims a stake in them, as they are in the open access regime, but rarely are willing to pay for this extractive use. These freshwater bodies are often subject to changes in land use in their catchments leading to reduction in inflows and deteriorating quality of the "runoff" traversing through agricultural fields and urban areas. On the other hand, many of them act as the "sink" for untreated effluents from urban centres and industries. Encroachment of reservoir area for urban development, excessive diversion of water for agriculture is yet another major problem (Verma et al., 2001). Lack of conformity among government policies in the areas of economics, environment, nature conservation, development planning is one reason for the deterioration of these water bodies (Turner et al., 2000). Lack of good governance and management are also major reasons (Kumar et al., 2013). Hence various legal and policy approaches should be adopted in India for their conservation and management.

THREATS TO WETLAND ECOSYSTEM

Freshwater wetland ecosystems are among the mostly heavily used, depended upon and exploited ecosystems for sustainability and well-being (Molur et al., 2011). More than 50% of specific types of wetlands in parts of North America, Europe, Australia, and New Zealand were converted during the twentieth century (MEA, 2005). In Asia alone, about 5000 km^2 of wetland area are lost annually to agriculture, dam construction, and other uses (McAllister et al., 2001). Further, dependence on water and other resources in this environment has placed enormous pressures on the ecosystem worldwide resulting in direct impacts to species diversity and populations (Molur et al., 2011). As a result many wetland dependent species including 21% of bird species; 37% of mammal species; and 20% of freshwater fish species are either extinct or globally threatened as shown in (Figure 3) (MEA, 2005). Loss in wetland area results in adverse impact on the key functions (ecosystem goods and services) performed by wetlands (Zedler and Kercher, 2005). Worldwide, the main causes of wetland loss have been: urbanization; land use changes; drainage to agricultural use; infrastructure development; pollution from industrial effluent and agricultural runoff; climate change and variability. Some of these factors which led to significant alterations in India's wetland ecosystems have been discussed below:

Figure 3. Wetland dependent species

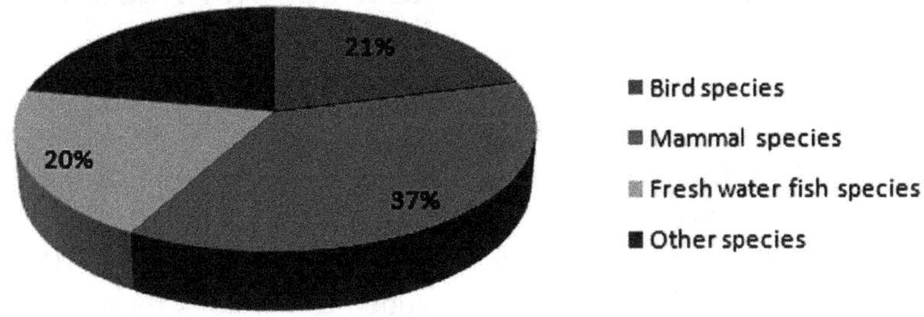

Urbanization

Between 1951 and 2011, total population in India increased from 0.4 billion to 1.2 billion with an average decadal growth rate of around 22%. During the 90 year period from 1901 to 1991, the number of urban centres doubled while urban population has increased eightfold (Bassi and Kumar, 2012). This magnitude of growth exerted tremendous pressure on wetlands and flood plain areas for meeting water and food demand of growing population. Between 1950–1951 and 2008–2009, total cultivated land in India increased from about 129 to 156 m ha. Also, area under non-agricultural uses (commercial or residential use) increased from 9 to 26 m ha (Data Source: India stat). In most of the major river basins of India, the increase in area for both agricultural and non-agricultural use was at the cost of conversion of flood plain areas, primary forests, grasslands and associated fresh water ecosystems to meet demands of growing population (Zhao et al., 2006). For instance, about 34,000 ha of the water spread area of the Kolleru lake (Andhra Pradesh) have been reclaimed for agriculture in recent years (MoEF, 1990). Further, there was a large scale development of irrigation and water supply infrastructure in the country which altered the inflows and water spread areas of many water bodies. Till 2007, about 276 major and 1000 medium irrigation projects were completed in India (Central Water Commission, 2009), with an estimated total water storage capacity of about 225 BCM (12% of total water resources potential of India). Though, the large reservoir projects have played a critical role in water supply; flood control; irrigation; and hydroelectric power production, the rapid proliferation of artificial water impounding structures without proper hydrological and economic planning (such as construction of small dams in semi-arid and arid regions where runoff potential is limited) has caused widespread loss and fragmentation of freshwater habitats (Zhao et al., 2006); and reduction in environmental flows (due to over allocation of water mainly for meeting agricultural and industrial water demands). Already, most of the river basins in southern and western India are experiencing environmental water scarcity, which means the discharge in these basins has already been reduced by water withdrawals to such levels that the amount of water left in the basin is less than that required by the freshwater dependent ecosystems (Smakhtin et al., 2004).Urbanization exerts significant influences on the structure and function of wetlands, mainly through modifying the hydrological and sedimentation regimes, and the dynamics of nutrients and chemical pollutants. Impact of urbanization is equally alarming on natural water bodies in the cities. A study found that out of 629 water bodies identified in the National Capital Territory of Delhi, as many as 232 cannot be revived on account of large scale encroachments (Khandekar, 2011). Similarly, between 1973 and 2007, Greater Bengaluru Region lost 66 wetlands with a water spread area of around 1100 ha due to urban sprawl (Ramachandra and Kumar, 2008). Further, poor management of water bodies, lack of concrete conservation plans, rising pollution, and rapid increase in localized demands for water are pushing these precious eco-balancers to extinction (Indian National Trust for Art and Cultural Heritage, 1998).

Pollution Due to Agricultural, Municipal, and Industrial Wastes

Water in most Asian rivers, lakes, streams and wetlands has been heavily degraded, mainly due to agricultural runoff of pesticides and fertilizers, and industrial and municipal wastewater discharges, all of which cause widespread eutrophication (Liu and Diamond, 2005; Prasad et al., 2002).As a result of intensification of agricultural activities over the past four decades, fertilizer consumption in India has increased from about 2.8 million tonne in 1973–1974 to 28.3 million tonne in 2010–2011 (Data Source: Indiastat). As per estimates, 10–15% of the nutrients added to the soils through fertilizers eventually find their way to the surface water system (Indian Institute of Technology, 2011). High nutrient contents stimulate algal growth, leading to eutrophication of surface water bodies. Studies indicate that 0.5 mg/l of inorganic Nitrogen and 0.01 mg/1 of organic Phosphorus in water usually stimulates undesirable algal growth in the surface water. Runoff from agricultural fields is the major source of non-point pollution for the Indian rivers flowing through Indo-Genetic plains (Jain et al., 2007a, b). Water from lakes that experience algal blooms is more expensive to purify for drinking or other industrial uses. Eutrophication can reduce or eliminate fish populations (Verhoeven et al., 2006) and can also result in loss of many of the cultural services provided by lakes. Along with runoff from agricultural fields, untreated wastewater also contributes significantly to pollution of water bodies. Less than 31% of the domestic wastewater from Indian urban centres is treated, compared to 80% in the developed world. In total of 35 metropolitan cities, treatment capacity exists for only 51% of the sewage generated. Conditions in smaller urban centres are even worse as treatment capacity exist for only about 18% of the sewage generated in Class I cities (population size of 100,000 or more but other than metropolitan cities) and 9% of the sewage generated in Class II towns (population between 50,000 and 100,000). Actual sewage treatment will be further low due to inadequacy of the sewage collection system and non-functional treatment plants. Thus, there is a huge gap in generation and treatment of wastewater in Indian urban centres and most of sewage is discharged without treatment in the natural water bodies such as streams and rivers (Central Pollution Control Board, 2009).Results from monitoring of Indian aquatic resources also show that water bodies, such as rivers and lakes, near to urban centres are becoming increasingly saprobic and eutrophicated due to the discharge of partly treated or untreated waste water (Central Pollution Control Board, 2010). River Yamuna, which passes through 6 Indian States, receives about 1789 MLD of untreated waste water from the capital city of Delhi alone. This is about 78% of the total pollution load that flows in to the river every day. As a result the water quality and hydrological character in the Delhi segment of the river is the most polluted as compared to other stretches in terms dissolved oxygen (DO) and biological oxygen demand (BOD). The DO level had decreased to 1.41 from 8.05 in the Himalayan segment and the BOD level has risen to 17.2 from 2.8.

Other Threats

Global climate change is expected to become an important driver of loss and change in wet-land ecosystem (MEA, 2005; UNESCO, 2007). These findings are important for Indian subcontinent where the mean atmospheric temperature and frequency of occurrence of intense rainfall events has increased, while the number of rainy days and total annual amount of precipitation have decreased due to increase in the concentration of greenhouse gases such as CO_2, CH_4 and N_2O in the atmosphere(Bates et al., 2008).Limited analysis on the impact of climate change on wetlands in India suggests that high altitude wetlands and coastal wetlands (including mangroves and coral reefs) are some of the most sensitive classes that will be affected by climate change (Patel et al., 2009). For instance, climate change induced rising level of glacial fed high altitude lakes, such as Tsomoriri in Ladakh, has submerged important breeding islands in the lake where endangered migratory birds like the Black-necked Crane and Bar-headed Goose would breed (Chandan et al., 2007). In case of the coastal wetlands such as Indian part of Sunderbans mangrove, rising sea surface temperature and sea level rise due to thermal expansion, could affect the fish distribution and lead to the destruction of significant portion of mangrove ecosystem. Further destruction of the Sundarbans mangroves would diminish their critical role as natural buffers against tropical cyclones resulting in loss of lives and livelihoods (Centre for Science and Environment, 2012; UNESCO, 2007).

The limited analysis also seems to suggest that the inland natural wetlands, especially those in arid and semi-arid regions, will be impacted through alteration in its hydrological regime due to changes in precipitation, runoff, temperature and evapotranspiration (Patel et al., 2009). Climate change induced rising temperature and declining rainfall pattern presents a potential danger to the already disappearing lakes in the Gangetic plains (Sinha, 2011). Decreased precipitation will exacerbate problems associated with already growing demands for water and hence alter the freshwater inflows to wetland ecosystems (Bates et al., 2008), whereas, rise in temperature can aggravate the problem of eutrophication, leading to algal blooms, fish kills, and dead zones in the surface water (Gopal et al.,2010). Also, seasonality of runoff in river basins (such as Ganges) will increase along with global warming, that is, wet seasons will become wetter and dry seasons will become drier (World Bank, 2012).This would have severe adverse impact on affected populations, especially if the seasonality of runoff change would be out of phase with that of demand. As per estimates, India will lose about 84% of coastal wetlands and 13% of saline wetlands with climate change induced sea water rise of 1 m (Blankespoor et al., 2012). As a result there will be adverse consequences on wetland species, especially those that cannot relocate to suitable habitats, as well as migratory species that rely on a variety of wetland types throughout their life cycle. However, it must be noted that projections about the extent of loss and degradation or decline of wetlands are not yet well established as climate models used for such predictions are not robust. It is not clear how the regions' temporal and spatial variability in rainfall gets captured by these models. Further, there is tendency to attribute hydrological regime changes in wetlands to climate change, rather than trying to find the real physical and socio-economic processes responsible for such changes (Kumar, 2013).

Pollution of wetlands by agricultural pesticides can cause different types of damage, from altering the growth of aquatic plants to reducing water fowl reproduction. This happens when broad spectrum pesticides directed at pests in cropland accidentally injure plant and animal species in nearby wetlands.

HERBICIDE TYPES AND DESCRIPTIONS

Herbicides are, by definition, toxic to plants. They can be separated into two groups based on their mode of action. Contact herbicides are applied in sufficient amounts to thoroughly cover stems and leaves of growing plants, affecting only the parts of the plant they actually "contact." Translocated (or systemic) herbicides "move" to other parts of a plant from the point of application (contact with foliage, soil incorporation, irrigation water uptake, etc.), where they alter normal plant functions such as growth, respiration, or photosynthesis. While some herbicides have a direct effect on non target vertebrate wildlife, most do not. By reducing plant cover and availability, however, herbicides may have indirect effects of populations.

INSECTICIDE TYPES AND DESCRIPTIONS

Insecticides are generally more toxic to animals than herbicides. Most modern insecticides are not as persistent as some used in the past, such as DDT. They also break down more quickly in the environment and generally do not accumulate to high concentrations in animal bodies. The two primary groups of insecticides in use today are organophosphates and carbamates, both of which kill pest insects by damaging their central nervous systems. Unfortunately, these insecticides have the same action on non target aquatic and terrestrial insects, fish, and wildlife. A third group of insecticides that has increasingly been used against pest insects in recent years is pyrethroids. These insecticides are synthetic formulations of naturally occurring insecticides that also damage the nervous systems of non target animals. However, toxic effects in birds and mammals are much less than those in fish, which have nervous systems that are more sensitive to these chemicals.

OTHER PESTICIDES

Other types of agricultural chemicals used to either reduce seed damage or increase the effectiveness of pesticide applications may also adversely affect plants and animals. These include fungicides and fumigants, surfactants, and drift retardants. Although used in smaller quantities than herbicides and insecticides, they must be used with caution to reduce effects on non target plants and animals.

PESTICIDE LETHAL AND SUB-LETHAL EFFECTS

The negative impacts of agricultural pesticides are more dramatic when plants and animals are observed dead after only a short period of exposure. This type of lethal effect is caused by direct, short term exposure of animals to a pesticide either by eating contaminated food, by drinking contaminated water, by breathing the pesticide, by absorbing the pesticide through the gills or skin, or by swallowing the pesticide while grooming. Plants are exposed by uptake of contaminated water through their roots or by direct pesticide deposits on leaf surfaces. Aquatic plants, like algae, are more sensitive than fish or invertebrates to contaminants such as herbicides. They form the base of the aquatic food web and impacts on them can cause adverse effects on all higher animal levels in a wetland ecosystem. However, the worst effects of pesticides on wetlands are those that harm a plant or animal in some way that cannot be observed immediately. These sublethal effects are due to pesticide-induced changes in the functions of enzymes, cells, or organs of plants and animals that in turn alter how a plant or animal competes for living space and food, avoids predators, reproduces, etc. Plants or animals harmed by pesticides may show these changes in two ways: altered population structure or altered community or ecosystem structure. For example, when a waterfowl population is altered, the flock may contain only adults and no juveniles because the youngsters were susceptible to pesticide poisoning. In a community or ecosystem structure harmed by sub-lethal pesticide exposure, a predator species may have been destroyed, resulting in excessively large populations of the animals who were once prey. When a wide variety of agricultural pesticides are used on croplands near wetlands, wetland plants and animals may be exposed to several herbicides and insecticides at the same time. This makes it difficult to determine if the wetland species were harmed from direct exposure to a single pesticide or indirectly through secondary changes in their population or ecosystem. In addition, pesticides break down in the environment, making it hard to tell whether toxic effects are linked to the original pesticide or its breakdown products.

IMPACT OF PESTICIDES ON BIRDS IN AND AROUND WETLANDS IN INDIA

Wetland birds or Water birds are under growing threat of pesticide toxicity as death of wetland birds due to pesticide poisoning. There are recent records of mortality of wetland birds such as Spot-billed Pelican, Painted Stork, Eurasian Spoonbill, Little Egret, Black-headed Ibis, and Black crowned Night Heron due to phosphamidon poisoning at Anna Zoological Park, Chennai; of the Red crested Pochard and Common Moorhen in Sitarganj Forest Range, Uttarakhand due to Chorpyrifos poisoning; of the Demoiselle Cranes in Amreli and Surendranagar districts of Gujarat due to phorate poisoning (as also rodenticide poisoning). Moreover, intentional pesticide poisoning of waterbirds (like Black-crowned Night Heron and Purple Heron) using carbofuran pesticides has been recorded at Virundunagar, Tamil Nadu. What is most disturbing is that what gets published about pesticide poisoning in birds of wetlands is a miniscule percentage of the reality Muralidharan et al., 2014. Pesticides not only threaten the existence of waterbirds; they endanger the existence of terrestrial birds too which live in the fields adjoining wetlands. Agricultural fields are often established on the banks of wetlands for easy and local irrigation water availability. But if such wetland waters are contaminated with pesticides, they can be the source of bioaccumulation and biomagnifications of toxic substances even in the body tissues of terrestrial birds like Indian Peafowl-the national bird of India and Schedule-I species as per Indian Wildlife (Protection) Act, 1972. Seeds and shoots of crops that are treated with pesticides can even cause direct toxicity in such birds. Since last 25 years, there have been reports of increasing large-scale mortalities due to increased use of insecticides/pesticides in agricultural lands Choudhury et al., 2007.

SUBLETHAL EFFECTS

Not all pesticide poisonings result in the immediate death of an animal. Small "sublethal" doses of some pesticides can lead to changes in behaviour, weight loss, impaired reproduction, inability to avoid predators, and lowered tolerance to extreme temperature.

Fish in streams flowing through croplands and orchards are likely to receive repeated low doses of pesticides if continuous pesticide applications run-off fields. Repeated exposure to certain pesticides can result in reduced fish egg production and hatching, nest and brood abandonment, lower resistance to disease, decreased body weight, hormonal changes, and reduced avoidance of predators. The overall consequences of sublethal doses of pesticides can be reduced adult survival and lowered population abundance. Sublethal Effects include:

Weight Loss, Low Diseases Resistance, Sterility, Reduced Egg Production, Loss of Attention and Low Predator Avoidance.

Pesticides can reduce the availability of plants and insects that serve as habitat and food for fish and other aquatic animals. Insect-eating fish can lose a portion of their food supply when pesticides are applied. A sudden, inadequate supply of insects can force fish to range farther in search of food, where they may risk greater exposure to predation. Spraying herbicides can also reduce reproductive success of fish and aquatic animals. The shallow, weedy nursery areas for many fish species provide abundant food and shelter for young fish. Spraying herbicides near weedy nurseries can reduce the amount of cover and shelter that young fish need in order to hide from predators and to feed. Most young fish depend on aquatic plants as refuge in their nursery areas. Aquatic plants provide as much as 80% of the dissolved oxygen necessary for aquatic life in ponds and lakes. Spraying herbicides to kill all aquatic plants can result in severely low oxygen levels and the suffocation of fish. Using herbicides to completely "clean up" a pond will significantly reduce fish habitat, food supply, dissolved oxygen, and fish productivity. The land owner who sprays a weedy fence line with herbicides may unintentionally kill the trumpet vine on which humming birds feed and the honeysuckle that nourish deer and quail. Similarly, the landowner who unnecessarily sprays his water plants kills the plants that fed the insects that fed the fish that fed the farmer. Casual use of herbicides for lake or farm pond "beautification" may reduce fish populations.

Toxic Effect of Pesticides in Fauna

Wild birds are of great importance to the ecosystem. Decline in the bird community serves as an indicator of environmental pollution. Continuous use of pesticides is one of the major causes for the reduction of birds. In many cases the impact is not direct, however repetitive use of pesticides like DDT in soil is taken up by earthworms which are then ingested by birds and thus their accumulation may result in a large loss in bird population (Fry, 1995). Subsequent research has also identified other pesticides and industrial chemicals that cause mortality and reproductive impairment, which affects both embryos and adult birds. The effects on embryos include mortality or reduced hatchability, wasting syndrome and teratological effects that produce skeletal abnormalities and impaired differentiation of the reproductive and nervous systems through mechanisms of hormonal mimicking of estrogens. The range of chemical effects on adult birds covers acute mortality, sub-lethal stress, reduced fertility, suppression of egg formation, eggshell thinning and impaired incubation and chick rearing behaviors (Gilman et al., 1979). Pesticides cause extinction, behavioral changes, loss of safe habitat and population decline in several birds. Prolonged use of pesticides causes a drastic decrease in birds like the peregrine falcon, sparrow hawk and bald eagle (Mitra et al., 2011). The levels of organochlorines in seabird eggs were indicated by forming a deposit of pollutants in the body, thus serving as a useful indicator of environmental contamination (Pearce et al., 1989).

Toxic Effect in Farm Animals

The prolonged use of pesticides in agriculture has caused serious health problems as these pesticides accumulate and affect the food chain. Organochlorine compounds are highly lipophilic and can accumulate in fat-rich food such as meat and milk (Hernandez et al., 1994). Pesticides are introduced into cattle mainly through fodder or contaminated water used for household and public purposes (Sabbah and Bouguerra 1997). Amphibians and insectivorous reptiles, like lizards, have an important function in linking invertebrates with vertebrates in the food chain. They serve as a food source for some organisms and are also a means by which chemical residues, especially residues of organochlorine pesticides taken in with contaminated prey, can enter food chains. Amphibians consume these pesticides by a number of ways, including inhalation, contact and through ingestion. Amphibians in open water bodies may also be exposed to pesticides due to run-off from adjacent agricultural land on which chemicals are used to control crop pests. Continuous exposure of honey bees to pesticides affects the quality of honey. The routes of honey contamination with pesticides are direct and indirect. The direct is treatment of beehives with pesticides (Tsipi et al., 1999). Wild animals, including the grass cutter (Thryonomys swinderianus), which are a good source of protein, are seriously affected by the use of pesticides. Grass cutters are a source of food for the people of Ghana in Africa (Blankson-Arthur et al., 2011). As pesticides have high effect on the animal and bird community, ultimately humans also take up pesticides as meat, milk and crops derived from these animals and plants are consumed by humans.

BIOTA AT RISK

Algae, vascular plants, invertebrates, fishes, and avian and mammalian wildlife all have some degree of ecological and social importance in estuarine habitats and each group has some degree of vulnerability to pesticide exposures. Exposures for aquatic species depend to a great extent on whether the biota are pelagic, epibenthic, or infaunal and whether the species or life stages are mobile or non motile forms. Exposures to relatively constant pesticide inputs can result in lasting changes in community structure and function, with only tolerant species prevailing. Exposures to pulse inputs associated with runoff from storm events or direct inputs following pesticide applications offer greater dynamic responses among the biota and fluxes of biotic viability. These dynamic effects can be imposed on natural, seasonal cycles of animal and plant movement in these systems, making exposure- response assessments difficult to quantify. Finally, the net effects of repeated pulse inputs over a growing season, and over several growing seasons, begin to define the total impact of pesticides on coastal wetlands. Pelagic organisms that are immobile or relatively non motile and become entrained in a contaminated water mass moving through the estuary may have exposure durations dependent on the rate of mixing and dispersion of the contaminated water with other site water. Mobile species, pelagic or epibenthic may actively avoid contamination. Avoidance may be especially important for pesticides associated with fresh or low-salinity waters, which become prevalent during a major storm runoff event. Sensing the changes in salinity, fish or shrimp may move out of coastal wetlands to deeper and more stable waters until the storm passes. Immobile species may experience pulses of pesticides as contaminated water and particles move through a system. Those species adapted to burrowing, reducing metabolism during stress, or some other means of minimizing exposure to external media may actively reduce their short-term exposures to the relatively high concentrations of pesticides pulsed through the system. If the input remains with a freshwater or low-salinity lens of water that moves through the immediate area, benthic species may not be exposed at all. Animals that spend only a portion of their life cycles within coastal wetlands and use them as nursery areas, feeding grounds, or migratory paths may be exposed to pesticides if visits coincide with inputs. Residual pesticides, trapped in sediments or persisting as body burdens in resident species as a result of prior exposures, may find their way to species that spend only portions of their lives in or near coastal wetlands. Persistence and bioaccumulation are major factors on which regulation and use restrictions are based in current regulatory processes (Urban and Cook, 1986). Historical examples of persistent, chlorinated pesticides that have resulted in extensive food-chain contamination serve as constant reminders that the potential for such problems must be continually evaluated.

EFFECTS ON MICROBIAL POPULATIONS

Effects on Microalgae

Two major effects of pesticides on rice field algae have been recorded:

- Selective toxicity which affects the composition of the algal population,
- Growth promoting effect of insecticides due to the decrease of invertebrate populations that grazes on algae.

Several reports indicates a preferential inhibitory effect of pesticides on green algae which results in the promotion of blue green algae (BGA) growth. This was observed with Benzene hexachloride (BHC) (Raghu & McRae 1967), pentachlorophenol (PCP) (Watanabe 1977), Symetryne (Yamagishi and Hashizume 1974), and algaedyn (Almazan & Robles 1956). Several insecticides have been reported to be harmless to BGA while they killed algal grazers, thus promoting BGA growth. This was observed in rice fields for parathion applied at 1 to 5 ppm in the irrigation water (Hirang et al. 1955). However, on a long term basis, insecticide application might be detrimental to BGA by decreasing species diversity and causing a rapid recruitment of ostracods. The relative acute lethal toxicity of carbofuran to the ostracod *Heterocypris luzonensis* was 2.4 $\mu g\ ml^{-1}$ and that of Lindane was 56.0 $\mu g\ ml^{-1}$ (Grant et al., 1983). Such resistance to conventional pesticides allows large densities of ostracods to develop after pesticide application (5,000 - 15,000 m^{-2}), particularly as the natural predators succumb first. Such populations may cause the disappearance of algal blooms in a few days. A field study by Takamura and Yasuno, (1986) reports the development of large populations of chironomids and ostracods in herbicide and insecticide treated fields. Simultaneously, the number of natural predators of chironomids and ostracods decreased. Benthic algae decreased in herbicide treated plots and did not increase in insecticide treated plots probably because of grazing by ostracods. There are also several reports indicating no significant effects of pesticides applied at recommended level on algal flora in the presence of soil (Megharaj et al., 1988).

Effects on Populations of Non-Photosynthetic Microorganisms

A non-exhaustive collection of the recorded effect of pesticides on populations of microorganisms other than algae and microbial activities in wetland rice soils is seen. On an average 17% of the trials report a decrease in bacterial population after a pesticide application, in 58% of the cases no significant change was observed, and in 25% of the cases an increase was recorded. Actinomycetes show a trend similar to that of bacterial counts. A population of fungi seems to be most sensitive to pesticides. However, it has to be kept in mind that the relative abundance of actinomycetes and fungi is much lower in wetland soils than in upland soils. Among the different groups of bacteria, N_2-fixers seem to be the less negatively affected, while other bacteria of the N cycle are relatively the most frequently inhibited. When considering original data, no trend is obvious within this last group as negative effects were recorded for ammonium oxidizers, denitrifiers, and nitrite oxydizers. The values presented for N_2-fixing bacteria are biased by a large number of tests with phyllospheric Azotobacter which have little agroecological implication. The absence of inhibitory effect on populations of nonrhizosperic N_2-fixers is in agreement with a very low number of records of negative effects of pesticide application on nitrogen fixation in soil but the absence of effect on populations of rhizosperic N_2-fixers do not with the relatively high frequency of inhibition of BNF in rhizosphere.

CONCLUSION

The wetlands in India are unique ecosystems that harbour diverse types of species and very important for balancing different food chains and food webs. These wetlands obviously provide livelihood to good number of people in India and have various resources that have increased the GDP of different states of the country. However, the intrusion of pesticides in these ecosystems have caused adverse effects not only to the species in these ecosystems, but have caused great damage to the wetland itself, as a result of that most of these ecosystems are under eutrophication. Hence, there is an urgent need to address the issues of pesticide contamination by making strict rules and regulations.

ACKNOWLEDGMENT

The authors would like to acknowledge Jiwaji University, Gwalior for all the necessary facilities.

REFERENCES

Almazan, L. L., & Robles, D. O. (1956). Exploratory study of the effect of algae upon lowland rice. *Araneta Journal of Agriculture (Philippines)*, *3*(4), 1-27.

Bassi, N., & Kumar, M. D. (2012). Addressing the civic challenges: Perspective on institutional change for sustainable urban water management in India. *Environment & Urbanization Asia*, *3*(1), 165–183. doi:10.1177/097542531200300109

Bates, B. C., Kundzewicz, Z. W., Wu, S., & Palutikof, J. P. (Eds.). (2008). Climate Change and Water. Inter governmental Panel on Climate Change, Geneva.

Blankespoor, B., Dasgupta, S., & Laplante, B. (2012). *Sea-level rise and coastal wetlands: impacts and costs*. Washington, DC: The World Bank. doi:10.1596/1813-9450-6277

Blankson-Arthur, S., Yeboah, P. O., Golow, A., Tutu, A. O., & Denutsui, D. (2011). Levels of Organochlorine Pesticide Residues in Grasscutter (Thryonomys swinderianus) Tissues. *Research Journal of Environmental and Earth Sciences*, *3*(4), 350–357.

Central Pollution Control Board (CPCB). (2008). *Status of Water Quality in India 2007*. New Delhi: Central Pollution Control Board, Ministry of Environment and Forests, Government of India.

Central Pollution Control Board (CPCB). (2009). *Status of Water Supply, Waste water Generation and Treatment in Class-I Cities and Class-II Towns of India*. New Delhi: Central Pollution Control Board, Ministry of Environment and Forests, Government of India.

Central Pollution Control Board (CPCB). (2010). *Status of Water Quality in India 2009*. New Delhi: Central Pollution Control Board, Ministry of Environment and Forests, Government of India.

Central Water Commission (CWC). (2009). *National Register of Large Dams 2009*. New Delhi: Central Water Commission.

Centre for Science and Environment (CSE). (2012). *Living with changing climate: Impact, Vulnerability and Adaptation Challenges in Indian Sundarbans*. New Delhi: Centre for Science and Environment.

Chandan, P., Chatterjee, A., & Gautam, P. (2007). Management planning of Himalayan high altitude wetlands. A case study of Tsomoriri and Tsokar Wetlands in Ladakh, India. In *Proceedings of Taal2007: The 12th World Lake Conference* (Vol. 1446, p. 1452). Academic Press.

Choudhury, B. C., & Sathyakumar, S., & Christi, S. (2007). An assessment of the current status of the Indian Peafowl (Pavo cristatus) in India based on questionnaire survey. *Galliformes of India, ENVIS Bulletin. Wildlife and Protected Areas, 10*, 53–60.

Cowardin, L. M., Carter, V., Golet, F. C., & LaRoe, E. T. (1979). *Classification of wetlands and deepwater habitats of the United States*. US Department of the Interior, US Fish and Wildlife Service.

Finlayson, C. M., & Davidson, N. C. (1999). Global review of wetland resources and priorities for wetland inventory. *Preface iv Summary Report*, 15.

Fry, D. M. (1995). Reproductive effects in birds exposed to pesticides and industrial chemicals. *Environmental Health Perspectives, 103*(Suppl 7), 165–171. doi:10.1289/ehp.95103s7165 PMID:8593865

Garg, J. K., Singh, T. S., & Murthy, T. V. R. (1998). *Wetlands of India. SAC*. Ahmadabad: Indian Space Research Organisation.

Gilman, A. P., Peakall, D. B., Hallett, D. J., Fox, G. A., & Norstrom, R. J. (1979). *Herring gulls (Larus argentatus) as monitors of contamination in the Great Lakes. Animals as monitors of environmental pollutants/sponsored by Northeastern Research Center for Wildlife Diseases, University of Connecticut, Registry of Comparative of Lab Animal Resources*. National Academy of Sciences.

Gopal, B., & Sah, M. (1995). Inventory and classification of wetlands in India. In *Classification and Inventory of the World's Wetlands* (pp. 39–48). Springer Netherlands. doi:10.1007/978-94-011-0427-2_5

Gopal, B., Shilpakar, R., & Sharma, E. (2010). *Functions and services of wetlands in the Eastern Himalayas: impacts of climate change. International Centre for Integrated Mountain Development*. ICIMOD.

Grant, I. F., Egan, E. A., & Alexander, M. (1983). Pesticides to control ostracods grazing on blue-green algae. *Soil Biology & Biochemistry, 15*(2), 193–197. doi:10.1016/0038-0717(83)90102-5

Hernández, L. M., Fernández, M. A., Jiménez, B., González, M. J., & Garcia, J. F. (1994). Organochlorine pollutants in meats and cow's milk from Madrid (Spain). *Bulletin of Environmental Contamination and Toxicology, 52*(2), 246–253. doi:10.1007/BF00198495 PMID:8123985

Hirang, T., Shiraisshi, K., & Naakano, K. (1955). Studies on the blue-green algae in lowland paddy soil, part 1. On some conditions for growth of blue-green algae in paddy soil and its effects on growth of paddy rice plant. *Shikoku Nogyo Shikenjo Hokoku, 2*, 121–137.

Indian Institute of Technology (IIT). (2011). Trends in Agriculture and Agricultural Practices in Ganga Basin: An Overview. Indian Institute of Technology Kanpur.

Indian National Trust for Art and Cultural Heritage (INTACH). (1998). *Delhi's Dying Water Bodies*. New Delhi: Indian National Trust for Artand Cultural Heritage.

Jain, C. K., Singhal, D. C., & Sharma, M. K. (2007a). Estimating nutrient loadings using chemical mass balance approach. *Environmental Monitoring and Assessment, 134*(1-3), 385–396. doi:10.100710661-007-9630-5 PMID:17616829

Jain, S. K., Agarwal, P. K., & Singh, V. P. (2007b). *Hydrology and Water Resources of India*. Springer Netherlands.

Kaul, S. (2007). National and International Initiatives in Wetland Management. *Kerala Environment Congress 2007: Thiruvananthapuram, Center for Environment Development.*

Khandekar, N. (2011). Delhi water bodies go under, almost. *Hindustan Times.*

Kumar, M. D. (2013). Climate in India: key features of the variables. *Climate Change Impacts on Water Resources Systems*, 104.

Kumar, M. D., Panda, R., Niranjan, V., & Bassi, N. (2013). Technology choices and institutions for improving economic and livelihood benefits from multiple uses tanks in western Orissa. Academic Press.

Lehner, B., & Döll, P. (2004). Development and validation of a global database of lakes, reservoirs and wetlands. *Journal of Hydrology (Amsterdam)*, *296*(1-4), 1–22. doi:10.1016/j.jhydrol.2004.03.028

Liu, J., & Diamond, J. (2005). China's environment in a globalizing world. *Nature*, *435*(7046), 1179–1186. doi:10.1038/4351179a PMID:15988514

McAllister, D. E., Craig, J. F., Davidson, N., Delany, S., & Seddon, M. (2001). *Biodiversity Impacts of Large Dams*. International Union for Conservation of Nature and United Nations Environmental Programme, Gland and Nairobi.

Megharaj, M., Venkateswarlu, K., & Rao, A. S. (1988). Tolerance of algal population in rice soil to carbofuran application. *Current Science*, *57*(2), 100–102.

Millennium Ecosystem Assessment (MEA). (2005). *Ecosystems and Human Well-being: Wetlands and Water Synthesis*. Washington, DC: World Resources Institute.

Ministry of Environment and Forests (MoEF). (1990). Wetlands of India: a directory. New Delhi: MoEF, Government of India.

Mitra, A., Chatterjee, C., & Mandal, F. B. (2011). Synthetic chemical pesticides and their effects on birds. *Res J Environ Toxicol*, *5*(2), 81–96. doi:10.3923/rjet.2011.81.96

Molur, S., Smith, K. G., Daniel, B. A., & Darwall, W. R. T. (2011). *The status and distribution of freshwater biodiversity in the Western Ghats, India*. Coimbatore, India: IUCN, Cambridge, UK and Glad, Switzerland and Zoo Outreach Organization.

Muralidharan, S., Ganesan, K., Nambirajan, K., Kirubhanandhini, V., & Dhanajayan, V. (2014). Wetland birds indicators of pesticide contamination- Current and future prospects for research in India. In G. V. Gopi & S. A. Hussain (Eds.), *Waterbirds & Protected Areas* (Vol. 16, pp. 314–323). Dehradun, India: Wildlife Institute of India.

Patel, J. G., Murthy, T. V. R., Singh, T. S., & Panigrahy, S. (2009, December). Analysis of the distribution pattern of wetlands in India in relation to climate change. In *Proceedings of the workshop on impact of climate change on agriculture* (pp. 17-18). Ahmedabad: International Society for Photogrammetry and Remote Sensing.

Pearce, P. A., Elliott, J. E., Peakall, D. B., & Norstrom, R. J. (1989). Organochlorine contaminants in eggs of seabirds in the Northwest Atlantic, 1968–1984. *Environmental Pollution*, *56*(3), 217–235. doi:10.1016/0269-7491(89)90039-0 PMID:15092478

Prasad, S. N., Ramachandra, T. V., Ahalya, N., Sengupta, T., Kumar, A., Tiwari, A. K., & Vijayan, L. (2002). Conservation of wetlands of India-a review. *Tropical Ecology*, *43*(1), 173–186.

Prigent, C., Matthews, E., Aires, F., & Rossow, W. B. (2001). Remote sensing of global wetland dynamics with multiple satellite data sets. *Geophysical Research Letters*, *28*(24), 4631–4634. doi:10.1029/2001GL013263

Raghu, K., & MacRae, I. C. (1967). The effect of the gamma isomer of benzene hexachloride upon the microflora of submerged rice soils: II. Effect upon nitrogen mineralization and fixation, and selected bacteria. *Canadian Journal of Microbiology*, *13*(6), 621–627. doi:10.1139/m67-082 PMID:4166540

Ramachandra, T. V., & Kumar, U. (2008). Wetlands of greater Bangalore, India: Automatic delineation through pattern classifiers. *Electronic Green Journal*, *1*(26), 1–22.

Ramsar. (2002). *Wetlands: Water, life, and culture*. 8th meeting of the conference of contracting parties to the convention of wetlands, Valencia, Spain.

Sabbah, S., & Bouguerra, M. L. (1997). Organochlorine pesticides in cow's milk from Tunisia. *Fresenius Environmental Bulletin*, *6*(7), 359–364.

Sinha, C. P. (2011). Climate change and its impacts on the wetlands of North Bihar, India. *Lakes and Reservoirs: Research and Management*, *16*(2), 109–111. doi:10.1111/j.1440-1770.2011.00457.x

Smakhtin, V., Revenga, C., & Döll, P. (2004). A pilot global assessment of environmental water requirements and scarcity. *Water International*, *29*(3), 307–317. doi:10.1080/02508060408691785

Space Applications Centre. (2011). *National Wetland Atlas. SAC*. Ahmedabad: Indian Space Research Organisation.

Takamura, K., & Yasuno, M. (1986). Effects of pesticide applications on chironomids, ostracods and other benthic organisms in rice fields. *Research Reports of the National Institute Environmental Studies of Japan*, *99*, 81–90.

Tsipi, D., Triantafyllou, M., & Hiskia, A. (1999). Determination of organochlorine pesticide residues in honey, applying solid phase extraction with RP-C18 material. *Analyst (London)*, *124*(4), 473–475. doi:10.1039/a809724k PMID:10605876

Turner, R. K., Van Den Bergh, J. C., Söderqvist, T., Barendregt, A., Van Der Straaten, J., Maltby, E., & Van Ierland, E. C. (2000). Ecological-economic analysis of wetlands: Scientific integration for management and policy. *Ecological Economics*, *35*(1), 7–23. doi:10.1016/S0921-8009(00)00164-6

United Nations Educational Scientific and Cultural Organization (UNESCO). (2007). *Case Studies on Climate Change and World Heritage*. UNESCO World Heritage Centre.

Urban, D. J., & Cook, N. J. (1986). Standard evaluation procedure ecological risk assessment. In *Standard evaluation procedure ecological risk assessment*. EPA.

Verhoeven, J. T., Arheimer, B., Yin, C., & Hefting, M. M. (2006). Regional and global concerns over wetlands and water quality. *Trends in Ecology & Evolution*, *21*(2), 96–103. doi:10.1016/j.tree.2005.11.015 PMID:16701481

Verma, M., Bakshi, N., & Nair, R. P. (2001). *Economic valuation of Bhoj Wetland for sustainable use.* Unpublished project report for World Bank assistance to Government of India, Environmental Management Capacity-Building. Bhopal: Indian Institute of Forest Management.

Watanabe, I. (1977). Pentachlorophenol-decomposing and PCP-tolerant bacteria in field soil treated with PCP. *Soil Biology & Biochemistry*, *9*(2), 99–103. doi:10.1016/0038-0717(77)90044-X

World Bank. (2012). *Turn Down the Heat: Why a 4oC Warmer World Must Be Avoided.* Washington, DC: The World Bank.

Yamagishi, A., & Hashizume, A. (1974). Ecology of green algae in paddy fields and their control with chemicals. *Weed Research.*

Zedler, J. B., & Kercher, S. (2005). Wetland resources: Status, trends, ecosystem services, and restorability. *Annual Review of Environment and Resources*, *30*(1), 39–74. doi:10.1146/annurev.energy.30.050504.144248

Zhao, S., Peng, C., Jiang, H., Tian, D., Lei, X., & Zhou, X. (2006). Land use change in Asia and the ecological consequences. *Ecological Research*, *21*(6), 890–896. doi:10.100711284-006-0048-2

ADDITIONAL READING

Agarwal, M. (2011). *Migratory birds in India: migratory birds dwindling*. Nature December.

Assessment, M. E. (2005). Ecosystems and human well-being: wetlands and water. World resources institute, Washington, DC, 5.

Bassi, N., Kumar, M. D., Sharma, A., & Pardha-Saradhi, P. (2014). Status of wetlands in India: A review of extent, ecosystem benefits, threats and management strategies. *Journal of Hydrology: Regional Studies*, *2*, 1–19.

Gopal, B., Shilpakar, R., & Sharma, E. (2010). *Functions and services of wetlands in the Eastern Himalayas: impacts of climate change. International Centre for Integrated Mountain Development.* ICIMOD.

KEY TERMS AND DEFINITIONS

Biota: Biota is defined as the animal and plant life of a particular region, habitat, or geological period.

Contamination: Is the presence of an unwanted constituent, contaminant or impurity in a material, physical body, natural environment, workplace, etc. Contaminants are biological, chemical, physical, or radiological substances.

Ecosystem: An ecosystem is a community of living organisms in conjunction with the nonliving components of their environment (like air, water, and mineral soil), interacting as a system.

Eutrophication: Is the enrichment of a water body with nutrients, usually with an excess amount of nutrients. This process induces growth of plants and algae and due to the biomass load, may result in oxygen depletion of the water body.

Food Chain: A food chain is a linear network of links in a food web starting from producer organisms (such as grass or trees which use radiation from the sun to make their food) and ending at apex predator species (like grizzly bears or killer whales), detritivores (like earthworms or woodlice), or decomposer species (such as fungi or bacteria).

Herbicide: A herbicide is a chemical substance used to control or manipulate undesirable vegetation, especially weeds. Herbicides are extensively used in gardening, farming, and landscape turf management.

Pesticide: A pesticide is a chemical that is used to control a pest. A pest can be an insect, weed, bacteria, fungus, rodent, fish, or any other troublesome organism.

Pollution: Pollution is the introduction of contaminants into the natural environment that causes adverse change. Pollution can take the form of chemical substances or energy, such as noise, heat, or light.

Section 2

Pesticides and Human Health

Chapter 5
Pesticide and Human Health: A Rising Concern of the 21st Century

Sonal Dixit
University of Lucknow, India

Madhu Prakash Srivastava
University of Lucknow, India

Yogesh Kumar Sharma
University of Lucknow, India

ABSTRACT

Pesticides are known to be one of the extremely useful and incredibly beneficial agents for preventing losses of crops as well as diseases in humans. They are used in a large number of conditions as in farms, orchards, gardens, parks, sports lawn, residences, industrial areas, shops, schools, hospitals, airports, railway lines, drains, on animals, and on people for control of diseases such as scabies and head lice. People are exposed to pesticides in their daily lives through multiple routes of exposure such as occupational or food, water, and air. Many pesticides can be used safely and effectively, but care must be taken while using them. Several pesticides are beneficial in agriculture for killing pests. Yet many times their injurious effects offset the positive ones. Uses of pesticides are apprehension for sustainability of environment and global stability. This chapter aims to discuss pesticides, their types, routes of their exposure, human health concerns related to them, methods to stop using them, and a future scenario of the world after eradicating pesticides.

INTRODUCTION

Pesticide is a substance or mixture of substances intended for preventing, destroying, repelling or mitigating any pest. Agricultural pesticides are those chemicals that are used by farmers to prevent the loss of growth and productivity of crops from pests. They have numerous beneficial effects including crop protection, preservation of food and materials and prevention of vector-borne diseases. For example pesticides may be used in the prevention of malaria, which kills up to 1 million children per year, and

DOI: 10.4018/978-1-5225-6111-8.ch005

for preventing other vector-borne diseases such as dengue, leishmaniasis and Japanese encephalitis. Pesticides are designed to kill the pest and they target systems or enzymes in the pests which may be identical or very similar to the systems or enzymes in human beings. The major drawback of pesticide is that their mode of action is not specific to one species; they often kill or harm organisms other than pests, including humans. Therefore, they pose risks to human health. The World Health Organization estimates that there are 3,000,000 cases of pesticide poisoning each year and up to 220,000 deaths, primarily in developing countries (Lah, 2011).

Pesticides are an important tool in modern agriculture, but the risks and benefits of using pesticides must be considered before an application takes place. Insects and pests are getting immune to the commercial pesticides due to over usage. Now a day's pesticides have been developed which target multiple species (Speck-Planche et al., 2012). Chemical pesticides and insecticides are becoming a dominant agent for eliminating pests. When these chemical pesticides are used in a combination of effective natural enemy then they result in enhanced integrated pest management and act as a comprehensive prophylactic and remedial treatment (Gentz et al., 2010).

Although pesticides are developed under very strict regulation processes to function with reasonable certainty and minimal impact on human health and the environment, serious concerns have been raised about health risks resulting from occupational exposure and from residues in food and drinking water. Pesticides have also posed a serious threat on biological integrity of marine and aquatic ecosystems. Non-regulated use of pesticides has led the environment into disastrous consequences. Serious concerns about human health and biodiversity are on rise due to overuse of pesticides (Agrawal et al., 2010). Pesticides are considered to be more water soluble, heat stable and polar which makes it very difficult to reduce their lethal nature. Pesticides are not only toxic to people related to agriculture, but they also cause toxicity to other people. Depending upon the target species, pesticides can cause toxicity in natural flora, natural fauna and aquatic life (Rashid et al., 2010).

Water pollution is on the rise due to these pesticides, even at low concentration, these pesticides have serious threat to the environment (Agrawal et al., 2010). There are organochlorines, which are used as pesticides. These pesticides are least biodegradable and their use is banned in many countries. Besides this fact, organochlorines are highly used in many places. This results in serious health hazards. The majority of farmers are unaware of the potential toxicities of pesticides. They have no information about types of pesticides, their level of poisoning, hazards and safety measures to be taken before use of those pesticides. Due to this reason, toxic and environmentally persistent chemicals are used to kill pests which can also lead to intentional, incidental or occupational exposure. These compounds have long term effects on human health. Awareness should be given for these farmers to reduce the use of toxic pesticides (Sharma et al., 2012).

ANCIENT TO CURRENT SCENARIO OF PESTICIDES

In Ancient times Romans started using burn sulphur for killing pests and salts, ashes and bitters for controlling weeds. Then a Roman naturalist advised to use arsenic for management of insects (History of pesticide use, 1998). In 1600s, mixture of honey and arsenic was used for controlling ants. There after the farmers in the USA started using certain chemicals such as sulphur, nicotine sulphate and calcium arsenate for field application in 1800s (Delaplane, 2000). The major breakthrough in pesticide development occurred in the period around and after World War-II, when several efficient and economical pesticides were produced like Aldrin, DDT in 1939, Dieldrin, β-Benzene Hexachloride (BHC), 2,4-Dichlorophenoxyacetic acid (2,4-D), Chlordane and Endrin (Jabbar and Mallick, 1994; Delaplane, 2000). Captan and Glyodin (fungicides) and Malathion (insecticide) were introduced in 1950–1955 followed by the discovery of triazine herbicides in the years 1955–1960 (Jabbar and Mallick, 1994). An herbicide Agent Orange was developed by Monsanto in 1961–1971 which was used during the War of Vietnam (History of pesticide use, 1998).

The use of pesticides reached its peak in 1961, however, after 1962, there was a marked decrease in the production of new pesticides, as the public awareness was drawn towards the environmental hazard of pesticide use. In 1962, an American scientist Rachel Carson highlighted in her book, Silent Spring, that spraying DDT in the field causes sudden death of non-target organisms (Jabbar and Mallick, 1994; Delaplane, 2000) either by direct or indirect toxicity. However, in the late 1960s "Integrated Pest Management" was introduced in which biological predators or parasites are used for controlling the pests, but unfortunately it could not became a substitute for chemical pesticides (Delaplane, 2000). In 1970–1980s, pyrethroids, sulfonylureas, synthetic fungicides triadimefron and metaxyl were introduced (History of pesticide use, 1998). In 1972, DDT was completely banned in the USA and use of Endosulfan, Dieldrin and Lindane was restricted. In 2001, an international treaty known as 'Stockholm Convention' was signed by 179 nations which were proposed to ban twelve Persistent Organic Pollutants (POP's) including DDT. Later in 2013, the European Union (EU) supported to ban the use of neonicotinoid pesticides (Jacobs, 2015). It has been observed that the overuse of pesticides affect primary producers and macroinvertebrates and also fishes including Salmon in aquatic ecosystems (Macneale et al., 2010).

The production of pesticides in India started in 1952 with the establishment of a plant for the production of BHC near Calcutta, and now India is the second largest manufacturer of pesticides in Asia after China and ranks twelfth globally (Mathur, 1999). There has been a steady growth in the production of technical grade pesticides in India, from 5,000 metric tons in 1958 to 102,240 metric tons in 1998. In 1996–97 the demand for pesticides in terms of value was estimated to be around Rs. 22 billion (USD 0.5 billion), which is about 2% of the total world market. However, the pattern of pesticide usage in India is different from that of the world. In India, 76% of the pesticide used as insecticides is against 44% globally (Mathur, 1999). The use of herbicides and fungicides is correspondingly less (Figure 1). The main use of pesticides in India is for cotton crops (45%), followed by paddy and wheat.

Figure 1. Comparison of Consumption pattern of pesticides in India and World

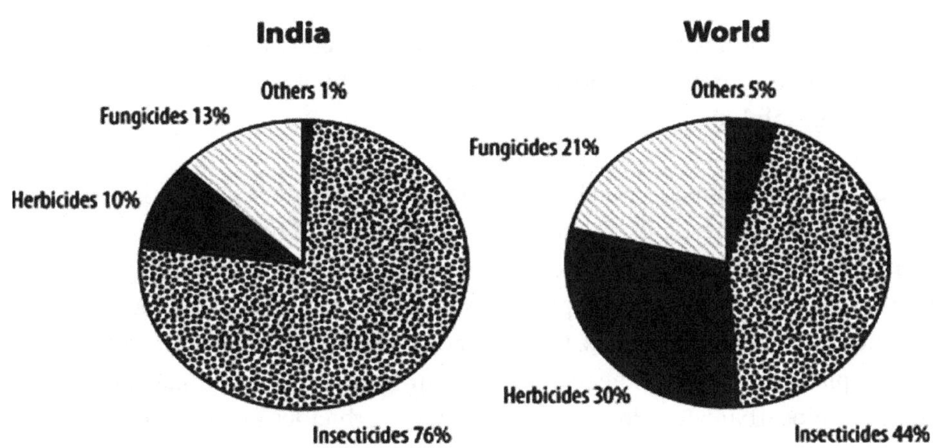

Table 1. Ancient to current scenario of pesticide use

Year	Events
1867	Paris Green (form of copper arsenite) was used to control Colorado potato beetle outbreak
1885	Introduction of a copper mixture by Professor Millardet to control mildew
1892	Potassium dinitro-2-cresylate was produced in Germany
1939	DDT discovered by Swiss chemist Paul Muller; organophosphate insecticides and phenoxyacetic herbicides were discovered
1950	Fungicides Captan and Glyodin and insecticide Malathion was discovered
1961-1971	Agent Orange was introduced
1972	DDT officially banned
2001	Stockholm Convention

Source: Mahmood et al., 2016.

Currently, preference is given to biological control of pests. This is a bioeffector method of controlling pests using biocontrolling agents including other living organisms. These biocontrolling agents are also known as bio-rational pesticides. An example of bio-rational pesticide is Insect Growth Regulators (IGRs) which are the hormones that regulate insect growth without affecting non-target organisms (Delaplane, 2000). Ancient to current scenario of pesticide use at a glance is shown in table 1.

CLASSIFICATION OF PESTICIDES

Based on the action, pesticides can be classified as destroying, repelling and mitigating agents. Pesticides are either restricted or unclassified. Restricted means it can cause harm to humans or the environment, and unclassified refers to all other pesticides. Pesticides are made up of active and inert ingredients. The active ones are those which cause damage to the pest. Federal law mandates that these ingredients should be clearly labelled on the packaging. Inert ingredients are not required to be labelled as in general they do not cause harm but not necessarily non-toxic, they are usually present as a solvent in the solution which may be toxic if inhaled or absorbed by the skin (Lah, 2011).

Worldwide pesticides are divided into different categories depending upon:-

1. Target organism e.g., herbicides, insecticides, fungicides, rodenticides, molluscicides and nematicides
2. Chemical structure e.g., organic, inorganic, synthetic, or biological
3. Physical state e.g. solid, liquid and gaseous

Many pesticides can be grouped into chemical families. Prominent insecticide families include organochlorines, organophosphates, and carbamates. Organochlorine hydrocarbons (e.g., DDT) could be separated into dichlorodiphenylethanes, cyclodiene compounds, and other related compounds (Kamrin, 1997). Organophosphate and carbamates operate through inhibiting the enzyme acetylcholinesterase, they have largely replaced organochlorines. Organophosphates are quite toxic to vertebrates and have in some cases been replaced by less toxic carbamates (Kamrin, 1997). Carbamates have two subclasses: thiocarbamate and dithiocarbamates. Prominent families of herbicides include phenoxy and benzoic acid

Table 2. Classification of pesticides based on target organism

S. No.	Pesticide Name	Target Pests
1	Algicides or algaecide	Algae
2	avicides	Birds
3	Bactericides	Bacteria
4	Fungicide	Fungi
5	Insecticide	Insects
6	herbicide	Plants
7	miticides	Mites
8	Milluscicides	Snails
9	Nematicides	Nematodes
10	Rodenticides	Rodents
11	Virucides	Viruses

Source: Mahmood et al., 2016.

herbicides (e.g. 2, 4-D), triazines (e.g., atrazine), ureas (e.g., diuron), and Chloroacetanilides (e.g., alachlor). Many commonly used pesticides are not included in these families, including glyphosate (Table 2).

Biopesticides include microbial pesticides and biochemical pesticides (EPA, 2009). Plant-derived pesticides, or "botanicals", have been developing quickly. These include the pyrethroids, rotenoids, nicotinoids, and a fourth group that includes strychnine and scilliroside (Kamrin, 1997). In 2009, the development of a new class of fungicides called paldoxins was announced which are believed to be safer and greener (Eurek, 2009)

HUMAN EXPOSURE TO PESTICIDES

Human exposure to pesticides may occur through professional activities or exposure to general population via contaminated food, water and air (Figure 2). In the case of agricultural workers in open fields and greenhouses, workers in the pesticide industry, and exterminators of house pests exposure of pesticides occurs from their workplace (Maroni et al., 2006; Tariq et al., 2007; Atreya, 2008; Martínez-Valenzuela et al., 2009; Soares and Porto, 2009). Workers who mix, load, transport and apply formulated pesticides are normally considered to be the group that will receive the greatest exposure because of the nature of their work and Therefore, at highest risk for possible acute intoxications (Fenske and Day, 2005). The exposure of workers increases in the case of not paying attention to the instructions on how to use the pesticides and particularly when they ignore basic safety guidelines on the use of personal protective equipment and fundamental sanitation practices such as washing hands after pesticide handling or before eating. The form of formulation of pesticide products may affect the extent of exposure. Liquids are prone to splashing and occasionally spillage, resulting in direct or indirect skin contact through clothing contamination. Solids may generate dust while being loaded into the application equipment, resulting in exposure to the face and the eyes and also respiratory hazards. General hygiene behaviour of workers during pesticide use can also have substantial impact on exposure. For example, workers who avoid mixing and spraying during windy conditions can reduce the exposure. Proper use and maintenance of protective clothing are considered important behaviours associated with reduced chemical exposures.

Figure 2. Routes of pesticide exposure to humans

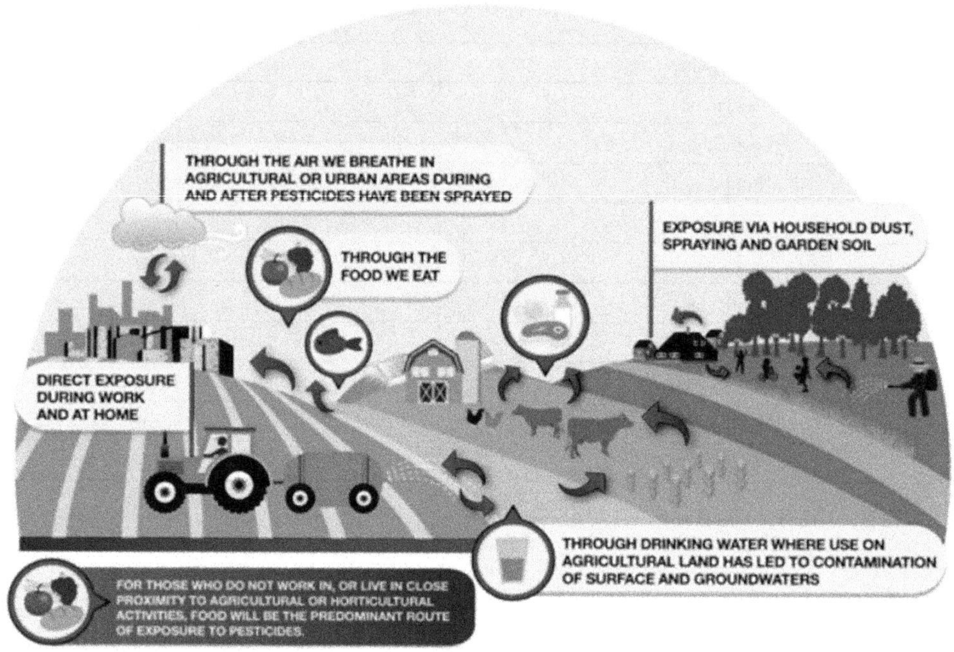

Exposure of the general population to pesticides occurs mainly through eating and drinking contaminated food and water, whereas substantial exposure to pesticides can also occur to the people residing in close vicinity to the workplace which uses pesticides (Davis et al., 1992; Jaga and Dharmani, 2003). Pesticide residues in food, air and drinking water generally involves low doses and is chronic (or semi-chronic). The actual acute exposure, however, may be higher than that anticipated due to certain food preferences residue variability between individual food items and greater than average consumption of a particular food item only at one sitting (Hamilton et al., 2004). Pesticide residues are commonly present in food that is grown through intensive industrial farming. Studies show that food often contains multiple residues and therefore pesticides are presented to us as mixtures or cocktails (Fenik et al., 2011). The toxic effect of these mixtures is particularly poorly understood, though it is recognised that some substances can interact synergistically and their combined effect is greater than that of the individual components (Reffstrup et al., 2010). Literature published between 2007 and 2014 suggested that legumes, leafy greens and fruits such as apples and grapes frequently contain the highest levels of pesticide residues (Bempah et al., 2012; Jardim et al., 2012; Fan et al., 2013; Yuan et al., 2014). Among many other pesticides, cypermethrin, chlorpyrifos, iprodione, boscalid, dithiocarbamates and acephate are regularly detected in our food (Claeys et al., 2011; Lozowicka et al., 2012; Yuan et al., 2014). Extensive research suggests that washing and cooking vegetables does reduce some of these residues that are on the surface of the plant, in some cases food preparation can actually concentrate their levels (Keikotlhaile et al., 2010). Concentrations of organotins are particularly high in the blood of those people who consume greater amounts of seafood and it has been suggested that regular monitoring of the levels of these substances be carried out for public health purposes (Yi et al., 2012).

Pesticide used in or around the residential area make the individuals vulnerable to be exposed during the preparation and application of pesticides or even after the applications. Delayed exposure can occur through inhalation of residual air or exposure to residues found on surfaces, clothing, bedding, food, dust, discarded pesticide containers, or application equipment (Davis et al., 1992). Accidental poisoning with pesticides used around the home and garden is also a possibility of contamination. Exposure is likely to occur from pesticide leak, improper use, poor storage, pesticide mishandling such as transferring the products from their original packages into household containers and also the lack of observance of instructions on the label can also be the sources of exposure (Jaga and Dharmani, 2003).

IMPACT OF PESTICIDES ON HUMAN HEALTH

Pesticides have improved the standard of human health by controlling vector-borne diseases; however, their long term and indiscriminate use have resulted in serious health effects. Human beings especially infants and children are at high risk to the detrimental effects of pesticides due to the non-specific nature of pesticides and poor application. About 2.2 million people, mainly belonging to developing countries are at increased risk of exposure to pesticides (Hicks, 2013). The impact of pesticides on human health are very extensive, affecting every part of the human body. Pesticides enter the human body through ingestion, inhalation or penetration via skin (Spear, 1991) but the majority of people get affected via the intake of pesticide contaminated food. Although human body has the mechanism for the excretion of toxins, however, in some cases, it retains them through absorption in the circulatory system (Jabbar and Mallick, 1994). Toxic effects are produced when the concentration of pesticide in the body increases far more than its initial concentration in the environment (Hayo and Werf, 1996). The effects of pesticides on human health are highly variable. They may appear in days and are immediate in nature called acute effect, or they may take months or years to manifest and hence are called chronic or long-term effects.

ACUTE EFFECTS OF PESTICIDES

Immediate effects of pesticide exposure include headache, stinging of the eyes and skin, irritation of the nose and throat, skin itching, appearance of the rash and blisters on the skin, dizziness, diarrhoea, abdominal pain, nausea and vomiting, blurred vision, blindness (Figure 3) and very rarely death. Acute effects of pesticide exposure are not severe enough for someone to seek medical help (Pesticides and Human Health, 2014), but Pyrethrins insecticides commonly used in common bug killers, can cause a potentially deadly condition if breathed in.

CHRONIC EFFECTS OF PESTICIDES

Chronic effects of pesticides are often lethal and may not appear even for years. These are long term effects that cause damage to multiple body organs. Ingestion of organochlorines causes hypersensitivity to light, sound, and touch, dizziness, tremors, seizures, vomiting, nausea, confusion and nervousness (Lah, 2011). Exposure to organophosphates and carbamates causes, symptoms similar to those of increased neurotransmitter-acetylcholine. These pesticides interfere with the normal nerve signal transduction,

Figure 3. Effects of acute pesticide poisoning on humans

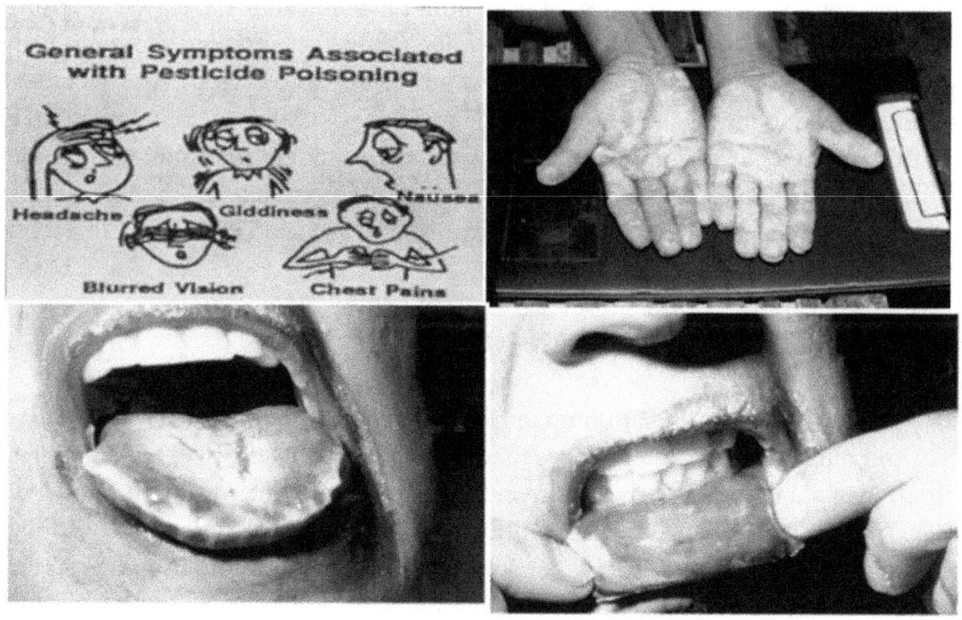

and exposure to them causes headaches, dizziness, confusion, nausea and vomiting, muscle and chest pain. Long-term pesticide exposure damages the immune system (Culliney et al., 1992) and can cause hypersensitivity, asthma and allergies. It can also damages liver, lungs, kidney and may cause blood diseases (Lah, 2011). Difficulty in breathing, convulsions, coma and even death may occur in severe cases. Pyrethroids can cause an allergic skin response, aggressiveness, hyper-excitation, reproductive or developmental effects in addition to causing tremors and seizures (Lah, 2011). It is also observed that there is a relationship between pesticides and Parkinson's disease and Alzheimer's disease (Casida and Durkin, 2013).

The risk assessment of the impact of pesticides on human health is not an easy and particularly accurate process because of differences in the periods and levels of exposure, the types of pesticides used, and the environmental characteristics of the areas where pesticides are usually applied. Also, the number of the criteria used and the method of their implementation to assess the adverse effects of pesticides on human health could affect risk assessment and would possibly affect the characterization of the already approved pesticides and the approval of the new compounds in near future. Pesticide exposure for prolonged periods of time results in following consequences:

Cancer

Cancer occur due to introduction of pesticide is the most researched issue associated with pesticides' toxicity. There is a significant number of epidemiological proofs which relate pesticides to cancer, and particularly with child cancer resultant from parental as well as direct childhood exposures. The common associated child cancers are leukaemia and brain cancer, other than that tumor in nerve tissues, bone tissues, kidney, liver, germ cell, eye etc. are also frequent (Infante-Rivard & Weichenthal, 2007; Carozza et al., 2008; Lyons & Watterson, 2010). An international study of seven countries recognized an association among brain tumours in children and maternal farm exposure to pesticides during the preceding five years of the diagnosis (Efird et al., 2003). A high rate of brain cancer was observed in children playing in orchards in Kashmir, India (Bhat et al., 2010). A significant increase in risk of cancer has been observed among farmers when compared with the risk of the common population, as they are highly exposed to pesticides (De Roos et al., 2005; Greenburg et al., 2008). The exposure to certain pesticides significantly increases the risk of cancer due to the nature of pesticides. The cancers in adults due to pesticides includes breast, lung, multiple myeloma, leukaemia, ovary, pancreas, prostate, kidney bladder, stomach, colon, rectal, lip, connective tissue, brain, and testicular. Out of these breast, prostate, and testicular cancers are considered to originated during early growth stages due to contact with environmental hormone disruptors (Waggoner et al., 2011; Alavanja and Bonner, 2012). Type of cancer associated with pesticides exposures are summarized (Table 3).

Neurobehavioral Problems

Various studies have revealed that many pesticides can act through neurotoxic mechanism that are related to human health, including carbamates, organochlorines, organophosphates, and pyrethroids (London et al., 2012). The exposures to pesticides have lasting adverse effects on the brain which has become a silent pandemic of developmental neurotoxicity (Harari et al., 2010). Pesticide exposure can cause a range of neurological health effects such as loss of coordination and memory, reduced visual ability, learning disorders, behavioural disorders and reduced motor signalling (Searles Nielsen et al., 2010; Lah, 2011; London et al., 2012). Suicidal cases and mental diseases are greater in children suffering from neurobehavioral disabilities; they are also in the greater risk of violence and criminal activities later in life (Spzir, 2006). High levels of exposure, such as with occupational exposures and poisonings, may result in increased risk of neuropsychiatric outcomes including increased anxiety, depression and suicide; and increased agricultural injury as a result of the depression (Beseler et al., 2008; London et al., 2012).

Depending upon the type of the pesticide used diverse results have been observed, some suggested that concerned neurological behaviour are linked with cumulative exposure of few fumigants and insecticides, and some researcher suggested that few pesticides does not show any neurological effect (Sathiakumar et al., 2004; Kamel et al., 2005). The effects of OP exposures during adolescence can manifest as mental and emotional disturbances (Jurewicz and Hanke, 2008). A relatively consistent pattern of neuro-behavioural deficits, including increased neuroses, have been observed in studies of pesticide applicators, greenhouse workers, agricultural workers and farm residents exposed repeatedly over months or years to low levels of OPs (Abdel Rasoul et al., 2008). A systematic meta-analysis of occupational exposure to pesticides by Van Maele-Fabry et al. (2012) concluded that there is a statistically significant increased risk of Parkinson's disease with occupational exposure of pesticides.

Teratogenic Effects

The presence of pesticides in the body for a relatively longer time affects reproductive capabilities by altering male and female reproductive hormones. As a result, it causes stillbirth, birth defects, spontaneous abortion and infertility. Studies regarding to the effect of pesticides causing birth defect in humans differ in various ways, some have not found any significant association with pesticides, but several have. Some reported that the high-quality birth defects found positive association with hypospadias, neural tube defects, and congenital diaphragmatic hernia (Sanborn et al., 2012). Various birth defects related with pesticides are missing or reduced limbs, anencephaly, cryptorchidism and micropenis, spina bifida and congenital heart disease (Rocheleau et al., 2009; Gaspari et al., 2011). Most pesticide affected person are the families of rural areas, pesticide appliers and due to maternal exposure (Schreinemachers, 2003; Brender et al., 2010). Assessments have made for organochlorine pesticides and it was observed that these are linked with a number of congenital defects for instance neural tube defects, undescended testicles, cryptorchidism, extra nipple in males, and cretinism (Brucker-Davis et al., 2008; Ren et al., 2011). Several studies demonstrate that herbicides 2,4-D, 2,4,5-T, MCPA, atrazine, and trifluralin, and pesticides endosulfan, chlorpyrifos, diclofop-methyl, cyanazine, dicamba, oxydemeton-methyl and metolachlor have been associated with teratogenicity.

Respiratory Problems

There are a number of studies defining correlation between respiratory diseases and pesticide exposure which suggested an increased risk of rhinitis, asthma, bronchitis, and wheeze among pesticide applicators (Slager et al., 2009; Hoppin et al., 2009). Asthma was found associated with occupational, domestic and environmental exposures particularly to organochlorine, organophosphate, biocide and fungicide. Herbicides and insecticides exposures to persons involve in farming, pesticide manufacturing, and pesticide spraying showed a subtle but persistent association with decreased lung functioning. There was found a strong relationship between a organophosphate pesticide chlorpyrifos and wheeze, chronic cough and shortness of breath. An evidence of Sarcoidosis, farmer's lung and chronic bronchitis were found to be associated with exposure to pesticides (Sanborn et al., 2012). Carbamate, organophosphate, and neonicotinoid insecticides are commonly found associated with restricted lung function.

Obesity, Diabetes, And Metabolic Diseases

Recently scientist has begun to focus on pesticides impact concerned with increasing rates of obesity, diabetes and metabolic disorders. Obesity is not a single disease; it is connected with hypertension, type 2 diabetes and cardiovascular diseases. Now there are a number of studies confirming the association of these conditions with pesticide exposure (Lee et al., 2006; Rignell-Hydbom et al., 2007; Montgomery et al., 2008; Valvi et al., 2012). Exposure of pesticides in prenatal and early childhood growth conditions are more risky because these stages are more susceptible, which lead to overweight condition and diabetes (Newbold et al., 2007).

An association between pesticide exposure and risk for diabetes has been observed with organochlorine and organophosphate compounds which are linked with increased incidences of diabetes (Cox et al., 2006; Montgomery et al., 2008). Pesticides are considered to cause obesity by interfering with the weight control mechanisms of the body. They alter weight controlling hormones as catecholamines, thyroxine, estrogens, testosterone, corticosteroids, insulin, growth hormones, and leptin; upsets the neurotransmitters dopamine, noradrenaline, and serotonin; hinder the metabolic processes; and damage nerve and muscle tissues (Baillie-Hamilton, 2002). A number of studies have associated insecticide organochlorine exposure to increased body mass index (Verhulst et al., 2009; Mendez et al., 2011; Valvi et al., 2012). The exposures to organophosphates results in pre-diabetes, abnormalities of lipid metabolism, and promotion of obesity in response to increased dietary fat. Sulfonylurea herbicides and imidazole fungicides have also been identified as potentially implicated in obesity and diabetes, because of their effects on weight gain, blood sugar levels, or the pancreas (Thayer et al., 2012).

Table 3. Type of cancer caused by Pesticides in individuals and WHO classification as stated in the IUPAC Pesticides Properties Database. WHO classification is denoted by U=unlikely to cause harm, O=obsolete SH=slightly hazardous, MH=moderately hazardous, EH=extremely hazardous.

Pesticide	Class	Type of Cancer	WHO Classification
Alachlor	OCP	All lymphohematopoietic*	MH
Aldicarb	Carbamate	Colon	EH
Carbaryl	Carbamate	Melanoma	MH
Diazinon	OPP	All lymphohematopoietic*, lung and leukemia	MH
Dicamba	Benzoic acid	Lung, colon	MH
Dieldrin	Chlorinated hydrocarbon	Lung	O
Chlordane	OCP	Rectal, leukemia	MH
Chlorpyrifos	OPP	All lymphohematopoietic*, lung, rectal, brain	MH
S-ethyl dipropylthiocarbamate (EPTC)	Thiocarbamate	Leukemia, colon, pancreatic	MH
Fonofos	OPP	Leukemia, prostate	O
Imazethapyr	Imidazolinone	Colon, bladder	U
Metolachlor	Chloroacetamide	Lung	SH
Pendimethalin	Dinitroanaline	Lung, rectal, pancreatic,	MH
Permethrin	Synthetic pyrethroid	Multiple myeloma	MH
Trifluralin	Dinitroanaline	Colon	U

Source: Weichenthal et al., 2012.
*Includes all lymphomas, leukemia and multiple myelomas.

ALLEVIATING THE HARMFUL IMPACT OF PESTICIDES

People have become increasingly concerned about pesticide use and particularly about their impacts on human health (Damalas, 2009). The easiest and simplest means to prevent the pesticide toxicity is through education and awareness. If farmers be acquainted about the risks associated with pesticides, they would be more careful about the instructions and basic safety guidelines while using the pesticides and also about the personal protection equipment and sanitation practices. Particular concern needs to be paid towards the people of the developing countries because major source of their earnings and economy is based on farming. Therefore, they are more vulnerable to the negative impacts of pesticides. In developed countries where wealth are not such big issues and people are more cautious for their health, consumers can decrease their risk of consuming pesticides by selecting organic products (which are costlier than normal ones) which means that no pesticides were used during the production of those products. There are some non government organizations (NGOs) which are working towards the prevention of pesticides exposure. They provide educational sessions, telephone hotlines and safety classes to the needy person.

Another way to minimize the negative effects of farming in which agricultural scientists are interested is Integrated Crop Management (ICM) systems. It includes the guidelines to be used by the farmer unions for production of safe agricultural products with simultaneous respect to the environment (Burger et al., 2008; Nwilene et al., 2008; Tsakiris et al., 2004; Baker et al., 2002). Moreover, ICM includes measures for implementation of good agricultural practices, the safety and hygiene of workers, the safety of the products, the full traceability of the measurements, and specific actions for the preservation of the environment (Chandler et al., 2008). Concerning pesticide use, ICM allows pesticide use only through an Integrated Pest Management (IPM) program (Nwilene et al., 2008; Chandler et al., 2008; Mariyono, 2008), which "emphasizes non chemical and cultural pest control strategies such as removal of diseased plant parts, crop rotation that may disrupt the life cycle of pests, and biological control such as the use of insect predators". Pesticides that are selected for use in IPM are biologically effective, user and environment friendly, and economically viable (Palacios Xutuc, 2010). The introduction of IPM system has contributed to a significant reduction of the pesticide impact on human health and the environment without affecting crop productivity.

New tools or techniques with greater reliability than those already existing are needed to predict the potential hazards of pesticides and thus contribute to reduction of the adverse effects on human health and the environment. The implementation of alternative cropping systems that are less dependent on pesticides, the development of new pesticides with novel modes of action and improved safety profiles, and the improvement of the already used pesticide formulations towards safer formulations could reduce the adverse effects of farming and particularly the toxic effects of pesticides. Moreover, the use of appropriate and well-maintained spraying equipment along with taking all the precautions required in all stages of pesticide handling could also reduce exposure to pesticides. The overall optimization of pesticide handling strictly according to the regulations and also considering the public concerns about pesticide residues in food and drinking water could contribute to reduction of the adverse effects of pesticides on human health and environment. The great scientific progress in many disciplines such as chemistry, biology, and molecular biology over the last years has improved considerably the way of searching for new agrochemicals and the re-assessment of safety for the already used pesticides (Bhattacharyya et al., 2009). The UN Food and Agriculture Organization and the convention on Persistent Organic Pollutants are making global efforts to decrease the presence and abundance of the harmful pesticides. Many states within the United States have installed a Clean Sweep Program which provides for proper cleanup and disposal of pesticides.

CONCLUSION

Pesticides have proved to be a boon for both agriculturist and common people by providing reliable supplies of agricultural produce at affordable prices, improving the quality of agricultural product, ensuring high profits to farmers and by decreasing diseases and infections caused by pests. Although pesticides are developed to function with reasonable certainty and minimal health risk, the issue of hazards caused by them to human health and the environment has raised concerns about their safety. Even though we cannot completely eliminate the hazards associated with pesticide use although we can evade them in one way or the other. The invention of new pesticides with harmless, improved, environment friendly formulations and novel modes of action could reduce the harmful effects associated with the pesticide usage. If the pesticides are used in appropriate quantities and used only when required or necessary, then pesticide risks can be minimised. In future chemical pesticides can be used in combination with natural treatments and remedies which result in more sustainable elimination of pests and insects. This combination not only promises environmental sustainability, but also has diverse applications in controlling of urban pests and invasive species. It is the need of time to integrate the studies of different disciplines including toxicology, environmental chemistry, population biology, community ecology, and conservation biology and landscape ecology to understand direct and indirect effects of pesticides on the environment. All these studies may sound difficult, but seem to be a promising way for maximizing the boon and minimizing the bane caused by pesticides.

ACKNOWLEDGMENT

The author SD is thankful to DSKPDF Cell, Pune, India, and University Grant Commission, New Delhi, India, for the award of the D.S. Kothari Postdoctoral Fellowship (F4-2/2006 (BSR)/BL/15-16/0156).

Authors MPS and YKS are thankful to the Center of Excellence Programme of UP state Government Letter No.: 1205/70-4-2013-46 (43) 2010 TC-11 for financial support.

REFERENCES

Abdel Rasoul, G. M., Abou Salem, M. E., Mechael, A. A., Hendy, O. M., Rohlman, D. S., & Ismail, A. A. (2008). Effects of occupational pesticide exposure on children applying pesticides. *Neurotoxicology*, *29*(5), 833–838. doi:10.1016/j.neuro.2008.06.009 PMID:18662718

Agrawal, A., Pandey, R. S., & Sharma, B. (2010). Water pollution with special reference to pesticide contamination in India. *Journal of Water Resource and Protection*, *2*(5), 432–448. doi:10.4236/jwarp.2010.25050

Aktar, M., Sengupta, D., & Chowdhury, A. (2009). Impact of pesticides use in agriculture: Their benefits and hazards. *Interdisciplinary Toxicology*, *2*(1), 1–12. doi:10.2478/v10102-009-0001-7 PMID:21217838

Alavanja, M. C. R., & Bonner, M. R. (2012). Occupational pesticide exposures and cancer risk: A review. *Journal of Toxicology and Environmental Health. Part B, Critical Reviews*, *15*(4), 238–263. doi:10.1080/10937404.2012.632358 PMID:22571220

Atreya, K. (2008). Health costs from short-term exposure to pesticides in Nepal. *Social Science & Medicine*, *67*(4), 511–519. doi:10.1016/j.socscimed.2008.04.005 PMID:18514373

Baillie-Hamilton, P. F. (2002). Chemical toxins: A hypothesis to explain the global obesity epidemic. *Journal of Alternative and Complementary Medicine (New York, N.Y.)*, *8*(2), 185–192. doi:10.1089/107555302317371479 PMID:12006126

Baker, B. P., Benbrook, C. M., Groth, E., & Lutz Benbrook, K. (2002). Pesticide residues in conventional, integrated pest management (IPM)-grown and organic foods: Insights from US data sets. *Food Addit Contam A*, *19*(5), 427–446. doi:10.1080/02652030110113799 PMID:12028642

Bempah, C. K., Buah-Kwofie, A., Enimil, E., Blewu, B., & Agyei-Martey, G. (2012). Residues of organochlorine pesticides in vegetables marketed in Greater Accra Region of Ghana. *Food Control*, *25*(2), 537–542. doi:10.1016/j.foodcont.2011.11.035

Beseler, C., Stallones, L., Hoppin, J., Alavanja, M., Blair, A., Keefe, T., & Kamel, F. (2006). Depression and Pesticide Exposures in Female Spouses of Licensed Pesticide Applicators in the Agricultural Health Study Cohort. *Journal of Occupational and Environmental Medicine*, *48*(10), 1005–1013. doi:10.1097/01.jom.0000235938.70212.dd PMID:17033500

Bhat, A. R., Wani, M. A., Kirmani, A. R., & Raina, T. H. (2010). Pesticides and brain cancer linked in orchard farmers of Kashmir. *Indian Journal of Medical and Paediatric Oncology: Official Journal of Indian Society of Medical & Paediatric Oncology*, *31*(4), 110–120. doi:10.4103/0971-5851.76191 PMID:21584215

Bhattacharyya, A., Barik, S. R., & Ganguly, P. (2009). New pesticide molecules, formulation technology and uses: Present status and future challenges. *Journal of Plant Protection Sciences*, *1*, 9–15.

Brender, J. D., Felkner, M., Suarez, L., Canfield, M. A., & Henry, J. P. (2010). Maternal pesticide exposure and neural tube defects in Mexican Americans. *Annals of Epidemiology*, *20*(1), 16–22. doi:10.1016/j.annepidem.2009.09.011 PMID:20006272

Brucker-Davis, F., Wagner-Mahler, K., Delattre, I., Ducot, B., Ferrari, P., Bongain, A., ... Fénichel, P. (2008). Cryptorchidism Study Group from Nice Area. Cryptorchidism at birth in Nice area (France) is associated with higher prenatal exposure to PCBs and DDE, as assessed by colostrums concentrations. *Human Reproduction (Oxford, England)*, *23*(8), 1708–1718. doi:10.1093/humrep/den186 PMID:18503055

Burger, J., Mol, F., & Gerowitt, B. (2008). The 'necessary extent' of pesticide use—Thoughts about a key term in German pesticide policy. *Crop Protection (Guildford, Surrey)*, *27*(3-5), 343–351. doi:10.1016/j.cropro.2007.06.006

Carozza, S. E., Li, B., Elgethun, K., & Whitworth, R. (2008). Risk of childhood cancers associated with residence in agriculturally intense areas in the United States. *Environmental Health Perspectives*, *116*(4), 559–565. doi:10.1289/ehp.9967 PMID:18414643

Casida, J. E., & Durkin, K. A. (2013). Neuroactive insecticides: Targets, selectivity, resistance, and secondary effects. *Annual Review of Entomology*, *58*(1), 99–117. doi:10.1146/annurev-ento-120811-153645 PMID:23317040

Chandler, D., Davidson, G., Grant, W. P., Greaves, J., & Tatchell, G. M. (2008). Microbial biopesticides for integrated crop management: An assessment of environmental and regulatory sustainability. *Trends in Food Science & Technology*, *19*(5), 275–283. doi:10.1016/j.tifs.2007.12.009

Claeys, W. L., Schmit, J.-F., Bragard, C., Maghuin-Rogister, G., Pussemier, L., & Schiffers, B. (2011). Exposure of several Belgian consumer groups to pesticide residues through fresh fruit and vegetable consumption. *Food Control*, *22*(3-4), 508–516. doi:10.1016/j.foodcont.2010.09.037

Cox, C., & Surgan, M. (2006). Unidentified inert ingredients in pesticides: Implications for human and environmental health. *Environmental Health Perspectives*, *114*(12), 1803–1806. PMID:17185266

Culliney, T. W., Pimentel, D., & Pimentel, M. H. (1992). Pesticides and natural toxicants in foods. *Agriculture, Ecosystems & Environment*, *41*(3-4), 297–320. doi:10.1016/0167-8809(92)90117-T

Damalas, C. A. (2009). Understanding benefits and risks of pesticide use. *Scientific Research and Essays*, *4*, 945–949.

Davis, J. R., Brownson, R. C., & Garcia, R. (1992). Family pesticide use in the home, garden, orchard, and yard. *Archives of Environmental Contamination and Toxicology*, *22*(3), 260–266. doi:10.1007/BF00212083 PMID:1616309

De Roos, A., Blair, A., Rusiecki, J., Hoppin, J., Svec, M., Dosemeci, M., ... Alavanja, M. (2005). Cancer Incidence among Glyphosate-Exposed Pesticide Applicators in the Agricultural Health Study'. *Environmental Health Perspectives*, *113*(1), 49–54. doi:10.1289/ehp.7340 PMID:15626647

Delaplane, K. S. (2000). *Pesticide usage in the United States: history, benefi ts, risks, and trends. Cooperative Extension Service.* The University of Georgia, College of Agricultural and Environmental Sciences. Bulletin, 1121. Retrieved from http://pubs.caes.uga.edu/ caespubs/pubs/PDF/B1121.pdf

Efird, J. T., Holly, E. A., Preston-Martin, S., Mueller, B. A., Lubin, F., Filippini, G., ... Bracci, P. M. (2003). Farm-related exposures and childhood brain tumours in seven countries: Results from the SEARCH International Brain Tumour Study. *Paediatric and Perinatal Epidemiology*, *17*(2), 201–211. doi:10.1046/j.1365-3016.2003.00484.x PMID:12675788

EPA. (2009). *Types of Pesticides*. EPA.

Eurek, A. (2009). *New 'green' pesticides are first to exploit plant defenses in battle of the fungi.* Academic Press.

Fan, S., Zhang, F., Deng, K., Yu, C., Liu, S. M., Zhao, P., & Pan, C. (2013). Spinach or Amaranth contains highest residue of metalaxyl, fluazifop-p-butyl, chlorfyrifos, and lamda-cyhalothrin on six leaf vegetables upon open field application. *Journal of Agricultural and Food Chemistry*, *61*(9), 2039–2044. doi:10.1021/jf304710u PMID:23387923

Fenik, J., Tankiewicz, M., & Biziuk, M. (2011). Properties and determination of pesticides in fruits and vegetables. *Trends in Analytical Chemistry*, *30*(6), 814–826. doi:10.1016/j.trac.2011.02.008

Fenske, R. A., & Day, E. W. (2005). Assessment of exposure for pesticide handlers in agricultural, residential and institutional environments. In C. A. Franklin & J. P. Worgan (Eds.), *Occupational and Residential Exposure Assessment for Pesticides. John Wiley & Sons; Chichester* (pp. 13–43). Academic Press. doi:10.1002/0470012218.ch1

Gaspari, L., Paris, F., Jandel, C., Kalfa, N., Orsini, M., Daurès, J. P., & Sultan, C. (2011). Prenatal environmental risk factors for genital malformations in a population of 1442 French male newborns: A nested casecontrol study. *Human Reproduction (Oxford, England)*, *26*(11), 3155–3162. doi:10.1093/humrep/der283 PMID:21868402

Gentz, M. C., Murdoch, G., & King, G. F. (2010). Tandem use of selective insecticides and natural enemies for effective, reduced-risk pest management. *Biological Control*, *52*(3), 208–215. doi:10.1016/j.biocontrol.2009.07.012

Greenburg, D., Rusiecki, J., Koutros, S., Dosemeci, M., Patel, R., Hines, C., ... Alavanja, M. (2008). Cancer incidence among pesticide applicators exposed to captan in the Agricultural Health Study. *Cancer Causes & Control*, *19*(10), 1401–1407. doi:10.100710552-008-9187-9 PMID:18587656

Hamilton, D. D., Ambrus, A., Dieterle, R., Felsot, A., Harris, C., Petersen, B., ... Tanaka, K. (2004). Pesticide residues in food-acute dietary exposure. *Pest Management Science*, *60*(4), 311–339. doi:10.1002/ps.865 PMID:15119595

Harari, R., Julvez, J., Murata, K., Barr, D., Bellinger, D. C., Debes, F., & Grandjean, P. (2010). Neurobehavioral deficits and increased blood pressure in school-age children prenatally exposed to pesticides. *Environmental Health Perspectives*, *118*(6), 890–896. doi:10.1289/ehp.0901582 PMID:20185383

Hayo, M. G., & Werf, V. D. (1996). Assessing the impact of pesticides on the environment. *Agriculture, Ecosystems & Environment*, *60*(2-3), 81–96. doi:10.1016/S0167-8809(96)01096-1

Hicks, B. (2013). *Agricultural pesticides and human health*. National Association of Geoscience Teachers. Available from http://serc.carleton.edu/NAGTWorkshops/ health / case_studies /pesticides.

History of pesticide use. (1998). Available from http://www2.mcdaniel.edu/Biology/eh01/pesticides/historyofpesticidesuse.html

Hoppin, J., Umbach, D., London, S., Henneberger, P., Kullman, G., Coble, J., ... Sandler, D. (2009). Pesticide use and adult-onset asthma among male farmers in the Agricultural Health Study. *The European Respiratory Journal*, *34*(6), 1296–1303. doi:10.1183/09031936.00005509 PMID:19541724

Infante-Rivard, C., & Weichenthal, S. (2007). Pesticides and childhood cancer: An update on Zahm and Ward's 1998 review. *Journal of Toxicology and Environmental Health. Part B, Critical Reviews*, *10*(1-2), 81–99. doi:10.1080/10937400601034589 PMID:18074305

Jabbar, A., & Mallick, S. (1994). *Pesticides and environment situation in Pakistan* (Working Paper Series No. 19). Sustainable Development Policy Institute (SDPI).

Jacobs, A. (2015). *Timeline-history of pesticides*. University of Oregon. Available from http://blogs.uoregon.edu/ajacobssu13gateway/timeline/

Jaga, K., & Dharmani, C. (2003). Sources of exposure to and public health implications of organophosphate pesticides. *Pan Am J Public Health*, *14*(3), 171–185. doi:10.1590/S1020-49892003000800004 PMID:14653904

Jardim, A. N. O., & Caldas, E. D. (2012). Brazilian monitoring programs for pesticide residues in food – results from 2001 to 2010. *Food Control*, *25*(2), 607–616. doi:10.1016/j.foodcont.2011.11.001

Jurewicz, J., & Hanke, W. (2008). Prenatal and childhood exposure to pesticides and neurobehavioural development: Review of epidemiological studies. *International Journal of Occupational Medicine and Environmental Health*, *21*(2), 121–132. doi:10.2478/v10001-008-0014-z PMID:18614459

Kamel, F., Engel, L., Gladen, B., Hoppin, J., Alavanja, M., & Sandler, D. (2005). Neurologic symptoms in Licensed Private Pesticide Applicators in the Agricultural Health Study. *Environmental Health Perspectives*, *113*(7), 877–882. doi:10.1289/ehp.7645 PMID:16002376

Kamrin, M. A. (1997). *Pesticide Profiles: toxicity, environmental impact, and fate*. CRC Press. doi:10.1201/9781420049220

Keikotlhaile, B. M., Spanoghe, P., & Steurbaut, W. (2010). Effects of food processing on pesticide residues in fruits and vegetables: A meta-analysi approach. *Food and Chemical Toxicology*, *48*(1), 1–6. doi:10.1016/j.fct.2009.10.031 PMID:19879312

Lah, K. (2011). Effects of pesticides on human health. *Toxipedia*. Available from http://www.toxipedia.org/display/toxipedia/Effects+of+Pesticides+on+Human+Health

Lee, D. H., Lee, I. K., Song, K., Steffes, M., Toscano, W., Baker, B. A., & Jacobs, D. R. Jr. (2006). A strong dose-response relation between serum concentrations of persistent organic pollutants and diabetes: Results from the National Health and Examination Survey 1999-2002. *Diabetes Care*, *29*(7), 1638–1644.

London, L., Beseler, C., Bouchard, M. F., Bellinger, D. C., Colosio, C., Grandjean, P., ... Stallones, L. (2012). Neurobehavioural and mneurodevelopmental effects of pesticide exposures. *Neurotoxicology*, *33*(4), 887–896. doi:10.1016/j.neuro.2012.01.004 PMID:22269431

Lozowicka, B., Jankowska, M., & Kaczyński, P. (2012). Pesticide residues in Brassica vegetables and exposure assessment of consumers. *Food Control*, *25*(2), 561–575. doi:10.1016/j.foodcont.2011.11.017

Lyons, G., & Watterson, A. (2010). *A Review of the Role Pesticides Play in some Cancers: Children, Farmers and Pesticide users at Risk?* CHEMTrust, UK. Retrieved from http://www.chemtrust.org.uk/Pesticidesandcancer.php

Macneale, K. H., Kiffney, P. M., & Scholz, N. L. (2010). Pesticides, aquatic food webs, and the conservation of Pacifi c salmon. *Frontiers in Ecology and the Environment*, *8*(9), 475–482. doi:10.1890/090142

Mahmood, I., Imadi, S. R., Shazadi,K., Gul, A. & Hakeem, K.R. (2016). *Effects of Pesticides on Environment.* Springer International Publishing.

Mariyono, J. (2008). Direct and indirect impacts of integrated pest management on pesticide use: A case of rice agriculture in Java, Indonesia. *Pest Management Science*, *64*(10), 1069–1073. doi:10.1002/ps.1602 PMID:18493927

Maroni, M., Fanetti, A. C., & Metruccio, F. (2006). Risk assessment and management of occupational exposure to pesticides in agriculture. *La Medicina del Lavoro*, *97*, 430–437. PMID:17017381

Martínez-Valenzuela, C., Gómez-Arroyo, S., Villalobos-Pietrini, R., Waliszewski, S., Calderón-Segura, M. E., Félix-Gastélum, R., & Álvarez-Torres, A. (2009). Genotoxic biomonitoring of agricultural workers exposed to pesticides in the north of Sinaloa State, Mexico. *Environment International*, *35*(8), 1155–1159. doi:10.1016/j.envint.2009.07.010 PMID:19665797

Mathur, S. C. (1999). Future of Indian pesticides industry in next millennium. *Pesticide Information*, *24*(4), 9–23.

Mendez, M. A., Garcia-Etseban, R., Guxens, M., Vrijheid, M., Kogevinas, M., Goñi, F., ... Sunyer, J. (2011). Prenatal organochlorine compound exposure, rapid weight gain, and overweight in infancy. *Environmental Health Perspectives*, *119*(2), 272–278. doi:10.1289/ehp.1002169 PMID:20923745

Montgomery, M. P., Kamel, F., Saldana, T. M., Alavanja, M. C., & Sandlerm, D. P. (2008). Incident diabetes and pesticide exposure among licensed pesticide applicators: Agricultural Health Study, 1993–2003. *American Journal of Epidemiology*, *167*(10), 1235–1246. doi:10.1093/aje/kwn028 PMID:18343878

Newbold, R. R., Padilla-Banks, E., Snyder, R. J., Phillips, T. M., & Jefferson, W. N. (2007). Developmental exposure to endocrine disruptors and the obesity epidemic. *Reproductive Toxicology (Elmsford, N.Y.), 23*(3), 290–296. doi:10.1016/j.reprotox.2006.12.010 PMID:17321108

Nwilene, F. E., Nwanze, K. F., & Youdeowei, A. (2008). Impact of integrated pest management on food and horticultural crops in Africa. *Entomologia Experimentalis et Applicata, 128*(3), 355–363. doi:10.1111/j.1570-7458.2008.00744.x

Palacios Xutuc, C. N. (2010). *Manual to Train Trainers on Safe and Correct Use of Plant Protection Products and Integrated Pest Management (IPM) CropLife Latin America*. Guatemala City, Guatemala: Pesticides and Human Health. 2014. Available from http://www. pesticidereform.org/section.php?id=38

Pimentel, D. (2005). Environmental and economic costs of the application of pesticides primarily in the United States. *Environment, Development and Sustainability*, *7*(2), 229–252. doi:10.100710668-005-7314-2

Rashid, B., Husnain, T., & Riazuddin, S. (2010). *Herbicides and pesticides as potential pollutants: a global problem. In Plant adaptation phytoremediation* (pp. 427–447). Dordrecht: Springer.

Reffstrup, T. K., Larsen, J. L., & Meyer, O. (2010). Risk assessment of mixtures of pesticides: Current approaches and future strategies. *Regulatory Toxicology and Pharmacology*, *56*(2), 174–192. doi:10.1016/j.yrtph.2009.09.013 PMID:19782118

Ren, A., Qiu, X., Jin, L., Ma, J., Li, Z., Zhang, L., ... Zhu, T. (2011). Association of selected persistent organic pollutants in the placenta with the risk of neural tube defects. *Proceedings of the National Academy of Sciences of the United States of America, 108*(31), 12770–12775. doi:10.1073/pnas.1105209108 PMID:21768370

Rignell-Hydbom, A., Rylander, L., & Hagmar, L. (2007). Exposure to persistent organochlorine pollutants and type 2 diabetes mellitus. *Human and Experimental Toxicology*, *26*(5), 447–452. doi:10.1177/0960327107076886 PMID:17623770

Rocheleau, C. M., Romitti, P. A., & Dennis, L. K. (2009). Pesticides and hypospadias: A meta-analysis. *Journal of Pediatric Urology*, *5*(1), 17–24. doi:10.1016/j.jpurol.2008.08.006 PMID:18848807

Sanborn, M., Bassil, K., Vakll, C., Kerr, K., & Ragan, K. (2012). *Systematic Review of Pesticide health Effects*. Toronto: Ontario College of Family Physicians. Retrieved from http://www.ocfp.on.ca/communications/newpublications/lists/new-publications/ocfp-2012-systematic-review-of-pesticide-health-effects

Sathiakumar, N., Delzell, E., MacLennan, P., Anne, M., Rosenberg, N., Cheng, H., & Myers, S. (2004). A crosssectional study of triallate exposure and neurological health among workers at a pesticide manufacturing and formulating facility'. *Occupational and Environmental Medicine*, *61*(1), 936–944. doi:10.1136/oem.2004.013128 PMID:15477288

Schreinemachers, D. M. (2003). Birth malformations and other adverse perinatal outcomes in four U.S. wheat-producing areas. *Environmental Health Perspectives*, *111*(9), 1259–1264. doi:10.1289/ehp.5830 PMID:12842783

Searles, N. S., McKean-Cowdin, R., Farin, F. M., Holly, E. A., Preston-Martin, S., & Mueller, B. A. (2010). Childhood brain tumors, residential insecticide exposure, and pesticide metabolism genes. *Environmental Health Perspectives*, *118*(1), 144–149. PMID:20056567

Sharma, D. R., Thapa, R. B., Manandhar, H. K., Shrestha, S. M., & Pradhan, S. B. (2012). Use of pesticides in Nepal and impacts on human health and environment. *J Agric Environ*, *13*, 67–72.

Slager, R., Poole, J., LeVan, T., Sandler, D., Alavanja, M., & Hoppin, J. (2009). Rhinitis Associated with Pesticide Exposure among Commercial Pesticide Applicators in the Agricultural Health Study. *Ocuppational and Environmental Medication*, *66*(11), 718–724. doi:10.1136/oem.2008.041798 PMID:19289390

Soares, W. L., & Porto, M. F. D. (2009). Estimating the social cost of pesticide use: An assessment from acute poisoning in Brazil. *Ecological Economics*, *68*(10), 2721–2728. doi:10.1016/j.ecolecon.2009.05.008

Spear, R. (1991). Recognised and possible exposure to pesticides. In W. J. Hayes & E. R. Laws (Eds.), *Handbook of pesticide toxicology* (pp. 245–274). San Diego, CA: Academic.

Speck-Planche, A., Kleandrova, V. V., & Scotti, M. T. (2012). Fragment-based approach for the in silic discovery of multi-target insecticides. *Chemometrics and Intelligent Laboratory Systems*, *111*(1), 39–45. doi:10.1016/j.chemolab.2011.11.010

Spzir, M. (2006). New thinking on neurodevelopment. *Environmental Health Perspectives*, *114*(2), 100–107. doi:10.1289/ehp.114-a100 PMID:16393665

Tariq, M. I., Afzal, S., Hussain, I., & Sultana, N. (2007). Pesticides exposure in Pakistan: A review. *Environment International*, *33*(8), 1107–1122. doi:10.1016/j.envint.2007.07.012 PMID:17765971

Thayer, K. A., Heindel, J. J., Bucher, J. R., & Gallo, M. A. (2012). Role of environmental chemicals in diabetes and obesity: A National Toxicology Programme workshop review. *Environmental Health Perspectives*, *120*(6), 779–789. doi:10.1289/ehp.1104597 PMID:22296744

Tsakiris, I. N., Danis, T. G., Stratis, I. A., Nikitovic, N., Dialyna, I. A., Alegakis, A. K., & Tsatsakis, A. M. (2004). Monitoring of pesticide residues in fresh peaches produced under conventional and integrated crop management cultivation. *Food Addit Contam A*, *21*(7), 670–677. doi:10.1080/02652030410001698715 PMID:15370840

Valvi, D., Mendez, M. A., Martinez, D., Grimalt, J. O., Torrent, M., Sunyer, J., & Vrijheid, M. (2012). Prenatal concentrations of PCBs, DDE, DDT and overweight in children: A prospective birth cohort study. *Environmental Health Perspectives*, *120*(3), 451–457. doi:10.1289/ehp.1103862 PMID:22027556

Van der Werf, H. M. G. (1996). Assessing the impact of pesticides on the environment. *Agriculture, Ecosystems & Environment*, *60*(2-3), 81–96. doi:10.1016/S0167-8809(96)01096-1

Van Maele-Fabry, G., Hoet, P., Vilain, F., & Lison, D. (2012). Occupational exposure to pesticides and Parkinson's disease: A systematic review and meta-analysis of cohort studies. *Environment International*, *46*, 30–33. doi:10.1016/j.envint.2012.05.004 PMID:22698719

Verhulst, S. L., Nelen, V., Den Hond, E., Koppen, G., Beunckens, C., Vael, C., ... Desager, K. (2009). Intrauterine exposure to environmental pollutants and body mass index during the first 3 years of life. *Environmental Health Perspectives*, *117*(1), 122–126. doi:10.1289/ehp.0800003 PMID:19165398

Waggoner, J. K., Kullman, G. J., Henneberger, P. K., Umbach, D. M., Blair, A., Alavanja, M. C., ... Hoppin, J. A. (2011). Mortality in the Agricultural Health Study. 1993-2007. *American Journal of Epidemiology*, *173*(1), 71–83. doi:10.1093/aje/kwq323 PMID:21084556

Weichenthal, S., Moase, C., & Chan, P. (2012). A review of pesticide exposure and cancer incidence in the agricultural health study cohort. *Environmental Health Perspectives*, *118*(8), 1117–1125. doi:10.1289/ehp.0901731 PMID:20444670

Wilson, C., & Tisdell, C. (2001). Why farmers continue to use pesticides despite environmental, health and sustainability costs. *Ecological Economics*, *39*(3), 449–462. doi:10.1016/S0921-8009(01)00238-5

Yi, A. X., Leung, K. M., Lam, M. H., Lee, J. S., & Giesy, J. P. (2012). Review of measured concentrations of triphenyltin compounds in marine ecosystems and meta-analysis of their risks to humans and the environment. *Chemosphere*, *89*(9), 1015–1025. doi:10.1016/j.chemosphere.2012.05.080 PMID:22704212

Yuan, Y., Chen, C., Zheng, C., Wang, X., Yang, G., Wang, Q., & Zhang, Z. (2014). Residue of chlorpyrifos and cypermethrin in vegetables and probabilistic exposure assessment for consumers in Zhejiang Province, China. *Food Control*, *36*(1), 63–68. doi:10.1016/j.foodcont.2013.08.008

KEY TERMS AND DEFINITIONS

Cancer: Cancer is a group of diseases involving abnormal cell growth with the potential to invade or spread to other parts of the body. These contrast with benign tumors, which do not spread to other parts of the body.

Chronic Effect: A chronic effect is the response to chemicals, biological agents, dusts, and vapors that occurs after a long period of exposure. Many chemicals and other substances in the workplace at allowable levels may not immediately provoke bad responses from the workers exposed to it.

Pesticide: Pesticide is a substance or mixture of substances intended for preventing, destroying, repelling, or mitigating any pest.

Teratogen: An agent that can produce a permanent alteration of structure or function in an organism exposed during embryonic or fetal life. The capability to cause malformations or defects to an embryo or fetus is called teratogenicity.

Chapter 6
Health Effects of Pesticides on Pregnant Women and Children

Mudasir Youssouf
Central University of Punjab, India

Arun Kalia
Central University of Punjab, India

Zahid Nabi
Central University of Jammu, India

Zubair A. Malik
Government HSS, India

ABSTRACT

Pesticides, along with hybrid seeds and fertilizers, are an integral part of the green revolution and are used to control and eradicate disease vectors for the improvement of agricultural production. Pesticides is an umbrella term for insecticides, nematocides, fungicides, herbicides, fumigants, repellents, and attractants. Pesticides are used against unwanted plants and animals to control diseases and losses. Efforts at different levels may help to reduce the impact of pesticides on newborn babies and on pregnant women. Different efforts can be considered at clinical, educational, and policymaking institutes. Use of risk assessment tools, encouragement of organic diets, educating parents working in agricultural fields from hazards of pesticides particularly in pregnancy and breast feeding, implementation of integrated pest management (IPM) programs, and encouraging policies supporting IPM can help in tackling the menace of pesticide hazards.

INTRODUCTION

Drinking water is classified among the most precious resources of the earth, however by anthropogenic activities both the quality and quantity of available water is continuously deteriorating (Benner *et al.*, 2013). A large part of world's population is forced to use contaminated drinking water (WHO, 2010). Millions of deaths mostly in developing countries could be prevented if people adhere to reliable safe

DOI: 10.4018/978-1-5225-6111-8.ch006

drinking water sources. Around 2.4 million deaths occur annually, mostly in developing countries by living in unhygienic conditions and having no access to potable water (Pruss-Ustun *et al.*, 2008). Among the two basic drinking water sources, surface water receives high extent of pollutants as compared to groundwater which is less exposed though groundwater can act as pollution source for decades due to higher residence times of pesticides and lower microbial activity as compared to surface water (Rodrigo *et al.*, 2014). With the varying pollutants and contaminants, the traditional water testing and monitoring processes and techniques (for microbial contamination) have also shifted to include the health risks of chemical contaminants, mostly when associated with chronic exposures (Fawell & Nieuwenhuijsen, 2003; Thompson *et al.*, 2007).

Due to widespread distribution, toxicity and persistence, pesticides are now the important class of water pollutants, even at very low concentrations pesticides can be hazardous to aquatic life because of bioconcentration process. Out of 22 identified POPs, 15 of them are pesticides mainly aldrin, dieldrin, endrin, chlordane, DDT, hexachlorobenzene, mirex, heptachlor, toxaphene, etc. Considering the severity of POPs a separate international environmental treaty (Stockholm conference) was signed in 2001 to eliminate or restrict the production and use of POPs (Xu *et al.*, 2013; Ali *et al.*, 2014). Properties like persistence in degradation process, ability to travel long distances, bioaccumulation, carcinogenic, hormone disruption and causing immunological and reproductive disorders has increased public concerns towards POPs (Vos *et al.*, 2000; Buccini, 2003; Sanpera *et al.*, 2003). Across the globe, 884 million people (13% of the world's population) depend on unprotected and distant water sources for drinking water collection and 3.6 billion people have well developed piped water system. However, in many low and middle-income countries, piped water system work for few hours and also are not safe, for example in Asian cities, more than one in five water supply schemes fail to meet national water quality standards (Bartram and Cairncross, 2010).

Pesticide contamination of surface water and groundwater can occur from both point sources (spill sites, disposal sites) and non-point sources which are the dominant source of pesticide pollution includes agricultural or urban runoff, infiltration from application sites, etc. (Fig. 1).

Figure 1. Schematic diagram depicting possible routes of pesticides into streams and groundwater
Thodal et al., 2009.

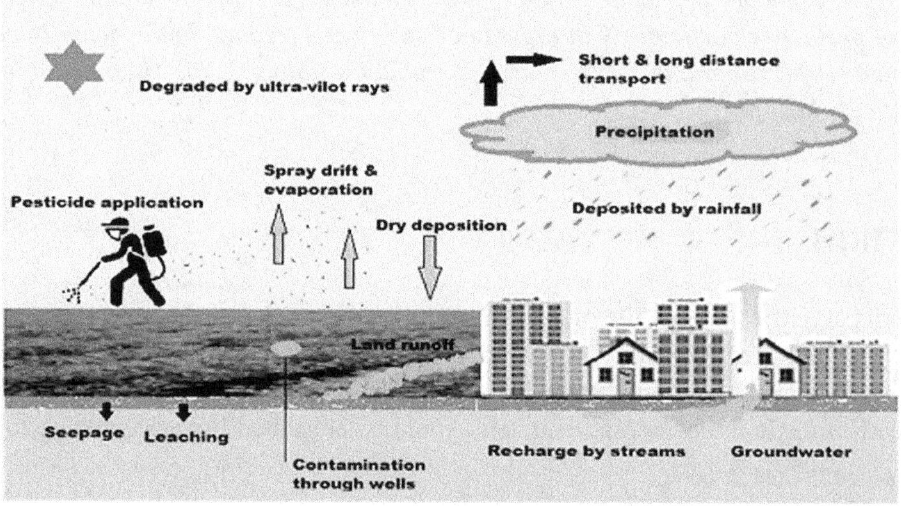

Table 1. Comparison of standards by different agencies for pesticide residues in drinking water

Pesticide	GV* µg/l (WHO)	USEPA		USA Health Advisory, Lifetime µg/l	BIS
		MCL** µg/l	MCLG*** µg/l		
Alachlor	20	2	0	-	Desirable limit for pesticides is given as "absent" In absence of alternate source the permissible limit is 0.001mg/l (1µg/l
Aldicarb	10	7	7	7	
Aldicarb sulfone	-	7	7	7	
Aldicarb sulfoxide	-	7	7	7	
Aldrin/ Dieldrin	0.03	-	-	-	
Atrazine	2	3	3	200	
Carbofuran	7	40	40	40	
Chlordane	0.2	2	0	-	
2,4-D	30	70	70	70	
DDT	2	-	-	-	
1,2-Dibromo-3-chloropane	1	0.2	0	-	
Diquat	10	20	20	-	
EDB	0.4-15	0.05	0	-	
Fenoprop (2,4,5-TP)	9	50	50	50	
Glyphosate	unnec	700	700	700	
Hexachlorobenzene	1	1	0	-	
Lindane	2	0.2	0.2	0.2	
Methoxychlor	20	40	40	40	
Molinate	6	-	-	-	
Pentachlorophenol	9	1	0	-	
Permethrin	20	-	-	-	
Picloram	-	500	500	500	
Propanil	20	-	-	-	
Simazine	2	4	4	4	
2,4,5-T	9	-	-	70	
Trifluralin	20	-	-	5	

Hamilton et al., 2003; BIS, 2012.

***GV:** Guideline Value, the concentration of a contaminant which doesn't show any significant risk to health over lifetime consumption.

****MCL:** Maximum Contaminant Level, allowed is drinking water and are enforceable standards set as closely possible to MCLGs taking technology and cost into consideration.

*****MCLG:** Maximum Contaminant Level Goal, level of the contaminant in drinking water below which there is no known or expected risk to health and are non-enforceable standards.

The importance and threat to freshwater ecosystem services were also highlighted in Millennium Ecosystem Assessment (MEA) report which identified heavy metals and pesticides as anthropogenic pollutants of extensive importance (Assessment, 2005).

For proper risk assessment and mitigation, information of the actual concentration of pesticides in the environment is fundamental. The most important matter of concern is the human exposure to pesticides by food and water (De Gerónimo *et al.*, 2014).The occurrence of pesticides in drinking water is related to high treatment costs, different toxicological incidences and hindrance in water usage. To restrict toxicological effects on human population and environmental pollution drinking water regulations are required (Lehmann *et al.*, 2017). There is a variation of regulatory limits for pesticide residues in drinking water from one regulating agency to another and it depends on different parameters like the type of water, type of residue, the analytical method followed and other environmental factors (Hamilton *et al.*, 2003). Each country must establish its own pesticide residue limits according to its actual environmental,

Table 2. List of pesticides which are banned and restricted in use in India: (As on 20th October 2015)

S. No.	Pesticides Banned for Manufacture, Import and Use	Pesticides Restricted for Use in the Country	Pesticides Refused Registration
1	Aldicarb	Aluminium Phosphide	2,4, 5-Trichloro phenoxy acetic acid
2	Aldrin	Captafol	Ammonium Sulphamate
3	Benzene Hexachloride	Cypermethrin	Azinphos Ethyl
4	Calcium Cyanide	Dazomet	Azinphos Methyl
5	Chlorbenzilate	Diazinon	Binapacryl
6	Chlordane	Dichloro Diphenyl Trichloroethane (DDT)	Calcium Arsenate
7	Chlorofenvinphos	Fenitrothion	Carbophenothion
8	Copper Acetoarsenite	Fenthion	Chinomethionate (Morestan)
9	Dibromochloropropane	Methoxy Ethyl Mercuric Chloride	Dicrotophos
10	Dieldrin		EPN
11	Endrin		Fentin Acetate
12	Ethyl Mercury Chloride		Fentin Hydroxide
13	Ethyl Parathion		Lead Arsenate
14	Ethylene Dibromide (EDB)		Leptophos (Phosvel)
15	Heptachlor		Mephosfolan
16	Lindane (Gamma-HCH)		Mevinphos (Phosdrin)
17	Maleic Hydrazide		Thiodemeton / Disulfoton
18	Menazon		Vamidothion
19	Metoxuron		
20	Nitrofen		
21	Paraquat Dimethyl Sulphate		
22	Pentachloro Nitrobenzene		
23	Pentachlorophenol		
24	Phenyl Mercury Acetate		
25	Sodium Methane Arsonate		
26	Tetradifon		
27	Toxaphene(Camphechlor)		
28	Trichloro acetic acid (TCA)		

economic and technological situation and shouldn't copy standards without considering the conditions under which they have been framed. The standard and guideline values for pesticide residues in drinking water by different agencies is given in Table 1

Also, there is an increasing need for the implementation of a scientifically efficient pesticide registration programme that filters out banned and damage causing pesticides (Teklu *et al.*, 2015). Sharing of information, cooperation and coordination between pesticide importing and exporting countries can help in accomplishing a single regional or international framework and compilation of a regional database (Islam *et al.*, 2017). Different countries have different regulations for pesticide consumption, for example, in India The Insecticide Act 1968 and Rules 1971 was enacted to regulate pesticide usage in the country and Central Insecticides Boards and registration Committee (CIB & RC), is responsible for granting or refusing registration to pesticides, also it gives detailed account of banned pesticides in the country, as is given in Table 2 (CIBRC, 2015).

In this chapter, we emphasized the possible routes of pesticide entry into drinking water and also discussed the effects of pesticide exposure to pregnant women and children. A very few studies has been carried out which discusses effects of pesticides on pregnant women and newborn children. Fetuses and children are the weakest age group to pesticide exposure and better understanding pesticide pollution and its control can help to lower health hazard incidences.

ENVIRONMENTAL FATE AND INPUT PATHWAYS OF PESTICIDES INTO DRINKING WATER

Once pesticides are accidentally or deliberately released into the environment, they either find their way directly into aquatic ecosystems or indirectly by atmospheric deposition during spray drift or by volatilization after their application (Schafer *et al.*, 2011). There are two sources by which pesticides enter water bodies which include diffused or by point sources. Leaching through the upper soil and vadose zone (unsaturated zone), and infiltration through river banks and beds are some of the examples of pesticide diffuse pathways into groundwater. Point source includes pesticide contamination from agricultural runoff, accidental spills, leaking or faulty equipment, storage canisters etc. (Reichenberger *et al.*, 2007).

The basic aim of drinking water schemes should be public health protection and can be delivered by constructing a regulatory framework encompassing health-based targets, proper and adequate treatment for every type of contaminant species and performing extensive monitoring (WHO, 2010). Once pesticides come in contact with the environment they are prone to degradation by biotic and abiotic sources. Pesticide degradation may start right from their application to their absorption/sink sites, however, degradation process may be slow due to the persistence of pesticides and variety of new compounds are formed as transformation products or degradates (Parsons *et al.*, 2008). The transformation products may be less or more toxic and stable than its base compound (Andreu and Pico, 2004). Pesticide and its degradation products follow different paths before they reach to major surface or groundwater sources, so there are optimal chances of pesticides contamination in water abstracted for drinking purposes (Damalas and Eleftherohorinos, 2011).

Conventional water treatment facilities (coagulation/flocculation, sedimentation, filtration and disinfection) are not able to remove or reduce pesticide residue concentration, so there should some specific treatment facilities for removal of entire pesticide classes (Rodrigo *et al.*, 2014). Over the past few decades, different water treatment technologies including chemical, physical and biologic processes

Figure 2. Common practices used for pesticide removal in drinking water

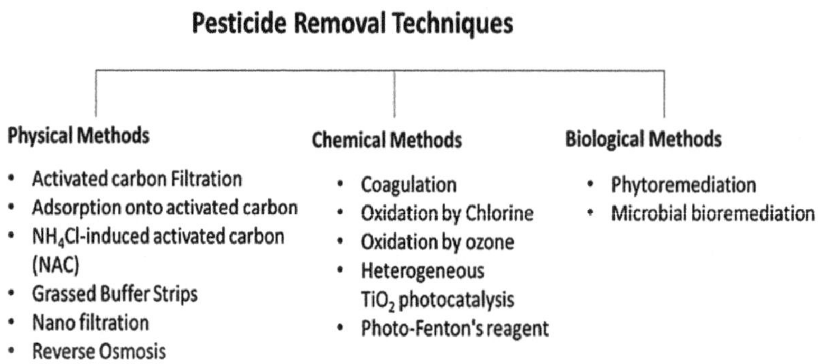

have been developed for pesticide removal. A brief account of different treatment processes and their limitations used in pesticide removal in drinking water is expressed in Figure 2.

PESTICIDE TOXICITY AND HEALTH PROBLEMS IN PREGNANT WOMEN AND NEWBORN BABIES

Pesticides were considered as a blessing for human life for enhancing agricultural productivity and wiping out infectious diseases but their large-scale use has caused numerous hazards to the ecosystem and human life (Mostafalou and Abdollahi, 2013). The extent or ability of pesticides to cause disease, injury or illness in living organisms is called the pesticide toxicity. The pesticide toxicity is studied by subjecting test animals to different doses of the active ingredient and at the same time each of its formulated products (Asghar *et al.*, 2016). On the basis of severity and time duration of toxicity symptoms two types of toxicity are acute and chronic (Gao and Lu, 2015).

There is no complete date available on toxicity of every pesticide, a brief account of acute oral and dermal toxicity of some pesticides is given in Table 3. However, chronic toxicity can't be identified at the same lethal doses (Criswell *et al.*, 2014).

Table 3. Acute oral and dermal lethal doses of different pesticides for humans

Common Name	Trade Name	Acute Oral LD50 mg/kg	Acute Dermal LD50 mg/kg
aldicarb	Temik	1	20
allethrin	(many)	480	11,200
azadirachtin	Aza-Direct, Ecozin, others	>5,000	>2,000
Carbaryl	Sevin	246-283	4,000
chlorfenapyr	Phantom, Pylon	560	>2,000
chlorpyrifos	Lorsban, Dursban, Durap	96-270	2,000
DDT	–	113	2,510
diazinon	Diazinon, Spectracide	300-400	3,600

continued on following page

Table 3. Continued

Common Name	Trade Name	Acute Oral LD50 mg/kg	Acute Dermal LD50 mg/kg
dichlorvos	DDVP, Vapona	80	105-107
dictotophos	Bidrin	17-22	224
diquat	Diquat, Reglone	215-235	>400
dienochlor	Pentac	3,160	>3,160
dimethoate	Dimethoate, Cygon	235	400
endosulfan	Thiodan, Phaser	160	359
hydroprene	Gen Trol	>34,000	5,100
lindane	Lindane, others	200	2,000
malathion	Cythion, Malathion	2,800	4,100
methoprene	Altosid, Precor, others	>34,000	>3,000
methoxfendozide	Intrepid	>5,000	>2,000
methyl bromide	(many)	214	–
permethrin	Ambush, Astro, others	2,215`	>2,000
phorate	Thimet, GX-118	4	6
phosphoric acid	Foray	1,530	2,740
potassium salts	M-Pede	>5,000	>2,000
propargite	Omite, Comite	4,029	2,940
pyrethrin	(many)	1,500	>1,800
pyriproxyfen	Distance	>5,000	>2,000
rotenone	(many)	350	940
sulfur	Microthiol, Thiodex	>2,000	2,000
sulfotepp	Bladafum	10	65
tebufenozide	Confirm	>5,000	>5,000
tebupirimphos	Aztec	132	>2,000
thiamethoxam	actara, Cruiser, others	>5,000	>2,000
thiodicarb	Larvin	166	>2,000
diuron	Karmex	3,40	2,000
fenac	Fenatrol	1,780	>3,160
glyphosate	Rodeo, Roundup	5,000	>5,000
hexazinone	Velpar	1,690	5,278
methazole	Probe	2,501	>12,500
metolachlor	Dual	2,780	>10,000
paraquat	Grmoxone, Cyclone	150	–
propachlor	Ramrod	500-1,700	–
simazine	Princep	>5,000	>3,100

Criswell et al., 2014.

Acute toxicity of pesticides results from mild to devastating and lethal consequences like a headache, vomiting, abdominal pain, respiratory problems, ulceration in the upper gastrointestinal tract, etc. (Roberts and Karr, 2012). Chronic effects (described by gradual progression and long-term pesticide contact) which are diagnosed in epidemiological studies are worse than acute effects and include adverse neurobehavioral and cognitive problems, cardiovascular diseases, respiratory problems like asthma, pediatric cancer, acute lymphocytic leukemia, preterm birth, low birth weight and hormonal imbalance etc (Souza *et al.* 2011; Mostafalou and Abdollahi 2012; Mostafalou and Abdollahi, 2013).

Although all age groups are vulnerable to pesticide toxicity but considering the stage of developing fetus and the children, they are more sensitive to chemicals as most of the organ systems are in developing stage, for example, their immune system may not be able to safeguard them against disease-causing agents or environmental threats (Weselak *et al.*, 2007). Children's behaviour and inability to understand their physical environment can place them at greater risk of exposure. Also, children for their weight, consume more food, drinks and air, increasing their possible dietary exposure. Therefore special attention should be paid to children's exposure to pesticides from food, water, inhalation, playground, etc. and the fetus can be exposed to pesticides across the placenta, amniotic fluid and ovarian follicular fluid (Colborn, 2006; Gilden *et al.*, 2010). Different studies have proved the vulnerability of fetuses to pesticides as they can pass through the placenta and blood-brain barrier and were also found in amniotic fluid (Bradman *et al.*, 2003). In comparison to pregnant women, children are more susceptible to pesticide poisoning due to their high respiratory and heart rate, higher metabolism, food consumption pattern and hand-to-mouth behaviour (Garry, 2004). Exposure to pesticides during fetal and early child development stages has been correlated with different diseases which will be briefly discussed in this chapter.

Pesticides and Birth Abnormalities

Exposure to toxicants during early stages of fetal development is associated with an increased risk of birth-related defects which are detectable at birth or within few years of childhood. The two common birth related deformities caused by pesticide exposure of fetuses are preterm birth or low birth weight (LBW), children born with these disorders experience higher rates of morbidity through the perinatal period as compared to normal ones (Stillerman *et al.*, 2008; Windham and Fenster, 2008). Household pesticide exposures through different sources on the time of pregnancy have adverse impacts on birth outcomes. Suppression of neurite outgrowth is associated with pesticide exposure as these inhibit acetylcholinesterase, down-regulate muscarinic receptors, inhibit the adenylate cyclase signalling cascade and leads to decrease brain DNA and RNA synthesis (Das and Barone, 1999; Yang *et al.*, 2008). Many studies have claimed a positive association between pesticide exposure with spontaneous abortions or fetal death (Petrelli *et al.*, 2000). Assessment of fetal growth is usually derived by surrogate measures at delivery, including gestation period, fetal size and birth weight. Dichlorodiphenyltrichloroethane (DDT) is an organochlorine pesticide and prototype persistent environmental chemical with endocrine disrupting effects. Longnecker *et al.* (2001) reported that dichlorodiphenyltrichloroethane use increases preterm births, small-for-gestational age which is a major contributor to infant mortality in a sampling subset of more than 44 000 eligible children born between 1959 and 1966 and measured the DDE concentration in their mothers' serum samples. Siddiqui *et al.*, (2003) examined the association between dichlorodiphenyltrichloroethane exposure and intra-uterine growth retardation and found that exposure of pregnant women to organochlorine pesticides elevate the risk of intra-uterine growth retardation, which is one of the factors for increased infant mortality in India. Curtis *et al.*, (1999) studied that application of pesticides was associated with a long time to pregnancy, resulting in a fecundability ratio of 0.46 and low spraying velocity resulted in a fecundability ratio of 0.47. Perera et al. (2005) reported that cord chlorpyrifos, and combined mixtures of cord chlorpyrifos, diazinon and propoxur-metabolite are associated with birth weight and length, indicating that prenatal chlorpyrifos exposures have impaired fetal growth among this minority cohort and that diazinon exposures may have contributed to the effects.

Neurologic and Neurobehavioral Effects

Susceptibility of fetuses and young children to potential neurotoxic effects of pesticides is higher than adults as their brains are developing rapidly (Eskenazi *et al.*, 2007). Different pesticides act as acetylcholinesterase (AChE) inhibitors and during developmental stages, pesticide exposure even at low levels can be detrimental to neurologic functioning (Munoz-Quezada *et al.*, 2013). Pesticides inhibit acetylcholinesterase enzyme (which is already lower during pregnancy) and prevent breakdown of the neurotransmitter acetylcholine, increasing its concentration and time in the neuronal junction. This acetylcholinesterase suppression disturbs synaptogenesis, and axonogenesis, cell replication and differentiation (Dam *et al.*, 2003; Tadeo, 2008). Eskenazi et al. (2007) reported that exposure of organophosphate (OP) pesticides is adversely impacting the mental development and pervasive developmental problems in newborns. Rauh *et al.* (2006) also studied the Prenatal Chlorpyrifos Exposure on Neurodevelopment by relating the maternal blood levels of a diethyl phosphate pesticide during pregnancy and performance on the Bayley Scales of Infant Development. They reported that in initial three years of life with high prenatal chlorpyrifos is directly related to delays in psychomotor and mental development and it also affected the mothers with symptoms of pervasive developmental problems. In another study, Rauh *et al.* (2011) highlighted the widespread use of chlorpyrifos and investigated that prenatal exposure of chlorpyrifos at 7 years of age is associated with deficits in Working Memory Index and Full-Scale IQ. Berkowitz *et al.* (2004) reported the relationship between paraoxonase (PON1) polymorphisms and enzyme activity and its effects on infant growth and neurodevelopment and reported that chlorpyrifos has deleterious effect on fetal neurodevelopment among mothers with low paraoxonase (PON1) activity.

Hormonal Imbalance and Endocrine Disruption Abnormalities

Reports of pesticides acting as endocrine disrupters were reported as early as 1949 when the aerial application of DDT was found as the main cause in low sperm count of men (Singer 1949). Pesticides interfere with the hormonal functioning of females, which lead to negative effects on the reproductive system through disordering the hormonal balance necessary for proper functioning (Bretveld et. al. 2006). Cell receptors become susceptible to the action of exogenous chemicals due to which imbalance of the endocrine system in female adversely affects the menstrual cycle and fertility (Figure 3) (Nicolopoulou-Stamati and Pitsos, 2001). Pesticides come under the class of endocrine-disrupting contaminants (EDC) that exert hormonal activity and when delivered during a specific period of embryonic development, have the potential to modify the organization of the reproductive, immune and nervous systems permanently (Guillette *et al.*, 1995). EDCs mimic the action of natural hormones by binding to different hormone receptors interfere with the synthesis, transport, metabolism and elimination of hormones. There has been enough research that reported certain pesticides acting as potential EDCs (Table 4). During the periods of gestation and breastfeeding fetuses and babies get larger doses of EDCs due to the mobilization of maternal fat reserves (Balabanic *et al.*, 2011). EDC's having estrogenic properties can also block ovula-

Figure 3. Potential effects of pesticides on female reproduction
Bretveld et al., 2006.

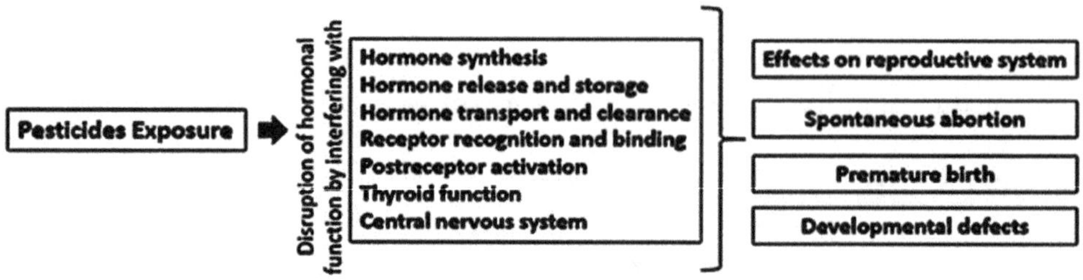

Table 4. Common endocrine-disrupting pesticide groups: their effects and modes of action

Pesticide Group	Hormones Affected	Mechanisms	References
Organochlorines	Oestrogens, androgens, prolactin	Inhibition of androgen receptor, oestrogen-sensitive reporter, binding to androgen receptors, interruption in induction of aromatase	Daxenberger, 2002; Lemaire *et al.*, 2004
Organophosphates	Oestrogens	Onset of oestrogen-related genomic activity	Gwinn *et al.*, 2005
Carbamates	Steroids, androgens, oestrogens	Oestrogen receptor interruption with cellular microtubule formation in oestrogen-sensitive cells	Lu *et al.*, 2006; Goad *et al.*, 2004
Triazines	Androgens	Inhibition of androgen receptors, binding to androgen-binding receptors, induction or inhibition of aromatase	Ishihara *et al.*, 2003
Pyrethrins	Progesterone, oestrogens	Inhibition or potentiation of oestrogen action by interfering progesterone action	Kim *et al.*, 2007

tion in females, by affecting the luteinizing hormone surge, like the contraceptive pills do (Ashby *et al.*, 2002). Different studies have suggested even adverse endocrine-disrupting effects, like cryptorchidism, hypospadias and different birth defects on children and fetuses exposed to pesticides (Carbone *et al.*, 2006). Farr *et al.* (2004) studied the relationship between pesticide exposure and menstrual function and reported that use of carbamate pesticides was linked with an increased chance of long cycles and with a lesser risk of irregular cycles.

Pediatric Cancer

Cancer is considered as the second leading cause for children's death after accidents and among the 12 different types of cancer Leukemia has the highest rate of incidence, whereas childhood brain tumours (CBT) and lymphomas are respectively on second and third most frequent groups (Nasterlack, 2007). Fetuses are more sensitive to harm from environmental toxicants and their prenatal exposures have been associated with adverse health impacts (Selevan, 2000). Various epidemiologic studies conducted all over the globe indicate that current pesticide exposures are associated with increased risks of childhood leukaemia, brain cancer, neuroblastoma, non-Hodgkin's lymphoma, Wilms' tumour, and Ewing's sarcoma. Pesticides of different groups including organochlorines, organophosphates, carbamates and pyrethroids are classified as possible carcinogens by USEPA and IARC (George and Shukla, 2011). Infante-Rivard and Weichenthal (2007) reported that parental occupational exposure to pesticides during pregnancy contribute to children's cumulative burden. Shim *et al.* (2009) diagnosed brain cancer cases at <10 years of age and reported the associations between parental exposure to pesticides and risk of astrocytoma in offspring from statewide cancer registries of four U.S. Atlantic Coast states. Turner *et al.* (2011) confirmed the positive association between unspecified residential pesticides, insecticides and herbicides exposures during pregnancy with childhood leukaemia in a meta-analysis of previous observational epidemiologic studies. Greenop *et al.* (2013) suggested that pesticide exposure during pregnancy was related to an increase in childhood brain tumours (CBT). Similarly, Chen *et al.* (2015) observed a positive relationship between exposure to pesticides and paediatric cancer and the strongest relation was found between from exposure to indoor pesticides and acute childhood leukaemia.

Many different diseases that are supposed to occur by the pesticide exposure, but lack of consistent evidence and ongoing investigations are major limitations to put those diseases on the list. We will give a brief account some of such astounding diseases. Pesticides as endocrine disrupters are believed to affect the development of the male reproductive system. Several studies revealed male children of women who were exposed to pesticides in early pregnancy showed signs of reduced genital size and change in hormone level (Andersen *et al.*, 2008; Wohlfahrt-Veje *et al.*, 2012a). Other scientists reported early breast development in female offspring of pregnant greenhouse workers who were exposed to non-persistent pesticides. Breast development started 1-1.5 years earlier in daughters of pesticide-exposed pregnant women as compared to daughters of unexposed pregnant women (Wohlfahrt-Veje *et al.*, 2012b). A very few reports showed a positive association of maternal pesticide exposure with Neural tube defects (NTDs), which occur during neurulation when neural tube remains open and is not closed in 21 to 28 days of postconception (Brender *et al.*, 2010; Yang *et al.*, 2014). But in other studies, the association of maternal pesticide exposure with NTDs has not consistently been ascertained (Lacasana *et al.*, 2006; Makelarski *et al.*, 2014). Shelton *et al.* (2012) reviewed that parental exposure to pesticides may or may not influence the trend of increasing Autism, which is composite behaviourally specified condition is usually found in children below the age of 3 years. Asthma, a paediatric widespread respiratory disease has been associated with foetal or early life exposure to pesticides by many researchers. Raanan *et al.* (2015) suggested pesticide exposure in early life stage is associated with different respiratory diseases and Salameh *et al.* (2006) revealed chronic respiratory symptoms and asthma were influenced by pesticide exposure and also found asthma in adult Lebanese caused by pesticide exposure.

CONCLUSION

In this study, we attempted to find the potential health effect of pesticide-contaminated drinking water on pregnant women (fetuses) and children. We tried to explore different routes of pesticide exposure into drinking water and also discussed different diseases caused in most risk-prone age groups. Stringent laws and regulations can help in tackling the pollution of drinking water with pesticides. There should be the monitoring and regulatory authorities to maintain safe and desirable drinking water supply. Also, utilization of biopesticides instead or chemical pesticide formulations and organic farming practices can lower the chemical pesticide load in the environment. Biopesticides are an eco-friendly alternative to synthetic chemical pesticides, which involves a broad array of biologically originated pesticides from microbes, nematodes, secondary metabolites of plants and also involves the exploitation of predatory or parasitic relation of organisms to control the pest population. Biopesticides are easily degradable as compared to conventional pesticides and also less hazardous to ecosystems. There is also a growing stress for the adoption of agricultural practices which depends on the use of biological inputs rather than synthetic chemical fertilizers and pesticides. Organic farming practices can provide better food quality without pesticide and nutrient pollution also the ecological soil quality can be maintained. In conclusion awareness about the pesticide contamination in drinking water and health hazards of these pesticide residues can help in minimizing the deleterious impacts of pesticides.

REFERENCES

Ali, U., Syed, J. H., Malik, R. N., Katsoyiannis, A., Li, J., Zhang, G., & Jones, K. C. (2014). Organochlorine pesticides (OCPs) in South Asian region: A review. *The Science of the Total Environment*, *476*, 705–717. doi:10.1016/j.scitotenv.2013.12.107 PMID:24522007

Andersen, H. R., Schmidt, I. M., Grandjean, P., Jensen, T. K., Budtz-Jørgensen, E., Kjaerstad, M. B., & Main, K. M. (2008). Impaired reproductive development in sons of women occupationally exposed to pesticides during pregnancy. *Environmental Health Perspectives*, *116*(4), 566–572. doi:10.1289/ehp.10790 PMID:18414644

Andreu, V., & Pico, Y. (2004). Determination of pesticides and their degradation products in soil: Critical review and comparison of methods. *Trends in Analytical Chemistry*, *23*(10), 772–789. doi:10.1016/j.trac.2004.07.008

Asghar, U., Malik, M. F., & Javed, A. (2016). Pesticide exposure and human health: A review. *Journal of Ecosystem and Ecography*, 5-12.

Ashby, J., Tinwell, H., Stevens, J., Pastoor, T., & Breckenridge, C. B. (2002). The effects of atrazine on the sexual maturation of female rats. *Regulatory Toxicology and Pharmacology*, *35*(3), 468–473. doi:10.1006/rtph.2002.1571 PMID:12202059

Assessment, M. E. (2005). *Ecosystems and human well-being: wetlands and water*. Washington, DC: World Resources Institute.

Balabanic, D., Rupnik, M., & Klemencic, A. K. (2011). Negative impact of endocrine-disrupting compounds on human reproductive health. *Reproduction, Fertility, and Development*, *23*(3), 403–416. PMID:21426858

Bartram, J., & Cairncross, S. (2010). Hygiene, sanitation, and water: Forgotten foundations of health. *PLoS Medicine*, *7*(11), e1000367. doi:10.1371/journal.pmed.1000367 PMID:21085694

Benner, J., Helbling, D. E., Kohler, H. P. E., Wittebol, J., Kaiser, E., Prasse, C., & Springael, D. (2013). Is biological treatment a viable alternative for micropollutant removal in drinking water treatment processes? *Water Research*, *47*(16), 5955–5976. doi:10.1016/j.watres.2013.07.015 PMID:24053940

Berkowitz, G. S., Wetmur, J. G., Birman-Deych, E., Obel, J., Lapinski, R. H., Godbold, J. H., & Wolff, M. S. (2004). In utero pesticide exposure, maternal paraoxonase activity, and head circumference. *Environmental Health Perspectives*, *112*(3), 388–391. doi:10.1289/ehp.6414 PMID:14998758

BIS IS10500. (2012). *Indian Standard Drinking Water–Specification (Second revision)*. Bureau of Indian Standards (BIS).

Bradman, A., Barr, D. B., Henn, B. G. C., Drumheller, T., Curry, C., & Eskenazi, B. (2003). Measurement of pesticides and other toxicants in amniotic fluid as a potential biomarker of prenatal exposure: A validation study. *Environmental Health Perspectives*, *111*(14), 1779–1782. doi:10.1289/ehp.6259 PMID:14594631

Brender, J. D., Felkner, M., Suarez, L., Canfield, M. A., & Henry, J. P. (2010). Maternal pesticide exposure and neural tube defects in Mexican Americans. *Annals of Epidemiology*, *20*(1), 16–22. doi:10.1016/j.annepidem.2009.09.011 PMID:20006272

Bretveld, R. W., Thomas, C. M., Scheepers, P. T., Zielhuis, G. A., & Roeleveld, N. (2006). Pesticide exposure: The hormonal function of the female reproductive system disrupted? *Reproductive Biology and Endocrinology*, *4*(1), 30. doi:10.1186/1477-7827-4-30 PMID:16737536

Buccini, J. (2003). The Development of a Global Treaty on Persistent Organic Pollutants (POPs). In Persistent Organic Pollutants. The Handbook of Environmental Chemistry (vol. 3, pp. 13-30). Springer.

Carbone, P., Giordano, F., Nori, F., Mantovani, A., Taruscio, D., Lauria, L., & Figa-Talamanca, I. (2006). Cryptorchidism and hypospadias in the Sicilian district of Ragusa and the use of pesticides. *Reproductive Toxicology (Elmsford, N.Y.)*, *22*(1), 8–12. doi:10.1016/j.reprotox.2006.01.006 PMID:16530380

Chen, M., Chang, C. H., Tao, L., & Lu, C. (2015). Residential exposure to pesticide during childhood and childhood cancers: A meta-analysis. *Pediatrics*, *136*(4), 719–729. doi:10.1542/peds.2015-0006 PMID:26371195

CIBRC. (2015). *List of pesticides which are banned, refused registration and restricted in use*. Retrieved March 8, 2018, from http://cibrc.nic.in/ibr2012.doc

Colborn, T. (2006). A case for revisiting the safety of pesticides: A closer look at neurodevelopment. *Environmental Health Perspectives*, *114*(1), 10–17. doi:10.1289/ehp.7940 PMID:16393651

Criswell, J. T., Shelton, K., & Luper, C. (2014). *Toxicity of pesticides. Pesticide Applicator Certificate Series*. Oklahoma Cooperative Extension Service EPP-7457-8. Available from: http://www.dasnr22.dasnr.okstate.edu/docushare/dsweb/get/.../EPP-7457web.pdf

Curtis, K. M., Savitz, D. A., Weinberg, C. R., & Arbuckle, T. E. (1999). The effect of pesticide exposure on time to pregnancy. *Epidemiology (Cambridge, Mass.)*, *10*(2), 112–117. doi:10.1097/00001648-199903000-00005 PMID:10069244

Dam, K., Seidler, F. J., & Slotkin, T. A. (2003). Transcriptional biomarkers distinguish between vulnerable periods for developmental neurotoxicity of chlorpyrifos: Implications for toxicogenomics. *Brain Research Bulletin*, *59*(4), 261–265. doi:10.1016/S0361-9230(02)00874-2 PMID:12464398

Damalas, C. A., & Eleftherohorinos, I. G. (2011). Pesticide exposure, safety issues, and risk assessment indicators. *International Journal of Environmental Research and Public Health*, *8*(5), 1402–1419. doi:10.3390/ijerph8051402 PMID:21655127

Das, K. P., & Barone, S. Jr. (1999). Neuronal differentiation in PC12 cells is inhibited by chlorpyrifos and its metabolites: Is acetylcholinesterase inhibition the site of action? *Toxicology and Applied Pharmacology*, *160*(3), 217–230. doi:10.1006/taap.1999.8767 PMID:10544056

Daxenberger, A. (2002). Pollutants with androgen-disrupting potency. *European Journal of Lipid Science and Technology*, *104*(2), 124–130. doi:10.1002/1438-9312(200202)104:2<124::AID-EJLT124>3.0.CO;2-T

De Gerónimo, E., Aparicio, V. C., Bárbaro, S., Portocarrero, R., Jaime, S., & Costa, J. L. (2014). Presence of pesticides in surface water from four sub-basins in Argentina. *Chemosphere*, *107*, 423–431. doi:10.1016/j.chemosphere.2014.01.039 PMID:24548646

Eskenazi, B., Marks, A. R., Bradman, A., Harley, K., Barr, D. B., Johnson, C., & Jewell, N. P. (2007). Organophosphate pesticide exposure and neurodevelopment in young Mexican-American children. *Environmental Health Perspectives*, *115*(5), 792–798. doi:10.1289/ehp.9828 PMID:17520070

Farr, S. L., Cooper, G. S., Cai, J., Savitz, D. A., & Sandler, D. P. (2004). Pesticide use and menstrual cycle characteristics among premenopausal women in the agricultural health study. *American Journal of Epidemiology*, *160*(12), 1194–1204. doi:10.1093/aje/kwi006 PMID:15583372

Fawell, J., & Nieuwenhuijsen, M. J. (2003). Contaminants in drinking waterEnvironmental pollution and health. *British Medical Bulletin*, *68*(1), 199–208. doi:10.1093/bmb/ldg027 PMID:14757718

Gao, F., & Lu, X. (2015). Detection of pesticides in foods by enzymatic biosensors. In *Improving and Tailoring Enzymes for Food Quality and Functionality*. Cambridge, UK: Woodhead publishing. doi:10.1016/B978-1-78242-285-3.00007-7

Garry, V. F. (2004). Pesticides and children. *Toxicology and Applied Pharmacology*, *198*(2), 152–163. doi:10.1016/j.taap.2003.11.027 PMID:15236951

George, J., & Shukla, Y. (2011). Pesticides and cancer: Insights into toxicoproteomic-based findings. *Journal of Proteomics*, *74*(12), 2713–2722. doi:10.1016/j.jprot.2011.09.024 PMID:21989265

Gilden, R. C., Huffling, K., & Sattler, B. (2010). Pesticides and health risks. *Journal of Obstetric, Gynecologic, and Neonatal Nursing*, *39*(1), 103–110. doi:10.1111/j.1552-6909.2009.01092.x PMID:20409108

Goad, R. T., Goad, J. T., Atieh, B. H., & Gupta, R. C. (2004). Carbofuran-induced endocrine disruption in adult male rats. *Toxicology Mechanisms and Methods*, *14*(4), 233–239. doi:10.1080/15376520490434476 PMID:20021136

Greenop, K. R., Peters, S., Bailey, H. D., Fritschi, L., Attia, J., Scott, R. J., & Milne, E. (2013). Exposure to pesticides and the risk of childhood brain tumors. *Cancer Causes & Control*, *24*(7), 1269–1278. doi:10.100710552-013-0205-1 PMID:23558445

Guillette, L. J. Jr, Crain, D. A., Rooney, A. A., & Pickford, D. B. (1995). Organization versus activation: The role of endocrine-disrupting contaminants (EDCs) during embryonic development in wildlife. *Environmental Health Perspectives*, *103*(Suppl 7), 157–164. doi:10.1289/ehp.95103s7157 PMID:8593864

Gwinn, M. R., Whipkey, D. L., Tennant, L. B., & Weston, A. (2005). Differential gene expression in normal human mammary epithelial cells treated with malathion monitored by DNA microarrays. *Environmental Health Perspectives*, *113*(8), 1046–1051. doi:10.1289/ehp.7311 PMID:16079077

Hamilton, D. J., Ambrus, A., Dieterle, R. M., Felsot, A. S., Harris, C. A., Holland, P. T., & Wong, S. S. (2003). Regulatory limits for pesticide residues in water (IUPAC Technical Report). *Pure and Applied Chemistry*, *75*(8), 1123–1155. doi:10.1351/pac200375081123

Infante-Rivard, C., & Weichenthal, S. (2007). Pesticides and childhood cancer: An update of Zahm and Ward's 1998 review. *Journal of Toxicology and Environmental Health. Part B*, *10*(1-2), 81–99.

Ishihara, A., Sawatsubashi, S., & Yamauchi, K. (2003). Endocrine disrupting chemicals: Interference of thyroid hormone binding to transthyretins and to thyroid hormone receptors. *Molecular and Cellular Endocrinology*, *199*(1), 105–117. doi:10.1016/S0303-7207(02)00302-7 PMID:12581883

Islam, M. N., Bint-E-Naser, S. F., & Khan, M. S. (2017). Pesticide Food Laws and Regulations. In *Pesticide Residue in Foods*. Cham, Switzerland: Springer. doi:10.1007/978-3-319-52683-6_3

Kim, S. D., Cho, J., Kim, I. S., Vanderford, B. J., & Snyder, S. A. (2007). Occurrence and removal of pharmaceuticals and endocrine disruptors in South Korean surface, drinking, and waste waters. *Water Research*, *41*(5), 1013–1021. doi:10.1016/j.watres.2006.06.034 PMID:16934312

Lacasana, M., Vazquez-Grameix, H., Borja-Aburto, V. H., Blanco-Muñoz, J., Romieu, I., Aguilar-Garduño, C., & García, A. M. (2006). Maternal and paternal occupational exposure to agricultural work and the risk of anencephaly. *Occupational and Environmental Medicine*, *63*(10), 649–656. doi:10.1136/oem.2005.023333 PMID:16873458

Lehmann, E., Turrero, N., Kolia, M., Konaté, Y., & De Alencastro, L. F. (2017). Dietary risk assessment of pesticides from vegetables and drinking water in gardening areas in Burkina Faso. *The Science of the Total Environment*, *601*, 1208–1216. doi:10.1016/j.scitotenv.2017.05.285 PMID:28605838

Lemaire, G., Terouanne, B., Mauvais, P., Michel, S., & Rahmani, R. (2004). Effect of organochlorine pesticides on human androgen receptor activation in vitro. *Toxicology and Applied Pharmacology*, *196*(2), 235–246. doi:10.1016/j.taap.2003.12.011 PMID:15081270

Longnecker, M. P., Klebanoff, M. A., Zhou, H., & Brock, J. W. (2001). Association between maternal serum concentration of the DDT metabolite DDE and preterm and small-for-gestational-age babies at birth. *Lancet*, *358*(9276), 110–114. doi:10.1016/S0140-6736(01)05329-6 PMID:11463412

Lu, C., Toepel, K., Irish, R., Fenske, R. A., Barr, D. B., & Bravo, R. (2006). Organic diets significantly lower children's dietary exposure to organophosphorus pesticides. *Environmental Health Perspectives*, *114*(2), 260–263. doi:10.1289/ehp.8418 PMID:16451864

Makelarski, J. A., Romitti, P. A., Rocheleau, C. M., Burns, T. L., Stewart, P. A., Waters, M. A., & Olney, R. S. (2014). Maternal periconceptional occupational pesticide exposure and neural tube defects. *Birth Defects Research. Part A, Clinical and Molecular Teratology*, *100*(11), 877–886. doi:10.1002/bdra.23293 PMID:25124525

Mostafalou, S., & Abdollahi, M. (2012). Concerns of environmental persistence of pesticides and human chronic diseases. *Journal of Clinical and Experimental Pharmacology*, *2*(3), 1–2.

Mostafalou, S., & Abdollahi, M. (2013). Pesticides and human chronic diseases: Evidences, mechanisms, and perspectives. *Toxicology and Applied Pharmacology*, *268*(2), 157–177. doi:10.1016/j.taap.2013.01.025 PMID:23402800

Munoz-Quezada, M. T., Lucero, B. A., Barr, D. B., Steenland, K., Levy, K., Ryan, P. B., & Vega, C. (2013). Neurodevelopmental effects in children associated with exposure to organophosphate pesticides: A systematic review. *Neurotoxicology*, *39*, 158–168. doi:10.1016/j.neuro.2013.09.003 PMID:24121005

Nasterlack, M. (2007). Pesticides and childhood cancer: An update. *International Journal of Hygiene and Environmental Health*, *210*(5), 645–657. doi:10.1016/j.ijheh.2007.03.001 PMID:17434797

Nicolopoulou-Stamati, P., & Pitsos, M. A. (2001). The impact of endocrine disrupters on the female reproductive system. *Human Reproduction Update*, *7*(3), 323–330. doi:10.1093/humupd/7.3.323 PMID:11392379

Parsons, S. A., Boxall, A. B. A., Sinclair, C. J., & Ramwell, C. (2008). *Pesticide degradates of concern to the drinking water community*. Denver, CO: AWWA Research Foundation.

Perera, F. P., Rauh, V., Whyatt, R. M., Tang, D., Tsai, W. Y., Bernert, J. T., & Diaz, D. (2005). A summary of recent findings on birth outcomes and developmental effects of prenatal ETS, PAH, and pesticide exposures. *Neurotoxicology*, *26*(4), 573–587. doi:10.1016/j.neuro.2004.07.007 PMID:16112323

Petrelli, G., Figa-Talamanca, I., Tropeano, R., Tangucci, M., Cini, C., Aquilani, S., & Meli, P. (2000). Reproductive male-mediated risk: Spontaneous abortion among wives of pesticide applicators. *European Journal of Epidemiology*, *16*(4), 391–393. doi:10.1023/A:1007630610911 PMID:10959949

Pruss-Ustun, A., Bos, R., Gore, F., & Bartram, J. (2008). *Safer water, better health: costs, benefits and sustainability of interventions to protect and promote health*. Geneva: World Health Organization.

Raanan, R., Harley, K. G., Balmes, J. R., Bradman, A., Lipsett, M., & Eskenazi, B. (2015). Early-life exposure to organophosphate pesticides and pediatric respiratory symptoms in the Chamacos cohort. *Environmental Health Perspectives*, *123*(2), 179. PMID:25369257

Rauh, V., Arunajadai, S., Horton, M., Perera, F., Hoepner, L., Barr, D. B., & Whyatt, R. (2011). Seven-year neurodevelopmental scores and prenatal exposure to chlorpyrifos, a common agricultural pesticide. *Environmental Health Perspectives*, *119*(8), 1196–1201. doi:10.1289/ehp.1003160 PMID:21507777

Rauh, V. A., Garfinkel, R., Perera, F. P., Andrews, H. F., Hoepner, L., Barr, D. B., & Whyatt, R. W. (2006). Impact of prenatal chlorpyrifos exposure on neurodevelopment in the first 3 years of life among inner-city children. *Pediatrics*, *118*(6), e1845–e1859. doi:10.1542/peds.2006-0338 PMID:17116700

Reichenberger, S., Bach, M., Skitschak, A., & Frede, H. G. (2007). Mitigation strategies to reduce pesticide inputs into ground-and surface water and their effectiveness; a review. *The Science of the Total Environment*, *384*(1), 1–35. doi:10.1016/j.scitotenv.2007.04.046 PMID:17588646

Roberts, J. R., & Karr, C. J. (2012). Pesticide exposure in children. *Pediatrics*, *130*(6), e1765–e1788. doi:10.1542/peds.2012-2758 PMID:23184105

Rodrigo, M. A., Oturan, N., & Oturan, M. A. (2014). Electrochemically assisted remediation of pesticides in soils and water: A review. *Chemical Reviews*, *114*(17), 8720–8745. doi:10.1021/cr500077e PMID:24983494

Salameh, P., Waked, M., Baldi, I., Brochard, P., & Saleh, B. A. (2006). Respiratory diseases and pesticide exposure: A case-control study in Lebanon. *Journal of Epidemiology and Community Health*, *60*(3), 256–261. doi:10.1136/jech.2005.039677 PMID:16476757

Sanpera, C., Ruiz, X., Jover, L., Llorente, G., Jabeen, R., Muhammad, A., & Fasola, M. (2003). Persistent organic pollutants in little egret eggs from selected wetlands in Pakistan. *Archives of Environmental Contamination and Toxicology*, *44*(3), 360–368. doi:10.100700244-002-2044-z PMID:12712296

Schafer, R. B., Van den Brink, P. J., & Liess, M. (2011). Impacts of pesticides on freshwater ecosystems. *Ecological Impacts of Toxic Chemicals*, 111-137.

Selevan, S. G., Kimmel, C. A., & Mendola, P. (2000). Identifying critical windows of exposure for children's health. *Environmental Health Perspectives*, *108*(Suppl 3), 451–455. doi:10.1289/ehp.00108s3451 PMID:10852844

Shelton, J. F., Hertz-Picciotto, I., & Pessah, I. N. (2012). Tipping the balance of autism risk: Potential mechanisms linking pesticides and autism. *Environmental Health Perspectives*, *120*(7), 944–951. doi:10.1289/ehp.1104553 PMID:22534084

Shim, Y. K., Mlynarek, S. P., & van Wijngaarden, E. (2009). Parental exposure to pesticides and childhood brain cancer: US Atlantic coast childhood brain cancer study. *Environmental Health Perspectives*, *117*(6), 1002–1006. doi:10.1289/ehp.0800209 PMID:19590697

Siddiqui, M., Srivastava, S., Mehrotra, P., Mathur, N., & Tandon, I. (2003). Persistent chlorinated pesticides and intra-uterine foetal growth retardation: A possible association. *International Archives of Occupational and Environmental Health*, *76*(1), 75–80. PMID:12592586

Singer, P. L. (1949). Occupational oligospermia. *Journal of the American Medical Association*, *140*, 1249.

Souza, A. D., Medeiros, A. D. R., Souza, A. C. D., Wink, M., Siqueira, I. R., Ferreira, M. B. C., & Torres, I. L. D. S. (2011). Evaluation of the impact of exposure to pesticides on the health of the rural population: Vale do Taquari, State of Rio Grande do Sul (Brazil). *Ciencia & Saude Coletiva*, *16*(8), 3519–3528. doi:10.1590/S1413-81232011000900020 PMID:21860952

Stillerman, K. P., Mattison, D. R., Giudice, L. C., & Woodruff, T. J. (2008). Environmental exposures and adverse pregnancy outcomes: A review of the science. *Reproductive Sciences (Thousand Oaks, Calif.)*, *15*(7), 631–650. doi:10.1177/1933719108322436 PMID:18836129

Tadeo, J. L. (Ed.). (2008). *Analysis of pesticides in food and environmental samples*. CRC Press. doi:10.1201/9781420007756

Teklu, B. M., Adriaanse, P. I., Ter Horst, M. M., Deneer, J. W., & Van den Brink, P. J. (2015). Surface water risk assessment of pesticides in Ethiopia. *The Science of the Total Environment*, *508*, 566–574. doi:10.1016/j.scitotenv.2014.11.049 PMID:25481716

Thodal, C. E., Carpenter, J., & Moses, C. W. (2009). *Monitoring for Pesticides in Groundwater and Surface Water in Nevada, 2008 (No. 2009-3093)*. US Geological Survey.

Thompson, T., Fawell, J., Kunikane, S., Jackson, D., Appleyard, S., & Callan, P. (2007). *Chemical safety of drinking water: Assessing priorities for risk management*. WHO.

Turner, M. C., Wigle, D. T., & Krewski, D. (2011). Residential pesticides and childhood leukemia: A systematic review and meta-analysis. *Ciencia & Saude Coletiva*, *16*(3), 1915–1931. doi:10.1590/S1413-81232011000300026 PMID:21519680

Vos, J. G., Dybing, E., Greim, H. A., Ladefoged, O., Lambre, C., Tarazona, J. V., & Vethaak, A. D. (2000). Health effects of endocrine-disrupting chemicals on wildlife, with special reference to the European situation. *Critical Reviews in Toxicology*, *30*(1), 71–133. doi:10.1080/10408440091159176 PMID:10680769

Weselak, M., Arbuckle, T. E., Wigle, D. T., & Krewski, D. (2007). In utero pesticide exposure and childhood morbidity. *Environmental Research*, *103*(1), 79–86. doi:10.1016/j.envres.2006.09.001 PMID:17084836

WHO. (2010). *Water for health. WHO guidelines for drinking water quality*. Retrieved from http://www.who.int/water_sanitation_health/WHS_WWD2010_guidelines_2010_6_en.pdf? ua=1

Windham, G., & Fenster, L. (2008). Environmental contaminants and pregnancy outcomes. *Fertility and Sterility*, *89*(2), 111–116. doi:10.1016/j.fertnstert.2007.12.041 PMID:18308050

Wohlfahrt-Veje, C., Andersen, H. R., Jensen, T. K., Grandjean, P., Skakkebaek, N. E., & Main, K. M. (2012a). Smaller genitals at school age in boys whose mothers were exposed to non-persistent pesticides in early pregnancy. *International Journal of Andrology*, *35*(3), 265–272. doi:10.1111/j.1365-2605.2012.01252.x PMID:22394112

Wohlfahrt-Veje, C., Andersen, H. R., Schmidt, I. M., Aksglaede, L., Sørensen, K., Juul, A., & Main, K. M. (2012b). Early breast development in girls after prenatal exposure to non-persistent pesticides. *International Journal of Andrology*, *35*(3), 273–282. doi:10.1111/j.1365-2605.2011.01244.x PMID:22404257

Xu, W., Wang, X., & Cai, Z. (2013). Analytical chemistry of the persistent organic pollutants identified in the Stockholm Convention: A review. *Analytica Chimica Acta*, *790*, 1–13. doi:10.1016/j.aca.2013.04.026 PMID:23870403

Yang, D., Howard, A., Bruun, D., Ajua-Alemanj, M., Pickart, C., & Lein, P. J. (2008). Chlorpyrifos and chlorpyrifos-oxon inhibit axonal growth by interfering with the morphogenic activity of acetylcholinesterase. *Toxicology and Applied Pharmacology*, *228*(1), 32–41. doi:10.1016/j.taap.2007.11.005 PMID:18076960

Yang, W., Carmichael, S. L., Roberts, E. M., Kegley, S. E., Padula, A. M., English, P. B., & Shaw, G. M. (2014). Residential agricultural pesticide exposures and risk of neural tube defects and orofacial clefts among offspring in the San Joaquin Valley of California. *American Journal of Epidemiology*, *179*(6), 740–748. doi:10.1093/aje/kwt324 PMID:24553680

Chapter 7
Pesticide Contaminated Drinking Water and Health Effects on Pregnant Women and Children

Sanjeevi Ramakrishnan
NIMS University, India

Anuradha Jayaraman
NIMS University, India

ABSTRACT

In the recent years, pesticide research and regulatory efforts have focused on the prevention of acute health effects from pesticide poisonings and pesticide residues on foods, but more attention is being given to the deleterious chronic health effects. Children and pregnant women's exposure to contaminated water in particular are at high risk for subsequent adverse health outcomes. The chapter summaries the health effects of water contamination.

INTRODUCTION

Pesticides have well-documented impacts on human health, and researchers have noted associations with cancer, neurological disease, respiratory impacts, birth outcomes, and other adverse conditions in human populations (Bonner & Alavanja, 2017). Children, in particular, are at greater risk for pesticide exposure and subsequent adverse health outcomes (Ferguson et al., 2017). Historically, pesticides research and regulatory efforts have focused on the prevention of acute health effects from pesticide poisonings and pesticide residues on foods, but more attention is being given to the deleterious chronic health effects resulting from low-level, ambient pesticide exposures.

For many years, the primary focus of public health surveillance has been on acute pesticide illnesses; yet, these acute outcomes are likely undercounted. Pesticide illness reports are often made only when the exposure is reported by the victim, the outcome is very serious and/or requires hospitalization, and

DOI: 10.4018/978-1-5225-6111-8.ch007

it is recognized as such by the attending medical provider. Furthermore, less policy attention has been paid to chronic health outcomes associated with pesticide exposures. With increasing knowledge about pesticides' impacts on cancer, diabetes, lung function, neurodegenerative disorders, intelligence quotient, birth defects, and reproductive impacts (Bonner & Alavanja, 2017), it will be important for public health research and surveillance efforts to improve the ability to identify and monitor chronic health impacts related to pesticide exposure.

Due to their size, physiology, and behavior, children are more vulnerable than adults to environmental hazards. Children on getting exposed to high toxic chemicals in proportion to their body mass index, suffers for more years of life ahead of them due to the earlier exposure in their life-time. Health at birth is predictive of important child outcomes, including educational attainment and adult earnings. Exposure to environmental pollution during pregnancy is a common source of potential fetal health shocks (Wang et al., 2016). Recent research shows that, even at levels below current air quality standards, air pollution can harm fetal health as measure d by the incidence of low birth weight and pre maturity. Drinking water contamination is another, potentially important, source of in utero exposure to pollution.

Special attention should therefore be paid to harmful substances to which pregnant women might be exposed. While a good deal of recent research focuses on the effects of air pollution on fetal health, there has been little attention paid to the potential harm caused by contaminated drinking water. First, the study shows that the women who live in areas with contaminated water supplies differ from other women in ways that one would expect to be correlated with worse fetal health. It is therefore important to control for these differences. Second, women may respond to contaminated water by moving elsewhere. We show that more educated women are more likely to vote with their feet. Third, there is a mechanical positive correlation between length of gestation and the probability of being exposed to most fetal health insults. It has been shown that correcting for the bias can have an important impact on the estimated magnitude of the effect. Fourth, there is good reason to expect effects to differ by socioeconomic stat us. Just as they are more likely to move, more educated women are more likely to take measures to protect themselves and their children from contaminated water (Currie *et al.,* 2013). The present chapter summaries that water contamination may be an important source of human pesticide exposure, and higher versus lower maternal residential water consumption has been associated with children's health and growing fetus.

PESTICIDE USE AND ITS EFFECT ON CHILDREN'S HEALTH AND PREGNANT WOMEN

More than 700 pesticide chemicals, including insecticides, herbicides, rodenticides, and fungicides, are currently registered with US EPA. These are chemicals deliberately engineered to kill or repel living things and thus have inherent toxic potential. Synthetic pesticides introduced after World War II are now ubiquitous. A 2000 US EPA survey demonstrated that 74% of US households use 1 or more pesticides around the home. Over time, insecticide use has evolved from chlorinated compounds such as dichlorodiphenyltrichloroethane (DDT) to organophosphates (OPs) and, more recently, pyrethroids. Though having low acute toxicity, organochlorine pesticides are persistent and associated with chronic health concerns. In making the switch to the OPs, we have substituted for the acute and neurodevelopmental toxicity of OPs.

Pesticides can be found in most of our everyday environments, including workplaces, homes, schools, in food, and in the community at large in our drinking and recreational waters, the air, soot, and soil. They are commonly used in a variety of ways including lawn sprays and household bug sprays and can be found in varying amounts in food, such as strawberries, blueberries, and apples. Children, in particular, are likely to be exposed when in their yards, on playgrounds, and on athletic fields (Galvez *et al.*, 2017). Exposure to pesticides can occur *via* inhalation, ingestion, dermal contact, or across the placenta. Biomarkers currently exist for some pesticides in blood serum, semen, ovarian follicular fluid, amniotic fluid, umbilical cord blood, breast milk, meconium, and urine. Biomonitoring results of the Centers for Disease Control and Prevention's (CDC) National Health and Nutrition Examination Survey (2005-2006) are published in a report titled Third Exposure to Environmental Chemicals Report (2005). This report stated that detectable body levels of about 50 pesticides have been found in a representative sample of the U.S. population. Although indications of health effects from pesticides are applicable to the general population, certain subpopulations warrant particular concern and special protections. During a critical period of development, conception to puberty, exposure to pesticides can result in increased risk for health outcomes (Weselak, *et al.*, 2007). Therefore, attention to women's exposure during pregnancy is especially important.

Due to regulatory actions based on demonstrated neurocognitive impacts, since the early 2000s, residential use of OPs has largely been replaced by pyrethroids. Initial concerns about pyrethroids being associated with allergic reactions have been accentuated with recent studies suggesting possible impacts on neurodevelopmental delay, autism, and male reproductive health (Saillenfait *et al.,* 2015; Shelton *et al.,* 2015). Newer, designer, pesticides have been introduced in recent years that target physiologic systems of insects and plants with the assumption that they would likely not have effects on human health. A prominent example is the neonicotinoid pesticides that target nicotinic acetylcholine receptors, which are different in insects than mammals (Casida *et al.,* 2013). There are few epidemiologic studies of the health impacts of these chemicals (used in agriculture and as flea and tick pesticides for pets) on humans though limited evidence has suggested a possible impact on neurodevelopment (Keil *et al.,* 2014). Glyphosate, the herbicide with greatest use worldwide, was long considered a relatively innocuous pesticide. But in 2015, based on new evidence of carcinogenicity from both animal studies and epidemiologic investigations, the IARC identified glyphosate as "probably carcinogenic to humans" (Guyton *et al.,* 2015). Understanding of the health hazards associated with chemical exposures changes with evolving science. Low-dose exposures to certain pesticides during pregnancy and childhood are associated with increased risk for attention problems, lower intelligence quotient (IQ), behavioral changes, and pervasive developmental delay, altered brain architecture, asthma and respiratory symptoms, and childhood leukemia. Though OP pesticides have largely been banned for residential use in the United States, they continue to be used in agriculture and their residues are found on food.

Pesticide residues (e.g. OPs and herbicides) on food contribute significantly to exposure in children, even those living in agricultural communities. Consumption of fresh fruits and vegetables is a priority for children, thus selection of organic products should be weighed against their cost and availability. Nonetheless, the evidence of health effects from low-dose exposures to some pesticides indicates a need to reduce developmental exposure to pesticides as much as is practical. In 2012, 4.7% of imported and 0.9% of domestic foods sampled contained residues greater than allowable in the United States. Some of the pesticides found on imported foods are not registered for use in the United States. Some products had high rates of unallowable residues. For example, more than 40% of imported Basmati rice was in violation. The Environmental Working Group uses federal agency data to determine fresh produce that has high (dirty dozen) and low (clean 15) pesticide loads. Those concerned about exposure can use these aids in determining which produce might be preferable. Residential use of pesticides is a major contributor to exposure. Integrated pest management (IPM) is an approach to minimizing pesticide use in residential, school, and agricultural settings. It integrates chemical and nonchemical methods to provide the least toxic control of pests. IPM has proven to be cost-effective and, at times, more effective at long-term control of pests while reducing pesticide exposure (Williams *et al.,* 2006; Brenner *et al.,* 2003).

Several billion pounds of pesticides are used annually worldwide, less than one quarter of that in the United States. The production and export of obsolete and persistent pesticides from developed to developing nations, though still a problem, has improved with international agreements and guidance. These documents provide information and tools useful for health care workers in addressing public health and are well summarized in International Tools for Preventing Local Pesticide Problems (EPA, 2011). Children have been exposed when parents bring contaminated clothing home, use agricultural pesticides from work at home, use empty containers to hold water or food, and when children work in fields or apply pesticides. Both acute and chronic low-level exposures have resulted from misuse as well as careless use and accidents (Naidoo *et al.,* 2010). Contamination of water, soil, housing, and food with current and obsolete pesticides is not rare. Often, old obsolete pesticides have been stockpiled and inadequately stored (at times degrading to more hazardous material) in developing nations.

Fetal development is a carefully orchestrated interplay of 'physical, electrical and chemical signaling among cells and organs' during which the single cell of the fertilized egg forms into the millions of cells that make up a newborn (Schettler *et al.,* 1999). Hormones play a vital role in this complex series of events, and any disruption in maternal or fetal hormone levels has the potential to negatively effect on fetal development. In the early stages of fetal development, cells have a flexibility that allows them to develop in numerous ways. For example, the cells in the middle mesoderm layer of the embryo have the potential to become the kidneys, skeleton, or muscle (Blackburn, 2007). As the cells develop more specialized characteristics called differentiation, their functionally decreases. On persistent exposure to such an environmental contaminant occurs to the embryo prior to this differentiation, normal fetal development may still occur as other cells are able to take over for those that have been injured (Schettler *et al.,* 1999). However, if the insult occurs after differentiation or during times of increased cell proliferation, abnormal development can result in structural or functional defects, altered growth, and even fetal death. These times of sensitivity to environmental contaminants are referred to as critical windows of susceptibility (Woodruff *et al.,* 008).

PRENATAL AND POSTNATAL EXPOSURE TO ORGANOPHOSPHATE PESTICIDES: A CASE STUDY AT SHANDONG, CHINA

The exposure of general population to organophosphate pesticides (OPs) through environmental pollution and consumption of the pesticide contaminated food and water has become a global issue. In China,> 300,000 tons of pesticides are used every year in agriculture, with OPs comprising approximately 70% of all pesticides used (Agriculture Information Network, 2006). Several studies have reported widespread exposure to OPs in susceptible populations (children and pregnant women), suggesting that OP exposure in children and pregnant women is a critical public health issue that deserves greater concern (Wang *et al.,* 2012). There is evidence that OPs could cross the placenta and blood–brain barrier, and they have been detected in amniotic fluid and meconium (Bradman *et al.,* 2003); this indicates that fetuses could be exposed to OPs.

Moreover, children and fetuses are thought to be highly vulnerable to OP toxicity, because the brain is in rapid development, and concentrations of protective enzymes that deactivate OPs are lower than those in adults (Holland *et al.,* 2006). Prenatal and early postnatal periods are critical for neurodevelopment. Evidence from studies in animals demonstrated that prenatal or early postnatal OP administration could adversely affect offspring's neurodevelopment (Dam *et al.,* 2000), such as producing changes in cognitive, motor, and sensory functions as well as the nervous system developmental processes. However, limited epidemiological investigations in human populations assessed childhood neurodevelopment after prenatal or early postnatal OP exposure, with controversial results. (Eskenazi *et al.* 2007) found that prenatal exposure to OPs was negatively associated with scores in the Mental Development Index (MDI), but postnatal exposure to OPs was positively associated with measurements in the MDI at 24 months of age. However, (Rauh *et al.* 2006) found lower Psychomotor Development Index (PDI) and MDI scores in children at 36 months of age exposed prenatally to higher chlorpyrifos concentrations. More recently, one study conducted in Jiangsu, China (known as an agricultural area with heavily used OPs) showed that prenatal OP exposure was related to neurodevelopmental delay in the adaptive domain, and postnatal OP exposure might be related to neurodevelopmental delay in the social and motor domains in 24-month-old children in the agricultural area (Liu *et al.,* 2016). These data suggested that prenatal OP exposure could adversely affect children's neurodevelopment at 24 months of age, especially among boys. The prenatal period might be a critical window of OP exposure. In view of the positive association with postnatal OP exposure, it is necessary to interpret findings with caution (Wang *et al.,* 2017).

PYRETHROIDS AND CARIBBEAN COUNTRY PREGNANT WOMEN'S CASE STUDY

Pyrethroid insecticides, synthetic versions of the natural compound pyrethrin produced by the Chrysanthemum flower, have been extensively used in agricultural and home formulations for more than 30 years and account for approximately one-fourth of the worldwide insecticide market (Casida and Quistad, 1998). Common human exposure to pyrethroids occurs primarily through the use of pyrethroid containing household insecticides and pet sprays, and through the ingestion of food and drinking water contaminated with pyrethroid residues. Like many other classes of pesticides, pyrethroids are neurotoxicants (Shafer *et al.,* 2005). Pyrethroid compounds can cross the placental barrier and are known to interfere with hormonal and neurological development, the immune system and other physiological functions (Doucet *et al.,* 2009). Bell *et al.*, (2001) reported an increased risk of fetal death due to congenital anomalies when synthetic pyrethroids were used in the same township, range, or section during the 3rd to 8th week of pregnancy. Hanke *et al.,* (2003) found a significant reduction in birth weight among the offspring of mothers potentially exposed to synthetic pyrethroids during the three months prior to conception and the first trimester of pregnancy. Conversely (Dabrowski *et al.*, 2003) found no significant reduction in birth weight after reported farm use of synthetic pyrethroids during the first or second trimester of pregnancy. Furthermore, the only study that measured biomarkers of exposure in maternal urine during the third trimester of pregnancy found no association with birth weight, birth length, or head circumference (Berkowitz *et al.,* 2004). Also, Horton *et al.*, (2011) did not observe a significant association between prenatal exposure to permethrin and adverse neurodevelopment. Finally, a study of 113 women using a pyrethroids cream to treat head lice did not show an increased risk for birth defects or pregnancy complications (Kennedy *et al.,* 2005). In the Caribbean region, pregnant and delivering mothers' fetuses are being exposed to various chemical substances. Dewailly *et al.*, (2014), states that part of a larger research project designed to address the question of prenatal exposure to persistent organic pollutants (POPs), heavy metals such as mercury and lead, and other classes of pesticides such as organophosphates, carbamates, and pyrethroids. The study showed that pyrethroids metabolite levels were higher in the 10 Caribbean countries than in Canada and the U.S. These higher levels may reflect the regular and consistent use of pyrethroids for residential pesticides in the Caribbean. The risks of exposure to this class of pesticides on the unborn child are currently unknown but suggestive of being adverse.

PESTICIDE EFFECT ON CHILDREN AND FETUS IN BRAZIL

Asmus *et al.*, (2016) presented a review by examining the effects of exposure to pesticides on the health of Brazilian children. Almost all of the studies described the occurrence of effects due to parental exposure. Exposure to pesticides before and during pregnancy, and during breastfeeding, was associated with higher risk for leukemia (Ferreira *et al.,* 2013), adverse pregnancy outcomes (Boccolini *et al.,* 2013), and congenital abnormalities (Gaspari *et al.,* 2012). Some studies investigated exposure through consumption of pesticides and others estimated the domestic and occupational use (Boccolini *et al.,* 2013). Measurement of concentrations of organochlorine pesticides (dichlorodiphenyltrichlorethane [DDT], and their metabolites, Hexachlorobenzene [HCB], and hexachlorocyclohexane [HCH]) was performed in breast milk, umbilical cord and maternal blood, and children's blood. The levels found in all studies were indicative of continued exposure, despite the fact that the use of organochlorine pesticides was discontinued in Brazil in 1998. Three studies were carried out in an organochlorine pesticide-contaminated area, located in Rio de Janeiro. A risk assessment determined that the estimated doses, for all the substances studied, exceeded the minimum risk levels for chronically exposed children living in the area. One study found that >60% of the children living in this area had detectable levels of most organochlorine pesticides. The authors determined the concentration of thyroid hormones and observed that the total triiodothyronine (T3) levels were above the reference range in 28% of children.

Results from another study reported an association between exposure to organochlorine pesticides and the occurrence of delayed puberty, for both boys and girls born in the contaminated area (Guimaraes *et al.,* 2013). Environmental pollution by pesticides is a common problem in Latin America. In Central America, 33 million tons of pesticides were imported each year from 1977 to 2006. Brazil is the number 1 consumer of pesticides in the world. The amount of pesticides consumed in Brazil increased by 700% in the past 40 years while the planted area increased 78% in this same time period. Annual consumption of pesticides has been >300,000 tons of commercial products or 130,000 tons of active compounds. Despite the heavy use of pesticides in Brazil, this review was only able to find a few studies on the effects of pesticide exposures on children's health. Most of these studies addressed reproductive effects or cancer (1 study). Additionally, half of these studies had an ecological design impairing the identification and establishment of cause-effect relationships. Even so, they all demonstrated that Brazil's children are exposed to pesticides. The paucity of information about the effects of pesticide exposure on Brazilian children's health is due to the lack of an efficient record system and of appropriate studies. According the PAHO, a positive and growing correlation is seen in Central America between the incidence rate of pesticide poisoning in children <15 years old and the import of pesticides into the region. Some pesticides can alter the action of hormones behaving as endocrine disruptors. Endocrine disruption may alter development and reproduction and induce birth defects. Pesticides also can have immune, genetic, and neurologic toxicity. Exposures during periods of rapid brain growth, especially in the intrauterine period and in early childhood, can produce subtle and permanent effects on the structure and function of the brain. The consequences can be chronic neurobehavioral and neurologic effects. Some studies have reported an association between prenatal and postnatal pesticide exposure and a higher risk for childhood cancer, mainly brain tumors, leukemia, and lymphomas (Ferreira *et al.,* 2013; Boccolini *et al.,* 2013; Gaspari *et al.,* 2012). Additionally, pesticide exposure before or during pregnancy has been associated with increased risk for infertility, prenatal death, spontaneous abortion, premature birth, fetal growth retardation, and congenital malformations (Eskenazi *et al.,* 1999; 2004).

In summary, the number and design of studies conducted to date are insufficient to examine the effect of pesticide exposures on the health of Brazilian children. According to Landrigan and Baker (2015) a "birth-cohort study with prenatal enrollment and long-term follow-up" is the best study design "to assess associations between early-life exposures and later disease." Large observational prospective studies are required for further evaluation. The monitoring of populations exposed to pesticide needs to consider not only adverse outcomes in early life, such as congenital malformations and low birth weight, but also late effects, such as cancer endocrine and neurodevelopmental effects.

TRATOGENIC EFFECTS OF PESTICIDE CONTAMINATED WATER CAUSING CONGENITAL HEART DEFECTS

Congenital heart defects (CHDs) are among the most common congenital malformations, occurring in nearly 1% of live births. CHDs account for approximately one-third of birth defects and are a leading cause of infant mortality due to birth defects (CDCP, 2010). The etiology of CHDs is poorly understood in the majority of non-syndromic CHD cases (i.e., those that do not occur due to a recognized genetic syndrome). The prevalence of these defects varies by subtype of CHD, geography, and season of conception (Van der Linde et al., 2011; Caton, (2012), which may suggest the involvement of environmental risk factors, such as pesticides, that also vary by geography and season. Water contamination may be an important source of human pesticide exposure, and higher versus lower maternal residential tap water consumption has been associated with CHDs in offspring. The pesticide atrazine is of particular interest because it is one of the most widely applied agricultural pesticides in the U.S., is present at substantial levels in ground water supplies (e.g., _0.003 mg/L, the U.S. Environmental Protection Agency (EPA) drinking water standard), and is suspected to have adverse reproductive effects (Shaw *et al.,* 1990; Brender & Weyer, 2016).

Currie (2013) evaluated the level of pesticide contamination in the mother in order to allow correlation for correlations between the siblings. The estimates suggest that exposure to chemicals in the drinking water during pregnancy raises the probability of low birth weight by 6.5%, while any water violation (including chemical contamination) increases the probability of low birth weight by 6.1%. Eskenazi et al., (2004) found clear decreases in gestational duration associated with two measures of in utero pesticide exposure: levels of metabolites of dim ethyl phosphate pesticide compounds and whole blood cholinesterase. Shortened gestational duration was most clearly related to increasing exposure levels in the latter part of pregnancy. In fact, we found increases in length and head circumference associated with some of these measures Perera et al., (2003), who reported decreased birth weight and length in association with blood measurements of the parent compound chlorpyrifos in pregnant residents of New York City. Disparities among study results may be due, in part, to differences in exposure measurements.

Andersen ., (2008) hypothesized that women who become pregnant while working in greenhouses where pesticides are applied have an increased risk of giving birth to a boy with abnormal development of the reproductive organs. Female greenhouse workers with confirmed pesticide exposure during pregnancy gave birth to boys with smaller penises and testicles, lower serum concentrations of testosterone and inhibit B, higher serum concentrations of SHBG and FSH, and higher LH: testosterone ratio than unexposed workers. These results suggest an adverse effect of pesticides on Leydig and Sertoli cells during testicular development. In addition, female greenhouse workers had a more than three-fold increased risk of delivering a boy with cryptorchidism compared with women from the urban area of Copenhagen.

As one of the most important organochlorine pesticides (OCPs), 1,1,1-trichloro-2,2′-bis (4-chlorophenyl)ethane (DDT) was widely employed in agricultural production and the vector-borne diseases control. Although it has been banned for N30 years, it is still categorized as a ubiquitous environmental pollutant by the United Nations Environment Programme (UNEP) (Stockholm Convention, 2001) because of its prolonged environmental persistence and toxicity on human beings. Developing neonates are at higher risks than adults to these contaminants since they are known to be more susceptible to the toxicological consequences of pollution. Adverse birth outcomes were previously observed in infants whose mothers were exposed to organic pollutants such as DDT during and before the pregnancy (Farhang et al., 2005). The ability of DDT to pass through the placental barrier has been demonstrated, and the high correlation coefficient between maternal and cord serum concentrations indicated that the cord serum levels of DDT compounds could be used to estimate the maternal burden (Vizcaino et al., 2010). Moreover, as a non-invasive human sample, cord serum is more convenient and safer to collect during the delivery process compared with neonatal blood or urine (Yin et al., 2016). The concentration of pollutants in the cord serum reflects the portion passed through the placental barrier, which is likely to be close to the actual concentration in newborns. Therefore, this type of low-fat matrix is believed to be an adequate bio-indicator to assess DDT levels in newborns and infants. Previous research based on cord serum samples indicated that exposure to DDT could affect the fetus and have continuing adverse effects on birth development (Falcon et al., 2004). A cross-sectional study suggested that low in utero DDT exposure led to a reduction in birth weight, birth height, crown-heel length and head circumference (Al-Saleh et al., 2012). Two birth cohort studies in Spain showed that cord serum DDT concentrations at birth were inversely associated with cognitive functioning at 4 years of age. Statistical analysis indicated that the exposure of these pollutants was correlated with maternal ages, pre-pregnant BMI, drinking water sources, parity and the number of abortions. The isomer ratios of different DDTs and/or its metabolites and enantiomeric patterns of o,p'-DDD and o,p'-DDT were used to identify the possible exposure sources of these pollutants. Using multivariable linear regressions, increasing p,p'-DDD and p,p'-DDT levels were found to be significantly associated with an increase in neonatal birth weight, which deserves additional attention to obesity risks.

TRATOGENIC EFFECTS OF PESTICIDE CONTAMINATED WATER CAUSING NERVOUS SYSTEM DISORDERS

The organochlorine pesticide DDT is known to cause excitation of the central nervous system and can lead to tremors, hyper excitability and convulsions. DDT's effect on the central nervous system is likely through interrupting the movement of ions through the neuronal membranes, which leads to the inappropriate release of neurotransmitters (ATSDR, 2002). In relation to in-utero exposure to organophosphates, both studies which examined neonates found a relation of abnormal reflexes and maternal organophosphate exposure. The studies suggest that there is reason to be cautious about exposure of pregnant women to DDT/DDE and organophosphates because of the potential effect on the neurodevelopment of their children (Rosas & Eskenazi, 2008). Organophosphate pesticides act through inhibition of acetyl cholinesterase in synaptic clefts, thereby deregulating the metabolism of acetylcholine. Some organophosphate pesticides also affect the nervous system by inhibiting neuropathy target esterase, which can result in a rare condition known as organophosphate induced delayed neuropathy (Vose et al., 2007).

CONCLUSION

Studies revealed that living in a water district with contaminated water during pregnancy is associated with an increase in low birth weight of 14.55% among less educated mothers. Similarly, potential exposure to contaminated water increases the incidence of prematurity by 10.3% among less educated mothers. Since not every mother in an affected water district is likely to be exposed to contaminated water, these estimates suggest large negative effects among those women who are actually exposed to water contaminated by chemicals during pregnancy. The pesticide like atrazine is of particular interest because it is one of the most widely applied agricultural pesticides. Its present substantial levels in ground water supplies (e.g., ≥0.003 mg/L), the U.S. Environmental Protection Agency (EPA) drinking water standard) and is suspected to have adverse reproductive effects to human beings (Kim et al., 2017). It is also found that small effects of drinking water contamination on all children, but large and statistically significant effects on birth weight and gestation of infants born to less educated mothers. Results had shown that those mothers who were most affected by contamination were the least likely to move between births in response to contamination. In addition, alteration in sex ratio of normal live births to fathers employed in pesticide application, deep sea divers, and carbon setters were also recorded (Mostafalou & Abdollahi, 2013). In general pesticide illness reports that on exposure, the victims face very serious chronic health outcomes impacting on cancer, diabetes, lung function, neurodegenerative disorders, intelligence quotient, birth defects, and reproductive disorders.

REFERENCES

Agency for Toxic Substances and Disease Registry. (2002). *Toxicological profile for DDT, DDE and DDD*. Atlanta, GA: US Department of Health and Human Services.

Agriculture Information Network. (2006). Analysis of pesticides demand in China [in Chinese]. *Plant Doctor.*, *19*, 16.

Al-Saleh, I., Al-Doush, I., Alsabbaheen, A., Mohamed, G. E. D., & Rabbah, A. (2012). Levels of DDT and its metabolites in placenta, maternal and cord blood and their potential influence on neonatal anthropometric measures. *The Science of the Total Environment*, *416*, 62–74. doi:10.1016/j.scitotenv.2011.11.020 PMID:22192892

Andersen, H. R., Schmidt, I. M., Grandjean, P., Jensen, T. K., Budtz-Jorgensen, E., Kjaerstad, M. B., ... Main, K. M. (2008). Impaired Reproductive Development in Sons of Women Occupationally Exposed to Pesticides during Pregnancy. *Environmental Health Perspectives*, *116*(4), 566–572. doi:10.1289/ehp.10790 PMID:18414644

Asmus, C. I. R. F., Camara, V. M., Landrigan, P. J., & Claudio, L. (2016). A Systematic Review of Children's Environmental Health in Brazil. *Annals of Global Health*, *82*(1), 132–148. doi:10.1016/j.aogh.2016.02.007 PMID:27325071

Bell, E. M., Hertz-Picciotto, I., & Beaumont, J. J. (2001). A case-control study of pesticides and fetal death due to congenital anomalies. *Epidemiology (Cambridge, Mass.)*, *12*(2), 148–156. doi:10.1097/00001648-200103000-00005 PMID:11246574

Berkowitz, G. S., Wetmur, J. G., Birman-Deych, E., Obel, J., Lapinski, R. H., Godbold, J. H., ... Wolff, M. S. (2004). In utero pesticide exposure, maternal paraoxonase activity, and head circumference. *Environmental Health Perspectives*, *112*(3), 388–391. doi:10.1289/ehp.6414 PMID:14998758

Blackburn, S. T. (2007). *Maternal, fetal, & neonatal physiology: A clinical perspective* (3rd ed.). St. Louis, MO: Saunders.

Boccolini, P. M., Boccolini, C. S., Meyer, A., Chrisman, J., & Guimarães, R. M. (2003). Pesticide exposure and low birth weight prevalence in Brazil. *International Journal of Hygiene and Environmental Health*, *216*, 2904. PMID:23021945

Bonner, M. R., & Alavanja, M. C. R. (2017). Pesticides, human health, and food security. *Food and Energy Security*, *6*(3), 89–93. doi:10.1002/fes3.112

Bradman, A., & Barr, D. B. (2003). Measurement of pesticides and other toxicants in amniotic fluid as a potential biomarker of prenatal exposure: A validation study. *Environmental Health Perspectives*, 111.

Bradman, A., Quiros-Alcala, L., Castorina, R., Schall, R. A., Camacho, J., Holland, N. T., ... Eskenazi, B. (2015). Effect of organic diet intervention on pesticide exposures in young children living in low-income urban and agricultural communities. *Environmental Health Perspectives*, *123*(10). doi:10.1289/ehp.1408660 PMID:25861095

Brender, J. D., & Weyer, P. J. (2016). Agricultural Compounds in Water and Birth Defects. *Current Environmental Health Reports*, *3*(2), 144–152. doi:10.100740572-016-0085-0 PMID:27007730

Brenner, B., Markowitz, S., Rivera, M., Romero, H., Weeks, M., Sanchez, E., ... Berkowitz, G. (2003). Integrated pest management in an urban community: A successful partnership for prevention. *Environmental Health Perspectives*, *111*(13), 1649–1653. doi:10.1289/ehp.6069 PMID:14527845

Casida, J. E., & Durkin, K. A. (2013). Neuroactive insecticides: Targets, selectivity, resistance and secondary effects. *Annual Review of Entomology*, *58*(1), 99–117. doi:10.1146/annurev-ento-120811-153645 PMID:23317040

Casida, J. E., & Quistad, G. B. (1998). Golden age of insecticide research: Past, present, or future? *Annual Review of Entomology*, *43*(1), 1–16. doi:10.1146/annurev.ento.43.1.1 PMID:9444749

Caton, A. R. (2012). Exploring the seasonality of birth defects in the New York State Congenital Malformations Registry. *Birth Defects Research. Part A, Clinical and Molecular Teratology*, *94*(6), 424–437. doi:10.1002/bdra.23006 PMID:22473628

Centers for Disease Control and Prevention. (2010). Racial differences by gestational age in neonatal deaths attributable to congenital heart defects—United States, 2003–2006. *MMWR. Morbidity and Mortality Weekly Report*, *59*, 1208–1211. PMID:20864921

Currie, J., Zivin, J. G., Meckel, K., Neidell, M., & Schlenker, W. (2013). Something in the water: Contaminated drinking water and infant health. *The Canadian Journal of Economics. Revue Canadienne d'Economique*, *46*(3), 791–810. doi:10.1111/caje.12039 PMID:27134285

Dabrowski, S., Hanke, W., Polanska, K., Makowiec-Dabrowska, T., & Sobala, W. (2003). Pesticide exposure and birthweight: An epidemiological study in Central Poland. *International Journal of Occupational Medicine and Environmental Health*, *16*(1), 31–39. PMID:12705715

Dam, K., Seidler, F. J., & Slotkin, T. A. (2000, June). Chlorpyrifos exposure during a critical neonatal period elicits gender-selective deficits in the development of coordination skills and locomotor activity. *Brain Research. Developmental Brain Research*, *121*(2), 179–187. doi:10.1016/S0165-3806(00)00044-4 PMID:10876030

Dewailly, E., Forde, M., Robertson, L., Kaddar, N., Sidi, E. A. L., Cote, S., ... Ayotte, P. (2014). Evaluation of pyrethroid exposures in pregnant women from 10 Caribbean countries. *Environment International*, *63*, 201–206. doi:10.1016/j.envint.2013.11.014 PMID:24317226

Doucet, J., Tague, B., Arnold, D. L., Cooke, G. M., Hayward, S., & Goodyer, C. G. (2009). Persistent organic pollutant residues in human fetal liver and placenta from Greater Montreal, Quebec: A longitudinal study from 1998 through 2006. *Environmental Health Perspectives*, *117*(4), 605–610. doi:10.1289/ehp.0800205 PMID:19440500

Eskenazi, B., Bradman, A., & Castorina, R. (1999). Exposures of children to organophosphate pesticides and their potential adverse health effects. *Environmental Health Perspectives*, *1*(Suppl 3), 409–419. doi:10.1289/ehp.99107s3409 PMID:10346990

Eskenazi, B., Harley, K., Bradman, A., Weltzien, E., Jewell, N. P., Barr, D. B., ... Holland, N. T. (2004). Association of in utero organophosphate pesticide exposure and fetal growth and length of gestation in an agricultural population. *Environmental Health Perspectives*, *112*(10), 1116–1124. doi:10.1289/ehp.6789 PMID:15238287

Eskenazi, B., Marks, A. R., Bradman, A., Harley, K., Barr, D. B., Johnson, C., ... Jewell, N. P. (2007). Organophosphate pesticide exposure and neurodevelopment in young Mexican-American children. *Environmental Health Perspectives*, *115*(5), 792–798. doi:10.1289/ehp.9828 PMID:17520070

Falcon, M., Oliva, J., Osuna, E., Barba, A., & Luna, A. (2004). HCH and DDT residues in human placentas in Murcia (Spain). *Toxicology*, *195*(2-3), 203–208. doi:10.1016/j.tox.2003.10.009 PMID:14751675

Farhang, L., Weintraub, J. M., Petreas, M., Eskenazi, B., & Bhatia, R. (2005). Association of DDT and DDE with birth weight and length of gestation in the child health and development studies, 1959–1967. *American Journal of Epidemiology*, *162*(8), 717–725. doi:10.1093/aje/kwi276 PMID:16120698

Ferguson, A., Penney, R., & Solo-Gabriele, H. (2017). A Review of the Field on Children's Exposure to Environmental Contaminants: A Risk Assessment Approach. *International Journal of Environmental Research and Public Health*, *14*(3), 265–289. doi:10.3390/ijerph14030265 PMID:28273865

Ferreira, J. D., & Couto, A. C. (2013). Pombo-de-Oliveira MS, Koifman S. Brazilian Collaborative Study Group of Infant Acute Leukemia. In utero pesticide exposure and leukemia in Brazilian children <2 years of age. *Environmental Health Perspectives*, *121*, 269–275. PMID:23092909

Galvez, M. P., & Balk, S. J. (2017). Environmental Risks to Children: Prioritizing Health Messages in Pediatric Practice. *Pediatrics*, *38*(6), 263–279. PMID:28572135

Gaspari, L., Sampaio, D. R., Paris, F., Audran, F., Orsini, M., Neto, J. B. S., & Sultan, C. (2012). High prevalence of micropenis in 2710 male newborns from an intensive-use pesticide area of northeastern Brazil. *International Journal of Andrology*, *35*(3), 253–264. doi:10.1111/j.1365-2605.2011.01241.x PMID:22372605

Guimaraes, R. M., & Froes Asmus, C. I. R. (2013). Organochlorine exposure and changes in primary and secondary sexual characters in Cidade dos Meninos. *Br Cad Saúde Colet*, *21*, 475–476.

Guyton, K. Z., Loomis, D., Grosse, Y., El Ghissassi, F., Benbrahim-Tallaa, L., Guha, N., ... Straif, K. (2015). Carcinogenicity of tetrachlorvinphos, parathion, malathion, diazinon, and glyphosate. *The Lancet Oncology*, *16*(5), 490–491. doi:10.1016/S1470-2045(15)70134-8 PMID:25801782

Hanke, W., Romitti, P., Fuortes, L., Sobala, W., & Mikulski, M. (2003). The use of pesticides in a Polish rural population and its effect on birth weight. *International Archives of Occupational and Environmental Health*, *76*(8), 614–620. doi:10.100700420-003-0471-4 PMID:12955525

Holland, N., Furlong, C., Bastaki, M., Richter, R., Bradman, A., Huen, K., ... Eskenazi, B. (2006). Paraoxonase polymorphisms, haplotypes, and enzyme activity in Latino mothers and newborns. *Environmental Health Perspectives*, *114*(7), 985–991. doi:10.1289/ehp.8540 PMID:16835048

Horton, M. K., Rundle, A., Camann, D. E., Boyd Barr, D., Rauh, V. A., & Whyatt, R. M. (2011). Impact of prenatal exposure to piperonyl butoxide and permethrin on 36-month neurodevelopment. *Pediatrics*, *127*(3), 699–706. doi:10.1542/peds.2010-0133 PMID:21300677

Keil, A. P., Daniels, J. L., & Hertz-Picciotto, I. (2014). Autism spectrum disorder, flea and tick medication, and adjustments for exposure misclassification: The CHARGE (Childhood Autism Risks from Genetics and Environment) case –control study. *Environmental Health*, *13*(1), 3. doi:10.1186/1476-069X-13-3 PMID:24456651

Kennedy, D., Hurst, V., Konradsdottir, E., & Einarson, A. (2005). Pregnancy outcome following exposure to permethrin and use of teratogen information. *American Journal of Perinatology*, *22*(2), 87–90. doi:10.1055-2005-837736 PMID:15731987

Kim, J., Swartz, M. D., Langlois, P. H., Romitti, P. A., Weyer, P., Mitchell, L. E., ... Agopian, A. (2017). The National Birth Defects Prevention Study. Estimated Maternal Pesticide Exposure from Drinking Water and Heart Defects in Offspring. *International Journal of Environmental Research and Public Health*, *14*(8), 889. doi:10.3390/ijerph14080889

Landrigan, P. J., & Baker, D. B. (2015). The National Children's Studydend or new beginning? *The New England Journal of Medicine*, *372*(16), 16. doi:10.1056/NEJMp1500276 PMID:25875255

Liu, P., Wu, C., Chang, X., Qi, X., Zheng, M., & Zhou, Z. (2016). Adverse associations of both prenatal and postnatal exposure to organophosphorous pesticides with infant neurodevelopment in an agricultural area of Jiangsu Province. China. *Environmental Health Perspectives*, *124*(10), 1637–1643. doi:10.1289/EHP196 PMID:27153333

Mostafalou, S., & Abdollahi, M. (2013). Pesticides and human chronic diseases: evidences, mechanisms, and perspectives. *Toxicol. Appl. Pharmacol.*, *268*(2), 157–77.

Naidoo, S., London, L., & Burdorf, A. (2010). Pesticide safety training and practices in women working in small-scale agriculture in South Africa. *Occupational and Environmental Medicine*, *67*(12), 823–828. doi:10.1136/oem.2010.055863 PMID:20576924

Perera, F. P., Rauh, V., Tsai, W. Y., Kinney, P., Camann, D., Barr, D., ... Whyatt, R. M. (2003). Effects of transplacental exposure to environmental pollutants on birth outcomes in a multiethnic population. *Environmental Health Perspectives*, *111*(2), 201–206. doi:10.1289/ehp.5742 PMID:12573906

Rauh, V. A., Garfinkel, R., Perera, F. P., Andrews, H. F., Hoepner, L., Barr, D. B., ... Whyatt, R. W. (2006). Impact of prenatal chlorpyrifos exposure on neurodevelopment in the first 3 years of life among inner-city children. *Pediatrics*, *118*(6), 1845–1859. doi:10.1542/peds.2006-0338 PMID:17116700

Rosas, L. G., & Eskenazi, B. (2008). Pesticides and child neurodevelopment. *Current Opinion in Pediatrics*, *20*(2), 191–197. doi:10.1097/MOP.0b013e3282f60a7d PMID:18332717

Saillenfait, A., Ndiaye, D., & Sabate, J. (2015). Pyrethroids: Exposure and health effects—an update. *International Journal of Hygiene and Environmental Health*, *218*(3), 281–292. doi:10.1016/j.ijheh.2015.01.002 PMID:25648288

Schettler, T., Solomon, G., Valenti, M., & Huddle, A. (1999). *Generations at risk*. Boston: Massachusetts Institute of Technology.

Shafer, T. J., Meyer, D. A., & Crofton, K. M. (2005). Developmental neurotoxicity of pyrethroid insecticides: Critical review and future research needs. *Environmental Health Perspectives*, *113*(2), 123–136. doi:10.1289/ehp.7254 PMID:15687048

Shaw, G. M., Swan, S. H., Harris, J. A., & Malcoe, L. H. (1990). Maternal water consumption during pregnancy and congenital cardiac anomalies. *Epidemiology (Cambridge, Mass.)*, *1*(3), 206–211. doi:10.1097/00001648-199005000-00005 PMID:2081254

Shelton, J. F., Geraghty, E. M., & Tancredi, D. J. (2015). Neurodevelopmental disorders and prenatal residential proximity to agricultural pesticides: The CHARGE study. *Environmental Health Perspectives*, *122*, 1103–1109. PMID:24954055

Stockholm Convention. (2001). *Stockholm Convention on Persistent Organic Pollutants*. Secretariat of the Stockholm Convention Clearing House.

US EPA. (2011). Pesticides Industry Sales and Usage: 2006 and 2007 Market Estimates. Washington, DC: Author.

Van der Linde, D., Konings, E. E., Slager, M. A., Witsenburg, M., Helbing, W. A., Takkenberg, J. J., & Roos-Hesselink, J. W. (2011). Birth prevalence of congenital heart disease worldwide: A systematic review and meta-analysis. *Journal of the American College of Cardiology*, *58*(21), 2241–2247. doi:10.1016/j.jacc.2011.08.025 PMID:22078432

Vizcaino, E., Grimalt, J. O., Lopez-Espinosa, M. J., Llop, S., Rebagliato, M., & Ballester, F. (2010). Maternal origin and other determinants of cord serum organochlorine compound concentrations in infants from the general population. *Environmental Science & Technology*, *44*(16), 6488–6495. doi:10.1021/es101397e PMID:20666461

Vose, S. C., Holland, N. T., Eskenazi, B., & Casida, J. (2007). Lysophosphatidylcholine hydrolases of human erythrocytes, lymphocytes, and brain: Sensitive targets of conserved specificity for organophosphorus delayed neurotoxicants. *Toxicology and Applied Pharmacology*, *224*(1), 98–104. doi:10.1016/j.taap.2007.06.008 PMID:17663017

Wang, A., Padula, A., Sirota, M., & Woodruff, T. J. (2016). Environmental influences on reproductive health: The importance of chemical exposures. *Fertility and Sterility*, *106*(4), 905–929. doi:10.1016/j.fertnstert.2016.07.1076 PMID:27513554

Wang, P., Tian, Y., Wang, X. J., Gao, Y., Shi, R., Wang, G. Q., ... Shen, X. M. (2012). Organophosphate pesticide exposure and perinatal outcomes in Shanghai. China. *Environment International*, *42*, 100–104. doi:10.1016/j.envint.2011.04.015 PMID:21601922

Wang, Y., Zhang, Y., Ji, L., Hu, Y., Zhang, J., Wang, C., ... Tian, Y. (2017). Prenatal and postnatal exposure to organophosphate pesticides and childhood neurodevelopment in Shandong, China. *Environment International*, *108*, 119–126. doi:10.1016/j.envint.2017.08.010 PMID:28843140

Weselak, M., Arbuckle, T. E., Wigle, D. T., & Krewski, D. (2007). In utero pesticide exposure and childhood morbidity. *Environmental Research*, *103*(1), 79–86. doi:10.1016/j.envres.2006.09.001 PMID:17084836

Williams, M. K., Barr, D. B., & Camann, D. E. (2006). An intervention to reduce residential insecticide exposure during pregnancy among an inner-city cohort. *Environmental Health Perspectives*, *114*, 1684–1689. PMID:17107853

Woodruff, T., Carlson, A., Schwartz, J., & Giudice, L. (2008). Proceedings of the summit on environmental challenges to reproductive health and fertility: Executive summary. *Fertility and Sterility*, *89*(2), 1–20. doi:10.1016/j.fertnstert.2008.01.065 PMID:18155200

Yin, S. S., Tang, M. L., Chen, F. F., Li, T. L., & Liu, W. P. (2016). Environmental exposure to polycyclic aromatic hydrocarbons (PAHs): The correlation with and impact on reproductive hormones in umbilical cord serum. *Environment and Pollution*. PMID:27838061

Chapter 8
Pesticide Contamination and Human Health

Jyoti Bisht
Shri Guru Ram Rai (PG) College, India

Lok Man S. Palni
Graphic Era University, India

Nirmal S. K. Harsh
Forest Research Institute Dehradun, India

ABSTRACT

Pesticides play a vital role in modern farming in order to meet the needs of growing population. However, due to their toxic effects, pesticides cause a serious threat to public health. Pesticides when used excessively and carelessly cause social conflict, as most of the workers are intoxicated by these chemicals. These chemicals not only affect farmers and applicators but also adversely affect surrounding communities, flora and fauna. During the present decade, there is an increased awareness among the people regarding pesticide poisoning. The present chapter highlighted the adverse effect of pesticides on environment and on human health. This review helps to seek the attention of researchers, government, and non-government organizations on health issues that have been associated with the exposure of harmful chemical pesticides and encourage research on finding the new concept in modern agriculture involving a reduction in the use of chemical pesticides.

INTRODUCTION

Ensuring food security for more than one billion Indians with diminishing cultivable land resources is a prodigious task. In order to meet the needs of a growing population, modern farming techniques play a crucial role and involved high yielding variety seeds, balanced fertilizers dose and an appropriate amount of quality pesticides along with farmers' education. The size of the Indian pesticide industry is $3.8 billion in the year 2011and expected to grow more in future. Over the years, India has emerged as the 4th largest producer of pesticides after USA, Japan and China and second largest producer and exporter of pesticides in Asia (Indian Chemical Industry, XIIth Five Year Plan: 2012-2017).

DOI: 10.4018/978-1-5225-6111-8.ch008

Any Substance or mixture of substances considered as a pesticide that alters biological processes of living organisms that supposed to be pests, whether these are insects, fungi and weeds. Pesticides are characterized by their uniqueness of their chemical structure or their patterns of use and their interaction with the environment. Depending upon the chemical nature, the major classes of pesticides include organochlorine, organophosphorus, carbamates, pyrethroids and neonicotinoids (US EPA, 2006). Among the various pesticides used in India, 40% of all the pesticides used are organochlorine. The other major category is organophosphate pesticides. Monocrotophos, chlorpyrifos, phorate, phosphamidon, methyl parathion, endosulfan and dimethoate are highly hazardous pesticides that are continually and indiscriminately used in India (Gupta, 2004).

ENVIRONMENTAL CONTAMINATION: A CAUSE OF CONCERN

Humans are continuously interacting with their environment, so it is logical to presume that humans' health is also affected by the environmental quality. It is truly said, "What goes around comes around". As many of the ways environment have been harmed by human activity come back to trouble future generations in the form of various health issues. The use of chemicals in modern agriculture has significantly increased productivity, but it has also significantly increased the concentration of pesticides in food and in our environment. The realization that the pesticide contaminated foods people eat, the smokestack-befouled air they breathe, and the petrochemical-based products they use, negatively affect the quality of life.

Until 1962, pesticide use in agriculture and public health was indiscriminate. Only after the publication of "Silent Spring" by Rachel Carson in 1962, peoples' awareness towards the ill effects of pesticides increased (Carson, 1962). The potential risk to human health makes the remediation of pesticide-contaminated sites a necessary and almost urgent undertaking. Most of the pesticides are banned under pesticide pollution and toxicity prevention act.

Pesticides can contaminate environment in many ways. Their fate in the environment depends on the quality of pesticide and medium in which it is transported. Chemical nature of pesticide and their interaction with environment decides the area it will move and their persistence. The longer the half-life resulted greater the potential for pesticide movement for longer distance. A pesticide with a half-life greater than 21 days may persist long enough to leach in soil or move with surface runoff before it degrades (Gavrilescu, 2005).

Soil constitutes a major environmental sink for many pesticides from which they are taken up by plants, moves into the bodies of invertebrates, pass into water or air, and are broken down (Khan, 1980). Heavy use of pesticides not only affects the surrounding communities but it also causes a serious threat to natural microflora and fauna of soil which are responsible for soil fertility (Sardar and Kole, 2005). Pesticides applied for the agricultural and nonagricultural purpose can move to the atmosphere by volatilization and by wind erosion of particles on which the pesticide is sorbed and cause pesticide drift (Seiber and Woodrow, 1995).

Pesticides can accumulate in the tissues of organisms. This process (the so-called bioaccumulation) leads to higher concentrations of pesticide. Pesticides that bioaccumulate in organisms are often very persistent in the environment. They do not break down easily and retain their form even when ingested and stored in the body. Among animals, tissues of the respiratory and digestive system are usually much more permeable than the skin (Fishel, 2005).

Recently, increased attention has been focused on chemical residues in food. It has been observed that a diet containing fruit and vegetables, crops treated with pesticides increased the potential risks of cancer, high blood pressure, heart disease, stroke, and other chronic infections (Dietary Guidelines, 2005). The first report of pesticide poisoning in India was from Kerala, where over 100 people died after consuming parathion contaminated wheat flour (Karunakaran, 1958).

Endosulfan Poisoning and Kasargod Tragedy

Endosulfan is particularly neurotoxic to both insects and mammals, including humans. It was classified by the United States Environmental Protection Agency (US EPA) as category: "highly toxic", based on an LD_{50} of 30 mg kg^{-1} for rats (US EPA, 2002). Endosulfan can easily be absorbed through skin, stomach, lungs and can pose hazards to different body parts. Exposure may result from drinking contaminated water, consuming contaminated food, breathing contaminated air, and even touching contaminated soil (ASTDR, 2013).

Endosulfan and endosulfan sulfate are well known endocrine disrupters (Jin *et al.*, 1997). Acute oral toxicity of endosulfan is higher than dermal toxicity (Silva and Gammon, 2009). Symptoms of poisoning include death, clinical signs, irritation of the stomach and small intestine, congestion of kidneys, lungs and adrenals, reddening of the small intestine, neurotoxicity, erythema, atonia, desquamation, hemorrhagic lung, granular livers, irritation of large intestine, congested kidneys, nausea, vomiting, seizures and dizziness. Long-term exposure is linked to immunosuppression, neurological disorders, congenital birth defects, chromosomal abnormalities, mental retardation, impaired learning and memory loss (Silva, 2007).

Swarga and other areas like Padre, Muliyar and Bellur in Kasaragod district of Kerala, India have become living examples of how the poison in pesticides could be lethal to our health when used excessively and carelessly. The villages of Kasargod experienced sustained exposure to endosulfan as a result of 20 years of aerial spraying of a nearby cashew nut plantation. Spraying had occurred at heights (up to 50 metres above canopy) which aided wind drift and widespread of pesticide into villages. Twelve streams used by the villagers originated in the plantation and were subsequently found to have contaminated sediment and water throughout the year (outside the spray season), as a result of endosulfan's persistence in soil and the soil being carried to the streams by storm runoff. Residues were still detected in the water and pond sediments 1.5 years after spraying ceased (Saiyed *et al.*, 2003). The villagers were also directly exposed to spray drift. Numerous congenital, reproductive, long-term neurological and other symptoms were experienced. Pesticide Action Network Asia and Pacific (PANAP) conduct Community Pesticide Monitoring led by Dr Romeo Quijano and experts from Community Health Cell, Thanal and local community leaders. A report confirms that "the occurrence of illnesses is due mainly to Endosulfan". Indian Council of Medical Research (ICMR) conducts detailed epidemiological study concludes on effects on male reproductive organs of children and mental impairment due to exposure to endosulfan (Sridhar, 2008). There were observations of similar effects in animals too: cows giving birth to deformed calves, cows and chickens dying inexplicably, domestic animals with miscarriages, bleeding, infertility, stunting of growth and deformities, as well as fish kills and dwindling populations of honeybees', frogs, and birds (NIOH, 2002). Endosulfan has subsequently been banned in the State of Kerala and compensation paid to some of the victims and/or their families by the State (Venugopal, 2008).

PESTICIDE AND HUMAN HEALTH

Pesticides are generally lethal to target species, but unfortunately their excessive use makes them harmful to non-target species including human being. The widespread use of these chemicals, under the adage, "if little is good, a lot more will be better" cause adverse effects on human health and other life forms (Aktar *et al.*, 2009; Nicolopoulou-Stamati *et al.*, 2016). Pesticide exposure is not only harmful to adults, but also young children and fetus during their developmental period are vulnerable to these pesticides due to their weak and inactive immune system and relatively more exposure effect on their small body mass (Darçın and Darçın, 2017).

The major risk of pesticide poisoning to those individuals who handle these chemicals directly, includes production workers, loaders, formulators, sprayers, agricultural farm workers, even children playing near to applied area. National Crime Records Bureau in India reported suicide cases by consuming pesticides. According to them, before 2001 there were on average 15,750 reported farmer suicides in a year. Since 2002, the annual reported average has risen to 17,366 in 2007. That is the equivalent of one suicide every 30 minutes. Among the worst-affected states figure the main cotton growing states of Maharashtra, Andhra Pradesh and Madhya Pradesh (WHO, 2009).

Organochlorine compounds are highly persistent in the environment and could pollute most of the life form. Due to highly persistent nature of organochlorine, these were replaced by organophosphates. But organophosphates are also associated with acute health problems such as dizziness, abdominal pain, nausea, vomiting, as well as skin and eye problems. Additionally, pesticide exposure for a longer period is also associated with other health problems such as depression, skin related issues, respiratory disorders, cancer, neurological defects, chances miscarriages in the pregnant female worker, and birth defects due to prenatal exposure (Hurley *et al.*, 1998).

Some time farmers combined different pesticides in a mixture for better results. Combining more than one pesticide together could be discouraged, because mixing of pesticides can alter their chemical properties, thereby increasing its harmful effects on health. Salameh *et al.* (2004) have already mentioned that the combined use of hazardous pesticides and the absence of appropriate precautions are detrimental to the farmers' health. Pesticides also act as an endocrine disruptor, sometime as a natural hormone in the body and linked to human health effects such as immune suppression, hormone disruption, diminished intelligence, reproductive abnormalities and cancer (Brouwer *et al.*, 1999).

Effect on Reproductive System

Cypermethrin, a synthetic pyrethroid insecticide and is used to control pests of a different variety of crops. Cypermethrin has been reported to induce gene mutations in male germ cells of Drosophila (Batiste–Alenton *et al.*, 1986) and genotoxicity and sperm abnormality in mice (Bhunya and Pati, 1988). Dibromochloropropane (IDBCP), a nematicide, resulted in the sterilization of around 1,500 banana plantation workers in Costa Rica. DBCP is manufactured in the United States, but in 1977 its application was banned in the United States after it was discovered to be the cause of the sterilization of 60 employees of the plant that manufactured it (Thrupp, 1991).

Carbamate pesticides also responsible for possible reproductive disorders (Jamal *et al.*, 2015), Triazines, such as atrazine, simazine, and ametryn, are another class of chemical pesticides that have been related to reproductive toxicity and delays in sexual maturation (Jin *et al.*, 2014). Neonicotinoid pesticides, such as imidacloprid, thiacloprid, and guadipyr, are extensively used insecticides and cause effects on the endocrine and reproductive systems of animals (Hoshi *et al.*, 2014).

Birth Defects

Birth defects are more common in families of pesticide applicators, families living near agriculture areas, and maternal exposure while working in orchards, greenhouses or grain farming (Andersen *et al.*, 2008). Maternal exposure to pesticides during the period from the month before conception and the first trimester increased the chance of multiple anomalies including nervous system defects and oral clefts (Garcia *et al.*,1998) use of insect repellents during the first trimester with hypospadias a condition with abnormally placed urinary opening on the penis (Dugas *et al.*, 2010) and paternal exposure in greenhouses producing vegetables and flowers during the 3 months prior to conception with cryptorchidism a condition with absence of one or both testes (Pierik *et al.*, 2004).

Arbuckle *et al.* (2001) suggested an increased risk of fetal deaths associated with pesticides and maternal employment in the agricultural industry. During pregnancy exposure to both DDT and dichlorodiphenyldichloroethylene (DDE) has been associated with neurodevelopmental effects in children (Eskenazi *et al.*, 2006). Chlorpyrifos and atrazine caused a significant reduction in head circumference of the foetus (Berkowitz *et al.*, 2004; Chevier *et al.*, 2011). Colborn and Carroll (2007) reported that exposure to atrazine, 2,4-D, and glyphosate cause foetal loss. Pesticides residues have also been detected in human breast milk samples, and cause of prenatal exposure and health effects in children (Pirsaheb *et al.*, 2015).

Endocrine Disruption

The endocrine, or hormonal system, is a delicately balanced system of glands and hormones that maintain homeostasis in the body, regulate metabolism, growth, function of the digestive, cardiovascular, renal and immune systems, sexual development and reproduction (Birnbaum 2010). DDT and its metabolite DDE also reported as potent endocrine-disruptor and carcinogen (Turusov *et al.*, 2002). Carbamate pesticides, such as aldicarb, carbofuran, and ziram, are another class of chemical pesticides also responsible for endocrine-disrupting activity (Goad *et al.*, 2004). Organophosphates, which were promoted as a more ecological alternative to organochlorines, include a great variety of pesticides, such as glyphosate, malathion, parathion, and dimethoate; some are known for their endocrine-disrupting potential (Gasnier *et al.*, 2009). Pesticides like phthalates, dioxins, polybrominated diphenyl ethers (PBDEs), and other halogenated organochlorines can destroy the thyroid gland and disturbed the normal thyroid function by affecting hormones production, transportation and metabolism (Patrick, 2009). Triazines (atrazine, simazine, and ametryn) also have been reported as endocrine disrupting pesticides (Jin *et al.*, 2014).

Cancer

Various epidemiological studies gave strong evidence of pesticide exposures and incidence of cancer. Application of pesticide on a commercial level and in houses will highly increase the risk of leukaemia, clone thyroid, brain and several other types of cancer. Dich *et al.* (1997) reviewed on cancer risk associated with pesticides exposure, links phenoxy herbicides with soft tissue sarcoma (STS), organochlorine insecticides with STS, non-Hodgkin's lymphoma (NHL), leukaemia and lung and breast cancer; organophosphorus are linked with NHL and leukaemia; and triazine with ovarian cancer. Kettles *et al.*, (1997) also reported that there is a possible statistical relationship between triazine and breast cancer incidence.

A study was conducted on cancer incidence in Costa Rica and the potential relation to pesticide exposure. Regions with a high concentration of banana plantations (and consequently the heavy use of DBCP, chlorothalonil, and mancozeb) had increased incidences of respiratory, ovary, and prostate cancer (Wesseling *et al.*, 2002). Aromatic amines used as pesticides also reported as carcinogenic and mainly responsible for the bladder cancer (Silverman *et al.*, 2006). Stella *et al.*, (2009) reported that person who is exposed to pesticides have increased the risk of bladder cancer.

The risk of brain cancer is increased in professional pesticide applicators which are involved in termite controlling activities (Kathryn *et al.*, 2013). *In vitro* study revealed the ability of carbamate pesticides to cause cytotoxic and genotoxic effects in hamster ovarian cells (Soloneski *et al.*, 2015) and to induce apoptosis and necrosis in human immune cells, natural killer cells (Li *et al.*, 2011), and increased risk for non-Hodgkin's lymphoma (Zheng *et al.*, 2001). Neonicotinoids were also reported to increase the expression of aromatase enzyme in females, which is responsible for increase breast (Caron-Beaudoin *et al.*, 2016).

Neurological Problems and Behavioral Disorders

Toxic pesticides like organophosphates, organochlorine and carbamates affect central and peripheral nervous system. Organophosphates exposure lead to a disorder called OP-induced delayed polyneuropathy which affects neurons and unable to produce the neuropathy target esterase enzyme cause overstimulation of postsynaptic cholinergic receptors (Lee *et al.*, 2003). Pesticides showed acute or chronic and long-term or short-term effects on the nervous system by the high or low-level exposure and may lead to the Alzheimer disease especially if exposed to person during their late life (Hayden *et al.*, 2010).

Some other herbicides (rotenone and paraquat) also disrupting the bioenergetic activities of mitochondria, oxygen metabolism and redox function which lead to Alzheimer disease (Thany *et al.*, 2013). Rotenone and paraquat also disrupt dopaminergic neuron of a brain and inhibit the production of dopamine. When dopamine is not produced by the brain, which leads to lack of coordination, trembling and loss of muscles control and resulted in Parkinson disease (Qi *et al*, 2014). Swanson *et al.* (2014) reported that exposure of glyphosate for longer period cause various health effects, such as autism, diabetes, hypertension, strokes, kidney failure, Parkinson's and Alzheimer's diseases.

Respiratory Problems

The association with chronic bronchitis, asthma was found for occupational, domestic and environmental exposures especially to paraquat, OC, OP, carbamate and pyrethroid insecticides. Chlorpyrifos exposure for longer period cause breathing difficulties like wheeze, chronic cough and shortness of breath (Sanborn *et al.*, 2012).

Metabolic Disorders and Other Adverse Effects

Fenvalerate, a third-generation synthetic pyrethroid has been reported to inhibit intracellular communication (Warngard and Flodstrom, 1989). Malathion is a commonly used organophosphorus insecticide has raised concern over its potential to cause genetic damage (Flessel *et al.*, 1993). Slight histopathological changes were observed at high doses of chlorpyrifos in the liver, adrenal glands and the kidneys. The observation from chronic and repeat dose studies included unstable heart rate and blood pressure, muscle cramps, skin flushing, diarrhoea, vomiting, increases in urinary frequency and epigastric cramping (NRA, 2000).

Organophosphate pesticides inhibit the function of cholinesterase enzymes resulted in problems related to the nervous and endocrine systems, decrease in insulin secretion, cause defects in the normal cellular metabolism of proteins, carbohydrates and fats, causing cellular oxidative stress (Karami-Mohajeri and Abdollahi, 2011). Sulfonylurea herbicides and imidazole fungicides have also been reported to cause obesity and diabetes, because of their effects on weight gain, blood sugar levels, or the pancreas (Thayer *et al.*, 2012). In vitro study revealed that Glyphosate can affect human erythrocytes (Kwiatkowska *et al.*, 2014) and in addition, cause extreme disruption in shikimate pathway. Disruption in this pathway may affect the synthesis of essential amino acids (Samsel and Seneff, 2015).

URGENT NEED FOR A NEW CONCEPT IN AGRICULTURE

Many of the pesticides have been associated with health and environmental issues. Presently no one has paid attention to the adequate regulatory process for assessing the effect of those endocrine disrupting pesticides. Nor do they adequately implement the precautionary principle, or the substitution of hazardous pesticides with less hazardous pesticides or non-chemical methods. The present review showed that most research focused on individuals with direct exposure to pesticides, e.g. farmers. Since research on people with indirect exposure was harder to find, thus more research should be conducted to analyze the potential health effects of pesticides on consumers or people with indirect contact with pesticides. Farmers should be informed about misuse, intentionally harmful use of non-purpose thread or application errors such as protective clothing use during application, comply with the rules of personal hygiene, overdose and unnecessary duplication, exposure and contacting to chemicals via educational programs. Safe use of pesticides should be provided. Chemicals are causing persistent organic pollutants should be banned.

REFERENCES

Aktar, W., Sengupta, D., & Chowdhury, A. (2009). Impact of pesticide use in agriculture: Their benefits and hazards. *Interdisciplinary Toxicology*, *2*(1), 1–12. doi:10.2478/v10102-009-0001-7 PMID:21217838

Andersen, H. R., Schmidt, I. M., Grandjean, P., Jensen, T. K., Budtz-Jørgensen, E., Kjaerstad, M. B., ... Main, K. M. (2008). Impaired reproductive development in sons of women occupationally exposed to pesticides during pregnancy. *Environmental Health Perspectives*, *116*(4), 566–572. doi:10.1289/ehp.10790 PMID:18414644

Arbuckle, T. E., Lin, Z., & Mery, L. S. (2001). An exploratory analysis of the effect of pesticide exposure on the risk of spontaneous abortion in an Ontario farm population. *Environmental Health Perspectives*, *109*(8), 851–857. doi:10.1289/ehp.01109851 PMID:11564623

ATSDR. (2013). *Draft toxicological profile for endosulfan*. U.S. Department of health and human services, Public Health Service Agency for Toxic Substances and Disease Registry. Retrieved from http://www.atsdr.cdc.gov/toxprofiles/tp41-p.pdf

Batiste-Alenton, X. M., Velazquez, N., Creus, A., & Marcos, R. (1986). Mutagenicity of the pyrethroid insecticide cypermethrin in Drisiphila. *Mutageneis*, *1*(5), 343–346. doi:10.1093/mutage/1.5.343

Berkowitz, G. S., Wetmur, J. G., Birman-Deych, E., Obel, J., Lapinski, R. H., Godbold, J. H., ... Wolff, M. S. (2004). In utero pesticide exposure, maternal paraoxonase activity, and head circumference. *Environmental Health Perspectives*, *112*(3), 388–391. doi:10.1289/ehp.6414 PMID:14998758

Bhunya, S. P., & Pati, P. C. (1988). Genotoxic effects of a synthetic pyrethroid insecticide cypermethrin in mice *in vivo*. *Toxicology Letters*, *41*(3), 223–230. doi:10.1016/0378-4274(88)90058-6 PMID:3376150

Birnbaum, L. (2012). Environmental chemicals: Evaluating low-dose effects. *Environmental Health Perspectives*, *120*(4), 143–144. doi:10.1289/ehp.1205179 PMID:22470049

Brouwer, A., Longnecker, M. P., Birnbaum, L. S., Cogliano, J., Kostyniak, P., Moore, J., ... Winneke, G. (1999). Characterization of potential endocrine related health effects at low dose levels of exposure to PCBs. *Environmental Health Perspectives*, *107*(Suppl 4), 639–649. doi:10.1289/ehp.99107s4639 PMID:10421775

Caron-Beaudoin, É., Denison, M. S., & Sanderson, J. T. (2016). Effects of neonicotinoids on promoter-specific expression and activity of aromatase (CYP19) in human adrenocortical carcinoma (H295R) and primary umbilical vein endothelial (HUVEC) cells. *Toxicological Sciences*, *149*(1), 134–144. doi:10.1093/toxsci/kfv220 PMID:26464060

Carson, R. (1962). Silent Spring. Boston: Houghton.

Chevrier, C., Limon, G., Monfort, C., Rouget, F., Garlantézec, R., Petit, C., ... Cordier, S. (2011). Urinary biomarkers of prenatal atrazine exposure and adverse birth outcomes in the PELAGIE birth cohort. *Environmental Health Perspectives*, *119*(7), 1034–1041. doi:10.1289/ehp.1002775 PMID:21367690

Colborn, T., & Carroll, L. E. (2007). Pesticides, sexual development, reproduction, and fertility: Current perspective and future direction. *Human and Ecological Risk Assessment*, *13*(5), 1078–1110. doi:10.1080/10807030701506405

Darçın, E. S., & Darçın, M. (2017). Health effects of agricultural pesticides. *Biomed Res- India; Special Section: Complex World of Neuroscience,* S13-17.

Dich, J., Zahm, S. H., Hanberg, A., Adami, H. O., & Adami, H.-O. (1997). Pesticides and cancer. *Cancer Causes & Control*, *8*(3), 420–443. doi:10.1023/A:1018413522959 PMID:9498903

Dietary Guidelines for Americans. (2005). U.S. Department of Health and Human Services U.S. Department of Agriculture.

Dugas, J., Nieuwenhuijsen, M. J., Martinez, D., Iszatt, N., Nelson, P., & Elliott, P. (2010). Use of biocides and insect repellents and risk of hypospadias. *Occupational and Environmental Medicine*, *67*(3), 196–200. doi:10.1136/oem.2009.047373 PMID:19951933

Environews Forum. Killer environment. (1999). *Environ Health Perspect, 107*, A62-A63. Retrieved from http://www.ncbi.nlm.nih.gov/pmc/articles/PMC1566330/?page=1

Eskenazi, B., Marks, A. R., Bradman, A., Fenster, L., Johnson, C., Barr, D. B., & Jewell, N. P. (2006). In utero exposure to dichlorodiphenyltrichloroethane (DDT) and dichlo-rodiphenyldichloroethylene (DDE) and neurodevelopment among young Mexican American children. *Pediatrics, 118*(1), 233–241. doi:10.1542/peds.2005-3117 PMID:16818570

Fishel, F. M. (2005). *43 Document. Pesticide-Organism Interactions, PI: Pesticide Information Office, Florida Cooperative Extension Service*. Institute of Food and Agricultural Sciences, University of Florida. Retrieved from http://edis.ifas.ufl.edu/pdffiles/PI/PI0800.pdf

Flessel, P., Quintana, P. J. E., & Hopper, K. (1993). Genetic toxicity of malathion: A review. *Environmental and Molecular Mutagenesis, 22*(1), 7–17. doi:10.1002/em.2850220104 PMID:8339727

Garcia, A. M., Benavides, F. G., Fletcher, T., & Orts, E. (1998). Paternal exposure to pesticides and congenital malformations. *Scandinavian Journal of Work, Environment & Health, 24*(6), 473–480. doi:10.5271jweh.371 PMID:9988089

Gasnier, C., Dumont, C., Benachour, N., Clair, E., Chagnon, M. C., & Séralini, G. E. (2009). Glyphosate-based herbicides are toxic and endocrine disruptors in human cell lines. *Toxicology, 262*(3), 184–191. doi:10.1016/j.tox.2009.06.006 PMID:19539684

Gavrilescu, M. (2005). Fate of pesticides in the environment and its bioremediation. *Engineering in Life Sciences, 5*(6), 497–526. doi:10.1002/elsc.200520098

Goad, E. R., Goad, J. T., Atieh, B. H., & Gupta, R. C. (2004). Carbofuran-induced endocrine disruption in adult male rats. *Toxicology Mechanisms and Methods, 14*(4), 233–239. doi:10.1080/15376520490434476 PMID:20021136

Gupta, P. K. (2004). Pesticide exposure: Indian scene. *Toxicology, 198*(1-3), 83–90. doi:10.1016/j.tox.2004.01.021 PMID:15138033

Hayden, K. M., Norton, M. C., Darcey, D., Ostbye, T., Zandi, J. C. S., Breitner, J. C. S., & Welsh-Bohmer, K. A. (2010). Occupational exposure to pesticides increases the risk of incident AD: The cache county study. *Neurology, 74*(19), 1524–1530. doi:10.1212/WNL.0b013e3181dd4423 PMID:20458069

Hoshi, N., Hirano, T., Omotehara, T., Tokumoto, J., Umemura, Y., Mantani, Y., ... Kitagawa, H. (2014). Insight into the mechanism of reproductive dysfunction caused by neon-icotinoid pesticides. *Biological & Pharmaceutical Bulletin, 37*(9), 1439–1443. doi:10.1248/bpb.b14-00359 PMID:25177026

Hurley, P. M., Hill, R. N., & Whiting, R. J. (1998). Mode of carcinogenic action of pesticides inducing thyroid follicular cell tumours in rodents. *Environmental Health Perspectives, 106*(8), 437–445. doi:10.1289/ehp.98106437 PMID:9681970

Indian Chemical Industry. (n.d.). *XI2Ith Five Year Plan: 2012-2017*. Planning Commission of India, Government of India. Retrieved from http://planningcommission.nic.in/aboutus/committee/wrkgrp12/wg_chem0203.pdf

Jamal, F., Haque, Q. S., Singh, S., & Rastogi, S. (2015). The influence of organophosphate and carbamate on sperm chromatin and reproductive hormones among pesticide sprayers. *Toxicology and Industrial Health*, 1–10. PMID:25647813

Jin, L., Trans, D. Q., Ide, C. F., McLachlan, J. A., & Arnold, S. F. (1997). Several synthetic chemicals inhibit progesterone receptor-mediated transactivation in yeast. *Biochemical and Biophysical Research Communications*, *233*(1), 13–146. doi:10.1006/bbrc.1997.6417 PMID:9144411

Jin, Y., Wang, L., Chen, G., Lin, X., Miao, W., & Fu, Z. (2014). Exposure of mice to atrazine and its metabolite diaminochlorotriazine elicits oxidative stress and endocrine disruption. *Environmental Toxicology and Pharmacology*, *37*(2), 782–790. doi:10.1016/j.etap.2014.02.014 PMID:24632104

Karami-Mohajeri, S., & Abdollahi, M. (2011). Toxic influence of organophosphate, carbamate, and organochlorine pesticides on cellular metabolism of lipids, proteins, and carbohydrates: A systematic review. *Human and Experimental Toxicology*, *30*(9), 1119–1140. doi:10.1177/0960327110388959 PMID:21071550

Karunakaran, C. O. (1958). The Kerala food poisoning. *Journal of the Indian Medical Association*, *31*, 204. PMID:13588025

Kathryn, R., Greenop, S., Peters, H., & Bailey, D. (2013). Exposure to pesticides and the risk of childhood brain tumors. *Cancer Causes & Control*, *7*, 1269–1278. PMID:23558445

Kettles, M. K., Browning, S. R., Prince, T. S., & Horstman, S. W. (1997). Triazine herbicide expo-sure and breast cancer incidence: An ecologic study of Kentucky counties. *Environmental Health Perspectives*, *105*(11), 1222–1227. doi:10.1289/ehp.971051222 PMID:9370519

Khan, S. U. (1980). Pesticides in the soil environment. Elsevier Scientific Publishing Company.

Kwiatkowska, M., Huras, B., & Bukowska, B. (2014). The effect of metabolites and impu-rities of glyphosate on human erythrocytes (*in vitro*). *Pesticide Biochemistry and Physiology*, *109*, 34–43. doi:10.1016/j.pestbp.2014.01.003 PMID:24581382

Lee, B. W., London, L., Paulauskis, J., Myers, J., & Christiani, D. C. (2003). Association between human paraoxonase gene polymorphism and chronic symptoms in pesticide-exposed workers. *Journal of Occupational and Environmental Medicine*, *45*(2), 118–122. doi:10.1097/01.jom.0000052953.59271.e1 PMID:12625227

Li, Q., Kobayashi, M., & Kawada, T. (2011). Ziram induces apoptosis and necrosis in human immune cells. *Archives of Toxicology*, *85*(4), 355–361. doi:10.100700204-010-0586-9 PMID:20842346

Nicolopoulou-Stamati, P., Maipas, S., Kotampasi, C., Stamatis, P., & Hens, L. (2016). Chemical Pesticides and Human Health: The Urgent Need for a New Concept in Agriculture. *Frontiers in Public Health*, *4*, 148. doi:10.3389/fpubh.2016.00148 PMID:27486573

NIOH (National Institute of Occupational Health). (2002). Final Report of the Investigation of Unusual Illnesses Allegedly Produced by Endosulfan Exposure. In *Padre Village of Kasargod District (N Kerala)*. Ahmedabad: National Institute of Occupational Health, Indian Council for Medical Research.

NRA (National Registration Authority). (2000). *Review of Chlorpyrifos*. Commonwealth of Australia. Retrieved from www.nra.gov.au/chemrev/chemrev.shtml

Patrick, L. (2009). Thyroid disruption: Mechanism and clinical implications in human health. *Alternative Medicine Review*, *14*, 326–346. PMID:20030460

Pierik, F. H., Burdorf, A., Deddens, J. A., Juttmann, R. E., & Weber, R. F. (2004). Maternal and paternal risk factors for cryptorchidism and hypospadias: A case-control study in newborn boys. *Environmental Health Perspectives*, *112*(15), 1570–1576. doi:10.1289/ehp.7243 PMID:15531444

Pirsaheb, M., Limoee, M., Namdari, F., & Khamutian, R. (2015). Organochlorine pesticides residue in breast milk: A systematic review. *Medical Journal of the Islamic Republic of Iran*, *29*, 228. PMID:26478886

Qi, Z., Miller, G. W., & Voit, E. O. (2014). Dopamine metabolism: A computational analysis of pesticide toxicity. *Toxicology*, *6*, 92–101. doi:10.1016/j.tox.2013.11.003 PMID:24269752

Saiyed, H., Dewan, A., Bhatnagar, V., Shenoy, U., Shenoy, R., Rajmohan, H., ... Lakkad, B. (2003). Effect of endosulfan on male reproductive development. *Environmental Health Perspectives*, *111*(16), 1958–1962. doi:10.1289/ehp.6271 PMID:14644673

Salameh, R. P., Daldi, I., Brochard, P., & Saleh, B. A. (2004). Pesticides in Lebanon: A knowledge, attitude and practise study. *Environmental Research*, *94*(1), 1–6. doi:10.1016/S0013-9351(03)00092-6 PMID:14643280

Samsel, A., & Seneff, S. (2015). Glyphosate, pathways to modern diseases III: Manganese, neurological diseases, and associated pathologies. *Surgical Neurology International*, *6*(1), 45. doi:10.4103/2152-7806.153876 PMID:25883837

Sanborn, M., Bassil, K., Vakll, C., Kerr, K., & Ragan, K. (2012). *Systematic Review of Pesticide health Effects*. Ontario College of Family Physicians, Toronto. Retrieved from www.ocfp.on.ca/communications/newpublications/lists/new-publications/ocfp-2012-systematic-review-of-pesticide-health-effects

Sardar, D., & Kole, R. K. (2005). Metabolism of Chlorpyriphos in relation to its effect on the availability of some plant nutrients in soil. *Chemosphere*, *61*(9), 1273–1280. doi:10.1016/j.chemosphere.2005.03.078 PMID:15894348

Seiber, J. N., & Woodrow, J. E. (1995). Origin and fate of pesticides in air. In *Eighth International Congress of Pesticide Chemistry–Options 2000*. Washington, DC: American Chemical Society.

Silva, M. H. (2007). *Endosulfan risk characterization document. Medical toxicology and worker health and safety branches department of pesticide regulation*. California environmental protection. Available from: www.cdpr.ca.gov/docs/emon/pubs/tac/.../endosulfan/endosulfan_doc.pdf

Silva, M. H., & Gammon, D. (2009). An assessment of the developmental, reproductive and neurotoxicity of endosulfan. *Birth Defects Research. Part B, Developmental and Reproductive Toxicology*, *86*(1), 1–28. doi:10.1002/bdrb.20183 PMID:19243027

Silverman, D. T., Devesa, S. S., Moore, L. E., & Rothman, N. (2006). *Cancer epidemiology and prevention*. New York: Oxford University Press.

Soloneski, S., Kujawski, M., Scuto, A., & Larramendy, M. L. (2015). Carbamates: A study on genotoxic, cytotoxic, and apoptotic effects induced in Chinese hamster ovary (CHO-K1) cells. *Toxicology In Vitro*, *29*(5), 834–844. doi:10.1016/j.tiv.2015.03.011 PMID:25820133

Sridhar, R. (2008). *Endosulfan Poisoning and the Struggle of the community in Kasaragod to regain life and the living land*. Retrieved from http://thanal.co.in/uploads/resource/document/endosulfan-poisoning-and-the-struggle-of-the-community-in-kasaragod-to-regain-life-and-the-living-land-a-presentation-at-the-cop3-of-rotterdam-conven-50269185.pdf

Stella, K., Charles, F., Lynch, X., & Won, M. A. (2009). Aromatic amine pesticide use and human cancer risk: Results from the U.S. Agricultural Health Study. *International Journal of Cancer*, *124*(5), 1206–1212. doi:10.1002/ijc.24020 PMID:19058219

Swanson, N. L., Leu, A., Abrahamson, J., & Wallet, B. (2014). Genetically engineered crops, glyphosate and the deterioration of health in the United States of America. *Journal of Organic Systems*, *9*, 6–37.

Thany, S. H., Reynier, P., & Lenaers, G. (2013). Neurotoxicity of pesticides: Its relationship with neurodegenerative diseases. *Medical Science*, *29*, 273–278. PMID:23544381

Thayer, K. A., Heindel, J. J., Bucher, J. R., & Gallo, M. A. (2012). Role of environmental chemicals in diabetes and obesity: A National Toxicology Programme workshop review. *Environmental Health Perspectives*, *120*(6), 779–789. doi:10.1289/ehp.1104597 PMID:22296744

Thrupp, L. A. (1991). Sterilization of Workers from Pesticide Exposure: The Causes and Consequences of DBCP-Induced Damage in Costa Rica and Beyond. *International Journal of Health Services*, *21*(4), 731–757. doi:10.2190/25B9-QFMV-Y28P-L736 PMID:1769760

Turusov, V., Rakitsky, V., & Tomatis, L. (2002). Dichlorodiphenyltrichloroethane (DDT): Ubiquity, persistence, and risks. *Environmental Health Perspectives*, *110*(2), 125–128. doi:10.1289/ehp.02110125 PMID:11836138

US EPA. (2002). *Reregistration Eligibility Decision for Endosulfan*. EPA, 738-R-02-013. Pollution, Pesticides and Toxic Substances (7508C), United States Environmental Protection Agency. Retrieved from http://www.epa.gov/oppsrrd1/REDs/endosulfan_red.pdf

US EPA. (2006). *About Pesticides*. Retrieved from http://www.epa.gov/pesticides/about/types.htm

Venugopal, P. N. (2008). *Endosulfan victims still suffer in India*. Retrieved from http://southasia.oneworld.net/article/view/157876/1

Warngard, L., & Flodstrom, S. (1989). Effects of tetradecanoye phorbol acetate, Pyrethroid and DDT. *Cell Biology and Toxicology*, *5*, 67–75. doi:10.1007/BF00141065 PMID:2920299

Wesseling, C., Keifer, M., Ahlbom, A., McConnell, R., Moon, J. D., Rosenstock, L., & Hogstedt, C. (2002). Long-term neurobehavioral effects of mild poisonings with organophos-phate and n-methyl carbamate pesticides among banana workers. *International Journal of Occupational and Environmental Health*, *8*(1), 27–34. doi:10.1179/oeh.2002.8.1.27 PMID:11843437

WHO. (2009). *Health implications from monocrotophos use: A review of the evidence in India*. WHO.

Zheng, T., Zahm, S. H., Cantor, K. P., Weisenburger, D. D., Zhang, Y., & Blair, A. (2001). Agricultural exposure to carbamate pesticides and risk of non-Hodgkin lymphoma. *Journal of Occupational and Environmental Medicine*, *43*(7), 641–649. doi:10.1097/00043764-200107000-00012 PMID:11464396

Section 3

Pesticides and Aquatic Life

Chapter 9
Pesticides and Their Impact on Aquatic Microorganisms

Riyes Un Aziz
Government (Model, Autonomous) Holkar Science College, India

Sameena
PMB Gujarati Science College, India

ABSTRACT

Microorganisms are the most dominant natives of aquatic ecosystems, where they fulfill very specific roles in primary productivity, decomposition, and nutrient cycling. Microorganisms get naturally exposed to pesticides in aquatic environments by the direct and indirect supply. The microorganism can make use of components entering the environment as feeding substrate for building material or a source of energy thus affecting balance in the ecosystem. Natural population possesses a number of responses to these contaminants, and quickest reaction has been reported from the microorganisms. Pesticides may affect the population dynamics by controlling individual reproduction, survival, and by changing sex ratio. The following patterns are recognized as effects of pesticides at the ecosystem and community levels like an increase of species richness reduction of energy transfer efficiency from primary producers to top predators. Thus, the purpose of this chapter is to review the significance of pesticides and their effects on aquatic microorganism and to study their ecological significance.

INTRODUCTION

Microorganisms are important natives of aquatic ecosystems, where they perform very important roles in primary productivity, decomposition and nutrient cycling. Aquatic ecosystems get indirect and direct pesticide supplies, imminently exposing microorganisms to pesticides. Moreover pesticides cause various chronic and acute toxic effects in microorganisms, microorganisms also have the ability to detoxify or metabolize and accumulate pesticides to some extent. Adverse effects of pesticides on the species of microorganisms may have proximate influence on top trophic levels .For example, shifts in community design of zooplankton grazers can affect their growth rate or lead to many changes in the macromolecular composition of species of phytoplankton (Ahlgren *et al.,* 1990). Aquatic environments serve as critical

DOI: 10.4018/978-1-5225-6111-8.ch009

nursery and feeding grounds for many aquatic organisms, including recreationally and commercially important shellfish and fish species. These diverse, productive environments are specifically sensitive to pesticidal pollution because they serve as storage places for pesticidal pollutants from upland sources. There is scarcity on the data of toxicity involving pesticides and microorganisms. Several investigations have targeted microbial degradation of pesticides rather than examining effects of pesticides on natural microorganism populations. Moreover, investigations of pesticidal effects on soil microbes are indeed far more common than investigations of those in aquatic environments.

Aquatic microorganisms include members of the protozoa, bacteria, plant kingdom and fungi. These aquatic microorganisms radically differ from each other but are similar only in their small size; most of them can be detected with the help of microscope, although colonies of many can be detected with the naked eye (Muturi *et al.,* 2017). Microorganisms are present in huge numbers everywhere and can survive even in unfavorable chemical and physical environmental conditions. Several aquatic microorganisms play basal roles in aquatic environments, trapping the light energy by the process of photosynthesis and moreover, they are considered to play an important role in liberating nutrients stored in organic tissue and also help in decomposition.

Bacteria

Most of the ancient and smallest organisms on the planet earth are bacteria; they are cosmopolitan in distribution i.e. present almost in every ecosystem and are far more abundant in all aquatic ecosystems. In streams and rivers, most of the bacteria penetrate in aquatic ecosystems from the surrounding land and their richness can increase adequately after a rainfall. The number of bacteria is measured in millions per milliliter (ml), and in the hundreds of millions per milliliter especially in polluted waters (Lydia, 2015).

If circumstances are favorable, bacteria divide rapidly by simple division to produce very large numbers of colonies in a very less time. Bacteria can be found emerged in the water, united with dead and decaying material (such as leaves or dead wood), or layered the surface of sand grains, stones and rocks as part of the biofilm (the lubricious layers on hard surfaces in rivers). They can make up a huge number of the living material in aquatic ecosystems (McArthur, 1992).

Bacteria exhibit the greatest magnitude in metabolic ability of any group of aquatic organisms. There are both heterotrophic and autotrophic bacteria. Autotrophic bacteria are primary producers in aquatic ecosystem as are true algae. So, for this reason, autotrophic bacteria (mostly cyanobacteria) are often referred as ‘algae’, although these aquatic microorganisms are by no means closely related to each other. Cyanobacteria used to be mistakenly known as ‘blue-green algae’. Actually ecologically, much of features what applies to algae is similar to autotrophic bacteria.

While as Heterotrophic bacteria act as decomposers and are considered a very important link in the decomposition of organic matter and the cycling of nutrients in aquatic environments.

Fungi

Fungi are present as a single cell or in filaments known as hyphae. Most aquatic fungi known as hyphomycetes are microscopic and are the most abundant in number and are very important (Lewin, 1974).

Fungi like heterotrophic bacteria are heterotrophic which attain their food by releasing exoenzymes into their existing environment, which helps to breakdown complex molecules into simpler and soluble substances that fungi can absorb. Fungi also act as decomposers as they are able to decompose plant matter in aquatic ecosystems. Moreover they are among the several microorganisms that can break down certain plant structural compounds such as lignin and cellulose.

Protozoa

Protozoa are single-celled microscopic organisms that often associate together to from colonies. There are both heterotrophic and autotrophic kinds of protozoa (Lee, 2001). Heterotrophic protozoa (such as the *Paramecium* and amoeba) feed on other organisms such as protists, bacteria or other algae. They are completely different from bacteria and fungi which absorb dissolved organic compounds from their surrounding environment. Protozoa together with other microorganisms make an association to form biofilm coated sediments and also make hard surfaces on the riverbeds although some protozoa are free-swimming microorganisms. Some of the protozoa are considered as parasites that cause diseases such as giardia (beaver fever).

Algae and Phytoplankton

Groups of several large autotrophic protists are known as algae. 'Microorganisms' term seems an unconventional term, used to describe microorganisms that do the process of photosynthesis e.g. the algae (cyanobacteria) are commonly included in it. The size of Algae varies from microscopic to large colonies that can be referred as macrophytes. Many types of algae like phytoplankton play a vital role in supplying the energy at the base of many aquatic food webs (Begon, 1990).

Phytoplankton is microscopically small plants that live emerged in the open water. Phytoplankton are generally more diverse and abundant in lakes than rivers and are completely absent from fast running streams, or where the rate at which they reproduce are lower than the rate at which the plants are washed downstream. Damming a river leads to lentic ecosystems which are more suitable for phytoplankton production and moreover nuisance algal blooms may also develop in reservoirs. Inputs of nutrients like phosphorus and nitrogen can also lead to algal blooms (Arora and Mehra, 2003).

Phytoplankton can live as colonies, in chains or as single cells. For many zooplankton and some fish, phytoplankton are direct food sources and are considered the base of the food webs in deep waters. The requirements of Phytoplankton vary for light, nutrients and other conditions. Aquatic ecosystems support a large complex mixture of phytoplankton's that can change with the change in environmental conditions (Bhatt *et al.,* 1999). The concentration of algal cells (number per unit volume) is generally highest when flows are lowest; in rivers containing significant amounts of phytoplankton while raised (particularly above the ground level) loads of suspended sediments during high water flows can lead to reduced photosynthesis and light. Some phytoplankton can lead to odour and taste problems in water, and also cause anoxic conditions that can kill fishes and other species. Some cyanobacteria are known to produce toxins that are lethal to various domestic species, fishes and wildlife.

Periphyton and Biofilm

Protozoa, bacteria, algae, fungi and the breakdown products of the cells that are dying form layers on submerged surfaces, including macrophytes, bottom sediments, submerged leaves and branches and rocks. Periphytonis a term that refers to a coat consisting mainly of algae, but the complete assemblage of layers is frequently called as *biofilm*. The rivers which possess a adequate light penetration like stony

Figure 1. An example of a food web

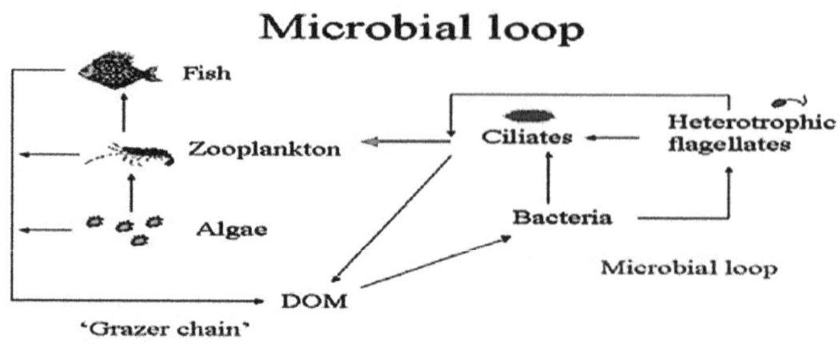

Table 1. Chronology of pesticide

Period	Example	Source	Characteristics
1800-1920s	Early organics, petroleum oils, chlorophenols, creosote, naphthalene, nitro-phenols.	By-products of coal gas production, Organic chemistry, etc.	Were toxic to user or un target organisms and lack specificity.
1945-1955	Chlorinated cyclodienes, HCCH, Chlorinated organics, DDT.	Organic synthesis	Long lasting, harmful ecological effects, good selectivity, no hazards to public health, resistance, good agricultural properties
1945-1970	Carbamates, organophosphorus compounds, Cholinesterase inhibitors.	Good use of structure-activity relationships, Organic synthesis	Some environmental problems, some user toxicity, lower persistence.
1970-1985	Biological pesticides, avermectins, juvenile hormone mimics, Synthetic pyrethroids.	New target systems, Refinement of structure activity relationships.	Costs and variable persistence, Some lack of selectivity, resistance.
1985- Till now	Genetically engineered organisms	Genetic alteration of plants to resist nontarget effects of pesticides, Transfer of genes for biological pesticides to other organisms and into beneficial plants and animals.	Disruption of microbiological ecology, Possible problems with mutations and escapes, monopoly on products.

Stephenson and Solomon, 1993.

and shallow rivers periphyton is an important and the main food source. Heterotrophic organisms, like some larger aquatic animals, such as fishes also feed on biofilm while as larger aquatic invertebrates such as insects and snails scrape the biofilm from surfaces. Biofilm can be essential in breaking down or absorbing chemical pollutants as well. Seasonal alterations in the abundance of periphyton reflect fluctuations in river discharge, as layers of algal cells build up in times of decreasing or low flowand wash away during flood periods (Saikia, 2011).

Pesticidescanbeclassifiedonthebasisofmechanismsofaction.Forexample,organophosphate,carbamate and organochlorineinsecticides that act primarily by disrupting and disfunctioning nervous system, while herbicides target mainly pathways of photosynthesis (Table 1). The mechanism of action of pesticide in aquatic microorganisms may not be the same as for the target organisms. In microorganisms, pesticides have been shown to interfere with biosynthetic reactions, photosynthesis and respirations well as molecular composition, division and cell growth.

AQUATIC TOXICOLOGY

Aquatic toxicology is the study of environmental contaminants and their effects on aquatic organisms such as pesticidal effect on the health of aquatic microorganism, fishes or other aquatic organisms. A pesticides ability to kill aquatic animals and fish is mainly a function of its (1) persistence in the environment (2) dose rate (3) exposure time and (4) toxicity.

How poisonous the pesticide is referred as Toxicity. Most of the pesticides are extremely far more toxic, whereas others are relatively not so toxic. Length of time, the animal is in contact with the pesticide is known as Exposure time. A longer exposure of pesticides may lead to harmful effects on aquatic fauna and flora while a brief exposure of pesticides (some chemicals) may have a very little effect on aquatic ecosystem including fishes (Christos and Spyridon, 2016).

The quantity of pesticide to which an animal is subjected through inhalation, orally and dermally is referred as dose rate. A little dose of high toxic chemicals can be more damaging than a high dose of low toxic chemicals. Dosage can be measured as the concentration of toxicant in the water or food supply (usually expressed as the weight of toxicant per unit (kilogram) or parts per million, ppm or parts per billion, ppb) or of body weight (expressed as mg pesticide/kg of body weight). A lethal dose (LD) is the amount of pesticide necessary to cause death or to kill organism. It is because of the fact that not all aquatic organisms or animals die at the same dose rate (most of them are more tolerant than others) so, for this a standard toxicity dose measurement called a lethal concentration 50 (LC50) is used. This is the concentration of a pesticide that kills 50% of a test population of organisms or animals within a set period of time usually ranging from 24 to 96 hours (Helfrich, 2009).

PERSISTENCE OF PESTICIDES

Some pesticides are so persistent that can remain in the ecosystem for many years and pass from one organism to another. Accumulation and Uptake of pesticides by aquatic organisms seem to be more likely a function of exchange equilibrium, habits, life cycle and habitat than of pharmacokinetics, size of organisms, food uptake and chemical and physical properties of the pesticide (Muirhead, 1971; Rosenberg, 1975).

The length of time a pesticide remains in the environment is known as persistence. Degradation of pesticide depends on environmental conditions and its chemical composition. They can be degraded by water or high air temperatures (thermal degradation), sunlight (photo decomposition), biological action (microbial decay), soil conditions (pH) and moisture conditions. Long lasting pesticides may be more available to aquatic animals and break down slowly (Deepa *et al.* 2006).

MECHANISM OF ACTION

From 500 BC up to 19th century, Pesticides came into existence and were defined as any mixture of substances used for destroying, preventing, mitigating or repelling any pest. According to US Environmental (2007) they included arsenic, sulphur, mercury and lead. Pesticides can be classified on the basis of their mechanisms of action. For example, insecticides like organophosphates,carbamates and organochlorinesact primarily by disrupting and disfunctioning nervous system, while herbicides frequently target pathways of photosynthesis (Ecobichon, 1991; DeLorenzo *et al.*, 2001).

Insecticides

Several insecticides disrupt the nervous system at different sites of target (figure1). The insecticides obstruct with the membrane transport of potassium, chlorine, sodium or calcium ions, inhibit activities of selective enzymes and may lead to the release and the persistence of chemical neurotransmitters at nerve endings (Ecobichon, 1991).

Organochlorines belong to one of the class of insecticide that affects the nervous system. They are practically stable unreactive compounds chemically and are characterized by their long lasting nature. Among them the most studied pesticide is DDT (Dicholoro di phenyl tri chloro ethane).According to US Environmental (2007) DDT was synthesized in 1874 and its insecticidal activity was discovered during the second half of World War II. It become the first synthetic organic pesticide and WHO (1979) reported in a report that it was used after the war for agricultural purpose. However, it was then alarmed to discontinue the use of organic synthetic pesticides, especially the toxic hydrocarbons that are chlorinated such as DDT and derivatives that have caused critical environmental pollution like water, air and soil, affecting non-target fauna (plants, animals and fish) and human health. This situation ultimately led to the ban of DDT in 2004 (UNEP, 2005). Younis *et al.* (2002) reported the recently DDT target protein in insects belonging to phylum arthropoda. The DDT target protein was found to be a subunit of the (ATPase) adenosine triphosphatase synthase, which is associated in the repolarization of neurons and thereby essential for normal functioning of nerves. It is noteworthy to mention here that DDT can also act through three more mechanisms at least (Matsumura, 1985). At the level of neuronal membrane, DDT reduces potassium transport across the membrane by altering the permeability to potassium ions. Moreover, during repolarization, DDT also disrupts the sodium channels, through which sodium ions can pass, thereby obstructing with the active transport of sodium out of the nerve axon. Finally, DDT also stops the activity of calmodulin, a calcium mediator in the nerve, to transport calcium ions that are important for the release of neurotransmitters. Other two main organochlorine insecticides are lindane and endrine, they are inhibitors of the γ -aminobutyric acid (GABA) receptors in both vertebrate central nervous system and insect in addition to the DDT-resembling effects (Wafford *et al.*, 1989). The GABA receptor functions as a receptor for the neurotransmitter GABA as it is an ion channel glycoprotein that traverses the cell membrane.

Organophosphorus insecticides, involving malathion and parathion replaced the organochlorines as DDT was withdrawn and banned from the market of many countries, in (Mulla and Mian, 1981). The groups of organophosphorus insecticides are also toxic to nervous system, because this group of insecticides acts on the nervous system by inhibiting the neurotransmitter enzyme acetyl cholinesterase (AChE). The reaction between the active site in the AChE protein and an organophosphorus insecticide leads in the formation of a intermediate, transient complex that hydrolyzes partially by leaving a stable, largely unreactive and phosphorylated inhibited enzyme that, under favorable conditions can be reactivated only at very slow rate (figure 2; Ecobichon, 1991). An irreversible inhibited enzyme is formed with many organophosphorus insecticides and the symptoms and signs of intoxication arepersistent and prolonged. Hence, most of the organophosphorus insecticides are considered to be more toxic only after metabolism by the cytochrome P-450 monooxygenase enzyme systems. This bioactivation process forms a metabolite that is a extremely strong inhibitor of the AChE than the parent compound (Belden and Lydy, 2000).

Figure 2. The reaction between organ phosphorus insecticides with the hydroxyl group in the active site of the enzyme acetyl cholinesterase (E-OH). The unstable, intermediate, complex formed before the release of the "leaving" group (ZH) is not shown. The dephosphorylation of the inhibited enzyme is the rate-limiting step to forming free enzyme
Ecobichon (1991).

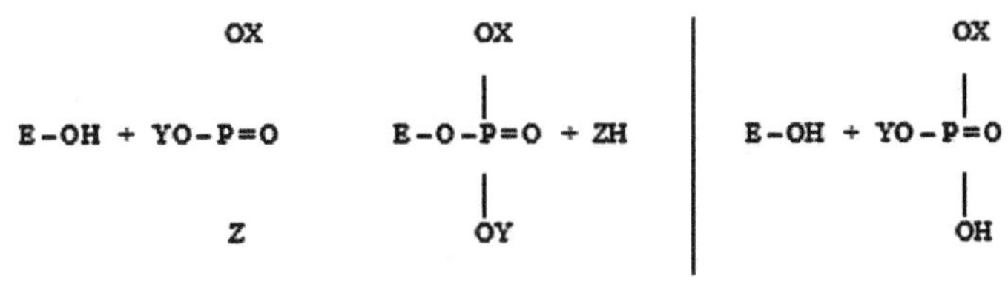

Carbamates are attach to the reactive site of AChE (acetylcholinesterase) enzyme and acts as an inhibitor (Ecobichon, 1991). Moreover, unlike organophosphorus insecicides, carbamates acts as a poor substrates for the cholinesterasetype enzymes, resulting in a reversible and short inhibition of AChE. Carbamates when attach to AChE (acetylcholinesterase) enzyme they undergo hydrolysis in two steps. The initial step is the formation of a carbamylated enzyme by removal of an alkyl group (e.g. $-CH_3$) or aryl (e.g. $-C_6H_5$). The second step is the the generation of once again a free and active enzyme by decarbamylation of the inhibited enzyme. The process is too much similiar as when organophosphorus pesticides bind to AChE enzyme, however the last stage occurs faster with the regeneration of a free enzyme.

Synthetic pyrethroids are categorized in the major classes of insecticides. The pyrethroids display two distant specific acidic portions, pyrethric or chrysanthemic acids resulting in type II syndrome and type I. In type I syndrome, peripheral nerves are also involved while type II syndrome implicate primarily an action in the central nervous system (Ecobichon, 1991). Both type II and I pyrethroid insecticides influence the sodium (Na^+) channels in the membranes of the nerves, leading to repetitive neuronal discharge, the effects quite similar to those produced by DDT. It results in the delay in the closing of the sodium activation gate by prolongation influx of sodium ions which results in prolonged and an increased sodium tail current (Narahashi, 1986; Bradbury and Coats, 1989; Ecobichon, 1991). Type II pyrethroids extended the sodium channel opentime more extremely than type I pyrethroids (Narahashi, 1986). Some other sites of action have been reported for the pyrethroid insecticides. Most of them are inhibitors of Mg^{2+}-ATPase and Ca^{2+}causing interferance with removal of calcium ions from the nerve endings, leading an increased release of neurotransmitter in the postsynaptic gap. Moreover, the protein calmodulin can be inhibited which is responsible for the intracellular binding of calcium ions to lower the spontaneous release of neurotransmitter. Type II Pyrethroids have also been found to bind to the GABA-receptor chloride channel complex leading blockage of chloride ion transport into the nerve cell. To conclude, there are many common characteristics between the mechanism of action of organochlorines and pyrethriod insecticides, subsequently there is a risk for synergistic or even additive effects.

Herbicides

Herbicides can be categorized in a number of ways. Herbicides are produced to injure or kill plants and therefore, affect number of mechanisms associated with e.g. nucleus division, photosynthesis, synthesis of proteins, carotenoids or lipids, growth and cellrespiration (Ecobichon, 1991).

Glyphosate, a herbicide which is the active component in the frequently used in preparation like Roundup™, and is essential for the control and management of weeds by acting as a inhibitor of a single plant enzyme, 5-enolpyruvoylshikimate 3-phosphate synthase (EPSPS) (Baylis, 2000; Sikorski and Gruys, 1997).5-enolpyruvoylshikimate 3-phosphate synthase is a key enzyme in the biosynthetic pathway of aromatic amino acid and inhibition of this enzyme causes severe effects on the synthesis of proteins. Moreover, the only known target enzyme of glyphosate is EPSPS or a herbicide that causes many physiological and physicochemical processes (Cole, 1985). Among the various processes, the degradation of chlorophyll, the reduction in photosynthesis as well as enhancement of auxin oxidation and inhibited transport of the plant growth hormone auxin are well known.

The most commonly used chlorophenoxy herbicides, like MCPA (4-chloro-*o*-toloxyacetic acid), 2,4,5-T (2,4,5-trichlorophenoxyacetic acid) and 2,4-D (2,4-dichlorophenoxyacetic acid) mimic the activity of auxin, a growth hormone in plants (Grossmann, 2000; Ecobichon, 1991).Elongation and growth by cell division is usually stimulated when herbicides are present in least concentrations at the cellular site of action. However, as the concentrations increases, different numbers of growth abnormalities are produced within 24 h of treatment (Grossmann, 2000). These involve the curvatures of stem and leaves, inhibition of growth of root and shoot and heavy intensified pigmentation of leaf. A large number of biochemical pathways are included in the growth and development of these abnormalities in the plants, like carbon assimilation inhibition, leading to stomata closure and overproduction of ethylene. Eventually, the exposure results in plant death, necrosis and desiccation.

Fungicides

As we know the breakdown of organic molecules by various metabolic processes forms energy for the survival of all living systems. In eukaryotes as well as in fungi, mitochondria is a site where a part of this catabolic process take place and results in the synthesis of the high energy intermediate known as ATP (Adenosine tri phosphate). Most of the groups of fungicides results in disturbance in energy supply in fungi and moreover, all such substances act as powerful inhibitors of germination of spores (Leroux, 1996). Among the powerful inhibitors, the R-S-CCl_3 compounds (e.g.dichlofluanid and captan) and dithiocarbamates (e.g. thiram and maneb) are some of the examples. These fungicides exhibit a multisite action by inhibiting many enzymes included in the respiratory processes.

Second group of fungicides, the phenylpyrroles, like iprodionel and fenpiclonil are known to exhibit spore germination inhibition and bring about various morphological alterations of germ tubes like inhibit elongation of germ-tube (Leroux *et al.*, 1992). Moreover, these fungicides are responsible for the uncoupling of oxidative phosphorylation and also leads to the inhibition of electron transport chain in respiration processes. Hence, more current studies reported that the main effect caused by fenpiclonil is to inhibit the biosynthesis of wall glycan and resulting in the accumulation of natural sugars. These effects show that the mechanism of action may be related to metabolism of glucose (Jespers and de Waard, 1995).The effect of the herbicide dichlobenil in plants is much similar to the effect of phenylpyrroles in fungi. However, the fungicide dichlobenil according to Delmer *et al.* (1987), may alter with a membrane-bound protein involved in the regulation of the synthesis of ß-glucan . Subsequently there is a risk for synergisticor even additive effects when the herbicide dichlobenil and phenylpyrroles fungicides are present at the same time in the environment.

Pesticide Toxicity to Aquatic Microorganisms

Microorganisms are most essential components or natives of aquatic environment, where they carry out various specific roles like nutrient cycling, decomposition and primary productivity. Microorganisms include three kingdoms, including > 50,000 different species of protozoa, fungi, algae and bacteria. They include a different range of size classes and morphologies, present in every environment and include a variety of growth rates and reproductive strategies, mobile and non mobile forms and multiple feeding types. Most of the available literatures on pesticide regarding aquatic microorganisms describe effects only on algae. Studies of herbicide like atrazine effects dominate. Therefore, very few pesticide studies exist for aquatic protozoa, bacteria and fungi. The mechanism of action of pesticide in aquatic microorganisms may not be similar as for the target organisms (Table 2). In aquatic microorganisms, pesticides have been reported to interfere with biosynthetic reactions, photosynthesis and respirations well as molecular composition, division and cell growth (DeLorenzo *et al.*, 2001).

Insecticides

Very few studies have been studied so far that showed the direct effects of insecticides causing toxicity to aquatic microorganisms. However, Peterson *et al.* (1994) studied the toxicity of the two carbamate insecticides like carbofuran and carbaryl to cyanobacteria and algae in EEC (expected environmental concentrations). The expected environmental concentrations (EEC) are an estimation of concentration for an aquatic environment in a lotic system which is based on the input of the maximum proposed application rate. Peterson *et al.* (1994) reported that the inhibition of growth, ranged from 35% to 86% by carbaryl exposure (3.7 mg/L) in *Anabaena inaequalis* and cyanobacteria *Pseudoanabaena* (being the most sensitive species tested). Actually, more than 50% inhibition was recorded in nine of the ten aquatic species involved in the study carried out by Peterson *et al.* (1994). Moreover, In contrast, carbofuran when applied at the EEC of 0.67 mg/L had relatively very low toxicity to most species tested. On the other side, the organophosphorus insecticides like pyridaphenthion and fenitrothion results in growth inhibition in cyanobacteria and algae (Sabater and Carrasco, 2001). The EC_{50}at 96 h for pyridaphenthion and fenitrothion for the green algae *Scenedesmus subspicatus, Scenedesmus acutus* and the *Pseudanabaena galeata* (cyanobacteria)ranged from 0.84 - 5.5 mg/L and for the green algae *Chlorella saccharophila* and *Chlorellavulgaris*it ranged from5.7 to 30.9mg/L. The most sensitive species for both insecticides was *Scenedesmusacutus* as its EC_{10}at96 h varied from 0.14to 0.32 mg/L, which is lower than the concentrations observed in natural waters (Sabater and Carrasco, 2001).

DeLorenzo *et al.* (1999) studied the effects of the agricultural insecticides like chlorpyrifos and endosulfan on an estuarine microbial food web in contrast to single species tests performed before. It was observed that Endosulfan was primarilyreported to target the phototropic portion of the community of bacteria including cyanobacteria. Endosulfan treatments of 1 and 10 µg/L resulted in reduction in total bacterial abundance, but not heterotrophic bacterial productivity. Cyanobacterias like *Oscillatoria*and *Melosira* were those cyanobacterias that were completely eliminated from the highendosulfan treatment. In the same study, chlorpyrifos were found affecting most of the phototrophic endpoints tested, e.g. in 1 and 10 µg/L of chlorpyrifos there was reduction in both chlorophyll *a*, and the phototrophic biovolume. In 10 µg/L treatment many diatom genera like *Tabellaria, Bacillaria, Asterionella*and *Gomphonema*) present in the controls were absent and moreover, heterotrophic flagellates and ciliates were reduced. In contrast, when chlorpyrifos was added, the bacterial abundance increased significantly as compared to the control. Hence, mainly in the high dose (10 μg/L), the alterations in the chlorpyrifos treatments occurred. At environmental concentrations (<0.02 μg/L) of chlorpyrifos (Miles and Pfeuffer, 1997), the aquatic micro organisms would automatically recover before any unfavorable effects are passed on to next higher trophic levels (DeLorenzo *et al.*, 2001). However, it was studied by Widenfalk *et al.* (2004) that communities of microbes of natural sediments can be even altered by very low concentrations of pesticides e.g. insecticides like pirimicarb and deltamethrin. Widenfalk *et al.* (2004) carried out microcosm experiments to prove that pirimicarb and deltamethrin inhibited the activity of bacteria while deltamethrin also affected the microbial biomass negatively.

Herbicides

Peterson *et al.* (1994) studied the inhibition of growth of diatoms (*Cyclotella meneghiana and Nitzschia* sp.),cyanobacteria (*Aphanizomenon flos-aquae, Oscillatoria* sp.,*Anabaenainaequalis, Pseudoanabaena* sp.and*Microcystis aeruginosa*) and green algae (*Scenedesmus capricornutumand Selenastrumquadricauda*) by giving treatment of 23 different pesticidesin EECs. Among the five triazine herbicides used for treatment (simazine, metribuzin, cyanazine, hexazinoneand atrazine (ranging from 2.7-2.9 mg/L) and diquat (0.7 mg/L) lead to more than 50% growth inhibition in all species tested. Studies conducted by Peterson *et al.* (1997) later on proved that 5% of the EEC of hexazinone (0.14 mg/L) caused inhibition of growth more than 80% in both green algae and diatoms. In addition to that 50% inhibition of growth of cyanobacteria and diatoms were caused by 0.7-14% of the EEC of diquat at 0.005-0.1 mg/L (Peterson *et al.*, 1997). The earlier reports by Peterson *et al.* (1994) also showed that glyphosate at a concentration of 2.8 mg/l was extremely toxic to nitrogen-fixing cyanobacterium *Anabaena flos-aquae and* diatoms at EECs, but relatively non-toxic to the other aquatic algal species. On the other side, Peterson *et al.* (1994) found thatat EECs ranging from 0.003-0.020 mg/L, four sulfonylurea herbicides (e.g. chlorsulfuron) caused no or very less growth inhibition of any of the algal species. Moreover, each sulfonylurea herbicides lead growth stimulation in either a cyanobacteria or diatom. The 2,4-D a phenoxyalkane herbicide at the concentration of 2.9 mg/L did not cause any inhibition of growth, whereas another phenoxyalkane MCPA at the concentration of 1.4 mg/L also caused growth inhibition in one species of the cyanobacteria, diatoms and green algae respectively (Peterson *et al.*, 1994).

Unlike Peterson *et al.* (1994), Nystrom *et al.* (1999) observed that two sulfonylurea herbicides like metsulfuronmethyl, and chlorsulfuron caused severe inhibition of growth in the *Synechococcus leopoliensis* a cyanobacteria at concentrations less than the EEC.

Moreover, the EC_{50}-values of two marine algae, *Nodularia harveyana* and *Amphidinium cararae* and cyano-bacteria *Anabaena flos-aquae* had equal or below to the EECvalue formetsulfuronmethyl (Nystrom *et al.*, 1999). Hence, the differences in species dependent sensitivity to sulfonylurea herbicides among taxa of algal seem to be extremely very high. Out of 20 marine and the 20 freshwater species recorded by Nystrom and coworkers (1999), the highest and the lowest EC_{50}value also varied. Dinoflagellates and Cyanobacteria appear to be particularly sensitive to the sulfonylurea herbicides, according to their study.

Even a large number of investigations have reported multiple effects of herbicide on single species of bacteria or algae (DeLorenzo *et al.*, 2001), very less investigations have concentrated on the interaction between natural sediment of microbial communities (like protozoans, fungi, bacteria and benthic algae) and pesticides. Nevertheless, DeLorenzo *et al.* (1999) also reported that the atrazine herbicide exhibited structural and functional changes of aquatic microorganism communities. The concentrations of Atrazine ranging from 40 to 160 μg/L, was reported to reduce simulating tidal creek environment, phototrophic biovolume in test chambers, chlorophyll *a*, and phototrophic carbon assimilation. In addition, Widenfalk *et al.* (2004) also reported the exposure of phenylurea herbicide isoproturon concentrations on microbial communities of natural sediment, that are consider to be environmentally safe. It was also recorded that isoproturon acts as a inhibitor and inhibited the activity of bacteria's and the microbial biomass was negatively affected. In spite of that, the reports did not show a inverse, consistent relationship with concentration of pesticides, showing the difficulty with proposing the low pesticide doses effects to microorganisms.

Fungicides

According to DeLorenzo *et al.* (2001) there is very insignificant information available about the toxicity caused by agricultural fungicide to aquatic microorganisms. There is even scarcity of literature about toxicity caused by fungicide in less concentration. Peterson *et al.* (1994) performed a single species test compared the sensitivity of 10 species of algae to the propiconazole, a triazole derivative fungicide. Inhibition of ^{14}C uptake was less than 20% to all species of algae at the EEC of 0.08 mg/L. Fungicide propicanazole, actually enhanced growth in the cyanobacteria to some extent. Moreover, The commercial formulation of propiconazole enhanced the growth of diatoms.

Fungicide fenpropimorph, reduced denitrification in a lake sediments at very low concentration I.e. 10 µg/kg in whole microbial community tests (Svensson and Leonardson, 1992).Hence, fungicide known as fenpropimorph has lipophilic nature it will obviously adsorb to organic sediment particles and create a persistent threat to microorganisms. Long lasting exposure also lead in an increased effect on the denitrifying bacteria (Svensson and Leonardson, 1992).Furthermore, Widenfalk *et al.* (2004) recorded that the bacterial activity in natural sediment was affected with the environmentally safe concentrations of the phtalimide fungicide captan. Unlike denitrification, respiration antimicrobial biomasses were not affected by phtalimide fungicide captan. Therefore, the negative effect on the activity of bacteria, studied at environmental safe concentrations, vanished at a higher test concentration which points out a non-linear relationship of the toxicity.

Table 2. Mechanisms of action of pesticides on target organisms

Pesticide Class	Groups Included	General Toxic Effect	Specific Site of Action
Organophosphates	Carbamates	Nervous system inhibition	Acetylcholinesterase
Organochlorines	Cyclodienes	Nervous system inhibition	GABA receptor
Herbicides	Ureas, cyclic ureas, triazines, acylanilides, phenylcarbamates, triazinones	Photosynthesis inhibition	Hill reaction of electron transport
	Bipyridiniums	Photosynthesis inhibition (lightreaction)	Reducing side of photosystem I
	Pyridazinones	Biosynthesis inhibition	Carotene accumulation
	Chloroacetamide	Biosynthesis inhibition	Fatty acid synthesis
	Dinitroanilines, phosphoric amides,chlorthaldimethyl, propyzamide,cholchicine, terbutol	Biosynthesis inhibition	Microtubule formation
Broad-spectrum biocides	Chlorophenols	Multiple inhibiting actions	Phophosphorylation, protein synthesis, lipid biosynthesis
	Tributyl tins, trialkyl tins	Respiratory system inhibition	Mitochonrial ATPase

ECOLOGICAL RELEVANCE OF PESTICIDE TOXICITY TO MICROORGANISMS

It is clear that there are significant differences in pesticide sensitivity among the microorganisms and that the uncertainty factor they use is necessary to provide an satisfactory, margin of safety in evaluating the hazard presented by these chemicals to aquatic ecosystems.

Using the environmental monitoring data of pesticide concentrations and microbial toxicity data, it can be predicted that which pesticides are most likely to cause an effect in aquatic organisms. Some of the pesticides which are described here are no longer used but are still usually traced in surface waters (e.g., DDT). These chemicals may inflict harm to microorganisms when they are resuspended in estuarine and marine sediments due to tidal action or dredging.

The lipophilic texture of various pesticides may exhibit a danger to higher estuarine organisms. For example, the pesticide endosulfan was not toxic to algae at levels likely to be present in the environment (1 mg/L), the compound may get assembled in large amount in algae and can be consumed by grazers in higher concentrations. Rao and Lal, (1987) reported active uptake of the agricultural pesticide used in general endosulfan in the cyanobacteria *Aulosira* and *Anabaena,* with concentrations reaching 700 times the exposed dose within 48 hours. The metabolite endosulfan sulfate gets bioaccumulated more and is more long lasting than the parent isomers (Callahan *et al.,* 1979). The accumulation of pesticide depends not only on adsorptive and transfer processes but also on the quantity continued to be released into the aquatic environment and conversion of the chemical (Heckman, 1994).Therefore there is, more potential for chronic effects, such as evolved sensitivity, bioaccumulation, and impacts on higher trophic levels. This is extremely true in estuaries, where contaminated sediments are constantly resuspended due to tidal action and dredging. Pesticides that get accumulated in prey species may then cause toxicity in the organisms that consume them or continue to bioaccumulation through the food web. Use of persistent pesticides in the environment should be constantly and carefully monitored.

Atrazine which is one of the most commonly used herbicides is constantly monitored in aquatic environments due to runoff from agricultural fields (Thurman *et al.,* 1992; Pereira and Hostettler, 1993). Out of the total pesticide volume applied to crops per year, atrazine accounts for 60% of the total pesticide with an estimate of 29 million kg of atrazine active ingredient applied in 1989 alone (Gianessi and Puffer, 1991).

Solomon *et al.* (1996) conducted an ecological risk assessment of atrazine and concluded that although atrazine thoroughly strengthens the populations of phytoplanktons at concentrations typically present in the environment, the ecological system usually restores rapidly from such exposures. Ecological effects were considered to occur only at concentrations of 50 mg/L or greater. It was moreover recommended that when atrazine is heavily used, site-specific risk assessments should be conducted for aquatic systems. The EC50s summarized for marine phytoplankton ranged from 20 to 600 mg/L. In a literature reviewed about the atrazine toxicity in aquatic ecosystems, Huber, (1993) revealed that 20 mg/L to be the no-observed-effect concentration. Several studies have shown that growth of population of phytoplankton can be stopped at concentrations as low as 0.1 mg/L (Bester *et al.,* 1995; Lampert *et al.*, 1989)

Atrazine contamination was found quite pervasive by monitoring estuarine for atrazine concentration along (Pennington, 1996). More than 97% of samples collected had a significant level of atrazine (February–October 1993). Furthermore, 52% of samples collected had atrazine levels exceeding 1 mg/ L, and over 5% of the samples collected along the mid-Texas coast had atrazine concentrations exceeding the 20-mg/L no-observed-effect-concentration value identified by Solomon *et al.*(1966) and Huber (1993). It was summarized that phytoplankton population, which is chronically exposed to atrazine along the mid-Texas coast, may not be able to recover because of the widespread nature of the contaminant (Pennington, 1996). Besides atrazine, eight other herbicides (hexazinone cyanazine, tebuthiuron, metribuzin, diquat, simazine,, carbaryl, and acrolein) were observed to be highly phytotoxic to algae when tested at the EECs (Peterson *et al.,* 1994). Based on the literature available, herbicides have maximum potential to cause serious effects on aquatic microorganisms at environment related concentrations.

The toxicity to microorganisms by pesticides may result in reduction of food resource for organisms occurring at top trophic levels. For example, 75% of a dominant zooplankton species in ponds was reduced when treated with 500 mg/L atrazine (Denoyelles *et al.,* 1982). Compositional changes in the phytoplanktonic population and decrease in photo-synthetic biomass and productivity have resulted in long-term reductions in the food resources of certain grazers (Ahlgren *et al.*, 1990). These alterations also affect the quality and quantity of nutrients cycled through the microbial food web and is at last received by top trophic levels (Goldman *et al.,* 1987).

Furthermore, microorganisms may be affected indirectly by pesticide effects existing at top trophic levels. For example, populations of phytoplankton enhanced after fenvalerate (a synthetic pyrethroid insecticide) exposure lead a significant reduction in the population of large-bodied cladocerans (Day, 1987). Likewise sub lethal concentrations of endosulfan suppressed filtration and ingestion rates of *Daphnia magna* fed the single celled alga *Nannochlorisoculata.* Filtration and ingestion rates were decreased to 50% of the control values at 0.44 and 0.61 mg/L endosulfan, respectively (Fernandez *et al.,* 1994). Large decline in rates of filtration and ingestion were also investigated with the rotifer *Brachionus ca-lyciflorus* (Fernandez *et al.,* 1992).

Lew *et al.* (2013) studied Impact of Pesticide Contamination on Aquatic Microorganism Populations in the Littoral Zone and according to their study, the quantity dynamics of aquatic microorganisms indicated that bacteria and fungi under the influence of long-term exposure to DDT can adapt to the presence of this pesticide in water. No modifying effect of DDT was observed on the quantity of microorganisms or the pattern of seasonal relationships in the eutrophic lake. Changes were shown in the percentage share of large groups of bacteria in the community of microorganisms as was an effect of contamination on the species diversity of fungi. Zachery *et al.* (2015) studied a Synthesis of the Effects of Pesticides on Microbial Persistence in Aquatic Ecosystems and found that Insecticides and fungicides also had deleterious direct effects in the majority of studies examining protozoa species, although herbicides were found to have inconsistent direct effects on protozoans. Our synthesis revealed mixed or no direct effects on bacterial species among all pesticide categories, with results highly dependent on the target species, chemical, and concentration used in the study. Examination of community-level, indirect effects revealed that all pesticide categories had a tendency to reduce higher trophic levels, thereby diminishing top-down pressures and favoring lower trophic levels. Often, indirect effects exerted greater influence than direct effects.

FUTURE SCOPE

Many questions still remain there regarding various pesticides like what levels of pesticides will be more toxic to aquatic microorganisms. Furthermore, there is scarcity in the pesticidal data for various aquatic microorganisms like protozoa. The pesticides causing toxicity essential aquatic microorganisms have been generally unnoticed and ignored. A lot of literature is available on bacterial studies. However, most of the studies deal with pesticidal degradation by soil bacteria rather than pesticidal toxicity to aquatic bacteria. Change of bacterial productivity and/or growth by pesticides has the capability to affect other microorganism population as well. Future research should be based on assessing the toxicity caused by pesticides to the microbial food web as a whole. Another point should be kept in mind regarding the influence of water quality on the toxicity caused by various pesticides to aquatic microorganisms.

CONCLUSION

Pesticides are usually considered a rapid, convenient solution for controlling and insect pests and weeds; however, their use comes at a cost that cause considerable damage to ecosystems. Pesticides may damage or harm no targeted species and diminish biodiversity along with disturbance to food chains and food webs that creates imbalance to ecosystems. Therefore, it is compulsory to decrease the after effects of pesticides through BMP practices and an overall reduction in reliance on chemical pest control.

REFERENCES

Ahlgren, G., Lundstedt, L., Brett, M., & Forsberg, C. (1990). Lipid com-position and food quality of some freshwater phytoplankton for cladoceran zooplankters. *J Phytoplankton Res*, *12*(4), 809–818. doi:10.1093/plankt/12.4.809

Arora, J., & Mehra, N. K. (2003). Species diversity of planktonic and epiphytic rotifers in the backwaters of the Delhi segment of the Yamuna River, with remarks on new records from India. *Zool. Stud. Taipei*, *42*, 239–247.

Baylis, A. D. (2000). Why glyphosate is a global herbicide: Strengths, weaknesses and prospects. *Pest Management Science*, *56*(4), 299–308. doi:10.1002/(SICI)1526-4998(200004)56:4<299::AID-PS144>3.0.CO;2-K

Begon, M., Harper, J. L., & Townsend, C. R. (1990). *Ecology: Individuals, Populations and Communities*. Oxford, UK: Blackwell Scientific Publications.

Belden, J. B., & Lydy, M. J. (2000). Impact of atrazine on organophosphate insecticide toxicity. *Environmental Toxicology and Chemistry*, *19*(9), 2266–2274. doi:10.1002/etc.5620190917

Bester, K., Hunerfuss, H., Brockmann, U., & Rick, H. J. (1995). Biological effects of triazine herbicide contamination on marine phyto-plankton. *Archives of Environmental Contamination and Toxicology*, *29*, 277–283. doi:10.1007/BF00212490

Bhatt, L. R., Lacoul, P., Lekhak, H. D., & Jha, P. K. (1999). Physico-chemical characteristics and phytoplanktons of Taudaha lake, Kathmandu. *Pollution Research*, *18*, 353–358.

Bradbury, S. P., & Coats, J. R. (1989). Comparative toxicology of the pyrethroid insecticides. *Reviews of Environmental Contamination and Toxicology*, *108*, 133–177. doi:10.1007/978-1-4613-8850-0_4 PMID:2646661

Callahan, M. A., Slimak, M. W., Gabel, N. W., May, I. P., Fowler, C. F., & Feed, J. R. (1979). Water-related environmental fate of 129 priority pol-lutants, Vol I. EPA 440/4-79-029B. U.S. Environmental Protection Agency.

Christos, A. D., & Spyridon, D. K. (2016). Farmers' Exposure to Pesticides: Toxicity Types and Ways of Prevention. *Toxics*, *4*(1), 1. doi:10.3390/toxics4010001 PMID:29051407

Cole, D. J. (1985). Mode of action of glyphosate - a literature analysis. In E. Grossbard & D. Atkinson (Eds.), *The herbicide glyphosate* (pp. 48–74). London: Butterworths.

Day, K. E., Kaushik, N. K., & Solomon, K. R. (1987). Impact of fenvalerate on enclosed freshwater planktonic communities and on in-situ rates of filtration of zooplankton. *Canadian Journal of Fisheries and Aquatic Sciences*, *44*(10), 1714–1728. doi:10.1139/f87-210

Deepa, T. V. G., Lakshmi, P. S., & Lakhsmi, S. K. (2006). *Effects of pesticides.* Retrieved from www.Interhopen.com

DeLorenzo, M. E., Lauth, J., Pennington, P. L., Scott, G. I., & Ross, P. E. (1999b). Atrazine effects on the microbial food web in tidal creek mesocosms. *Aquatic Toxicology (Amsterdam, Netherlands)*, *46*(3-4), 241–251. doi:10.1016/S0166-445X(98)00132-5

DeLorenzo, M. E., Scott, G. I., & Ross, P. E. (1999). Effects of the agricultural pesticides atrazine, deethylatrazine, endosulfan, and chlorpyrifos on an estuarine microbial food web. *Environmental Toxicology and Chemistry*, *18*(12), 2824–2835. doi:10.1002/etc.5620181224

DeLorenzo, M. E., Scott, G. I., & Ross, P. E. (2001). Toxicity of pesticides to aquatic microorganisms: A review. *Environmental Toxicology and Chemistry*, *20*(1), 84–98. doi:10.1002/etc.5620200108 PMID:11351418

DeNoyelles, F., Kettle, W. D., & Sinn, D. E. (1982). The response of plank-ton communities in experimental ponds to atrazine, the most heavily used pesticide in the United States. *Ecology*, *63*(5), 1285–1293. doi:10.2307/1938856

Ecobichon, D. J. (1991). Toxic effects of pesticides. In M. O. Amdur & ... (Eds.), *Casarett and Doull's Toxicology- The basic science of poisons* (4th ed.; pp. 565–622). New York: McGraw-Hill, Inc.

US Environmental. (2007). *What is a pesticide?* Epa.gov.

Fernandez-Casalderrey, A., Ferrando, M. D., & Andreu-Moliner, E. (1992). Filtration and ingestion rates of *Brachionus calyciflorus* after exposure to endosulfan and diazinon. *Comparative Biochemistry and Physiology*, *C*, 103, 357–361.

Fernandez-Casalderrey, A., Ferrando, M. D., & Andreu-Moliner, E. (1994). Effect of sublethal concentrations of pesticides on the feeding behavior of *Daphnia magna. Ecotoxicology and Environmental Safety*, *27*(1), 82–89. doi:10.1006/eesa.1994.1008 PMID:7525207

Gianessi, L. P., & Puffer, C. (1991). *Herbicide Use in the United States.* Washington, DC: Resources for the Future.

Goldman, J. G., Caron, D. A., & Dennett, M. R. (1987). Nutrient cycling in a microflagellate food chain. IV. Phytoplankton-microflagel-late interactions. *Marine Ecology Progress Series*, *38*, 75–87. doi:10.3354/meps038075

Grossmann, K. (2000). Mode of action of auxin herbicides: A new ending to a long, drawn out story. *Trends in Plant Science*, *5*(12), 506–508. doi:10.1016/S1360-1385(00)01791-X PMID:11120460

Heckman, C. W. (1994). Pesticide chemistry and toxicity to algae. In L. C. Rai, J. P. Gaur, & C. J. Soeder (Eds.), *Algae and Water Pollution* (pp. 205–234). Stuttgart, Germany: E. Schweitzerbart.

Helfrich, L.A. (2009). *Pesticides and Aquatic animals: A guide to reducing impacts on aquatic system.* Virginia, Dept of game and inland fisheries.

Huber, W. (1993). Ecotoxicological relevance of atrazine in aquatic systems. *Environmental Toxicology and Chemistry*, *12*(10), 1865–1881. doi:10.1002/etc.5620121014

Jespers, A. B. K., & De Waard, M. A. (1995). Effect of fenpiclonil on phosphorylation of glucose in *Fusarium sulphureum. Pesticide Science*, *44*(2), 167–175. doi:10.1002/ps.2780440210

Lampert, W., Fleckner, W., Pott, E., Schober, U., & Storkel, K. U. (1989). Herbicide effects on planktonic systems of different complexity. *Hydrobiologia*, *188/189*(1), 415–424. doi:10.1007/BF00027809

Lee, J. J., Seymour, H. H., & Eugene, C. B. (2001). *An Illustrated Guide to the Protozoa* (2nd ed.). Lawrence, KS: Society of Protozoologists.

Leroux, P. (1996). Recent developments in the mode of action of fungicides. *Pesticide Science*, *47*(2), 191–197. doi:10.1002/(SICI)1096-9063(199606)47:2<191::AID-PS415>3.0.CO;2-I

Leroux, P., Lanen, C., & Fritz, R. (1992). Similarities in the antifungal activities of fenpiclonil, iprodione and tolclofos-methyl against *Botrytis cinerea* and *Fusarium nivale. Pesticide Science*, *36*(3), 255–261. doi:10.1002/ps.2780360312

Lew, S., Lew, M., Biedunkiewicz, A., & Szarek, J. (2013). Impact of Pesticide Contamination on Aquatic Microorganism Populations in the Littoral Zone. *Archives of Environmental Contamination and Toxicology*, *64*(3), 399–409. doi:10.100700244-012-9852-6 PMID:23229198

Lewin, R. A. (1974). Biochemical taxonomy. *Blackwell Scientific Publications.*

Lydia, H. Z. (2015). Stream microbial diversity in response to environmental changes: Review and synthesis of existing research. *Frontiers in Microbiology*, *6*, 454. PMID:26042102

Matsumura, F. (1985). *Toxicology of insecticides* (2nd ed.). New York: Plenum Press. doi:10.1007/978-1-4613-2491-1

McArthur, J. V., Leff, L. G., & Smith, M. H. (1992). Genetic diversity of bacteria along a stream continuum. *Journal of the North American Benthological Society*, *11*(3), 269–277. doi:10.2307/1467647

Miles, C. J., & Pfeuffer, R. J. (1997). Pesticides in canals of South Florida. *Archives of Environmental Contamination and Toxicology*, *32*(4), 337–345. doi:10.1007002449900194 PMID:9175497

Muirhead, T. R. C. (1971). Pesticides and freshwater fauna. Academic Press.

Mulla, M. S., & Mian, L. S. (1981). Biological and environmental impacts of the insecticides malathion and parathion on nontarget biota in aquatic ecosystems. *Residue Reviews*, *78*, 100–135. PMID:7017857

Muturi, E. J., Donthu, R. K., Fields, C. J., Moise, I. K., & Kim, C. H. (2017). Effect of pesticides on microbial communities in container aquatic habitats. *Scientific Reports*, *16*(7), 44565. doi:10.1038rep44565 PMID:28300212

Narahashi, T. (1986). Nerve membrane ionic channels as the target of toxicants. *Archives of Toxicology. Supplement. Archiv fur Toxikologie. Supplement*, *9*, 3–13. doi:10.1007/978-3-642-71248-7_1 PMID:2434060

Nystrom, B., Bjornsater, B., & Blanck, H. (1999). Effects of sulfonylurea herbicides on non-target aquatic micro- organisms: Growth inhibition of micro-algae and short-term inhibition of adenine and thymidine incorporation in periphyton communities. *Aquatic Toxicology (Amsterdam, Netherlands)*, *47*(1), 9–22. doi:10.1016/S0166-445X(99)00007-7

Pennington, P. L. (1996). *The toxicity of the herbicides atrazine and alachlor on the estuarine phytoplankter* Pavlova sp. *(*Prymne-siophyceae*) with an emphasis on acute toxicity testing of in-dividual herbicides, herbicide mixtures and multi-generational chronic bioassays* (MS thesis). University of Charleston, Charleston, SC.

Pereira, W. E., & Hostettler, F. D. (1993). Nonpoint source contamina-tion of the Mississippi River and its tributaries by herbicides. *Environmental Science & Technology*, *27*(8), 1542–1552. doi:10.1021/es00045a008

Peterson, H. G., Boutin, C., Freemark, K. E., & Martin, P. A. (1997). Toxicity of hexazinone and diquat to green algae, diatoms, cyanobacteria and duckweed. *Aquatic Toxicology (Amsterdam, Netherlands)*, *39*(2), 111–134. doi:10.1016/S0166-445X(97)00022-2

Peterson, H. G., Boutin, C., Martin, P. A., Freemark, K. E., Ruecker, N. J., & Moody, M. J. (1994). Aquatic phyto-toxicity of 23 pesticides applied at expected environmental concentrations. *Aquatic Toxicology (Amsterdam, Netherlands)*, *28*(3-4), 275–292. doi:10.1016/0166-445X(94)90038-8

Rao, V. V. S. N., & Lal, R. (1987). Uptake and metabolism of insecticides by blue-green algae *Anabaena* and *Aulosira fertilissima. Mi-crobios Lett*, *36*, 143–147.

Rosenberg, D. M. (1975). Food chains concentrations of chlorinated hydrocarbons pesticides in invertebrate communities: A reevaluation. *Quaestiones Entomologicae*, *11*, 97–110.

Sabater, C., & Carrasco, J. M. (2001a). Effects of pyridaphenthion on growth of five freshwater species of phytoplankton. A laboratory study. *Chemosphere*, *44*(8), 1775–1781. doi:10.1016/S0045-6535(00)00575-0 PMID:11534908

Sabater, C., & Carrasco, J. M. (2001b). Effects of the organophosphorus insecticide fenitrothion on growth in five freshwater species of phytoplankton. *Environmental Toxicology*, *16*(4), 314–320. doi:10.1002/tox.1038 PMID:11501280

Saikia, S. K. (2011). Review on Periphyton as Mediator of nutrient transfer in aquatic ecosystems. *Ecologia Balkanica*, *3*, 65–78.

Sikorski, J. A., & Gruys, K. J. (1997). Understanding Glyphosate's molecular mode of action with EPSP synthase: Evidence favoring an allosteric inhibitor model. *Accounts of Chemical Research*, *30*(1), 2–8. doi:10.1021/ar950122

Solomon, K. R., Baker, D. B., Richards, R. P., Dixon, K. R., Klaine, S. J., La Point, T. W., ... Williams, W. M. (1996). Ecological risk assessment of atrazine in North American surface waters. *Environmental Toxicology and Chemistry*, *15*(1), 31–76. doi:10.1002/etc.5620150105 PMID:23147529

Stephenson, G. A., & Solomon, K. R. (1993). Pesticides and the environment. Department of Environmental Biology, University of Guelph.

Thurman, E. M., Goolsby, D. A., Meyer, M. T., & Kolpin, D. W. (1992). Her-bicides in surface waters of the midwestern United States: The effect of spring flush. *Environmental Science & Technology*, *25*(10), 1794–1796. doi:10.1021/es00022a018

Thurman, E. M., Goolsby, D. A., Meyer, M. T., Mills, M. S., Pomes, M. L., & Kolpin, D. W. (1992). A reconnaissance study of herbicides and their metabolites in surface water of the midwestern United States using immunoassay and gas chromatography/mass spec-trometry. *Environmental Science & Technology*, *26*(12), 2440–2447. doi:10.1021/es00036a016

UNEP. (2005). Ridding the World of POPs. In A guide to the Stockholm Convention on Persistent Organic Pollutants. UNEP.

Wafford, K. A., Sattelle, D. B., Gant, D. B., Eldefrawi, A. T., & Eldefrawi, M. E. (1989). Noncompetitive inhibition of GABA receptors in insect and vertebrate CNS by endrin and lindane. *Pesticide Biochemistry and Physiology*, *33*(3), 213–219. doi:10.1016/0048-3575(89)90119-3

WHO. (1979). *Environmental Health Criteria: DDT and its derivatives*. WHO.

Widenfalk, A., Svensson, J. M., & Goedkoop, W. (2004). Effects of the pesticides captan, deltamethrin, isoproturon and pirimicarb on the microbial community of a freshwater sediment. *Environmental Toxicology and Chemistry*, *23*(8), 1920–1927. doi:10.1897/03-345 PMID:15352481

Younis, H. M., Abo-El-Saad, M. M., Abdel-Razik, R. K., & Abo-Seda, S. A. (2002). Resolving the DDT target protein in insects as a subunit of the ATP synthase. *Biotechnology and Applied Biochemistry*, *35*, 9–17. PMID:11834125

Zachery, R. S., Valerie, J. H., & Jason, R. R. (2015). A Synthesis of the Effects of Pesticides on Microbial Persistence in Aquatic Ecosystems. *Critical Reviews in Toxicology*, *45*(10), 813–836. doi:10.3109/10408444.2015.1065471 PMID:26565685

ADDITIONAL READING

Lozano, R. B., & Prant, J. R. (1993). Interaction of toxicants and communities: The role of nutrients. *Environmental Toxicology and Chemistry*, *13*(3), 361–368. doi:10.1002/etc.5620130302

Melendez, A. L., Kepner, R. L. Jr, Balczon, J. M., & Pratt, J. R. (1993). Effects of diquat on freshwater microbial communities. *Archives of Environmental Contamination and Toxicology*, *25*(1), 95–101. doi:10.1007/BF00230718

Rajendran, R., Rajendran, N., & Venugopalan, V. K. (1990). Effect of organochlorine pesticides on the bacterial population of a tropical estuary. *Microbios Lett*, 44, 57-63. https://www.cabdirect.org/cabdirect/abstract/19920505200

Vanden Brink, P. J., Van Dank, E., Gylstra, R., Crum, S. J. H., & Brock, T. C. M. (1995). Effects of chronic low concentartions of the pesticides chlorpyrifos and atrazine in indoor freshwater microcosms. *Chemosphere*, *31*(5), 3181–3200. doi:10.1016/0045-6535(95)00180-G

KEY TERMS AND DEFINITIONS

EC_{50}: Half maximal effective concentration (EC_{50}) refers to the concentration of a drug, antibody or toxicant which induces a response halfway between the baseline and maximum after a specified exposure time. It is commonly used as a measure of a drug's potency.

Exoenzyme: An exoenzyme, or extracellular enzyme, is an enzyme that is secreted by a cell and functions outside of that cell. Exoenzymes are produced by both prokaryotic and eukaryotic cells and have been shown to be a crucial component of many biological processes.

LC_{50}: Lethal concentration 50 is the lethal concentration required to kill 50% of the population.

Neurotransmitters: They also known as chemical messengers are endogenous chemicals that enable neurotransmission. They transmit signals across a chemical synapse, such as a neuromuscular junction, from one neuron (nerve cell) to another "target" neuron, muscle cell, or gland cell.

Pharmacokinetics: Sometimes abbreviated as PK, is a branch of pharmacology dedicated to determining the fate of substances administered to a living organism. The substances of interest include any chemical xenobiotic such as: pharmaceutical drugs, pesticides, food additives, cosmetics, etc.

Species Richness: The number of different species represented in an ecological community, landscape, or region. Species richness is simply a count of species, and it does not take into account the abundances of the species or their relative abundance distributions.

Chapter 10
Impact of Pesticides on Aquatic Life

Zahid Nabi
Central University of Jammu, India

Mudasir Youssouf
Central University of Punjab Bathinda, India

Javid Manzoor
Jiwaji University Gwalior, India

ABSTRACT

Humans made use of pesticides to kill pests infesting crops. This was done to increase agricultural yields and improve public health. Pesticides however turn out to be damaging for the environment, causing many harmful impacts. Certain pesticides after being applied to the environment show long-term residual effects while others show acute fatal effects particularly to aquatic life. For example, organochlorine pesticides are persistent in the environment; as a result of this, these pesticides find their way to contaminate ground water, surface water, food products, air, soil, and may also affect human beings through direct contact. Pesticide exposure to humans has been found to be an important cause of some diseases such as cancer, respiratory diseases, skin diseases, endocrine disruption, and reproduction disorders. It is this aspect of pesticides in the environment that has raised concern among environmental scientists to study their behavior in the environment and then come out with a sound alternative so as to rescue the human population from their adverse effects. Fifty years (half a century) after Rachel Carson's warning to the world about the devastating effect pesticides have on birds and beneficial insects, pesticides continue to be in use. Continued usage of pesticides can be described as a massive chemical assault on our environment which threatens the survival of many birds, fish, insects, and small aquatic organisms that form the basis of the food web. More generally, pesticides reduce species diversity in the animal kingdom and contribute to population decline in animals and plants by destroying habitats, reducing food supplies, and impairing reproduction. Organisms in ecosystems exist in complex interdependent associations such that losses of one keystone species as a result of pesticides (or other causes) can have far reaching and unpredictable effects. A keystone species is a species that is disproportionately connected to more species in the food-web. The many connections that a keystone species holds mean that it maintains the organization and structure of entire communities. The loss of a keystone species

DOI: 10.4018/978-1-5225-6111-8.ch010

results in a range of dramatic effects that alters trophic structure, other food-web connections, and can cause the extinction of other species in the community. A pesticide may eliminate a species essential to the functioning of the entire community, or it may promote the dominance of undesired species or it may simply decrease the number and variety of species present in the community. This may disrupt the dynamics of the food webs in the community by breaking the existing dietary linkages between species.

INTRODUCTION

Water can be described as one of the most priceless gifts of nature; as a result it is also termed as the lifeline of earth. Evolution of life and beginning of human civilization would not have been possible without it. Throughout the history of humans, social and economic development of civilization has been closely related with the availability of sources of water. The world population has shown a tremendous growth with the volume of water remaining the same, this increase in population has lead to over exploitation of water resources particularly post industrialization. Due to the impact of human activities, environmental disturbances on the water cycle have also increased (McMichael, 2009). Many factors such as unplanned urbanization, population explosion, and deforestation have caused pollution crisis on earth. Industrialization results in problems of pollution to lithosphere, hydrosphere and atmosphere based on the type of industry, the nature of raw materials, processes involved and types of equipments used (Hodges *et al.*, 1973). Most of the water bodies that are located in and around human settlements have been polluted by industrial effulent, which come out of different factories. All the chemicals of the industrial water are toxic to all forms of aquatic life ranging from minute organisms to giant fishes. Among the most prominent pollutants that have rendered aquatic resources polluted and caused huge damage to aquatic organisms are pesticides.

Pesticides may be described as a mixture of substances which can be of chemical or biological origin, used by human society to mitigate or repel pests such as bacteria, nematodes, insects, mites, mollusks, birds, rodents, and other organisms that affect food production or interfere with human welfare. They usually act by disrupting some component of the pest's life processes to kill or inactivate it. Pesticides also include substances such as insect attractants, herbicides, plant defoliants, desiccants, and plant growth regulators.

The concept of pesticides can be traced to 1000 B.C, when Homer referred to the use of sulfur to fumigate homes. By 900 A.D, the Chinese started using arsenic to control garden pests. Although there were outbreaks of pests, such as potato blight, which destroyed most of the potato crop in Ireland during the middle of nineteenth century, not until later that century were pesticides such as arsenic, pyrethrum, lime sulfur, and mercuric chloride put into use. Between this period and World War II, inorganic and biological substances, such as Paris green, lead arsenate, calcium arsenate, selenium compounds, lime–sulfur, pyrethrum, mercury, copper sulfate and nicotine were used, but the amount and frequency of use were limited, and most pest control employed cultural methods such as rotations, tillage, and manipulation of sowing dates. After World War 2nd the use of pesticides increased tremendously, and there are currently more than 1,600 pesticides available and about 4.4 million tons used annually, at a cost of more than $20 billion. The United States accounts for more than 25 percent of pesticide market. The first synthetic insecticide, DDT (dichlorodiphenyl-trichloroethane), was manufactured in Switzerland in 1939 and it was very effective and used extensively to control head and body lice, human disease vectors and agricultural pests. In the decades leading up to the 1970s, Benzene hexachloride (BHC) and chlordane were discovered and toxaphene (and heptachlor) slightly later. Shortly thereafter, two cyclodiene organochlorines, aldrin and dieldrin, were introduced, followed by endrin, endosulfan, and isobenzan. All these insecticides acted by blocking an insect's nervous system, causing malfunction, tremors, and death. All organochlorines are relatively insoluble, persist in soil and aquatic sediments, can bioconcentrate in the tissues of invertebrates and vertebrates from their food, move up trophic chains, and affect top predators. Pesticide residues have also been found in rain and groundwater. Studies by the UK government showed that pesticide concentrations exceeded those allowable for drinking water in some samples of river water and groundwater

ECOLOGICAL IMPACTS OF PESTICIDES

Pesticides can reach the water by four major routes. They may drift outside of the area when it is sprayed, they may percolate, or leach, through the soil, and they may be carried to the water as runoff or may be spilled accidentally or through negligence. They may also be carried to water by eroding soil. Factors that affect a pesticide's ability to contaminate water include its water solubility, the distance from an application site to a body of water, weather, soil type, presence of a growing crop, and the method used to apply the chemical. Pesticides can have considerable adverse environmental effects, which may be extremely diverse: sometimes relatively simple but often extremely complex. Some pesticides are highly specific and others broad spectrum; both types can affect terrestrial wildlife, soil, water systems, and humans (Matson et al., 1997).

Pesticides have exhibited some of their most striking effects on birds, particularly those in the higher tropic levels, such as bald eagles, hawks, and owls. These birds are often rare, endangered, and susceptible to pesticide residues such as those occurring from the bioconcentration of organochlorine insecticides through terrestrial food chains. Pesticides may kill plant-feeding birds, and the elimination of many rare species of ducks and geese has been reported. Populations of insect-eating birds such as partridges and pheasants have decreased due to the loss of their food (certain insect species) in agricultural fields through the use of insecticides. Bees, which are highly important in the pollination of crops and wild plants, have been killed by pesticides, resulting in the considerably reduced yield of crops dependent on bee pollination (Goulson et al., 2015).

The literature on pest control lists many examples of new pest species that have been developed when their natural enemies are killed by pesticides. This has created a further dependence on pesticides much similar to drug dependence. Finally, the effects of pesticides on the biodiversity of plants and animals in agricultural landscapes, whether caused directly or indirectly by pesticides, constitute a major adverse environmental impact of pesticides (Edwards, 1993).

The impacts of pesticide on aquatic systems are often studied using a hydrology transport model to study movement and fate of chemicals in rivers and streams. As early as the 1970s quantitative analysis of pesticide runoff was conducted in order to predict amounts of pesticide that would reach surface waters. Pesticides have been found to exert their impacts at multiple levels i.e. molecules, tissues, organs, individuals, populations and communities. A variety of tests have also been designed to assess these impacts (Cairns & Niederlehner, 1995).

After the application of pesticides for the protection of crop from pests and diseases, only around fifteen percent of the preparation hits the target and rest is distributed into the soil, water and air; which will finally reach the nearby water bodies through runoff (Wyman et al., 1985; Richards & Baker, 1993).

These pesticides present in water body can cause damage to the structure and function of the aquatic ecosystem (Relyea & Jason, 2008). Thus, pesticides are considered a main source of contaminants for aquatic bodies. Pesticides result in the reduction of the zooplankton diversity and biomass of predatory insects (Relyea, 2005). Some species of aquatic communities are highly impacted even by very low concentration of pesticides. Compared to individual pesticides, mixtures of pesticides pose greater effect and even result in 99% mortality of aquatic species (Relyea, 2009). Most of the pesticides possess enzyme and hormone disrupting mechanisms which contribute to the decline of zooplankton, macro invertebrates and vertebrate populations in aquatic biomes. This can in turn result in blooming of individual species (especially primary consumers) that can be damaging (Roger *et al.,* 1994). Excessive inflow of both phosphate and nitrogen is known to enhance the algal growth indirectly altering the aquatic community (Cathleen *et al.*, 2002). Pesticide pollution from agricultural activities causes adverse effect the environment. Fertilizers are considered one of the main pollutants and agricultural runoff that leads to contamination of streams, lakes and surface water bodies resulting in enhancement of the algal growth, richness of aquatic organisms and eutrophication. Fertilizers also have greater effects on the physiology of non-target species when combined with pesticides (Vijayavel, 2006). Even low concentrations of pesticide (0.13 to 0.46 mg malathion L^{-1}) can affect the food web and predators by reducing zooplankton diversity and abundance, subsequently increasing the phytoplankton community (Relyea and Jason, 2008). However, level of toxicity, time of exposure, quantity of dose and persistence capacity in environment is the key factors to determine the impacts of pesticides and fertilizers on aquatic organisms and productivity (Surendra, 2010). Many studies have revealed that individual effect of pesticides negatively influence on population of zooplankton, while fertilizers increase the population of phytoplanktons.

Aquatic life Fish and other aquatic biota are greatly impacted by water contaminated with pesticides. Pesticide surface runoff into rivers and streams can be extremely lethal to aquatic life, sometimes resulting in mass killing of all the fish. Application of herbicides to water bodies can kill fish when the dead plants decay and consume the water's oxygen, resulting in suffocation of the fish. Pesticides such as copper sulfite that are applied to kill unwanted plants are toxic to fish and other water animals at concentrations similar to those used to kill the plants. Continued exposure to sub lethal concentrations of some pesticides can reduce fish populations by causing physiological and behavioral changes, decrease in immunity to disease and lesser predator avoidance. Application of herbicides to bodies of water can also result in killing of plants on which fish depend for their habitat. Pesticides can accumulate in aquatic ecosystems to levels that kill off zooplankton, the main source of food for young fish. Pesticides can also mass kill insects on which some fish feed, causing the fish to travel farther in search of food and exposing them to greater risk from their predators. The faster a given pesticide breaks down in the environment, the less threat it poses to aquatic life. Insecticides are typically more toxic to aquatic life than herbicides and fungicides. In the past several decades, amphibian populations have declined across the world, reasons which are still under scientific investigation and for which pesticides may be a part (Werner & Moran, 2008).

Pesticide appears to have a cumulative toxic effect on frogs. Tadpoles from ponds containing multiple pesticides take longer to metamorphose and are smaller when they normally do, this decreases their ability to catch hold of prey and avoid predators. Exposures of tadpoles to the organochloride pesticide endosulfan at levels that are likely to be found in habitats near fields that receive sprays chemical kill the tadpoles and causes behavioral and growth abnormalities. The herbicide atrazine can turn male frogs into hermaphrodites, decreasing their ability to reproduce. Embryonic exposure in turtles to various pesticides causes a sex reversal. Across the United States and Canada disorders, in such aquatic organisms, decreased hatching success, feminization, skin lesions, and other developmental abnormalities have been reported (Hayes et al., 2006).

CRITERIA TO EVALUATE IMPACTS OF PESTICIDES

It is difficult to generalize the impact of pesticides as different categories have different types of effects on aquatic organisms, although terrestrial impacts by pesticides do occur, the principal pathway that causes ecological impacts is that of water contaminated by pesticide runoff. The two principal mechanisms are bioconcentration and biomagnification.

Bioconcentration

Bioconcentration is the movement of a chemical from the surrounding medium into an organism. The primary "sink" for some pesticides is fatty tissue (lipids). Some pesticides, such as DDT, are "lipophilic", it means that they are soluble in, and accumulate in, fatty tissue such as edible fish tissue and human fatty tissue (Johnson et al, 2006).

Biomagnification

The term Biomagnification describes the increasing concentration of a chemical as we move up the food chain. When smaller organisms are eaten by larger ones, the concentration of pesticides and other chemicals increases in tissues of other organs. Highest concentrations can be observed in top predators, including man. Effects at the organism or ecological level are usually considered to be an early warning indicator of potential human health impacts. The major types of effects are listed below and will vary depending on the organism under investigation and the type of pesticide. Different pesticides have markedly different effects on aquatic life which makes generalization very difficult. The important point is that many of these effects are chronic (not lethal), are often not noticed by casual observers, yet have consequences for the entire food chain (Margni et al., 2002.).

- Cancers, tumors and lesions on fish and animals.
- Cellular and DNA damage.
- Death of the organism.
- Disruption of endocrine (hormonal) system.
- Intergenerational effects (effects are not apparent until subsequent generations of the organism).

Other physiological effects such as egg shell thinning. These effects are not necessarily caused only by exposure to pesticides or other organic contaminants, but may be associated with a combination of environmental stresses such as eutrophication and pathogens. These associated stresses need not be large to have a synergistic effect with organic micro pollutants. Ecological effects of pesticides extend beyond individual organisms and can extend to ecosystems. Swedish work has found that application of pesticides is thought to be one of the most significant factors affecting biodiversity. Jonsson *et al.,* (1990) report that the continued decline of the Swedish partridge population is related with changes in land use and the use of chemical weed control. Chemical weed control which implies use of pesticides has the

Figure 1. Impacts of pesticides on aquatic life
http://www.fao.org

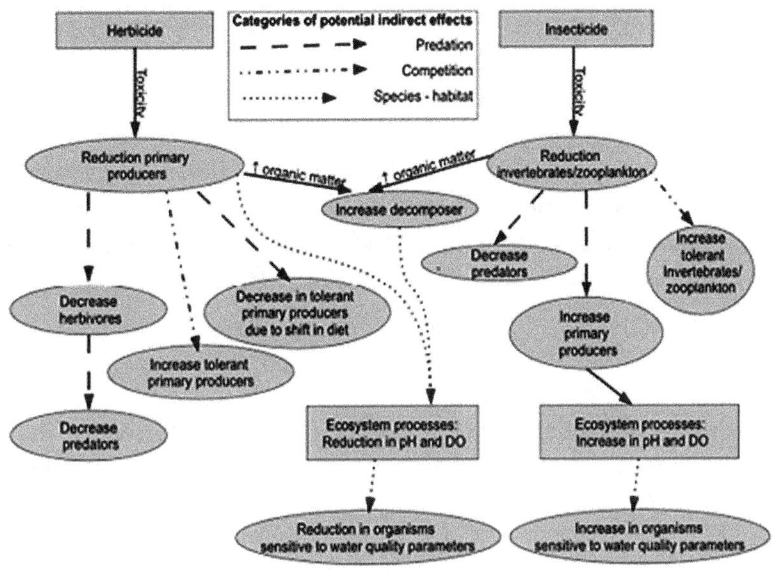

effect of reducing habitat, decreasing the number of weed species, and of shifting the balance of species in the plant community. Swedish studies also show the impact of pesticides on soil fertility, including inhibition of nitrification with reduced uptake of nitrogen by plants (Torstensson, 1990). These studies also suggest that pesticides adversely affect soil microorganisms which are responsible for microbial degradation of plant matter (and of some pesticides), and for soil structure.

- Poor fish health marked by low red to white blood cell ratio, excessive slime on fish scales and gills, etc.
- Reproductive inhibition or failure.
- Suppression of immune system.
- Teratogenic effects (physical deformities such as hooked beaks on birds).

Impacts of pesticides on aquatic life have been depicted in Figure 1.

FACTORS AFFECTING PESTICIDE TOXICITY IN AQUATIC SYSTEMS

Following factors determine the ecological impacts of pesticides in water:

Toxicity

Toxicity of a chemical or a pesticide for that case is usually expressed as LD50 (Lethal Dose). The lower the LD50, the greater the toxicity; values of 0-10 are highly toxic. Drinking water and food guidelines are determined using a risk based assessment. Generally, Risk = Exposure (amount and/or duration)

× Toxicity. Response to toxic substances can be (death) or chronic (an effect that does not cause death over the test period but which causes observable effects in the test organism such as cancers and tumors, reproductive failure, growth inhibition, teratogenic effects, etc.). (Gaines, 1969).

Persistence

Persistence of a pesticide is measured by means of its half life. Half-life is determined by biotic and abiotic degradation processes. Biotic processes are biodegradation and metabolism; abiotic processes are mainly hydrolysis, photolysis, and oxidation. Modern pesticides tend to have short half lives that reflect the period over which the pest needs to be controlled. (Hornsby et al, 1995).

Degradation

On degradation pesticides undergo chemical changes and may lead to formation of "degradates" which may have greater, equal or lesser toxicity than the parent compound. As an example, DDT degrades to DDD and DDE.

POLLUTION DUE TO PESTICIDES

Pesticides can further increase existing problem of air pollution. Pesticide drift takes place when wind carries pesticides suspended in the air as particles are to other areas, making them contaminated. When pesticides are applied to crops, they can volatilize and are occasionally blown by winds into nearby areas, potentially posing a threat to wildlife. Spread of the pesticide in the air is determined by weather conditions at the time of application including temperature and humidity (Cohen & Pinkerton, 1996).

As wind velocity increases so does the spray drift and exposure. Less relative humidity and high temperature usually cause more spray evaporation. Quite understandably droplets of sprayed pesticides from pesticides applied may travel as dust along with particles to other areas, or pesticides may stick to particles that blow in the wind, such as dust particles. Ground spraying also produces less pesticide drift than aerial spraying does. In order to prevent pesticides from reaching atmosphere, farmers can employ a buffer zone around their crop, consisting of empty land or non-crop plants such as evergreen trees to serve as windbreaks and absorb the pesticides, preventing drift into other areas. In countries like Netherlands, pesticides that are sprayed on to fields and used to fumigate soil can give off chemicals called volatile organic compounds, which can react with other chemicals and form a pollutant called tropospheric ozone. Pesticide use accounts for about six percent of total tropospheric ozone production (Alloway & Ayres, 1997).

Most of the chemicals used in pesticides are soil contaminants that persist for a very long time, whose impact may remain for decades and adversely affect soil fertility. Use of pesticides results in decrease in soil quality and lack of organic matter due to pesticides may result in less water retention in the soil. This decreases yields for farms in drought years. The use of pesticides decreases the general biodiversity in the soil. A smaller content of organic matter in the soil increases the amount of pesticide that will leave the area of application, because organic matter binds to and helps break down the pesticides. Degradation and sorption are two factors which influence the persistence of pesticides in soil. Depending on the chemical nature of the pesticide, such processes take control of the transportation from soil to water, and in turn to air and our food. Breaking down organic substances, degradation, involves interactions among microorganisms in the soil. Sorption affects bioaccumulation of pesticides which are dependent on organic matter in the soil. Nitrogen fixation, on the other hand which is essential for the growth of higher plants, is hindered by pesticides present in soil. The insecticides DDT, methyl parathion, and can be lethal to nitrogen fixing bacteria (Kalia & Gosal, 2011).

A CASE OF PESTICIDE POLLUTION IN INDIA

One of the most shocking cases of pesticide contamination of groundwater was reported when pesticide residues were found in bottled drinking water. During the months of July and December 2002, the Pollution Monitoring Laboratory of the New Delhi Based Centre for Science and Environment (CSE) tested seventeen brands of bottled water; both packaged drinking water and packaged natural mineral water, usually sold in areas that fall within the national capital region of Delhi. Pesticide residues of organochlorine and organophosphorus pesticides, which are most commonly used in India, were found in all the samples. Among the organochlorines, gamma-hexachlorocyclohexane (lindane) and DDT were prevalent, while among organophosphorus pesticides, Malathion and Chlorpyrifos were the most common. All these were present above the permissible limits specified by the European Economic Community (EEC). One wonders as to how these pesticide residues made its way into bottled water which is manufactured by big companies (Kaprow, 1985). This can be due to several reasons. Lack of regulation, that the bottled water Industry must be located in 'clean' zones. Currently, the manufacturing plants of most brands are situated in the dirtiest industrial estates or in the midst of agricultural fields. Most companies use bore-wells to pump out water from the ground from depths varying from 24-152 meters below the ground. The raw water samples collected from the plants also revealed the presence of pesticide residues. Thus, the fault obviously lies in the treatment methods used. These plants use membrane technology, where the water is filtered using membrane with ultra-small pores to remove fine suspended solids and all bacteria and protozoa and even viruses (Cameotra & Dhanjal, 2010). While nano filtration can remove insecticides and herbicides but it is expensive and thus rarely used. Most industries also use an activated charcoal adsorption process, which is effective in removing organic pesticides but not heavy metals. To remove pesticides, the plants use reverse osmosis and granular activated charcoal methods. So even though the manufacturers claim to use these processes, the presence of pesticide residues points to the fact that either the manufacturers do not use the treatment process effectively or only treat a part of the raw water. The low concentrations of pesticide residues in bottled water do not cause acute or immediate effects. However, repeated exposure to even miniscule amounts can result in chronic effects like cancer, liver and kidney damage, disorders of the nervous system, damage to the immune system and birth defects. CSE reported pesticide residues in bottled water as well as in popular cold drink brands sold across the country. This is because the main ingredient in a cold drink or a carbonated non-alcoholic beverage is water and there with standards specified for water to be used in these beverages in India. There were no standards for bottled water in India till September 29, 2000,when the Union Ministry of Health and Family Welfare issued a notification (No. 759(E)) amending the Prevention of Food Adulteration Rules, 1954 (Handford et al., 2015).

The BIS (Bureau of Indian Standards) certification mark became mandatory for bottled water after March 29, 2001. However, the parameters for pesticide residues remained ambiguous. A series of Committees were established and eventually on 18th July 2003, amendments were made in the Prevention of

Food Adulteration Rules stating that pesticide residues considered individually should not exceed 0.0001 mg/L and that the total pesticide residues should not be more than 0.0005 mg/L and that the analysis shall be conducted by using internationally established test methods meeting the residue limits specified herein. This notification came into force from January 1st, 2004. (Gruère & Rao, 2007).

MITIGATING IMPACTS OF PESTICIDE POLLUTION

There exist many alternatives to reduce the effects of pesticides. Alternatives include applying heat, covering weeds with plastic, manual removal, placing traps, removing pest breeding sites, maintaining healthy soils that breed healthy, more resistant plants, cropping native species that are naturally more resistant to native pests and supporting biocontrol agents such as birds and other pest predators. Biological control of pesticides such as resistant plant varieties and the use of pheromones have been successful and have at times permanently resolved a pest problem. Integrated Pest Management employs chemical use only when other alternatives are ineffective (Chandler et al, 2011). IPM causes less harm to humans and the environment. The focus is broader than on a specific pest, considering a range of pest control alternatives. Strains of crops can be genetically modified (GM) to increase their resistance to pesticides (Su & Scheffrahn, 1998).

CONCLUSION

The use of pesticides was initiated to enhance agricultural produce by eliminating pests that cause diseases to crops. However, their negative impacts caused environmental problems ranging from water contamination to soil infertility. Although, off late humans have learned to minimize the use of pesticides, by incorporating organic form of pest killers that are environmentally sustainable, but this has not been fully practiced at field level at large scale. Hence, there is a dire need to develop pesticides that are environmentally safe and can be used at large scale at cost effective rates.

REFERENCES

Alloway, B., & Ayres, D. C. (1997). *Chemical principles of environmental pollution*. CRC press.

Cairns, J., & Neiderlehner, B. R. (Eds.). (1985). Ecological Toxicity Testing. Lewis Publishers.

Cameotra, S. S., & Dhanjal, S. (2010). Environmental nanotechnology: nanoparticles for bioremediation of toxic pollutants. In *Bioremediation Technology* (pp. 348–374). Dordrecht: Springer. doi:10.1007/978-90-481-3678-0_13

Chandler, D., Bailey, A. S., Tatchell, G. M., Davidson, G., Greaves, J., & Grant, W. P. (2011). The development, regulation and use of biopesticides for integrated pest management. *Philosophical Transactions of the Royal Society of London. Series B, Biological Sciences*, *366*(1573), 1987–1998. doi:10.1098/rstb.2010.0390 PMID:21624919

Cohen, J. M., & Pinkerton, C. (1966). *Widespread translocation of pesticides by air transport and rain-out*. Academic Press.

Edwards, C. A. (1993). The impact of pesticides on the environment. In *The pesticide question* (pp. 13–46). Boston, MA: Springer. doi:10.1007/978-0-585-36973-0_2

Gaines, T. B. (1969). Acute toxicity of pesticides. *Toxicology and Applied Pharmacology, 14*(3), 515–534. doi:10.1016/0041-008X(69)90013-1 PMID:5787520

Goulson, D., Nicholls, E., Botías, C., & Rotheray, E. L. (2015). Bee declines driven by combined stress from parasites, pesticides, and lack of flowers. *Science, 347*(6229), 1255957. doi:10.1126cience.1255957 PMID:25721506

Gruère, G. P., & Rao, S. R. (2007). *A review of international labeling policies of genetically modified food to evaluate India's proposed rule*. Academic Press.

Handford, C. E., Elliott, C. T., & Campbell, K. (2015). A review of the global pesticide legislation and the scale of challenge in reaching the global harmonization of food safety standards. *Integrated Environmental Assessment and Management, 11*(4), 525–536. doi:10.1002/ieam.1635 PMID:25765969

Hapeman, C. J., Dionigi, C. P., Zimba, P. V., & McConnell, L. L. (2002). Agrochemical and nutrient impacts on estuaries and other aquatic systems. *Journal of Agricultural and Food Chemistry, 50*(15), 4382–4384. doi:10.1021/jf020457n PMID:12105974

Hayes, T. B., Case, P., Chui, S., Chung, D., Haeffele, C., Haston, K., & Tsui, M. (2006). Pesticide mixtures, endocrine disruption, and amphibian declines: Are we underestimating the impact? *Environmental Health Perspectives, 114*(Suppl 1), 40–50. doi:10.1289/ehp.8051 PMID:16818245

Hodges, R. R., Jr., Hoffman, J. H., Johnson, F. S., & Evans, D. E. (1973). Composition and dynamics of lunar atmosphere. In *Lunar and Planetary Science Conference Proceedings* (*Vol. 4*, p. 2855). Academic Press.

Hornsby, A. G., Wauchope, R. D., & Herner, A. (1995). *Pesticide properties in the environment* (Vol. 1). Springer Science and Business Media.

Johnson-Restrepo, B., Kannan, K., Rapaport, D. P., & Rodan, B. D. (2005). Polybrominated diphenyl ethers and polychlorinated biphenyls in human adipose tissue from New York. *Environmental Science & Technology, 39*(14), 5177–5182. doi:10.1021/es050399x PMID:16082945

Kalia, A., & Gosal, S. K. (2011). Effect of pesticide application on soil microorganisms. *Archives of Agronomy and Soil Science, 57*(6), 569–596. doi:10.1080/03650341003787582

Kaprow, M. (1985). Manufacturing danger: Fear and pollution in industrial society. *American Anthropologist, 87*(2), 342–356. doi:10.1525/aa.1985.87.2.02a00070

Margni, M., Rossier, D., Crettaz, P., & Jolliet, O. (2002). Life cycle impact assessment of pesticides on human health and ecosystems. *Agriculture, Ecosystems & Environment, 93*(1-3), 379–392. doi:10.1016/S0167-8809(01)00336-X

McMichael, P. (2009). Contemporary contradictions of the global development project: Geopolitics, global ecology and the 'development climate'. *Third World Quarterly*, *30*(1), 247–262. doi:10.1080/01436590802622987

Relyea, R. A. (2005). The impact of insecticides and herbicides on the biodiversity and productivity of aquatic communities. *Ecological Applications*, *15*(2), 618–627. doi:10.1890/03-5342

Relyea, R. A. (2009). A cocktail of contaminants: How mixtures of pesticides at low concentrations affect aquatic communities. *Oecologia*, *159*(2), 363–376. doi:10.100700442-008-1213-9 PMID:19002502

Relyea, R. A., & Hoverman, J. T. (2008). Interactive effects of predators and a pesticide on aquatic communities. *Oikos*, *117*(11), 1647–1658. doi:10.1111/j.1600-0706.2008.16933.x

Richards, R. P., & Baker, D. B. (1993). Pesticide concentration patterns in agricultural drainage networks in the Lake Erie basin. *Environmental Toxicology and Chemistry*, *12*(1), 13–26. doi:10.1002/etc.5620120104

Roger, P. A., Simpson, I., Oficial, R., Ardales, S., & Jimenez, R. (1994). Effects of pesticides on soil and water microflora and mesofauna in wetland ricefields: A summary of current knowledge and extrapolation to temperate environments. *Australian Journal of Experimental Agriculture*, *34*(7), 1057–1068. doi:10.1071/EA9941057

Su, N. Y., & Scheffrahn, R. H. (1998). A review of subterranean termite control practices and prospects for integrated pest management programmes. *Integrated Pest Management Reviews*, *3*(1), 1–13. doi:10.1023/A:1009684821954

Vijayavel, K., Balasubramanian, S. E., & Balasubramanian, M. P. (2006). Effect of urea and ammonium chloride on Aspartate and Alanine transaminases of Oreochromis mossambicus (Peters). *Indian Journal of Fisheries*, *53*(1), 19–22.

Werner, I., and Moran, K. (2008). *Effects ofpyrethroid insecticides on aquatic organisms*. Academic Press.

Wyman, J. A., Jensen, J. O., Curwen, D., Jones, R. L., & Marquardt, T. E. (1985). Effects of application procedures and irrigation on degradation and movement of aldicarb residues in soil. *Environmental Toxicology and Chemistry*, *4*(5), 641–651. doi:10.1002/etc.5620040508

ADDITIONAL READING

Edwards, C. A. (Ed.). (2013). *Environmental pollution by pesticides* (Vol. 3). Springer Science & Business Media.

Khan, M. (Ed.). (2013). *Pesticides in aquatic environments* (Vol. 10). Springer Science & Business Media.

Li, M. (1977). Pollution in nation's estuaries originating from the agricultural use of pesticides. *In Proc. Conf. Estuarine Pollut. Control Assess* (Vol. 1977, pp. 451-466).

Matson, P. A., Parton, W. J., Power, A. G., & Swift, M. J. (1997). Agricultural intensification and ecosystem properties. *Science*, *277*(5325), 504–509. doi:10.1126cience.277.5325.504 PMID:20662149

Meena, R. P., Meena, S. S., & Meena, S. R. (2008). Use of agrochemicals in agriculture, their consequences and possible solution. *Green Farming*, *1*(4), 54–55.

Rasmussen, J. J., Wiberg-Larsen, P., Baattrup-Pedersen, A., Cedergreen, N., McKnight, U. S., Kreuger, J., & Friberg, N. (2015). The legacy of pesticide pollution: An overlooked factor in current risk assessments of freshwater systems. *Water Research*, *84*, 25–32. doi:10.1016/j.watres.2015.07.021 PMID:26207877

KEY TERMS AND DEFINITIONS

Aquatic Animal: An aquatic animal is an animal that lives in water for most or all of its life. Aquatic animals may breathe air or extract oxygen from that dissolved in water through specialized organs called gills, or directly through the skin.

Pesticide: A pesticide is any substance used to kill, repel, or control certain forms of plant or animal life that are considered to be pests.

Phytoplanktons: Phytoplanktons are minute organisms of lakes, streams, and oceans that make their own food from sunlight through photosynthesis. Phytoplanktons occur almost anywhere there is water and sunlight.

Zooplankton: Zooplankton is the common name given to many small species of animals found in fresh and marine waters throughout the world. The word *zooplankton*, derived from Greek, means "wandering animals." They float in the water column and drift with the currents.

Chapter 11
Impact of Pesticides on Invertebrates in Aquatic Ecosystem

Azad Gull
Jiwaji University Gwalior, India

Ashaq Ahmad Dar
Jiwaji University Gwalior, India

Jaya Chaturvedi
Jiwaji University Gwalior, India

ABSTRACT

Aquatic ecosystems do not contain more than a fragment of the global water resources, but they are exclusive and complex habitats due to the extremely close association between terrestrial and aquatic habitats. The important fish stocks and a unique set of organisms that provides priceless consumer services, such as chemical water purification and organic matter processing, are affected. The pollution of aquatic ecosystems with pesticides applied in agricultural production is widely acknowledged as one of the greatest anthropogenic stressors to stream ecosystems, and agricultural pesticides are known to cause a threat to all living organisms in stream ecosystems. The general objective of this chapter is to study the effects of agricultural pesticides on invertebrates. There are only a few evaluating effects of pesticide contamination resulting from normal agricultural practice on invertebrates, and there is a lack of studies focusing on the indirect effects of pesticides. The importance of physical habitat degradation in the assessment and mitigation of pesticide risk in agricultural streams will be discussed.

DOI: 10.4018/978-1-5225-6111-8.ch011

Figure 1. Major transport routes into receiving streams for pesticides that are applied to agricultural fields

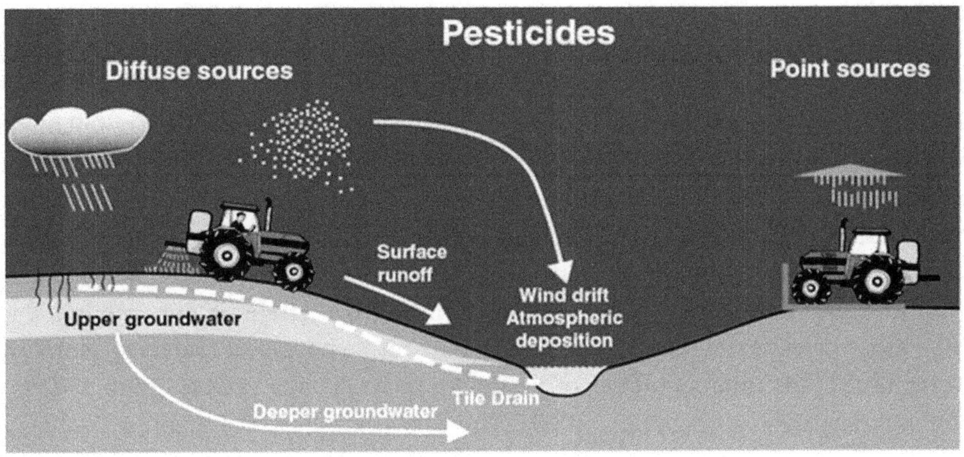

INTRODUCTION

The main routes through which the pesticides are transported to water bodies include surface runoff, tile drains, groundwater and leakage from point sources (Schulz, 2004). The transport pathways are strongly governed by climatic and geological conditions of the area and also by the physicochemical properties of the pesticide compound (Figure 1).

Consequently, peak pesticide concentration lasts for short period and occurs during heavy precipitation. The maximum exposure concentration is important for estimating ecological effects (Schulz & Liess, 2000). The combination of extensive agricultural activities and the close connectivity between land and stream makes them more vulnerable to pesticides. Overall, many environmental factors that act on stream ecosystems are directly or indirectly correlated. Pollutants like pesticides have potential impact on agricultural streams and are effected by other local stressors like channelisation and dredging (MacArthur & Wilson, 1967). The stream ecosystem is also influenced by the local factors which are believed to be more important for aquatic biodiversity (Pedersen & Friberg, 2009). The aquatic biodiversity is very sensitive to different xenobiotic substances like pesticides. Pesticides are believed to have more impact on invertebrates as compared to other pollutants (Beketov & Liess, 2012).

Some pesticides have severe impact on organisms even at low concentration while some are less toxic even at high concentration although it depends on the chemistry and properties of the pesticide. Synergetic effects are generally shown by different compounds with similar mode of action, and mostly trigger the effect on the exposed organism (Belden & Lydy, 2000). Few studies have shown that pesticides possess joint toxicity on aquatic invertebrates (Belden & Lydy, 2000). Bailey and coworkers (1997) showed that two organophosphorus insecticides, diazinon and chlorpyriphos, exhibit additive toxicity to Ceriodaphniadubia. Additivity between diazinon and chlorpyriphos is reasonable given that both are metabolically activated organophosphorus insecticides and act similarly with respect to binding with acetylcholine acetrase (AChE) (Ecobichon, 1991).

Indiscriminate use of pesticide spraying may cause ecological disturbance in natural aquatic environments. Lack of information about pesticide is the serious concern for ecosystems. Although mode of action of several pesticides is known (Ecobichon, 1991), however very little is known about their toxicity. Liess & Schulz (1999) and Schulz & Liess (1995) observed runoff-related insecticide input on stream invertebrate dynamics. It is confirmed from different studies that different species disappeared in macroinvertibrate community due to toxicity of pesticides (Liess & Schulz, 1999; Schulz & Liess, 1999). Hydraulic stress during runoff events was reduced in microcosms operated in the bypass to the stream (Liess & Schulz, 1999). Moreover, hydraulic stress plays synergetic role in population reduction, especially whereas the agricultural insecticide input change the dynamics of the total macroinvertebrate community and cause a serious threat to the stream ecosystem.

Sometimes the direct effect of pesticide exposure is very difficult to detect. However, the mechanism of indirect effects are even more complex to describe. Woin (1998) while studying microcosm showed that the pesticides cause indirect structural changes in a macroinvertebrate community. Nonpersistent pesticides may cause detrimental effects resulting in long term changes at the ecosystem level of organization (Woin, 1998). The indirect effects of pesticides in freshwater ecosystem include proliferation of algae in the microcosms (Friberg *et al.*, 2003). This is an indication that exposure to pesticides may shift an aquatic ecosystem to an algal dominated turbid state. The exposure to pesticides may trigger a competitive advantage for algae in relation to macrophytes, it is possible that the indirect effects of the pesticide groups can act synergistically at the ecosystem level, even though their direct toxic effects are exerted through very different modes of action. However, the rate of pesticides, toxicity of pesticides and its dosage is a critical factor to determine its effect on microorganisms.

Some pesticides like atrizine is less toxic to larvae of midge chironomus tentans (Pape-Lindstrom & Lydy, 1997). However, atrazine in binary combination with the organophosphorus insecticides exhibit synergistic toxicity in case of fourth instar of the midge. While exposure of the organophosphorus mevinophos or the organochlorine methoxychlor together with atrazine results in less additive (Pape-Lindstrom & Lydy, 1997) This can be explained due to:

- The penetration of the insecticides through the midge cuticula or the cellular permeability may increase due to atrizine and
- Atrazine may enhance the biotransformation of the organophosphorus insecticides.

Belden & Lydy (2000), reported synergistic toxicity to Chironomustentans larvae at lower concentrations in combination with the organophosphorus chlorpyrifos and methyl parathion. Exposure to the organophosphorus diazinon showed similar results. Atrazine enhances the uptake rate of the organophosphorus insecticides, however, the following increase in toxicity would be minimal as compared to the nearly 50% decrease in the EC50-value. Body residue analyses of exposed larvae indicates that higher amounts of metabolites were generated in atrazine treated larvae in contrast to the controls. Further atrazine-treated larvae in vitro assay generated an increase in a specific toxic chlorpyrifos metabolite. Therefore, the increase in toxicity occurs due to an increase in biotransformation rates of the organophosphorus insecticides and results in the formation of synergetic metabolites (Belden & Lydy, 2000). Some pesticides have low synergetic toxicity due to chemical structure (Pape-Lindstrom & Lydy, 1997). The difference in pesticide structure may result in slightly different biotransformation pathways that reduce the influence of pesticide unless it is present at very high levels. These studies indicate that more commonly used pesticides, may constitute synergistic toxicity towards aquatic invertebrates.

Pesticides applied to the fields have a wide range of impacts on organisms from all trophic levels ranging from algae to fish (Liess *et al.*, 2005). However, benthic macroinvertebrates in particular strongly respond to pesticide exposure, especially insecticides (Norum *et al.*, 2010). Further, macroinvertebrate are widely used as indicator to determine the ecological quality of streams (Metcalfe-Smith, 1996). That is why, macroinvertebrate ecology is well researched, and benthic macroinvertebrates are keystone organisms for detecting the impacts of various anthropogenic stressors.

OBJECTIVES AND APPROACH

The general objective of this book chapter was to investigate the effects of pesticides on aquatic ecosystem structure and function with specific focus on benthic macroinvertebrates and ecosystem processes related to this group of organisms.

More specifically, we aimed to:

- Characterise the potential problem of pesticide contamination in aquatic ecosystem for the macroinvertebrate communities
- Review the pesticide effects on selected ecosystem processes with relevance to macroinvertebrates
- Review the (indirect) effects of changes in microorganism communities, that is the result of pesticide exposure, on the foraging behaviour of macroinvertebrates that use this food resource.

PESTICIDE TOXICITY TO AQUATIC INVERTEBRATE

Benthic communities have a significant role in transferring of different nutrients and environmental pollutants, sediment energy, to higher tropic levels (Burton, Nelson & Ingersoll, 1992. Horne and Goldman, 1994). The pesticides may accumulate in the sediments that might cause a serious damage to the ecosystem and their exposure to aquatic invertebrates through body walls, across respiratory surfaces, and ingestion of contaminated sediment particles (Power & Chapman, 1992). The behavior of the pesticide depends on the bioavailability and toxicity of the pesticide with interaction to a receptor site, which may cause significant biological effects.

The sediment toxicity tests with benthic macroinvertebrates are main tools for risk assessment of pestisides (Streloke & Kopp, 1995; U.S. Environmental Protection Agency, 2000). It is important to consider that there are numerous variations are associated with tests performed in laboratory. A study by Goedkoop and Peterson (2003) shows that both larval burrowing activity and sediment organic matter strongly modify test conditions. Earlier, confounding effects of food additions on the toxicity test has been investigated in various studies (Ristola *et al.*, 1999). The toxicity of currently used pesticides to aquatic invertebrates have been addressed in various studies (Conrad, Fleming, & Crane, 1999).

The organochlorine pesticide Lindane is frequently found pesticide in aquatic environments (Ulen; Kreuger & Sundin, 2002). Schultz & Liess (1995) collected toxicity data of lindane and parathion from several aquatic macroinvertebrates indicated that lindane acutely toxic. Hirthe *et al.* (2001) observed that short-term exposures to lindane to 4th instar larvae of Chironomus riparius not only affect the organisms during the exposure, but also after the exposure as adults. The normal larval behavior was affected, the emergence was delayed for both sexes, and as adults the midges got shorter wings and lower fecundity. Similar results was found when *C. riparius* larvae were exposed to much lower concentrations of lindane as 2nd instars over 10 days (Maund *et al.*, 1992). The emergence time was increased by 20% and 45%, respectively, in larvae exposed to 0.8 and 2.0 μg/L in comparison with emergence of the control.

Further the larval growth was decreased at 1 µg/L lindane. The caddisfly Limnephilus lunatus being most sensitive species was also affected and results in delayed emergence.

Similarly, parathion did not alter the emergence rate of Limnephilus lunatus (Schulz & Liess, 1995). In contrast, it reduces the emergence rate of Limnephilusbipunctatus Curtis at 1 ng/L. Biological and environmental impacts of parathion and malathion on the aquatic ecosystem indicates that non-target insects are very sensitivity to these pesticides (Mulla & Mian 1981). Several modern insecticides such as pyrethriod deltamethrin and fenvalerate are highly toxic to non-target aquatic invertebrates (Schulz and Liess, 2001; Hedlund., 2002; Friberg-Jensen *et al.*, 2003). The toxicity ranges from altered interspecific relationship and life history cycles to lethal effects.

Even though, it is difficult to compare the toxicity that single pesticide will cause deleterious effects to aquatic invertebrates at low concentrations. It is important to mention that pesticide exposure in the field is a single compound exposure, but rather an exposure of a mix of substances can have synergistic effects.

AQUATIC FLORA AND FAUNA AND THEIR HABITAT

Biological diversity in aquatic ecosystem is highly depends on the habitat diversity of the streams. Stable stream systems enhance diversity and availability of habitats. That is why the stream channel stability and the restoration of natural functions is always considered in stream corridor restoration activities. Moreover, the aquatic habitat is affected by the cross-sectional shape and dimensions, slope and confinement, the distribution of bed sediments, and even plan form of stream. Stream biota are often classified in seven groups—bacteria, algae, macrophytes (shallow rooted plants), protists, microinvertebrates (invertebrates less than 0.02 inch in length), macroinvertebrates (invertebrates greater than 0.02 inch in length) and vertebrates.

Aquatic plants are usually attached to different types of stream substrates. Rooted aquatic vegetation occurs where substrate is suitable and high currents do not wash the away. vascular plants grow in clear water, with substrates rich in nutrient value, and in slow water velocities. Large bedrock or stones are often covered by mosses and algae and various forms of micro- and macroinvertebrates (Ruttner 1963). Planktons are usually limited but can also occur where the watershed contains lakes, ponds, floodplain waters, or slow current areas (Odum, 1957).

The benthic invertebrate community of streams has a variety of biota and are found in or on a multitude of microhabitats in streams. Unicellular organisms and microinvertebrates are the most abundant biota in streams. Larger macroinvertebrates plays a significant role in determining the community structure (Morin & Nadon, 1991; Bourgmann & Morin 1995). Furthermore, they often play important roles in determining community composition of other components of the ecosystem

Invertebrates of ricefield ecosystem are derived from adjoining water bodies and spent their wholelife as freshwater fauna (Fernando *et al.,* 1980). The commonly found groups are crustaceans, microcrustaceans, aquatic insect larvae, aquatic insects, molluscs, annelids, nematodes and rotifers (Heckman, 1979; Clement *et al.*, 1977). Microorganisms and benthic invertebrates together breakdown the organic material that enters the stream from external sources. The macroinvertebrates are divided down into functional feeding groups. Invertebrates act as shredders, collectors, filter smaller organic material from the water or grazers scrape material off surfaces, or feed on material deposited on the substrate (Moss, 1988). These activities result in the breakdown to different types of organic matter and elaboration of invertebrate tissue on which other consumers like other predators groups feed on.

Benthic macroinvertebrates are widely used as indicators of stream health and condition mostly for pollution estimation. They also acts as a food source for many fish species either by direct browsing on the benthos or by catching benthic organisms (Walburg, 1971).

PESTICIDES FLUCTUATION IN AQUATIC INVERTEBRATES

Aquatic invertebrates is an important role in nutrient cycling in various ways. Grazers and detritivores perform important roles in the decomposition of the photosynthetic aquatic biomass (PAB). Aquatic oligochaetes perform translocation of accumulated organic matter in the detritus layer at the soil-water interface and also nutrient cycling through their bioperturbations, which release native minerals from the soil (Grant *et al.*, 1986) and-through the decomposition and mineralization of their body tissues. Primary productivity encourages the proliferation of grazer populations, which may inhibit algal growth and decreases biological N, fixation (Wilson *et al.*, 1980). Factors that changes the relationship between primary production and grazers can have consequences for nutrient availability. The aquatic oligochaetes that contribute to nutrient cycling and modifying biological activities in rice fields have gained more attention than other invertebrate groups. Furthermore, research conducted on these organisms in other freshwater environments could be applicable to flooded rice fields. The rapid use of pesticides decreases the number of invertebrates in aquatic ecosystem, however, it also increase their resistance with the span of time which results their increase in number (Lim, 1980; Ishibashi & ltoh, 1981; Roger, Heong *et al.*, 1991). Invertebrates were reported to be abundant in insecticide-treated ricefields (IRRI, 1986). This was due to insecticide resistance, reduced competition and predation, high tolerance of juveniles relative to adults, parthenogenetic reproduction, and increased fecundity (Wong, 1979; Grant *et al.*, 1983). Use of benthiocarb to rice fields in reduces the population of snails, cladocerans, odonatans, midges, and mosquito larvae. Regeneration of midges, cladocerans, and mosquito larvae occurred rapidly to densities exceeding those of the controls (Ishibashi and Itoh, 1981). According to Lim (1980) in nontreated fields nematodes, hemipterans, and dipterans dominated in contrast to ostracods, dipterans, and conchostracans dominated in fields when pesticides were applied at the recommended dose. Zooplankton reduced from 1,500 per litre to 400 per litre after the use of carbofuran. Applications of endosulfan and carbaryl produced indecisive results and no apparent impact, respectively. Carbaryl which is applied as a spray had no impacts on invertibrates because the rice plants obstruct it and prevent it from entering the aquatic system. The regeneration of chironomid larvae and ostracods was reported in ricefields when predatory invertebrates were decreased following the application of a mixture of propoxur, thiobencarb, and simetryne (Takamura & Yasuno, 1986). Grigarick *et al.*, (1990) observed that the fungicide triphenyltin hydroxide (TPTH) affected a wider range of microcrustaceans than a chitin synthesis inhibitor Benzoylphenyl urea (BPU) and that mosquito larvae increased due to reduction in predation. Gorbach, Haaring, Knauf, and Werner (1971) found that mortality of Coleoptera and Tipulidae larvae after the use of endosulfan (0.5 kg ai/ha) in a ricefield and the use of fenitrothion decreases the population of the zooplankton Moina species. (Takaku *et al.*, 1979). The use of pesticides at prescribed rates was not acutely toxic to invertebrate rice field, but populations were indirectly reduced later in the growing season (Lim *et al.*, 1984). Simpson *et al.* (1994) investigated the effect of realistic carbofumn and butachlor use on the population change of invertebrates in ricefields and significant effects were observed on some invertebrates like ostracod, copepod, cladoceran, and chironomid, and mosquito larvae although the impacts were small, transient, and inconsistent. It was concluded that at realistic application rates of carbofuran and butachlor did not affect crop cycle population dynamics invertebrates. After endosulfan use gastropod population showed no signs of mortality (Gorbach, Haaring, Knauf & Werner, 1971). Some pesticides like benzene hexachloride treated plants molluses favour a population increase (Rogers, Grant & Reddy, 1985). Ishibashi and Itoh (1981) reported larger snail populations in fields initially treated with the herbicide benthiocarb than in the control after harvesting. Simpson, Roger, Oficial, and Grant *et al.* (1994) found little evidence that snail populations were affected by carbofuran or butachlor applications. The impact of pesticide on non-target mollusks species is likely to enhance with the elevated use of molluscicides to control the golden snail problem. Aquatic invertebrate groups found in the floodwater of rice fields are similar to those found in shallow freshwater ecosystems. Due to the lack of information available on in situ toxicity of pesticides in the floodwater of rice fields, useful information about impacts of pesticide can be obtained from field and laboratory experiments related to these habitats. Crosby and Mabury (1992) reviewed the potential impacts of pesticides commonly used in Californian ricefields. The impact of pesticides reported on aquatic invertebrates in other shallow fresh- water ecosystems include differential mortality and recovery patterns between species and life stages, feeding effects, changes in population density and community structure (Hurlbert *et al.*, 1970; Hurlbert *et al.*, 1972). Many studies observed the toxicity of pesticide to aquatic invertebrates found in rice fields. Barrion and Litsinger (1982) observed that carbofuran was acutely toxic to ostracods, chironomid larvae, corixids, and some predatory insects. Filtration and assimilation rates of algae by cladocerans and copepods were affected at low concentrations of the pyrethroid fenvalerate (Day, Kaushik, & Solomon, 1987). Pyrethroids were acutely toxic to species of cladocerans and copepods even at low concentration by reducing reproduction and filtering rates (Day, 1989). The growth and egg production in cladoceran species is reduced by carbryl and endosulfan (Krishnan & Chockalingam., 1989).

EFFECT OF PESTICIDES ON ABIOTIC SYSTEM

Diversity of stream biota reflects variations in both abiotic and biotic factors. These factors affect the growth, survival, and reproduction of aquatic organisms. The mechanism for transfering food and nutrients limit the ability of organisms to remain in a stream segment. Some organisms also respond to temporal variations in flow, which can change the physical structure of the stream channel, as well as increase mortality, modify available resources, and disrupt interactions among species (Resh *et al.*, 1988, Bayley & Li, 1992).

The flow velocity determines the planktonic forms in streams. The slower the currents the more closely the composition and configuration of biota at the shore (Ruttner, 1963). High flows are responsible for migration and spawning of some fishes, cleanse the streambed materials and also scour the pools which may limit young fish production (Kohler & Hubert, 1993).

Water Temperature

According to Arrihenius and Von't Hoff's equation (Francisco *et al.*, 1988) the effect of pesticides on temperature variation in an aquifer is dependent upon saturation and unsaturated phase, the temperature of the water body will increase and may not cause significant increase with decrease in water temperature in an aquifer. Water temperature varies with air temperature, altitude, latitude, origin of the water, and solar radiation (Ward 1985; Sweeney 1993). Temperature controls many biochemical and physiological processes in poikilothermal aquatic organisms thus, water temperature plays a significant role in determining growth, development, and behavioral patterns. The entanglement between temperature and growth, development, and behavior can be strong enough to affect geographic ranges of some species. Water temperature is one of the most important factors determining the distribution of aquatic organisms. Many aquatic organisms can tolerate a narrow temperature range such as salmonids and sculpins in contrast to species like largemouth bass, small-mouth bass, suckers, minnows, sun-fishes and catfishes are present in warmer streams (Walburg, 1971).

Toxicity of xenobiotic chemicals is greatly influenced by temperature depending on the species and the chemical nature of the toxicant. The study shows that pesticides exert more toxic effect at high temperature. Gupta and Rajbanshi (1991) reported that fishes were more susceptible to the metals at high temperature.

Dissolved Oxygen

The dissolved oxygen is an important parameter that decides the diversity of micro organisms in the water body. Oxygen enters the water by absorption directly from the atmosphere and by plant photosynthesis (Mackenthun, 1969). Due to the shallow depth, large surface exposure to air and constant motion, streams contain abundant dissolved oxygen. Oxygen content of the water containing pesticide medium decreases at high temperature resulting in the death of aquatic organisms due to decrease in oxygen consumption at high temperature also noticed by Sultana and Devi (1995).

pH

The properties of water influence the effectiveness of some pesticides. One of the most important parameter is pH of the water. Many pesticides undergo a chemical reaction in alkaline water that decreases their effectiveness. This process is known as alkaline hydrolysis. Greater the alkalinity of water, the more quickly the pesticide breaks down. The severity of the pesticides is determined by:

- Susceptibility of pesticide towards alkaline hydrolysis,
- Duration for which the pesticide in contact with the alkaline water,
- The temperature of the diluted pesticide mixture, and
- Alkalinity (pH) of the water.

The hydrolysis is fast when the pH of the water is more than 8 or 9. For every unit in pH increase, the rate of hydrolysis increases 10 times. Some pesticides undergo rapid break down especially when the pH of the water is very high. Aquatic organisms from a wide range of taxa exist and thrive in aquatic systems with nearly neutral hydrogen ion activity (pH 7). Deviations in pH increase chronic stress levels and eventually decrease species diversity and abundance. One of the most widely recognized impacts of change in pH has been attributed to increased acidity of rainfall, especially areas downwind of industrial and urban emissions. Acidic rainfall can be especially harmful to environments where runoff already tends to be slightly acidic as a result of natural conditions.

EUTROPHICATION AND PESTICIDES

The primary productivity of streams varies depends on geographic location, stream size, and season (Odum, 1957, Minshall, 1978). According to river continuum concept (Vannote *et al.* 1980) the primary productivity has minimal importance in shaded head-water streams but its significance increases as stream size increases and vegetation on banks has no longer restrict the entry of light in stream. In certain ecosystems, such as streams, grassland and desert ecosystems the primary productivity is of great importance. Flora can vary from diatoms in high mountain streams to macrophytes in low gradient streams.

Enrichment of streams with a pesticides that are composed of nitrogen and phosphorus can increase the rate of algae and aquatic plant growth, a process known as eutrophication. Organic matter decomposition can deplete oxygen reserves and result in killing of fishes and other aesthetic problems in water bodies.

Eutrophication is indirectly measured as standing crops of phytoplankton biomass. In smaller streams phytoplankton biomass is usually not the dominant portion due to periods of energetic flow. Eutrophication may result in algal blooms and oxygen depletion and excessive plant growth can occur in streams at low concentrations of nitrogen and phosphorus because the stream currents promote nutrient exchange.

Shading or turbidity may limit the light for algal growth in various water bodies and biota depend on organic matter cab be transported from elsewhere. Allochthonous materials undergo rapid changes in streams (Cummins, 1974) and soluble organic compounds are removed through leaching. Microorganisms colonize the leaf material and utilize them as carbon source. The microbial biomass enhances the protein content of the leaves, which becomes a food resource for invertebrates.

Microbial decomposition and invertebrate shredding/ scraping reduces the particle size of the organic matter. These finer particles become the energy inputs to the down- stream portions of the stream. In lotic systems this unidirectional movement of nutrients and organic matter is reduced by the temporary retention, storage, and utilization of nutrients.

Organic matter processing has been shown to have nutrient-dependent relationships like that of primary productivity. Decomposition can be limited by either nitrogen or phosphorus, with predictive N:P ratios being similar to those for growth of algae and periphyton. Leaves and organic matter are generally low in protein value. However, the colonization of organic matter by bacteria and fungi increases the concentration of nitrogen and phosphorus due to the accumulation of proteins and lipids contained in microbial biomass. These compounds are a major nutritive source for aquatic invertebrates. Decaying organic matter represents a major storage component for nutrients in streams, as well as a primary pathway of energy and nutrient transfer within the food web (Fig II). Ultimately, the efficiency of retention and utilization is reflected at the top of the food web in the form of fish biomass. Organisms usually respond to variations in the availability of autochthonous, allochthonous, and upstream sources.

Aquatic macroinvertebrates are consumers that play a significant role in transferring energy and nutrients to higher trophic levels of the food chains of aquatic ecosystems. These animals feed upon submerged and emergent macrophytes, plankton, and suspended organic material in the water column. Burrowing and feeding activities help in the decomposition of plant and animal matter and ultimately the recycling of nutrients. Furthermore, these organisms prey each other and serve as food for fishes, certain birds, and other animals. But, they will be killed by high concentration of pesticides and the process of eutrophication will

Figure 2. Typical stream food web

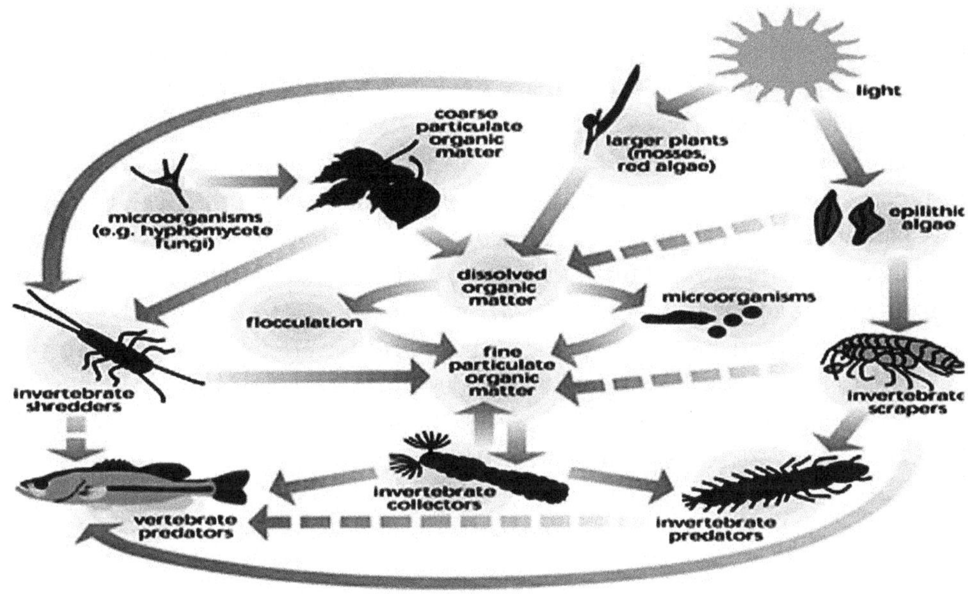

MACROINVERTIBRATES AS BIOLOGICAL SENSORS

The aquatic organisms like macroinvertebrates of streams and rivers acts as biological monitoring. To determine the richness of the biological community scientists observe different changes that occur in organisms present in particular stream system and also observe the total number of organisms in an area, or the density of the community. If community richness and community density change with respect to time, it indicates the effects of human activity on the stream. Biological stream monitoring is based on the fact that different species react to the pollution in different ways. Pollution-sensitive organisms like mayflies, stoneflies, and caddis flies are susceptible to the effects of physical or chemical changes in a stream and serve as pollution indicators. Pollution tolerant organisms are less susceptible to these changes. The presence or absence of indicator organisms is an indirect measure of pollution. When a stream becomes polluted, pollution-sensitive organisms and pollution tolerant organisms decrease and increase in number respectively. Benthic macroinvertibrates being sensitive to changes in the stream's over all ecological integrity offer other advantages to scientists looking for indications of stream pollution. They are sessile, easy to sample, abundant and can be easily collected and identified by skilled volunteers. Unlike, fish can escape toxic spills or degraded habitats by swimming away and migratory animals may spend only a small portion of their life cycles in a particular stream but most macroinvertebrates spend a large part of their life cycle in stream. The composition of a macroinvertebrate community in a stream reflects that stream's physical and chemical conditions of stream over time. Monitoring for certain water quality parameters describe the condition of the water. Benthic macroinvertebrates are a important part of the aquatic food web. They form an important link in the food chain and also reflects the stability and diversity of the larger aquatic food web.

The fauna living on the streambed react to changes in water quality (Marneffe *et al.*, 1996), and the subsequent species organization is probably going to reflect both the physical (e.g., particulate material, shading) and compound (e.g., supplements) nature of the waterway all in all. A few gatherings of macroinvertebrates are exceptionally delicate to the oxygen consumption that emerges as a result of any natural/supplement advancement thus won't get by under these conditions, while different gatherings flourish just when the opposition has been expelled by enhancement (Crawford *et al.*, 1992) Recently a noteworthy report surveying the effect of different cultivating frameworks (e.g., arable, meat, dairy, blended arable-hamburger) on catchment water quality was attempted in southwest and upper east of Scotland. The outcomes demonstrated that yields from even a solitary ranch created perceptible increments in stream water nitrate (Hooda *et al.*, 1997a) and phosphate (Hooda *et al.*, 1997b) focuses. The work exhibited in this paper included a study of benthic macroinvertebrates and related water quality/ecological segments in three streams depleting ranch estimated catchments, and framed a piece of a similar report. Mechanical and metropolitan waste contributions to waterways are known to significantly affect the benthic macroinvertebrates (Griffiths, 1991; Crawford *et al.*, 1992). It is generally seen that grassland farming frameworks are earth kind, however the effect of overflow from such frameworks on macroinvertebrates in catchment waters is moderately obscure.

CONCLUSION

The pesticides present in streams have undesirable effects on water quality parameters viz., temperature, dissolved oxygen, pH, may cause eutrophication that inturn effects the diversity of invertebrates in these ecosystems. Sometimes, the impacts are so significant that it may erode the chains of invertebrates from the streams.

Studies has shown that only few pesticides contribute to a major part of summed toxicity in streams, despite the large number of pesticides that occur in the samples and also the effects on algal growth indicating that pesticide effects on algae occur under in situ conditions. Effects on the macroinvertebrate communities were primarily explained by physiochemical conditions in the aquatic ecosystems, while pesticide toxicity did not contribute. This could be due to a low toxicity for macroinvertebrates or due to a bias caused by other stressors such as eutrophication or low habitat complexity. However, some organisms decreased (Asellus aquaticus) or increased (Oligochaeta) with pesticide exposure. It has been also found that physical water parameters such as temperature, pH etc plays a significant role in the toxicity of the pesticides. It is the need of time to integrate the studies of different disciplines including toxicology, environmental chemistry, population biology, community ecology, conservation biology and landscape ecology to understand direct and indirect effects of pesticides on the aquatic organisms.

Since there are limited studies that have studies the bioaccumulation of pesticides in different food chains and food web along with ecological assessment. There is a need to encourage the research towards this direction for better understanding.

ACKNOWLEDGMENT

We are highly thankful to Head, School of Studies in Zoology and Director, CSR& TI, Central Silk Board, Pampore for the necessary facilities.

REFERENCES

Bailey, H. C., Miller, J. L., Miller, M. J., Wiborg, L. C., Deanovic, L., & Shed, T. (1997). Joint acute toxicity of diazinon and chlorpyrifos to Ceriodaphnia dubia. *Environmental Toxicology and Chemistry*, *16*(11), 2304–2308. doi:10.1002/etc.5620161115

Barrion, A. T., & Litsinger, J. A. (1982). *The aquatic invertebrate fauna of IRRI ricefields*. Paper presented at the thirteenth National Conference of the Pest Control Council of the Philippines, Baguio City, Philippines.

Bayley, P. B., & Li, H. W. (1992). Riverine fishes. In The rivers handbook: hydrological and ecological principles (vol. 1.). Oxford, UK: Blackwell Scientific Publications.

Beketov, M., & Liess, M. (2012). Ecotoxicology and macroecology - Time for integration. *Environmental Pollution*, *162*, 247–254. doi:10.1016/j.envpol.2011.11.011 PMID:22243871

Belden, J. B., & Lydy, M. J. (2000). Impact of atrazine on organophosphate insecticide toxicity. *Environmental Toxicology and Chemistry*, *19*(9), 2266–2274. doi:10.1002/etc.5620190917

Bourgmann, N., & Morin, A. (1995). Relationship between size structure of invertebrate assemblages and trophy and substrate composition in streams. *Journal of the North American Benthological Society*, *14*(3), 393–403. doi:10.2307/1467205

Burton, G. A., Nelson, M. K., & Ingersoll, C. G. (1992). Freshwater benthic toxicity tests. In G. A. Burton Jr., (Ed.), *Sediment toxicity assessment* (pp. 213–240). Chelsea, MA: Lewis Publishers.

Clement, S. L., Grigarick, A. A., & Way, M. O. (1977). The colonization of California rice paddies by chironomid midges. *Journal of Applied Ecology*, *14*(2), 379–389. doi:10.2307/2402551

Conrad, A. U., Fleming, R. J., & Crane, M. (1999). Laboratory and field response of Chironomus riparius to a pyrethroid insecticide. *Water Research*, *33*(7), 1603–1610. doi:10.1016/S0043-1354(98)00397-2

Crawford, C. G., Wangsness, D. J., & Martin, J. D. (1992). Recovery of benthic invertebrate communities in the White River near Indianapolis, following advance treatment of municipal wastewater. *Archiv für Hydrobiologie*, *126*, 67–84.

Crosby, D. G., & Mabury, S. (1992). *Management of pesticide residues in ricefield soil water-a report to the Soil Conservation Service of the US Department of Agriculture*. Department of Environmental Toxicology, University of California-Davis.

Cummins, K. W. (1974). Structure and function of stream ecosystem. *Bioscience*, *24*(11), 631–641. doi:10.2307/1296676

Day, K. E. (1989). Acute, chronic and sublethal effects of synthetic pyrethroids on freshwater zooplankton. *Environmental Toxicology and Chemistry*, *8*(5), 411–416. doi:10.1002/etc.5620080507

Day, K. E., Kaushik, N. K., & Solomon, K. R. (1987). Impact of fenvalerate on enclosed planktonic communities and on in situ rates of filtration of zooplankton. *Canadian Journal of Fisheries and Aquatic Sciences*, *44*(10), 1714–1728. doi:10.1139/f87-210

Ecobichon, D. J. (1991). Toxic effects of pesticides. In M. O. Amdur & ... (Eds.), *Casarett and Doull's Toxicology- The basic science of poisons* (4th ed.; pp. 565–622). New York: McGraw-Hill, Inc.

Fernando, C. H., Furtado, J. I., & Lim, R. P. (1980). The ecology of rice fields with special reference to aquatic fauna. In J.I. Furtado (Ed.), *Tropical Ecology and Development: Proceedings of the International Society of Tropical Ecology* (pp. 943-951). Academic Press.

Francisco, P., Pierre, L., Claude, R., & Jean-P, V. (1988). Modeling the transport and the fate of pesticides in the unsaturated zone considering temperature effects. *Ecological Modelling*, *44*(1-2), 73–88. doi:10.1016/0304-3800(88)90083-X

Friberg, N., Lindstrom, M., Kronvang, B., & Larsen, S. E. (2003). Macroinvertebrate/sediment relationships along a pesticide gradient in Danish streams. *Hydrobiologia*, *494*(1-3), 103–110. doi:10.1023/A:1025493627272

Friberg-Jensen, U., Wendt-Rasch, L., Woin, P., & Christoffersen, K. (2003). Effects of the pyrethroid insecticide, cypermethrin, on a freshwater community studied under field conditions. I. Direct and indirect effects on abundance measures of organisms at different trophic levels. *Aquatic Toxicology (Amsterdam, Netherlands)*, *63*(4), 357–371. doi:10.1016/S0166-445X(02)00201-1 PMID:12758002

Goedkoop, W., & Peterson, M. (2003). The fate, distribution, and toxicity of lindane in tests with Chironomus riparius: Effects of bioturbation and sediment organic matter content. *Environmental Toxicology and Chemistry*, *22*(1), 67–76. doi:10.1002/etc.5620220109 PMID:12503748

Gorbach, S., Haaring, R., Knauf, W., & Werner, H. J. (1971). Residue analyses and biotests in rice-fields of East Java treated with thiodan. *Bulletin of Environmental Contamination and Toxicology*, *6*(3), 163–199. doi:10.1007/BF01539926

Grant, I. F., Roger, P. A., & Watanabe, I. (1986). Ecosystem manipulation for increasing biological nitrogen fixation by blue-green algae (Cyanobacteria) in lowland ricefields. *Biological Agriculture and Horticulture*, *3*(2-3), 299–315. doi:10.1080/01448765.1986.9754477

Grant, I. F., Tirol, A. C., Aziz, T., & Watanabe, I. (1983). Regulation of invertebrate grazers as a means to enhance biomass and nitrogen fixation of cyanophyceae in wetland ricefields. *Soil Science Society of America Journal*, *47*(4), 669–675. doi:10.2136ssaj1983.03615995004700040013x

Griffiths, R. W. (1991). Environmental-quality assessment of the St Clair River as refiected by the distribution of benthic macroinvertebrates in 1985. *Hydrobiologia*, *219*(1), 143–164. doi:10.1007/BF00024752

Grigarick, A. A., Webster, R. K., Meyer, R. P., Zalom, F. G., & Smith, K. A. (1990). Effects of pesticide treatments on non-target organisms in California rice paddies. *Hilgardia*, *58*, 1-36. doi:10.3733/hilg.v58n01p036

Gupta, A. K., & Rajbanshi, V. K. (1991). Toxicity of Copper and Cadmium to Heteropneustes fossils (Bloch). *Acta Hydrochimica et Hydrobiologica*, *19*(3), 331–340. doi:10.1002/aheh.19910190313

Heckman, C. W. (1979). Ricefield ecology in northeastern Thailand: The effect of wet and dry seasons on a cultivated aquatic ecosystem. In J. Illies (Ed.), Monographia Biologicae (Vol. 34). The Hague, The Netherlands: Junk.

Hedlund, M. (2002). *Deltametrins toxiska effekter för Chironomus riparius - en jämförande studie mellan kontaminerat sediment och vatten. Examensarbete, Sveriges Lantbruksuniversitet.* Uppsala, Sweden: SLU. (in Swedish)

Hirthe, G., Fisher, T. C., Crane, M., & Callaghan, A. (2001). Short-term exposure to sub-lethal doses of lindane affects developmental parameters in Chironomus riparius Meigen, but has no effect on larval glutathione-S-transferase activity. *Chemosphere, 44*(4), 583–589. doi:10.1016/S0045-6535(00)00471-9 PMID:11482645

Hooda, P. S., Moynagh, M., Svoboda, I. F., Thurlow, M., Stewart, M., Thomson, M., & Anderson, H. A. (1997a). Streamwater nitrate concentrations in six agricultural catchments in Scotland. *The Science of the Total Environment, 201*(1), 63–78. doi:10.1016/S0048-9697(97)84053-3 PMID:9232025

Hooda, P. S., Moynagh, M., Svoboda, I. F., Thurlow, M., Stewart, M., Thomson, M., & Anderson, H. A. (1997b). Soil and land use effects on phosphorus in six streams draining small agricultural catchments in Scotland. *Soil Use and Management, 13*(4), 196–204. doi:10.1111/j.1475-2743.1997.tb00586.x

Horne, A. J., & Goldman, C. R. (1994). Limnology (2nd ed.). McGraw-Hill.

Hurlbert, Mulla, Kieth, Westlake, & Dusch. (1970). Biological effects and persistence of Dursban in freshwater ponds. *Journal of Economics Entomology, 63*, 43-51.

Hurlbert, S. H., Mulla, M. S., & Wilson, H. R. (1972). Effects of an orthophosphorus insecticide on the phytoplankton, zooplankton and insect populations of freshwater ponds. *Ecological Monographs, 42*(3), 269–299. doi:10.2307/1942211

International Rice Research Institute (IRRI). (1986). *Annual reportfor 1985*. Manila, Philippines: IRRI.

Ishibashi, N., & Itoh, S. (1981). Effects of herbicide benthiocarb on fauna in paddy field [in Japanese]. *Proceedings of the Association for Plant Protection of Kyushu, 27*, 90–93. doi:10.4241/kyubyochu.27.90

Kohler, C. C., & Hubert, W. A. (1993). *Inland fisheries management in North America*. Bethesda, MD: American Fisheries Society.

Krishnan & Chockalingam. (1989). Toxic and sublethal effects of endosulfan and carbaryl on growth and egg production of Moina micruru Kurz (Cladocera: Moinidae). *Environmental Pollution, 56*, 319-326.

Liess, M., Brown, C. D., Dohmen, G. P., Duquesne, S., Hart, A., Heimbach, F., ... Tarazona, J. V. (2005). *Effects of pesticides in the field*. Berlin, Germany: SETAC press.

Liess, M., & Schulz, R. (1999). Linking insecticide contamination and population response in an agricultural stream. *Environmental Toxicology and Chemistry, 18*(9), 1948–1955. doi:10.1002/etc.5620180913

Lim, R. P. (1980). Population changes of some aquatic invertebrates in ricefields. In J. I. Furtado (Ed.), *Tropical Ecology and Development* (pp. 971–980). Kuala Lumpur, Malaysia: Proceedings International Society of Tropical Ecology.

Lim, R. P., Abdullah, M. F., & Fernando, C. H. (1984). Ecological studies of cladocera in ricefields of Tanjung Karang, Malaysia, subject to pesticide treatment. *Hydrobiologi'u, 113*(1), 99–103. doi:10.1007/BF00026596

MacArthur, R. H., & Wilson, E. O. (1967). *The theory of Island Biogeography*. Princeton, NJ: Princeton University Press.

Mackenthun, K. M. (1969). The practice of water pollution biology. U.S. Dept. of Interior, Federal Water Pollution Control Administration.

Marneffe, Y., Comblin, S., Bussers, J. C., & Thome, J. P. (1996). Biomonitoring of the water quality in the River Warche (Belgium): Impact of tributaries and sewage effluents. *Netherlands Journal of Zoology*, *46*(3), 337–350. doi:10.1163/156854295X00294

Maund, S. J., Peither, A., Taylor, E. J., Jüttner, I., Beyerle-Pfnür, R., Lay, J. P., & Pascoe, D. (1992). Toxicity of lindane to freshwater insect larvae in compartments of an experimental pond. *Ecotoxicology and Environmental Safety*, *23*(1), 76–88. doi:10.1016/0147-6513(92)90023-V PMID:1375150

Metcalfe-Smith, J. L. (1996). Biological water-quality assessment of rivers: use of macroinvertebrate communities. In G. Petts & P. Calow (Eds.), *River Restoration*. Oxford, UK: Blackwell Science.

Minshall, G. W. (1978). Autotrophy in stream ecosystems. *Bioscience*, *28*(12), 767–771. doi:10.2307/1307250

Morin, A., & Nadon, D. (1991). Size distribution of epilithic lotic invertebrates and implication for community metabolism. *Journal of the North American Benthological Society*, *10*(3), 300–308. doi:10.2307/1467603

Moss, B. (1988). *Ecology of Fresh waters, Man and Medium* (2nd ed.). Blackwell.

Mulla, M. S., & Mian, L. S. (1981). Biological and environmental impacts of the insecticides malathion and parathion on nontarget biota in aquatic ecosystems. *Residue Reviews*, *78*, 100–135. PMID:7017857

Norum, U., Friberg, N., Jensen, M. R., Pedersen, J. M., & Bjerregaard, P. (2010). Behavioral changes in three species of freshwater macroinvertebrates exposed to the pyrethroid lambda-cyhalothrin: Laboratory and stream microcosm studies. *Aquatic Toxicology (Amsterdam, Netherlands)*, *98*(4), 328–335. doi:10.1016/j.aquatox.2010.03.004 PMID:20362345

Odum, H. T. (1957). Tropic structure and productivity of Silver Springs, Florida. *Ecological Monographs, 27*, 55-112.

Pape-Lindstrom, P. A., & Lydy, M. J. (1997). Synergistic toxicity of atrazine and organophosphate insecticides contravenes the response addition mixture model. *Environmental Toxicology and Chemistry*, *16*(11), 2415–2420. doi:10.1002/etc.5620161130

Pedersen, M. L., & Friberg, N. (2009). Influence of disturbance on habitats and biological communities in lowland streams. *Fundamental and Applied Limnology; Official Journal of the International Association of Theoretical and Applied Limnology*, *174*(1), 27–41. doi:10.1127/1863-9135/2009/0174-0027

Power, E. A., & Chapman, P. M. (1992). Assessing sediment quality. In G. A. Burton Jr., (Ed.), *Sediment toxicity assessment* (pp. 1–18). Chelsea, MA: Lewis Publishers.

Resh, V. H., Brown, A. V., Covich, A. P., Gurtz, M. E., Li, H. W., Minshall, G. W., ... Wissmar, R. C. (1988). The role of disturbance in stream ecology. *Journal of the North American Benthological Society*, *74*(4), 433–455. doi:10.2307/1467300

Ristola, T., Pellinen, J., Ruokolainen, M., Kostamo, A., & Kukkonen, J. V. K. (1999). Effect of sediment type, feeding level, and larval density on growth and development of a midge (Chironomus riparius). *Environmental Toxicology and Chemistry*, *18*(4), 756–764. doi:10.1002/etc.5620180423

Roger, P. A., Grant, I. F., & Reddy, P. M. (1985). *Blue-green algae in India: A trip report.* Los Bailos, Philippines: IRRI.

Roger, P. A., Heong, K. L., & Teng, P. S. (1991). Biodiversity and sustainability of wetland rice production: role and potential of microorganisms and invertebrates. In D. L. Hawksworth (Ed.), *The Biodiversity of Microorganisms and Invertebrates: Its Role in Sustainable Agriculture* (pp. 117–136). Wallingford: CABI.

Ruttner, F. (1963). *Fundamentals of Limnology* (3rd ed.). Toronto: University Press Toronto.

Schulz, R. (2004). Field studies on exposure, effects, and risk mitigation of aquatic nonpoint-source insecticide pollution: A review. *Journal of Environmental Quality*, *33*(2), 419–448. doi:10.2134/jeq2004.4190 PMID:15074794

Schulz, R., & Liess, M. (1995). Chronic effects of low insecticide concentrations on freshwater caddisfly larvae. *Hydrobiologia*, *299*(2), 103–113. doi:10.1007/BF00017562

Schulz, R., & Liess, M. (2000). Toxicity of fenvalerate to caddisfly larvae: Chronic effects of 1-vs 10-h pulse-exposure with constant doses. *Chemosphere*, *41*(10), 1511–1517. doi:10.1016/S0045-6535(00)00107-7 PMID:11057675

Schulz, R., & Liess, M. (2001). Toxicity of aqueous-phase and suspended particle-associated fenvalerate: Chronic effects after pulse-dosed exposure of Limnephilus lunatus (Trichoptera). *Environmental Toxicology and Chemistry*, *20*(1), 185–190. doi:10.1002/etc.5620200121 PMID:11351407

Simpson, I. C., Roger, P. A., Oficial, R., & Grant, I. F. (1994). Effects of nitrogen fertilizer and pesticide management on floodwater ecology in a wetland ricefield: I. Experimental design and dynamics of the photosynthetic aquatic biomass. *Biology and Fertility of Soils*, *17*(2), 129–137. doi:10.1007/BF00337745

Streloke, M., & Kopp, H. (1995). Long-term toxicity tests with Chironomus riparius: development and validation of a new system Heft 315. Biologische Bundesanstalt für Land und Forstwirtschaft (BBA), Blackwell Wissenschafts-Verlag.

Sultana, R., & Devi, U. (1995). Oxygen consumption in a Cat fish Hystus gulio (Ham) exposed to heavy metals. *Journal of Environmental Biology*, *16*, 207–210.

Sweeney, B. W. (1993). Effects of streamside vegetation on microinvertebrate communities of White Clay Creek in Eastern North America. *Proceedings. Academy of Natural Sciences of Philadelphia*, *144*, 291–340.

Takaku, T., Takahashi, M., & Otsuki, A. (1979). Dispersion of an organophosphorus insecticide, fenitrothion, in paddy fields and its effect on microorganisms. *Japanese Journal of Limnology*, *40*(3), 137–144. doi:10.3739/rikusui.40.137

Takamura, K., & Yasuno, M. (1986). Effects of pesticide application on chironomid larvae and ostracods in ricefields. *Applied Entomology and Zoology*, *21*(3), 370–376. doi:10.1303/aez.21.370

Ulen, B., Kreuger, J., & Sundin, P. (2002). *Undersökning av bekämpningsmedel i vatten från jordbruk och samhälle år 2001 Rapport 2002:4*. Institutionen för Miljöanalys, Sveriges lantbruksuniversitet. (in Swedish)

U.S. Environmental Protection Agency. (2000). Methods for measuring the toxicity and bioaccumulation of sediment-associated contaminants with freshwater invertebrates (2nd ed.). U.S. Environmental Protection Agency, Office of Research and Development EPA/600/R-99/064.

Vannote, R. L., Minnshall, W. G., Cummins, K. W., Sedell, J. R., & Cushing, C. E. (1980). The river continuum concept. *Can J Aquat Sci*, *37*(1), 130–137. doi:10.1139/f80-017

Walburg, C. H. (1971). Loss of young fish in reservoir discharge and year class survival Lewis and Clark Lake, Missouri River. American Fisheries Society.

Ward, J. (1985). Thermal characteristics of running waters. *Hydrobiol*, *125*(1), 31–46. doi:10.1007/BF00045924

Wilson, J. T., Greene, S., & Alexander, M. (1980). Effects of microcrustaceans on bluegreen algae in flooded soil. *Soil Biology & Biochemistry*, *12*(3), 237–240. doi:10.1016/0038-0717(80)90067-X

Woin, P. (1998). Short- and long-term effects of the pyrethroid insecticide fenvalerate on an invertebrate pond community. *Ecotoxicology and Environmental Safety*, *41*(2), 137–156. doi:10.1006/eesa.1998.1694 PMID:9756702

Wong, M. C. (1979). *The ecology of the ostracod, Stenocypris major Baird 1859 with reference to the application of insecticides, carbofuran and endosulfan* (B.S. honour thesis). University of Malaya, Malaysia.

ADDITIONAL READING

Ansara-Ross, T. M., Wepener, V., van den Brink, P. J., & Ross, M. J. (2012). Pesticides in South African fresh waters. *African Journal of Aquatic Science*, *37*(1), 1–16. doi:10.2989/16085914.2012.666336

Baptista, D., Buss, D., Egler, M., Giovanelli, A., Silveira, M., & Nessimian, J. L. (2007). A multimetric index based on benthic macroinvertebrates for evaluation of Atlantic Forest streams at Rio de Janeiro State, Brazil. *Hydrobiologia*, *575*(1), 83–94. doi:10.100710750-006-0286-x

Beketov, M., Schäfer, R. B., Marwitz, A., Paschke, A., & Liess, M. (2008). Long-term stream invertebrate community alterations induced by the insecticide thiacloprid: Effect concentrations and recovery dynamics. *The Science of the Total Environment*, *405*(1-3), 96–108. doi:10.1016/j.scitotenv.2008.07.001 PMID:18760829

Rasmussen, J. J., Wiberg-Larsen, P., Baattrup-Pedersen, A., Monberg, R. J., & Kronvang, B. (2012). Impacts of pesticides and natural stressors on leaf litter decomposition in agricultural streams. *The Science of the Total Environment*, *416*, 148–155. doi:10.1016/j.scitotenv.2011.11.057 PMID:22177030

Smith, V. H., Tilman, G. D., & Nekola, J. C. (1999). Eutrophication: Impacts of excess nutrient inputs on freshwater, marine and terrestrial ecosystems. *Environmental Pollution*, *100*(1-3), 179–196. doi:10.1016/S0269-7491(99)00091-3 PMID:15093117

KEY TERMS AND DEFINITIONS

Abiotic Component: The nonliving component of environment.

Biotic Component: The living component of environment.

Ecosystem: It is a segment of nature consisting of biotic and abiotic components, both interacting with each other and exchanging materials between them.

Eutropication: The enrichment of a water body by nitrate and phosphate containing organic material leading to increase in growth of aquatic plants and often to algal blooms.

Invertebrate: Any animal that lacks a vertebral column or backbone.

Macrophytes: Large aquatic plants.

Organophosphorus Insecticides: These are synthetic pesticides widely used all over the world. They belong to the group of chlorinated hydrocarbon derivatives.

Pesticide: Any toxic substance used to kill animals or plants that cause damage to living organisms.

Synergism: Relating to the interaction of two or more substances to produce a combined effect greater than the sum of their separate effects.

Toxicity: Sum of adverse effects or the degree of danger posed by a substance to living organism.

Xenobiotic Compounds: Xenobiotic compounds are chemicals which are foreign to the biosphere.

Section 4

Pesticide Hazards, Analysis, and Purification

Chapter 12
Pesticides as an Occupational Hazard Facts and Figures

Nazia Tarannum
Chaudhary Charan Singh University, India

Meenakshi Singh
MMV BHU Varanasi, India

Ranjit Hawaldar
Centre for Materials for Electronics Technology, India

ABSTRACT

The chapter gives insight into the harmful use of pesticides in different professional environments. It portrays the use of pesticides as the potential risks to the health of users and third parties and a danger to the environment. The use of pesticides has increased at a phenomenal rate. Pesticides and their threat to the biological world have reached almost hysterical proportions. Their residues are found everywhere, particularly those of the so-called "hard pesticides" or organochlorine compounds, DDT. Herein, an attempt has been made to reflect pesticide exposure in different occupational settings and their harmful effects on humans. Excess use of pesticide in agriculture has placed workers in this industry at risk of lethal exposure. Personnel working in domestic pest control service is also from continuous exposure to the pesticide. Further, the chapter highlights various corrective measures to be taken by the people working in different occupational settings to combat the dangerous effects of pesticides in everyday life.

INTRODUCTION

Fungicide, Federal Insecticide and Rodenticide Act (FIFRA) (US EPA, 1947) define "Pesticides are substances or mixtures intended for preventing, destroying, repelling, any insects, rodents, nematodes, fungi or any other forms of life declared to be pests; designed for use as a plant regulator." Pesticides consist of various categories of toxins. Pesticide-related health effects can be acute or chronic. The duration of exposure, the dose of exposure and the route helps in determination of the severity of health effects. The intention of use defines pesticide validity as per the FIFRA. For example, bleach is used differently in

DOI: 10.4018/978-1-5225-6111-8.ch012

different places, viz as a (disinfectant) in bathrooms (Figure1), floors or tissue culture room (Figure 2) but it is not considered a pesticide when it is used for whitening of clothes. The United States uses over 1/4th of the world's total pesticide. It is the biggest pesticide hub for 4.5 billion pounds of chemicals per year required for the manufacture of 890 active ingredients, which are packed commercially into 30,000 formulations. 75% of pesticides used in agricultural settings and the rest 25% of structural fumigation, homes, gardens, hospitals, etc. In England, indiscriminate use of pesticides caused a reduction in the number of birds in gardens and farmland. An Indian Press Report (http://www.organicconsumers.org/Toxic/pepsi_coke_pesticides.cfm) confirmed the presence of high level of pesticides and insecticides such as lindane, DDT, malathion, chlorpyrifos in twelve popular soft-drink brands. The Food and Agriculture team of Greenpeace claimed that tea sold by some biggest tea brands in India contained pesticides. The team tested 49 samples, out of which 46-tested positive for one or more pesticide. The statistical data collected from vegetable samples of local vendors from different parts of the city Pune, India were inspected. A few samples revealed the presence of residues of banned pesticides such as DDT, chlordane, captafol, and carbofuran. The demands for imported chemicals by developing countries have extensively increased the value of pesticide purchase by Third World countries (World resources, 1986). Pesticide use in public health care and agriculture has impacted the lives of human. Reports from Central America suggested that large-scale spraying of DDT on cotton developed resistance to the pesticide in Anopheles, a malaria vector (Chapin & Wasserstrom, 1981). Further, in recent years, the number of reports of mortality and human poisonings due to pesticides are growing (Foo, 1985). The most glaring problem faced today in developing countries is the presence of organochlorine residues in foodstuffs like in poultry, red meat, and vegetables in Nigeria (Atuma, 1985), eggs in Kenya (Mugambi et al., 1989), and potatoes in Egypt (Lakwah et al., 1989). These residues get detected in human milk. A survey in Third World countries revealed levels of organochlorine residues in milk samples of nursing infants were greater than the acceptable daily intakes proposed by the Food and Agriculture Organization (FAO/WHO, 1988). Apart from the health disaster imposed by pesticide exposure, an environmental disaster is also

Figure 1. Use of bleach as a (disinfectant) in bathrooms

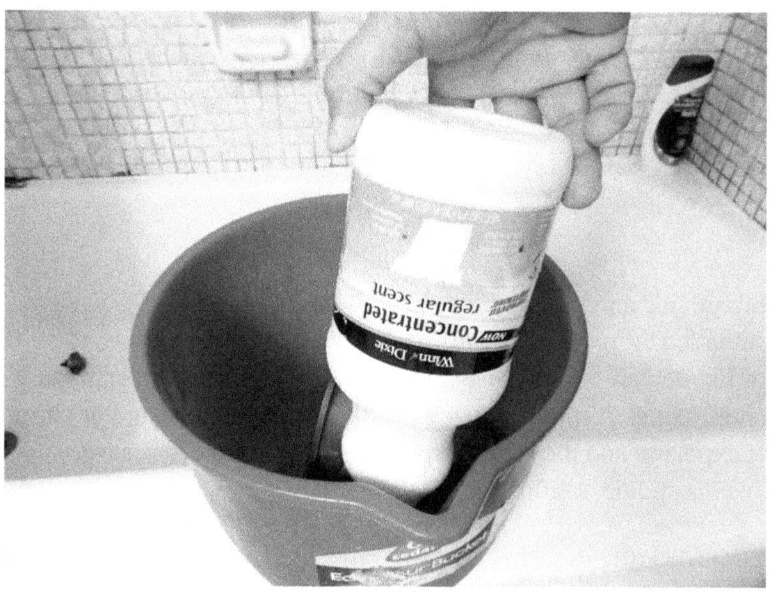

Figure 2. Use of bleach as a disinfectant in cell culture and tissue culture room

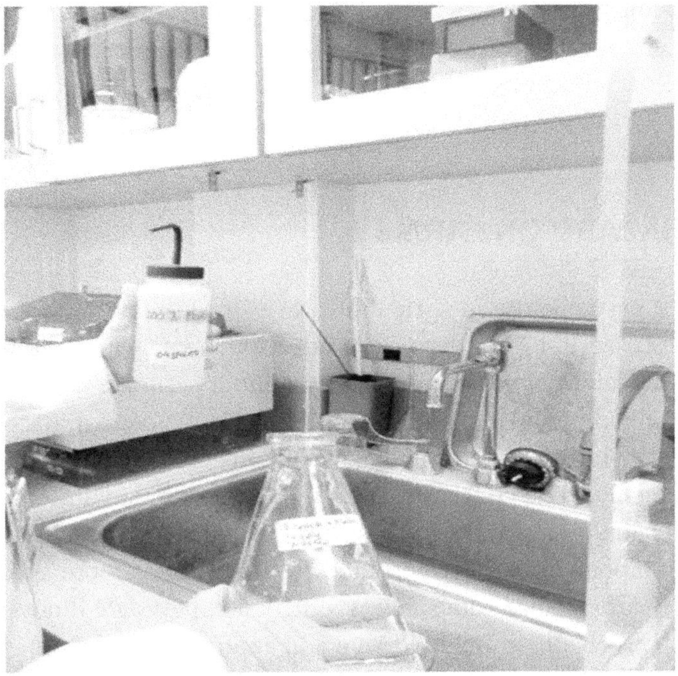

a matter of great concern. In rural areas, pesticide application equipment or empty pesticide containers are discarded and disposed in fields, on river banks and water bodies. Pesticide contamination reduces the soil organic, matter, changes soil chemistry, and pH. Hence, excessive use of a pesticide also impacts our natural resources like agricultural products, drinking water, air, and land.

Pesticide Exposure-Occupational Settings

- Multiple industries
    - Agriculture/Farms
    - Manufacturer and Transportation
    - Domestic Pest Control
- Variety of workers
    - Applicators, Field Workers, Firefighters, Scientific Research

Pesticides often spread to vicinity areas beyond their target due to the physical movement of pesticide via air, water, and aerosols at the time of application or afterward, a process known as drift. The drift of pesticide has resulted in 1/4th of pesticide-related illness reported in California in 2000. Schools, hospitals and other settings with sensitive subpopulations could be adversely affected by pesticide drift. Workers working in the field should be cautious as they may bring home pesticide from work through contaminated clothing or containers and expose family members. Secondary contamination of cars, furniture, and carpets at home happens by wearing contaminated clothing from work. Domestic exposure to bleach and other household disinfectants also result in pesticide exposure. Finally, these harmful pesticides enter our food cycle through food and may lead to low-level ingestion exposure, which may be dangerous especially to children. US EPA has set a maximum allowable reference dose for residues of pesticide in food as per the data provided by manufacturers. The recent regulatory decisions regarding concerns for infant exposure to pesticides include the elimination of the use of aldicarb on bananas (Wilson *et al.*, 2003).

OCCUPATIONAL HAZARD ON FARMS

In the US, the pattern of pesticide use in agriculture varies from region to region and by commodity to commodity. Pesticides used in high volume in production areas, which require hand labor, like orchards, nurseries, vineyards and raw vegetables. The amount of pesticide use is lower where production is mechanized like livestock dipping and grain agriculture. The workers in the fields are at a high-risk occupation with exposure of chemicals on a daily basis. Many of these pesticides may produce a carcinogenic effect, and some may cause congenital disabilities. California witnessed high rates of organophosphate pesticide poisoning in the agricultural sector during 1982 to 1990. According to the survey, the workers who loaded, applied or mixed pesticides were 35 times more susceptible to illness than the average rate. Further, surveillance data showed remarkable pesticide illness rates among people in the agricultural industry, especially in field workers and applicators (crop protection workers). As per the annual data of the United States, an estimated 5.1 billion pounds of pesticides were invested into plants and adverse health impacts of these chemicals (Figure 1) affected approximately 1-2 million farm workers. Further, the annual data by the Federal government also suggested acute pesticide poisonings among 10,000–20,000 employees in the agricultural industry. Short-term effects of pesticide exposures include nausea, rashes, blisters, stinging eyes, respiratory problems, and headaches. Cumulative long-term exposures may increase the threat of chronic health problems such as neurological impairments, congenital disabilities, cancer and Parkinson's disease for farmworkers, their families, and children. Manifestations of illness include detrimental effects on the central nervous system, causing impaired mental alertness, anxiety, and nervousness, which provoke sleep difficulties with nightmares, impairment of memory, apathy, and depression (Grob & Harvey, 1953; Durhami *et al.*, 1965). A medical history of organophosphorus substance exposure in population showed complaints about visual difficulty, forget-

Figure 3. Farm workers at highest risk of direct exposure to pesticides.

fulness, and persistent muscular pains along with a loss of interest in work (Metcalf & Holmes, 1969). Dermatitis and other skin diseases affect farmers handling pesticides (Spencer, 1966). Most challenging protocols in some countries are unrefined Worker Protection Standard. As a result, farm workers lack knowledge of hazardous pesticides used in professional setting and potential exposure to their families. One common occupational risks associated with farm workers include the unsanitary condition and lack of potable water. The farm workers, therefore, are at potential risk of exposure to harmful chemicals and waterborne parasites (Figure 3).

OCCUPATIONAL HAZARD IN DOMESTIC PEST CONTROL

In the domestic setting, there are various pests like beetles, lice, garden moths, ants, mites, mosquitoes that harbor home for their survival. Hence, an integrated pest management program strives to minimize pest risk through proofing, hygiene, and environmental management. The use of pesticide in home, gardens, hospitals, schools, offices, etc. leads to chemical exposure. Pest control management team to evacuate the place before fumigating the indented area. But residues of chemicals are left or are being carried by drift in the vicinity area which may contaminate their body, clothes or food they take. Further, the pest control personnel are also at the risk of endangerment due to constant exposure to hazardous chemicals (Figure 2). Several cases of the use of pesticides as a choice for suicide-homicide attempts got reported. Pesticides used in veterinary products for the treatment of fleas and ticks, which results in indirect exposure to the family. A pesticide may recklessly treat in homes for protection against mosquitoes, cockroaches or rodents. Some pesticides are used in schools or playgrounds to destroy weeds. Domestically, certain topical pharmaceuticals directly applied to children's skin or scalp, which may contain an insecticide to control scabies or lice. High doses or long-term exposure to such products may result in acute poisoning. The pharmaceuticals in International Pharmacopeia to treat head lice contain 2% of organophosphorus (Malathion) and organochlorine (Lindane) pesticides. Boric acid, also called hydrogen borate, often used as an antiseptic, insecticide, in house holds. Powdered roach-killing products that contain boric acid is hazardous and could cause acute or chronic poisoning if swallowed by someone (Figure 5).

Figure 4. Pest Control activity carried out by professionals inside home

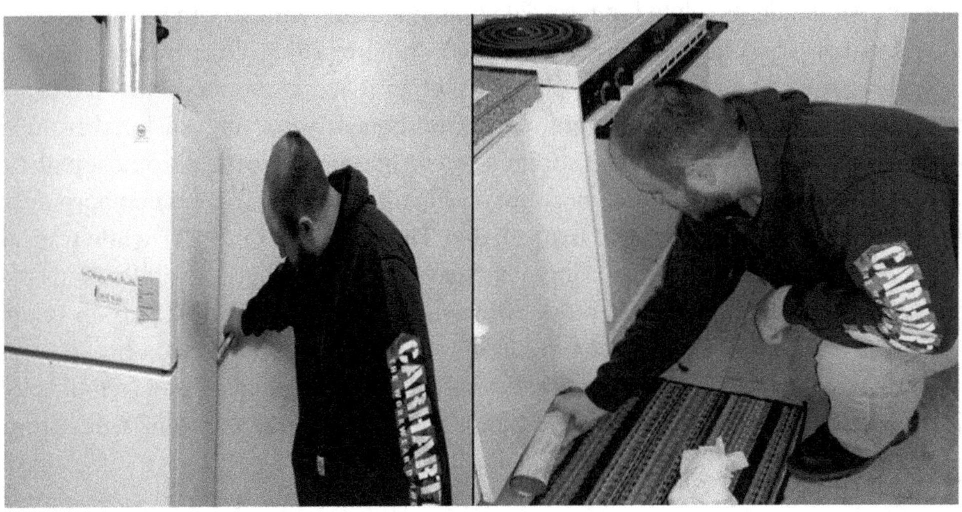

Figure 5. Roach control activity at home by using powdered roach-killing products containing boric acid

OCCUATIONAL HAZARD IN MANUFACTURE AND TRANSPORTATION OF PESTICIDES

The manufacturing and transporting of pesticides have resulted into an alarming occupational hazard. The pesticides manufactured in bulk are given by different means to dealers and distributors and finally reach the users. This whole procedure involves many people who directly or indirectly encounter harmful chemicals. Pesticides should be transported, stored and secured properly to protect people and environment. Cases of occupational deaths related to ammonia transportation were reported. Workers are exposed to various harmful chemicals for eight to ten hours daily in certain pesticide industries. They become a source of pesticides fragment carriers through their contaminated clothes, shoes, bags, and body to their families. The transportation of synthesized pesticides to the final destination is another challenge and a threat to the society. Accidents and damage by highly flammable or highly toxic pesticides during transportation and storage is alarming. Spills and leaks, which occur while transporting, lead to environmental pollution and menace to residential areas (Figure 6).

One of the terrible incidents related to pesticide manufacturing industry in history is the accident at Union Carbide's carbaryl and aldicarb plant in Bhopal, India in 1985. Over 60,000 pounds of methyl isocyanate, a toxic chemical and over 30,000 pounds of other reaction products spread like a toxic cloud over 15 square miles distance adjacent to the plant. Consequently, victims suffered from various ailments such as eye injury, lung damage, and suppression of immune system, chromosomal damage, and changes in blood chemistry, spontaneous abortions, congenital disabilities and newborn deaths (Mehta *et al.*, 1990). Another unfortunate incident happened in The Northwest in 1991 while transportation of toxic pesticides in bulk between manufacturing plant and final users where several gallons of carbamate soil fumigant spilled into Sacramento River, California from a derailed freight train (Garcia, 1992). Similarly, reports reveal a freeway accident, where ammonia discharged from a broken loading valve on a truck. The death caused much panic, and the residents were evacuated, and schools closed for the day. An episode happened during anhydrous ammonia transfer into delivery tank truck from a railroad

Figure 6. Schematic route of the pesticides from manufacture to its disposal

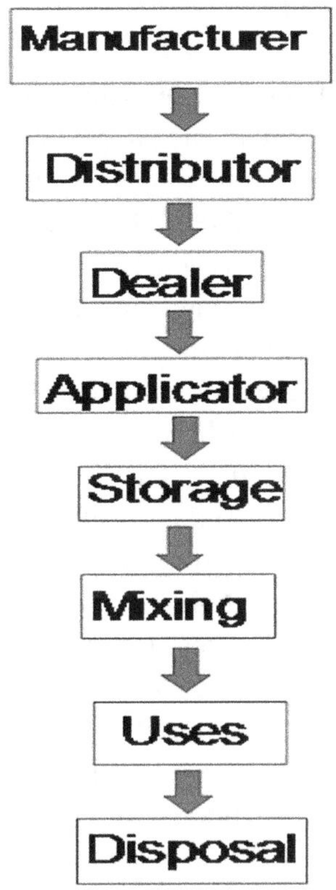

tank car in a fertilizer supply operation. The death occurred by inhalation when ammonia tank exploded at a chemical plant. Investigations revealed repeated exposure to methyl bromide while manufacturing or transportation led to many deaths. Methyl bromide exposure shows symptoms like nausea, vomiting, chest pain for several hours. The repeated exposure damage lungs, nervous system, and kidneys. Accidental contact with eyes or skin causes skin rashes and severe burns (West, 1964).

OCCUPATIONAL HAZARD VIA VARIETY OF WORKERS

Pesticide exposure occurs in various occupational settings. Besides the enormous use of pesticides in agriculture, farmers in the field or people involved in the transportation of pesticides are subjected to high exposure. Fieldworkers are exposed while pesticides application or picking the treated crops. Emergency response personnel viz firefighters may get exposure to victims soiled with pesticides at an accidental site. The exposure develops through body fluids, such as vomitus or stool, viscera or stained clothing of severely exposed patients. Some airlines treat aircraft with pesticides to control international vector transport. Flight attendants and passengers in cases exposed to pesticides within an enclosed area. Thus, subsequent exposure to pesticides may occur by contaminated victims to health care staff and emergency personnel. Thus, a corrective measure should be taken to control such exposure.

PESTICIDE EXPOSURE IN OCCUPATIONAL SETTINGS: A GROWING HAZARD TO HEALTH

Literature survey carried out so far, report a consistent pattern of health risk on the farmers who are prone to pesticide exposure (Antle *et al.*, 1995). The exposure to pesticides may result in acute (Ambridge, 1988; WHO/UNEP, 1990) and chronic health effects in humans. Health impairments of high concern include acute or chronic neurotoxicity (Tayler, 1976; Abou-Donia & Preissin, 1976; Abou-Donia & Preissig 1976; Abou-Doma & Preissig, 1976; Xintaras *et al.*, 1978; Savage, *et al.*, 1988; Eskenazi & Maizlish 1988; Robinson *et al.*, 1993; Kishi *et al.*, 1995), lung damage, respiratory failure (Ames, Brown, Mengle, Kahn, Stratton & Jackson, 1989) and infertility in men (Restrepo *et al.*, 1990; Rupa *et al.*, 1991). Various reports of cancers emerged due to pesticide exposure, such as hematopoietic types of cancer (Guzelian, 1980) and soft-tissue sarcomas (IARC, 1988). Chronic effects such as malfunction of the neurological system, cancer, reproductive disability, reduced growth or congenital disabilities in neonates have been reported due to devastating exposures. Cases of endocrine disruption (Rutherford, 1996; Colborn, Myers & Dumanoski, 1996) and allergic dermatitis sensitization (Adams, 1983) are also observed. Inhibition of cholinesterase results by organophosphates and N-methyl carbamate pesticides depress parasympathetic nervous system (Lopez-Carillo & Lopez-Cervantes, 1993; Miller & Shah, 1982).

Long-term pesticide exposure in different occupational settings leads to a challenging hike in early-onset of Parkinson's disease (Butterfield *et al.*, 1993; Gorell *et al.*, 1998). Studies on animal brains revealed some pesticides cause lasting behavioral aberrations during a critical period of cell division (Chanda & Pope, 1996; Eriksson, 1996). Agriculture workers are vulnerable to zoonotic infectious disease like methicillin-resistant Staphylococcus aureus (MRSA), avian flu, and West Nile virus etc. Farmworkers face chronic respiratory hazard by bioaerosols, such as organic specks of dust, endotoxins, and chemical toxicants from field agrochemicals (Kirkhorn & Garry, 2000). National Institutes of Health (NIH) investigated that workers using chlorinated pesticides exceeding hundred days developed the risk of diabetes (Montgomery, Kamel, Saldana, Alavanja & Sandler, 2008). Researchers involved with pesticide studies collaborates with industries like Imperial Cancer Research Fund in England and Hans-Olov Adami of the Karolinska Institute, Sweden for health and safety reasons (Hardell *et al.*, 2007). Intense exposure to fumigants may produce confusion, sleepiness, poor coordination or unconsciousness. Pyrethrin is extracted naturally from the flowers of chrysanthemum whereas pyrethroids obtained synthetically in pesticide laboratories. Pyrethrins and synthetic pyrethroids affect the central nervous system and may cause convulsions and poor coordination. Paraquat is a pyrylium herbicide, which if accidentally swallowed or inhaled may cause lung fibrosis. Asthma is another common respiratory disorder caused by pesticide exposures in multiple occupational settings (Selgrade *et al.*, 2006; Hernandez *et al.*, 2011). A case of persistent asthma due to inhalation of the organophosphate insecticide Dichlorvos reported in France (Deschamps *et al.*, 1994) while two instances of occupational asthma due to fungicides Fluazinam and Chlorothalonil (Draper *et al.*, 2003) observed in the UK. Further, in Belgium, a case of occupational asthma was found due to the chronic exposure to tetramethrin, a pyrethroid insecticide (Vandenplas *et al.*, 2000).

CORRECTIVE MEASURES TO CONTROL THE HAZARD

The common pesticides or metabolites detected in the general population are organophosphates, organochlorines, carbamates, herbicides, pest repellents and disinfectants. National Health and Nutrition Examination Survey (NHANES) in 1999-2000 examined blood and urine samples of selected participants for chemicals or their metabolites. The national report on human exposure to environmental chemicals showed bio monitoring exposure data on environmental chemicals for non-institutionalized US civilian population over two year periods. During 1993 to 1996, pesticides graded among main items involved in human exposures. Another common type of pesticide exposure was disinfectant. International treaties on "prior informed consent" and "persistent organic pollutants" aimed to reduce the global burden of hazardous or lethal chemicals that affected US exports. Inadequate law enforcement and lack of regulations in developing countries increased pesticide issues in the world (Calvert, *et al.*, 2001; Smith, 2001).

The Environmental Protection Agency (EPA) keeps a check on pesticide use as per Worker Protection Standard (WPS). WPS targets agricultural workers or pesticides handlers to control pesticide poisonings and injuries. WPS protocol compiles importance of pesticide safety training, personal protective equipment beside material and safety notification of pesticide's applications, decontamination supplies, and emergency medical assistance. The US EPA regulates and checks the sale and use of pesticides, and is responsible for health and safety of agricultural workers. A pesticide may be for a general or restricted use, and licensed applicators may use restricted pesticides. USEPA regulates pesticide label information such as pesticide toxicity, registration number, and recommended first aid measures. A pesticide label shows the toxicity category, highlighting "Danger" and "Caution" for high and low toxic pesticides respectively, along with essential precautionary statements (Rosenberg, & O'Malley, 1997). Many commercial pesticide formulations contain hazardous compounds, like piperonyl butoxide, synergist, a constituent in most pyrethrin or pyrethroid formulations. A significant portion of pesticide formulations comprised of "Inert" ingredients. "Inert" ingredients may not be active against the target pest but may cause some toxic effect on exposed personnel. Pesticide inventory is maintained to organize redundant storage. Proper disposal and controlling spillage of pesticide are also significant. Pesticide safety is necessary to limit access to pesticide storage facility to authorized personnel only. Security also includes not leaving pesticide containers or sprayers unattended anytime. Safety is critical, especially during transportation. Accidental pesticide spill can be observed by using specially equipped vehicles during transit. Side rails on trucks keep a check on falling cargo while making turns. Application equipment and bulk tanks, fittings, hoses should be checked periodically for leaks, cracks, or defects. Proper precaution should be taken in the event of pesticide spills, and drivers should be trained accordingly to deal with such a mishap. According to the U.S. Department of Transportation, specialized driver training and a license are mandatory.

Pest control service in residential areas should be used with caution. Items of use especially edibles must be covered or removed from areas to be treated. Pets should be displaced during treatment and until the pesticide is dry. It is critical to sterilize patients with pesticide exposure to prevent any chances of subsequent exposure. The detrimental effect of pesticide exposure depends on the degree of contamination and dose of pesticide. In the case of pesticide contact with skin, the individual should take a shower with soap and shampoo immediately. Spaces under fingernails, scalp, ear, etc. may act as a reservoir for the pesticide. Therefore, careful scrubbing of this area is required. (Reigart & Roberts, 1999) In a severely contaminated patient, body fluids, such as urine and stool, can be saved for residue analysis of pesticide. The infected body fluids of patients should be put in airtight containers in a secure place. Healthcare staff and forensic scientist may be exposed to passing pesticides during analysis. The symptoms of such exposure may vary from short-term eye irritation, headaches, rashes and odor-related symptoms to symptoms that are more persistent.

CONCLUSION

The objective of discussion of this chapter is to assess the health hazards during pesticides handling, storage and use to create awareness to control its exposure. Researchers advise biological monitoring techniques to reduce the intensity of absorption of selected pesticides. The common pesticides or metabolites detected in the general population are organophosphates, organochlorines, carbamates, herbicides, pest repellents, and disinfectants. Indiscriminate use of pesticidal chemicals is responsible for many complications in humans. Pesticides are poisons and should be used restrictively as per instructions. The exposure to pesticides in occupational settings may produce acute and chronic diseases. Employees dealing with pesticides must be aware of chemicals composition, first aid, and medical help in case of pesticide chemical accident. Therefore, taking necessary precautions during the use of pesticide may avoid the risk of diseases in human, plants and animals.

REFERENCES

Abou-Doma, M. B., & Preissig, S. H. (1976). Delayed neurotoxicity from continuous low-dose oral administration of leptophos to hens. *Toxicology and Applied Pharmacology*, *38*(3), 595–608. doi:10.1016/0041-008X(76)90190-3 PMID:65030

Abou-Donia, M. B., & Preissig, S. H. (1976). Delayed neurotoxicity of leptophos; toxic effects on the nervous system of hens. *Toxicology and Applied Pharmacology*, *35*(2), 269–282. doi:10.1016/0041-008X(76)90287-8 PMID:57652

Adams, R. M. (1983). *Occupational skin diseases*. New York: Grune and Stratton.

Ambridge, E. M. (1988). Pesticide management in relation to user safety. In R. T. Prinsley & P. J. Terry (Eds.), *Crop protection for small-scale farms in East and Central Africa'; A review* (pp. 98–105). Commonwealth Science Council.

Ames, R. G., Brown, S. K., Mengle, D. C., Kahn, E., Stratton, J., & Jackson, R. J. (1989). Cholinesterase activity depression among California agricultural pesticide applicators. *The American Journal of Emergency Medicine*, *15*, 143–150. PMID:2729280

Antle, J. M., Pingali, P. L., & Pingali, P. L. (Eds.). (1995). Pesticides, productivity and farmer health: a Philippine case study. Impact of pesticides on farmer health and the rice environment. Bulletin Environmental Contamination Toxicology, 35, 735-738.

Butterfield, P. G., Valanis, B. G., Spencer, P. S., Lindeman, C. A., & Nutt, J. G. (1993). Environmental antecedents of young-onset Parkinson's disease. *Neurology*, *43*(6), 1150–1158. doi:10.1212/WNL.43.6.1150 PMID:8170560

California Dept. of Health Services. Environmental Epidemiology and Toxicology Program. (1992, June). *Acute health effects of the Cantara metam sodium spill.* Author.

Calvert, G. (2001). Surveillance of pesticide-related illness and injury in humans. In R. Krieger (Ed.), *Handbook of Pesticide Toxicology* (2nd ed.; pp. 603–641). New York: Academy Press. doi:10.1016/B978-012426260-7.50030-6

Chanda, S. M., & Pope, C. N. (1996). Neurochemical and neurobehavioral effects of repeated gestational exposure to chlorpyrifos in maternal and developing rats. *Pharmacology, Biochemistry, and Behavior, 53*(4), 771–776. doi:10.1016/0091-3057(95)02105-1 PMID:8801577

Chapin, G., & Wasserstrom, R. (1981). Agricultural production and malaria resurgence in Central America and India. *Nature, 292*(5829), 181–185. doi:10.1038/293181a0 PMID:7278974

Colborn, T., Myers, J. P., & Dumanoski, D. (1996). *Our Stolen Future. Penguin Books.*

Deschamps, D., Questel, F., Baud, F. J., Gervais, P., & Dally, S. (1994). Persistent asthma after acute inhalation of organophosphate insecticide. *Lancet*, 344–352. PMID:7996993

Draper, A., Cullinan, P., Campbell, C., Jones, M., & Newman Taylor, A. (2003). Occupational asthma from fungicides fluazinam and chlorothalonil. *Occupational and Environmental Medicine, 60*(1), 76–77. doi:10.1136/oem.60.1.76 PMID:12499462

Durhami, W. F., & .... (1965). Organophosphorus insecticides and mental alertness. *Archives of Environmental & Occupational Health, 10*(1), 55–66. doi:10.1080/00039896.1965.10663953 PMID:14219489

El Lakwah, F., Meuser, F., Suckow, P., Abdel Gawaad, A. A., & Abdel Karim, K. (1989). Organochlorine insecticide residues in potatoes used for consumption in Egypt. In *Abstracts of the 3rd World Conference on Environmental and Health Hazards of Pesticides*. Academic Press.

Eriksson, P. (1996). Developmental neurotoxicology in the neonate-effects of pesticides and polychlorinated organic substances. *Archives of Toxicology. Supplement. Archiv fur Toxikologie. Supplement, 18*, 81–88.

Eskenazi, B., & Maizlish, N. (1988). Effects of occupational exposure to chemicals on neurobehaviourial functioning. In R. E. Tarter (Ed.), *Neuropsychological disorder in neural illness* (pp. 223–264). New York: Plenum Press.

FAO/WHO. (1988). Guidelines for predicting the dietary intake of pesticide residues. *Bulletin of the World Health Organization, 66*, 429–434. PMID:3262445

Foo, G. S. (1985). *The pesticide poisoning report: A survey of some Asian countries*. Kuala Lumpur, Malaysia: International Organization of Consumers' Unions.

Garcia, D. (1991, July 17). Spill threatening Shasta: Response to accident was slow. *San Francisco Chronicle,* pp. A1-4.

Gorell, J. M., Johnson, C. C., Rybicki, B. A., Peterson, E. L., & Richardson, R. J. (1998). The risk of Parkinson's disease with exposure to pesticides, farming, well water, and rural living. *Neurology, 50*(5), 1346–1350. doi:10.1212/WNL.50.5.1346 PMID:9595985

Grob, D., & Harvey, A. M. (1953). Effects and treatment of nerve gas poisoning. *The American Journal of Medicine, 14*(1), 52–58. doi:10.1016/0002-9343(53)90358-1 PMID:13016586

Guzelian, P. S. (1980). Liver structure and function in patients poisoned with chlordecone. *Gastroenterology*, *78*, 206–213. PMID:6153073

Hardell, L., Walker, M. J., Walhjalt, B., Friedman, L. S., & Richter, E. D. (2007). Secret ties to industry and conflicting interests in cancer research. *American Journal of Industrial Medicine*, *50*(3), 227–233. doi:10.1002/ajim.20357 PMID:17086516

Hernandez, A. F., Parron, T., & Alarcon, R. (2011). Pesticides and asthma. *Current Opinion in Allergy and Clinical Immunology*, *11*(2), 90–96. doi:10.1097/ACI.0b013e3283445939 PMID:21368619

IARC. (1988). *Overall evaluations of earcinogenicity: An updating of IARC monographs* (Vol. 1-42). International Agency for Research on Cancer.

Kirkhorn, S. R., & Garry, V. F. (2000). Agricultural Lung Diseases. *Environmental Health Perspectives*, *108*(4), 705–712.

Kishi, M., Hirschhorn, N., Djajadisastra, M., Saterlee, L. N., Strowman, S., & Dilts, R. (1995). Physiological effcts of work stress and pesticide exposure in tree planting by British Columbia silviculture workers. *Ergonomics*, *36*, 951–961.

Lopez-Carillo, L., & Lopez-Cervantes, M. (1993). Effect of exposure to organophosphate pesticides on serum cholinesterase levels. *Archives of Environmental Health*, *48*(5), 359–363. doi:10.1080/00039896.1993.9936726 PMID:8215602

Mehta, P. S., Mehta, A. S., Mehta, S. J., & Makhijani, A. B. (1990). Bhopal Tragedy's Health Effects: A review of Methyl Isocyanate Toxicity. *Journal of the American Medical Association*, *264*(21), 2781–2787. doi:10.1001/jama.1990.03450210081037 PMID:2232065

Metcalf, D. H., & Holmes, H. (1969). Toxicology and physiology; EEG, psychological, and neurological alterations in humans with organophosphorus exposure. *Annals of the New York Academy of Sciences*, *160*(1 Biological Ef), 357–365. doi:10.1111/j.1749-6632.1969.tb15857.x PMID:5257403

Miller, S., & Shah, M. A. (1982). Cholinesterase activities of workers exposed to organophosphorus insecticides in Pakistan and Haiti and an evaluation of the tintometric method. *Journal of Environmental Science and Health*, *17*(2), 125–130. doi:10.1080/03601238209372307 PMID:7077053

Montgomery, M. P., Kamel, F., Saldana, T. M., Alavanja, M. C., & Sandler, D. P. (2008). Incident diabetes and pesticide exposure among licensed pesticide applicators. *Agricultural Health Study, 1993–2003.*

Mugambi, J., Kanja, L., Maitho, T. E., Skaare, J. U., & Lokken, P. (1989). Organochlorine pesticide residues in domestic fowl (Gallus domesticus) eggs from central Kenya. *Journal of the Science of Food and Agriculture*, *48*(2), 165–176. doi:10.1002/jsfa.2740480205

Reigart, J. R., & Roberts, J. R. (1999). *Recognition and Management of Pesticide Poisonings* (5th ed.). US EPA.

Restrepo, M., Munoz, N., Day, N. E., Parra, J. E., De Romero, L., & Nguyen-Dinh, X. (1990). Testicular function in DBCP exposed pesticide workers. *Journal of Occupational Medicine and Toxicology (London, England)*, *21*, 161–166.

Robinson, D. G., Trites, D. G., & Banister, E. W. (1993). Neurotoxic concerns of human pesticide exposures. *American Journal of Industrial Medicine*, *18*, 327–331. PMID:2220839

Rosenberg, J., & O'Malley, M. (1997). Pesticides. In Occupational and Environmental Medicine (2nd ed.). Stamford, CT: Appleton Lange.

Rupa, D. S., Reddy, P. P., & Reddi, O. S. (1991). Prevalence of adverse reproduction outcomes in a population occupationally exposed to pesticides in Columbia. *Scandinavian Journal of Work, Environment & Health*, *16*, 232–238.

Rutherford, B. (1996). The synthetic chemical threat - endocrine disruptors. *The Journal of the Pesticide trust. Pesticide News*, *31*, 6–7.

Savage, E. P., Keefe, T. J., Mounce, L. M., Heaton, R. K., Lewis, J. A., & Bucar, P. J. (1988). Chronic neurological sequelae of acute organophosphate pesticide poisoning. *Archives of Environmental Health*, *43*(1), 38–45. doi:10.1080/00039896.1988.9934372 PMID:3355242

Selgrade, M. K., Lemanske, R. F. Jr, Gilmour, M. I., Neas, L. M., Ward, M. D., Henneberger, P. K., ... Yeatts, K. B. (2006). Induction of asthma and the environment: What we know and need to know. *Environmental Health Perspectives*, *114*(4), 615–619. doi:10.1289/ehp.8376 PMID:16581555

Smith, C. (2001). Pesticide exports from US ports, 1997-2000. *International Journal of Occupational and Environmental Health*, *7*(4), 266–274. doi:10.1179/oeh.2001.7.4.266 PMID:11783856

Spencer, M. C. (1966). Herbicide dermatitis. *Journal of the American Medical Association*, *198*(12), 1307–1308. doi:10.1001/jama.1966.03110250121037 PMID:5953527

Taylor, J. R. (1976). Neurologic disorder induced by Kepone. A preliminary report. *Neurology*, *26*, 358–363.

U.S. Environmental Protection Agency. (1947). *The Federal Insecticide, Fungicide, and Rodenticide Act of 1947*. Author.

Vandenplas, O., Delwiche, J. P., Auverdin, J., Caroyer, U. M., & Cangh, F. B. (2000). Asthma to tetramethrin. *Allergy*, *55*(4), 417–418. doi:10.1034/j.1398-9995.2000.00599.x PMID:10782541

West, I. (1964). Occupational hazards to pest control operators with special reference to pesticides. *Proceedings of the 2nd Vertebrate Pest Control Conference*, 3-6.

WHO/UNEP. (1990). *Public Health impact of pesticides used in agriculture*. Geneva: WHO.

Wilson, N. K., Chuang, J. C., Lyu, C., Menton, R., & Morgan, M. K. (2003). Aggregate exposures of nine preschool children to persistent organic pollutants at day care and at home. *Journal of Exposure Analysis and Environmental Epidemiology*, *13*(3), 187–202. doi:10.1038j.jea.7500270 PMID:12743613

World Resources Institute and the International Institute for Environment and Development. (1986). *World Resources*. New York, NY: Basic Books.

Xintaras, C., Burg, J. R., Tanaka, S., Lee, S. T., Johnson, B. L., Cottrill, C. A., & Bender, J. (1978). Occupational exposure to leptophos and other chemicals. Cincinnati, OH: DHEW (NIOSH).

ADDITIONAL READING

Ecobichon, D. J. (1998). *Occupational Hazards of Pesticide Exposure: Sampling, Monitoring*. Taylor and Francis.

Kumari, D., & John, S. (2017) Safety and Occupational Health Hazards of Agricultural Workers Handling Pesticides: A Case Study Conference paper *Advances in Health and Environment Safety*, 75-82

Long, K. R. (1971). Pesticides: An Occupational Hazard on Farms. *The American Journal of Nursing*, *71*(4), 740–743. doi:10.2307/3421761 PMID:5206465

KEY TERMS AND DEFINITIONS

Hazards: A hazard is an agent that has the potential to cause harm to a vulnerable target.

Pest Control: Pest control is the regulation or management of a species defined as a pest, a member of the animal kingdom that impacts adversely on human activities.

Pesticide Drift: Pesticide drift refers to the unintentional diffusion of pesticides and the potential negative effects of pesticide application.

Pesticide Handling: The use of pesticides in farm, forest, nursery, or greenhouse.

Chapter 13
Advances and Evolution of Techniques for Pesticide Estimation

Krishnasamy Lakshmi
Bharathiar University, India

S. Kanagasubbulakshmi
Bharathiar University, India

Krishna Kadirvelu
Bharathiar University, India

Venkatramanan Varadharajan
National Institute of Technology, India

ABSTRACT

Pesticides that are meant for the demolition of pests have become a nuisance for public health and environmental safety. Hence, the detection of pesticides in various types of samples such as food, environment, and even bodily fluids from various living entities is essential. Even though spectroscopic techniques are widely used for routine detection and quantification of pesticides, these techniques cannot determine pesticides with accuracy assured by chromatographic techniques. This chapter deals with various pesticide analysis techniques and the limitations associated with each technique.

INTRODUCTION

Pesticides are chemical compounds used for the demolition of various types of pests in agriculture. Pesticides can be classified according to the type of pests targeted as herbicides or weedicides (unwanted vegetation), insecticides (insects), rodenticides (rodents like rat), bactericides (bacteria), larvicides (insect larvae) and fungicides (fungus). As agricultural practices had seen major developments in last few decades owning to the increased requirement of food, the production and usage of pesticides got increased. Between 1960 and 2000, the global production of cereal grains had increased from 0.8 to

DOI: 10.4018/978-1-5225-6111-8.ch013

$2.0x10^9$ megatonnes. Similarly, pesticide production in global scale also got increased from 0.4 to $3.0x10^3$ tonnes within the same time period (Tilman et al., 2002). This exponential raise in pesticide production demonstrated the use of pesticides in almost all agricultural lands to improve productivity and vice versa. Increasing resistance of pests towards pesticides demands for higher application quantity per unit area and thereby requirement of more pesticides for agricultural fields. Alternatively, new potent pesticides that can eliminate pests at much lower doses are developed (Carvalho, 2006). Currently, natural and environmentally safe organic farming methods are gaining momentum although the expenses associated with it is high. As organic farming methods show far less productivity compared to conventional farming, it is less likely that organic farming can overhaul conventional farming practices. Hence, the production of pesticides is also less likely to get reduced drastically over the course of time.

Studies have found compelling evidences that prove detrimental effects of pesticides on human health and environment. In France, pesticide contamination of river streams have shown to expunged certain sensitive species and caused a 2.5 fold decrease in leaf-litter breakdown rate (Schäfer et al., 2006). Chlorothalonil, an organochlorine pesticide is correlated positively with an increase in mortality rate of various aquatic organisms such as amphibians, zooplanktons and algae. As a direct result of pesticide contamination, deregulations in various aquatic ecosystems such as decreased rate of organic decomposition, reduction in water clarity and increased dissolved oxygen were observed (McMahon et al. 2012). Apart from environmental toxicity, pesticide residues present in various consumer products affects general public health at large. Even though pesticides are designed specific for pests in agricultural fields, many off-target organisms that are exposed to pesticides undergo significant health related changes. Numerous reports have claimed various health-related issues associated with the mere exposure of toxic pesticides. One of the common biological effects associated with pesticide exposure is the inhibition of enzyme acetylcholine esterase (AChE) (Fulton & Key, 2001). This enzyme is responsible for the rapid clearance of acetyl choline, a neurotransmitter, from synaptic cleft. When inhibited, the absence of AChE leads to the accumulation of acetyl choline and thereby affecting the normal functioning of central nervous system. Apart from this, studies have correlated the pesticide exposure with a myriad of disorders including neurodegenerative conditions such as Parkinson's (Baltazar et al., 2014) and Alzheimer's disease (Singh et al., 2014), Dermatological conditions such as dermatitis (Sanborn et al., 2007), Reproductive dysfunctions (Sharpe & Irvine, 2004) such as fetal abortion and infertility and various types of cancers (Abdi et al., 2017).

Increasing pesticide contamination of various natural sources is considered one of the common factors for public exposure to pesticides. Natural sources such as rivers, lakes and ground water are easily contaminated by pesticides present in agricultural runoff water. Processing of pesticide contaminated water is absolutely necessary before releasing it to the natural sources or to direct public consumption. Many municipal wastewater treatment plants are inefficient in processing pesticide containing wastewater. Studies have found the trace levels of pesticides in both influent and effluent from wastewater treatment plants (Sadaria et al., 2017). Hence, unique processing and treatment methods specific to pesticides should be used prior to the release of water from treatment plants. Ground water contamination with pesticides is one of widely reported events that can result in direct public exposure to pesticides. In India, ground water contamination with organophosphorous and organochlorine pesticides has resulted in the presence of pesticide residues in packaged drinking water. Ground water is widely used for the manufacturing of packaged waters only after proper purification. As ground water contamination was evident and presence of pesticides in drinking water was confirmed, industries that process drinking water were asked to increase the stringency of their purification methods. Government of India has established the mandate requirement of high quality standards in bottle drinking water pertaining to the issue of pesticide residues in packaged water (Sharma, 2018).

Another major route of public exposure to pesticides is from the agricultural products. Various agricultural products such as grains, fruits and vegetable are found to be laced with trace levels of pesticides (Sun et al., 2018). Apart from this, consumer products that make use of agriculture based raw materials also contain pesticide residues and it includes cosmetic products that contain essential oils (Fillâtre, Gray, & Roy, 2017). As pesticide contamination remains a nuisance to public health and environment, detection of pesticides in various sources is needed prior to the determination of techniques to degrade or process it.

Pesticide analysis falls under general chemical analysis, where numerous techniques can be employed for accurate detection and estimation of pesticide quantity in various products. Some techniques allow for direct measurement of pesticides in the sample while others require pre-processing of the samples prior to the estimation. Currently, analytical techniques that employ chromatographic or spectroscopic methods are used for the detection and quantification of pesticides (Vidal et al., 2009). Novel sensors that can detect trace levels of pesticides in various samples are also developed. These sensors have nano-based materials or enzymes to detect the pesticides in water samples or agricultural products (Vakurov et al., 2004). This chapter is a comprehensive review about various pesticide analysis techniques and it also provides insight into pre-processing steps required for the analysis of pesticides.

SPECTROSCOPIC METHODS FOR PESTICIDE ANALYSIS

Spectrophotometric methods for estimation of pesticides rely on the principles of Beer-Lambert law. According to Beer-Lambert law, absorbance of light by the sample is directly proportional to the molar concentration of analyte present in the sample.

$$A = \varepsilon bc \tag{1}$$

Where, A is the absorbance of sample, ε is the molar absorptivity, b is the path length and c is the molar concentration of the analyte (Calloway, 1997). Usually, the absorbance of the sample is around the visible range. A color changing reaction accompanied by the addition of analyte is required for qualitative determination of pesticides. If the changing color intensifies with the addition of higher concentration of analyte, then it can be quantified using spectrophotometer. A few examples of pesticide estimation with spectrophotometry is given below.

Carboxin Estimation

Swarna et al., (2011) has proposed a colorimetric method for rapid estimation of fungicide carboxinusing spectrophotometer (Swarna et al., 2009). The authors used two different methods for quantification. The first one is based on the oxidative coupling of 1,10-phenonthrolin with carboxin to produce orange colored complex with maximum absorbance (λ_{max}) at 510 nm. The second method gave rise to similar colored complex upon reaction with 2,2′-bipyridine and it has λ_{max} value at 522 nm. These methods have limitations with respect to the concentration of samples. Both of these methods followed beer-lambert law at the concentration range of 10-100 μg/mL and 5-50 μg/mL respectively.

Oxyfluorfen Estimation

Oxyfluorfen is a diphenyl ether herbicide widely employed for control of monocotyledonous weeds. Alrahman et al. (2015) has given a spectroscopic method for the determination of oxyfluorfen. According to this method, an orange colored charge transfer complex will be formed upon treatment of oxyfluorfen with an electron acceptor (1,2- naphthoquinone-4-sulphonate) as oxyflurofen acts as an electron donor. The resulting complex exhibited λ_{max} at 460 nm. There are several factors to be considered before setting up of the reaction and it includes pH (optimum value at pH 13), buffer solution and reaction time. This method can analyze oxyfluorfen only within the concentration range of 0.4-4 µg/mL.

Carbofuran Estimation

Carbofuran is a carbomate pesticide used to control a variety of insects in various food crops. Jan et al. 2003 has employed an indirect method for quantitative estimation of carbofuran in various commercial products. Upon hydrolysis, carbofuran forms methylamine and it reacts with sodium nitroprusside to form a purple colored complex. The maximum aborbance of this colored complex was 530 nm. Instead of carbofuran, authors have used methylamine as standard to construct calibration plot. The measurement limitation was around 0.8-2.68 ppm.

Glyposate Estimation

A spectrophotometric method for estimation of potent herbicide glyposate in soil samples is conferred by Sharma et al. (2012). In this method, the amino group in glyphosate is transformed to dithiocarbomate derivative in acetonitrile with the aid of carbon disulphide and sodium bicarbonate. After reaction with copper perchlorate, a yellow colored complex with λ_{max} of 392 nmwas formed. The limitation of detection was in the range of 3.38-33.8 µg/mL.

Aminocarb and Carbaryl Estimation

Ni et al. (2009) has described a procedure for simultaneous estimation of aminocarb and carbaryl in the vegetable and water samples. This method make use of differential oxidation rates of these pesticides upon treatment with a strong oxidizing agent (potassium ferricyanide) in alkaline conditions. The lambda max value for the resulting oxidized complex was around 420 nm. Estimation limit for aminocarb and carbaryl using this method was in the range of 0.05-0.6 mg/L and 0.1-1.2 mg/L respectively. Using different chemometric methods such as partial least square and artificial neural networks, the presence of these pesticides in different samples can be predicted.

Paraquat Estimation

One of the commonly used herbicide paraquat (N,N′-dimethyl-4,4′-bipyridinium dichloride) was estimated spectrophotometrically by Ganesan et al. (1979). Upon reaction with teraiodomercurate (II), paraquat form a yellow colloidal solution with composition of paraquatHgI_4. This colloidal solution can be stabilized using stabilizing agents such as starch solution prior to measurement of absorbance at 400-420 nm. The linearity of calibration can be maintained till 3µg/mL.

Endosulfan Estimation

Endosulfan is an organochlorine, sulfur containing insecticide and acaricide which is banned by several countries. Raju et al. (1990) have given a simple protocol for estimation of endosulfan in soil and water samples using spectrophotometer. In this method, Sulphur dioxide from endosulfan was liberated and then absorbed by a suitable absorbant (malonyldihydrazide). This complex upon treatment with p-aminoazobenzene and formaldehyde forms a pink colored complex with λ_{max} of 505 nm. The concentration range at which beer-lambert law is obeyed was around 1-6 ppm.

Porpoxur Estimation

Porpoxur is a carbamate insecticide which is known for its long term effects in controlling various types of insect pests. Sastry et al. 1986 has provided a spectrophotometric method for measurement of propoxur and carbaryl in food samples like grains and pulses. The colorimetric reaction involves the addition of propoxur and carbaryl with p-dimethylaminocinnamaldehyde and p-dimethylaminobenzaldehyde to form colored complexes with λ_{max} values of 480 and 560 nm respectively. The beers limits for propoxur and carbaryl were 1.2-10 and 2-12 μg/mL.

Malathion and Dimethoate Estimation

Two organophosphorous pesticides malathion and dimethoate were estimated using three different methods by Gouda et al. (2010). The first two methods were based on the quantificantion of oxidant that remained in the reaction mixture containing pesticides in acidic medium (H_2SO_4) and excess of Ce^{4+}. The absorbance was measured at 523 nm if chromotrope 2R was used or 525 nm if rhodamine 6G was used. The third method was based on the oxidation of pesticides by N-bromosuccinamide and detection of unoxidized oxidant with amaranth dye and simultaneous measurement at 520 nm. The beer's detection limit was 0.1-4.2 μg/mL.

Chlorpyrifos Estimation

Chlorpyrifos is an organophosphate pesticides marketed under different brand names is widely used to demolish large variety of insects and worms. Shahabadi et al. (2013) has used a simultaneous method for quantification of chlorpyrifos and carbaryl in agricultural samples and boosted it with partial least square (PLS) method. In this method the pesticide containing samples were diluted with dimethylformamide and then mixed with citrate buffer to maintain pH at 5. The samples were then subjected to wavelength scan in the range of 250-375 nm. A calibration can be built with standards measured at several points within this wavelength range. The linearity of calibration can be maintained within 1.6 to 45 μg/mL for carbaryl and 1.5 to 50 μg/mL for chlorpyrifos.

Methyl Parathion Estimation

Methyl parathion is an organophosphorous insecticide with potent neurodegenerative effects on off-target organism. Gaur et al. (2008) has conferred a simple method for quantification of methyl parathion in various aqueous samples. Upon coupling of methyl parathion with alkaline medium, a greenish yellow complex was formed. This complex has maximum absorbance at 395 nm. The limit of calibration ranges from 0.1 to 1.5 μg/mL.Another method was provided by Tiwari et al. 2013 for measurement of methyl parathion in vegetable samples. In this method, an inhibition of redox reaction between bromate and hydrochloric acid takes place in the presence of methyl parathion. The reduction in neutral red during the course of the reaction was determined spectrophotometrically at 530 nm. The beer's law limit observed in this method was in the range of 0.025-0.3 μg/mL of methyl parathion.

Cypermethrin Estimation

A simple method for detection and estimation of widely used insecticide cypermethrin was reported by Janghel et al. 2007. In this method, the cyanide ions obtained from hydrolysis of cypermethrin in alkaline medium reacts with potassium iodide and leuco crystal violet to form a crystal violet complex with maximum absorbance at 595 nm. The linearity of calibration was maintained within the concentration range of 0.12-0.68 ppm. The structure and properties of pesticides described above are given in Table 1.

Table 1. Structure and properties of various pesticides

Pesticide	Chemical Formula	Structure	Molar Mass (g/mol)
Carboxin	$C_{12}H_{13}NO_2S$	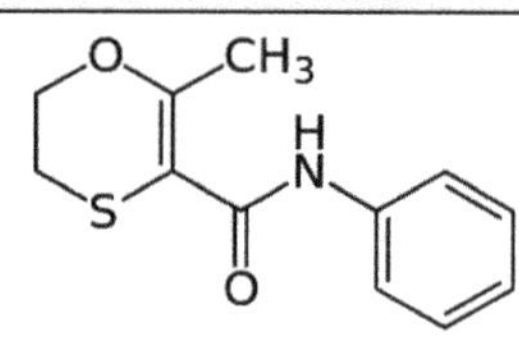	235.30
Oxyfluorfen	$C_{15}H_{11}ClF_3NO_4$	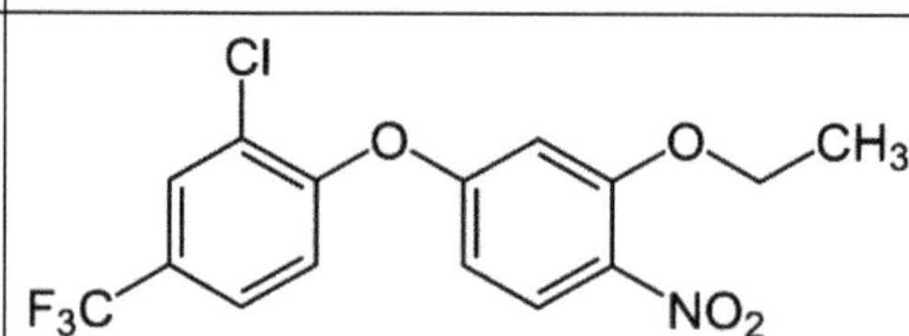	361.70
Carbofuran	$C_{12}H_{15}NO_3$	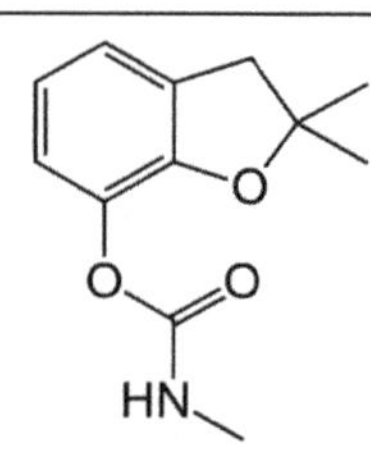	221.26
Glyposate	$C_3H_8NO_5P$	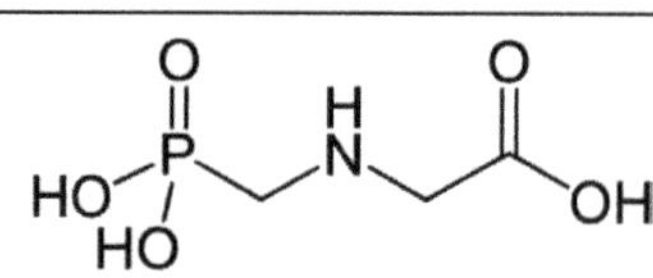	169.07
Aminocarb	$C_{11}H_{16}N_2O_2$	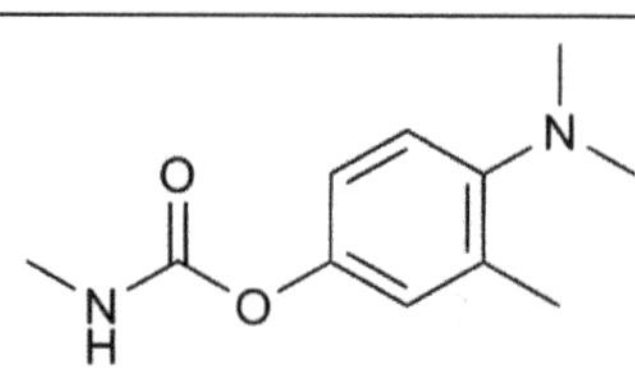	208.26

continued on following page

Table 1. Continued

Pesticide	Chemical Formula	Structure	Molar Mass (g/mol)
Carbaryl	$C_{12}H_{11}NO_2$	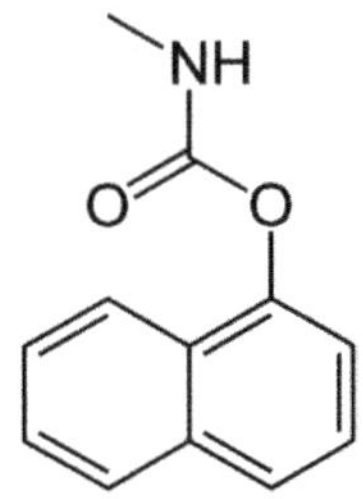	201.23
Paraquat	$C_{12}H_{14}Cl_2N_2$	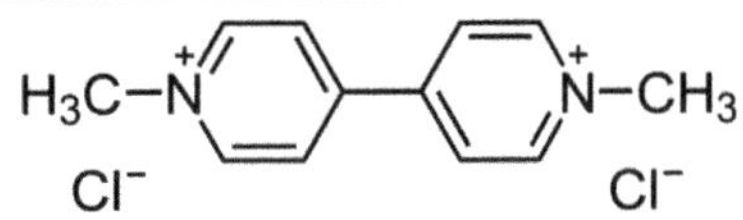	257.16
Endosulfan	$C_9H_6Cl_6O_3S$	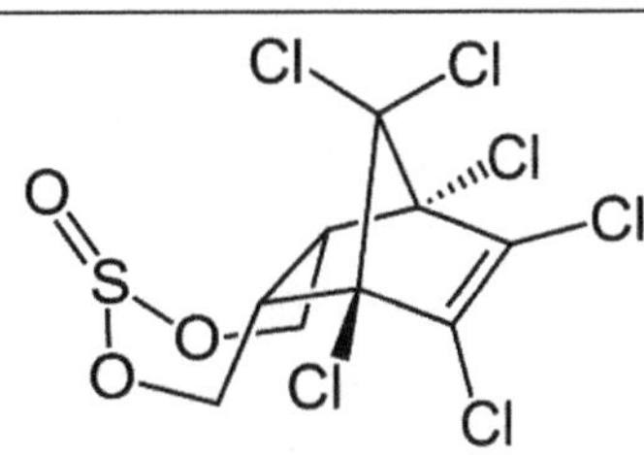	406.90
Porpoxur	$C_{11}H_{15}NO_3$	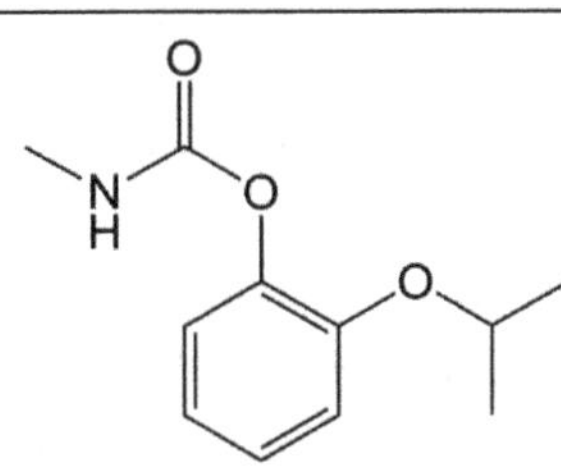	209.25
Malathion	$C_{10}H_{19}O_6PS_2$	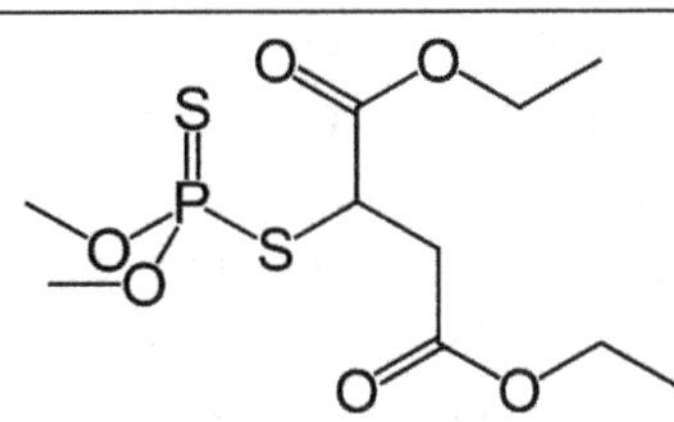	330.36
Dimethoate	$C_5H_{12}NO_3PS_2$	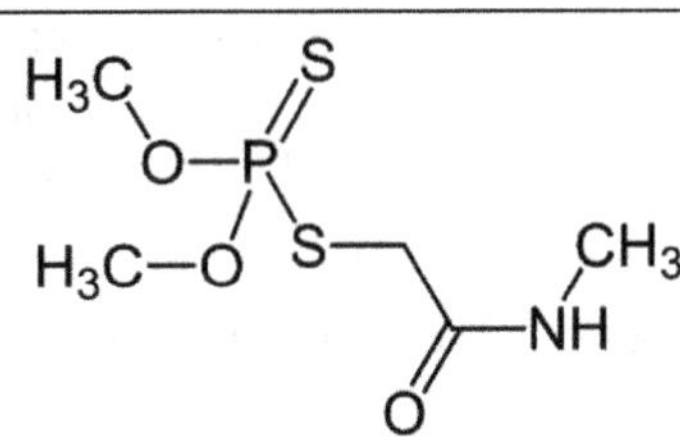	229.26
Chlorpyrifos	$C_9H_{11}Cl_3NO_3PS$	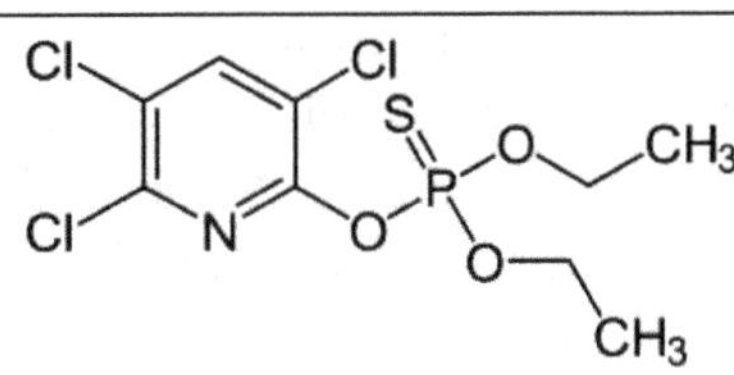	350.57

continued on following page

Table 1. Continued

Pesticide	Chemical Formula	Structure	Molar Mass (g/mol)
Methyl parathion	$(CH_3O)_2P(S)OC_6H_4NO_2$	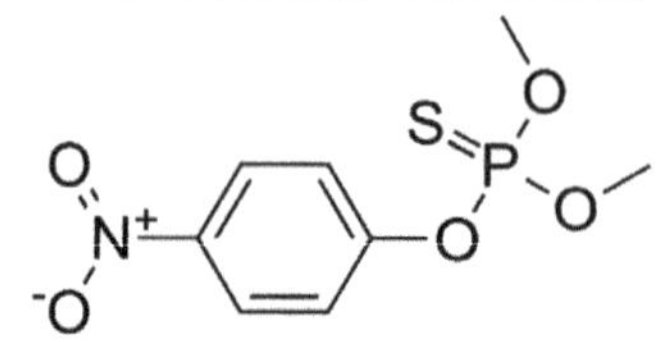	263.20
Cypermethrin	$C_{22}H_{19}Cl_2NO_3$	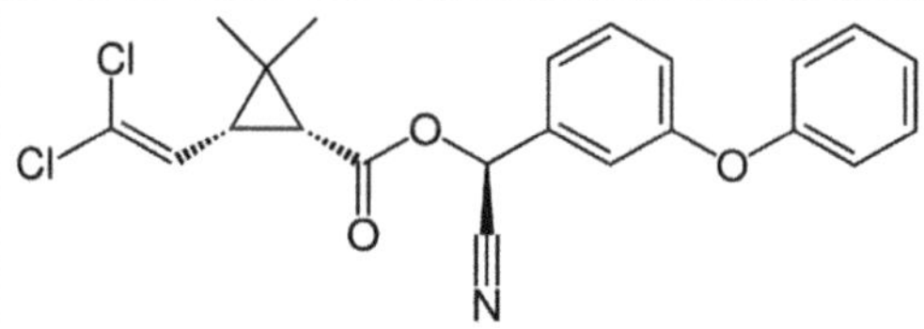	416.30

OTHER SPECTROSCOPIC METHODS

Apart from spectrophotometric method, various other spectroscopic methods such as raman spectroscopy and infrared spectroscopy are widely used for measurement of pesticides. Raman spectroscopy is used for the detection of various frequency modes of a chemical system including vibrational and rotational frequencies (Bowley et al., 2012). Identification of molecular structure can be done using the raman spectrum. Fan et al. (2015) had used surface-enhanced raman spectroscopy (SERS) coupled with multivariant analysis for the estimation of carbaryl in fuji apple samples. Suface enhanced raman spectroscopy has tremendous advantages over conventional spectroscopy as integration of analyte to metallic nanoparticles can enhance the raman scattering of analyte. Hence, detection of chemical species can be made at ease with surface enhancement (Nie & Emery, 1997). The authors have identified the lowest possible detection to be 0.5 μg/g. Further, quantification of carbaryl was done using PLS and vector regression. One of the important subjects of interest when it comes to the estimation using SERS is the processing of the samples prior to the detection. Since food sample are mostly analyzed by SERS, the complexity of food matrices play a major role in determining the accuracy of detection. Another important factor to be considered is the stability of the pesticide in the samples as most pesticides degrade to form secondary compounds that are difficult to detect using raman spectroscopy (Xu et al., 2017). Yang et al. (2016) has used SERS bases scattering mapping of pesticide penetration in edible leaf. In that study, the authors have utilized the penetrative ability of gold nanoparticles (AuNPs) (SERS substrate) for real time monitoring of penetration of thiabendazole and ferbam. A SERS mapping image of ferbam penetrated on spinach leaf is illustrated in Figure 1a. Raman spectroscopy can be coupled with different techniques for increasing the pesticide detection potential. Zhu et al. (2017) has integrated SERS with paper based microfluidics for detection of thiram residues. Paper based microfluidic devices act as highly portable capillary devices to facilitate the mixing of chemicals added to it at different locations. The authors have constructed a microfluidic device out of printed filter paper with hydrophilic sample addition chamber and hydrophobic mixing chamber. Upon on addition of AuNPs and thiram, they are carried to the mixing chamber by capillary action. The mixing chamber was directly subjected to raman spectroscopy for detection of thiram. The detection limit was found to be around 10^{-9} mol/L. Another

Figure 1. a) SERS based detection of pesticide penetration in spinach leaf and b) IR spectra of various pesticides present in algal samples
a) Reprinted with permission from Yang et al. (2016); b) Reprinted with permission from Shao et al. 2016.

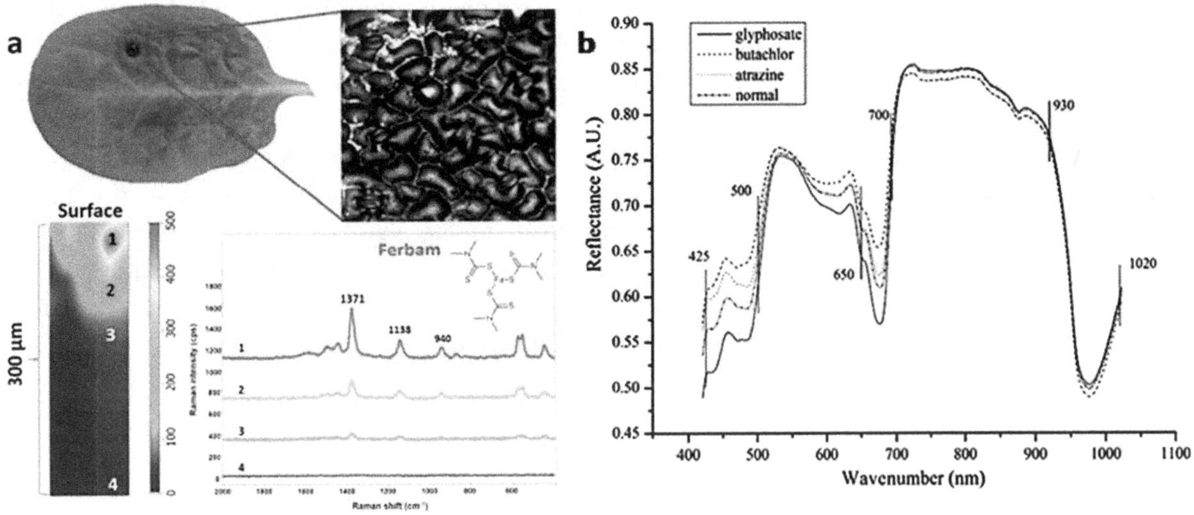

important spectroscopic method used for detection of pesticides is infrared spectroscopy. Martin et al. (2017) has established a protocol for detection of triadimefon using near infrared spectroscopy (NIR) with fiber optic probe. The quantification of triadimefon was carried out using PLS method. The authors have quantified different pesticide residues in the samples using Gas chromatography coupled with mass spectroscopy (GC-MS) prior to the measurement with NIR. Even though NIR is not as accurate in detecting the chemical components as GC-MS, the root mean square error was negligible (0.36). Shao et al. (2016) have used near/visible IR spectra for differentiating the pesticide variety present in algae samples. The authors have compared three different models to construct hyperspectral images for pesticides in algae. Among these models, linear discrimination analysis with regression coefficients was found best to produce hyperspectral image at classification rate of 97%. The IR spectrum of various pesticides in algal samples is depicted in Figure 1b. Mass spectroscopy (MS) is the highly used spectroscopic method for accurate detection of pesticide residues. As it is coupled with chromatographic techniques, it is described in detail in next section.

CHROMATOGRAPHIC TECHNIQUES FOR PESTICIDE ANALYSIS

Chromatographic techniques are widely employed for the accurate detection and quantification of pesticides in various samples. Even though direct surface measurement is not possible with chromatographic techniques. Samples are needed to be dissolved in suitable solvents prior to the estimation. In chromatography, the separation of constituents takes place by the principles of partitioning of chemical species in two different phases based on their relative affinities. Different types of chromatographic techniques are employed for the detection of chemical species including pesticides. The commonly used techniques include thin layer chromatography (TLC), Gas chromatography (GC) and liquid chromatography (LC and HPLC). Usually, GC and LC are coupled with MS for detection of pesticides.

PRINCIPLES OF CHROMATOGRAPHIC TECHNIQUES

There are three basic components of chromatography. They are stationary phase, mobile phase and separated molecules (Hansen & Reubsaet, 2015). Stationary phase comprises of solid phase or a liquid adsorbed solid phase (thin layer of liquid coating). Mobile phase is either liquid or gas phase. Molecular separation takes place based on the affinity of the molecule to stationary or mobile phase. Usually, analyte in suitable solvent is injected to the mobile phase of the chromatographic set up where it is carried to a column packed with stationary phase. If the compound exhibits higher affinity towards mobile phase, it will be the first to get detected by a detector system fixed with chromatographic set up. If it has higher affinity towards stationary phase, it will remain in stationary phase for a certain time and exit the system at end. The results of chromatographic experiments can be visualized as a chromatogram that comprises of a graph drawn between the detector response and elution time. From this chromatogram, the following information (Corradini, 2016) can be extrapolated.

Retention Time (t_R)

It is the time taken by the chemical species after injection till it is detected by the detector.

Retention Volume (V_R)

It is defines as the volume of mobile phase required for the complete elution of particular component of the sample injected to the chromatographic set up. It is given by the formula,

$$V_R = t_R \times F \quad (2)$$

Where, F is the volumetric flow rate of mobile phase.

Dead Time (t_m)

It is the time required for the unretained mobile phase to exit the chromatographic set up.

Adjusted Retention Time (t_R')

It is the excess in time required by the analyte to cross through the column beyond the time taken by unretained mobile phase. It is given by the equation,

$$t_R' = t_R - t_m \quad (3)$$

Relative Retention (α)

It is ratio of adjusted retension times of two different analytes in the sample. It is given by,

$$\alpha = t_{R\,1}^{'}/t_{R\,2}^{'} \quad (4)$$

Where, $t_{R\,1}^{'} > t_{R\,2}^{'}$. So, α is greater than 1.

Capacity or Retention Factor (k')

It is defined as the ratio of time spent by the analyte in stationary phase to the mobile phase. It is given by,

$$k^{'} = (t_{R} - t_{m})/t_{m} \quad (5)$$

THIN LAYER CHROMATOGRAPHY

Thin layer chromatography (TLC) is one of the planer chromatographic techniques used for the determination of pesticides in different types of samples (Sherma, 2016). A simple TLC set up consists of a thin sheet of silica gel coated matrix which act as stationary phase. The sample of analysis is usually placed in the corner of the silica coated plate and then allowed to dry. The sample containing TLC plates were kept in TLC chamber containing mobile phase (usually a solvent of the analyte). The mobile phase moves through the stationary phase along with the analytes due to capillary action. TLC plates will be removed from the chambers as mobile phase moves up to 2/3 distance of the length of the plates. After removal, the plates will be sprayed with reagents or kept in iodine or UV chambers for visualization of analyte spots. The determination of pesticide can be made by keeping an internal standard or measuring the retention factor, which is given by,

Retention factor (R_f) = Distance travelled by solute/Distance travelled by solvent (6)

A typical assembly of TLC is given in figure 2. TLC can be successfully integrated with other techniques such as MS. A variant of TLC called HPTLC (High performance thin layer chromatography) is an enhanced form of TLC with higher resolution and it is used for accurate quantitative estimations. Sample preparation for TLC based separations requires knowledge of suitable solvent systems. Generally, sample preparations follow stirring, soaking or shaking of samples with suitable solvents followed by liquid-liquid extraction of suitable components from the sample mixture. The solvent system is selected on the basis of polarity of the pesticide that needs to be separated out of the sample mixture.

Detection of pesticide by TLC based separation requires the spraying of reactive chemicals that can react with pesticide of interest to form colored complexes. Chromogenic or fluorogenic reagents are used for the detection of pesticides. In case of organopesticides containing nitrophenyl ring such as ethyl, methyl parathion and fenitrothion, 5% of stannous chlorine in acidic medium, 5% sodium nitrate and 0.5% iminodibenzyl in ethanol produces purple colored spots. Furfural spraying is widely employed for the detection of nitrobenxene insecticide (Sherma, 2017). Fluorescent silica gel can also be used to detect the separated components without spraying any chromogenic reagents. In such case, dark spots can be visualized agains bright fluorescent background of the silica gel. In HPTLC, the spots were analyzed using UV-Visible spectra and each of the spots is detected with MS.

Figure 2. A sketch of TLC experimental set up

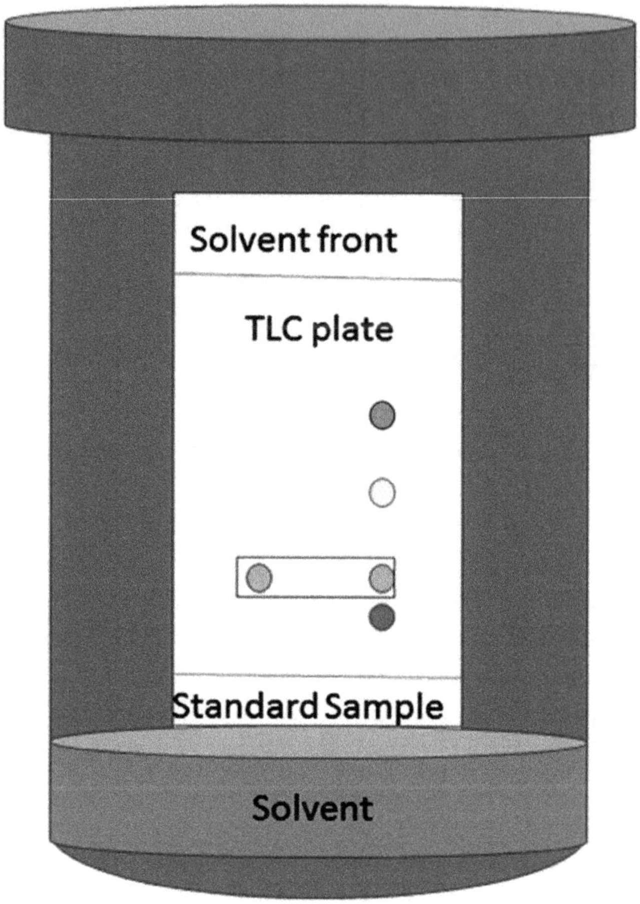

GAS CHROMATOGRAPHY

In gas chromatography, the mobile phase is always a gaseous system and usually an inert gas such as helium or nitrogen is used. About 90% of the GC system in current use operates with helium gas. But, higher resolution can be obtained with the use of nitrogen. Gas chromatography is usually suitable for volatile chemicals and chemicals that can resist thermal degradation during vaporization of samples during separation. A typical GC system is outlined in figure 3. It consists of a sample injector system and the common types of sample injection systems include split/splitless injector system, PTV injector or purge and trap system. In split/splitless injector system, a small chamber is located next to the injection assembly, which in turn heats up the sample and volatilizes it. Usually a portion of the volatilized sample or in some cases whole of the sample will be introduced to the mobile phase. PTV injector is the programmed temperature vaporizer,that uses low initial temperatures to maintain the samples below its boiling point on the contrary to hot injection systems. Finally, purge and trap system works by purging inert gas into samples to separate volatile insoluble materials to get trapped on the absorbent material. The volatile constituents are further discharged to the mobile phase (Dettmer-Wilde & Engewal,d 2016). The volume of sample injection will be in some micro liters (usually 100 μL).

Figure 3. A schematic representation of gas chromatography

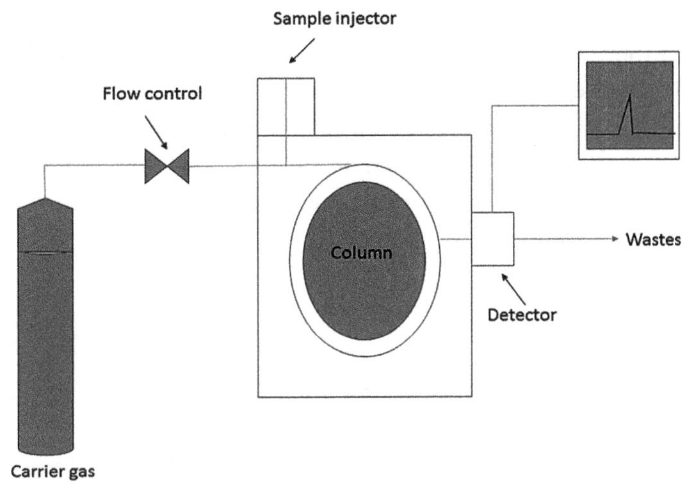

Upon introduction of the sample into the mobile phase, it reaches the packed column kept inside an electric oven. The selection of column is based on the nature of the sample such as polarity, functional group of the analyte to be measured and so on. Three types of columns are used for GC and it includes wall coated open tubular colunms (WCOT), surface coated open tubular (SCOT) and fused silica open tubular (FCOT). In WCOT, a thin flim of liquid is coated to the wall of the capillary tube. In SCOT, a thin solid matrix with adsorbed liquid is coated on the walls of the capillary tube. The wall of FCOT consists of a thin solid coating ossified by polyamide.

As the sample reaches the packed columns, the temperature of the oven laid a significant role in separating the constituents effectively. Isothermal conditions in the oven can be maintained throughout the analysis or the temperature inside oven can be varied over time by a temperature control. This temperature based control is termed as "Ramp" oven temperature program. It has two advantages including effective separation at the beginning of the elution and quicker elution of final component from the GC system[44].

After the separation of sample constituents, the detection of each of these constituents takes place due to the presence of suitable detector systems. A myriad of detecting platforms can be integrated to the GC system and it includes UV detectors, thermal conductivity detector (TCD), flame ionization detector (FID) and electron capture detectors (ECD). In TCD, the thermal conductivity of the sample constituents was measure using tungsten-rhenium filament kept in helium or nitrogen atmosphere where thermal conductivity is very. Upon interaction of analyte with filament, its thermal conductivity changes and gets recorded. In case of FID, current generated by the pyrolysis of carbonaceous compounds that release cations is measured. This method has very low detection limit. ECD detectors work by measuring the electric current deviation that occurs when halogen or nitro group containing compounds capture electron from carrier gases that generate electric current upon excitation with ionizing radiation (Bai et al., 2015; Cinelli et al., 2014).

Shamsipur et al., (2016) have used GC-MS method for detection of pesticide residues in water, milk, honey and fruit juices. The authors have used methanol and water for extraction of pesticides from the samples. Methanol was used further for purification and the optimal flowrate was found to be 15 mL/min. A total of 19 pesticides from dichlorovos to fenvalarate with retention time of 14.73 to 48.81 min was detected in the samples with detection limit of 1 – 10000 ng/Kg of samples. Naksen et al. (2016) have conferred a GC-FID based detection of organophosphorous pesticide in human plasma and breast milk. The authors have used acetone and methylene chloride for the extraction of pesticide from the samples and then purified it with the aid of aminopropyl extraction cartridges. Samples showed the presence of ethion and chlorpyrifos in the range of 54% and 94%. The detection limit of this method was in the range of 0.8 – 10 ng/L. Apart from this, many different kind of samples including consumer products like wine (dos Anjos and de Andrade 2015) and environmental samples such as ground water (Lafuente et al., 2016) can be tested for pesticide residues using GC.

LIQUID CHROMATOGRAPY

Liquid chromatography is widely used for the separation, detection and quantification of chemical compounds in variety of samples. It is even used in the manufacturing of certain drugs and biopharmaceuticals as a purification step (Gerber et al., 2004). In liquid chromatography, the mobile phase used is always a liquid. The instrumentation follows similar set up as GC but does not contain gas carrier or oven. A solvent reservoir system is placed and fitted with pumping system instead of the gas carrier. Currently, high pressure pumps that can pump mobile phase at the pressure of 100 – 350 bar is used in liquid chromatography. Hence, it is termed as high pressure liquid chromatography or high performance liquid chromatography (HPLC). A scheme of HPLC fitted with MS is depicted in figure 4. The mobile phase of the HPLC system can be pumped in different modes. One is isocratic mode in which the ratio of solvents used in mobile phase is pumped at constant rate throughout the separation process. The second is the gradient mode in which the solvent ratio is varied at different time intervals to ease the separation process at the end of the elution (MacNair, Patel, & Jorgenson, 1999). The sample with mobile phase gets separated at the stationary packed column that contains very fine sized particles of package materials to increase separation. HPLC can be operated in normal phase or reverse phase based on the polarity of the mobile phase and stationary phase. If the mobile phase consists of less polar solvents and stationary phase of polar package material like silica, it is termed as normal phase HPLC. If the mobile phase is highly polar and stationary phase is less polar (like C18 column), it is termed as reverse phase HPLC. The separated constituents can be collected on the waste or elution collector terminal in order of elution. It is widely applicable for purification process. If MS is coupled to HPLC system, the elution constituents will be passed through MS equipment. The detection and quantification of pesticides in HPLC is possible if pure standards are used prior to the estimation of sample. A linear calibration graph can be constructed based on the detector response of the standards and concentration of the standards, from which unknown quantity of pesticides in sample can be extrapolated. The detectors used in HPLC includes UV detectors, TCD and refractive index detectors (Chávez-Servın, Castellote, & López-Sabater, 2004).

Mass spectrometer consists of four components such as ionizer, electromagnet, ion selector and detector. The sample constitutions are ionized by an ionizing source such as electron beam which leads to the fragmentation of the molecules. Each of these fragments are sorted based on their charge to mass ratio using an ion selection assembly typically TOF (time of flight) or ion trap (orbitrap) systems. The selected ions are detected through various type of charge detectors such as electron multipliers (Hu et al., 2005). The fragmentation pattern based on charge to mass ratio is unique for every molecules and the pattern of unknown molecules can be checked against the MS spectral library to identify it, if the MS spectral of the unknown is available with the library.

Figure 4. Identification of pesticides through HPLC coupled with mass spectroscopy
Reprinted with permission from Niihori et al. (2018).

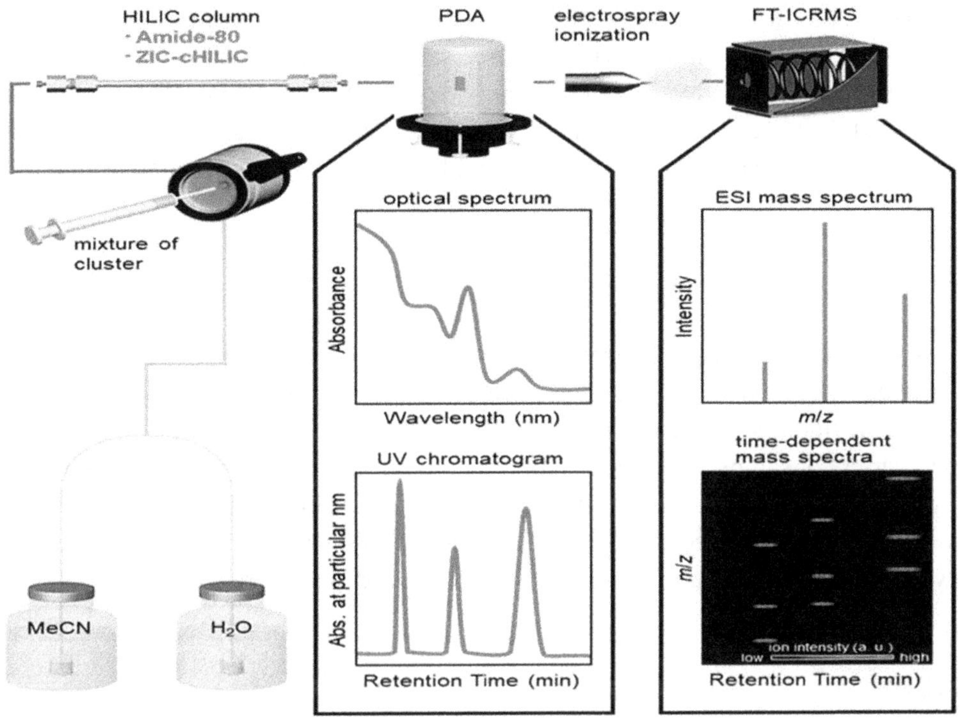

Table 2. Mass spectroscopy parameters for analysis of various pesticides

Pesticide	Chemical Group	t_R (min)	Precursor Ion, m/z	Product Ions, m/z (CE, eV)	LOD/LOQ (mg kg^{-1})	Recovery (%) mg kg^{-1}
2,4-D	phenoxy	2.52	219.2	125.0 (27) 161.0 (15)	0.001/0.004	104.2
atrazine	triazine	4.80	216.1	96.1 (23) 174.1 (18)	0.001/0.005	93.0
benfuracarb	carbamate	8.63	411.1	190.0 (13) 195.0 (23)	0.001/0.003	81.6
bromoxynil	nitrile	1.87	275.8	78.7 (17) 80.6 (17)	0.003/0.010	83.6
carbaryl	carbamate	4.19	202.0	117.0 (28) 145.0 (22)	0.001/0.002	99.2
diafenthiuron	thiourea	9.68	385.4	278.3 (33) 329.3 (19)	0.001/0.003	94.0
etoxazole	diphenyl oxazoline	9.41	360.5	141.2 (28) 177.3 (20)	0.001/0.003	93.8
mecarbam	organophosphate	6.78	330.0	97.0 (35) 227.1 (8)	0.001/0.003	94.0
propyzamide	benzamide	6.37	256.1	173.0 (23) 190.0 (16)	0.001/0.004	103.4
triadimenol	triazole	6.75	296.1	70.2 (10) 99.1 (15)	0.001/0.004	106.2

Golge & Kabak, 2018.

Alder et al. (2006) have compared the GC-MS and LC-MS techniques for the detection of large number of pesticides. The group had detection around 500 priority pesticides with GC and LC systems and found LC-MS as best for detection of almost all classes of pesticides except organochlorines based on the sensitivity and detection limit. A wide range of samples cam be used in LC based determination and estimation such as fruits and vegetables (Kmellár et al., 2010), grains (Pareja et al., 2007), milk (Mortensen et al., 2005), honey (Kasiotis et al., 2014) and environmental samples (Cappiello et al., 2002). A list of MS parameters for analysis of various pesticides is tabulated in Table 2.

CONCLUSION

A wide range of pesticides are currently used in the field of pest control that pose a severe threat to public and environmental wellbeing. In order to determine the trace levels of pesticides in various samples, adequate detection methods should be readily available. Different techniques are employed for detection of pesticides in variety of samples. Although a best method of detection is one that can detect nanogram quantities of pesticides in short period of time, numerous techniques lack either one of these advantages. From very routinely employed spectrometric techniques to complex chromatographic techniques efforts are taken towards the improvement of accuracy with broader detection range.

REFERENCES

Abdi, H., Lee, J., Ellison, G., Lai, G., & Lam, T. (2017). *Pesticides and primary liver cancer: A systematic review and meta-analysis*. Academic Press.

Adamovics, J. A. (Ed.). (2017). *Chromatographic analysis of pharmaceuticals*. Routledge.

Alder, L., Greulich, K., Kempe, G., & Vieth, B. (2006). Residue analysis of 500 high priority pesticides: Better by GC–MS or LC–MS/MS? *Mass Spectrometry Reviews*, *25*(6), 838–865. doi:10.1002/mas.20091 PMID:16755599

Alrahman, K. F., Elbashir, A. A., & El-obid, A. H. (2015). Development and Validation of Spectrophotometric Method for Determination of Oxyfluorfen Herbicide Residues. *Medicinal Chemistry (Shariqah, United Arab Emirates)*, *5*, 383–387.

Bai, L., Smuts, J., Walsh, P., Fan, H., Hildenbrand, Z., Wong, D., & Schug, K. A. (2015). Permanent gas analysis using gas chromatography with vacuum ultraviolet detection. *Journal of Chromatography. A*, *1388*, 244–250. doi:10.1016/j.chroma.2015.02.007 PMID:25724098

Baltazar, M. T., Dinis-Oliveira, R. J., de Lourdes Bastos, M., Tsatsakis, A. M., Duarte, J. A., & Carvalho, F. (2014). Pesticides exposure as etiological factors of Parkinson's disease and other neurodegenerative diseases—a mechanistic approach. *Toxicology Letters*, *230*(2), 85–103. doi:10.1016/j.toxlet.2014.01.039 PMID:24503016

Bowley, H. J., Gerrard, D. L., Louden, J. D., & Turrell, G. (2012). *Practical Raman spectroscopy*. Springer Science & Business Media.

Calloway, D. (1997). Beer-lambert law. *Journal of Chemical Education*, *74*(7), 744. doi:10.1021/ed074p744.3

Cappiello, A., Famiglini, G., Palma, P., & Mangani, F. (2002). Trace level determination of organophosphorus pesticides in water with the new direct-electron ionization LC/MS interface. *Analytical Chemistry*, *74*(14), 3547–3554. doi:10.1021/ac015685f PMID:12139067

Carvalho, F. P. (2006). Agriculture, pesticides, food security and food safety. *Environmental Science & Policy*, *9*(7-8), 685–692. doi:10.1016/j.envsci.2006.08.002

Chávez-Servın, J. L., Castellote, A. I., & López-Sabater, M. C. (2004). Analysis of mono-and disaccharides in milk-based formulae by high-performance liquid chromatography with refractive index detection. *Journal of Chromatography. A*, *1043*(2), 211–215. doi:10.1016/j.chroma.2004.06.002 PMID:15330094

Cinelli, G., Avino, P., Notardonato, I., Centola, A., & Russo, M. V. (2014). Study of XAD-2 adsorbent for the enrichment of trace levels of phthalate esters in hydroalcoholic food beverages and analysis by gas chromatography coupled with flame ionization and ion-trap mass spectrometry detectors. *Food Chemistry*, *146*, 181–187. doi:10.1016/j.foodchem.2013.09.064 PMID:24176330

Corradini, D. (Ed.). (2016). *Handbook of HPLC*. CRC Press.

Dettmer-Wilde, K., & Engewald, W. (2016). *Practical Gas Chromatography*. Springer-Verlag Berlin An.

dosAnjos, J. P., & de Andrade, J. B. (2015). Simultaneous determination of pesticide multiresidues in white wine and rosé wine by SDME/GC-MS. *Microchemical Journal*, *120*, 69–76. doi:10.1016/j.microc.2015.01.009

Fan, Y., Lai, K., Rasco, B. A., & Huang, Y. (2015). Determination of carbaryl pesticide in Fuji apples using surface-enhanced Raman spectroscopy coupled with multivariate analysis. *Lebensmittel-Wissenschaft + Technologie*, *60*(1), 352–357. doi:10.1016/j.lwt.2014.08.011

Fillâtre, Y., Gray, F. X., & Roy, C. (2017). Pesticides in essential oils: Occurrence and concentration in organic and conventional orange essential oils from eleven geographical origins. *Analytica Chimica Acta*, *992*, 55–66. doi:10.1016/j.aca.2017.08.039 PMID:29054150

Fulton, M. H., & Key, P. B. (2001). Acetylcholinesterase inhibition in estuarine fish and invertebrates as an indicator of organophosphorus insecticide exposure and effects. *Environmental Toxicology and Chemistry*, *20*(1), 37–45. doi:10.1002/etc.5620200104 PMID:11351414

Ganesan, M., Natesan, S., & Ranganathan, V. (1979). Spectrophotometric method for the determination of paraquat. *Analyst (London)*, *104*(1236), 258–261. doi:10.1039/an9790400258

Gaur, M. S., Sharma, A. K., Sharma, P., Mishra, V., & Tiwari, R. K. (2008). Spectrophotometric assessments of methyl parathion in water samples. *Can J Pure ApplSci*, *2*, 581–587.

Gerber, F., Krummen, M., Potgeter, H., Roth, A., Siffrin, C., & Spoendlin, C. (2004). Practical aspects of fast reversed-phase high-performance liquid chromatography using 3 µm particle packed columns and monolithic columns in pharmaceutical development and production working under current good manufacturing practice. *Journal of Chromatography. A*, *1036*(2), 127–133. doi:10.1016/j.chroma.2004.02.056 PMID:15146913

Golge, O., & Kabak, B. (2018). Pesticide residues in table grapes and exposure assessment. *Journal of Agricultural and Food Chemistry*, *66*(7), 1701–1713. doi:10.1021/acs.jafc.7b05707 PMID:29364655

González-Martin, M. I., Revilla, I., Vivar-Quintana, A. M., & Salcedo, E. B. (2017). Pesticide residues in propolis from Spain and Chile. An approach using near infrared spectroscopy. *Talanta*, *165*, 533–539. doi:10.1016/j.talanta.2016.12.061 PMID:28153295

Gouda, A. A., Amin, A. S., El-Sheikh, R., & Akl, M. A. (2010). Sensitive spectrophotometric methods for determination of some organophosphorus pesticides in vegetable samples. *Chemical Industry & Chemical Engineering Quarterly*, *16*(1), 11–18. doi:10.2298/CICEQ090417003G

Hansen, S. H., &Reubsaet, L. (2015). High-Performance Liquid Chromatography (HPLC) and High-Performance Liquid Chromatography–Mass Spectrometry (LC-MS). *Bioanalysis of Pharmaceuticals: Sample Preparation, Separation Techniques, and Mass Spectrometry*, 123-172.

Hu, Q., Noll, R. J., Li, H., Makarov, A., Hardman, M., & Graham Cooks, R. (2005). The Orbitrap: A new mass spectrometer. *Journal of Mass Spectrometry*, *40*(4), 430–443. doi:10.1002/jms.856 PMID:15838939

Jan, M. R., Shah, J., & Khan, H. (2003). Investigation of new indirect spectrophotometric method for the determination of carbofuran in carbamate pesticides. *Chemosphere*, *52*(9), 1623–1626. doi:10.1016/S0045-6535(03)00480-6 PMID:12867195

Janghel, E. K., Rai, J. K., Rai, M. K., & Gupta, V. K. (2007). A new sensitive spectrophotometric determination of cypermethrin insecticide in environmental and biological samples. *Journal of the Brazilian Chemical Society*, *18*(3), 590–594. doi:10.1590/S0103-50532007000300015

Kasiotis, K. M., Anagnostopoulos, C., Anastasiadou, P., & Machera, K. (2014). Pesticide residues in honeybees, honey and bee pollen by LC–MS/MS screening: Reported death incidents in honeybees. *The Science of the Total Environment*, *485*, 633–642. doi:10.1016/j.scitotenv.2014.03.042 PMID:24747255

Kmellár, B., Abrankó, L., Fodor, P., & Lehotay, S. J. (2010). Routine approach to qualitatively screening 300 pesticides and quantification of those frequently detected in fruit and vegetables using liquid chromatography tandem mass spectrometry (LC-MS/MS). *Food Additives and Contaminants*, *27*(10), 1415–1430. doi:10.1080/19440049.2010.490791 PMID:20582778

MacNair, J. E., Patel, K. D., & Jorgenson, J. W. (1999). Ultrahigh-pressure reversed-phase capillary liquid chromatography: Isocratic and gradient elution using columns packed with 1.0-μm particles. *Analytical Chemistry*, *71*(3), 700–708. doi:10.1021/ac9807013 PMID:9989386

McMahon, T. A., Halstead, N. T., Johnson, S., Raffel, T. R., Romansic, J. M., Crumrine, P. W., & Rohr, J. R. (2012). Fungicide-induced declines of freshwater biodiversity modify ecosystem functions and services. *Ecology Letters*, *15*(7), 714–722. doi:10.1111/j.1461-0248.2012.01790.x PMID:22587750

Mortensen, G. K., Main, K. M., Andersson, A. M., Leffers, H., & Skakkebæk, N. E. (2005). Determination of phthalate monoesters in human milk, consumer milk, and infant formula by tandem mass spectrometry (LC–MS–MS). *Analytical and Bioanalytical Chemistry*, *382*(4), 1084–1092. doi:10.100700216-005-3218-0 PMID:15933851

Naksen, W., Prapamontol, T., Mangklabruks, A., Chantara, S., Thavornyutikarn, P., Robson, M. G., & Panuwet, P. (2016). A single method for detecting 11 organophosphate pesticides in human plasma and breastmilk using GC-FPD. *Journal of Chromatography. B, Analytical Technologies in the Biomedical and Life Sciences*, *1025*, 92–104. doi:10.1016/j.jchromb.2016.04.045 PMID:27232054

Ni, Y., Xiao, W., & Kokot, S. (2009). Application of chemometrics methods for the simultaneous kinetic spectrophotometric determination of aminocarb and carbaryl in vegetable and water samples. *Journal of Hazardous Materials*, *168*(2-3), 1239–1245. doi:10.1016/j.jhazmat.2009.03.003 PMID:19349118

Nie, S. M., & Emery, S. R. (1997). Probing single molecules and single nanoparticles by surface-enhanced Raman scattering. *Science*, *275*(5303), 1102–1106. doi:10.1126cience.275.5303.1102 PMID:9027306

Niihori, Y., Shima, D., Yoshida, K., Hamada, K., Nair, L. V., Hossain, S., & Negishi, Y. (2018). High-performance liquid chromatography mass spectrometry of gold and alloy clusters protected by hydrophilic thiolates. *Nanoscale*, *10*(4), 1641–1649. doi:10.1039/C7NR07840D PMID:29192295

Pareja, L., Cesio, V., Heinzen, H., & Fernández-Alba, A. R. (2011). Evaluation of various QuEChERS based methods for the analysis of herbicides and other commonly used pesticides in polished rice by LC–MS/MS. *Talanta*, *83*(5), 1613–1622. doi:10.1016/j.talanta.2010.11.052 PMID:21238760

Raju, J., & Gupta, V. K. (1991). A simple spectrophotometric determination of endosulfan in river water and soil. *Fresenius' Journal of Analytical Chemistry*, *339*(6), 431–433. doi:10.1007/BF00322365

Rodriguez-Lafuente, A., Piri-Moghadam, H., Lord, H. L., Obal, T., & Pawliszyn, J. (2016). Inter-laboratory validation of automated SPME-GC/MS for determination of pesticides in surface and ground water samples: Sensitive and green alternative to liquid–liquid extraction. *Water Quality Research Journal*, *51*(4), 331–343. doi:10.2166/wqrjc.2016.011

Sadaria, A. M., Sutton, R., Moran, K. D., Teerlink, J., Brown, J. V., & Halden, R. U. (2017). Passage of fiproles and imidacloprid from urban pest control uses through wastewater treatment plants in northern California, USA. *Environmental Toxicology and Chemistry*, *36*(6), 1473–1482. doi:10.1002/etc.3673 PMID:27808432

Sanborn, M., Kerr, K. J., Sanin, L. H., Cole, D. C., Bassil, K. L., & Vakil, C. (2007). Non-cancer health effects of pesticides: Systematic review and implications for family doctors. *Canadian Family Physician Medecin de Famille Canadien*, *53*(10), 1712–1720. PMID:17934035

Sastry, C. S. P., Vijaya, D., & Rao, K. E. (1986). Sensitive spectrophotometric determinations of carbaryl and propoxur in formulations, water, grains and pulses. *Food Chemistry*, *20*(2), 157–162. doi:10.1016/0308-8146(86)90152-4

Schäfer, R. B., Caquet, T., Siimes, K., Mueller, R., Lagadic, L., & Liess, M. (2007). Effects of pesticides on community structure and ecosystem functions in agricultural streams of three biogeographical regions in Europe. *The Science of the Total Environment*, *382*(2-3), 272–285. doi:10.1016/j.scitotenv.2007.04.040 PMID:17555800

Shamsipur, M., Yazdanfar, N., & Ghambarian, M. (2016). Combination of solid-phase extraction with dispersive liquid–liquid microextraction followed by GC–MS for determination of pesticide residues from water, milk, honey and fruit juice. *Food Chemistry*, *204*, 289–297. doi:10.1016/j.foodchem.2016.02.090 PMID:26988504

Shao, Y., Jiang, L., Zhou, H., Pan, J., & He, Y. (2016). Identification of pesticide varieties by testing microalgae using Visible/Near Infrared Hyperspectral Imaging technology. *Scientific Reports*, *6*(1), 24221. doi:10.1038rep24221 PMID:27071456

Sharma, A. (2018). Science-based mandatory standards and the implementation gap: the case of bottled water regulations in India. *Current Science, 114*(1).

Sharma, D. K., Gupta, A., Kashyap, R., & Kumar, N. (2012). Spectrophotometric method for the determination of Glyphosate in relation to its environmental and toxicological analysis. *Arch. Environ. Sci*, *6*, 42–49.

Sharpe, R. M., & Irvine, D. S. (2004). How strong is the evidence of a link between environmental chemicals and adverse effects on human reproductive health? *BMJ (Clinical Research Ed.)*, *328*(7437), 447–451. doi:10.1136/bmj.328.7437.447 PMID:14976101

Sherma, J. (2016). Thin layer chromatography: Recent advances. *Thin-Layer and Liquid Chromatography and Pesticides of International Importance: Analytical Methods for Pesticides and Plant Growth Regulators*, *7*, 1.

Sherma, J. (2017). Review of thin-layer chromatography in pesticide analysis: 2014–2016. *Journal of Liquid Chromatography & Related Technologies*, *40*(5-6), 226–238. doi:10.1080/10826076.2017.1298024

Singh, N. K., Banerjee, B. D., Bala, K., Basu, M., & Chhillar, N. (2014). Polymorphism in cytochrome P450 2D6, glutathione S-transferases Pi 1 genes, and organochlorine pesticides in Alzheimer disease: A case–control study in North Indian population. *Journal of Geriatric Psychiatry and Neurology*, *27*(2), 119–127. doi:10.1177/0891988714522698 PMID:24584466

Sun, J., Pan, L., Tsang, D. C., Li, Z., Zhu, L., & Li, X. (2018). Phthalate esters and organochlorine pesticides in agricultural soils and vegetables from fast-growing regions: A case study from eastern China. *Environmental Science and Pollution Research International*, *25*(1), 34–42. doi:10.100711356-016-7725-7 PMID:27738865

Swarna, C., Naidu, K. P., Nagendrudu, G., Naidu, N. V. S., & Saraswathi, K. (2011). Spectrophotometric Method for the Determination of Carboxin in its Formulations and Environmental Samples. *Journal of Chemistry*, *8*(4), 1680–1685.

Tilman, D., Cassman, K. G., Matson, P. A., Naylor, R., & Polasky, S. (2002). Agricultural sustainability and intensive production practices. *Nature*, *418*(6898), 671–677. doi:10.1038/nature01014 PMID:12167873

Tiwari, N., Asthana, A., & Upadhyay, K. (2013). Kinetic-spectrophotometric determination of methyl parathion in water and vegetable samples. *Spectrochimica Acta. Part A: Molecular and Biomolecular Spectroscopy*, *101*, 54–58. doi:10.1016/j.saa.2012.09.084 PMID:23099160

Vakurov, A., Simpson, C. E., Daly, C. L., Gibson, T. D., & Millner, P. A. (2004). Acetylcholinesterase-based biosensor electrodes for organophosphate pesticide detection: I. Modification of carbon surface for immobilization of acetylcholinesterase. *Biosensors & Bioelectronics*, *20*(6), 1118–1125. PMID:15556357

Vidal, J. M., Plaza-Bolanos, P., Romero-González, R., & Frenich, A. G. (2009). Determination of pesticide transformation products: A review of extraction and detection methods. *Journal of Chromatography. A*, *1216*(40), 6767–6788. doi:10.1016/j.chroma.2009.08.013 PMID:19720377

Xu, M. L., Gao, Y., Han, X. X., & Zhao, B. (2017). Detection of Pesticide Residues in Food Using Surface-Enhanced Raman Spectroscopy: A Review. *Journal of Agricultural and Food Chemistry*, *65*(32), 6719–6726. doi:10.1021/acs.jafc.7b02504 PMID:28726388

Yang, T., Zhang, Z., Zhao, B., Hou, R., Kinchla, A., Clark, J. M., & He, L. (2016). Real-time and in situ monitoring of pesticide penetration in edible leaves by surface-enhanced Raman scattering mapping. *Analytical Chemistry*, *88*(10), 5243–5250. doi:10.1021/acs.analchem.6b00320 PMID:27099952

Zare-Shahabadi, V., Abbasitabar, F., Akhond, M., & Shamsipour, M. (2013). Simultaneous determination of chlorpyrifos and carbaryl by spectrophotometry and boosting partial least squares. *Journal of the Brazilian Chemical Society*, *24*(10), 1561–1569.

Zhu, J., Chen, Q., Kutsanedzie, F. Y., Yang, M., Ouyang, Q., & Jiang, H. (2017). Highly sensitive and label-free determination of thiram residue using surface-enhanced Raman spectroscopy (SERS) coupled with paper-based microfluidics. *Analytical Methods*, *9*(43), 6186–6193. doi:10.1039/C7AY01637A

Chapter 14
Cost-Effective Methods of Monitoring Pesticide Pollution in Water

S. Kanagasubbulakshmi
Bharathiar University, India

Krishnasamy Lakshmi
Bharathiar University, India

K. Kadirvelu
Bharathiar University, India

ABSTRACT

Environmental protection efforts require numerous advanced technologies to prevent and monitor the health and ecological effects associated with abiotic and biotic systems. Development of innovative tools and methodologies with the help of multidisciplinary approach to assess the transport, accumulation, and impact of pesticides will avoid the long-term effects in the environment. The lack of information about the pesticides hampers the labeling requirements that lead to misuse and discharge of pesticide-contaminated effluents into the water resources. This chapter covers the information on major sources of pesticides, chronic impacts, labeling of pesticides, multidisciplinary approach for monitoring, current cost-effective technologies, pros and cons of current technologies, and future perspectives of the pesticide monitoring technologies.

BACKGROUND INFORMATION

Pesticides are commonly known as inhibiting agents of pests and hostile plants that are harmful in agriculture practices. In recent past, the development of new technologies has given many target based pesticides that protect beneficiary crops and increases the production. Pesticides usages in highly populated countries like India have aided the food production to meet out the increasing demands. Pesticides also reduce the harmful diseases that occur in plants system throughout the growth phases

DOI: 10.4018/978-1-5225-6111-8.ch014

and enhance the yield of production. However, enormous increase in pesticides usage have alarmed the environmental regulatory bodies to monitor the ecological and health impacts of pesticides in living system.(Samsidar *et al.,* 2018)

The fate of pesticides in ecosystem depends on the following factors such as chemical composition, reactivity, toxicity, mode of action, secondary metabolites and degradation capabilities. These factors decide the persistent level of pesticides in the environment that can cause many harmful impacts. Major sources of pesticides not only depend on agricultural runoff that also includes the sectors of health maintenance (vector borne diseases), building maintenance (to control insects), domestic managements and industries. Pesticides contaminated water runoff from major sources finally end up in water resources and wastewater treatment plants. Current treatment technologies to remove pesticides contamination in water resources require additional micro pollutant removal supports.(Rousis *et al.,* 2017)

World Health Organization says that there are three million poisoning cases occur every year in particular Organophosphorous (OP) category possess main role to cause severe health impacts. The OP pesticide's mode of action mainly depends on acetylcholine esterase (AChE) enzyme, which has a main role in central nervous system. OP pesticides inhibits AChE and causes death in insects in the same way if it get accumulated in human system that lead to several health issues and finally end up in disordering of nervous system.(Kanagasubbulakshmi *et al.,* 2017) According to current reports, the rigorous usage of pesticides automatically will laid platform to develop resistant in pests against pesticides. Therefore, findings of new composition for pesticidal effect again will end up in harmful effects to humans and to ecosystem as well. Moreover, the secondary metabolites can cause more harmful effects than the primary pesticides. A regulation to pesticide management is an essential tool to monitor and prevent lethal impacts on the society. United States Department of Agriculture (USDA) has amended the Federal Insecticide, Fungicide, and Rodenticide Act (FIFRA) on 1972, which provides for federal regulation of pesticide distribution, sale, and use. In India, the Insecticides Act, 1968 and Insecticides Rules, 1971 regulate the import, registration process, manufacture, sale, transport, distribution and use of insecticides (pesticides) with a view to prevent risk to human beings or animals and for all connected matters, throughout India. All insecticides (pesticides) have to undergo necessarily the registration process with the Central Insecticides Board & Registration Committee (CIB & RC) before they can be made available for use or sale.(Act *et al.,* 1968).

Pesticides monitoring, determination and detection methodologies have attracted many researches to develop field deployable monitoring systems to safe guard the environment. The conventional detection systems that include high-end analytical instruments have many pros and cons. Gas chromatography (GC) and liquid chromatography (LC) are considered as very laborious, time consuming, require skilled persons and not cost effective(Zhang *et al.,* 2017). Therefore, the detection and monitoring of pesticides still a fascinated field among the multi-disciplinary researchers. Several technologies were developed for the detection and monitoring for pesticides by using signature innovations in nano and multidisciplinary researches. Electrochemistry, fluorescent on – off technologies, immuno assays and lab on chip methods have many advantages than the traditional techniques as well as the disadvantages.(Wang *et al.,* 2017a)

Therefore, this book chapter covers the classification of pesticides, list of banned pesticides, need of detection, cost effective technologies in current scenario, mode of mechanism of the detection systems, pros and cons of existing technologies and research gap in the pesticide monitoring system.

Pesticides and Classifications

Pesticides have been classified according to their mode of action, chemical composition and which type of organism that inhibits.(Samsidar *et al.,* 2018) In this contest, the classification is as follows,

Classification by Target

- **Insecticides:** Insects
- **Herbicides:** Plants
- **Rodenticides:** Rodents (rats and mice)
- **Bactericides:** Bacteria
- **Fungicides:** Fungi
- **Larvicides:** Larvae

Classification by Chemistry

Based on their chemical composition, it can be categorized as organic and inorganic pesticides.(Foo & Hameed, 2010) Organic pesticides are complex in chemical structure and can be obtained as naturally as well as through synthetic methods. Inorganic pesticides are simple compounds easily soluble in water persist in the environment for longer time.

- Organochlorines
- Organophosphates
- Pyrethrum
- Pyrethroids
- Biorational pesticides or biopesticides
- Microbial pesticides

The chemical composition of the pesticides decides the persistent nature in the environment, toxicity and harmful effects to the living systems.

Banned Pesticides in India

Pesticides banning procedure falls under different categorization as follows,

- Pesticides Banned for manufacture, import and use: e.g. Aldicarb and Ethyl Parathion
- Pesticide / Pesticide formulations banned for use but their manufacture is allowed for export: e.g. Nicotin Sulfate
- Pesticide formulations banned for import, manufacture and use: e.g. Methomyl and Carbofuron
- Pesticides withdrawn: e.g. Warfarin and Simazine
- List of pesticides refused registration: e.g. Vamidothion and Azinphos Methyl
- Pesticides restricted for use in India: e.g. Methyl parathion and Endosulfan

Figure 1. Nano-based pesticide formulation increases bioavailability and efficiency
Reprinted with permission from the publisher, (Huang et al., 2018).

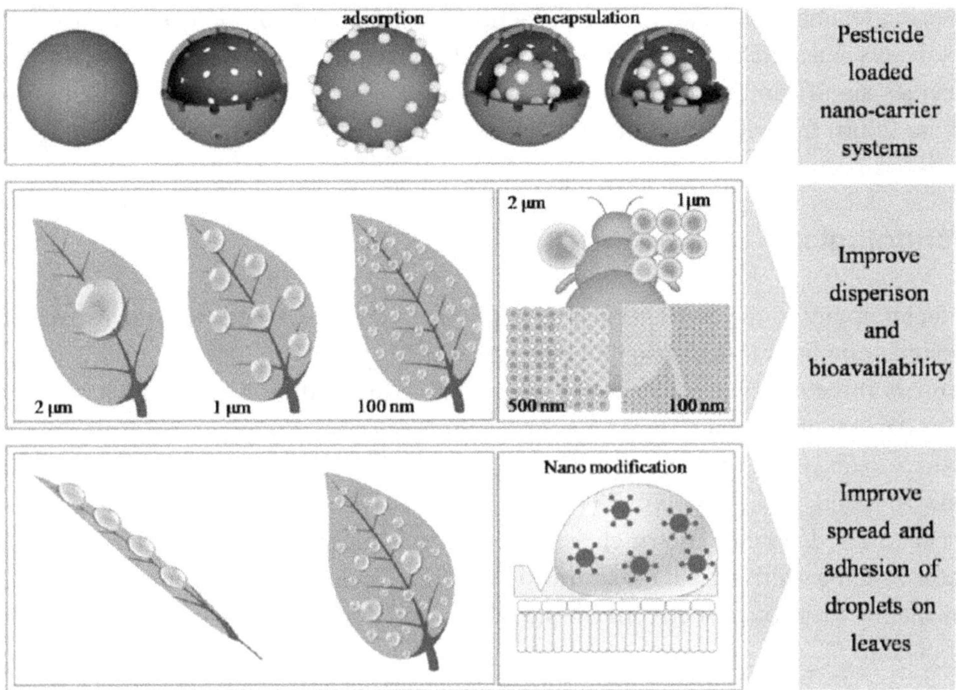

The rules and regulations as mentioned Indian government paved the way to avoid harmful chemical contaminants in the biotic system and safeguard the environment. (Insecticidal act, 1968)

Nanotechnology in Controlled Release of Pesticides

Development of new technologies for the controlled release of pesticides vastly depends on nonmaterial's and nanofromulations. That can effectively control the projected release and maintain the concentration of pesticides according to the requirements. These nanoformulations can be applied in the environment directly devoid of environmental factors. The pH, salinity, temperature and other environmental factors were found to be causing less impact on these nano pesticides. In particular, nanoformulated pesticides have wide advantages in terms of wettability, adsorption, reactivity, targeted delivery and high surface area to control the pests. Nanobiotechnology also reduces the risk factors associated with the environment and effectively target the controlled released of pesticides.(Damalas, 2018)

Conventional Pesticides Monitoring Technologies

Conventional monitoring technologies for pesticides includes gas and liquid chromatography, high performance liquid chromatography, (Guan *et al.*, 2010) enzyme-linked immunoabsorbant assays, (Rekha *et al.*, 2000)and capillary electrophoresis.(Hsu & Whang 2009)Even though the method is very sensitive and selective towards the analytes, still it has own drawbacks in terms of highly skilled laboratory persons, costly instrument and complicated procedures. Pesticides come under micro pollutants category, as it is available in very low concentration in the polluted sources. The sample preparation for conventional monitoring technologies starts from extraction from the mother sources that also need to focus on suitable solvents and suitable method selection.(Schenck *et al.,* 2002) The development in extraction methods receives significant attention as it is devoid of solvents and complex extraction procedures. (Leandro *et al.,* 2006) In terms of cost effectiveness, these techniques still require additional supports from multi-disciplinary subjects.

Smart Technologies

Alternative technologies to the conventional technologies were developed with the use of different materials that avoid complicated extraction procedures.

In Biosensors there are two components to determine an analyte namely, bioactive material and transducers to develop the signals from an analyte. In recent past, the bioactive material development that corresponds to an analyte draws a significant attention. The bioactive material includes antibodies, aptamers, enzymes and binding proteins. Transducers work as a signal developer that can be amperometry, fluorescence based, potentiometry, chemiluminescence, piezo - electric or fiber-optic principle, and semiconductor or ion-sensitive field effect transistor (ISFET) techniques. (Kumar *et al.,* 2015) Current technologies mainly focus on cost effective point of care and lab on chip methodologies that can be used by common person.

Biosensing: Aptamers, Antibodies and Enzyme Based Technologies

The problems in conventional technologies lead to the development of immunoassays for the detection of pesticides. In the last decade development of detection technologies using antibodies and enzymes had a significant attentions among the researchers. In Biosensors, the bioactive material can be aptamers, antibodies and specific enzymes to bind with targeted pesticides. On other hand the transducer can be nanomaterials (nanoparticles, quantum dots, nanofibers) and polymer based substrates to develop the signals to be sensed effectively.

Enzymes

ELISA (Enzyme linked immunosobent assay) has a main role in the recent technologies to develop the sensing platforms that is based on enzyme inhibition. Few neurotransmitter enzymes can be used in ELISA platform to detect the pesticides effectively. Acetylcholine esterase and Organophosphorous hydrolase that can bind with Organophosphorous based pesticides. Variety of nanomaterials can be used as a signal producer for the sensing of pesticides e.g. CdTe quantum dots and carbon nanotubes.(Meng *et al.,* 2013; Gothwal *et al.,* 2014) Apart from the development of nanomaterial and enzyme based complexes there are many other studies mainly focus on colorimetry based detection platforms for pesticides. On the basis of the facts of inhibition of AChE activity by organophosphrous and carbamates, inexpensive and commercially available H_2O_2 or thiocholine sensing methodology was developed in colorimetric sensor array. Array base sensor system not only useful in the selective sensing approach as well as it helps in the discrimination of cross reactivity of closely related pesticides.(Qian and Lin, 2015)

Figure 2. Design of the dual-readout (colorimetric and fluorometric) assay for pesticides
Xia et al., 2015, Reprinted with permission from the publisher.

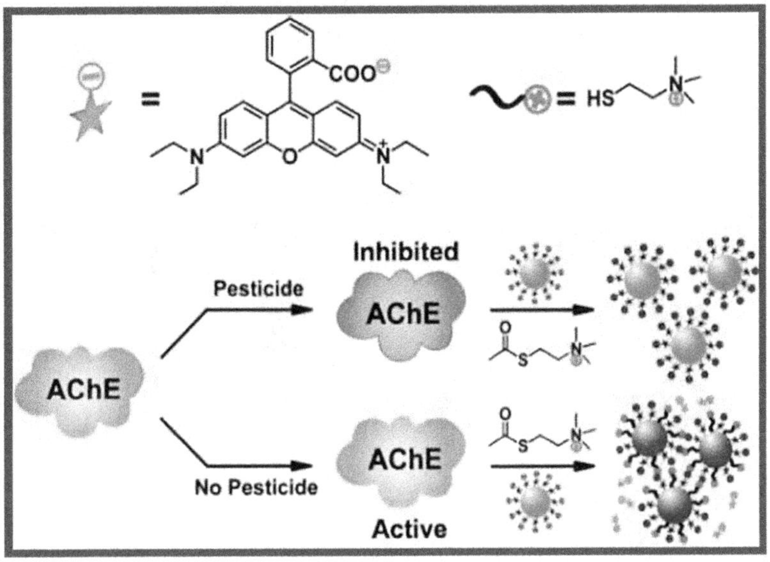

Figure 3. Schematic Illustration of the Detection Principle of the Developed Colorimetric Sensor Array for Organophosphorus and Carbamate Pesticides
Qian and Lin 2015, Reprinted with permission from the publisher.

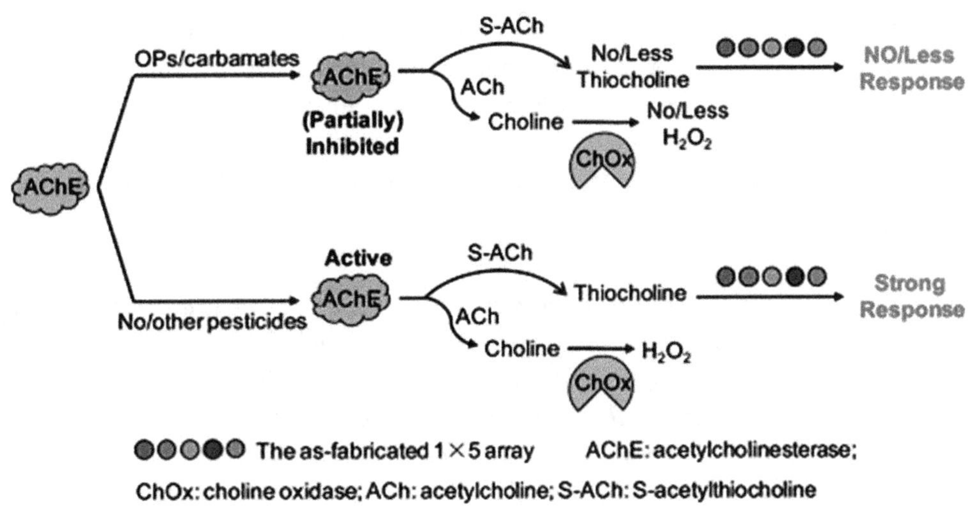

Antibodies

Conventional monoclonal and polyclonal antibodies have been used as recognizing elements in recent sensing technologies. The broad specific antibodies attract its own attention due to the capability in single analogue assays. The non immunogen substances should be conjugated with proteins to make it immunogenic. Other antibodies like a binder and novel nanobodies can effectively detect the contaminants

Figure 4. (a) Color difference maps of the array for 10-6 g/L methomyl in the absence (I) and presence of 1% apple juice (II) and5% green tea drinks (III). For purposes of visualization, the color range of these difference maps of methomyl was expanded from 3 to 8bits per color (RGB range of 3-10 expanded to 0-255). (b)Responses of the developed sensor array to 10-6 g/L methomyl in the absence (I) and presence of 1% apple juice (II) and 5% green tea drinks (III). The error bars showed the standard deviation of triplicate experiments
Qian & Lin 2015, Reprinted with permission from the publisher.

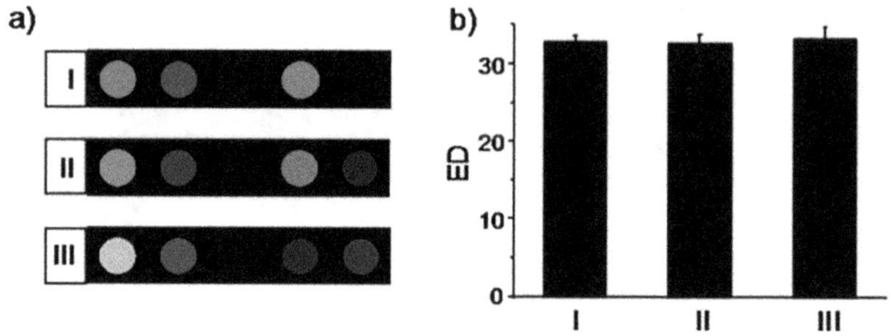

easily. Recombinant antibodies can be designed to capture the antigens in the development of detection platforms for pesticides.(Zhang *et al.*, 2017)

Aptamers

Aptamers have significant advantages than the antibodies in electrochemical sensing, fluorescence based sensing, chemiluminescence and colorimetric detection system. Aptamers can be generated against target based molecules and used to incorporate with nanomaterials to develop sensing platforms. Aptamers are highly specific, stable and sensitive towards the analytes. Surface Plasmon resonance phenomena of nanoparticles help in the detection of targets precisely in colorimetric detection. Binding of targeted pesticides on the surface of aptmers lead to aggregation of nanoparticles to give visual interpretation.(Bala *et al.*, 2016)

Figure 5. Schematic representation of the reversible inhibition of the nanonozyme activity of gold nanoparticles (GNPs) using an acetamiprid-specificS-18 single strand DNA aptamer. Step A shows intrinsic peroxidase-like activity of GNPs that gets inhibited after shielding of GNP surface through conjugation of S 18 aptamer molecules (Step B). In the presence of acetamiprid target, aptamer undergoes target responsive structural changes and forms a supra molecular complex with acetamiprid resulting in free GNP to resume its peroxidase like activity (Step C)
Weerathunge et al., 2014, Reprinted with permission from the publisher.

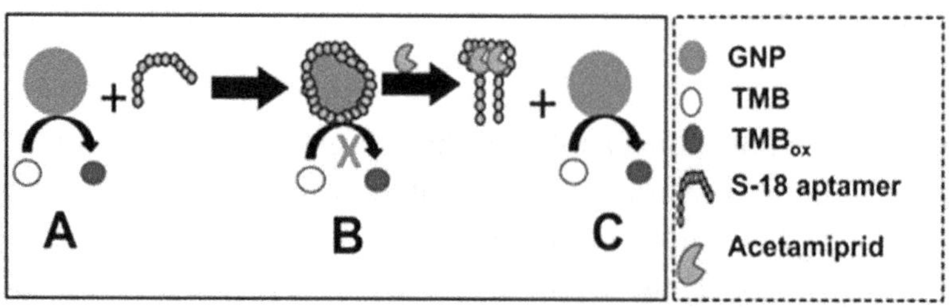

Figure 6. Schematic Illustrations of the Principle of Immunosensing Platform Based on (A) Immunodetection of Enzyme Activity and (B) Immunoassay of Total Amount of Enzyme Simultaneously for Biomonitoring of OP Exposure
Reprinted with permission from the publisher, Du et al., 2011.

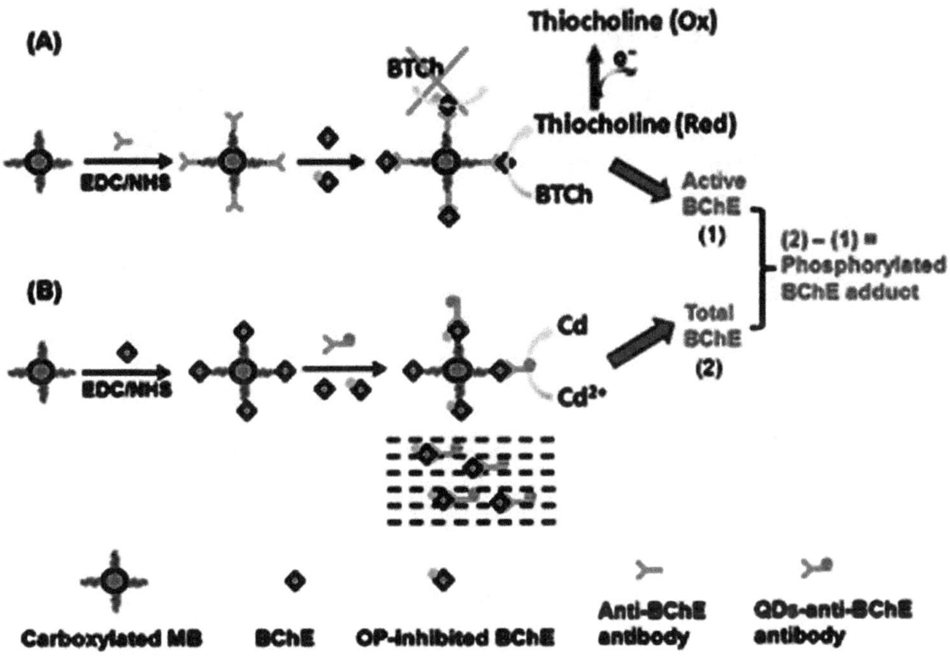

Electrochemical Sensing

Now days the electrochemical sensing methodology is a fascinating research area because of its efficiency in the detection of OPs pesticides in terms of accuracy and selectivity. Here, the sensor system is mainly based on the active groups those are used to design the electrode's property to obtain the specific signals. (Huo *et al.,* 2014) Over the past, the interaction mechanism of OPs pesticides with AChE enzyme have been used to develop the best alternative sensor systems.(Liang *et al.,* 2014)Also the nanomaterials and enzyme linked hybrids are being used to create an accurate signal for specific detection and quantification by AChE activity. Gothwal *et al.,* 2014 On those nonmaterial's, quantum dots (QDs) have gained more interest due to its excellent photophysical properties and have been used as a fluorescent probe in the sensing systems. The fluorescent properties can be tuned by providing enzyme functionalities on the surface of QDs that also can provide turn of visual detection platforms for the detection of OPs pesticides.(Meng *et al.,* 2013)

Surface-Enhanced Raman Scattering (SERS)

Label free Surface-Enhanced Raman Scattering (SERS) sensing method using nanomaterials were developed for the signal based detection of organic and inorganic molecules.(Yang *et al.,* 2014) Since SERS based techniques mainly depend on the excitation of molecules at higher energy states that corresponds to the Raman Effect, it can precisely detect the pesticides. Raman detection strategies mainly

Figure 7. SERS spectra of an apple surface containing (a) omethoate pesticide and gold nanoparticles, (b) chlorpyrifos pesticide and gold nanoparticles, (c) omethoate pesticide only (d) chlorpyrifos pesticide only, (e) gold nanoparticles only, (f) a clean apple surface
Chen et al., 2018, Reprinted with permission from the publisher.

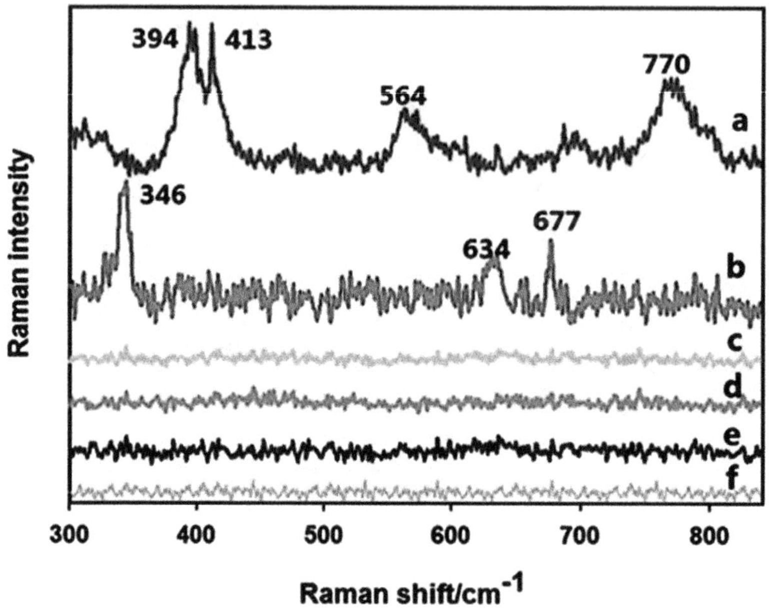

focus on metal based platforms to enhance the signal from the substrates. The substrates development in Raman based technologies require special attention in the field of nano and material chemistry. From the discovery of SERS based detection strategies, it has been exploited for the detection of pesticides in vegetables and food samples for continuous monitoring. Several methodologies for the detection of pesticides were developed using different nanomaterials like Au, Ag and semiconductor based quantum dots. Nanofibers have main role in the field of detection of pesticides as solid substrate to enhance the signals. Recently Raman based detection system was developed for the detection of dithiocarbamates, chlorpyrifos and omethoate.(Chen *et al.,* 2018; Fateixa *et al.,* 2018)In those studies the SERS based microscopic imaging was successfully demonstrated to detect the pesticides on the surface of the samples.

Photo Induced Electron Transfer Process (PET) Based Techniques

The molecular imprinting based turn on sensing methodology for the detection of 2,4-dichlorophenoxyacetic acid(2,4-D), 4-Nitrophenol and other organic molecules was developed using photo induced electron transfer process (PET). (Wang *et al.,* 2016a, 2018; Yu *et al.,* 2017)PET based techniques require special attention due to the specific attraction towards the molecule that effectively sense an analyte. Here, an electron donor and acceptor place a significant role in the signal development. For example, in the detection of 2,4-D the nitrobenzoxadiazole and red quantum dots were used for signal enhancement based detection of pesticides. The strategy for designing fluorescence sensor was based on the occurrence of PET between NBD and the amine groups of APTES, since the amine group with nitrogen lone-pair electron can quench a nearby NBD fluorophore through an N-to-NBD electron transfer. Binding of amine

Figure 8. Schematic Illustration of Sensing Strategy for OPs Detection Based on CDs-MnO2 Nanosheets
Reprinted with permission from the publisher.

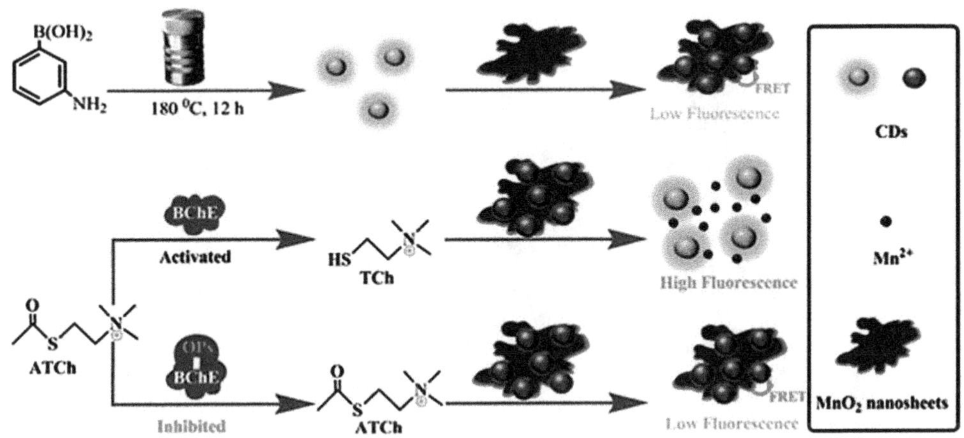

groups with 2,4-D through hydrogen bond enhances the fluorescence intensity and denotes that the PET process was inhibited, while the fluorescence intensity of red QDs maintained constant. Increase in the concentration of 2,4- D laid platform for the visual detection.(Wang *et al.,* 2016b)

Fluorescence Resonance Energy Transfer (FRET) Based Techniques

Fluorescent quantum dots are being considered as an excellent alternative to the conventional organic dyes to develop optical sensing technologies for pesticides. Semiconductor QDs like CdTe, CdS, CdSe and core shell QDs have been exploited for FRET based sensing for many pesticides. QDs based Biosensors were developed by utilizing its prominent photophysical properties. In recent studies(Zhang *et al.,* 2010; Zheng *et al.,* 2011)AChE enzyme was used to coat multilayer on the QDs to achieve an efficient sensing platforms for the pesticides. Layer by layer techniques was used to assemble the bi-enzymes on the QDs to obtain lowest limit of detection and sensitivity towards an analyte. Thus, the developed sensor was capable enough to detect ultralow level detection of pesticides in fruit samples. To avoid the biomolecules in FRET based techniques the ligand replacement based turn on detection methodology was developed using QDs. The selective binding of fluorescent ligands with the pesticides was obtained by turn on FRET based mechanisms in many studies.(Shi *et al.,* 2015; Wu *et al.,* 2017) Carbon dots (CDs) have been holding its significant role in the field of sensing technologies as a potential replacement to the conventional semiconductor QDs. Compared to conventional dye molecules, quantum dots and metal clusters, CDs possess distinct advantages, such as excellent biocompatibility, low photo bleaching, and good water solubility. Recently, FRET based probes for the detection of pesticides was developed by using MnO_2 and CDs for pesticides detection that gave comparatively low detection limit.

Other Cost Effective Smart Technologies

Detection and monitoring of pesticides have been widely focused by the researchers to develop cost effective technologies. Recently, the chemosensor based detection platforms draw a significant atten-

Figure 9. Schematic illustration of Benzimidazolium based fluorescent cations and their fluorescent aggregates for the detection of OPs pesticides
Singh et al., 2016a, Reprinted with permission from the publisher.

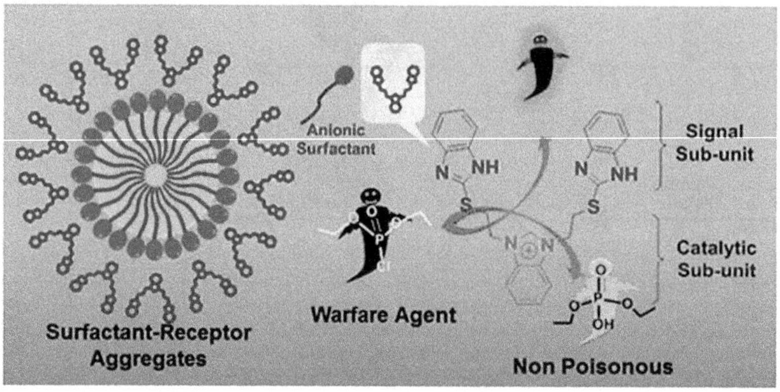

tion to detect the organic and inorganic molecules. New colorimetric based technologies to provide the visual detection platforms have been developed by using exact chemical composition of the particular pesticides. For example the AZO dye was used to develop detection platform for carbaryl pesticides. The chemistry behind this detection platform is in pretreatment of carbaryl pesticides with an organic base yielded 1-naphthol that can react with a diazonium salt and gives visual color change. Further azo dye can change the color by reacting with organic base that was used in the pretreatment, Thus developed detection platform can be used to detect the contaminants directly from the source with simple dye based techniques.(Lee *et al.,* 2018) Benzimidazolium based fluorescent cations and their fluorescent aggregates can be used in the detection platform for Organophosphorous (OPs) based pesticides. Aggregation of Benzimidazolium particles can enhance the fluorescent emission spectra but addition of OPs pesticides in the detection system did not allow the enhancement and quenches the fluorescence of Benzimidazolium particles. That particular phenomenon was taken for the detection of OPs pesticides directly from the contaminant sources.(Singh *et al.,* 2016a)

Metal ion complex systems were used to develop ligand formation based sensing technology for the detection of pesticides. In this study (Raj *et al.,* 2016) hexadentate ligands H2L1-L3 with mixed S, N, O donor sites and possessing substituents having either "no" or electron releasing/withdrawing nature at terminal ends have been developed and tested for binding with library of metal ions. The ligands were found to be capable with the binding of Ni^{2+} ions and used as a chemosensor for OPs pesticides.

The recent developments in the detection and monitoring technologies for pesticides mainly deal with multidisciplinary approach with wide variety of nanomaterials. Nickel cobaltite ($NiCo_2S_4$) based nanostructure was generated for voltammetry based sensing for methyl parathion. That sensor was able to give lowest limit of detection from the contaminant sources.(Justino *et al.,* 2017)

The paper based sensing technology was developed recently for the visual detection of pesicides by simple colour change as a confirmation of an analyte presence in the source sample. Portable fluorescent sensor that integrates animmuno chromatographic test strip assay (ITSA) with a quantum dot (QD) label and a test strip reader for the detection of chlorpyrifos that was based on the biomarker binding with targeted analyte. (Zou *et al.,* 2010) The lateral flow assay based bioactive paper strip was developed for the onsite detection of pesticides using indophenyl acetate (IPA) and AChE enzyme. The bioactive strip

Figure 10. Possible mechanism of interaction of (A) phosmet with dinuclear nickel complex and (B) chlorpyrifos with mononuclear nickel complex
Reprinted with permission from the publisher, Raj et al., 2016.

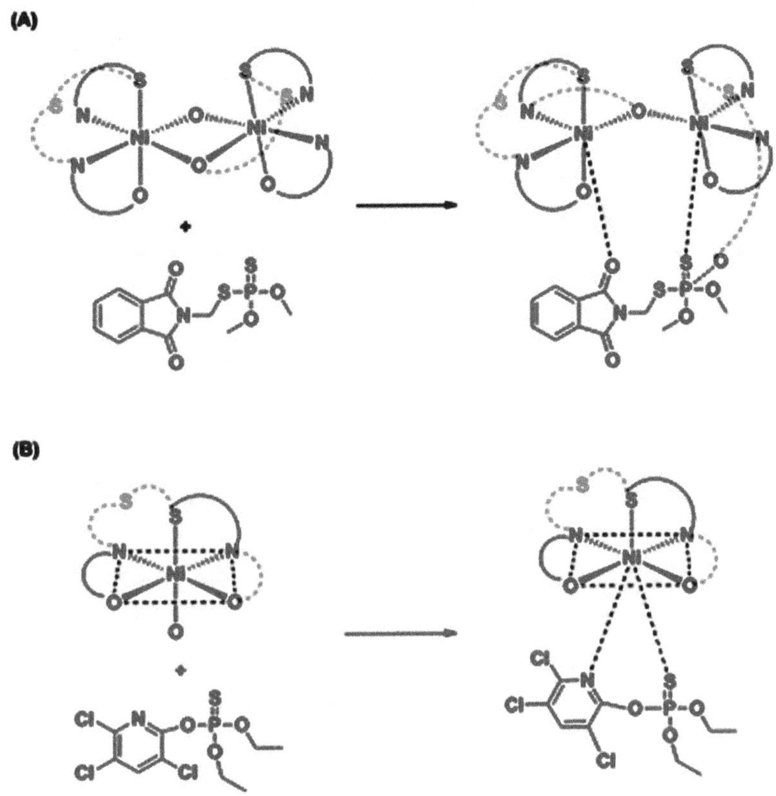

was able detect paraxon, carbaryl and malathion effectively within the span of five minutes.(Hossain *et al.*, 2009)

In contrast to fluorescence based detection technologies the chemiluminescence resonance energy transfer (CRET) was developed by employing hapten functionalized QDs in a competitive immune assay for detection of sulfamethazine (SMZ). The CRET based immuno assay was able to give a pico molar level limit of detection for SMZ.(Ma *et al.*, 2016)

Nanobiosensor based on an atomic force microscopy(AFM) tip functionalized with the acetolactate synthase (ALS) enzyme was developed for the detection of herbicide metsulfuron-methyl, The detection strategy was mainly based on acquisition of force curves from the biosensor. This developed methodology has shown promising application in future for possible detection of pesticides in the environment. (da Silva *et al.*, 2013)

The technology developments in smartphone integration have paved the platform to develop sensing technologies for variety of organic and inorganic molecules. In that contest, the smartphone-based technology was developed for the detection of pesticides. The sensor was capable of screening 8samples in a microplate single-stripe. The device has so many advantages in terms of size, ease of usage and cost effective. The detection of 2,4-Dichlorophenoxyacetic acid was achieved efficiently up to ppb level. (Wang *et al.*, 2017b)

Figure 11. Schematic illustration of the principle of fluorescent ITSA. (A) Aqueous sample containing analytes was applied to the sample zone.(B) Analytes migrate together with QD-conjugated competitors toward the other end of test strip by capillary force.(C) Competitive immunoreactions among analytes, conjugated competitors, and antibodies. Excess conjugated competitors and analytes continue to migrate toward the absorption pad. In a control assay (no analyte), the QD-conjugated competitors are fully binding to the antibodies in the test zone. (D) Fluorescence detection on the test strip reader
Zou et al., 2010, Reprinted with permission from the publisher.

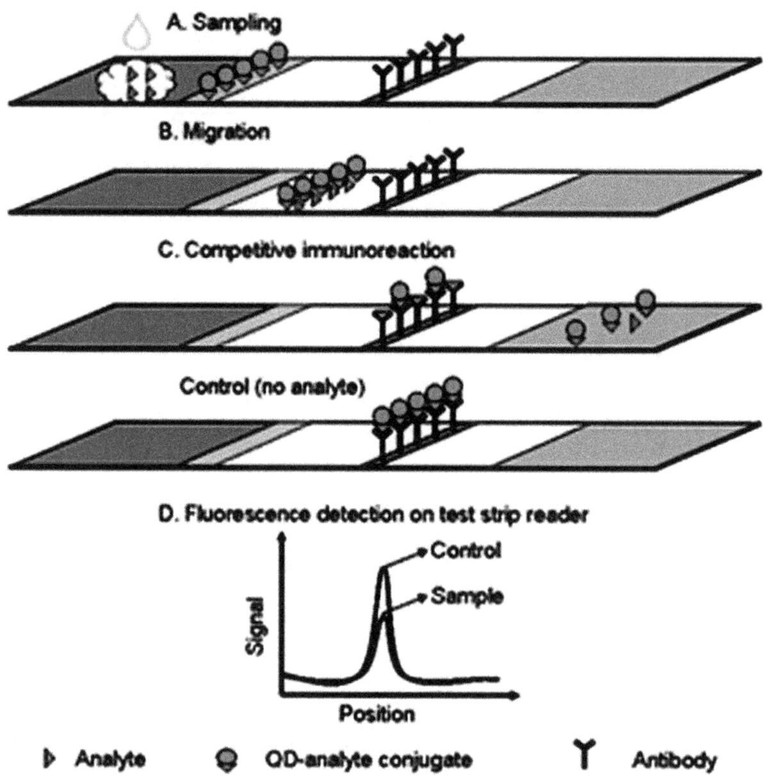

Advantages of Smart Technologies

Developments of new technologies have many advantages than the conventional methodologies as mentioned below,

1. Cost effective
2. Field deployable
3. Onsite detection methodologies
4. Visual interpretations
5. Target can be tailored towards an analyte
6. Multidisciplinary approaches provide the platform for update as per the requirements.
7. Ease of usage

Figure 12. (a) Effects of cationic polyvinyl amine PVAm on entrapment of indophenoxide anion in a lateral flow based paper sensor system. Color intensity (CI) due to elution of IPA (3 mM, final concentration) in the lateral flow-based platform. The areas within the dashed boxes were printed/over spotted without (control) and with PVAm (0.5 wt. %) followed by printing/over spotting of AChE (500 U/mL). PVAm concentrates the reaction product, the blue indophenoxide anion, while in the control experiment; the blue indophenoxide anion is dispersed over a large area. (b) Proof of concept for the development of the reagent less bioactive paper-based lateral flow platform, in which the sensor was dipped into ddH2O to move the IPA reagent into the sensing region for the generation of the blue color
Hossain et al., 2009, Reprinted with permission from the publisher.

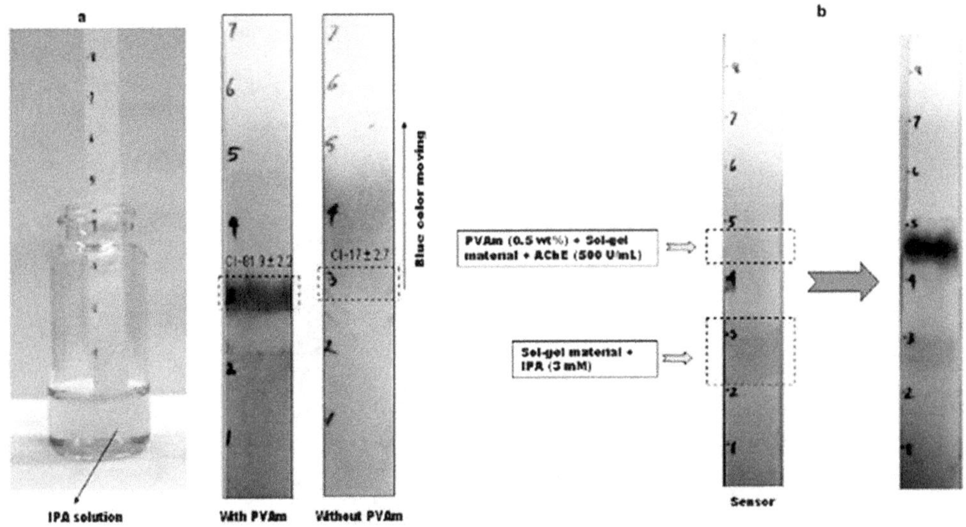

Figure 13. (A) Schematic illustration of the CRET process between the donor, luminol, and the accepter, QDs, based on the immune reaction between mAb-HRP and BS-QDs; and (B) Schematic illustration of the CRET-based competitive immunoassay for SMZ
Ma et al., 2016, Reprinted with permission from the publisher.

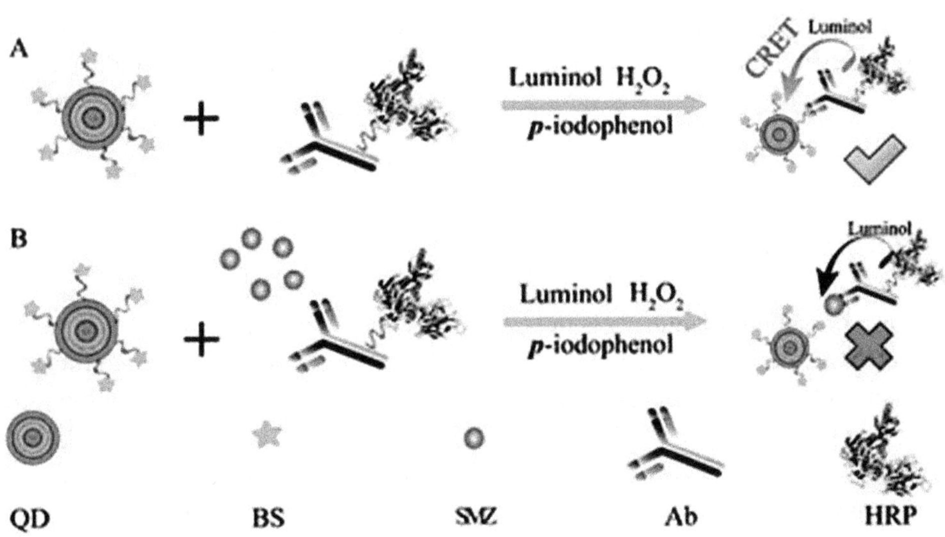

Figure 14. Schematic illustrating the functionalization of the silicon nitride tip and the detection of metsulfuron-methyl herbicide on the mica surface
da Silva et al., 2013, Reprinted with permission from the publisher.

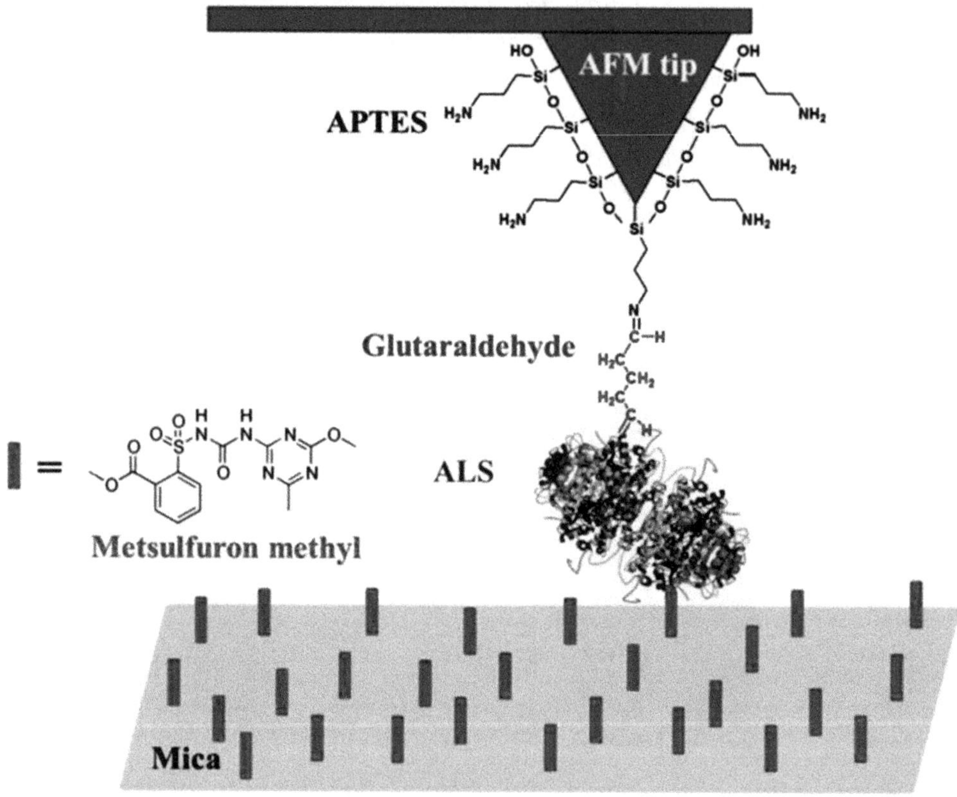

Disadvantages of Smart Technologies

In terms of disadvantages, the technology has a draw back in the requirements for material development and designing. Though, the nanomaterial can provide sensitive and selective approach towards an analyte the field level application still need collaboration and input from the industrial sectors. The possible interferences in the environment should be addressed when the smart technologies come to the pilot scale development.

FUTURE PROSPECTS

The health and environmental monitoring is the main areas that require huge number of developments in all aspects. The harmful impacts of the potential contaminant could be detectable on the site to provide significant precautions. Point of care systems using multidisciplinary platforms can give fruitful outcomes in the sensing and monitoring technologies. The monitoring of pesticides from the contaminated sources should provide immediate accessible platforms for users. Upcoming technologies should focus on these factors to provide a beneficial technology to the common person aspects.

Figure 15. (A) configuration of the aperture arrays, microplate stripe, and prism array, (B) structure of the prism array, (C) Three-dimensional structure of the smartphone based single stripe microplate reader, (D) actual device with an iPhone 6s for image acquisition (The strip was partially slide out in Figure 1D to indicate the entrance for the strip, thus not all 8 wells were shown in the smartphone screen)
Wang et al., 2017b, Reprinted with permission from the publisher.

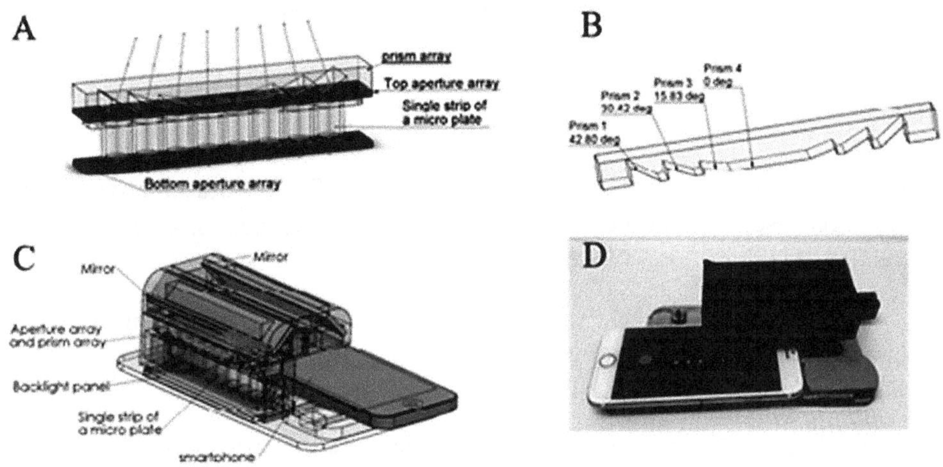

CONCLUSION

Over all point of view about the cost effective technologies for the detection and monitoring of pesticides

Table 1. Analytical performances of smart technologies for pesticides sensing

Sensor System	Response	Limit of Detection	Selectivity	Reference
Magnetic nanoparticles + AChE	Electrochemical	6.66×10^{-3}mM	None	(Dzudzevic Cancar *et al.*, 2016)
Benzimidazolium cations	PL quenching	15 nM	Moderate	(Singh *et al.*, 2016b)
Aptamer-Gold nanoparticles	Colorimetric	0.06 pM	Moderate	(Bala *et al.*, 2016)
AChE-CdTe	PL Quenching	3.20×10^{-8}mol/L	Moderate	(Sun *et al.*, 2011)
Competitive fluorescence-linked immunosorbent assay	Colorimetric	16.2 ng m/L	Excellent*	(Chen *et al.*, 2010)
Immunochromatographic test strip assay	Fluorescence	1ng/mL	Excellent	(Zou *et al.*, 2010)
Organophosphate hydrolase+CNTs	Electrochemical	322.58 μM	Moderate	(Gothwal *et al.*, 2014)
Dithizone+CdTe QDs	FRET	0.1nM	Moderate	(Zhang *et al.*, 2010)
TGA +CdTe QDs	PL Quenching	0.68nM	Excellent**	(Kanagasubbulakshmi *et al.*, 2018)

have driven the new innovation in the field of nanotechnology. Upcoming technologies namely, biosensing, electrochemical sensing, FRET, PET, SERS and paper based bioactive strips already have proved its own efficiency in terms of field deployable detection methodologies. Next step to bring out the lab level technologies to the market requires essential output from the researchers, industries and funders. The successful implementation of multidisciplinary efforts would bring the point of care detection monitoring technologies for the common person's aspects that is field deployable without laborious methodologies.

REFERENCES

Bala, R., Kumar, M., Bansal, K., Sharma, R. K., & Wangoo, N. (2016). Ultrasensitive aptamer biosensor for malathion detection based on cationic polymer and gold nanoparticles. *Biosensors & Bioelectronics*, *85*, 445–449. doi:10.1016/j.bios.2016.05.042 PMID:27208476

Chen, J., Dong, D., & Ye, S. (2018). Detection of pesticide residue distribution on fruit surfaces using surface-enhanced Raman spectroscopy imaging. *RSC Advances*, *8*(9), 4726–4730. doi:10.1039/C7RA11927E

Chen, Y., Ren, H. L., Liu, N., Sai, N., Liu, X., Liu, Z., ... Ning, B. (2010). A fluoroimmunoassay based on quantum dot-streptavidin conjugate for the detection of chlorpyrifos. *Journal of Agricultural and Food Chemistry*, *58*(16), 8895–8903. doi:10.1021/jf101778t PMID:23654227

Da Silva, A. C. N., Deda, D. K., da Róz, A. L., Prado, R., Carvalho, C., Viviani, V., & Leite, F. (2013). NanoBiosensors based on chemically modified AFM probes: A useful tool for metsulfuron-methyl detection. *Sensors (Switzerland)*, *13*(2), 1477–1489. doi:10.3390130201477 PMID:23348034

Damalas, C. A. (2018). Current Status and Recent Developments in Biopesticide Use. *Agriculture*, *8*(1), 13. doi:10.3390/agriculture8010013

Du, D., Wang, J., Wang, L., Lu, D., Smith, J. N., Timchalk, C., & Lin, Y. (2011). Magnetic electrochemical sensing platform for biomonitoring of exposure to organophosphorus pesticides and nerve agents based on simultaneous measurement of total enzyme amount and enzyme activity. *Analytical Chemistry*, *83*(10), 3770–3777. doi:10.1021/ac200217d PMID:21462919

Dzudzevic Cancar, H., Soylemez, S., Akpinar, Y., Kesik, M., Göker, S., Gunbas, G., ... Toppare, L. (2016). A Novel Acetylcholinesterase Biosensor: Core-Shell Magnetic Nanoparticles Incorporating a Conjugated Polymer for the Detection of Organophosphorus Pesticides. *ACS Applied Materials & Interfaces*, *8*(12), 8058–8067. doi:10.1021/acsami.5b12383 PMID:26956086

Fateixa, S., Raposo, M., Nogueira, H. I. S., & Trindade, T. (2018). A general strategy to prepare SERS active filter membranes for extraction and detection of pesticides in water. *Talanta*, *182*, 558–566. doi:10.1016/j.talanta.2018.02.014 PMID:29501193

Foo, K.Y., & Hameed, B.H. (2010). *Detoxification of pesticide waste via activated carbon adsorption process*. Academic Press.

Gothwal, A., Beniwal, P., Dhull, V., & Hooda, V. (2014). Preparation of electrochemical biosensor for detection of organophosphorus pesticides. *International Journal of Analytical Chemistry*. PMID:25667593

Guan, H., Brewer, W. E., Garris, S. T., & Morgan, S. L. (2010). Disposable pipette extraction for the analysis of pesticides in fruit and vegetables using gas chromatography/mass spectrometry. *Journal of Chromatography. A, 1217*(12), 1867–1874. doi:10.1016/j.chroma.2010.01.047 PMID:20144461

Hossain, S. M. Z., Luckham, R. E., McFadden, M. J., & Brennan, J. D. (2009). Reagentless bidirectional lateral flow bioactive paper sensors for detection of pesticides in beverage and food samples. *Analytical Chemistry, 81*(21), 9055–9064. doi:10.1021/ac901714h PMID:19788278

Hsu, C. C., & Whang, C. W. (2009). Microscale solid phase extraction of glyphosate and aminomethylphosphonic acid in water and guava fruit extract using alumina-coated iron oxide nanoparticles followed by capillary electrophoresis and electrochemiluminescence detection. *Journal of Chromatography. A, 1216*(49), 8575–8580. doi:10.1016/j.chroma.2009.10.023 PMID:19853856

Huang, B., Chen, F., Shen, Y., Qian, K., Wang, Y., Sun, C., ... Cui, H. (2018). Advances in Targeted Pesticides with Environmentally Responsive Controlled Release by Nanotechnology. *Nanomaterials (Basel, Switzerland), 8*(2), 102. doi:10.3390/nano8020102 PMID:29439498

Huo, D., Li, Q., Zhang, Y., Hou, C., & Lei, Y. (2014). A highly efficient organophosphorus pesticides sensor based on CuO nanowires-SWCNTs hybrid nanocomposite. *Sensors and Actuators. B, Chemical, 199*, 410–417. doi:10.1016/j.snb.2014.04.016

Justino, C., Duarte, A., & Rocha-Santos, T. (2017). Recent Progress in Biosensors for Environmental Monitoring: A Review. *Sensors (Basel), 17*(12), 2918. doi:10.339017122918 PMID:29244756

Kanagasubbulakshmi, S., Gopinath, A., & Kadirvelu, K. (2017). Enzyme free Thiol capped CdS Quantum dots based sensing method for the detection of Malathion. *Materials Today: Proceedings, 4*(13), 12448–12456. doi:10.1016/j.matpr.2017.10.015

Kanagasubbulakshmi, S., Kathiresan, R., & Kadirvelu, K. (2018). Structure and physiochemical properties based interaction patterns of organophosphorous pesticides with quantum dots: Experimental and theoretical studies. *Colloids and Surfaces. A, Physicochemical and Engineering Aspects, 549*, 155–163. doi:10.1016/j.colsurfa.2018.04.007

Kumar, P., Kim, K.-H., & Deep, A. (2015). Recent advancements in sensing techniques based on functional materials for organophosphate pesticides. *Biosensors & Bioelectronics, 70*, 469–481. doi:10.1016/j.bios.2015.03.066 PMID:25864041

Leandro, C. C., Bishop, D. A., Fussell, R. J., Smith, F. D., & Keely, B. J. (2006). Semiautomated determination of pesticides in water using solid phase extraction disks and gas chromatography-mass spectrometry. *Journal of Agricultural and Food Chemistry, 54*(3), 645–649. doi:10.1021/jf051874d PMID:16448162

Lee, M. G., Patil, V., Na, Y. C., Lee, D. S., Lim, S. H., & Yi, G.-R. (2018). Highly stable, rapid colorimetric detection of carbaryl pesticides by azo coupling reaction with chemical pre-treatment. *Sensors and Actuators. B, Chemical, 261*, 489–496. doi:10.1016/j.snb.2018.01.151

Liang, H., Song, D., & Gong, J. (2014). Signal-on electrochemiluminescence of biofunctional CdTe quantum dots for biosensing of organophosphate pesticides. *Biosensors & Bioelectronics, 53*, 363–369. doi:10.1016/j.bios.2013.10.011 PMID:24184599

Ma, M., Wen, K., Beier, R. C., Eremin, S. A., Li, C., Zhang, S., ... Wang, Z. (2016). Chemiluminescence Resonance Energy Transfer Competitive Immunoassay Employing Hapten-Functionalized Quantum Dots for the Detection of Sulfamethazine. *ACS Applied Materials & Interfaces*, *8*(28), 17745–17750. doi:10.1021/acsami.6b04171 PMID:27362827

Meng, X., Wei, J., Ren, X., & ... . (2013). Biosensors and Bioelectronics A simple and sensitive fl uorescence biosensor for detection of organophosphorus pesticides using H 2 O 2 -sensitive quantum dots / bi-enzyme. *Biosensors & Bioelectronics*, *47*, 402–407. doi:10.1016/j.bios.2013.03.053 PMID:23612061

Qian, S., & Lin, H. (2015). Colorimetric sensor array for detection and identification of organophosphorus and carbamate pesticides. *Analytical Chemistry*, *87*(10), 5395–5400. doi:10.1021/acs.analchem.5b00738 PMID:25913282

Raj, P., Singh, A., Kaur, K., Aree, T., Singh, A., & Singh, N. (2016). Fluorescent Chemosensors for Selective and Sensitive Detection of Phosmet/Chlorpyrifos with Octahedral Ni2+ Complexes. *Inorganic Chemistry*, *55*(10), 4874–4883. doi:10.1021/acs.inorgchem.6b00332 PMID:27115348

Rekha, K., Thakur, M. S., & Karanth, N. G. (2000). Biosensors for the detection of organophosphorous pesticides. *Critical Reviews in Biotechnology*, *20*(3), 213–235. doi:10.1080/07388550008984170 PMID:11039330

Rousis, N. I., Bade, R., Bijlsma, L., Zuccato, E., Sancho, J. V., Hernandez, F., & Castiglioni, S. (2017). Monitoring a large number of pesticides and transformation products in water samples from Spain and Italy. *Environmental Research*, *156*, 31–38. doi:10.1016/j.envres.2017.03.013 PMID:28314152

Samsidar, A., Siddiquee, S., & Shaarani, S. M. (2018). A review of extraction, analytical and advanced methods for determination of pesticides in environment and foodstuffs. *Trends in Food Science & Technology*, *71*, 188–201. doi:10.1016/j.tifs.2017.11.011

Schenck, F. J., Lehotay, S. J., & Vega, V. (2002). Comparison of solid-phase extraction sorbents for cleanup in pesticide residue analysis of fresh fruits and vegetables. *Journal of Separation Science*, *25*(14), 883–890. doi:10.1002/1615-9314(20021001)25:14<883::AID-JSSC883>3.0.CO;2-7

Shi, J., Tian, F., Lyu, J., & Yang, M. (2015). Nanoparticle based fluorescence resonance energy transfer (FRET) for biosensing applications. *Journal of Materials Chemistry. B, Materials for Biology and Medicine*, *3*(35), 6989–7005. doi:10.1039/C5TB00885A

Singh, A., Raj, P., & Singh, N. (2016a). Benzimidazolium-Based Self-Assembled Fluorescent Aggregates for Sensing and Catalytic Degradation of Diethylchlorophosphate. *ACS Applied Materials & Interfaces*, *8*(42), 28641–28651. doi:10.1021/acsami.6b09983 PMID:27731616

Singh, A., Raj, P., & Singh, N. (2016b). *Benzimidazolium-Based Self-Assembled Fluorescent Aggregates for Sensing and Catalytic Degradation of Diethylchlorophosphate*. Academic Press.

Sun, X., Liu, B., & Xia, K. (2011). A sensitive and regenerable biosensor for organophosphate pesticide based on self-assembled multilayer film with CdTe as fluorescence probe. *Luminescence*, *26*(6), 616–621. doi:10.1002/bio.1284 PMID:21567885

The Insecticides Act, 1968.

Wang, X., Yu, J., Kang, Q., Shen, D., Li, J., & Chen, L. (2016a). Molecular imprinting ratiometric fluorescence sensor for highly selective and sensitive detection of phycocyanin. *Biosensors & Bioelectronics, 77*, 624–630. doi:10.1016/j.bios.2015.10.019 PMID:26485176

Wang, X., Yu, J., Wu, X., Fu, J., Kang, Q., Shen, D., ... Chen, L. (2016b). A molecular imprinting-based turn-on Ratiometric fluorescence sensor for highly selective and sensitive detection of 2,4-dichlorophenoxyacetic acid (2,4-D). *Biosensors & Bioelectronics, 81*, 438–444. doi:10.1016/j.bios.2016.03.031 PMID:27015146

Wang, X., Yu, S., & Liu, W. (2018). *Molecular Imprinting Based Hybrid Ratiometric Fluorescence Sensor for the Visual Determination of Bovine Hemoglobin*. ACS Sensors acssensors.7b00804.

Wang, Y., Hu, Q., Tian, T., & Yu, L. (2017a). Simple and sensitive detection of pesticides using the liquid crystal droplet patterns platform. *Sensors and Actuators. B, Chemical, 238*, 676–682. doi:10.1016/j.snb.2016.07.114

Wang, Y., Zeinhom, M. M. A., Yang, M., Sun, R., Wang, S., Smith, J. N., ... Du, D. (2017b). A 3D-Printed, Portable, Optical-Sensing Platform for Smartphones Capable of Detecting the Herbicide 2,4-Dichlorophenoxyacetic Acid. *Analytical Chemistry, 89*(17), 9339–9346. doi:10.1021/acs.analchem.7b02139 PMID:28727917

Weerathunge, P., Ramanathan, R., Shukla, R., Sharma, T. K., & Bansal, V. (2014). Aptamer-controlled reversible inhibition of gold nanozyme activity for pesticide sensing. *Analytical Chemistry, 86*(24), 11937–11941. doi:10.1021/ac5028726 PMID:25340286

Wu, X., Song, Y., Yan, X., Zhu, C., Ma, Y., Du, D., & Lin, Y. (2017). Carbon quantum dots as fluorescence resonance energy transfer sensors for organophosphate pesticides determination. *Biosensors & Bioelectronics, 94*, 292–297. doi:10.1016/j.bios.2017.03.010 PMID:28315592

Xia, N., Wang, Q., & Liu, L. (2015). Nanomaterials-based optical techniques for the detection of acetylcholinesterase and pesticides. *Sensors (Switzerland), 15*(1), 499–514. doi:10.3390150100499 PMID:25558991

Yang, J. K., Kang, H., Lee, H., Jo, A., Jeong, S., Jeon, S.-J., ... Lee, Y.-S. (2014). Single-step and rapid growth of silver nanoshells as SERS-active nanostructures for label-free detection of pesticides. *ACS Applied Materials & Interfaces, 6*(15), 12541–12549. doi:10.1021/am502435x PMID:24988366

Yu, J., Wang, X., Kang, Q., Li, J., Shen, D., & Chen, L. (2017). One-pot synthesis of a quantum dot-based molecular imprinting nanosensor for highly selective and sensitive fluorescence detection of 4-nitrophenol in environmental waters. *Environmental Science. Nano, 4*(2), 493–502. doi:10.1039/C6EN00395H

Zhang, K., Mei, Q., Guan, G., Liu, B., Wang, S., & Zhang, Z. (2010). Ligand replacement-induced fluorescence switch of quantum dots for ultrasensitive detection of organophosphorothioate pesticides. *Analytical Chemistry, 82*(22), 9579–9586. doi:10.1021/ac102531z PMID:20973515

Zhang, Z., Zeng, K., & Liu, J. (2017). Immunochemical detection of emerging organic contaminants in environmental waters. TrAC -. *Trends in Analytical Chemistry, 87*, 49–57. doi:10.1016/j.trac.2016.12.002

Zheng, Z., Li, X., Dai, Z., Liu, S., & Tang, Z. (2011). Detection of mixed organophosphorus pesticides in real samples using quantum dots/bi-enzyme assembly multilayers. *Journal of Materials Chemistry*, *21*(42), 16955. doi:10.1039/c1jm11631b

Zou, Z., Du, D., Wang, J., Smith, J. N., Timchalk, C., Li, Y., & Lin, Y. (2010). Quantum dot-based immunochromatographic fluorescent biosensor for biomonitoring trichloropyridinol, a biomarker of exposure to chlorpyrifos. *Analytical Chemistry*, *82*(12), 5125–5133. doi:10.1021/ac100260m PMID:20507134

KEY TERMS AND DEFINITIONS

Label-Free Technology: Label-free technologies encompasses a wide range of methods that can eliminate the need for labeling agents used for detection of specific chemical species. The major label-free methods include physical, electrochemical, mass-based, and optical techniques. This technology is widely used in drug discovery and high-end analysis of toxic agents.

Nanomaterials: Nanomaterials are materials that are made up of nanoscale structure in the range of 1 – 100 nm. Currently, nanomaterials are employed in various applications including non-stick coatings, scaffolds for coverage of wounds, nanoscale electronic circuits, etc.

Optical Detection: Optical detection is the process of using visible light to detect different chemical species. The optical detection system usually comprises of a light source, which is made to pass through the sample and then absorption or emission characteristics is detected further by a light detector such as photo-multiplier tube.

Pesticides: Pesticides are chemical compounds used to control various types of pests. It is extensively used in agriculture and households to eradicate pest infestations. Several pesticides are proven to be toxic even at very low doses. Hence, advances are being made to use natural products to control pests including bio-control agents such as plant extracts and microbes.

Chapter 15
Occurrence of Pesticides and Their Removal From Aquatic Medium by Adsorption

Irfan Rashid Sofi
Jiwaji University Gwalior, India

Rayees Ahmad Bhat
Jiwaji University Gwalior, India

Rabia Quadir
Jiwaji University Gwalior, India

Javid Manzoor
Jiwaji University Gwalior, India

ABSTRACT

Large amounts of pesticides are used annually, and in some cases, a part of the pesticide enters the water bodies by surface runoff to form long-term residues. In the recent past, the adverse effects of pesticides on the environment and human health received serious attention by the public and the competent authorities. Various conventional methods are used to remove these pesticides from water, but those methods are either costly or typical in operation. Therefore, adsorption is considered as an ecofriendly method. The adsorbent derived from biomaterial is considered an encouraging adsorbent due to its cost-effective and high adsorption capacity. In this chapter, detailed information on different types of pesticides, their metabolites, environmental concerns, and present status on degradation methods using adsorbents will be reviewed. This chapter presents a comprehensive overview on the recent advancement in the utilization of different adsorbents for the removal of pesticides. Overall, this study assists researchers to move forward in exploring a simple and economically viable technique to produce adsorbents with outstanding physiochemical properties and excellent adsorption capacity, so that the pesticides can be removed from aquatic ecosystem.

DOI: 10.4018/978-1-5225-6111-8.ch015

INTRODUCTION

Different types of pesticides have been extensively used by modern agriculture practices. The major pathways thought which pesticides reach into the environment are by application, disposal, runoff or a spill (Adachi et al., 2001). Environmental contamination due to the extreme use of pesticides has become a great opposing effect on human health and environment (Bakour et al., 2009). The unwarranted use of pesticides to control the crop destroying insects has gained momentum in last two to three decades. Due to their extensive usage, they easily contaminate different media i.e., air, soil and ground water. Their deposits may remain in soil which decreases the biodiversity in the soil and they may also pass into the ground waters by percolating through soil (Shukla et al., 2006). Many Organophosphate compounds are currently being used as pesticides are of concern due to their diligence, bioaccumulation as well as their toxicological effects on human health and the environment (Monirith et al., 2003). Pesticides are hazardous and toxic in nature and persist in the aquatic environment for many years. (Radan et al., 1999).

Pesticides are the substances that are used to kill, destroy or control pests which harm the agricultural produce. In the recent time the increasing population results in increasing demand of agricultural products and the quality of these products is controlled by using different kinds of pesticides and these pesticides may enter the human body through food chain. Pesticides used for agriculture purpose are released into the water bodies by surface runoff from agricultural landscape and also through municipal and industrial discharge which results in deterioration of water quality (Luo et al., 2008). The presence of pesticides in water bodies affect the aquatic biodiversity and other associated living organisms (Simazaki et al., 2015, Gou et al., 2016). It is important to understand the impact of pesticides on the aquatic animals and their ecosystem and also on humans which are sensitive to pesticides which may suffer from serious acute and chronic diseases ((Maund et al.,1997; Liess et al., 2003; Noyes et al.,2009).The exposure of human to those pesticides results in many disorders like infant mortality (Cremonse, 2014), Alzheimer's disease (Hayden, 2010), Carcinogenicity (Vopham, 2015), Neurotoxicity (Kiefer, 2007), reproductive toxicity (Chevier, 2013), and metabolic toxicity (Evanglou, 2016).

Pesticides are the kind of substances which cause major stress in aquatic ecosystem because they are persistent, bio-accumulative and toxic and can affect all kind of aquatic biodiversity ((Malaj et al., 2014;) Kohler &Triebskorn, 2013). The water bodies that are small in size and surrounded by landscape are mostly affected by pesticides and other pollutants. These water bodies have high proportion of bio-diversity and lower dilution potential and the pesticides are adsorbed on the sediments which results in strong ecological effects as compared to the large water bodies (Lorenz et al., 2016, Munz et al., 2017, Neale et al., 2017, Szocs et al., 2017). Therefore, the small water bodies are given the more attention because they provide variety of habitat to the aquatic biodiversity (Downing, 2010; Biggs et al., 2016; Hill et al, 2016).

The different kinds of pesticides that are used to kill pests and some of them are banned in some developing countries and irrespective of the ban they are still producing these toxic pesticides in bulk quantity in order to increase their agricultural yield (Luo et al., 2016). The improper handling, lack of knowledge about toxic pesticides and poverty status of users becomes the major cause of deaths by pesticides in developing countries (Kesavachandran et al., 2009) and also the pesticides and their metabolites were not recognized as environmental toxicants. Pesticides are great threat to the environment globally because of their low polarity, high solubility in water, bioaccumulation in food chain (Afful et al., 2010). The pesticides in water bodies in a concentration which exceeds the limit damages the aquatic biodiversity (Kole, Banerjee, & Bhattacharyya, 2001).

Figure 1. Top ten pesticides consuming countries (millions of kilograms)

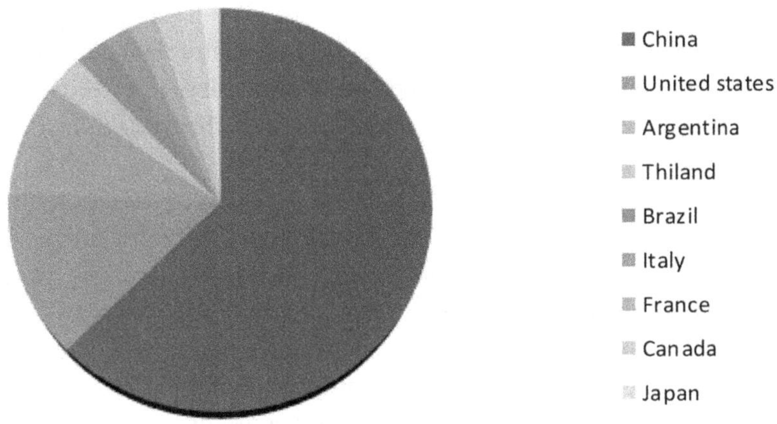

Now the research has begun to assess the persistence, resistance, fate and transport of these pesticides from different sources which ends up in the human body through food chain (Schaefer et al., 2013). Different researchers have reported the transport of pesticides to water bodies through different routes (Squillace et al., 1993; Kolpin et al., 2001). Moreover, it is also important to understand the eco toxicological impact of pesticides dose over successive generations in aquatic environment (Artigas et al., 2012). Nowadays genetic engineering is used in preparation of pesticides and these pesticides are safer than traditional pesticides (Uneke, 2007).

Now the need of the hour is to remove these persistent pesticides from the water by the ecofriendly methods. Adsorption is believed to be the ecofriendly method among all the conventional methods by using adsorbents based on agricultural wastes, which are used to remove various pollutants from waste water. The adsorbents includes rice husk, fruit seed, nut shells, palm fruit shell and fruit peels by adsorption (Johan et al. 2011; Lim et al. 2012; Olu-owolabi et al. 2012; Omri and Benzina 2013; Sugashini and Begum 2013). These adsorbents are abundant in supply and are known as low cost based adsorbents (Bailey et al. 1999; Jain et al. 2013). The purpose of this study is to give insight about the pesticide pollution and various potential low cost materials which can be utilized as adsorbents for the conventional pesticide recovery from water in a cost effective way.

INITIAL DISTRIBUTION

The fate of a pesticide is influenced by the initial pesticide distribution and proportion of application. The determination of amount depends upon the type of vegetation, weather condition and topography. Pesticides can move to the non-target site and may cause water pollution. They may be brokendown by sunlight or leached away to the water before reaching their target. In soil, microorganisms broke down the pesticide and are absorbed by plants through their roots. The properties of pesticide, their use and characteristics also govern the fate of pesticides in the environment and also influence its persistence and mobility.

PERSISTENCE AND DEGRADATION

The fate of the pesticides can be predicted by half-life period. Pesticides are categorized into three types based on their half-life period: non-persistent with half-life period less than 30days, moderately persistent with half-life 30 to100 days and persistent with half-life more than 100 days. Pesticides degrade by microbial activity, chemical activity, or sunlight.

- **Microbial Activity:** Involves break down of pesticides by microorganisms when they use the pesticides as food along with other substances. Microbial degradation is favoured by the warm, moist, and well aerated soil with neutral pH.
- **Chemical Degradation:** Involves the reaction with water, oxygen with other chemicals in the soil. Rapid chemical degradation is supported by low microbial activity.
- **Photo Degradation:** The breakdown of pesticides by sunlight is called photo degradation. The rate of photodegradation is governed by the intensity of light, length of exposureand pesticide properties.

Pesticides are degraded by chemical and photochemical reaction and also by biological mechanism though microbiological digestion and metabolism of pesticides by organisms. All these processes help in reducing the toxicity of pesticides. The degradation of pesticides in the soil and water may vary greatly depending upon local topographical conditions. The degradation products include water, carbon dioxide and minerals and many other products andmore ever its degradation have severe impact on environment and human health.The various factors which are of prime importance in the degradation of pesticides include pesticide chemistry, environmental conditions, microbial activity, chemical activity and sunlight and pH.The rate of degradation is also mediated by the enzymes and is doubled by increasing the temperature by 10°C. Pesticides tend to be more persistent in soil rather than canopy foliage. The pesticides are taken as food by microbes. The microbial activities in degradation of pesticides are influenced by moisture, temperature, aeration and pH. The warmest soil with neutral pH helps the microbes to degrade the pesticides to greater extent.

Many pesticides are rapidly converted into simpler compounds through the process known as mineralization which includes chemical reactions such as hydrolysis, photolysis, microbial catabolism and metabolism and these minerals are readily dissipated into soils. Some pesticides are easily broken down while some are very typical and remain persistent in soil (Stephenson, Harris and Solomon, 1993).

PROCESS OF METABOLISM

The toxic effect of xenobiotics entering through food supply are protected by organisms through the process of metabolism, where toxic chemicals are converted into less toxic simpler compounds with the help of certain enzymes.

MOBILITY

Environmental toxicology is well understood by the mobility of pesticides. Pesticides should be mobile so that they can reach the target site. Pesticides with more mobility get dissipated immediately after application and result in contamination of water and sediments. As mentioned earlier, mobility also depends on soil and water environment, soil characteristics, pesticides and timing of application. The mobility of some pesticides may result in the redistribution at application site. They may get adsorbed to the soil particles, and vegetation near the site of application and sometimes are eroded with the along with the soil particle by surface runoff and are also absorbed by the plant roots. Some pesticides are volatile and get eroded from foliage by the wind and spread into the air. Mobility of pesticides is also influenced by the characteristics of pesticides, water solubility and vapor pressure. Furthermore, environmental characteristics are also responsible for the mobility of pesticides which includes topography, ground cover, canopy, organic matter and texture.

Pesticides when adsorbed by the soil particles are subject to the plant uptake or microbial or chemical degradation. However, plant uptake and microbial degradation of pesticides are very less when they are adsorbed strongly by the soil particles and weakly adsorbed pesticides are percolated through the soil by infiltrating water. Adsorption of pesticides is also influenced by the pH of the soil, moisture content, their hydrophobic attraction and hydrogen bonding. Pesticides are easily adsorbed by dry soil as compared to wet soil because of competition between water and pesticides in the moist soil. More ever adsorption potential of pesticides is also increased by high clay content and organic matter in soil and reduces their mobility.

ECOLOGICAL EFFECTS OF PESTICIDES

Pesticides are among the micro pollutants which have great impact in environment. Different pesticides effects the different living entity through different ways and it is therefore difficult to generalize them. Pesticides also have impact on terrestrial environment and pesticide runoff is believed to be one of the major pathways of water contamination and hence cause the ecological impacts. The two principle mechanisms are bio magnification and bio concentration.

- **Bioconcentration:** It is process by which concentration of pesticides increases in the aquatic medium and is transferred from the surrounding medium to organize. Some pesticides are lipophilic and the lipids are the primary sink for pesticides and may get accumulated in the fatty tissue.
- **Biomagnification:** It may be defined as the increase in concentration of any chemical inside the body of any organism. When the pesticides enter the food chain they get magnified in the fatty tissue and their highest concentration is observed in case of top predators.

The effect of micro pollutants is often inter-related and their impact on human health is considered as potential indicator of effects on environment. Different pesticides have different impacts depending upon the type of pesticide and type of organism under investigation. The consequences of pesticide pollution are given below:

1. It causes tumor or lesions in fish and animals and may result in death of organisms.
2. It causes weakening of immune system, disruption of endocrine glands, failure of reproductive ability and also may damage the DNA,
3. It can also affect the birds and may result in deformities and is also responsible for the poor health of fishes by lowering down their blood cells.

These effects may be caused by the micro pollutants including pesticidesand also may result by environmental stresses. Pesticides can have impact on whole ecosystem rather than any individuals of the ecosystem. It is indicated by Swedish work that biodiversity is mostly affected by the pesticides. Pesticides used to control weeds and also alter the habitat and may result in imbalance among the community. Moreover, the pesticides also affectthe fertility of soil by inhibiting the process of nitrification and also effect the soil microbes responsible for degradation of organic matter and pesticides (Torstensson and stenstrom, 1990).

In urban areas the common pollutant found in the soil, water and air are pesticides. According to the US Geological Survey 1999, the most common pesticide in urban streams is chloropyrifos which is highly toxic to the aquatic life. Other pesticides including Trifluralin are also toxic for all cold and warm fishes. Glyphosate also causes the erratic swimming and labored breathing in fishes and result in lethal effects in their lung etc. 2,4-D herbicide have physiological stress on sockeye salmon (Mc Bride, Tyler &Hovde 1981) and also reduce the capability of rainbow trout in gathering of food. The water mammal Dolphin also has been affected by the pesticide poisoning. Being the higher trophic level in the food chainand having low drug metabolism activities, dolphins receive high concentration of persistent pollutants. River dolphins are among the world's most seriously endangered species because riverine and estuarine ecosystems are mostly reducing the pollution because they are vulnerable to human activities. The river dolphins are decreasing in number and are facing the threat of extinction. Other factors which are the possible threat to dolphins include their habitat destruction, boat traffic, fishing, incidental and chemical pollution. The continuous use of organochlorine pesticides end polychlorinated biphenyls also impacts the marine or fresh water animals including Ganga river dolphin (Kannan et al., 2005). The exposure to the high concentration of pesticides which are bio accumulating, toxic and persistent including DDT and PCB affect the reproductive and immune system of aquatic animals. Some aquatic animals are very sensitive to the chemical contamination are quickly affected which includes otters and Mink. A pesticide when transported to their final destination affects all the members in the community from invertebrates to birds and humans. These pesticides are responsible for the death of birds, fishes and zooplanktons. The breeding sites of amphibians when contaminated by pesticides result in long term effect on their survival pattern. The frogs and late wood frogs are affected after 24 days exposure to atrazine (Storrs&Kiesecker, 2004).

PESTICIDE MONITORING IN SURFACE WATER

The level of pesticides in the surface water is increasing rapidly day by day in much of the world mostly in developing countries. The key pesticides are used to develop a set by including them in monitoring schedule but cost of analysis and sampling at different time of year, inadequate facility, financial support and impure reagents makes it impossible. Nowadays new techniques reduce the cost and increase reliability in detecting the presence or absence of pesticides.

BIOADSORBENTS USED IN REMOVAL OF PESTICIDES

Chemical pesticides like insecticides, herbicides and fungicides are frequently used in agricultural fields to protect the agricultural produce from pests. However, the excessive use of these agrochemicals has adverse effects on environment result in reduction in population of insect pollinators, alteration of habitat of birds, and threat to the endangered species. Upon consumption; these pesticides may cause various health related issues which includes chronic disease such as skin, eye and nervous disorder and cancer upon prolonged exposure. Various techniques in the past have been developed on the basis of surface adsorption, membrane filtration and biological degradation to remove the content of pesticides. However, these techniques show slow response, less specificity and sensitivity and were cost effective. In recent times, Nanotechnology has emerged as a helping tool for the sensing and remediation of pesticides. Various nanomaterials have been synthesized by various techniques and were classified in various groups such as nano-particles, nano-tubes and nano-composites that are used for detection, degradation and removal of pesticides (Rawtani, 2018).

Different researchers have prepared different potential adsorbents from different materials and used them for the removal of different pollutants from water which shows greater efficiency for their removal. In this study we tried to give insight about the adsorbents used for the removal of pesticides from the aqueous solution and which are validated by certain isotherm and kinetic models. A detailed summary about the different adsorbents used for the removal of pesticides is give below:

Alahabadi and Moussavi (2017) synthesized Carbonate-induced activated biochar (CAB) and used it in the removal of atrazine. This material shows good adsorption for atrazine with the capacity of 714mg/g and at temperature of 40°Cat neutral pH with Adsorbent BET. The adsorption was validated by various techniques which include pseudo-second order model and thermodynamics and it was found that the adsorption is spontaneous and exothermic.

Ignatowicz 2011 experimentally evaluated that adsorption of HCH on coconut shell based activated carbon type NP-5 and found that HCH is adsorbed as monolayer and the competition between pesticides and and water is negligible. The adsorption was better fitted by isotherm model. Similarly adsorption of glyphosate, s-metolachlor and epoxiconazole was investigated by Aslam 2013 using decomposed maize mulch residues as an adsorbent. During the study it was revealed that adsorption of s-metolachlor and epoxiconazole increased with decomposed mulch whereas glyphosate adsorption was less as expected.

Rodríguez-Liébana (2016) collected nine natural clay samples and used them in the removal of metalaxyl and fludioxonil. The clay contains low organic content and having high Ca content and low surface area. The retention of both the pesticides was dependent upon the organic content and it was observed that metalaxyl was more adsorbed as compared to fludioxonil. Metalaxyl was better explained by Fruendlich equation while fludoxionil was better fitted by both Fruendlich and Langmuir. Long (2015) has developed up to fluidic bio sensing platform by combining with advance photonics and microfluidic technology for rapid and sensitive detection and adsorption kinetics. EWOB helps prediction of adsorption of atrazine on soil with great robustness, reusability and accuracy. The adsorption process follows the pseudo second order kinetic model. Soil plays important role in the adsorption of pesticides as it favours the transportation, degradation and mobility and bioaccumulation.

Lui, Cai& Chen 2011 synthesized seven spherical porous cyclodextrin based polymer in order to determine their adsorption potential for pesticides from water. It was observed that adsorbent can adsorb the pesticides from mixture with open homogenous network through different interaction. Khoshnood, &Azizian (2012) studied removal of pesticide (2,4-dichlorophenoxyacetic acid) from aqueous solution by magnetic and graphitic carbon nanostructures from both equilibrium and kinetic point of view. Taha (2014) used biochars and charcoal for the adsorption of pesticides from mixture of fifteen pesticides. The individual concentration of each pesticides were 400mg/l. the adsorbent were characterized by some valid techniques which includes SEM, FTIR, BET. It was observed that biochars shows more adsorption as compared to charcoal. Biochars synthesized from rice straw reduced the value of pesticides to 0.005mg/l just in two hours.

Derylo-Marczewska (2017) studied the role of porous carbon adsorbent in removal of chlorophenoxy herbicides. The carbon was used in three fractions. The equilibrium and kinetics were taken into account to describe the adsorption data and moreover the adsorbate property and its effectiveness on the rate of adsorption were also determined. Alfonso (2017) found four organophosporus pesticides in the ground water whose concentration were more than the 5 μg/l and evaluated the process of adsorption for these pesticides in soil. The study was performed in batch process techniques. The adsorption process was validated by Henry, Langmuir and Freundlich models. The results indicate that the Freundlich model provides the best correlation of the experimental data. Freundlich adsorption coefficients K_f were in the range of 1.62–2.35 for sulfotep, 2.43 to 3.25 for dimethoate, from 5.54 to 9.27 for methyl parathion, and 3.22 to 5.17 for diazinon.

Salman, Njoku, &Hameed (2011) prepared the activated carbon from banana stalk by potassium hydroxide activated in presence of Co_2 gas. The same carbon was characterized by FTIR and utilized to remove the 2,4-dichlorophenoxyacetic acid. Two kinetic models were used to analyze the adsorption and it was found that pseudo second order fits best. Higher adsorption was obtained due to the –cl groups and smaller molecular groups of 2,4-dichlorophenoxyacetic acid. It is therefore found that banana stalk activated carbon is good adsorbent for the removal of 2,4-dichlorophenoxyacetic acid from aqueous solution.

Agarwal (2016) synthesized Graphene quantum dots through microwave-assisted hydrothermal route and used the graphene quantum dots as an adsorbent for the removal of noxious pesticide compound i.e. oxamyl from aqueous solutions. Batch adsorption was adopted to optimize the parameters such as agitation speed, pH, adsorbent dose, contact time, temperature and initial concentration on sorption of pesticides. The optimized ph, dose and concentration were 8.0, 0.6 g and 25 minute respectively, which favours maximum adsorption of oxamyl. It was evident for the results increase in oxamyl concentration considerably favours the removal percentage of oxamyl. The isotherm and kinetic model were in good agreement with the adsorption data. Rojas et al., (2015) elucidated the adsorption potential of ricehusk sunflower and composite sewage sludge. It was observed that trifluralin and chlorfenvinphos were adsorbed significantly by all kinds of adsorbents used as shown by adsorption kinetics. Simazine was best adsorbed by rice husk among all adsorbents. It was revealed that organic waste can be good source for the synthesis of adsorbents and because of their natural, economical and high adsorption capacity. Behloul et al., (2017) used biomass of pluerotusmutilus biomass for the adsorption of metribuzan. The biomass was physically treated and the characterized for it structural changes and more ever the influence of other parameters like pH, agitation, and particle size was also evaluated. From the results it was revealed that adsorption of metribuzin was convincing and the rate of adsorption was about 70%.

Mandal, Singh, &Purakayastha (2017), Investigated biochars of different agricultural wastes, these agricultural wastes include corn cob, eucalyptus bark, rice straw, bamboo chips, and rice husk and acid treated RSBC (T-RSBC). Kinetics study of these biochars was best explained by the pseudo second order and modified Elovich model. All the five biochars showed good adsorption phenomena but the rice straw showed maximum adsorption for atrazine (37.5 – 70.7%) and imidacloprid (39.9 – 77.8%). The Adsorption of both pesticides was enhanced by phosphoric acid treatment of rice straw. The factors that affected the adsorption of pesticides on biochars were their weak acid fraction, pH, pore diameter, aromaticity and polarity. From this study, it is clear that biochars of rice straw have significant potential for environmental implications and it can be best utilized as an adsorbent for waste water purification in pesticide industry.The high use of pesticides in agricultural activities, the pesticides concentration in aquatic environments has increased. In this study, low cost bioadsorbent chestnut shells were used for the pesticide removal. The pesticides removed from this bioadsorbent were pirimicarb, imidacloprid, acetamiprid and thi-amethoxam. These pesticides are widely used in agricultural activities. To enhance the removal efficiency of this bioadssorbent, the economic acid pretreatments were done. Citric acid pretreatment of chestnut shells showed highest removal than raw chestnut shells increasing a 15% of adsorption capacity (Cobas et al., 2016).

Haung, Ling, & Zhang (2017) explained that engineered nanoparticles (ENPs) in soil have high potential for the adsorption of organic pollutants. They are believed to affect the transport of pesticides in soils and also affect the uptake and transformation of pesticides. In presence of nano-SiO_2, the adsorption pattern of racemic-metalaxyl on agricultural soils including kinetics and isotherms changes. When the soil and SiO_2 were mixed, the adsorption of racemic-metalaxyl on SiO_2 decreased to some extent, and the absolute decrease was dependent on soil properties.

Wang et al., (2017) synthesized low cost adsorbent, novel phenyl-modified magnetic graphene/mesoporous silica (MG-MS-Ph) composites with hierarchical bridge-pore structure. The synthesized MG-MS-Ph exhibited large surface area (446.5 m2g/1), highly ordered mesoporous with uniform pore size (2.8 nm) and pore volume (0.32 $cm^3g/1$) and higher saturation magnetization (25 emu g^{-1}) makes it prominent adsorbent for the removal of pesticides. It was found through batch adsorption studies that the novel composites unveil good adsorption capacity of pesticides as compared to other adsorbents like Activated carbon, Multi-walled carbon nanotube and Single-walled carbon nanotube. Therefore, it was clear from the investigations that MG-MS-Ph as low cost and effective adsorbents for the removal of toxic pesticides would be economically and technically feasible.

Pinto (2016) synthesized the porous carbonaceous material (PCM) by pyrolysis at 850 ^{0}C of a composite made of Laponite and Cassava starch. That material synthesized has almost twice the specific surface area and up to 20 times more mesoporous volume than the carbonaceous material obtained without the clay. The adsorbent (PCM) exhibits a high removal efficiency of Dicamba, with a maximum adsorption capacity of 251.9 mg g^{-1} related to a pseudo-second-order kinetic adsorption model with a strong pH dependence. Liu (2017) prepared a novel magnetic copper-based metal–organic framework by using Fe_4O_3–graphene oxide (GO)–cyclodextrin (-CD) nanocomposite and used it for adsorption and removal of neonicotinoid insecticide pollutants from aqueous solution. During the study it was observed that adsorption of thiacloprid fitted with the Langmuir monolayer adsorption other insecticides tested showed Freundlich bimolecular layer adsorption. From the results it is evident that M-MOF is good adsorbent with efficient adsorbent capacity and is a favorable hybrid adsorbent for expeditious removal of insecticide pollutants from environmental waters.

CONCLUSION

Pesticides are very often used to the control the pests in the urban landscapes. However frequent use of these pesticides, their persistence and pesticides residues which are found in every sphere of environment result in the severe impact on human health. Environmental contamination due to pesticides poses a greater risk to the microorganism which includes insects, plants, fish, and birds. To reduce the environmental contamination by the pesticides, the best way is to use non chemical pesticides. Furthermore to avoid any kind environmental pollution by the pesticides constant monitoring and appropriate analysis is to be adopted by the concerned agencies. It is also important to make the people aware about the consequences of pesticides use. This can be done by public awareness and implementation of strict rules by the government agencies about the use of harmful pesticides. Although some countries have banned the use of some pesticides while some developing countries are still using these harmful chemicals to increase their agriculture yield without thinking about its consequence on the environment. Furthermore some researchers have tried some methods to remove these pesticides from the water. Adsorption is one of the methods used to remove the pesticides from waste water. Many agricultural based or many synthetic adsorbents have been prepared in order to remove pesticides from water. It has been observed that the use of these adsorbent is the alternative methods for the removal of pesticides from the water.

REFERENCES

Adachi, A., Ikedia, C., Takagi, S., & Fukao, N. (2001). Efficiency of rice bran for removal of Organochlorine compounds and benzene from industrial waste water. *Journal of Agricultural and Food Chemistry*, *49*(3), 1309–1314. doi:10.1021/jf001147c PMID:11312856

Afful, S., Anim, A. K., & Serfor-Armah, Y. (2010). Spectrum of organochlorine pesticide residues in fish samples from the Densu Basin. *Res J Environ Earth Sci*, *2*(3), 133–138.

Agarwal, S., Sadeghi, N., Tyagi, I., Gupta, V. K., & Fakhri, A. (2016). Adsorption of toxic carbamate pesticide oxamyl from liquid phase by newly synthesized and characterized graphene quantum dots nanomaterials. *Journal of Colloid and Interface Science*, *478*, 430–438. doi:10.1016/j.jcis.2016.06.029 PMID:27362399

Akhtar, M., Hasany, S. M., Bhanger, M. I., & Iqbal, S. (2007). Low cost sorbents for the removal of methyl parathion pesticide from aqueous solutions. *Chemosphere*, *66*(10), 1829–1838. doi:10.1016/j.chemosphere.2006.09.006 PMID:17109916

Akhtar, M., Iqbal, S., Bhanger, M. I., Zia-Ul-Haq, M., & Moazzam, M. (2009). Sorption of organophosphorous pesticides onto chickpea husk from aqueous solutions. *Colloids and Surfaces. B, Biointerfaces*, *69*(1), 63–70. doi:10.1016/j.colsurfb.2008.10.019 PMID:19091530

Aksu, Z., & Kabasakal, E. (2005). Adsorption characteristics of 2, 4-dichlorophenoxyacetic acid (2, 4-D) from aqueous solution on powdered activated carbon. *Journal of Environmental Science and Health. Part. B, Pesticides, Food Contaminants, and Agricultural Wastes*, *40*(4), 545–570. doi:10.1081/PFC-200061533 PMID:16047879

Alahabadi, A., & Moussavi, G. (2017). Preparation, characterization and atrazine adsorption potential of mesoporous carbonate-induced activated biochar (CAB) from CalligonumComosum biomass: Parametric experiments and kinetics, equilibrium and thermodynamic modeling. *Journal of Molecular Liquids*, *242*, 40–52. doi:10.1016/j.molliq.2017.06.116

Alfonso, L. F., Germán, G. V., del Carmen, P. C. M., & Hossein, G. (2017). Adsorption of organophosphorus pesticides in tropical soils: The case of karst landscape of northwestern Yucatan. *Chemosphere*, *166*, 292–299. doi:10.1016/j.chemosphere.2016.09.109 PMID:27700995

Armaghan, M., & Amini, M. M. (2017). Adsorption of diazinon and fenitrothion on nanocrystalline magnesium oxides. *Arabian Journal of Chemistry*, *10*(1), 91–99. doi:10.1016/j.arabjc.2014.01.002

Artigas, J., Arts, G., Babut, M., Caracciolo, A. B., Charles, S., Chaumot, A., & Friberg, N. (2012). Towards a renewed research agenda in ecotoxicology. *Environmental Pollution*, *160*, 201–206. doi:10.1016/j.envpol.2011.08.011 PMID:22015334

Aslam, S., Garnier, P., Rumpel, C., Parent, S. E., & Benoit, P. (2013). Adsorption and desorption behavior of selected pesticides as influenced by decomposition of maize mulch. *Chemosphere*, *91*(11), 1447–1455. doi:10.1016/j.chemosphere.2012.12.005 PMID:23434076

Ayranci, E., & Hoda, N. (2005). Adsorption kinetics and isotherms of pesticides onto activated carbon-cloth. *Chemosphere*, *60*(11), 1600–1607. doi:10.1016/j.chemosphere.2005.02.040 PMID:16083766

Bailey, R., Barrie, L. A., Halsall, C. J., Fellin, P., & Muir, D. C. G. (2000). Atmospheric organochlorine pesticides in the western Canadian Arctic: Evidence of transpacific transport. *Journal of Geophysical Research, D, Atmospheres*, *105*(D9), 11805–11811. doi:10.1029/1999JD901180

Bakour, E. I., & Morillo, H. (2009). usero J and ooassini A, Natural attenuation of pesticide water contamination by using ecological adsorbents, application for chlorinated pesticide included in European water framework directive. *J. of Hydro.*, *364*, 175–181. doi:10.1016/j.jhydrol.2008.10.012

Behloul, M., Lounici, H., Abdi, N., Drouiche, N., & Mameri, N. (2017). Adsorption study of metribuzin pesticide on fungus Pleurotusmutilus. *International Biodeterioration & Biodegradation*, *119*, 687–695. doi:10.1016/j.ibiod.2016.07.005

Biggs, J., Von Fumetti, S., & Kelly-Quinn, M. (2017). The importance of small waterbodies for biodiversity and ecosystem services: Implications for policy makers. *Hydrobiologia*, *793*(1), 3–39. doi:10.100710750-016-3007-0

Chang, K. L., Lin, J. H., & Chen, S. T. (2011). Adsorption studies on the removal of pesticides (Carbofuran) using activated carbon from rice straw agricultural waste. *World AcadSciEngTechnol, 5*, 4-21.

Chatterjee, S., Das, S. K., Chakravarty, R., Chakrabarti, A., Ghosh, S., & Guha, A. K. (2010). Interaction of malathion, an organophosphorus pesticide with Rhizopusoryzae biomass. *Journal of Hazardous Materials*, *174*(1-3), 47–53. doi:10.1016/j.jhazmat.2009.09.014 PMID:19783095

Chevrier, C., Warembourg, C., Gaudreau, E., Monfort, C., Le Blanc, A., Guldner, L., & Cordier, S. (2013). Organochlorine pesticides, polychlorinated biphenyls, seafood consumption, and time-to-pregnancy. *Epidemiology (Cambridge, Mass.)*, *24*(2), 251–260. doi:10.1097/EDE.0b013e31827f53ec PMID:23348067

Cobas, M., Meijide, J., Sanromán, M. A., & Pazos, M. (2016). Chestnut shells to mitigate pesticide contamination. *Journal of the Taiwan Institute of Chemical Engineers*, *61*, 166–173. doi:10.1016/j.jtice.2015.11.026

Council, E. P. A. (2009). Directive 2009/128/EC of the European Parliament and of the Council of 21 October 2009 establishing a framework for Community action to achieve the sustainable use of pesticides. *Official JL*, *309*, 71–86.

Cremonese, C., Freire, C., Camargo, A., Lima, J., Koifman, S., & Meyer, A. (2014). Pesticide consumption, central nervous system and cardiovascular congenital malformations in the South and Southeast region of Brazil. *International Journal of Occupational Medicine and Environmental Health*, *27*(3), 474–486. doi:10.247813382-014-0269-5 PMID:24847732

Daneshvar, N., Aber, S., Khani, A., & Khataee, A. R. (2007). Study of imidaclopride removal from aqueous solution by adsorption onto granular activated carbon using an on-line spectrophotometric analysis system. *Journal of Hazardous Materials*, *144*(1), 47–51. doi:10.1016/j.jhazmat.2006.09.081 PMID:17084525

Dehghani, M. H., Niasar, Z. S., Mehrnia, M. R., Shayeghi, M., Al-Ghouti, M. A., Heibati, B., ... Yetilmezsoy, K. (2017). Optimizing the removal of organophosphorus pesticide malathion from water using multi-walled carbon nanotubes. *Chemical Engineering Journal*, *310*, 22–32. doi:10.1016/j.cej.2016.10.057

Derylo-Marczewska, A., Blachnio, M., Marczewski, A. W., Swiatkowski, A., & Buczek, B. (2017). Adsorption of chlorophenoxy pesticides on activated carbon with gradually removed external particle layers. *Chemical Engineering Journal*, *308*, 408–418. doi:10.1016/j.cej.2016.09.082

El Bakouri, H., Usero, J., Morillo, J., & Ouassini, A. (2009). Adsorptive features of acid-treated olive stones for drin pesticides: Equilibrium, kinetic and thermodynamic modeling studies. *Bioresource Technology*, *100*(18), 4147–4155. doi:10.1016/j.biortech.2009.04.003 PMID:19409776

El Bakouri, H., Usero, J., Morillo, J., Rojas, R., & Ouassini, A. (2009). Drin pesticides removal from aqueous solutions using acid-treated date stones. *Bioresource Technology*, *100*(10), 2676–2684. doi:10.1016/j.biortech.2008.12.051 PMID:19186054

Evangelou, E., Ntritsos, G., Chondrogiorgi, M., Kavvoura, F. K., Hernández, A. F., Ntzani, E. E., & Tzoulaki, I. (2016). Exposure to pesticides and diabetes: A systematic review and meta-analysis. *Environment International*, *91*, 60–68. doi:10.1016/j.envint.2016.02.013 PMID:26909814

Furini, L. N., Constantino, C. J. L., Sanchez-Cortes, S., Otero, J. C., & López-Tocón, I. (2016). Adsorption of carbendazim pesticide on plasmonic nanoparticles studied by surface-enhanced Raman scattering. *Journal of Colloid and Interface Science*, *465*, 183–189. doi:10.1016/j.jcis.2015.11.045 PMID:26674234

Gou, Y. Y., Lin, S., Que, D. E., Tayo, L. L., Lin, D. Y., Chen, K. C., ... Chuang, K. P. (2016). Estrogenic effects in the influents and effluents of the drinking water treatment plants. *Environmental Science and Pollution Research International*, *23*(9), 8518–8528. doi:10.100711356-015-5946-9 PMID:26791027

Gupta, V. K., & Ali, I. (2001). Removal of DDD and DDE from wastewater using bagasse fly ash, a sugar industry waste. *Water Research*, *35*(1), 33–40. doi:10.1016/S0043-1354(00)00232-3 PMID:11257890

Gupta, V. K., Jain, C. K., Ali, I., Chandra, S., & Agarwal, S. (2002). Removal of lindane and malathion from wastewater using bagasse fly ash—a sugar industry waste. *Water Research*, *36*(10), 2483–2490. doi:10.1016/S0043-1354(01)00474-2 PMID:12153014

Hameed, B. H., Salman, J. M., & Ahmad, A. L. (2009). Adsorption isotherm and kinetic modeling of 2, 4-D pesticide on activated carbon derived from date stones. *Journal of Hazardous Materials*, *163*(1), 121–126. doi:10.1016/j.jhazmat.2008.06.069 PMID:18667269

Hayden, K. M., Norton, M. C., Darcey, D., Østbye, T., Zandi, P. P., Breitner, J. C. S., & Welsh-Bohmer, K. A.Cache County Study Investigators. (2010). Occupational exposure to pesticides increases the risk of incident AD The Cache County Study. *Neurology*, *74*(19), 1524–1530. doi:10.1212/WNL.0b013e3181dd4423 PMID:20458069

Hill, M. J., Death, R. G., Mathers, K. L., Ryves, D. B., White, J. C., & Wood, P. J. (2017). Macroinvertebrate community composition and diversity in ephemeral and perennial ponds on unregulated floodplain meadows in the UK. *Hydrobiologia*, *793*(1), 95–108. doi:10.100710750-016-2856-x

Huang, J., Liang, C., & Zhang, X. (2017). Effects of nano-SiO2 on the adsorption of chiral metalaxyl to agricultural soils. *Environmental Pollution*, *225*, 201–210. doi:10.1016/j.envpol.2017.03.065 PMID:28388518

Huguenot, D., Bois, P., Jézéquel, K., Cornu, J. Y., & Lebeau, T. (2010). Selection of low cost materials for the sorption of copper and herbicides as single or mixed compounds in increasing complexity matrices. *Journal of Hazardous Materials*, *182*(1), 18–26. doi:10.1016/j.jhazmat.2010.05.062 PMID:20594640

Ignatowicz, K. (2011). A mass transfer model for the adsorption of pesticide on coconut shell based activated carbon. *International Journal of Heat and Mass Transfer*, *54*(23-24), 4931–4938. doi:10.1016/j.ijheatmasstransfer.2011.07.005

Ioannidou, O. A., Zabaniotou, A. A., Stavropoulos, G. G., Islam, M. A., & Albanis, T. A. (2010). Preparation of activated carbons from agricultural residues for pesticide adsorption. *Chemosphere*, *80*(11), 1328–1336. doi:10.1016/j.chemosphere.2010.06.044 PMID:20598734

Islam, M. A., Sakkas, V., & Albanis, T. A. (2009). Application of statistical design of experiment with desirability function for the removal of organophosphorus pesticide from aqueous solution by low-cost material. *Journal of Hazardous Materials*, *170*(1), 230–238. doi:10.1016/j.jhazmat.2009.04.106 PMID:19477587

Johan, N. A., Kutty, S. R. M., Isa, M. H., Muhamad, N. S., & Hashim, H. (2011). Adsorption of copper by using microwave incinerated rice husk ash (MIRHA). *International Journal of Civil and Environmental Engineering*, *3*(4), 211–215.

Jusoh, A., Hartini, W. J. H., & Endut, A. (2011). Study on the removal of pesticide in agricultural run off by granular activated carbon. *Bioresource Technology*, *102*(9), 5312–5318. doi:10.1016/j.biortech.2010.12.074 PMID:21232934

Kannan, K., Ramu, K., Kajiwara, N., Sinha, R. K., & Tanabe, S. (2005). Organochlorine pesticides, polychlorinated biphenyls, and polybrominateddiphenyl ethers in Irrawaddy dolphins from India. *Archives of Environmental Contamination and Toxicology*, *49*(3), 415–420. doi:10.100700244-005-7078-6 PMID:16170447

Keifer, M. C., & Firestone, J. (2007). Neurotoxicity of pesticides. *Journal of Agromedicine*, *12*(1), 17–25. doi:10.1300/J096v12n01_03 PMID:18032333

Khoshnood, M., & Azizian, S. (2012). Adsorption of 2, 4-dichlorophenoxyacetic acid pesticide by graphitic carbon nanostructures prepared from biomasses. *Journal of Industrial and Engineering Chemistry, 18*(5), 1796–1800. doi:10.1016/j.jiec.2012.04.007

Khoshnood, M., & Azizian, S. (2012). Adsorption of 2, 4-dichlorophenoxyacetic acid pesticide by graphitic carbon nanostructures prepared from biomasses. *Journal of Industrial and Engineering Chemistry, 18*(5), 1796–1800. doi:10.1016/j.jiec.2012.04.007

Köhler, H. R., & Triebskorn, R. (2013). Wildlife ecotoxicology of pesticides: Can we track effects to the population level and beyond? *Science, 341*(6147), 759–765. doi:10.1126cience.1237591 PMID:23950533

Kole, R. K., Banerjee, H., & Bhattacharyya, A. (2001). Monitoring of market fish samples for Endosulfan and Hexachlorocyclohexane residues in and around Calcutta. *Bulletin of Environmental Contamination and Toxicology, 67*(4), 554–559. doi:10.1007001280159 PMID:11779071

Kolpin, D. W., Thurman, E. M., & Linhart, S. M. (2001). Occurrence of cyanazine compounds in groundwater: Degradates more prevalent than the parent compound. *Environmental Science & Technology, 35*(6), 1217–1222. doi:10.1021/es001520x PMID:11347936

Kuo, W. S., & Regan, R. W. Sr. (1999). Removal of pesticides from rinsate by adsorption using agricultural residuals as medium. *Journal of Environmental Science and Health. Part. B, Pesticides, Food Contaminants, and Agricultural Wastes, 34*(3), 431–447. doi:10.1080/03601239909373207

Liu, G., Li, L., Xu, D., Huang, X., Xu, X., Zheng, S., ... Lin, H. (2017). Metal–organic framework preparation using magnetic graphene oxide–β-cyclodextrin for neonicotinoid pesticide adsorption and removal. *Carbohydrate Polymers, 175*, 584–591. doi:10.1016/j.carbpol.2017.06.074 PMID:28917904

Liu, H., Cai, X., Wang, Y., & Chen, J. (2011). Adsorption mechanism-based screening of cyclodextrin polymers for adsorption and separation of pesticides from water. *Water Research, 45*(11), 3499–3511. doi:10.1016/j.watres.2011.04.004 PMID:21529879

Lizano-Fallas, V., Masís-Mora, M., Espinoza-Villalobos, D., Lizano-Brenes, M., & Rodríguez-Rodríguez, C. E. (2017). Removal of pesticides and ecotoxicological changes during the simultaneous treatment of triazines and chlorpyrifos in biomixtures. *Chemosphere, 182*, 106–113. doi:10.1016/j.chemosphere.2017.04.147 PMID:28494353

Long, F., Zhu, A., Shi, H., Sheng, J., & Zhao, Z. (2015). Adsorption kinetics of pesticide in soil assessed by optofluidics-based biosensing platform. *Chemosphere, 120*, 615–620. doi:10.1016/j.chemosphere.2014.09.072 PMID:25462305

Lorenz, S., Rasmussen, J. J., Süß, A., Kalettka, T., Golla, B., Horney, P., ... Schäfer, R. B. (2017). Specifics and challenges of assessing exposure and effects of pesticides in small water bodies. *Hydrobiologia, 793*(1), 213–224. doi:10.100710750-016-2973-6

Ludvík, J., & Zuman, P. (2000). Adsorption of 1, 2, 4-triazine pesticides metamitron and metribuzin on lignin. *Microchemical Journal, 64*(1), 15–20. doi:10.1016/S0026-265X(99)00015-6

Luo, D., Pu, Y., Tian, H., Cheng, J., Zhou, T., Tao, Y., ... Mei, S. (2016). Concentrations of organochlorine pesticides in umbilical cord blood and related lifestyle and dietary intake factors among pregnant women of the Huaihe River Basin in China. *Environment International*, *92*, 276–283. doi:10.1016/j.envint.2016.04.017 PMID:27123771

Luo, Y., Zhang, X., Liu, X., Ficklin, D., & Zhang, M. (2008). Dynamic modeling of organophosphate pesticide load in surface water in the northern San Joaquin Valley watershed of California. *Environmental Pollution*, *156*(3), 1171–1181. doi:10.1016/j.envpol.2008.04.005 PMID:18457909

Ma, L., He, F., Huang, T., Zhou, Q., Zhang, Y., & Wu, Z. (2016). Nitrogen and phosphorus transformations and balance in a pond-ditch circulation system for rural polluted water treatment. *Ecological Engineering*, *94*, 117–126. doi:10.1016/j.ecoleng.2016.05.051

Malaj, E., Peter, C., Grote, M., Kühne, R., Mondy, C. P., Usseglio-Polatera, P., ... Schäfer, R. B. (2014). Organic chemicals jeopardize the health of freshwater ecosystems on the continental scale. *Proceedings of the National Academy of Sciences of the United States of America*, *111*(26), 9549–9554. doi:10.1073/pnas.1321082111 PMID:24979762

Mandal, A., & Singh, N. (2017). Optimization of atrazine and imidacloprid removal from water using biochars: Designing single or multi-staged batch adsorption systems. *International Journal of Hygiene and Environmental Health*, *220*(3), 637–645. doi:10.1016/j.ijheh.2017.02.010 PMID:28433639

Mandal, A., Singh, N., & Purakayastha, T. J. (2017). Characterization of pesticide sorption behaviour of slow pyrolysis biochars as low cost adsorbent for atrazine and imidacloprid removal. *The Science of the Total Environment*, *577*, 376–385. doi:10.1016/j.scitotenv.2016.10.204 PMID:27817922

McBride, M. B., Tyler, L. D., & Hovde, D. A. (1981). Cadmium Adsorption by Soils and Uptake by Plants as Affected by Soil Chemical Properties 1. *Soil Science Society of America Journal*, *45*(4), 739–744. doi:10.2136ssaj1981.03615995004500040013x

Memon, G. Z., Bhanger, M. I., Akhtar, M., Talpur, F. N., & Memon, J. R. (2008). Adsorption of methyl parathion pesticide from water using watermelon peels as a low cost adsorbent. *Chemical Engineering Journal*, *138*(1), 616–621. doi:10.1016/j.cej.2007.09.027

Memon, G. Z., Moghal, M., Memon, J. R., Memon, N. N., & Bhanger, M. I. (2014). Adsorption of selected pesticides from aqueous solutions using cost effective walnut shells. *Adsorption*, *4*(10).

Mohammad, S. G., Ahmed, S. M., & Badawi, A. F. M. (2015). A comparative adsorption study with different agricultural waste adsorbents for removal of oxamyl pesticide. *Desalination and Water Treatment*, *55*(8), 2109–2120. doi:10.1080/19443994.2014.930797

Monirith, I., Uneno, D., Takahashi, S., & Nakata, H. (2003). sudryanto A, Subramanian A, karupiah S, ismail A, Muchtar M, Zhehg J, Richardson B.J., Prudente M, hue N.D., Jana T.s., Tkalin A.V and Tanabe S, Asia –pacific mussel Watch: Monitoring contamination of persistent Organochlorine compounds in coastal water of Asian countries, Mar. pollu. *Bull.*, *46*, 281–300.

Mostafalou, S., & Abdollahi, M. (2017). Pesticides: An update of human exposure and toxicity. *Archives of Toxicology*, *91*(2), 549–599. doi:10.100700204-016-1849-x PMID:27722929

Moussavi, G., Hosseini, H., & Alahabadi, A. (2013). The investigation of diazinon pesticide removal from contaminated water by adsorption onto NH 4 Cl-induced activated carbon. *Chemical Engineering Journal, 214*, 172–179. doi:10.1016/j.cej.2012.10.034

Munz, N. A., Burdon, F. J., De Zwart, D., Junghans, M., Melo, L., Reyes, M., ... Stamm, C. (2017). Pesticides drive risk of micropollutants in wastewater-impacted streams during low flow conditions. *Water Research, 110*, 366–377. doi:10.1016/j.watres.2016.11.001 PMID:27919541

Naushad, M., ALOthman, Z. A., Khan, M. R., ALQahtani, N. J., & ALSohaimi, I. H. (2014). Equilibrium, kinetics and thermodynamic studies for the removal of organophosphorus pesticide using Amberlyst-15 resin: Quantitative analysis by liquid chromatography–mass spectrometry. *Journal of Industrial and Engineering Chemistry, 20*(6), 4393–4400. doi:10.1016/j.jiec.2014.02.006

Njoku, V. O., Islam, M. A., Asif, M., & Hameed, B. H. (2014). Utilization of sky fruit husk agricultural waste to produce high quality activated carbon for the herbicide bentazon adsorption. *Chemical Engineering Journal, 251*, 183–191. doi:10.1016/j.cej.2014.04.015

Olu-Owolabi, B. I., Oputu, O. U., Adebowale, K. O., Ogonsolu, O., & Olujimi, O. O. (2012). Biosorption of Cd2+ and Pb2+ ions onto mango stone and cocoa pod waste: Kinetic and equilibrium studies. *Scientific Research and Essays, 7*(15), 1614–1629. doi:10.5897/SRE11.2248

Omri, A., & Benzina, M. (2013). Adsorption characteristics of silver ions onto activated carbon prepared from almond shell. *Desalination and Water Treatment, 51*(10-12), 2317–2326. doi:10.1080/19443994.2012.734585

Otero, R., Fernández, J. M., González, M. A., Pavlovic, I., & Ulibarri, M. A. (2013). Pesticides adsorption–desorption on Mg–Al mixed oxides. Kinetic modeling, competing factors and recyclability. *Chemical Engineering Journal, 221*, 214–221. doi:10.1016/j.cej.2013.02.007

Pal, O. R., & Vanjara, A. K. (2001). Removal of malathion and butachlor from aqueous solution by clays and organoclays. *Separation and Purification Technology, 24*(1), 167–172. doi:10.1016/S1383-5866(00)00226-4

Pavlovic, I., Barriga, C., Hermosín, M. C., Cornejo, J., & Ulibarri, M. A. (2005). Adsorption of acidic pesticides 2, 4-D, Clopyralid and Picloram on calcinedhydrotalcite. *Applied Clay Science, 30*(2), 125–133. doi:10.1016/j.clay.2005.04.004

Pinto, M. D. C. E., Gonçalves, R. G. L., dos Santos, R. M. M., Araújo, E. A., Perotti, G. F., dos Santos Macedo, R., ... Tronto, J. (2016). Mesoporous carbon derived from a biopolymer and a clay: Preparation, characterization and application for an organochlorine pesticide adsorption. *Microporous and Mesoporous Materials, 225*, 342–354. doi:10.1016/j.micromeso.2016.01.012

Radan, B. D., Pennigton, D. W., Eckley, N., & Boethling, A. (1999).. . *Environmental Science & Technology*, 33–3482.

Ratola, N., Botelho, C., & Alves, A. (2003). The use of pine bark as a natural adsorbent for persistent organic pollutants–study of lindane and heptachlor adsorption. *Journal of Chemical Technology and Biotechnology (Oxford, Oxfordshire), 78*(2-3), 347–351. doi:10.1002/jctb.784

Rawtani, D., Khatri, N., Tyagi, S., & Pandey, G. (2018). Nanotechnology-based recent approaches for sensing and remediation of pesticides. *Journal of Environmental Management*, *206*, 749–762. doi:10.1016/j.jenvman.2017.11.037 PMID:29161677

Rodríguez-Cruz, M. S., Andrades, M. S., Parada, A. M., & Sánchez-Martín, M. J. (2008). Effect of different wood pretreatments on the sorption– desorption of linuron and metalaxyl by woods. *Journal of Agricultural and Food Chemistry*, *56*(16), 7339–7346. doi:10.1021/jf800980w PMID:18642844

Rodriguez-Cruz, S., Andrades, M. S., Sanchez-Camazano, M., & Sanchez-Martin, M. J. (2007). Relationship between the adsorption capacity of pesticides by wood residues and the properties of woods and pesticides. *Environmental Science & Technology*, *41*(10), 3613–3619. doi:10.1021/es062616f PMID:17547186

Rodríguez-Liébana, J. A., López-Galindo, A., de Cisneros, C. J., Gálvez, A., Rozalén, M., Sánchez-Espejo, R., ... Peña, A. (2016). Adsorption/desorption of fungicides in natural clays from Southeastern Spain. *Applied Clay Science*, *132*, 402–411. doi:10.1016/j.clay.2016.07.006

Rojas, R., Morillo, J., Usero, J., Vanderlinden, E., & El Bakouri, H. (2015). Adsorption study of low-cost and locally available organic substances and a soil to remove pesticides from aqueous solutions. *Journal of Hydrology (Amsterdam)*, *520*, 461–472. doi:10.1016/j.jhydrol.2014.10.046

Salman, J. M., & Hameed, B. H. (2010). Adsorption of 2, 4-dichlorophenoxyacetic acid and carbofuran pesticides onto granular activated carbon. *Desalination*, *256*(1), 129–135. doi:10.1016/j.desal.2010.02.002

Salman, J. M., Njoku, V. O., & Hameed, B. H. (2011). Adsorption of pesticides from aqueous solution onto banana stalk activated carbon. *Chemical Engineering Journal*, *174*(1), 41–48. doi:10.1016/j.cej.2011.08.026

Salman, J. M., Njoku, V. O., & Hameed, B. H. (2011). Bentazon and carbofuran adsorption onto date seed activated carbon: Kinetics and equilibrium. *Chemical Engineering Journal*, *173*(2), 361–368. doi:10.1016/j.cej.2011.07.066

Sathishkumar, M., Binupriya, A. R., Kavitha, D., Selvakumar, R., Jayabalan, R., Choi, J. G., & Yun, S. E. (2009). Adsorption potential of maize cob carbon for 2, 4-dichlorophenol removal from aqueous solutions: Equilibrium, kinetics and thermodynamics modeling. *Chemical Engineering Journal*, *147*(2), 265–271. doi:10.1016/j.cej.2008.07.020

Schäfer, R. B., Caquet, T., Siimes, K., Mueller, R., Lagadic, L., & Liess, M. (2007). Effects of pesticides on community structure and ecosystem functions in agricultural streams of three biogeographical regions in Europe. *The Science of the Total Environment*, *382*(2), 272–285. doi:10.1016/j.scitotenv.2007.04.040 PMID:17555800

Senthilkumaar, S., Krishna, S. K., Kalaamani, P., Subburamaan, C. V., & Subramaniam, N. G. (2010). Adsorption of organophosphorous pesticide from aqueous solution using “waste” jute fiber carbon. *Modern Applied Science*, *4*(6), 67. doi:10.5539/mas.v4n6p67

Seyhi, B., Drogui, P., Gortares-Moroyoqui, P., Estrada-Alvarado, M. I., & Alvarez, L. H. (2014). Adsorption of an organochlorine pesticide using activated carbon produced from an agro-waste material. *Journal of Chemical Technology and Biotechnology (Oxford, Oxfordshire)*, *89*(12), 1811–1816. doi:10.1002/jctb.4256

Shukla, G., Kumar, A., Bhanti, M., Joseph, P.E., & Taneja, A. (2006). Organochlorine pesticide contamination of ground water in the city of Hyderabad. *Environ. Inter., 32*, 244-247.

Simazaki, D., Kubota, R., Suzuki, T., Akiba, M., Nishimura, T., & Kunikane, S. (2015). Occurrence of selected pharmaceuticals at drinking water purification plants in Japan and implications for human health. *Water Research*, *76*, 187–200. doi:10.1016/j.watres.2015.02.059 PMID:25835589

Solomon, K. R., Harris, S. A., & Stephenson, G. R. (1993). *Applicator and bystander exposure to home garden and landscape pesticides*. Academic Press.

Squillace, P. J., Thurman, E. M., & Furlong, E. T. (1993). Groundwater as a nonpoint source of atrazine and deethylatrazine in a river during base flow conditions. *Water Resources Research*, *29*(6), 1719–1729. doi:10.1029/93WR00290

Storrs, S. I., & Kiesecker, J. M. (2004). Survivorship patterns of larval amphibians exposed to low concentrations of atrazine. *Environmental Health Perspectives*, *112*(10), 1054–1057. doi:10.1289/ehp.6821 PMID:15238276

Streat, M., & Horner, D. J. (2000). Adsorption of highly soluble herbicides from water using activated carbon and hypercrosslinked polymers. *Process Safety and Environmental Protection*, *78*(5), 363–382. doi:10.1205/095758200530880

Sugashini, S., & Begum, K. M. S. (2013). Optimization using central composite design (CCD) for the biosorption of Cr (VI) ions by cross linked chitosan carbonized rice husk (CCACR). *Clean Technologies and Environmental Policy*, *15*(2), 293–302. doi:10.100710098-012-0512-3

Taha, S. M., Amer, M. E., Elmarsafy, A. E., & Elkady, M. Y. (2014). Adsorption of 15 different pesticides on untreated and phosphoric acid treated biochar and charcoal from water. *Journal of Environmental Chemical Engineering*, *2*(4), 2013–2025. doi:10.1016/j.jece.2014.09.001

Torstensson, L., & Stenström, J. (1990). Persistence of herbicides in forest nursery soils. *Scandinavian Journal of Forest Research*, *5*(1-4), 457–469. doi:10.1080/02827589009382628

Uneke, C. J. (2007). *Integrated pest management for developing countries: a systemic overview*. Nova Publishers.

VoPham, T., Brooks, M. M., Yuan, J.-M., Talbott, E. O., Ruddell, D., Hart, J. E., ... Weissfeld, J. L. (2015). Pesticide exposure and hepatocellular carcinoma risk: A case-control study using a geographic information system (GIS) to link SEER-Medicare and California pesticide data. *Environmental Research*, *143*, 68–82. doi:10.1016/j.envres.2015.09.027 PMID:26451881

Vukčević, M. M., Kalijadis, A. M., Vasiljević, T. M., Babić, B. M., Laušević, Z. V., & Laušević, M. D. (2015). Production of activated carbon derived from waste hemp (Cannabis sativa) fibers and its performance in pesticide adsorption. *Microporous and Mesoporous Materials*, *214*, 156–165. doi:10.1016/j.micromeso.2015.05.012

Wang, X., Wang, H., Lu, M., Teng, R., & Du, X. (2017). Facile synthesis of phenyl-modified magnetic graphene/mesoporous silica with hierarchical bridge-pore structure for efficient adsorption of pesticides. *Materials Chemistry and Physics*, *198*, 393–400. doi:10.1016/j.matchemphys.2016.12.017

Zheng, W., Guo, M., Chow, T., Bennett, D. N., & Rajagopalan, N. (2010). Sorption properties of greenwastebiochar for two triazine pesticides. *Journal of Hazardous Materials*, *181*(1), 121–126. doi:10.1016/j.jhazmat.2010.04.103 PMID:20510513

ADDITIONAL READING

Adeboyejo, O. A., Clarke, E. O., & Olarinmoye, M. O. (2011). Organochlorine pesticides residues in water, sediments, fin and shell-fish samples from Lagos lagoon complex Nigeria. *Researcher*, *3*, 38–45.

Adeyemi, D., Anyakora, C., Ukpo, G., Adedayo, A., & Darko, G. (2011). Evaluation of the levels of organochlorine pesticide residues in water samples of Lagos Lagoon using solid phase extraction method. *Journal of Environmental Chemistry and Ecotoxicology*, *3*, 160–166.

Aguilar, A., Borrell, A., & Reijnders, P. J. H. (2002). Geographical and temporal variation in levels of organochlorine contaminants in marine mammals. *Marine Environmental Research*, *53*(5), 25–52. doi:10.1016/S0141-1136(01)00128-3 PMID:12054104

Azarkan, S., Peña, A., Draoui, K., & Sainz-Díaz, C. I. (2016). Adsorption of two fungicides on natural clays of Morocco. *Applied Clay Science*, *123*, 37–46. doi:10.1016/j.clay.2015.12.036

Ipek, I. Y. (2014). Application of kinetic, isotherm, and thermodynamic models for atrazine adsorption on nanoporous polymeric adsorbents. *Separation Science and Technology*, *49*(15), 2358–2365. doi:10.1080/01496395.2014.915415

Long, E. R., MacDonald, D. D., Smith, S. L., & Calder, F. D. (1995). Incidence of adverse biological effects within ranges of chemical concentration in marine and estuarine sediments. *Environmental Management*, *19*(1), 81–97. doi:10.1007/BF02472006

Nowell, L. H., Caple, P. D., & Dileanis, P. D. (1999). *Pesticides in sediments and aquatic biota.Distribution, trends, and governing factors.Pesticides in hydrologic system series*. Boca Raton: CRC. doi:10.1201/9781439822708

Upadhi, F., & Wokoma, O. A. F. (2012). Examination of some pesticide residues in surface water, sediment and fish tissue of Elechi Creek, Niger Delta. *Nigeria Research Journal of Environmental and Earth Sciences*, *4*(11), 939–944.

Zhou, R., Zhu, L., Yang, K., & Chang, Y. (2006). Distribution of organochlorine pesticide in surface water and sediment from Qiantang River, East China. *Journal of Hazardous Materials. A*, *137*(1), 68–78. doi:10.1016/j.jhazmat.2006.02.005 PMID:16540236

KEY TERMS AND DEFINITIONS

Activated Carbon: This is a form of carbon processed to have small, low-volume pores that increase the surface area available for adsorption or chemical reactions. Activated carbon is usually derived from charcoal and is sometimes utilized as biochar. Those derived from coal and coke is referred as activated coal and activated coke respectively.

Adsorption: This refers to the adhesion of atoms, ions, or molecules from a gas, liquid, or dissolved solid to a surface and creates a film of the adsorbate on the surface of the adsorbent. Adsorption is a surface-based process while absorption involves the whole volume of the material.

Agricultural Chemistry: The study of both chemistry and biochemistry which are important in agricultural production, the processing of raw products into foods and beverages, and in environmental monitoring and remediation. These studies emphasize the relationships between plants, animals, and bacteria and their environment.

Animal Repellents: These are products designed to keep certain animals away from objects, areas, people, plants, or other animals.

Bioaccumulation: This is the accumulation of substances, such as pesticides, or other chemicals in an organism. Bioaccumulation occurs when an organism absorbs a substance at a rate faster than that at which the substance is lost by catabolism and excretion.

Disinfectants: The antimicrobial agent used to destroy the microorganisms on the nonliving surface are called disinfectants. Disinfection does not necessarily kill all microorganisms, especially resistant bacterial spores.

Ecotoxicology: Ecotoxicology is a multidisciplinary field, which integrates toxicology and ecology and it involves the study of effect of toxic chemicals at the population, community, ecosystem, and biosphere levels.

Insecticides: The substance used to kill insects is called insecticide. They include ovicides and larvicides used against insect eggs and larvae, respectively. Insecticides are used in agriculture, medicine, industry.

Pesticide Drift: When the pesticides diffuse to the off target is caused by the runoff or spray drift from plant or soil is called pesticide drift and can have potential effect on human health.

Pests: A pest is a plant or animal detrimental to humans or human concerns including crops, livestock, and forestry. The term is also used of organisms that cause a nuisance, such as in the home. In its broadest sense, a pest is a competitor of humanity.

APPENDIX

Table 1. Adsorbents and their role in removal of pesticides

S. No.	Adsorbent	Pesticides	References
1	Calligonum Comosum biomass	Atrazine	Alahabadi, &Moussavi, (2017)
3	Coconut shell	Hexachlorocyclohexane	Ignatowicz, 2011
4	Maize mulch	gly-phosate, s-metolachlor and Gpoxiconazole	Aslam et al., 2013
5	Natural clays	Metalaxyl and fludioxonil	Rodríguez-Liébana et al., 2016
6	Activated carbon-cloth	Ametryn, aldicarb, dinoseb and diuron	Ayranci,&Hoda 2005.
7	Soil	Atrazine	Long et al., 2015
9	Graphitic carbon nanostructures	2,4-dichlorophenoxyacetic acid	Khoshnood, &Azizian 2012
10	Biochars and charcoal	Azinphos-methyl, Phosmet, Boscalid, Chlorfenvinphos, Flutolanil, Diazinon, Carbaryl, Malathion	Taha et al., 2014
11	Plasmonic nanoparticles	Carbendazim pesticide	Furini et al., 2016
13	Soil	Organophosphorus pesticide	Alfonso et al., 2017
15	Organohydrotalcite	Carbetamide&Metamitron	Bruna et al., 2006
19	Graphene quantum dots nanomaterials	Oxamyl	Agarwal, 2016
20	Sunflower seed shells, rice husk, composted sewage sludge and soil	Chlorfenvinphos, chlorpyrifos, simazine and trifluralin	Rojas et al., 2015
21	Pleurotusmutilus	Metribuzin	Behloul et al.,2017
22	Eucalyptus bark (EBBC), corn cob (CCBC), bamboo chips (BCBC), rice husk (RHBC) and rice straw (RSBC)	Atrazine and imidacloprid	Mandal, Singh & Purakayastha, 2017
23	Chestnut shells	Pirimicarb, imidacloprid, acetamiprid and thi-amethoxam	Cobas, 2016
24	Nano-SiO2	Metalaxyl	Huang, Liang, & Zhang 2017
25	Phenyl-modified magnetic graphene/mesoporous silica	Avemictin, Imidacloprid, Acetamaprid, pyridabin, phoxim, Dursban, Isocarbophos, Dichlorvos	Wang et al., 2017
26	Magnetic copper-based metal–organic	Neonicotinoid insecticide	Liu et al., 2017
27	Mg–Al mixed oxides	Nicosulfuron&Mecoprop-P	Otero et al.,2013
28	Olive kernel, corn cobs, rapeseed stalks and soya stalks	Bromopropylate	Ioannidou et al., 2017
29	Cannabis sativa	Acetamiprid, Dimethoate, Nicosulfuron, Carbofuran And Atrazine	Vukcevic et al., 2015
33	Humic acids graft copolymer	Parathion-methyl, carbaryl and carbofuran	Yang et al., 2014
34	Magnetoliposomes	α-chlordane, chlordecone, hexachlorobenzene (HCB), lindane, p,p'-dichlorodiphenyltrichloroethane (DDT),	Wang et al. 2018
35	Activated carbon	chlorophenoxy pesticides	Derylo-Marczewska et al., 2017

continued on following page

Table 1. Continued

S. No.	Adsorbent	Pesticides	References
36	Chemviron F-400 activated carbon and Purolite MN-200 hypercrosslinked polymer	Benazolin, Bentazone, Imazapyr And Triclopyr	Streat& Horner, 2000
37	Carbon derived from a biopolymer and a clay	Dicamba	Pinto et al., 2016
38	*Trametesversicolor*	Atrazine/terbuthylazine/terbutryn, chlorpyrifos	Lizano-Fallas et al., 2017
39	*Plasmonic nanoparticles*	Carbendazim	Furini et al., 2016
40	*Rice straw biochars*	Atrazine and imidacloprid	Mandal& Singh, 2017
41	*Calcinedhydrotalcite*	2,4-D, Clopyralid and Picloram	Pavlovic et al., 2005
42	*Graphitic carbon nanostructures*	,4-dichlorophenoxyacetic acid	Khoshnood,&Azizian, 2012
43	*Lignin*	1,2,4-triazine	Ludvic, 2000
45	*NH4Cl-induced activated carbon*	Diazinon	Moussavi, Hosseini, &Alahabadi (2014)
46	*Nanocrystalline magnesium oxides*	Diazinon and fenitrothion	Armaghan,&Amini 2014
47	*Pomegranate Peel Banana Peel*	oxamyl pesticide	Mohammad, Ahmad &Badawi, 2015
48	*Wheat straw*	Heptachlor	Seyhi et al., 2013
49	*Walnut Shells*	carbofuran and Chloropyriphos	Menon et al., 2014
50	*Rice Straw*	Carbofuran)	Chang, Lin,& Chen 2011
51	Montmorillonite clays, dimethyl-dialkylamine, aminopropyltriethoxysilane,	Alachlor, Metolachlor, Chlorpyriphos, Fipronil, α-endosulfan, β-endosulfan, p,p'-DDT	Saha et al., 2013
52	Sunflower seed shell, rice husk, composted sewage sludge and soil	atrazine, alachlor, endosulfan sulfate and trifluralin	Rojas et al., 2014
53	sky fruit husk	bentazon	Njoku et al., 2014
54	Jute Fiber	Malathion	Senthilkumaar et al., 2010
55	Powdered activated carbon (PAC)	2,4-dichlorophenoxyacetic acid (2,4-D),	Aksu, &Kabasakal, (2005).
56	Date stones	aldrin, dieldrin and endrin	ELBakouri et al., 2009
57	*Rhizopusoryzae* biomass (ROB)	Malathion	SubhankarChatterjee et al, 2010.
58	Wood	linuron and metalaxyl	Rodrı́guez-Cruz et al, 2008.
59	Wood Residues	linuron, alachlor, metalaxyl, and chlorpyrifos	Rodriguez-Cruz, S et al., 2007.
60	Bagasse fly ash	DDD [2,2-Bis(4-chlorophenyl)-1,1-dichloroethane] and DDE [2,2-Bis(4-chlorophenyl)-1,1-dichloroethene]	Gupta, & Ali, 2001
61	Activated carbon derived from date stones (DSAC)	2,4-dichlorophenoxyacetic acid (2,4-D)	Hameed, Salman & Ahmad, 2009
62	Acid-treated olive stones	Drin	El Bakouri, 2009
63	Tea leaves	quinalphos [QP: *O,O-diethyl O-2-quinoxalinyl phosphorothioate*]	Islam, Sakkas, &Albanis, 2009
64	Activated carbon fiber prepared from cotton stalk	*p*-nitroaniline (PNA)	Li, et al., 2009

continued on following page

Table 1. Continued

S. No.	Adsorbent	Pesticides	References
65	Rice, bagasse fly ash, *Moringa,* rice husk	methyl parathion pesticide (MP	Akhtar, et al., 2007
66	kaolin, montmorillonite, bentonite clays	malathion and butachlor	Pal, &Vanjara, (2001).
67	Bagasse fly ash	lindane and malathion	Gupta, et al, 2002
68	Chickpea husk of black gram	triazophos (TAP) and methyl parathion (MP)	Akhtar et al., 2009
69	watermelon peels	methyl parathion (MP)	Memon et al., 2008
70	granular activated carbon; coconut shell and palm shells	malathion	Jusoh, Hartini, &Endut, 2011
71	Sugar beet pulp, corncob, corncob char, perlite, vermiculite, sand, sediment	glyphosate, diuron and 3,4-dichloroaniline (3,4-DCA)	Huguenot et al., 2010
72	Amberlyst-15 resin	malathion.	Naushad et al., 2014
73	Greenwastebiochar	trazine and simazine	Zheng et al., 2010
74	Maize cob	2,4-dichlorophenol (2,4-DCP)	Sathishkumar et al., 2009
75	Granular activated carbon (GAC)	imidaclopride	Daneshvar et al., 2007
76	Pine bark	lindane and heptachlor	Ratola, Botelho&Alves, 2003
77	Date seed	bentazon and carbofuran	Salman,Njoku, &Hameed, 2011
78	Mushroom compost	carbaryl, carbofuran, and aldicarb	Kuo, & Regan Sr, 1999
79	Banana stalk	2,4-dichlorophenoxyacetic acid (2,4-D) and bentazon	Salman, Njoku, &Hameed, 2011
80	Rice straw	carbofuran	Chang, Lin & Chin, 2011
81	Granular activated carbon,	2,4-D and carbofuran	Salman, &Hameed, 2010

Chapter 16
Residual Analysis of Pesticides in Surface Water of Nagpur, India:
An Approach to Water Pollution Control

Binota Thokchom
IIT Guwahati, India

Neeta Thacker
NEERI, India

ABSTRACT

Seventy-five percent of India's economy depends on agriculture with statewide pesticide consumption of 0.5 kg/h. The highest pesticide consuming states are Tamil Nadu and Andhra Pradesh in between 0.8 to 2 kg/ha. Maharashtra is the topmost consumer of pesticides with over 23.5% share. Nagpur city (the present study area) of Maharashtra has high population density with intensive farming practices. Organochlorine and organophorous pesticide residues were measured in surface water collected from major lakes and rivers located in and around this city. A comparative study with previous records has also been discussed. Monitoring experiments conducted during pre-monsoon, monsoon, and post-monsoon seasons allowed the different samples to show their susceptibility for the above-mentioned pesticide residues.

INTRODUCTION

Pesticides that are applied to soil or sprayed over crop fields are finally released to the environment. Some of them causes serious contamination and threaten human health. Monitoring of pesticide residues is one of the most important aspects in minimizing the potential hazards to human health (Ntow, 2005). Among the most prominent pesticides, organochlorines are highly toxic because of their persistence in the environment and their ability to bioaccumulate in food chain (Ballesteros & Parrado, 2004). In an effort to substitute persistent organochlorines, agricultural sectors have shifted towards organophosphate pesticides. However, organophosphates are generally much more toxic to vertebrates compared to other

DOI: 10.4018/978-1-5225-6111-8.ch016

classes of insecticides even though they rapidly degrade in the environment (Chambers *et al.*, 2001). The usage of both the groups were banned and restricted in developed countries during 1970s and 1980s but some developing countries, including India are still using them for agricultural and public health purposes because of their low cost and versatility in controlling pest (Iwata *et al.*, 1993). With the applications of pesticides in agriculture, runoff wastewater and the surrounding ecology, including the wells located in the middle of the agricultural fields are also contaminated. On consuming such commodities, bioaccumulation occurs, with human as the final pathway. Moreover, if more than one pesticide is used, the cocktail of multiple compounds synergize and antagonistic effects occurs, including the complexity in developing treatment system for such cocktailed contaminated water. The synergistic impact has posed the common discharge limit set for each individual pesticides (1-10 mg l^{-1}) as haphazard (Fernández-Alba *et al*, 2001). Besides, new generation pesticides, including the ones reported in this study, are highly soluble in water. Their low-sorption affinity to soils has also rapidly lead to extensive groundwater and surface water contamination. These compounds, often toxic, chemically stable and difficult to mineralize, have been proven recalcitrant to biological treatment. Due to their persistence and bio-accumulation, and consequent long-term toxicity, they have been designated as priority substances in EU legislation (Loos, 2012).

CLASSIFICATION OF PESTICIDES

1. **Based on the Target Organism (Yadav & Devi, 2017):** Chemical classification of pesticides (insecticides) (see Figures 1 and 2).
2. **Based on Applications:** These are generally sprays, dusts, aerosols, emulsifiable concentrations etc.

Originally, these chemicals were classified on the basis of their mode of entry in the bodies of insects, viz. stomach poisons, contact poisons and fumigants. However, this classification has become outdated, because most of the organic insecticides act both as stomach and contact poisons and some have also the added fumigant action.

Figure 1.

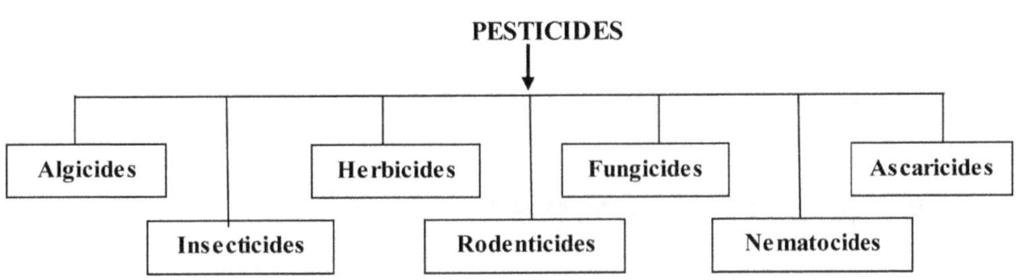

Figure 2.

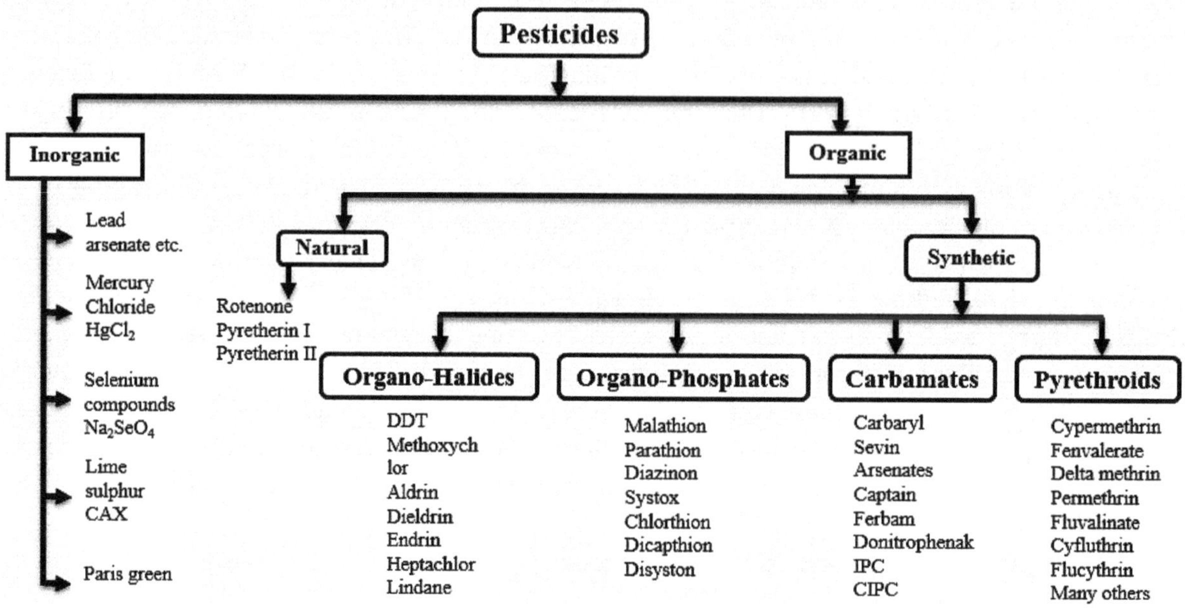

TYPES AND PROPERTIES OF SOME PESTICIDES

1. **Organochlorines:** Members of this class of pesticides have given rise to the greatest environmental concern because of their great stability in the aquatic ecosystem. These pesticides are usually inert to hydrolysis, however if exposed to photo-chemical reaction they tend to form more toxic compounds. It has also been reported that they don't easily dissolve in water, thereby creating the problem of bio-accumulation (Malini *et al.*, 2016). Biologically, reductive dechlorination and dehydrochlorination in the DDT type pesticides and the BHC isomers are the major metabolic routes for those compounds, resulting in highly stable products (Matsumara, 1973).
2. **Organophosphates:** These compounds are toxic at localized level as compared to organochlorines. They are also less persistent; however their toxicity is quite higher than the chlorinated alternatives. Besides, they tend to bioaccumulate more due to their higher solubility in water (Maheswari & Ramesh, 2012).

Three major types of this class of pesticides and common examples are shown in Figure 3.

PESTICIDES IN AQUATIC ENVIRONMENT

Pesticides enter the water from various sources such as (Edward, 1973):

1. Runoff from agricultural lands
2. Direct entry from spray operations
3. Industrial effluents

Figure 3.

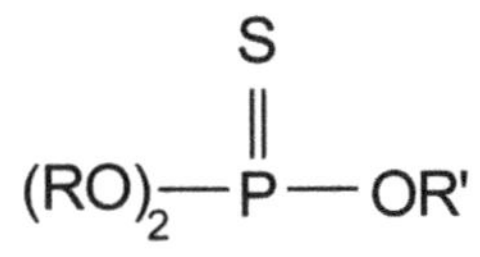

Phosphorothionates (parathions, fenitrothion, diazinon)

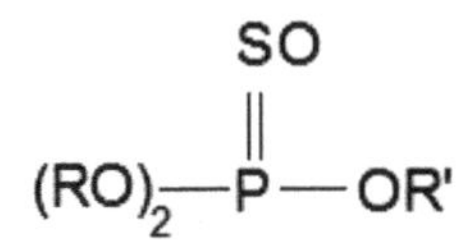

Phosphothiolates (demeton)

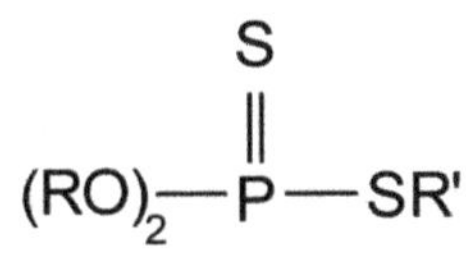

Phosphorodithioates (malathion, azinphosmethyl)

4. Sewage effluents
5. Spraying of cattle
6. Dust and rainfall

In water, pesticides and their intermediate products are distributed among the soluble category and those adsorbed into sediments, benthic invertebrates, aquatic plants, plankton, aquatic invertebrates, suspended detritus, fish etc. They leave water either by evaporation or co-distillation into the environment. They reaches human, birds and animals through consumption of fish or aquatic foods (Maurya & Malik, 2016).

PESTICIDES IN GROUNDWATER

The groundwaters in India are contaminated with pesticide residues as a result of excessive use of pesticides by farmers. Organochlorine pesticides contamination of groundwater in the city of Hyderabad was reported by Shukla *et al.*, 2006. Groundwater samples analyzed were found to be contaminated with four pesticides i.e. DDT, β-Endosulfan, α-Endosulfan and Lindane. DDT was found to range between 0.15 and 0.19 μg L^{-1}, β-Endosulfan ranges between 0.21 and 0.87 μg L^{-1}, α-Endosulfan ranges between 1.34 and 2.14 μg L^{-1} and Lindane ranges between 0.68 and 1.38 μg L^{-1}respectively. These concentrations of pesticides in the water samples were found to be above their respective Acceptable Daily Intake (ADI) values for Humans. In other study undertaken by Jayashree & coworker, 2006, organochlorine pesticide residues in ground water of Thiruvallur district were conducted. The samples were highly contaminated with DDT, HCH, endosulfan and their derivatives. Among the HCH derivatives, Gamma HCH residues were found maximum of 9.8 μg L^{-1} in Arumbakkam open wells. Concentrations of pp-DDT and op-DDT were 14.3 μg L^{-1} and 0.8 μg L^{-1}. The maximum residue (15.9 μg L^{-1}) of endosulfan sulfate was recorded in Kandigai village bore well. The study showed that the ground water samples were highly contaminated with organochlorine residues.

Sharma *et al.*, 2008 have reported the presence of persistent pesticide residues in ground water of Agra region using solid phase extraction and Gas Chromatography. Both organochlorine and organophosphorous pesticides viz., malathion, chloropyriphos, γ-HCH, α-HCH, endosulfan (α and β isomers) and isomers of

DDT (p'-DDE and p'-DDT) were detected in concentrations much above the prescribed limits. Seasonal variations were observed in the level of pesticide residues. A study of different multi-residue methods for the determination of pesticides in groundwater samples of Castellon province (Spain) was made by comparing several extraction procedures which included liquid-liquid partitioning and solid-phase extraction. Ground- water samples were found to contaminate with organochlorine pesticides (lindane, dicofol, chlorfenson and tetradifon), organophosphorous (dimethoate, fenitrothion and methidathion) and phenoxyacid herbicides (MCPA and 2, 4-D). Assessment of priority pesticides, degradation products, and pesticide adjuvants in groundwaters and top soils from agricultural areas of the Ebro river basin were conducted by Hildebrandt *et al.*, (2007). Gas chromatography-mass spectrometry (GC/MS) was employed for the determination of 30 widely used pesticides including various transformation products and alkylphenols in water and agricultural soils with the aim of assessing the impact of these compounds in agricultural soils and the underlying aquifer. The multi-residue extraction procedures were applied to the analysis of groundwaters and agricultural soils from the Ebro river basin (NE Spain). Most ubiquitous herbicides detected were triazines but some acetanilides and organophosphorus pesticides were also found; the pesticide additive tributyl phosphate was found in all water samples. A study on the presence of carbamate pesticides, namely aldicarb, aldicarb sulfoxide, baygon, benthiocarb, carbofuran, 3-hydroxycarbofuran, carbaryl, desmedipham, methiocarb, methomyl, thiodicarb, oxamyl, and propham in ground and surface waters from an agricultural zone of the Yaqui Valley located in northwest Mexico was conducted by Llasera *et al.*, (2001).

PESTICIDES IN SURFACE WATER

Phillips *et al*, (2000) reported a survey on occurrence of pesticides and their Metabolites in three small public water-supply reservoir systems of Western New York, which was initiated by the U.S. Geological Survey (USGS) and the New York State Department of Environmental Conservation (NYSDEC) as a combined project to assess the occurrence of pesticides in groundwater and surfacewater at various states of New York. Among the investigated 60 pesticides and its metabolites, 23 were detected in the samples studied with maximum concentration between 10 and 50% of the lowest applicable water quality standard. These exceeded limit were from tributary sites and they were mostly herbicides atrazine, alachlor, and cyanazine, and the insecticide p,p'-DDE. It was found that the concentration of pesticides detected were in co-ordination with the amount of cropland within the watersheds.

PESTICIDE APPLICATION AND MANAGEMENT

Production of pesticides worldwide increases almost four times within 20 years (1954 to 1973). It increased from 6000 million pounds to 24,000 million pounds within this short span. Although DDT saved 5 million lives and prevented 100 million serious illnesses due to malaria, typhus, dysentery within the first 10 years, however, 600 million pounds produced were organochlorines alone. Again, even after these huge production, the annual loss of agricultural sector was Rs. 1,00,000 crores approx. on a global scale. (De, 2003).

Although these chemicals control insects, weeds, and other pests and hence increase agricultural products and minimize diseases to humans and animals, some can also remain active in the environment for long periods of time, and some can affect the nontarget organisms such as fish and wildlife. Surprisingly, pesticides and its residues were detected in Polar Regions and (Edwards, 2013).The bio-accumulative tendency and nontarget side effects of these chemicals could pose a hazard to health and to the environmental ecosystem. Therefore, the monitoring and surveillance of these chemicals in food and in the environment are very necessary and are basic steps for health protection, environmental assessment, and pollution control.

Recently the number and type of pesticides required has already increased. The use of pesticides for public health programme and for controlling household pests is rather limited but for agriculture their number is large because our crops are affected by over 200 major pests, about 100 plant diseases, hundreds of weeds and other pests like nematodes, rodents. Legal control exists for the consideration of the importance and need to control manufacture, labeling, transportation and use of drugs and other chemicals including pesticides by ordinances, regulations and status of the city, state and federal levels of the Government. The various legislations are strictly enforced in several developed countries of the world. The use of pesticides in India commenced around 1948-49, the first insecticide act of 1968 was initiated to regulate the import, manufacture, sale, transport, distribution and use of insecticides with a view of preventing risk to human beings or animals and for matters connected therewith. The Act was enacted in the year 1972 with the introduction of insecticide rules which came into force on the 1st of August 1971(https://www.pesticidelaw.in/acts-rules/the-insecticides-act-1968/). At the International level, the need to exert some control over chemicals was recognized by the European Economic Community (EEC), and in 1997, a directive (67/548/EEC) was issued on the approximation of the laws and regulations and administrative provisions relating to the classification, packaging and labeling of dangerous substances (Roy, 2015).

With respect to the quality of water intended for human consumption, the Drinking Water Directive (98/83/EC) sets a limit of 0.1 μg/L for a single active ingredient of pesticides, and 0.5 μg/L for the sum of all individual active ingredients detected and quantified through monitoring, regardless of hazard or risk (Narita, 2014).

Table 1 shows the important pesticides being banned in India after the introduction of various laws as stated above.

HEALTH IMPACTS OF PESTICIDES

Some of the suspected chronic effects from exposure to certain pesticides include birth defects, production of tumors, blood disorders, and neurotoxic effects (nerve disorders). Organochlorines alter the permeability of the axonial membrane by delaying the closing of some of the sodium channels, therefore acts as a nerve poison by upsetting the sodium balance in nerve membranes. The general symptoms of organochlorines poisoning in insects and vertebrates are violent tremors, loss of movement followed by convulsions, and death indicating that organochlorines act on the nervous system and produces damage in nervous tissue at much lower concentrations than induce toxic effects in other tissues and enzyme systems (Kirk, 1991).

The organophosphorous and carbamate owe their insecticidal properties because they phosphorylate or carbomylate the enzyme acetyl cholinesterase. In the absence of effective cholinesterase, there is an accumulation of acetylcholine which violates the function of the nervous system which results in giv-

Table 1. Pesticides/Pesticides Formulations Banned in India

(A) Pesticides Banned for Manufacture, Import and Use (24 Nos.)		(B) Pesticides Restricted for Use in India	
1	Aldrin	1	Aluminium phosphide
2	Benzene Hexachloride	2	DDT
3	Calcium cyanide	3	Lindane
4	Chlordane	4	Methyl bromide
5	Copper acetoarsenite	5	Methyl parathion
6	CIbromochloropropane	6	Sodium cyanide
7	Endrin	7	Methoxy ethyl mercuric chloride
8	Ethyl mercury chloride	8	Monocrotophos (banned for vegetables)
9	Ethyl parathion	9	Endosulfan (banned in Kerala)
10	Heptachlor	10	Fenitrothion
11	Menazone		
12	Nitrofen		
13	Paraquat dimethyl sulphate		
14	Pentachloro nitrobenzene		
15	Pentachlorophenol		
16	Sodium methane arsonate		
17	Tetradifon		
18	Toxafen		
19	Aldicarb		
20	Chlorobenzilate		
21	Dieldrine		
22	Maleic hydrazide		
23	Ethylene dibromide		
24	Trichloro acetic acid		

Source: DPPQ & S, Faridabad (National Centre for Integrated Pest Management)
http://cibrc.nic.in/list_pest_bann.htm

ing rise to the typical cholinergic symptoms associated in insects in poisoning hyper activity, tremors, convulsions, paralysis and death (Fukuto, 1990). More detailed impacts can be known from Table 2.

PESTICIDES IN INDIAN SCENERIO

The use of pesticides in Indian scenario began with the urgent need to control malaria and locust invasion. In 1948, DDT and BHC were imported to control them. Later, their production started in 1952 at the scale of around 5000 metric tonnes. At present, nearly 145 pesticides are registered with increased production rate of around 85,000 metric tonnes. This increased production is mainly due to the introduction of Green Revolution in 1960. Mention can be made that Indians are predominantly agrarian and

Table 2. Health Impacts of Pesticides

Sl. No.	Pesticides	Uses	Health Impacts
1	DDT	Effective against wide variety of insects, including domestic insects and mosquitoes	Chronic liver damage cirrhosis and chronic hepatitis, endocrine and reproductive disorders, immuno suppression, cytogenic effects, breast cancer, Non hodkins lymphoma, polyneuritis.
2	Endosulfan	It is used as a broad spectrum non systemic, contact and stomach insecticide, and acaricide against insect pests on various crops	Effects kidneys, developing foetus, and liver Immuno-suppression, decrease in the quality of semen, increase in testicular and prostate cancer, increase in defects in male sex organs, and increased incidence of breast cancer. It is also mutatagenic
3.	Aldrin	Effective against wireworms and to control termites	Lung cancer, liver diseases
4.	Dieldrin	Used against ectoparasites such as blowflies, ticks, lice and widely employed in cattle and sheep dips. Also used to protect fabrics from moths, beetles and against carrot and cabbage root flies/ Also used as seed dressing against wheat and bulbfly	Liver diseases, Parkinson's & Alzheimer's diseases
5.	Heptachlor	It controls soil inhibiting pests.	Reproductive disorders, blood dyscariasis
6.	Chlordane	It is a contact, stomach and respiratory poison suitable for the control of soil pests, white grubs and termites.	Reproductive disorders, blood dyscariasis, brain cancer, Non Hodkins lymphoma
7.	Lindane	It is used against sucking and biting pest and as smoke for control of pests in grain sores. It is used as dust to control various soil pests.such as flea beetles and mushroom flies. It is effective as soil dressing against the attack of soil insects	Chronic liver damage-cirrhosis and chronic hepatitis, endocrine and reproductive disorders, allergic dermatitis, breast cancer, Non hodkins lymphoma, polyneuritis.

http://www.cseindia.org/node/653

Table 3. Summary of some of the reported incidences of pesticide poisonings in India.

Place	Year	Numbers Affected	Deaths	Causative Agent
Kerala	1958	360	100	Parathion
Madhya Pradesh	1967	35	5	Malathion
Madhya Pradesh	1967-68	12	12	Aldrin and HCH
Uttar Pradesh	1977	8	-	HCH
Uttar Pradesh	1978	250	4	HCH
Madhya Pradesh	1978	?	6	Phosgene
Madhya Pradesh	1984	250000	8000	Methyl isocyanate
Uttar Pradesh	1992	8	6	Aluminium phosphide
Andhra Pradesh*	1997-2000	8040	1817	Monochrotophos & Endosulfan
Yavatmal (Maharashtra)**	2017	1000	23	Organophosphorus
Maharashtra***	2017		52	Monochrotophos & Mixed pesticides

Source: Pesticide exposure: Indian scene (Gupta, 2004); *Rao et al, 2005; ** The Hindu, 2017; *** Times of India, February, 2018.

the average consumption of vegetables is about 150–250 g per day per person. Therefore, there is need for sufficient production of green vegetables. Currently, India is the largest producer of pesticides in Asia and ranks twelfth in the world for the use of pesticides. 70% of the pesticides consumed in India are mainly gamma-HCH, DDT and malathion. These pesticides are considered as farmer friendly due to low cost, availability and widely applicable. India predominantly use organochlorine more than other pesticides. Again, India consumes 80% insecticides, 15% herbicides, 1.46% fungicide against the worldwide consumption 47.5% herbicides, 29.5% insecticides and 17.5% fungicides. Countrywide, Europe consumes the largest with 45%, followed by US with 24% and the rest of the countries consume only 25% (Pathak, 2012). In total the consumption is around 2 million/year. The consumption rate in India is 0.5 kg/ha. However, careless handling and installation of low quality plants have led to many unavoidable accidents, claiming number of lives. On the other hand, more than 50% of food commodities are still contaminated, with almost 20% above the permissible limit. Even after that, the crop losses range in between 10–30% due to pests alone, accounting to around Rs. 2,90,000 million/year (Yadav *et al.*, 2015). Table 2 shows the negative impacts related to pesticides production and usage in various Indian states.

The first case of pesticide poisoning in India was reported in Kerala (1958). As many as 100 people died. It was caused due to the contamination of wheat flour by Parathion (Gupta, 2004). Later, in December 1984, the infamous Bhopal tragedy eventuated (Table 3). The installation of pesticide plant at Bhopal which produced sevin brand carbaryl insecticide and temik brand aldicarb pesticide was actually under the provision of Green Revolution. It was one of the greatest tragedies in the world related to pesticides with an estimation of around 3,800 immediate death, 15,000 death follow up and 15,000 to 20,000 premature deaths reportedly occurring in the subsequent two decades. Ironically, the incident hasn't brought many changes in agricultural practice patterns of India. Still today, around 3 million people suffer pesticide poisoning per year with at least 22,000 deaths (Edward, 2005). One very example of direct ingestion of contaminated food can be known from the case of Haryana. In 2001, Kumari *et al* tested vegetable samples available in Haryana state. It was found that all the vegetables the team tested were contaminated. Among them 23% of Organophosphate compounds were above Maximum Residue Limit. Besides, Organochlorines, hexachlorocyclohaxane, DDT, endosulfan and aldrin were also detected. It can be mentioned here that DDT and aldrin have already been banned in India from April 1993, and BHC since April 1997. Apart from it, our team has already studied on the contamination of surface water and drinking water of Vidarbha region (Nagpur and Amravati Division) of Maharashtra. Nearly 65% Vidarbha region is known to depend on agriculture and its allied activities for its livelihood. Lari *et al* (2014), from NEERI, Nagpur studied the contamination of some of the ground water and surface water which are considered to be sources of drinking water in this region. Upon study, surface water was found to be more contaminated than ground water. Again, organophosphate was more abundant than organochlorines. Throughout the monitoring study, α–HCH (0.39μg/L in Amravati region), α–endosulphan (0.78 μg/L in Yavatmal region), chlorpyrifos (0.25 μg/L in Bhandara region) and parathion-methyl (0.09 μg/L in Amravati region) were frequently found pesticide in ground water, whereas α,β,γ-HCH (0.39μg/L in Amravati region), α,β–endosulphan (0.42μg/L in Amravati region), dichlorovos (0.25 μg/L in Yavatmal region), parathion-methyl (0.42 μg/L in Bhandara region), phorate (0.33 μg/L in Yavatmal region) were found in surface water. Surprisingly, pesticides detected in the surface water samples from Bhandara and Yavatmal region exceeded the European Union limit of 1.0 μg/L. Another team of ours (Chavan *et al.*, 2013) examined the residual contamination of surface water located in the intensive agricultural area of Nagpur. Analysis was carried out to identify aldrin, HCH, endosulphan, endosulphan sulfate, DDT, DDE, Dicofol in surface water of that area. The highest concentration of HCH was detected in the sample from Ukkarwahi Lake (0.292 μg/L) in November. The lowest concentration of HCH was obtained in Umred River (0.0107 ug/L) in the monsoon season. Though Endosulphan was found prominent in Makardhokada dam and Saiki Lake almost in all the seasons, the concentration of endosulphan was

Figure 4. (A) India map showing Maharastra, (B) Maharashtra map showing Vidarbha region (dark yellow on the upper right).

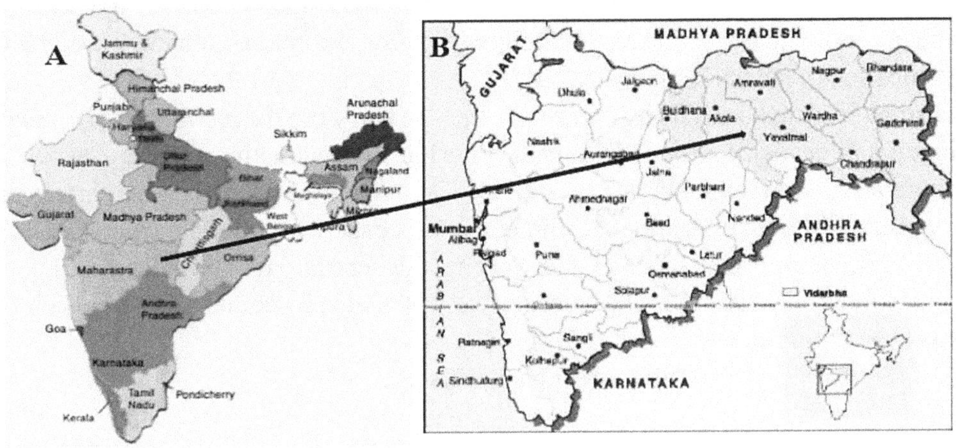

found highest (0.198 ug/L) in Paradgaon Lake in June. The highest concentration of DDT was observed in Paradgaon Lake followed by Undari Lake and Makardhokada dam especially in pre-monsoon season. Aldrin was detected in October, almost in all the sampling sites. The concentration of total pesticide was found maximum in Paradgaon Lake at the range of 0.016-0.826 ug/L. The highest concentration of total pesticide was recorded in November, 2012 i.e. winter season. The lowest concentration of total pesticide was recorded in August, 2012 (monsoon). Both the studies revealed the certainty of more contamination in future if no restriction or remediation measure has been imposed in this region.

On the other hand, Times of India (Oct. 2017) reported 18 deaths in Yavatmal district of which 12 persons were using power spray (Chinese spray) and six used regular sprayer. Again, according to the Times of India City published on February 7th, 2018, 52 accidental deaths due to inhalation of pesticides were reported during July to September, 2017 in Vidarbha region. The main cause was expected to be the use of monocrotophos in pure or mixed form (Times od India, 2018). Although Nagpur is a very important city of India with India's zero mile located at its center and being a hub for orange plantation, till today, limited information has been reported on pesticide contamination of this region, except for the intensive studies done at National Environmental Engineering Research Institute, Nagpur under our team. Hence, the lack of enough information has given rise to the present study to monitor and assess various persistent pesticide residues in surface water of major lakes and rivers in and around Nagpur city, which are also the major source of drinking water for Vidarbha region (Figure 4).

Target analytes in the determinations include potentially water contaminating and commonly used pesticide active ingredients such as herbicides, insecticides, fungicides etc namely:

- **Organochlorines:** α-HCH, β-HCH, γ-HCH, δ-HCH, Heptachlor, Aldrin, Dicofol, α-endosulfan, p,p' – DDE, p,p' – DDD, p,p' –DDT, β-endosulfan, Endosulfan sulfate
- **Organophosphate:** Acephate, Monocrotophos, Dimethoate, Phosphamidon, Parathion methyl, Malathion, Chlorpyriphos, Quinalphos, Profenophos, Ethion, Triazophos, Fenamiphos

MATERIALS AND METHOD

The pesticides that are most commonly related with water include organochlorine, and organophosphate pesticides. The liquid-liquid extraction method followed by gas chromatography was used for the separation and identification of pesticides in water.

Before starting the actual analysis it is very necessary to carry out the recovery experiments to check out the accuracy and pesticide losses. The main criterion for conducting recovery experiments is to standardize suitable method for analysis of pesticides. Detailed investigations were carried out to study the efficiency of liquid-liquid extraction techniques. In this experiment distilled water was spiked with known amount of pesticides standards and with varying concentrations. The spiked samples were then extracted, concentrated and analyzed. The recovery was calculated by comparing sample chromatogram with standard chromatogram.

The general sequence for the pesticide residue analysis:

1. **Sampling Sites:** Nagpur is located at the Earth's geographical region of 79.0882° E longitude and 21.1458° N latitude (Wikipedia). It is situated at a height of 312.42 meters above sea level. The most important river that flows across this city is Kanhan. Other than it, Wardha, Wainganga and Pench are the rivers that flow in this industrial city. River water samples were collected from three of these streams (Wainganga, Kanhan and Wardha rivers). Lake samples were collected from Ambazari, Futala, Gorewada Khindsi, Sakkardara, Sukrawari (Nagpur Region). These surface water bodies are major sources for household and industrial water supply in this region.
   Ambazari lake water is supplied after filteration to Maharastra Industrial Development Corporation area at Nagpur. Raw water from Gorewada is supplied to the city after conventional treatment and

Figure 5.

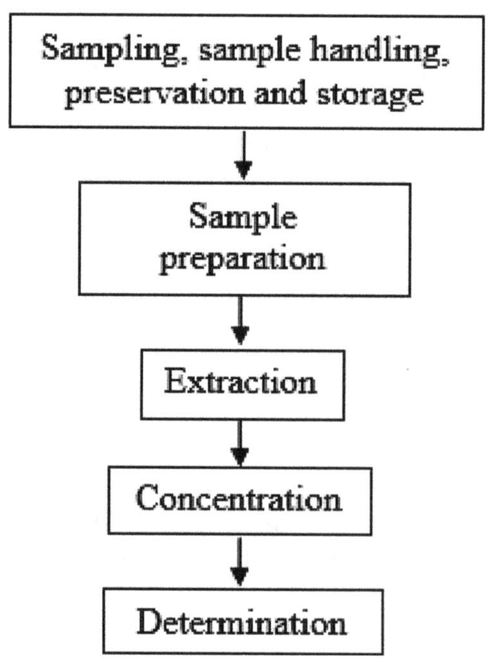

serves 85% of the city's population. The Kanhan River is 18 kilometers away from the city and the Kanhan water treatment plant is located on the bank of this river [6].

2. **Sampling, Collection and Storage:** Samples were collected in April 2008 (pre monsoon), July 2008 (monsoon) and September 2008 (post monsoon). Due to the instability of many of the pesticides in water, samples were extracted within a week and analyzed within 40 days. Samples were collected in 1 liter amber colored glass bottles through grab sampling. The bottles were first rinsed with the water to be taken as samples and then they were carefully filled to over-flowing, such that no air bubbles are trapped. The samples were transported from the sites using ice cooled opaque boxes. Ultrapure water (Millipore) was used throughout the experiment and glasswares were air-dried. Inside the laboratory, samples were stored at 4°C.
3. **Preparation of Standard and Stock Solutions:** The standard mixtures of organochlorine and organophosphorus were prepared in the range of 0.8-3.52 ppm and 5-45 ppm respectively. Individual pesticide stock standard solutions were prepared using high-purity substances in 10 mL volumetric flasks and filled up with n-hexane. The standard solutions were stocked in a deep freezer below −20 °C, away from light. Working standards of desired concentrations were prepared from it.
4. **Preparation of Glass Column:** Glass columns were prepared by placing glass wool (A.R. grade) at the bottom of the columns. Anhydrous sodium sulphates were filled upto 3/4th of the length of each column. Before passing the solvent extracts, the columns were pre-wetted with n-hexane.
5. **Extraction of Pesticides From Water Samples Using Dichloromethane (DCM):** The pesticides in the water were extracted by liquid-liquid extraction as in Figure 6. Measured volumes of 800 mL of samples were taken into 1 L separatory funnels equipped with Teflon stopcocks. 80 g of sodium chloride were added into them. Extractions were carried out for three times with different volumes of organic solvent. In the first extraction 80 mL of DCM were added in each sample and they were shaken vigorously for 1-2 minutes with frequent removal of pressure. The mixtures were kept undisturbed for 3-5 minutes allowing the organic solvents to separate from water. The separated organic layers were then passed into 250 mL flat-bottom flasks through the prewetted columns. The same steps were proceeded for next the two extractions using 40 mL of DCM. Finally the anhydrous sodium sulphate columns were rinsed with n-hexane.
6. **Concentration:** Extractions were followed by concentrations. The process of concentration was carried out in 2 steps. The first step was of concentrating the solvent upto 1-2 ml using heating mantle as shown in Figure 7. Glass beads were added & ideal temperatures were maintained at 30-35 °C to avoid bumping. In the 2nd step the concentrated samples from the round bottom flask were rinsed with n-hexane & transferred into the KDs. They were again heated in water bath at 45-50 °C till the final volume was 0.5-1.0 mL.
7. **Determination/Analysis:** Initially solvent and standard were run on GC. The run of the solvent is necessary for rinsing of column. Then 0.8 μL concentrated samples were injected into GC for analysis. Chromatograms were observed with retention time against the peak area. Since different pesticides have different retention time, distinguished peaks were obtained according to the concentration of pesticides. The concentration of the pesticides can be calculated by comparing the sample chromatogram with that of standard chromatogram.

Figure 6. Extraction Assembly

Figure 7. Concentration Assembly

Box 1. Protocol for extraction

```
800 ml water sample
↓
Add 80 g NaCl
↓
Shake vigorously to dissolve (1-2 min)
↓
Rinse the anhydrous sodium sulphate column and round bottom flask with n-Hexane
(HPLC grade)
↓
Add 80 ml Dichloromethane (DCM) in water sample
↓
Shake vigorously & remove pressure
↓
Allow to separate two layers
↓
Pass the separated organic layer through the sodium sulphate column
↓
Add 40 ml DCM, shake vigorously with removing pressure
↓
Pass the organic layer through the sodium sulphate column
↓
Again add 40 ml DCM, and repeat the procedure as above.
↓
Finally rinse column with n-Hexane and collect the sample in round bottom flask.
```

The pesticide residues were analyzed by gas chromatograph equipped with Electron Capture Detector for organochlorines and Flame Ionization detector for organophosphates. The analytical procedure was carried out following the 20th edition of Standard Methods of Water and Wastewater by APHA/AWWA/WEF (1998). The specification of the instruments used along with its operating conditions is given in Table 4. The quantification of pesticide was performed by injecting 1 μL of each sample in gas chromatograph.

8. 8. **Calculation for Concentration:** The concentration of the analyzed pesticide was calculated using the following equation:

$$Conc.\,of\;pesticide = \frac{Conc.\,of\;standard}{Peak\;area\;of\;standard} \times \frac{Peak\;area\;of\;sample}{Volume\;injected} \times \frac{Volume\;in\;KD}{Sample\;taken}$$

Table 4. Gas chromatograph conditions for analysis of organochlorine pesticides

Item	Condition	
Gas Chromatograph	Make: Perkin Elmer, Clarus-500	
Column	Equity™-5 (30mX0.25mm, 0.25µm film)	
	Organochlorine	Organophosphorus
Oven temperature (°C)	Initial 100 for 2 mins, 160 °C at 15 ml/min for 2 min and finally 270 for 18 min at 5 ml/min	150°C for 1 min to 225°C @ 5°C/min for 10 min.
Injector temperature	250 °C	220 °C
Injection volume	1 µL	1 µL
Detector	Electron Capture Detector	Flame Thermionic Detector
Detector temperature	300 °C	270 °C
Carrier gas	Nitrogen	
Carrier gas flow	1.23 ml/min	0.9 ml/min
Total run-time	48 min	26 min

RESULTS AND DISCUSSION

The major food crops and vegetables cultivated during the sampling period in and around Nagpur city were paddy, soyabean, cotton, citrus, chili, tomato, brinjal, cauliflower, lady's finger, turmeric and sugarcane. Common important organochlorines found to be used in these regions are Deltamethrin, Endosulfan, Lindane, Cypermethrin, Dicofol etc. Common organophosphates are Triazophos, Chlorpyriphos, Monocrotophos, Phosphomidon, Acephate, Phorate, Dichlorvos, Propenfos and Fenvalerate. Other pesticides such as Buprofenzin, Acetamipirid and bio-fertilizers have also been reported to be used.

1. **Recovery of the Spiked Samples:** To analyze the pesticide samples standardization of suitable method is very important. Thus to study the efficiency of liquid-liquid extraction techniques recovery experiments were carried out. The recovery can be checked out by spiking the known amount of pesticide standard in distilled water. In this experiment sample was spiked with concentration of 2µg/mL. Each spiked sample was homogenized (for even distribution) and stored (for equilibrium) before extraction, clean-up, concentration, and analysis as described. The recovery percentages were calculated using peak heights on the chromatograms with and without the standard. Recovery experiments are conducted to optimize the method for reproduction of a rational and valid analysis of pesticides in water samples.
2. **Quality Assurance and Quality Control:** The quality assurance and quality control (QA/QC) of the method were established. The percent recovery and standard deviation are shown in the Table 5 and 6. The chromatogram showing retention times for organo-chlorines shown in Figure 6. Recovery studies were performed by adding known amount of standard mixture of pesticides in ultra-pure water. Validation studies for each concentration were done in triplicates. Detection limits for organochlorine pesticides varied between No GuidelineValue (well below health concern in drinking water) to 30 µg/L (WHO, 1967).

Table 5. Percent recovery of organochlorine pesticides in the spiked sample.

Sl. No.	Pesticides	Recovery
1	α-HCH	71.63%
2	γ-HCH	79.23%
3	δ-HCH	86%
4	Heptachlor	70%
5	Aldrin	71%
6	α-Endosulfan	85.78%
7	p,p'-DDE	82.47%
8	β-Endosulfan	85.69%
9	p,p'-DDD	88.32%
10	p,p'-DDT	86.28%
11	Fenpropathrin	85.31%
12	λ-Cyhalothrin	84.6%

Table 6. Percent recovery, SD and RSD of Lindane, Endosulfan, DDT.

Sl. No.	Pesticides	Recovery	Standard Deviation (SD)	Relative Standard Deviation (RSD)
1	γ-HCH(lindane)	79.23%	0.13	9.2
2	Endosulfan	85.73%	0.09	6.08
3	DDT	86.28%	0.2	17.72

Despite the official ban and restriction on the usage of some organochlorines, pesticide residues were detected in the analysis by Thacker *et. al.* (1997) in and around Nagpur including the above given sampling sites though they were well below the guideline values given by WHO. However, the detection of these pesticides shows its origin from various pesticide rich sources, mainly agricultural areas and household. The surface water sources from the lakes and rivers of Nagpur region were analyzed for more number of persistent pesticides in comparison to the organochlorine residues monitored by Thacker and her group in 1997.

It can be observed from Figure 8, 9 and 10 that the pesticides, particularly Lindane, Endosulfan and DDT which were present in significant concentration at these sites have come down to a very low level. In this context it can also be mentioned that the pesticides DDT and Lindane have been banned in India since 1989. Results of monitoring conducted on various specified water bodies from the different zones of the city during pre-monsoon, monsoon and post-monsoon seasons allowed the different samples to show their susceptibility for the above mentioned pesticide residues. The previous study in 1997 clearly showed that DDT was highest above 1 μgL^{-1} at Futala Lake in both the years. Again, Endosulfan and Lindane have been recorded with 0.3 and 0.8 μgL^{-1}, respectively in Sakkardara and Futala Lakes.

Figure 8. Lindane levels in Surface water at Nagpur Region (WHO guideline values for lindane is 2 μL^{-10})

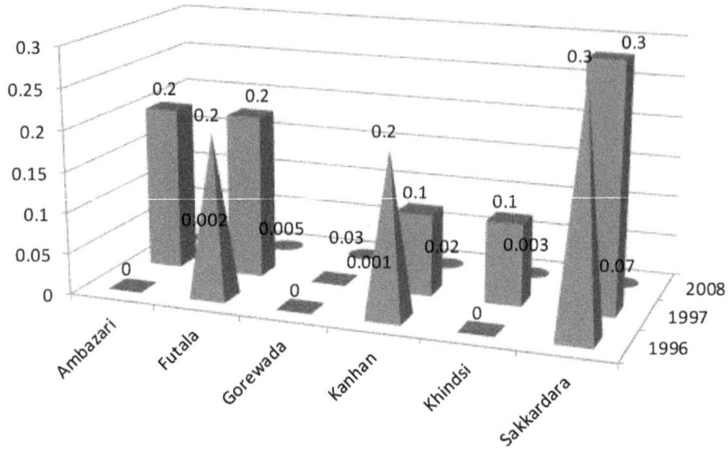

Figure 9. Endosulfan levels in Surface water at Nagpur Region (WHO guideline value for endosulfan not available).

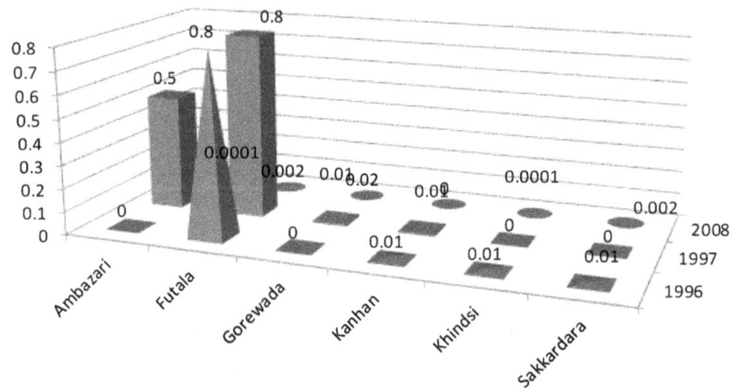

Figure 10. Total DDT levels in Surface water at Nagpur Region (WHO guideline values for Total DDT is 1 $\mu g\ L^{-1}$)

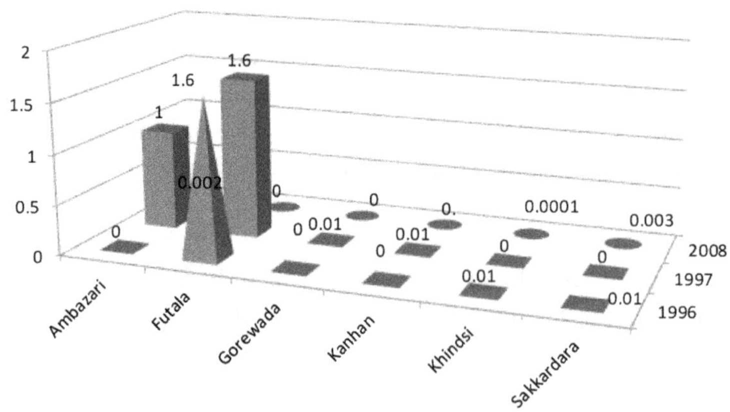

CONCLUSION

Increased use of pesticides has resulted in increased crop production, lower-cost maintenance, and control of public health hazards. However, concerns and public awareness about the potential adverse effects of pesticides on the environment and human health have also grown simultaneously. In many respect the greatest potential for unintended adverse effects of pesticides is through contamination of the hydrologic system, which supports aquatic life and related food chains and is used for recreation, drinking water, and many other purposes. Water is one of the primary media in which pesticides are transported from targeted application sites to other parts of the environment.

To minimize health risks, sensitive and accurate detection and reproducible quantification of these organics are of paramount importance. Thus number of experiments has been conducted to optimize the analytical methods for monitoring concentration of pesticides in the water. Samples were collected from various sampling sites and were subjected to liquid-liquid extraction and analysed for the presence of organoclorines and organophophates pesticides. The most suitable and accurate method for analysis of these pesticides was Gas Chromatography equipped with different detector systems. The conditions are accurately optimized for analysis and results are recorded. The important pesticides that are found as contaminants are chlorinated hydrocarbons and their derivatives. Regarding the present project the pesticides monitored were all found to be under permissible limit given by WHO. Some were not even detected. These indicate that the water is safe for use. The results of our study show that the use of pesticide has been reduced and thereby its concentration in water bodies too. The marginal variation in three seasons with the result showing a decreasing trend of the residual detection reflects that although there exists some changes but these are not statistically significant. Further, from the monitoring status various agricultural areas including the rice and cotton fields around the sampling sites have been switched more to comparatively less harmful and less persistent synthetic pyrethroids and organophosphorus and with the analysis almost all of them have been found below the detectable limits.

However, how much less it be but pesticides are noted contaminants and are defined as poisons harmful to human health as per the Provisions of the Prevention of Food Adulteration Act (PFA Act) which governs the quality of food and essential water consumed by the country's population.

ACKNOWLEDGMENT

Authors are grateful to the Director, National Environmental Engineering Research Institute (NEERI), Nagpur for granting permission to publish this work. Acknowledgments are also due to the scientists and staff of the Instrumentation Division of the Institute for their kind assistance and cooperation during sample analysis.

REFERENCES

American Public Health Association (APHA). (1998). *American Water Works Association (AWWA)* (20th ed.). Washington, DC: Standard Methods for the Contamination of Water and Wastewater.

Ballesteros, E., & Parrado, M. J. (2004). Continuous solid-phase extraction and gas chromatographic determination of organophosphorus pesticides in natural and drinking waters. *Journal of Chromatography. A*, *12*(1-2), 267–273. doi:10.1016/j.chroma.2003.12.009 PMID:15032372

Chambers, H. W., Boone, J. S., Carr, R. L., & Chanbers, J. E. (2001). *Handbook of Pesticide Toxicology*. Academic Press. doi:10.1016/B978-012426260-7.50047-1

Chavan, M., Thacker, N., Tara, J. L., & Gandhi, K. (2013). Monitoring the water quality for organochlorine pesticides in the area of intensive agriculture activity of Nagpur District, India. *Proceedings of the 13th International Conference on Environmental Science and Technology*.

De, A. K. (2003). New Age International. *Environmental Chemistry*, 406.

Edward, B. (2005). The Bhopal disaster and its aftermath: A review. *Environmental Health*, *4*(1), 6. doi:10.1186/1476-069X-4-6 PMID:15882472

Edwards, C. (2013). *Environmental Pollution by Pesticides*. Springer Science & Business Media.

Edwards, C. A. (1973). *Persistent Pesticides in the Environment* (2nd ed.). CRC Press.

Fernández-Alba, A. R., Guil, L. H., López, G. D., & Chisti, Y. (2001). Toxicity of pesticides in wastewater: A comparative assessment of rapid bioassays. *Analytica Chimica Acta*, *426*(2), 289–301. doi:10.1016/S0003-2670(00)00874-6

Gangesh, S., Anoop, K., Mayank, B., Joseph, P. E., & Ajay, T. (2006). Organochlorine pesticide contamination of ground water in the city of Hyderabad. *Environment International*, *32*(2), 244–247. doi:10.1016/j.envint.2005.08.027 PMID:16183122

García de Llasera, M. P., & Bernal-González, M. (2001). Presence of carbamate pesticides in environmental waters from the northwest of Mexico: Determination by liquid chromatography. *Water Research*, *35*(8), 1933–1940. doi:10.1016/S0043-1354(00)00478-4 PMID:11337839

Gupta, P. K. (2004). Pesticide exposure—Indian scene. *Toxicology*, *198*(1-3), 83–90. doi:10.1016/j.tox.2004.01.021 PMID:15138033

Hildebrandt, A., Lacorte, S., & Barceló, D. (2007). Assessment of priority pesticides, degradation products, and pesticide adjuvants in groundwaters and top soils from agricultural areas of the Ebro river basin. *Analytical and Bioanalytical Chemistry*, *87*(4), 1459–1468. doi:10.100700216-006-1015-z PMID:17211597

Iwata, H., Tanabe, S., Sakai, N., & Tatsukawa, R. (1993). Distribution of persistent organochlorines in the oceanic air and surface seawater and the role of ocean on their global transport and fate. *Environmental Science & Technology*, *27*(6), 1080–1098. doi:10.1021/es00043a007

Jayashree, R., & Vasudevan, N. (2007). Organochlorine pesticide residues in ground water of Thiruvallur district, India. *Environmental Monitoring and Assessment*, *128*(1-3), 209–215. doi:10.100710661-006-9306-6 PMID:17016752

Kirk, K. L. (1991). Persistent Polyhalogenated Compounds: Biochemistry, Toxicology, Medical Applications, and Associated Environmental Issues. Biochemistry of the Elements, 9, 245–254.

Kumari, B., Madan, V. K., Kumar, R., & Kathpal, T. S. (2002). Monitoring of seasonal vegetables for pesticide residues. *Environmental Monitoring and Assessment*, *74*(3), 263–270. doi:10.1023/A:1014248827898 PMID:11944799

Lari, S. Z., Khan, N. A., Gandhi, K. N., Meshram, T. J., & Thacker, N. P. (2014). Comparison of pesticide residues in surface water and ground water of agriculture intensive areas. *Journal of Environmental Health Science & Engineering*, *12*(1), 11. doi:10.1186/2052-336X-12-11 PMID:24398360

Loos, R. (2012). *Analytical Methods for the new proposed Priority Substances of the European Water Framework Directive (WFD)*. Revision of the Priority Substance List.

Maheswari, S. T., & Ramesh, A. (2012). Adsorption and Degradation of Anilofos in Different Soils and its Environmental Impact in Soils. *International Journal of Environmental of Research*, *6*, 451–456.

Malini, T. P., Ramesh, A., Selvi, J. A., Arthanareeswari, M., & Kamaraj, P. (2016). Kinetic Modeling of Photocatalytic Degradation of Alachlor using TiO_2 (Degussa P25) in Aqueous Solution. *Oriental Journal of Chemistry*, *32*(6), 3165–3173. doi:10.13005/ojc/320639

Matsumura, F. (1973). Degradation of Pesticide Residues in the Environment. In C. A. Edwards (Ed.), *Environmental Pollution by Pesticides. Environmental Science Research*. doi:10.1007/978-1-4615-8942-6_14

Maurya, P. K., & Malik, D. S. (2016). Bioaccumulation of xenobiotics compound of pesticides in riverine system and its control technique: A critical review. *Journal of Industrial Pollution Control*, *32*, 580–594.

Narita, K., Matsui, Y., Iwao, K., Kamata, M., Matsushita, T., & Shirasaki, N. (2014). Selecting pesticides for inclusion in drinking water quality guidelines on the basis of detection probability and ranking. *Environment International*, *63*, 114–120. doi:10.1016/j.envint.2013.10.019 PMID:24286868

Niti, S., & Alka, P. (2008). Determination of Persistent Pesticide Residues in Ground Water of Agra Region Using Solid Phase Extraction and Gas Chromatography. *Asian Journal of Water, Environment and Pollution*, *5*, 91–94.

Ntow, W. J. (2005). Pesticide residues in Volta Lake, Ghana. *Lakes and Reservoirs: Research and Management*, *10*(4), 243–248. doi:10.1111/j.1440-1770.2005.00278.x

Pathak, R. K., & Dikshit, A. K. (2012). Atrazine and its Use. *International Journal of Research in Chemistry and Environment*, *2*, 1–6.

Phillips, P. J., Eckhardt, D. A., & Rosenmann, L. (2000). *Pesticides and Their Metabolites in Three Small Public Water-Supply Reservoir Systems, Western New York, 1998-99*. U.S. Department of the Interior, U.S. Geological Survey.

Rao, C. S., Venkateswarlu, V., Surender, T., Eddleston, M., & Buckley, N. A. (2005, June). Pesticide poisoning in south India: Opportunities for prevention and improved medical management. *Tropical Medicine & International Health*, *10*(6), 581–588. doi:10.1111/j.1365-3156.2005.01412.x PMID:15941422

Roy, R. (2015). Risk Assessment in the European Union (EU). In J. Torres & S. Bobst (Eds.), *Toxicological Risk Assessment for Beginners*. Cham: Springer. doi:10.1007/978-3-319-12751-4_9

Thacker, N. P., Kaur, P., & Rudra, A. (1997). Halogenated compounds in and around Nagpur City water resources, India. *Environmental Education and Information*, *16*, 367.

Thacker, N. P., Vaidya, M. V., Sipani, M., & Kalra, A. (1997). Removal technology for pesticide contaminants in potable water. *Journal of Environmental Science and Health. Part. B, Pesticides, Food Contaminants, and Agricultural Wastes*, *B32*(4), 483–496. doi:10.1080/03601239709373099

WHO. (1967). *Specifications for Pesticides Used in Public Health* (3rd ed.). Geneva: WHO.

Yadav, I. C., Devi, N. L., Syed, J. H., Cheng, Z., Li, J., Zhang, G., & Jones, K. C. (2015). Current status of persistent organic pesticides residues in air, water, and soil, and their possible effect on neighboring countries: A comprehensive review of India. *The Science of the Total Environment*, *511*, 123–137. doi:10.1016/j.scitotenv.2014.12.041 PMID:25540847

Yadav, I. S., & Devi, N. L. (2017). Pesticides Classification and Its Impact on Human and Environment. *Environment Science and Engineering*, 6.

Yadav, I. S., & Devi, N. L. (2017). Pesticides Classification and Its Impact on Human and Environment. *Environment Science and Engineering*, 6.

Chapter 17
Pesticide Analysis Techniques, Limitations, and Applications

Ebru Kafkas
University of Cukurova, Turkey

Nebile Daglioglu
University of Cukurova, Turkey.

Nigar Yarpuz-Bozdogan
University of Cukurova, Turkey

Mozhgan Zarifikhosroshahi
University of Cukurova, Turkey

ABSTRACT

Natural or synthetic chemical compounds in pesticides are commonly used to kill pests or weeds. In general, pesticides are potentially toxic to not only organisms but also the environment and should be used safely disposed of expediently. Pesticide residues in foods may cause various crucial diseases in the body. The damage of pesticides can be changed depending on the application dose or type of active compounds. For this reason, identification and quantification of pesticides via highly selective, sensitive, accurate, and renewable techniques are of vital importance due to the large amounts of possible interfering substances during the extraction stages. Analysis of pesticide residues by analytical methods can be fluctuate based on the pesticide types. For food and health safety, maximum residue limits (MRL) of pesticides in foods were determined by the European Community. There are many analytical methods developed for identification and quantification of pesticides. Although there are some limitations, the multi-residue methods sensible for analyzing a great number of pesticides in one single run is the fastest, the most favorite, and efficient choice. However, some of the pesticides need specific methodologies and single-residue methods apply as compulsory for them. In this chapter, recent advances in the various analysis of pesticide residues in crops and their applications and limitations are discussed.

DOI: 10.4018/978-1-5225-6111-8.ch017

INTRODUCTION

Pesticides are generally used for obtaining high quality and yield of agricultural crops by farmers. They can prevent agricultural crops from quality and quantity losses. Pesticides are chemicals and widely used all over the world for many years. Known as an essential nutrient for plants, sulfur was the oldest pesticide that kills insects, mites, fungi, and rodents. Until the 1940s, chemicals which were derived from plants and inorganic compounds were used for pest control (Fishel, 2009). Then chemicals including plant growth regulators and synthetic substances started to be used for plant protection. However, most of the chemicals are poisonous to the environment and living organisms. These chemicals can be classified on the basis of their use for insect killers as insecticides, weed killers as herbicides, fungi killers as fungicides, rodent killers as rodenticides, etc. (Aktar et al., 2009). Classification of pesticides are given in Table 1 (Jayaraj et al., 2016)

Applied pesticides can be absorbed by various parts of plants such as roots and leaves and also can move and translocate capability from one part to other parts of the plant. Due to having several benefits particularly protection of yield and crop reduction and improving fruit quality, pesticides are widely used for agricultural production for many years. However, pesticides are dangerous and have unwanted side

Table 1. Classification of pesticides

Chemical Group	Chemical Names
Organochlorines	DDT,DDD, Dicofol, Aldrin, Dieldrin, Chlorobenziate, Lindane, BHC, Methoxychloroaldrin, Chlordane, Heptaclor, Endosufan, Isodrin, Isobenzan, Toxaphene.
Organophosphates	Dimefox, Mipafox, methyl parathion, Ronnel, enitrothion, Bidrin, Phorate, Fenthion, caumphos, Abate, Dichlorovas, Diptrex, Phosphomidon, Demetox, Oxydemeton-methyl, Malathion, Dimethoate, Trichlorofan
Carbamates	**Methyl:** Carbaryl, Carbanolate, Prupoxur, Dimethan, Dimetilan, Isolan, Carbofuran, Pyrolan, Aminocarb, Aldicarb **Thio:** Vernolate, Pebulate, Diallate, Monilate, Butylate, Cycloate, Trillate, Thiourea **Dithio:** Methan, Thiram, Ferban, Amoban, Naban, Zineb, Maneb, Zirampolyran, Dithane M- 45
Pyrethroids	Allethrin, Bonthrin, Dimethrin, Tetramethrin, Ptrethrin, Cyclethrin, Furethrin, Fenevelerate, Alphamethrin, Decamethrin, Cypermethrin
Phenylamides	**Carbanilates:** Barban, Carbetamide, Chlororprofan, Prophan, Phenylurea, Fenuron, Monuron, Diuron, Flumeturon, Chloroxuron, Neburon, Bromuron. **Acylanalide:** Propanil, Solan, Dicryl, Karsil, Propachlor, Alachlor, Butachlor. **Toluidines:** Trifluralin, Dipropanil, Benefin, Oryzalin, Isopropanil, Nitralin. **Acetamide:** Diphenamid
Phenoxyalkonates	2,4-D(2,4 Dichlorophenoxyaceticacid), 2,4 5 T(2,4 5 Trichlorophenoxyaceticacid), Dichloroprop, Mecoprop, Erbin, Sesone
Trazines	Atrazine, Simazine, Ametryn, Atratone, Chlorazine, Cynazine, Cyprazine, Metribuzin, Propazine, Turbutryn, Simetryn
Benzoicacid	Dicamba, Dichlorobenil, Chloroambin, Tricamba, Neptalan, Bromoxynil
Phtalimides	Captan, Diflotan, Folpet
Dipyrids	Paraquat, Diaquat
Others	Pentachlorophenol, Floroacetate, Phenylmercuricacetate, Ethylmercuricphosphate, Methylmercuricchloride, Sodiumarsenate, Calciumarsenate, Leadarsenate, Cacodylicacid, Aluminiumphosphide, Zincphosphide

Jayaraj et al., 2016.

effects for the environment, living organisms and humans, especially the implementing workers due to high contamination risks for soil and groundwater (Yarpuz-Bozdogan et al., 2017; Yarpuz-Bozdogan & Bozdogan, 2016; Bozdogan & Yarpuz-Bozdogan, 2015; Bozdogan, 2014). Groundwater contamination with pesticides is also one of the vital worldwide problems and once water is polluted with pesticides, it is so difficult and may take long periods or years to get rid of contamination. Sometimes it may turn into a complex phenomenon which is impossible solve. (Waskom, 1994; Oneil, 1998; USEPA, 2001). Similarly, soil impurity with pesticides has also poison risks for soil microorganisms. In pesticide applications, overdose and incorrect pesticide applications can cause serious problems on human health and environment (Palis et al., 2006; Bozdogan & Yarpuz-Bozdogan, 2008a ; 2008b; Yarpuz-Bozdogan & Bozdogan, 2009; Damalas and Eleftherohorinos, 2011). In recent years, one of the most important threats to human health is pesticide residues on food. The most important effect of these residues cancer cases which are very difficult to treat. Another effect is the difficulty of giving the right decision about harvest time. It is essential to take great care of pre-harvest interval (PHI) recommended on the pesticide label by the manufacturer. If it pays no attention to PHI, food is presented to the consumer with no regard to pesticide residue. Therefore, human health is negatively affected by pesticide residue on food.

In recent years, pesticide residues have been noticed as agricultural products are exported. Exports of products above the Maximum Residue Limits (MRL) value are prohibited and these products are destroyed. The pesticides licensed by the Ministry should be applied at the recommended dose and the time between the last application and the harvest must be considered. Due to the Rapid Alert System, insecticide residues are analyzed in products coming to European Union countries and products with insecticide residues are banned from entering EU countries. With this system, products containing pesticide residues are reported to all EU countries and entry to the country is banned.

Pesticide hazard can be changed based on the active ingredient of the chemical component and with regard to toxicity degree, it can be toxic or relatively toxic and non-toxic and those can be either acute or chronic. Acute toxicity of a pesticide can be exposed by inhaling, dermally, orally or contact of eyes and it is measured and expressed as LD_{50} (lethal dose). However, chronic toxicity does not result in poisoning in the traditional sense and is evident in the chronic or acute effects on the body. Kole et al., (2001) reported that generally insecticides and herbicides groups are known as the acutest toxic groups and may also affect negatively non-target organisms. Dr. Elaine Ingham reported that using excessive chemical pesticides are similar to antibiotics overuse on human (Savonen,1997). Based on the previously published papers, sulfur (s) compound was the first known pesticide for controlling insects and mites 4500 years ago by Sumerians (Anonymus, 2008). However, in 1975, recognition of pesticides hazard was primarily published and pesticides were classified based on their acute toxicity levels. In 2006, the FAO Council endorsed FAO participation in the Strategic Approach to International Chemicals Management (SAICM) and especially on the progressive banning of Highly Hazardous Pesticides (HHPs) and this organizations have annual meetings internationally and then publishes guidelines and documents (Durovic & Dordevic, 2011; Anonymous, 2016). Since the awareness of pesticides hazards for living organisms, identification and quantification of pesticides using accurate and sensitive techniques has been substantial. In this chapter, advanced techniques, various methodologies of pesticides and advantages, limitations and applications in agricultural crops are discussed.

BIOMONITORING OF EXPOSURE TO PESTICIDE IN CUKUROVA REGION, TURKEY

Çukurova is a region that is located in the Mediterranean area in Turkey where intense agriculture is done and 32% of agricultural chemicals consumed in our country are used in this region. Acute poisonings due to pesticides are still among the causes of death in our country. In the retrospective study that we carried out between 2006 and 2008 in the Adana Forensic Medicine Institution, 4199 autopsy cases were analyzed in the forensic toxicology laboratory with the request of pesticide analysis and positive pesticides were found in 72 cases. Among these pesticides, the most abundant is found is endosulfan with the rate of 47.2% followed by dichlorvos (Daglioglu et al., 2011). In the conventional forensic analysis, blood, body fluids and organs are studied in deaths due to intoxication. Different biological specimens can be studied in cases where it is putrefied, buried without autopsy or determined that it is necessary to re-examine by removing from the grave. In our study in 2009, we analyzed pesticides by GC-MS in putrefication structures and bone marrow of rabbits acutely exposed to pesticides (Akcan et al., 2009).

Persistent organic pollutants (POPs) are chemicals that are consciously produced or produced as by-products, which are used as pesticides in many countries or formed as a result of industrial activities. Many countries in the world have come together to sign the Stockholm Convention, and it is aimed to control, track and ban these pollutants (Figure 1).

POPs are environmental chemicals that are toxic, persistent and tend to accumulate in the nutritional chain. They are transported across borders by diffusing in the air and the water and may threaten ecosystems and living organisms by accumulating both in terrestrial and aquatic ecosystems (Figure 2).

The use of OCPs pesticides were banned in developed countries in the 1970s. In our country, the prohibition of the use of some organochlorine pesticides were implanted he middle of 1980's. After 1985, these compounds were prohibited except for endosulfan and tossofen due to the length of their half-lives and accumulation characteristics. While restriction and prohibition decisions on PCBs compounds were taken in the USA in the early 1970s, it was made in our country in 1996. It is estimated that there are approximately 11 tons of stock DDT and 6.5 tons of PCBs in Turkey. Approximately 213 tons of PCBs are used in Turkish Electricity Generation and Conduction. Unfortunately, there is not enough data on other organochlorine compounds (Erdogrul et al., 2004).

Figure 1. Persistent Organic Pollutants (POPs)

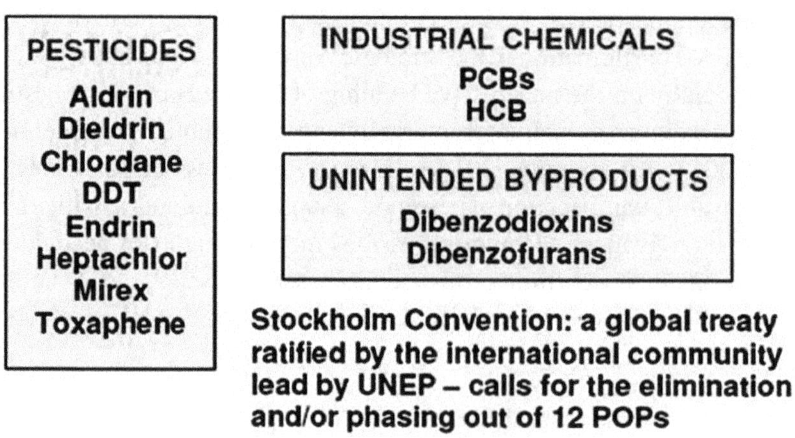

Figure 2. POPs in Environment

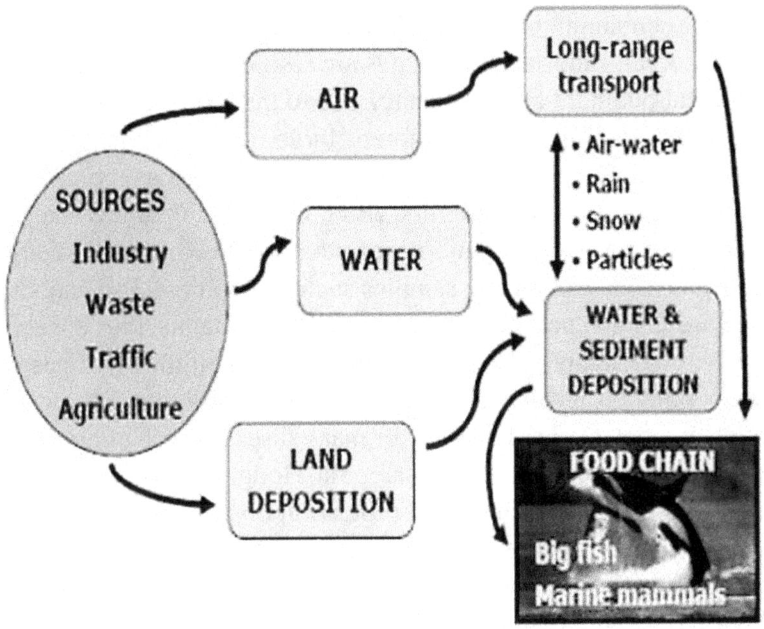

Figure 3. Structural formulas of organochlorine pesticide

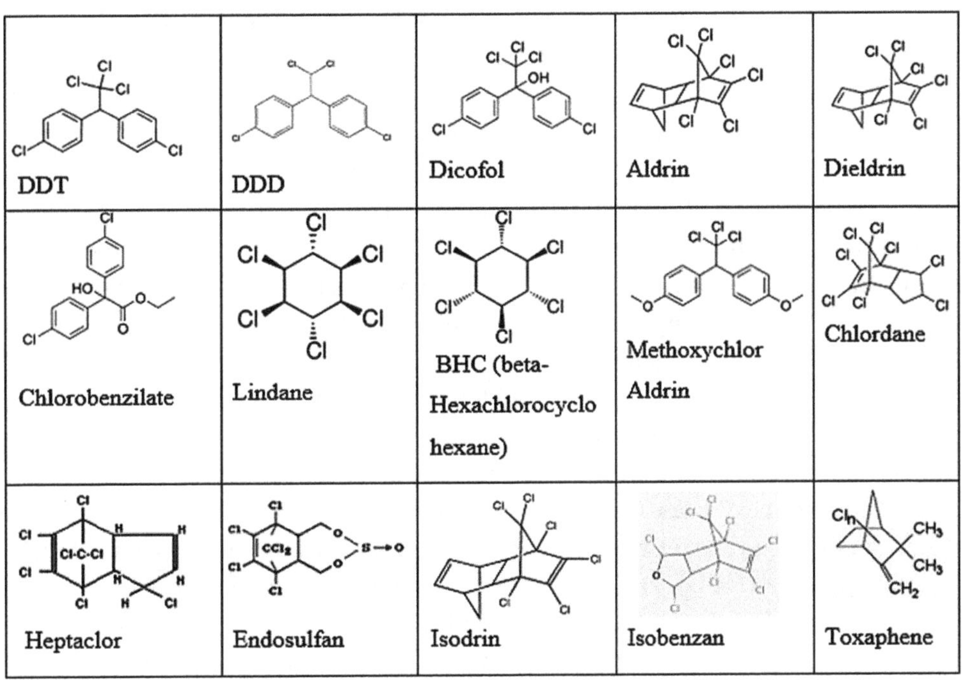

Organochlorine pesticides (Figure 3) cause heavy pollution due to their usage in agriculture. Humans are exposed to these lipophilic compounds in different ways. These ways are; 1. Consumption of contaminated food, especially that containing high fat (fish, meat, chicken etc.) 2. Local use 3. Environmental Contamination. Although organochlorine compounds are restricted and prohibited by law, they are still an important environmental pollutant for our country due to the bioconcentration factor.

Organochlorine pollutants accumulate in different biological specimens; Blood, Adipose tissue, Breastmilk, Hair, and Meconium e.g. Biological monitoring, defined as the detection of the amounts of environmental contaminants in body tissues and fluid, is now being undertaken to identify residues in people living in various regions in the majority of developed countries. When bioconcentration is taken into account, analysis of environmental samples such as water and soil cannot be sufficient in the determination of the environmental pollution caused by these pollutants, thus the necessity of biological monitoring emerges. There are many studies on the biological monitoring of pesticides and their metabolites in the general population and occupational exposures. Studies have shown that these pesticides are endocrine disrupting and potential risk factors for many diseases such as non-Hodgkin's lymphoma, leukemia, brain cancer, uterine cancer, soft-tissue sarcoma, Hodgkin's disease and low sperm concentration. There are different health and environmental protection organizations working in this area. These include the International Agency for Research on Cancer (IARC) and the Environmental Protection Agency (EPA). These organizations have determined that OCPs are as well arcinogenic to humans (2B) and PCBs as probably carcinogenic to humans (2A). In another study we performed, we studied HCH and DDT isomers and metabolites and HCB in adipose tissue of 82 postmortem cases. 100% of adipose tissues have been found positive for *p,p'*–DDE (a metabolite of DDT). Organochlorine pesticides were significantly higher in adipose tissue of females than males ($p < 0.05$) (Daglioglu et al., 2010). In our region, these pesticides are still used illegally. In some pesticides, unwanted impurities are very high. For example, the ratio of the impurity of Dicofol formulation in DDT is about 14.3% (Turgut et al., 2009). Perinatal exposure to POPs occurs in different ways, and biological monitoring of them involves the use of different biological samples (amniotic fluid, cord blood, breast milk, etc).

POPs accumulate in the adipose tissue during pregnancy and are first transported to the placenta with blood and then the fetus (Sala et al., 2001). When compared with adults, fetuses and children are more vulnerable to these compounds these compounds, which may result in low detoxification levels, physiologic immaturity and short life after exposure. There happen various side effects in fetal development, as a result of exposure to OCs compounds in utero such as asthma, reproductive disorders, diabetes, growth delay and neurobehavioral disorders.

In another study we conducted from July to September 2006, milk samples were collected from 59 mothers in their 1st to 30th postpartum days. These samples were analyzed in our forensic toxicology laboratory and OCPs were detected in 62.7% of the participating mothers (Aytaç et al., 2010). In our another study, organochlorine pesticides (OCPs) and polychlorinated biphenyls (PCBs) levels were screened in the amniotic fluid of 200 pregnant women living in the Çukurova Region. OCPs (hexachlorohexane (HCH), Hexachlorobenzene (HCB), DDT and various metabolites) and different PCB compounds (28, 52, 101, 118, 138, 153, and 180) were identified by Gas Chromatography-Electron Capture Detector. While organochlorine compounds could not be detected in only 5% of pregnant women, one or more compounds were found at the ratio of 80%. This study is important in terms of the initial exposure of the fetus to the organochlorine compounds (Daglioglu et al.,2013). These findings indicate that our region is a region with intense pollution. In this regard, these compounds need to be biologically monitored in the general population and occupational exposure to produce the necessary data.

PESTICIDE ANALYSIS: SAMPLE PREPARATION, EXTRACTION, AND PURIFICATION

Pesticides are chemical compounds and applied for the protection of plants against insects and pathogens. Due to having active substances, pesticides are potentially toxic or poisonous to other living organisms and environments. For this reason, pesticide applications or treatments should be planned safely and with lowest or non-residue on crop, other living organisms as well as the environment. The traces of pesticides on plant materials have a vital importance and are called "residues" and maximum residue level is important. A term maximum residue level (MRL) can be explained as the highest level of a residue that is tolerated in crops legally and those limits were set by several organizations such as Codex Alimentarius, European Commission (EC), and some national governments, also countries such as Japan and Canada (Colborn et al., 1993; Lintelmann et al., 2003). For this reason, pesticide analysis and their sensitivity, reliability, and reproducibility are important. There are many analytical methods which are published periodically in Laboratory Guidebook such as Food and Drug Administration (FDA), Association of Analytical Communities (AOAC), Pesticide Analytical Manual Volumes I, Volumes II (PAM I and PAM II), and Food Safety and Inspection Service in the USA (USDA-FSIS).

Sampling and labeling technique is the first and the most critical step for residue analysis and accurate sampling and a good cross-section of the material being sampled are extremely important which generally takes more than half of the total time required for the analysis. In addition, some pesticides need special handling and for this reason, such samples should be frozen immediately. The condition and duration of sample shipping are also vital and should be considered. The physical properties of foods also have an importance to test active ingredients, especially their chemical properties such as acidic or basic characteristics, solubility, and stability situation, etc. To obtain a uniform matrix, blending or homogenization is needed (Durovic and Dordevic, 2011). The second step is the extraction which means disparity of pesticide residues from the matrix with solvents or non-solvents. To obtain accurate and reproducible results, the separation should be done using efficient and fruitful extraction procedures that remove the target chemicals from ample matrix (Tan and Cahi, 2011). There are many previously reported soxhlet and liquid-liquid extraction techniques. A liquid-liquid one was used as a basic one and most favoured. A principle of this technique is to separate compounds according to solubility degrees. The extraction efficiency or achievement depends on the ratio and diffusion or passing of the solute through the liquid border layer and suitable and fruitful solvent needed (Sapkale et al., 2010). However, those traditional techniques have some disadvantages such as too much time consumption, labor requirement, requirement of great amounts of solvents and loss of some analytic quantity. The first analytical methods of pesticides were evolved in the 1960s, and acetone was used for extraction as the primary solvent, followed by insertion of a sodium chlorur and nonpolar solvent and then cleanup steps that are employed to remove interfering matrix components. As an organic solvent, dichloromethane has strong extraction efficiency and suitable for concurrent analysis due to the having low boiling point and easy reconcentration after extraction and easy seperation from water and non-flammable characteristic. However, dichloromethane is harmful due to being carcinogenic. Addition to these solvents, particularly ethyl acetate and acetonitrile, were also commonly used in pesticide analysis. Some other solvents such as hexane, toluene, and isooctane, are also used as exchange solvents before a GC analysis. Moreover, some of the pesticides can be unstable in solvent extraction method. In the 1990s, robotic technology provided to reduce manual sample preparation techniques and fast sample preparation facilities. These innovations initially were very promising but not efficient for pesticides analysis due to the high cost.

It was also inconvenient to the extraction of various pesticide compounds in foods and for this reason optimization was needed based on the analysis (Cunha et al. 2011). Recently, modern solvent extraction techniques such as accelerated solvent extraction (ASE), supercritical fluid extraction (SFE),microwave assisted extraction (MAE),solid phase extraction (SPE), solid phase microextraction (SPME),matrix solid-phase dispersion (MSPD) extraction and QuEChERS (quick, easy, cheap, effective, rugged and safe) have been rectified to overcome the weakness of the conventional extraction procedures (Durovic and Dendovic, 2011).

ASE has almost same fundamental to the Soxhlet extraction procedure, but extraction duration is short due to the using elevated temperature and pressure and with a less amount of solvent and also reduced waste levels (Gan et al., 1999). Cervera et al. (2010) reported that ASE also provides desiccation of high moisture samples before the extraction step.

SFE principle is based on the process that separates the extractant from the matrix using a CO_2 supercritical solvent which is known as the king of solvents for extraction utilization. In addition, its harmlessness and low cost and provide extraction especially and trouble or impetus to achieve in traditional solvents. This technique is an express business in which extraction can be completed in 10-60 minutes and inert characteristics (Sapkale et al., 2010). However, SFE has some difficulties especially cannot match the range possible using traditional liquid solvents using most commercial instruments.

MAE is a technique that needs microwave energy it can be used successfully in a wide range of chemicals in many matrices and provides high selectivity and fast technique. Lopez-Avilla and Young (1994) reported that this technique reduced solvent usage and provided shorter extraction period (such as 10 min) as well as up to 12 samples, which makes this technique much more efficient than Soxhlet technique. The authors also implied that this energy enhances the efficiency of extraction from various solid matrices from food to soils and much more efficient technique compares to the soxhlet, particularly polar ones. MAE is also a suitable technique for soil samples (Ranz et al., 2008) However, the main disadvantage of MAE is lack of selectivity resulting in the co-extraction of significant amounts of interfering compounds. For this reason additional cleanup is required before injection to chromatography. Another disadvantage is the weak efficiency once non-polar or volatile solvent used and flammable solvents can be caused by serious hazards and needed much more attentiveness.

SPE is a useful technique that provides pesticides isolation by mini-column including special sorbent This helps to remove matrix interferences easily and then enhance the isolation efficiency and enrichment of the target compounds. SPE technique needs very small amount of solvent and its cartridges are less expensive. Sandstrom et al. (1992), Sandstrom (1995), Zaugg et al. (1995), Lindley et al. (1996), and Werner et al. (1996) reported previously that solid phase Extraction (SPE) is a fast and modern alternative method to liquid-liquid extraction technique.

SPME is a solvent-less extraction technique that provides both extraction and concentration steps into a single step (Arthur and Pawliszyn et al., 1990). The principle of this technique is based on absorption of analytes onto a polymeric-coated with a fused silica fiber (Tan and Chai, 2011). SPME can be applied through either Immersion directly (Im-SPME) or Headspace (HS-SPME) extraction techniques to detect pesticide residue in food and various materials both gas and liquid samples (Beltran et al., 2003; Blanco et al., 2002; Tomkins and Barnard, 2002; Hernandez et al., 2000).

MSPD is a technique that was introduced in 1989 and its principle is similar with SPE, to extract solid and/or semi-solid samples, which may or may not be viscous. In this technique, a small scale of sample or solvent can be used and extraction and cleanup can be done in a single step (Barker, 2000). This technique has recently been used for the determination of a new generation of pesticides.

QuEChERS is a new and fast dispersive solid phase extraction cleanup (dSPE) method for extraction of pesticides using a small amount solvent. In addition materials compare to traditional SPE methods in terms of the low cost, fast, easy application and simple and the most popular methods for routine pesticide analysis and developed by Anastassiades et al.,(2003). This mini-multi residue method has been standardized as the AOAC 2007.01, EN 15662. Previous extraction methods were difficult, took time and needed a considerable amount of solvent. It is possible to analyze various pesticides such as non-polar, polar and planary ones and also an efficient method for extraction from interfering matrix compounds such as phenolics, sugars, etc. However, the analysis of QuEChERS extracts, especially in acetonitrile by GC-MS, is not totally tangible due to the degradation of the GC column by the polar solvent, vapor overload of the insert liner due to the high thermal expansion coefficient, contamination of the system by co-extractives (Hetmanski et al., 2010), and reduced enrichment factors (Chiesa et al., 2017). More recently, Bedassa et al. (2017) have reported salting-out assisted liquid-liquid extraction (SALLE) as selective and efficient sample preparation procedure for the determination of multi-residue pesticides in alcoholic beverages by High-Performance Liquid Chromatography. The same authors reported that SALLE is fast, cheap and environmentally friendly. According to the procedure, appropriate amount of salts (MgSO4, $(NH4)_2SO4$, (NaCl), ($CaCl_2$), (K_2CO_3) ($CaSO_4$) reduces the mutual miscibility between liquids. The same authors also reported that SALLE applied previously for extraction of various organic compounds including α-dicarbonyl compounds, biogenic amines, alcoholic beverages N-nitrosamines, trimetazidine, methoxetamine, sulfonamides, temozolomide, benzimidazole fungicides, sulfonylurea, triazole, chlorophenols, etc with high efficiency.

The Gas Chromatography (GC), Liquid Chromatography (LC) and high-performance liquid chromatography (HPLC) techniques with different types of detector systems can provide identification and quantification of maximum residue levels (MRL) at trace levels. When MS analyzer is used, the acquisition modes mainly selected are single ion monitoring (SIM) and full scan m/z 50–600. In order to achieve a valuable identification and quantification of the analyte, at least 2–3 ions are selected in SIM mode. As an alternative to the single quadrupole, the ion trap (IT) has also been applied, in which a scan acquisition mode allows the ion selection to be monitored post-acquisition. Nowadays, combinations of most MS analyzers are possible, allowing tandem-MS (QqQ) to be performed.GC–MS with electron impact (EI) ionization and the combination of LC with tandem mass spectrometers(LC-MS/MS) with electrospray ionization (ESI) are techniques using identification and quantification in multi-residue methods for pesticides at the present. However, some of the pesticides, such as the organochlorine pesticides, can be detected by GC-MS more efficiently. Even though various methods, such as GC, HPLC, and combined with mass spectrometry (MS), have been developed for pesticide analysis, most of them cannot handle the sample matrices directly. GC can be used for detection of pesticide chemical classes which do not require derivatization include organochlorines (OCs), pyrethroids, organophosphorus pesticides (OPs), triazines, and chloroacetanilides. However, some of the polar chemicals such as phenoxy acid herbicides and carbamates need derivatization to detect by GC. Furthermore, the pesticides such as dicarboximides (vinclozin, iprodione), dinitroaniline (trifluralin, ethalfluralin), dinitrophenol (dinoseb), and dithiocarbamate (triallate). have been analyzed by GC/MS usually included in multi-residue methods but these methods have not tackled the entire range of compounds. LC coupled with tandem mass spectrometry (MS/MS) minimizes the derivatization before GC analysis. Mass spectrometry in selected reaction monitoring (SRM) mode is now more frequently used for LC rather than selected ion monitoring (SIM) with LC/MS as the ionization process for LC/MS is a softer process than that of GC/MS ion sources such as EI and CI. For atmospheric pressure ionization (API) sources most frequently used in

LC/MS/MS, most pesticides have only one ion formed during ionization (the protonated or deprotonated molecular ion or sometimes an adduct ion and consequently there is little confirmation ability. The main chemical classes of pesticides that have been more recently analyzed by LC/MS/MS or LC/MS methods and include phenoxyacid herbicides and a related nitrile herbicide often used in formulations with phenoxyacid herbicides, phenylureas, sulfonylureas, carbamates, pyrethroids, azoles, and a more extensive list of dithiocarbamates. Phenylureas, sulfonylureas, and most dithiocarbamates are not GC amenable and many azoles have significantly lower detection limits with LC/MS/MS.

The selection of ionization mode often depends upon whether the analysis is targeted for specific chemical classes or is a multi-residue analysis method for determination of hundreds of pesticides in a sample extract. A comparison of GC/MS or GC/MS/MS with EI to LC/MS/MS has been reviewed for a large number of compounds and suggests for most pesticides other than organochlorines that LC/MS/MS can provide lower detection limits (Alder et al., 2006; Pihlström et al., 2007; Paya et al., 2007; Lambropoulou and Albanis; 2007). However, lower or comparable detection limits have also been found for chloroacetanilides (metolachlor, acetochlor, alachlor) and selected triazines by GC/MS or GC/MS/MS with EI relative to LC/APCI-MS/MS (Dagnac et al., 2005) or LC/ESI-MS/MS (Freitas et al., 2004). GC/MS of a wider range of triazines has also been done by GC-EI/MS (Nagaraju and Huang, 2007; Zambonin and Palmisano, 2000; Jiang et al., 2005; Gonçalves et al., 2006; Albanis et al., 1998). Chemical ionization is often not considered in comparisons of GC and LC mass spectrometry methods.

LIMITATIONS

Pesticide analysis has some limitations when the analysis is performed such as unsuitable laboratory conditions, inexperienced staff, unsuitable analytical methodologies, etc. In addition, laboratory safety is one of the most important issues and the stringent working conditions must be recognized. In addition, smoking, eating, drinking or application of cosmetics should not be permitted in the working area. The use of highly toxic solvents and reagents should be minimized. Moreover all the waste solvent should be stored safely and disposed of both safely and in an environmentally friendly manner taking into account specific national regulations where available. Extraction, clean-up and concentration steps should be carried out in a well-ventilated area and safety screens should be used when glassware is used under vacuum or pressure. Residue analysis of pesticides consists of a chain of procedures. For his reason appropriate professional and experienced staff must be fully trained and experienced in the correct use of apparatus, evaluation and interpretation in appropriate laboratory skills andshould be aware of Analytical Quality Assurance (AQA) systems. In the laboratory adequate and reliable supplies of electricity, water, reagents, solvents, gas, glassware, chromatographic materials etc. of suitable quality are required. Furthermore, chromatographic equipment, balances, spectrophotometers etc. must be serviced and calibrated regularly and a record of all servicing/repairs must be maintained for every item of equipment. Labeling is of great importance and must immediately be assigned a sample identification code which should accompany it from through all stages of the analysis to the reporting of the results. If possible, the samples should be subjected to an appropriate disposal review system and records should be kept. If samples cannot be analyzed immediately but are to be analyzed quickly, they should be stored at (1-5°C) away from direct sunlight and analyzed within a few days. However, samples received from deep-frozen must be kept at -20°C.

Evaluation of Maximum Residue Limits (MRLs) is also important to determine the residues in a crop or an environmental sample and to measure residues at very low levels usually become very expensive and difficult. Lowest Calibrated level (LCL) has an advantage of reducing the technical difficulty of obtaining the data and would also reduce costs. Residues with agreed MRLs, the LCL can be specified as a fraction of the MRL. It is extremely important in pesticide residue studies that analytical results be reported in a consistent and unambiguous manner. The analytical methods used in the analysis of food samples must be as sensitive as possible. The sensitivity of the overall analytical procedure is usually defined in terms of Limit of Detection (LOD) and Limit of Quantitation (LOQ) or determination. The Limit of Quantitation (LOQ) is the minimum concentration of a contaminant in a food sample that can be determined quantitatively with an acceptable accuracy and consistency. The Codex Committee on Pesticide Residues employs the term limit of determination which is defined as the lowest practical concentration of a pesticide residue on the contaminant that can be quantitatively measured and identified in the specified food commodity or animal feedstuff with an acceptable degree of certainty by current regulatory methods of analysis.

Lower Limit of Detection (LLD) The smallest amount of sample activity which will yield a net count for which there is confidence at a predetermined level that activity is present.

It is extremely important in pesticide residue studies that analytical results be reported in a consistent and unambiguous manner. Often, national or international organizations that summarize the analytical results and calculate dietary intakes must evaluate and interpret data obtained from different laboratories, each reporting results in a different format. To facilitate these evaluations, laboratories should report enough details about the detection and quantitation limits of the analytical method to enable correct interpretations of the data to be made. Laboratories should ensure consistent detection and quantitation limits throughout a study. Results of recovery tests for the different contaminants should also be reported. However, analytical findings should be reported as measured, without the use of correction factors that take recovery into account.

REFERENCES

Akcan, R., Hilal, A., Daglioglu, N., Cekin, N., & Gulmen, M. K. (2009). Determination of pesticides in postmortem blood and bone marrow of pesticide-treated rabbits. *Forensic Sci Int., 189*(1-3), 82-7.

Aktar, W., Sengupta, D., & Chowdhury, A. (2009). Impact of pesticides use in agriculture: Their benefits and hazards. *Interdisciplinary Toxicology, 2*(1), 1–12. doi:10.2478/v10102-009-0001-7 PMID:21217838

Albanis, T. A., Hela, D. G., Sakellarides, T. M., & Konstantinou, I. K. (1998). Monitoring of pesticide residues and their metabolites in surface and underground waters of Imathia (N. Greece) by means of solid-phase extraction disks and gas chromatography. *Journal of Chromatography. A, 823*(1–2), 59–71. doi:10.1016/S0021-9673(98)00304-5 PMID:9818393

Alder, L., Greulich, K., Günther, K., Kempe, G., Bärbel, V., & Vieth, B. (2006). Residue analysis of 500 high priority pesticides: Better by GC-MS or LC-MS/MS. *Mass Spectrometry Reviews, 25*(6), 838–865. doi:10.1002/mas.20091 PMID:16755599

Anastassiades, M., Lehotay, S. J., Stajnbaher, D., & Schenck, F. J. (2003). Fast and Easy Multiresidue Method Employing Acetonitrile Extraction/Partitioning and "Dispersive Solid-Phase Extraction" for the Determination of Pesticide Residues in Produce. *Journal of AOAC International, 86*(2), 412–431. PMID:12723926

Anonymous. (2008). *The History of Pesticides, Organic Pesticides.* Retrieved from http://blog.ecosmart.com/index.php/2008/09/19/the-history-of-pesticides/

Anonymous. (2016). *International Code of Conduct on Pesticide Management Guidelines on Highly Hazardous Pesticides.* Food and Agriculture Organization of the United Nations Rome.

Arthur, C., & Pawliszyn, L. (1990). Solid-phase microextraction with thermal-desorption using fused-silica optical fibers. *Journal of Analytical Chemistry, 62*(19), 2145–2148. doi:10.1021/ac00218a019

Aytaç, N., Hilal, A., Yapicioglu, A. B., Daglioglu, N., Gülmen, M. K., & Tanir, F. (2010). Organochlorine Pesticide Level in Breast Milk. *Turkiye Klinikleri J Med Sci, 30*(1).

Barker, S. A. (2000). Matrix solid-phase dispersion. *Journal of Chromatography. A, 885*(1-2), 115–127. doi:10.1016/S0021-9673(00)00249-1 PMID:10941670

Bedassa, T., Negussie, M., & Abera, G. (2017). Salting-out Assisted Liquid-Liquid Extraction for the Determination of Multiresidue Pesticides in Alcoholic Beverages by High Performance Liquid Chromatography. *Science Journal of Analytical Chemistry, 5*(3), 38–45. doi:10.11648/j.sjac.20170503.11

Beltran, J., Peruga, A., Pitarch, E., Lopez, F. J., & Hernandex, F. (2003). Application of solid-phase microextraction for the determination of pyrethroid residues in vegetable samples by GC-MS. *Analytical and Bioanalytical Chemistry, 376*, 502–511. doi:10.100700216-003-1916-z PMID:12750867

Blanco, M. C., Rodriguez, B. R., Grande, B. C., & Gandara, J. S. (2002). Optimization of solid-phase extraction and solid-phase microextraction for the determination of α- and β- endosulfan in water by gas chromatographyelectron capture detection. *Journal of Chromatography. A, 976*(1-2), 293–299. doi:10.1016/S0021-9673(02)01232-3 PMID:12462621

Bozdogan, A. M. (2014). Assessment of total risk on non-target organisms in fungicide application for agricultural sustainability. *Sustainability, 6*(2), 1046–1058. doi:10.3390u6021046

Bozdogan, A. M., & Yarpuz-Bozdogan, N. (2008a). Determination of Dermal Bystander Exposure of Malathion For Different Application Techniques. *Fresenius Environmental Bulletin, 17*, 2103–2210.

Bozdogan, A. M., & Yarpuz-Bozdogan, N. (2008b). Assessment of Buffer Zones to Ditches of Dicofol For Different Applied Doses and Replication Numbers In Pesticide Applications In Adana Province, Turkey. *Fresenius Environmental Bulletin, 17*, 275–281.

Bozdogan, A. M., & Yarpuz-Bozdogan, N. (2015). Determination of potential risks of pesticide applications in peanut cultivation on human health and environment. *Fresenius Environmental Bulletin, 24*, 683–689.

Cervera, M. I., Medina, C., Portolés, T., Pitarch, E., Beltrán, J., Serrahima, E., ... Hernández, F. (2010). Multi-Residue Determination of 130 Multiclass Pesticides in Fruits and Vegetables by Gas Chromatography Coupled to Triple Quadrupole Tandem Mass Spectrometry. *Analytical and Bioanalytical Chemistry, 397*(7), 2873–2891. doi:10.100700216-010-3597-8 PMID:20237919

Chiesa, L. M., Labella, G. F., Panseri, S., Britti, D., Galbiati, F., Villa, R., & Arioli, F. (2017). Accelerated solvent extraction by using an 'in-line' clean-up approach for multiresidue analysis of pesticides in organic honey. *Food Additives & Contaminants. Part A, Chemistry, Analysis, Control, Exposure & Risk Assessment, 34*(5), 809–818. PMID:28277176

Colborn, T., VomSaal, F. S., & Soto, A. M. (1993). Developmental effects of endocrine-disrupting chemicals in wildlife and humans. *Environmental Health Perspectives, 101*(5), 378–384. doi:10.1289/ehp.93101378 PMID:8080506

Cunha, S. C., Fernandes, J. O., Oliveira, B. M., & Oliveira, P.P. (2011). Current Trends in Liquid-Liquid Microextraction for Analysis of Pesticide Residues in Food and Water. *Pesticides Strategies for Pesticides Analysis.*

Daglioglu, N., Akcan, R., Gulmen, M. K., Yener, F., & Efeoglu, P. (2011). Pesticide intoxications in Cukurova, Turkey: Three years analysis. *Human and Experimental Toxicology, 30*(12), 1892–1895. doi:10.1177/0960327111402241 PMID:21393348

Daglioglu, N., Efeoglu, P., Akcan, R., Inandiklioglu, N., Gulmen, M. K., & Demirhan, O. (2013). Polychlorinated biphenyls and organochlorine pesticides in amniotic fluids of pregnant women in South-central Turkey. *Toxicological and Environmental Chemistry, 95*(6), 954–961. doi:10.1080/02772248.2013.840370

Daglioglu, N., Gulmen, M. K., Akcan, R., Efeoglu, P., Yener, F., & Unal, I. (2010). Determination of organochlorine pesticides residues in human adipose tissue, data from Cukurova, Turkey. *Bulletin of Environmental Contamination and Toxicology, 85*(1), 97–102. doi:10.100700128-010-0057-2 PMID:20585752

Dagnac, T., Bristeau, S., Jeannot, R., Mouvet, C., & Baran, N. (2005). Determination of chloroacetanilides, triazines and phenylureas and some of their metabolites in soils by pressurised liquid extraction, GC–MS/MS, LC-MS and LC-MS/MS. *Journal of Chromatography. A, 1067*(1–2), 225–233. doi:10.1016/j.chroma.2004.11.058 PMID:15844528

Damalas, C. A., & Eleftherohorinos, I. G. (2011). Pesticide exposure, safety issues, and risk assessment indicators. *International Journal of Environmental Research and Public Health, 8*(5), 1402–1419. doi:10.3390/ijerph8051402 PMID:21655127

Durovic, R., & Dordevic, T. (2011). Modern Extraction Techniques for Pesticide Residues Determination in Plant and Soil Samples. In M. Stoytcheva (Ed.), *Pesticides in the Modern World Trends in Pesticides Analysis* (pp. 221–246). Rijeka, Croatia: InTech. doi:10.5772/17312

Erdogrul, O., Covaci, A., Kurtul, N., & Schepens, P. (2004). Levels of organohalogenated persistent pollutants in human milk from Kahramanmaraş region, Turkey. *Environment International, 30*(5), 659–666. doi:10.1016/j.envint.2003.12.004 PMID:15051242

Fishel, M. (2009). *Pest Management and Pesticides: A Historical Perspective*. UF/IFAS Extension. Retrieved from http://edis.ifas.ufl.edu/

Freitas, L. G., Gotz, C. W., Ruff, M., Singer, H. P., & Muller, S. R. (2004). Quantification of the new triketone herbicides, sulcotrione and mesotrione, and other important herbicides and metabolites, at the ng/l level in surface waters using liquid chromatography-tandem mass spectrometry. *Journal of Chromatography. A*, *1028*(2), 277–286. doi:10.1016/j.chroma.2003.11.094 PMID:14989481

Gan, J., Papiernik, S. K., Koskinen, W. C., & Yates, S. R. (1999). Evaluation of Accelerated Solvent Extraction (ASE) for Analysis of Pesticide Residues in Soil. *Environmental Science & Technology*, *33*(18), 3249–3253. doi:10.1021/es990145+

Gonçalves, C., Esteves da Silva, J. C. G., & Alpendurada, M. F. (2006). Chemometric interpretation of pesticide occurrence in soil samples from an intensive horticulture area in north Portugal. *Analytica Chimica Acta*, *560*(1-2), 164–171. doi:10.1016/j.aca.2005.12.021

Hernandez, F., Pitarch, E., Beltran, J., & Lopez, F. J. (2000). Headspace solid-phase microextraction in combination with gas chromatography and tandem mass spectrometry for the determination of organochlorine and organophosphorus pesticides in whole human blood. *Journal of Chromatography. B, Biomedical Sciences and Applications*, *769*, 65–77.

Hetmanski, M. T., Fussell, R., Godula, M., & Hübschmann, H. J. (2010, July). Rapid analysis of pesticides in difficult matrices using GC-MS-MS. *LC GC Europe*, 14–15.

Jayaraj, R., Megha, P., & Sreedev, P. (2016). Organochlorine pesticides, their toxic effects on living organisms and their fate in the environment. *Interdisciplinary Toxicology*, *9*(3-4), 90–100. doi:10.1515/intox-2016-0012 PMID:28652852

Jiang, D. G., Fang, C. R., Yang, D. J., & Wang, Z. T. (2005). Determination of 27 organochlorine and pyrethroid pesticide multi-residues in teas by gas chromatography. *Zhongguo Shipin Weisheng Zazhi*.

Kole, R. K., Banerjee, H., & Bhattacharyya, A. (2001). Monitoring of market fish samples for Endosulfan and Hexachlorocyclohexane residues in and around Calcutta. *Bulletin of Environmental Contamination and Toxicology*, *67*(4), 554–559. doi:10.1007001280159 PMID:11779071

Lambropoulou, D. A., & Albanis, T. A. (2007). Liquid-phase micro-extraction techniques in pesticide residue analysis. *Journal of Biochemical and Biophysical Methods*, *70*(2), 195–228. doi:10.1016/j.jbbm.2006.10.004 PMID:17161462

Lindley, C. E., Stewart, J. T., & Sandstrom, M. W. (1996). Determination of low concentrations of acetochlor in water by automated solid-phase extraction and gas chromatography with mass-selective detection. *Journal of AOAC International*, *79*(4), 962–966.

Lintelmann, J., Katayama, A., Kurihara, N., Shore, L., & Wenzel, A. (2003). Endocrine disruptors in the environment. *Pure and Applied Chemistry*, *75*(5), 631–681. doi:10.1351/pac200375050631

Lopez-Avilla, V., & Young, R. (1994). Microwave-assisted extraction of organic compounds from standards reference soils and sediments. *Analytical Chemistry*, *66*(7), 1, 1097–1106.

Nagaraju, D., & Huang, S. D. (2007). Determination of triazine herbicides in aqueous samples by dispersive liquid-liquid microextraction with gas chromatography-ion trap mass spectrometry. *Journal of Chromatography. A*, *1161*(1–2), 89–97. doi:10.1016/j.chroma.2007.05.065 PMID:17574561

O'Neil, W., Raucher, R., & Wayzata, M. N. (1998). *Groundwater Policy Education Project; Groundwater public policy leaflet series #4: The costs of groundwater contamination*. Retrieved from http://www.dnr.state.wi.us/org/water/dwg/gw/costofgw.htm

Palis, F. G., Flor, R. J., Warburton, H., & Hossain, M. (2006). Our farmers at risk: Behavior and belief system in pesticide safety. *J. Public Health*, *28*(1), 43–48. doi:10.1093/pubmed/fdi066 PMID:16436451

Payá, P., Anastassiades, M., Mack, D., Sigalova, I., Tasdelen, B., Oliva, J., & Barba, A. (2007). Analysis of pesticide residues using the Quick Easy Cheap Effective Rugged and Safe (QuEChERS) pesticide multiresidue method in combination with gas and liquid chromatography and tandem mass spectrometric detection. *Analytical and Bioanalytical Chemistry*, *38*(6), 1697–1714. doi:10.100700216-007-1610-7 PMID:17909760

Pihlström, T., Blomkvist, G., Friman, P., Pagard, U., & Österdahl, B. (2007). Analysis of pesticide residues in fruit and vegetables with ethyl acetate extraction using gas and liquid chromatography with tandem mass spectrometric detection. *Analytical and Bioanalytical Chemistry*, *389*(6), 1773–1789. doi:10.100700216-007-1425-6 PMID:17609934

Ranz, A., Eberl, A., Maier, E., & Lankmayr, E. (2008). Microwave-assisted derivatization of acidic herbicides for gas chromatography-mass spectrometry. *Journal of Chromatography. A*, *1192*(2), 282–288. doi:10.1016/j.chroma.2008.03.076 PMID:18405907

Sala, M., Ribas-Fitó, N., Cardo, E., de Muga, M. E., Marco, E., Mazón, C., ... Sunyer, J. (2001). Levels of hexachlorobenzene and other organochlorine compounds in cord blood: Exposure across placenta. *Chemosphere*, *43*(4-7), 895–901. doi:10.1016/S0045-6535(00)00450-1 PMID:11372882

Sandstrom, M. W. (1995). U.S. Geological Survey Water-Resources Investigations Report: Vol. 95-4105. Filtration of water-sediment samples for the determination of organic compounds. Academic Press.

Sandstrom, M. W., Jelinski, J. C., & Tai, D. Y. (1992). *Apparatus for sampling pesticide residues in run-off with control of sample pump and distributor valve.* U.S. Patent 5, 167, 802.

Sapkale, G. N., Patil, S. M., Surwase, U. S., & Bhatbhage, P. K. (2010). Supercritical fluid extraction. *International Journal of Chemical Sciences*, *8*(2), 729–743.

Savonen, C. (1997). Soil microorganisms object of new OSU service. *Good Fruit Grower*. Retrieved from http://www.goodfruit.com/archive/1995/6other.html

Tan, G. H., & Chai, M. K. (2011). Sample preparation in the analysis of pesticides residue in food for chromatographic techniques. In M. Stoytcheva (Ed.), *Pesticides—Strategies for Pesticides Analysis* (pp. 28–59). Rijeka, Croatia: InTech.

Tomkins, B. A., & Barnard, A. R. (2002). Determination of organochlorine pesticides in ground water using solid phase microextraction followed by dual-column gas chromatography with electron-capture detection. *Journal of Chromatography. A*, *964*(1-2), 21–33. doi:10.1016/S0021-9673(02)00592-7 PMID:12198850

Turgut, C., Gokbulut, C., & Cutright, T. J. (2009). Contents and sources of DDT impurities in dicofol formulations in Turkey. *Environmental Science and Pollution Research International*, *16*(2), 214–217. doi:10.100711356-008-0083-3 PMID:19052792

US EPA. (2001). Water protection practices bulletin. Washington, DC: Office of Water.

Waskom, R. (1994). *Best management practices for private well protection*. Colorado State Univ. Cooperative Extension. Retrieved from http://hermes.ecn.purdue.edu:8001/cgi/convertwq?7488

Werner, S. L., Burkhardt, M. R., & DeRusseau, S. N. (1996). Methods of Analysis by the U.S. Geological Survey National Water Quality Laboratory: Determination of Pesticides in a Water by CarboPak-B Solid-Phase Extraction and High-Performance Liquid Chromatography. U.S. Geological Survey.

Yarpuz-Bozdogan, N., & Bozdogan, A. M. (2009). Comparison of Field and Model Percentage Drift Using Different Types of Hydraulic Nozzles in Pesticide Applications. *International Journal of Environmental Science and Technology*, *6*, 191–196.

Yarpuz-Bozdogan, N., & Bozdogan, A. M. (2016). Pesticide exposure risk on occupational health in herbicide application. *Fresenius Environmental Bulletin, 25*, 3720-3727.

Yarpuz-Bozdogan, N., Bozdogan, A. M., Daglioglu, N., & Erdem, T. (2017). Determination of dermal exposure of operator in greenhouse spraying. *AMA. Agricultural Mechanization in Asia, Africa and Latin America*, *48*, 33–38.

Zambonin, C. G., & Palmisano, F. (2000). Determination of triazines in soil leachates by solid-phase microextraction coupled to gas chromatography-mass spectrometry. *Journal of Chromatography A.*, *874*(2,7), 247-255.

Zaugg, S. D., Sandstrom, M. W., Smith, S. G., & Fehlberg, K. M. (1995). *Methods of Analysis by the U.S. Geological Survey National Water Quality Laboratory Determination of Pesticides in Water by C-18 Solid-Phase Extraction and Capillary-Column Gas Chromatography/Mass Spectrometry With Selected-Ion Monitoring*. Open-File Rep. U.S. Geol. Surv. 95-181.

KEY TERMS AND DEFINITIONS

Biomonitoring: Biological monitoring, or biomonitoring, is the use of biological responses to assess changes in the environment, generally changes due to anthropogenic causes. In analytical chemistry, biomonitoring is the measurement of the body burden of toxic chemical compounds, elements, or their metabolites in biological substances. Biomonitoring programs may be qualitative, semi-quantitative, or quantitative.

Chromatography: Chromatography is a laboratory technique to separate the components of a mixture. The mixture is dissolved in a fluid called the mobile phase. Mobile phase carries the mixture through a stationary phase called column. The components are separated based on their molecular weight. Generally, in the analytical analysis, there are two kinds of chromatography, liquid chromatography and gas chromatography.

Forensic Toxicology: Forensic toxicology deals with the investigation of toxic substances, environmental chemicals, or poisonous products. In other words, forensic toxicology is the use of toxicology and other disciplines such as analytical chemistry, pharmacology, and clinical chemistry to aid medical or legal investigation of the death, poisoning, and drug use. It also shares ties with some of the environmental sciences.

Pesticide: Pesticides are chemical compounds that are used to kill or control pests, including insects, rodents, fungi, and unwanted plants (weeds). Pesticides are used in public health to kill vectors of disease, such as mosquitoes, and in agriculture to kill pests that damage crops. By their nature, pesticides are potentially toxic to other organisms, including humans, and need to be used safely and disposed of properly.

Solid-Phase Microextraction: Solid-phase microextraction (SPME) is an innovative, solvent-free sample prep technology that is fast, economical, and versatile. SPME uses a fiber coated with a liquid (polymer), a solid (sorbent), or a combination of both. The fiber coating extracts the compounds from your sample by absorption in the case of liquid coatings or absorption in the case of solid coatings. The SPME fiber is then inserted directly into the chromatograph for desorption and analysis. SPME has gained widespread acceptance as the technique of preference for many applications including flavors and fragrances, forensics and toxicology, environmental and biological matrices, and product testing to name a few.

Chapter 18
Computational Tools and Techniques to Predict Aquatic Toxicity of Some Halogenated Pollutants

Raghunath Satpathy
Majhighariani Institute of Technology and Science, India

ABSTRACT

Halogenated organic compounds are usually xenobiotic in nature and used as ingredients for the synthesis of pesticides, solvents, surfactants, and plastics. However, their introduction to the aquatic ecosystems resulted in ecological danger due to their toxic effects. The usual method of toxicity assessment is by performing the experimental approach by considering some model organism. In this aspect the computational techniques such as QSAR (quantitative structure activity relationship) is considered an effective method. By computing several molecular features and the experimental activity, the toxic effect of a compound can be correlated. This chapter describes the aquatic toxicity of the compounds. The information about different computational resources (databases, tools, and modeling tools) have been given. Also, the application of QSAR to predict aquatic toxicity of different halogenated compounds available in the literature has been reviewed.

INTRODUCTION

From the industry and agriculture sector, huge amount of halogenated organic compounds are produced (Gribble, 1994; Song *et al.*, 2000). Due to their continuous entry into the environment, the propensity for the accumulation of these compounds in the habitats represents a global threatening (Perocco *et al.*, 1983; Damstra, 2002; Dewan *et al.*, 2013). Due to their persistency, many of these compounds are banned. However due to their versatility, long history of formulation and use as major industrial chemicals; these have been detected in soil, sediments as well as in water ecosystems (Persistent, 2000). Another feature in case of the halogenated pollutants is their toxicity increases with an increasing number of halogen

DOI: 10.4018/978-1-5225-6111-8.ch018

Table 1. Showing the example of halogenated substances causes aquatic toxicity

S. No.	Common Sources of Halogenated Compounds	Remark
1	Persistent Halogenated compounds	Persistent organic pollutants (POPs) include halogenated compounds as a major component. Important examples are polychlorinated biphenyls (PCBs), dioxins (e.g. TCDD), furans, and organochlorine insecticides.
2	Paper- and pulp-mill effluents	Since chlorinated compounds have disappeared from effluents, the major toxic compounds are natural compounds of trees, such as resin acids from coniferous trees and phenolics from deciduous trees
3	Endocrine-disrupting compounds	These include several types of compounds with various modes of action. Although several different types of hormonal pathways could be targeted, the term is most commonly used for compounds that disturb reproductive hormone cycles
4	Pesticides	Pesticides contain several different types of compounds, including herbicides, insecticides, and fungicides

atoms and number of aromatic rings present in the molecule (Nikinmaa, 2014). Aquatic organisms are frequently faced with periods of exposure to various environmental pollutants, often as the result of the release of chemicals from agricultural and/or industrial activities. Among them, paper and pulp mills were the major sources of halogenated compounds, especially chlorinated compounds because of chlorine bleaching (Walker & Peterson, 1994; Ali & Sreekrishnan, 2001).Different categories of halogenated compounds that are persistent and having ecotoxicological effects are polychlorinated biphenyls (PCBs), dioxins and many organochlorine insecticides (Table 1).

The ecotoxic information is gathered by the study about the effect of pollutant molecules on fish and other aquatic organisms. Most often these results are based on materials tested independently under laboratory conditions in field studies. Therefore, this information may not replicate the realistic measure. The reason is that several toxic substances often occur together in significant amounts in polluted water and are likely to interfere with their actions. For example, the interactions among the substances like dieldrin, DDT, and methoxychlor can cause adverse effects when these substances occur in combination rather than individual effect (Jana, 1994).

The basic application of these toxicity predictions is

1. To study the change the modification of water quality that may affect the aquatic taxa
2. To study genetic impact (mutagenic potential) of aquatic pollutants on aquatic organisms
3. Risk assessment of toxic compounds

Since the experimental methods are difficult to perform for toxicity prediction, as alternative methods, the computational techniques are used. This chapter will basically focus on the importance of computational analytical methods for the prediction analysis of aquatic toxicity of special focus to halogenated substances along with application and challenges exist. In addition to that, different computational tools databases associated with this type of study are to be highlighted by thorough literature survey.

CAUSE OF AQUATIC CONTAMINATION AND TEST MEASUREMENT FOR AQUATIC TOXIC COMPOUNDS

A chemical compound that is released into the aquatic environment and generates the toxic effect and the degree of toxicity is dependent upon the following factors (Deneer, 2000; Arnot & Gobas, 2006).

1. The amount of the toxic compound released into the ecosystem.
2. The water solubility of the compound. (related to the uptake route and the bio-magnification of the compound)
3. The fugacity (the tendency of the compound to escape from the material) of the compound.
4. The transformation of the compound. (compounds that are transformed from the original into daughter compounds)
5. Complex formation by the other compounds in the aquatic environment.

Usually, the common approach to access the aquatic toxic effect is the analysis of the degree of toxicity in the level of (aquatic) organism or at the cellular level. More conveniently, this is performed in the laboratory that uses a single organisms and single pure toxic compounds with the limited the duration of exposure. Although this type of experimentation provides valuable information, it fails to impart the essential knowledge about the interaction in between toxicants or between the organisms. Knowledge of both is required and it is a challenging aspect to understand fully about the functioning of a contaminated ecosystem. This approach is widely used by considering by using different model organisms. Table 2 describes the different types of model systems used for measurement of aquatic toxicity.

Aquatic toxicity tests serve to measure the effects of chemical exposure on a variety of endpoints, including survival, reproduction, and physiologic and biochemical responses. The major objective of the toxicity testing is to evaluate toxicity using standard methods to derive water quality criteria. In this section, the testing criteria followed by some of the important organizations have been summarized.

Table 2. Showing the model organisms used for aquatic toxicity predictions

S. No.	Name	Remark
1	Zebrafish (*Daniorerio*)	Used in biomedicine research and toxicity testing
2	Rainbow trout (*Oncorhynchusmykiss*)	Most commonly used big fish in aquatic toxicology
3	Fathead minnow (*Pimephalespromelas*)	Used to study the reproductive toxicity and life-cycle toxicity
4	Guppy (*Poeciliareticulata*)	Used to study of bioaccumulation
5	*Chironomusriparius*	The chironomids have been studied to evaluate bioaccumulation and toxicity of metals
6	Oyster (*Crassostreagigas*)	To study reproductive toxicity
7	Water flea (*Daphnia* sp.)	Used in aquatic environmental toxicity studies
8	*Nereis* sp. (particularly virens)	Used for study particularly in sediment toxicology
9	*Tetrahymena* sp. (pyriformis and thermophila)	Used to study aquatic toxicity
10	Duckweed (*Lemna* sp.)	They are common models for aquatic toxicology studies on green plants
11	*Chlamydomonas reinhardii*	One of the most important model algae to study aquatic toxicity
12	*Vibrio fischeri*	The bioluminescent marine bacterium is commonly used in in toxicity testing (inhibition of bioluminescence by contaminants)

Table 3. Some selected ISO toxicity test standards for aquatic toxicity testing

Sl. No.	Name of the Test	Test No.	Basis of the Test
1	Water quality	ISO 13641-2:2003	test for low biomass concentrations
2		ISO 13829:2000	determination of the genotoxicity of water and wastewater
3		ISO 14371:2012	determination of freshwater sediment toxicity to Crustacea
4		ISO 14380:2011	determination of the acute toxicity to *Thamnocephalus platyurus*
5		ISO 14442:2006	guidelines for algal growth inhibition tests with poorly soluble materials, volatile compounds, and wastewater
6		ISO 15088:2007	determination of the acute toxicity of waste water to zebrafish eggs
7		ISO/DIS 16303	determination of toxicity of freshwater sediments by using *Hyalella azteca*
8		ISO 21427-1:2006	evaluation of genotoxicity using amphibian larvae
9		ISO 20666:2008	determination of the chronic toxicity to *Brachionus calyciflorus* in 48 hours
10		ISO 20079:2005	determination of the toxic effect of water constituents and wastewater on duckweed (*Lemna* minor growth inhibition test

Table 4. Showing some OECD tests relevant for aquatic toxicology

Sl. No.	Name of the Test	Test No.	Basis of the Test
1	Fish-embryo acute toxicity (FET) test	236	Evaluation of the acute toxicity of chemicals to zebrafish embryos by exposing it to the test chemical for a period of 96 hours
2	*Daphnia magna* reproduction test	211	Analysing the effect of chemicals on the reproduction of Daphnia
3	Freshwater alga and cyanobacteria growth inhibition test	201	Determines the effects of a chemical on the growth of freshwater microalgae
4	Amphibian metamorphosis assay	231	Screens substances that may interfere with the normal functioning of the hypothalamo–pituitary–thyroid axis with the help of amphibian metamorphosis
5	*Lemna sp.* growth inhibition test	221	Evaluates the toxicity of substances to freshwater aquatic plants of the genus *Lemna*
6	*Daphnia sp.* acute immobilization test	202	Evaluates the acute effects of chemicals on daphnids
7	Fish acute toxicity test	203	Fish are exposed to the test toxic substance and mortalities recorded at 24, 48, 72, and 96 hours
8	Fish, juvenile growth test	215	Evaluates how prolonged exposure to chemicals affects the growth of juvenile fish
9	Determination of the inhibition of the activity of anaerobic bacteria	224	Predicts the likely effect of a test toxic substance on gas production in anaerobic digesters

One way of testing is by using the International Organization for Standardization (ISO) tests methods. The accepted ISO toxicity test standards can be bought from http://www.iso.org, and a selection of them (either accepted or under development) is given in Table 3. Another major source of testing of toxic compounds is of international toxicity test standards is the Organisation for Economic Co-operation and Development (OECD) and the protocol is available freely. Table 4 summarizes some of the methods used in the aquatic toxicology. In addition to this; the US Environmental Protection Agency (EPA) has an extensive compilation of aquatic toxicity testing methods, which can be found at http://water.epa.gov/scitech/methods/cwa/.

COMPUTATIONAL METHODS AND TOXICITY PREDICTION OF HALOGENATED COMPOUNDS

As it is discussed in the above section about the aquatic toxic effect of the halogenated compounds so an effective way to study the toxicity can be studied by Bioinformatics based methods. The objective of the Bioinformatics is for the effective storing, processing and analyzing the biological data with the help of the computer. In this case, by using the bioinformatics-based study, the relationships between the structure of a compound and its toxicity effect can be inferred by computing the toxicological effect of halogenated chemicals on aquatic based on several factors (Barron, 2003; La Farre et al., 2008). Another bioinformatics approach is also utilized for toxicity prediction is by the analysis of toxico kinetic modeling. This technique basically integrates the uptake of pollutants, metabolism, and excretion of chemicals in an individual organism of study (Landrum *et al.,* 1992). In this chapter, the discussion will be mainly about the theoretical basis of toxicity prediction includes the structure-activity analysis (SAR) analysis of toxic pollutants in the form of a statistical model.

Usually, the SAR (structure-activity relationship) models are obtained by regression or classification methods by relating one or more descriptors of the chemical structure of a substance. The descriptors may rely on the topology (such as the number of hydrogen atoms or the number of double bonds between two carbon atoms), electronic properties (e.g., the charge of the molecule), geometry (e.g., the surface and the volume of the molecule) or the physicochemical properties (e.g. lipophilicity, ionisation). QSAR models are based on the principle that molecules possessing similar chemical structures will possess similar

Figure 1. Showing factors responsible for aquatic toxicity of halogenated compounds

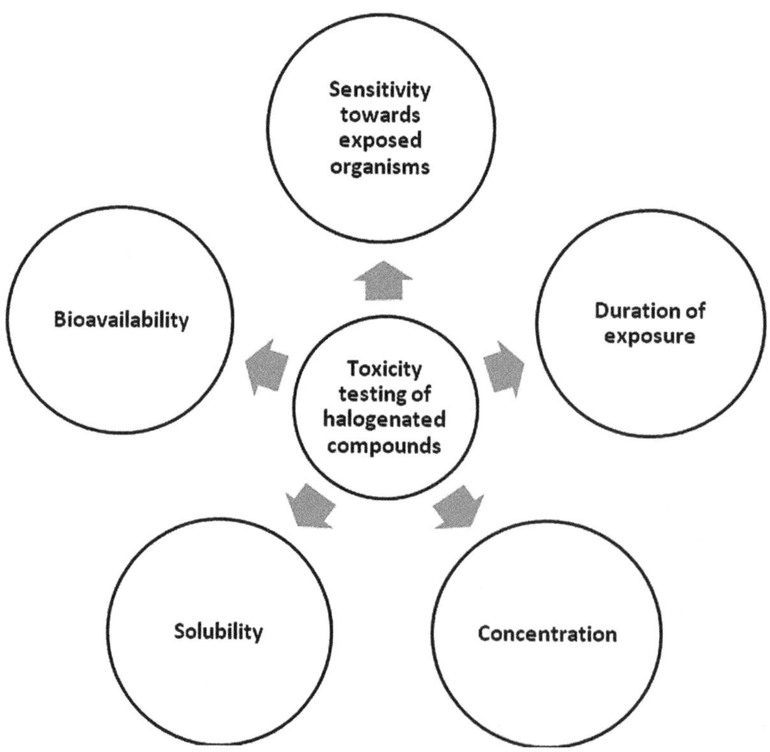

Figure 2. Showing the general outline for bioinformatics based modeling methods for prediction of toxicity

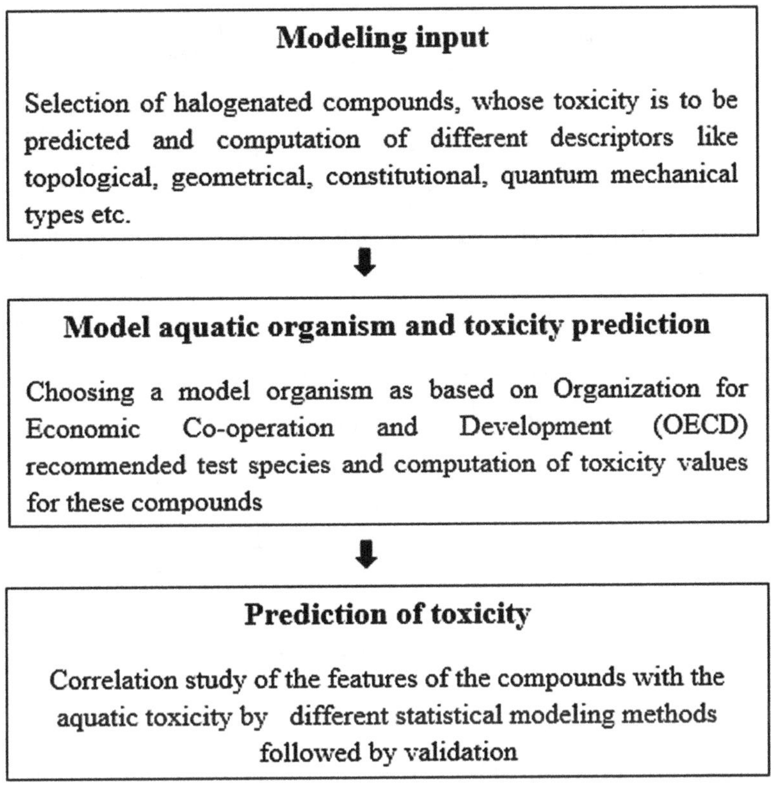

activities or properties. The robustness of the model is then tested and mainly depends on the amount, the range and the quality of the data available for their construction (Bradbury, 1995; Akers *et al.*, 1999).

Developing a QSAR model is a process that involves 3 basic steps.

- First of all, a high-quality dataset comprising relevant (biological) activity data, associated with a given chemical structure, needs to be identified.
- Molecular descriptors for these chemicals can then be generated – those most relevant to the activity should be considered.
- Finally, a function describing the relationship between the activity and the selected descriptors is to be established and validated.

The detailed process of QSAR methods has been discussed in below sections:

Dataset Preparation for QSAR Modeling

A typical dataset used in QSAR modeling consists of a list of compounds with corresponding experimental measurements obtained for a specific endpoint. The variety of experimental values can be used in QSAR, usually originated from may be of biological data and obtained from in vitro or in vivo studies. Many of such variables and their explanation are shown in Table 5.

Table 5. Major variables to predict the toxicity of different types of toxic chemicals

Sl No.	Type of Toxic Compounds	Toxicity Variable Parameter	References
1	phenols and thiophenols	EC50	(Ghamali et al., 2016)
2	nitroaromatics compounds	IGC_{50}	(Artemenko et al., 2011)
3	non-polar narcotic	EC50	(Aruoja et al., 2014)
4	1-(3,4-dichlorophenyl)-3-methlyurea (DCPMU), 3-(3-chlorophenyl)-1,1-dimethylurea (MCPDMU), and 1-(3,4-dichlorophenyl)urea (DCPU)	EC50	(Neuwoehner et al., 2010)
5	poly-substituted benzenes	EC50	(Netzeva et al., 2004)
6	nitrobenzene	EC50	(Altenburger et al., 2005)
7	alcohol ethoxylate surfactants	EC50	(Wong et al., 1997)
8	normal and branched alkanes, alkylbenzenes, polyaromatic hydrocarbons, alkanols, polyols, phenylcarbinols, aliphatic primary amines etc	P_{OW}	(Chicu et al., 2000)
9	Linear alkyl benzene sulphonates and ester sulphonates	EC50	(Hodges et al., 2006)
10	carboxylic acids	IGC50	(Seward & Schultz, 1999)
11	Anilines and phenols	Kow	(Damborsky & Schultz, 1997)
13	organophosphorus pesticides	LC_{50}	(deBruijn & Hermens, 1993)

Where, EC50= effective concentration 50, IGC50 = 50% inhibition growth concentration, Kow and Pow= n-octanol-water partition coefficients, LC 50= lethal concentration 50

Molecular Descriptors Calculation

Molecular descriptors are obtained through the process of converting molecular structures into a set of continuous numerical or binary values describing specific molecular (Dudek *et al.* 2006). Basically, two families of descriptors for the compounds can be computed are described in following sections

2D (Two Dimensional) Descriptors

This family of descriptors is calculated without information relating to the 3D spatial arrangement of the molecule (Katritzky and Gordeeva 1993; Todeschini and Consonni 2009). 2D descriptors can be further classified as:

1. **Constitutional Descriptors:** These category describe simple properties related to the presence of elements that constitute a molecular structure. Typical examples of such descriptors include the number of specific atoms, the number of single, double and triple bonds or the number of aromatic rings.
2. **Topological Descriptors:** These are the result of the application of graph theory to describe the way the atoms are connected. Atoms are represented by vertices and bonds are represented by edges. Different computed properties such as Wiener index, Balaban's J index are derived from the distance and adjacency matrices. Similarly the properties like Kier and Hall indices consider the electro-topological state of the atoms within the molecule.

3. **Electronic Descriptors:** These are related to electronic properties of molecules. Polarisability and hydrogen bonding are among such descriptors with the latter especially important in ligand-receptor interactions. This category also encompasses quantum mechanical descriptors such as the energy of the highest occupied molecular orbital (EHOMO) and the energy of the lowest unoccupied molecular orbital (ELUMO). These two descriptors are especially useful for studying chemical reactivity.
4. **Steric Descriptors:** These describe size and shape of a molecule. Molecular volume (a sum of the van der Waals volumes), molecular surface area or molar refractivity, as a measure of the size of a molecule, are commonly used in QSAR studies.
5. **Hydrophobicity Descriptors:** This is an important group of descriptors that are widely used in drug design and discovery as they can be applied to modeling both pharmacodynamics (receptor binding) and pharmacokinetic properties (e.g. the uptake and distribution of a xenobiotic relying on partitioning through biological membranes). Partition coefficient (log P), distribution coefficient (log D) and aqueous solubility (log S) are important hydrophobicity/ hydrophilicity descriptors.
6. **Molecular Fingerprints and Fragment-Based Descriptors:** These are very useful for screening purposes and can also be used in QSAR modeling. A molecular fingerprint is a string of bits representing the presence or lack of specific molecular features and as such, it can be used to quantify the similarity between molecules. In cases where only a specific substructure is to be sought within a molecule.

3D (Three Dimensional) Molecular Descriptors

These three dimensional descriptors are more computationally expensive than 2D features. To obtain 3D descriptors the conformation of the structures has to be established either experimentally or by molecular mechanics calculations. Alignment-dependent and alignment-independent 3D descriptors exist. The former relies on aligning structures in space and therefore depends on knowledge of ligand-receptor complexes. Comparative molecular field analysis (CoMFA) is a common method. It samples compounds, which are represented by their steric and electrostatic fields, in a three-dimensional lattice and then it analyses the data using the partial least square (PLS) technique to determine which parts of the molecule are important for the biological activity (Cramer *et al.,* 1998).

Statistical Modeling Methods

The relationship between the biological activities (such as toxicity) or any other properties of a molecule and its chemical structure can be established by utilizing the statistical analysis. As an example, the linear regression statistics can be used to predict a relationship between toxicity and a particular property of the chemicals, such as log P. In such cases, the relationship is referred to as quantitative structure-activity relationship (QSAR) analysis (Livingstone & Manallack, 1993; Papa et al., 2005; Hewitt et al., 2010). Classification of toxic compounds by using QSAR modeling is a common method for the toxicity predictions of compounds. By applying several data mining techniques, the toxic compounds can be placed into a specific category (e.g. toxic or non-toxic). In addition to this another techniques such as artificial intelligence methods have also increased the application of machine learning to QSAR modeling intended to provide improved and accurate predictions. Machine learning provides sophisticated algorithms that can learn from data and then use this knowledge to build both linear and non-linear models. A variety of machine learning techniques are available such as: multiple linear regression (MLR), partial least squares (PLS), decision trees, k-nearest neighbors (kNN), random forest (RF), multilayer perceptron (MLP) and support vector machines (SVM) and also unsupervised learning techniques such as: self-organizing maps and principal component analysis (PCA). Supervised learning uses labeled data (e.g. inhibitor/non-inhibitor) to build a model that can predict a value or class for new data points (compounds).

QSAR Validation

In order to be acceptable for regulatory purposes, QSAR models need to be scientifically valid, the organization for Economic Cooperation and Development (OECD) introduced a set of guidelines known as the OECD principles for the validation of QSAR. The following general rules were introduced to guide the development of QSAR models but are applicable to a broader range of computational methods (Cherkasov et al., 2014; Gramatica, 2007). The following general rules were introduced to guide the development of QSAR models:

1. A defined endpoint: to ensure that the modeled data were obtained using suitable experimental protocols and conditions.
2. An unambiguous algorithm: methods used for building a QSAR model should be transparent.
3. A defined domain of applicability: models are based on specific subsets of chemical space, for example, the unreliable predictions may be obtained for compounds that do not belong to this subset of chemical space must be avoid.
4. Appropriate measures of goodness-of-fit, robustness and productivity: the quality of models should be assessed by both internal and external validation by using different parameters such as Coefficient of determination (R2), adjusted determination coefficient (R2adj), standard error of estimate (s), F-statistics and t-values were indicated to represent statistical parameters describing the goodness-of-fit for multilinear regression models. The use of external datasets is the most appropriate procedure to assess predictively. However, in practical terms, the availability of external datasets may be limited. Therefore, it is common to use internal validation by following such simulation procedures as cross-validation (leave-one-out and leave-many-out), bootstrapping, Y-scrambling or training/test set splitting.
5. A mechanistic interpretation: the explanation of the QSAR in terms of chemistry or toxicology should be provided.

Interpretation of QSAR Study

A general scheme that encompasses aquatic toxicity in general, i.e. it is not related to a specific species, as defined by Verhaar et al., (1992). The scheme assigns chemicals to one of four identified toxicity categories, which are associated with estimated ranges of effective concentrations. Hence, the scheme can be used for preliminary screening or prioritization for testing. The identified classes are outlined below.

- **Inert Chemicals:** The chemicals are not reactive so the toxic potency of these chemicals depends only on their *lipophilicity*.
- **Less Inert Chemicals:** These chemicals are also not reactive but their toxicity is slightly higher than that of inert chemicals, so the toxicity value depends upon the presence of *polar groups*.

- **Reactive Chemicals:** These group of chemical show an excess toxicity and can react with biological structures by following different modes of action.
- **Specifically Acting Chemicals:** This is a category that includes diverse chemical classes that are known to react with the specific biological target.

However, the interpretation depends upon the refined data, proper selection of descriptors and modeling methods used for statistics generation and validation.

APPLICATION OF COMPUTATIONAL METHODS IN AQUATIC TOXICITY PREDICTION

The degree of toxic action of any chemicals is the measure of the adverse effect of the chemicals. Specific types of these adverse effects are called toxicity endpoints, such as carcinogenicity and nontoxicity and so on. These actions can be quantified as different quantitative parameters such as LD50 (lethal dose to 50% of tested individuals) or qualitatively as binary (e.g., toxic or non-toxic) or ordinary (e.g., low, moderate, or high toxicity) (Rowe, 2010; Deeb & Goodarzi, 2012). The computational methods now becomes a popular aid to correlate both in *vitro* and *in vivo* toxicity tests. The major application is to minimize the need for animal testing, reduce the cost and time of toxicity tests, as well as to improve toxicity prediction and safety assessment process. Another recent implementation of the computational methods is to estimate the toxicity of any chemicals even prior to their synthesized. Several computational resources are used for prediction of toxic substances requires a wide variety of features as given in below section (Zhu, 2013).

- Software tools and databases for storing and effective retrieval of data about chemicals, structure, and properties (both physical and chemical), toxicity
- Software for computing the molecular descriptors
- Simulation tools for molecular docking and molecular dynamics simulation
- Modeling methods and implementation of important tools for the effective generation of prediction models
- Application of visualization software tools

The most commonly used of this computational resources is given in Table 6, Table 7 and Table 8. In this way, the knowledge of molecular properties that will minimize the unintended biological activity of the industrial chemicals at their synthetic stage presents an opportunity to rationally develop less hazardous and thus greener chemicals (Connors et al. 2014).

IMPLEMENTATION OF QSAR TECHNIQUES IN PREDICTION OF AQUATIC TOXICITY OF SOME HALOGENATED COMPOUNDS

Prediction and modeling techniques, such as quantitative structure-activity relationships (QSAR) are applied to fill data gaps and to predict, assess and extrapolate the toxicity of halogenated chemicals (McKim *et al.*, 1987; Satpathy, 2016, 2018). In one of the work, the QSAR study was performed for the

Table 6. Specific database for toxic chemical data

S. No.	Name of Database	Availability	Remark
1	ECOTOX	https://cfpub.epa.gov/ecotox/	Publicly available database providing the single chemical environmental toxicity data on aquatic life, terrestrial plants and wildlife.
2	AQUIRE (AQUatic toxicity Information REtrieval)	http://www.epa.gov/med/databases/aquire.html	Quick access to a comprehensive, systematic, computerized compilation of aquatic toxic effects data.
3	Oak Ridge National Laboratory (ORNL) Benchmark	http://www.hsrd.ornl.gov/ecorisk/	The basic eco-toxicity measuring criteria have been developed for a wide range of contaminants including metals, and organochlorine and organic compounds.
4	Acute Toxicity Database	https://www.cerc.usgs.gov/data/acute/acute.html	Database summarizes the results from aquatic acute toxicity tests conducted by the USGS CERC located in Columbia, Missouri.
5	MOAtox	http://epa.gov/ceampubl/fchain/webice/	Comprehensive mode of action and acute aquatic toxicity database for predictive model development
6	Chemical Aquatic Fate and Effects (CAFE)	https://response.restoration.noaa.gov/cafe	Constitutes a software program enable the user to estimate the fate and effects of thousands of chemicals.
7	PAN pesticide database	http://www.pesticideinfo.org/Search_Ecotoxicity.jsp	The Aquatic Ecotoxicity section includes 223,853 aquatic toxicity results from U.S. EPA's AQUIRE database.
8	Environmental Health and Safety freeware	http://www.ehsfreeware.com/	A free virtual library contains hundreds of environmental, health and safety software, apps other resources.
9	The ARS Pesticide Properties Database	https://www.ars.usda.gov/northeast-area/beltsville-md/beltsville-agricultural-research-center/adaptive-cropping-systems-laboratory/docs/ppd/pesticide-properties-database/	A database of chemical and physical properties of 334 widely used pesticides and their degradation characteristics.
10	Toxicology Data Network (Toxnet) Developmental and Reproductive Toxicology Database (DART)	http://toxnet.nlm.nih.gov/cgibin/sis/htmlgen?DARTETIC	Contains references to the aspects of developmental and reproductive toxicology.
11	Endocrine Disruptor Knowledge Base (EDKB)	http://www.fda.gov/ScienceResearch/BioinformaticsTools/EndocrineDisruptorKnowledgebase	Contains in vitro and in vivo experimental data for more than 3,000 chemicals
12	Endocrine Active Substances information system (EASIC)	https://eurl-ecvam.jrc.ec.europa.eu/databases/eas_database	Searchable database giving information chemical structure, toxicity
13	NureXbase	http://www.nursa.org	Information on chemical structure, crystal structure, physical descriptors, nuclear receptors and mechanism of endocrine action
14	OECD (Q)SAR Toolbox	http://www.oecd.org/env/ehs/riskassessment/theoecdqsartoolbox.htm	Contains several databases, including reproductive toxicity data
15	Acute Toxicity Database	https://www.cerc.usgs.gov/data/acute/acute.html	database of aquatic acute toxicity test results for thousands of chemicals
16	Distributed Structure-Searchable Toxicity (DSSTox) Database	https://www.epa.gov/chemical-research/distributed-structure-searchable-toxicity-dsstox-database	Resource for supporting improved predictive toxicology.
17	TerraBase	http://www.terrabase-inc.com/	provide for the quick search of compounds with specific biological effects and properties
18	EXTOXNET	http://extoxnet.orst.edu/pips/ghindex.html	Pesticide information including experimental toxicity value

Table 7. Software tools and servers details for calculating variables in toxicity prediction in case of compounds

Sl. No.	Software	Availability	Application
1	ADMET Predictor	http://www.simulations-plus.com/	Quantitative prediction of oestrogen receptor toxicity.
2	ACD ToxSuite (ToxBoxes)	http://www.acdlabs.com/products/admet/tox/	Prediction of Endoplasmic Reticulum (ER) binding affinity prediction.
3	CAESAR	http://www.lhasalimited.org/	Two classification models for developmental toxicity
4	Derek	https://www.lhasalimited.org/	Different levels of classification models (based on developmental toxicity
5	Leadscope	http://www.leadscope.com/	Classification models for developmental toxicity in the rodent fetus
6	MolCode Toolbox	http://molcode.com/	Quantitative prediction of rat ER binding affinity and AhR binding affinity
7	MultiCASE	http://www.multicase.com/	Classification models for developmental toxicity associated with drugs
8	OSIRIS property explorer	http://www.organic-chemistry.org/prog/peo/	predicts mutagenicity, tumorigenicity, irritating effects and reproductive effects
9	PASS	http://ibmc.p450.ru/PASS//	Classification models giving the probability of reproductive toxic effects.
10	T.E.S.T.: The Toxicity Estimation Software Tool	http://oasis-lmc.org/	Developmental toxicity estimation.
11	TOPKAT	http://www.accelrys.com	Classification model for developmental toxicity of pesticides, industrial chemical
12	Toxboxespharma algorithms	http://pharma-algorithms.com/tox_boxes.htm	A classification model for the prediction of ER binding.
13	VirtualToxLa	http://www.biograf.ch	Classification model for endocrine disrupting potential
14	HAZARD EXPERT	http://www.compudrug.com/hazardexpertpro	Human carcinogenicity and genotoxicity prediction
15	Toxline	https://toxnet.nlm.nih.gov/cgi-bin/sis/htmlgen?TOXLINE	Human neurotoxicity prediction
16	BCABAF	https://www.epa.gov/tsca-screening-tools/epi-suitetm-estimation-program-interface	Prediction of bio- concentration of toxic substances
17	PCKOCWIN	cpas.mtu.edu/cencitt/oppt/tsld019.htm	Prediction of soil sorption with the toxic chemicals
18	BIOWIN	envirosim.com/products/biowin	Prediction of biodegradability of toxic substances
19	KOWWIN	https://www.epa.gov/tsca-screening-tools/epi-suitetm-estimation-program-interface	estimates octanol-water partition coefficient of toxic chemicals
20	AMBIT	http://ambit.sourceforge.net/intro.html	Chemical structure search, experimental data and predictive model can be obtained
21	PreADMET	https://preadmet.bmdrc.kr/	A web-based application for predicting ADME data and also toxicity prediction

Table 8. Standard list of software tools details used for QSAR analysis. The programs are basically data mining tools involves in classification, clustering, modeling, validation of the model

Sl. No.	Simulation Software	Availability
1	Molecular Operating Environment (MOE)	https://www.chemcomp.com/MOE-Cheminformatics_and_QSAR.htm
2	BIOVIA QSAR Workbench	http://accelrys.com/products/collaborative-science/biovia-qsar-workbench/
3	VEGAHUB	https://www.vegahub.eu/
4	WEKA	http://www.cs.waikato.ac.nz/ml/weka/
5	KNIME	https://www.knime.org
6	BuildQSAR	http://www.profanderson.net/files/buildqsar.php
7	Orange	http://orange.biolab.si/
8	Rapid Miner	https://rapidminer.com/
9	MALLET	http://mallet.cs.umass.edu/
10	R	www.r-project.org

potential persistent organic pollutants per fluorinated for its importance and possible risk in the aquatic environment (Chen *et al.*, 2013).Similarly, the chronic toxicity of 12 compounds of parabens and their chlorinated by-products was investigated using 7-day *Ceriodaphnia dubia* survival test to generate information on how to disinfect by-products of preservatives that are discharged in aquatic systems (Terasaki *et al.,* 2015).The chronic aquatic toxicities of an antimicrobial halogenated substance 2,2-dibromo-3-nitrilopropionamide (DBNPA) in *Daphnia magna* and rainbow trout were evaluated and observed about its significant effect on the reproduction and survival of *D. Magna* (Chen, 2012). An extensive work by Agatonovic-Kustrin *et al.* (2014) has resulted an in-silico QSAR model, that predict the aquatic toxicity of pesticides by considering experimentally calculated lethal dose (LD50) parameter for fish, without use of in vivo testing procedure (Agatonovic-Kustrin*et al.,* 2014). Since the variable selection is important in QSAR process, to address this a hierarchical approach was implemented for the prediction of acute aquatic toxicity and the model was developed. In the model a hybrid form (counter propagation neural network that is coupled with genetic algorithms) of algorithm was used successfully for the variable selection (Mazzatorta *et al.*, 2005). During the risk assessment process of pesticides, a fragment-based QSAR approach was used by taking LC 50 calculated experimentally during 96 h in case of rainbow trout. By considering the dependent variable the acute molecular toxicity was predicted (Casalegno *et al.,* 2006). A similar method was implemented for the prediction of based on the experimental LC50 values of 258 chemicals (including halogenated substances) available in the literature. The toxicity prediction accuracy was predicted and validated based on the value correlation coefficient of 0.9495 (Tao *et al.,* 2002).In an another experiment, the cytotoxicity of a series of small chlorinated alkanes was analysed by considering in case Chinese hamster ovary (CHO) cells as well as for the fish comparatively by QSAR methods (Zvinavashe *et al.,* 2008). The acute toxicity of the compounds such as aniline, 4-chloroaniline, 3,5-dichloroaniline and 2,3,4-chloroaniline was tested in one prokaryote species (*Escherichia coli*) and three eukaryote aquatic species (*Pseudokirchneriella subcapitata*, *Daphnia magna*, and *Daniorerio*) by using SAR methods (Dom *et al.,* 2010). The diverse biological activities of substituted phenols (X-phenols) in organisms such as from protozoa to animals are comparatively analyzed. Correlations between biological activities and physicochemical attributes of X-phenols reveal strong consistencies about the parameters specifically, hydrophobicity and electronic terms dominate the extent of interactions between these chemical entities and their molecular targets leads to interpreting about molecular interaction involved in the binding of toxic substances (Selassie & Verma, 2015).The models for acute to the chronic estimation of the *Daphnia magna* reproductive toxicities of chemical substances from their *Daphnia magna* acute immobilization toxicities was analyzed by quantitative structure–activity–activity relationship (QSAAR) models. The models combined the acute toxicities with structural and physicochemical descriptors (Furuhama *et al.,* 2016). Many currently existing QSAR models can predict the effects of a wide range of substances to biota, particularly aquatic biota. The difficulty for regulatory programs is in choosing the appropriate QSAR model or models for application in their new and existing substances programs. The performance of six QSAR modeling packages has been comparatively analysed such as: Ecological Structure Activity Relationship (ECOSAR), TOPKAT, a Probabilistic Neural Network (PNN), a Computational Neural Network (CNN), the QSAR components of the Assessment Tools for the Evaluation of Risk (ASTER) system, and the Optimized Approach Based on Structural Indices Set (OASIS) system. By using a testing data set of 130 substances the 96-h median lethal concentrations (LC50s) to fathead minnows was used as dependent variable. From the analysis, it was analyzed that, the TOPKAT based analysis was obtained as the best during accessing the performance of the model building task (Moore *et al.,* 2003). Another study was performed to analyse the chronic toxicity predictions and extrapolations for a set of chlorinated compounds such as aniline, 4-chloroaniline, 3,5-dichloroanilineand 2,3,4-trichloroaniline (Dom *et al.,* 2012).A review that provides a brief overview of four databases created for the purpose of developing QSARs for estimating the toxicity of chemicals to aquatic organisms in terms of chemical-specific mode of action classification and associated QSAR selection for estimating potential toxicological effects of organic chemicals is presented (Bradbury *et al.,* 2003).The establishment of toxicologically-credible techniques such as QSAR has been utilized to assess the mode of toxic action from the chemical structure by using toxicodynamic knowledge have been established for use in predictive aquatic toxicology applications (Bradbury, 1994).

CONCLUSION

The uncontrolled human activities have introduced thousands of halogenated substances into aquatic systems throughout the world, which significantly affect the water quality. The number of these chemicals used in industry is increasing, and as a consequence workers in chemical industries are thought to have many opportunities for being exposed to chemicals. The use of quantitative structure-activity relationships (QSARs) is used as an alternative to the difficult experimental methods to accesses the potential toxic effects of organic chemicals on aquatic organisms. Therefore, the regulatory agencies have used the QSAR based approach as an essential tools to help prioritize the risk assessments when empirical data are not available to evaluate toxicological effects. Especially in the aquatic toxicology, quantitative structure-activity relationships (QSARs) have developed as scientifically-credible tools for predicting the toxicity of chemicals when little or no empirical data are available. This chapter specifically reviewed about the basic approach to measure to represent the aquatic toxicity of some halogenated compounds along with the computational resources used to predict the same.

REFERENCES

Agatonovic-Kustrin, S. W., Morton, D., & Razic, S. (2014). In silico modelling of pesticide aquatic toxicity. *Combinatorial Chemistry & High Throughput Screening*, *17*(9), 808–818. doi:10.2174/13862 07317666141021110738 PMID:25335880

Akers, K. S., Sinks, G. D., & Schultz, T. W. (1999). Structure–toxicity relationships for selected halogenated aliphatic chemicals. *Environmental Toxicology and Pharmacology*, *7*(1), 33–39. doi:10.1016/S1382-6689(98)00048-9 PMID:21781907

Ali, M., & Sreekrishnan, T. R. (2001). Aquatic toxicity from pulp and paper mill effluents: A review. *Advances in Environmental Research*, *5*(2), 175–196. doi:10.1016/S1093-0191(00)00055-1

Altenburger, R., Schmitt, H., & Schüürmann, G. (2005). Algal toxicity of nitrobenzenes: Combined effect analysis as a pharmacological probe for similar modes of interaction. *Environmental Toxicology and Chemistry*, *24*(2), 324–333. doi:10.1897/04-032R.1 PMID:15719992

Arnot, J. A., & Gobas, F. A. (2006). A review of bioconcentration factor (BCF) and bioaccumulation factor (BAF) assessments for organic chemicals in aquatic organisms. *Environmental Reviews*, *14*(4), 257–297. doi:10.1139/a06-005

Artemenko, A. G., Muratov, E. N., Kuz'min, V. E., Muratov, N. N., Varlamova, E. V., Kuz'mina, A. V., ... Tropsha, A. (2011). QSAR analysis of the toxicity of nitroaromatics in Tetrahymena pyriformis: Structural factors and possible modes of action. *SAR and QSAR in Environmental Research*, *22*(5-6), 575–601. doi:10.1080/1062936X.2011.569950 PMID:21714735

Aruoja, V., Moosus, M., Kahru, A., Sihtmäe, M., & Maran, U. (2014). Measurement of baseline toxicity and QSAR analysis of 50 non-polar and 58 polar narcotic chemicals for the alga Pseudokirchneriella subcapitata. *Chemosphere*, *96*, 23–32. doi:10.1016/j.chemosphere.2013.06.088 PMID:23895738

Barron, M. G. (2003). Bioaccumulation and bioconcentration in aquatic organisms. In Handbook of ecotoxicology (2nd ed.). Lewis Publishers.

Bradbury, S. P. (1994). Predicting modes of toxic action from chemical structure: An overview. *SAR and QSAR in Environmental Research*, *2*(1-2), 89–104. doi:10.1080/10629369408028842 PMID:8790641

Bradbury, S. P. (1995). Quantitative structure-activity relationships and ecological risk assessment: An overview of predictive aquatic toxicology research. *Toxicology Letters*, *79*(1-3), 229–237. doi:10.1016/0378-4274(95)03374-T PMID:7570660

Bradbury, S. P., Russom, C. L., Ankley, G. T., Schultz, T. W., & Walker, J. D. (2003). Overview of data and conceptual approaches for derivation of quantitative structure-activity relationships for ecotoxicological effects of organic chemicals. *Environmental Toxicology and Chemistry*, *22*(8), 1789–1798. doi:10.1897/01-234 PMID:12924578

Casalegno, M., Sello, G., & Benfenati, E. (2006). Top-priority fragment QSAR approach in predicting pesticide aquatic toxicity. *Chemical Research in Toxicology*, *19*(11), 1533–1539. doi:10.1021/tx0601814 PMID:17112242

Chen, F. (2012). The chronic aquatic toxicity of a microbicide dibromo nitrilopropionamide. *Toxicology and Industrial Health*, *28*(2), 181–185. doi:10.1177/0748233711410904 PMID:21937531

Chen, H., Yao, J., Wang, F., Cai, M., & Liu, H. (2013). Toxicity of perfluorooctanoic acid to Pseudomonas putida in the aquatic environment. *Journal of Hazardous Materials*, *262*, 726–731. doi:10.1016/j.jhazmat.2013.09.046 PMID:24140521

Cherkasov, A., Muratov, E. N., Fourches, D., Varnek, A., Baskin, I. I., Cronin, M., & Consonni, V. (2014). QSAR modeling: Where have you been? Where are you going to? *Journal of Medicinal Chemistry*, *57*(12), 4977–5010. doi:10.1021/jm4004285 PMID:24351051

Chicu, S. A., Herrmann, K., & Berking, S. (2000). An approach to calculate the toxicity of simple organic molecules on the basis of QSAR analysis in Hydractinia echinata (Hydrozoa, Cnidaria). *Molecular Informatics*, *19*(3), 227–236.

Connors, K. A., Voutchkova-Kostal, A. M., Kostal, J., Anastas, P., Zimmerman, J. B., & Brooks, B. W. (2014). Reducing aquatic hazards of industrial chemicals: Probabilistic assessment of sustainable molecular design guidelines. *Environmental Toxicology and Chemistry*, *33*(8), 1894–1902. doi:10.1002/etc.2614 PMID:24839109

Cramer, R. D., Patterson, D. E., & Bunce, J. D. (1988). Comparative molecular field analysis (CoMFA). 1. Effect of shape on binding of steroids to carrier proteins. *Journal of the American Chemical Society*, *110*(18), 5959–5967. doi:10.1021/ja00226a005 PMID:22148765

Damborsky, J., & Schultz, T. W. (1997). Comparison of the QSAR models for toxicity and biodegradability of anilines and phenols. *Chemosphere*, *34*(2), 429–446. doi:10.1016/S0045-6535(96)00361-X PMID:9057301

Damstra, T. (2002). Potential effects of certain persistent organic pollutants and endocrine disrupting chemicals on the health of children. *Journal of Toxicology. Clinical Toxicology*, *40*(4), 457–465. doi:10.1081/CLT-120006748 PMID:12216998

de Bruijn, J., & Hermens, J. (1993). Inhibition of acetylcholinesterase and acute toxicity of organophosphorous compounds to fish: A preliminary structure-activity analysis. *Aquatic Toxicology (Amsterdam, Netherlands)*, *24*(3-4), 257–274. doi:10.1016/0166-445X(93)90075-C

Deeb, O., & Goodarzi, M. (2012). In silico quantitative structure toxicity relationship of chemical compounds: Some case studies. *Current Drug Safety*, *7*(4), 289–297. doi:10.2174/157488612804096533 PMID:23062241

Deneer, J. W. (2000). Toxicity of mixtures of pesticides in aquatic systems. *Pest Management Science*, *56*(6), 516–520. doi:10.1002/(SICI)1526-4998(200006)56:6<516::AID-PS163>3.0.CO;2-0

Dewan, P., Jain, V., Gupta, P., & Banerjee, B. D. (2013). Organochlorine pesticide residues in maternal blood, cord blood, placenta, and breast milk and their relation to birth size. *Chemosphere*, *90*(5), 1704–1710. doi:10.1016/j.chemosphere.2012.09.083 PMID:23141556

Dom, N., Knapen, D., Benoot, D., Nobels, I., & Blust, R. (2010). Aquatic multi-species acute toxicity of (chlorinated) anilines: Experimental versus predicted data. *Chemosphere*, *81*(2), 177–186. doi:10.1016/j.chemosphere.2010.06.059 PMID:20637490

Dom, N., Knapen, D., & Blust, R. (2012). Assessment of aquatic experimental versus predicted and extrapolated chronic toxicity data of four structural analogues. *Chemosphere*, *86*(1), 56–64. doi:10.1016/j.chemosphere.2011.08.050 PMID:21944038

Dudek, A. Z., Arodz, T., & Gálvez, J. (2006). Computational methods in developing quantitative structure-activity relationships (QSAR): A review. *Combinatorial Chemistry & High Throughput Screening*, *9*(3), 213–228. doi:10.2174/138620706776055539 PMID:16533155

Furuhama, A., Hayashi, T. I., & Tatarazako, N. (2016). Acute to chronic estimation of Daphnia magna toxicity within the QSAAR framework. *SAR and QSAR in Environmental Research*, *27*(10), 833–850. doi:10.1080/1062936X.2016.1243151 PMID:27767343

Ghamali, M., Chtita, S., Ousaa, A., Elidrissi, B., Bouachrine, M., & Lakhlifi, T. (2017). QSAR analysis of the toxicity of phenols and thiophenols using MLR and ANN. *Journal of Taibah University for Science*, *11*(1), 1–10. doi:10.1016/j.jtusci.2016.03.002

Gramatica, P. (2007). Principles of QSAR models validation: Internal and external. *Molecular Informatics*, *26*(5), 694–701.

Gribble, G. W. (1994). The natural production of chlorinated compounds. *Environmental Science & Technology*, *28*(7), 310A–319A. doi:10.1021/es00056a712 PMID:22662801

Hewitt, M., Ellison, C. M., Enoch, S. J., Madden, J. C., & Cronin, M. T. D. (2010). Integrating (Q) SAR models, expert systems and read-across approaches for the prediction of developmental toxicity. *Reproductive Toxicology (Elmsford, N.Y.)*, *30*(1), 147–160. doi:10.1016/j.reprotox.2009.12.003 PMID:20006701

Hodges, G., Roberts, D. W., Marshall, S. J., & Dearden, J. C. (2006). The aquatic toxicity of anionic surfactants to Daphnia magna—a comparative QSAR study of linear alkylbenzene sulphonates and ester sulphonates. *Chemosphere*, *63*(9), 1443–1450. doi:10.1016/j.chemosphere.2005.10.001 PMID:16293293

Jana, B. B. (1994). Deposition of pollutants and their impacts on fisheries. *Environmental Monitoring, 2*.

Katritzky, A. R., & Gordeeva, E. V. (1993). Traditional topological indexes vs electronic, geometrical, and combined molecular descriptors in QSAR/QSPR research. *Journal of Chemical Information and Computer Sciences*, *33*(6), 835–857. doi:10.1021/ci00016a005 PMID:8113335

La Farre, M., Pérez, S., Kantiani, L., & Barceló, D. (2008). Fate and toxicity of emerging pollutants, their metabolites and transformation products in the aquatic environment. *TrAC Trends in Analytical Chemistry*, *27*(11), 991–1007. doi:10.1016/j.trac.2008.09.010

Landrum, P. F., Lydy, M. J., & Lee, H. II. (1992). Toxicokinetics in aquatic systems: Model comparisons and use in hazard assessment. *Environmental Toxicology and Chemistry*, *11*(12), 1709–1725. doi:10.1002/etc.5620111205

Livingstone, D. J., & Manallack, D. T. (1993). Statistics using neural networks: Chance effects. *Journal of Medicinal Chemistry*, *36*(9), 1295–1297. doi:10.1021/jm00061a023 PMID:8487267

Mazzatorta, P., Smiesko, M., Lo Piparo, E., & Benfenati, E. (2005). QSAR model for predicting pesticide aquatic toxicity. *Journal of Chemical Information and Modeling*, *45*(6), 1767–1774. doi:10.1021/ci0502471 PMID:16309283

McKim, J. M., Bradbury, S. P., & Niemi, G. J. (1987). Fish acute toxicity syndromes and their use in the QSAR approach to hazard assessment. *Environmental Health Perspectives*, *71*, 171–186. doi:10.1289/ehp.8771171 PMID:3297660

Moore, D. R., Breton, R. L., & MacDonald, D. B. (2003). A comparison of model performance for six quantitative structure-activity relationship packages that predict acute toxicity to fish. *Environmental Toxicology and Chemistry*, *22*(8), 1799–1809. doi:10.1897/00-361 PMID:12924579

Netzeva, T. I., Dearden, J. C., Edwards, R., Worgan, A. D., & Cronin, M. T. (2004). QSAR Analysis of the Toxicity of Aromatic Compounds to Chlorella v ulgaris in a Novel Short-Term Assay. *Journal of Chemical Information and Computer Sciences*, *44*(1), 258–265. doi:10.1021/ci034195g PMID:14741035

Neuwoehner, J., Zilberman, T., Fenner, K., & Escher, B. I. (2010). QSAR-analysis and mixture toxicity as diagnostic tools: Influence of degradation on the toxicity and mode of action of diuron in algae and daphnids. *Aquatic Toxicology (Amsterdam, Netherlands)*, *97*(1), 58–67. doi:10.1016/j.aquatox.2009.12.005 PMID:20044149

Nikinmaa, M. (2014). *An introduction to aquatic toxicology*. Elsevier. doi:10.1016/B978-0-12-411574-3.00002-5

Papa, E., Villa, F., & Gramatica, P. (2005). Statistically Validated QSARs, Based on Theoretical Descriptors, for Modeling Aquatic Toxicity of Organic Chemicals in Pimephales p romelas (Fathead Minnow). *Journal of Chemical Information and Modeling*, *45*(5), 1256–1266. doi:10.1021/ci050212l PMID:16180902

Perocco, P., Bolognesi, S., & Alberghini, W. (1983). Toxic activity of seventeen industrial solvents and halogenated compounds on human lymphocytes cultured in vitro. *Toxicology Letters*, *16*(1), 69–75. doi:10.1016/0378-4274(83)90012-7 PMID:6836616

Persistent, P. B. (2000). Environmental Medicine, Part 4: Pesticides–Biologically Persistent and Ubiquitous Toxins. *Alternative Medicine Review*, *5*(5), 432–447. PMID:11056413

Rowe, P. H. (2010). *Statistical methods for continuous measured endpoints in in silico toxicology. In Silico Toxicology: Principles and Applications* (pp. 228–251). Cambridge, UK: The Royal Society of Chemistry.

Satpathy, R. (2016). Quantitative structure activity relationship (QSAR) study of some haloalkane compound properties with their aquatic toxicity feature. In Advanced engineering: Current prospective (pp. 111-119). Research India Publication.

Satpathy, R. (2018). Quantitative Structure-Activity Modelling of Toxic Compounds. In *Nanotechnology, Food Security and Water Treatment* (pp. 313–331). Cham: Springer. doi:10.1007/978-3-319-70166-0_10

Selassie, C., & Verma, R. P. (2015). QSAR of toxicology of substituted phenols. *Journal of Pesticide Science*, *40*(1), 1–12. doi:10.1584/jpestics.D14-097

Seward, J. R., & Schultz, T. W. (1999). QSAR analyses of the toxicity of aliphatic carboxylic acids and salts to Tetrahymena pyriformis. *SAR and QSAR in Environmental Research, 10*(6), 557–567. doi:10.1080/10629369908033224 PMID:10674293

Song, B., Palleroni, N. J., & Häggblom, M. M. (2000). Isolation and characterization of diverse halobenzoate-degrading denitrifying bacteria from soils and sediments. *Applied and Environmental Microbiology, 66*(8), 3446–3453. doi:10.1128/AEM.66.8.3446-3453.2000 PMID:10919805

Tao, S., Xi, X., Xu, F., & Dawson, R. (2002). A QSAR model for predicting toxicity (LC 50) to rainbow trout. *Water Research, 36*(11), 2926–2930. doi:10.1016/S0043-1354(01)00514-0 PMID:12146883

Terasaki, M., Abe, R., Makino, M., & Tatarazako, N. (2015). Chronic toxicity of parabens and their chlorinated by-products in Ceriodaphniadubia. *Environmental Toxicology, 30*(6), 664–673. doi:10.1002/tox.21944 PMID:24376163

Todeschini, R., & Consonni, V. (2009). Molecular descriptors for chemoinformatics: Vol. 41. *2 volume set.* John Wiley & Sons.

Verhaar, H. J., Van Leeuwen, C. J., & Hermens, J. L. (1992). Classifying environmental pollutants. *Chemosphere, 25*(4), 471–491. doi:10.1016/0045-6535(92)90280-5 PMID:10718581

Walker, M. K., & Peterson, R. E. (1994). Aquatic toxicity of dioxins and related chemicals. In Dioxins and Health (pp. 347-387). Springer US. doi:10.1007/978-1-4899-1462-0_11

Wong, D. C., Dorn, P. B., & Chai, E. Y. (1997). Acute toxicity and structure-activity relationships of nine alcohol ethoxylate surfactants to fathead minnow and Daphnia magna. *Environmental Toxicology and Chemistry, 16*(9), 1970–1976.

Zhu, H. (2013). From QSAR to QSIIR: searching for enhanced computational toxicology models In Computational Toxicology (vol. 930, pp. 53-65). New York: Humana Press. doi:10.1007/978-1-62703-059-5_3

Zvinavashe, E., van den Berg, H., Soffers, A. E., Vervoort, J., Freidig, A., Murk, A. J., & Rietjens, I. M. (2008). QSAR models for predicting in vivo aquatic toxicity of chlorinated alkanes to fish. *Chemical Research in Toxicology, 21*(3), 739–745. doi:10.1021/tx700367c PMID:18254607

ADDITIONAL READING

Bedard, D. L. (2003). Dehalogenation Microbial Processes And Environmental Applications. *Polychlorinated Biphenyls In Aquatic Sediments: Environmental Fate And Outlook For Biological Treatment.* DOI: 10.1007/b101852

Nantasenamat, C., Isarankura-Na-Ayudhya, C., Naenna, T., & Prachayasittikul, V. (2009). A Practical Overview of Quantitative Structure-Activity Relationship. *EXCLI Journal, 8*, 74–88.

Puzyn, T., Leszczynski, J., & Cronin, M. T. (Eds.). (2010). Recent advances in qsar studies: methods and applications (vol. 8). springer science & business media. DOI: doi:10.1007/978-1-4020-9783-6

KEY TERMS AND DEFINITIONS

Aquatic Toxicity: Refers to the effects of a compound to organisms living in the water and is usually determined on organisms representing, usually considering in the three trophic levels, such as vertebrates (fish), invertebrates (crustaceans as *Daphnia* spp.), and algae.

Computational Techniques: Computational tools are the implemented techniques in computers to solve problems by either step-wise, repeated, and iterative solution methods; also known as in-silico methods.

EC50: (Effective Concentration 50): This refers to the concentration of a toxic substances that indices 50% of mortality in cells after a specified exposure time.

Halogenated Compounds: Halogenated compounds are normally man-made (xenobiotic).Their toxicity increases with an increasing number of halogen moieties as well as the increasing in number of aromatic rings in the molecule.

IGC50 (50% Inhibition Growth Concentration): IGC 50 inhibits the growth of cells by 50%.

Kow and Pow (N-Octanol-Water Partition Coefficients): This is widely used property for assessing the partitioning behavior of chemicals in the environment to estimate the fate, behavior, and effects of toxic chemicals in the environment.

LC 50 (Lethal Concentration 50): LC50 value is the concentration of a material in air that will kill 50% of the test subjects (animals, typically mice or rats) when administered as a single exposure (typically 1 or 4 hours).

Machine Learning Methods: Obtained from the study of pattern recognition and computational learning theory in artificial intelligence.

Model Validation: Model validation is verification of the generated model under a guided frame work to access the performance to a desired level.

Molecular Descriptor: The molecular descriptor is quantitative chemical information encoded within a symbolic representation which is the result of some standardized experiment.

QSAR Models: Quantitative structure-activity relationship models are regression or classification models used in the chemical and biological sciences and engineering to predict the effect by considering suitable dependent and independent variable.

Statistical Modeling: A statistical model is a class of mathematical model that provides a set of assumptions concerning the generation of some sample data.

Toxic Substances: A substance that can be lethal or cause health hazards when uptake by the cell or body.

Chapter 19
Water Purification Using Different Chemical Treatment

Rupali Rastogi
ITM University, India

ABSTRACT

Water from surface sources is often contaminated by microbes, whereas groundwater is normally safer, but even groundwater can be contaminated by harmful chemicals from human activities or from the natural environment. The purification process of water may reduce the concentration of particulate matter including suspended particles, parasites, bacteria, algae, viruses, fungi, and a range of dissolved and particulate material derived from the surfaces. Water purification is the process of removing undesirable chemicals, materials, and biological contaminants from contaminated water. Most water is purified for human consumption (drinking water), but water purification may also be designed for a variety of other purposes, such as medical, pharmacology, chemical, and industrial applications. In general, the methods used include physical processes such as filtration and sedimentation, biological processes such as slow sand filters or activated sludge, chemical processes such as flocculation and chlorination, and the use of electromagnetic radiation such as ultraviolet light.

INTRODUCTION

Even though over 70% of the Earth is covered with water, only 3% is fit for human consumption, of which two thirds is comprised of frozen and largely uninhabited ice caps and glaciers, leaving 1% available for consumption. Remaining 97% is saltwater, which cannot be used for agriculture or drinking. Unique qualities and properties of water makes it so important and basic need of life. The cells in our bodies are full of water. The excellent ability of water is to dissolve so many substances allows our cells to use valuable nutrients, minerals, and chemicals in biological processes. Water's "stickiness" (from surface tension) plays a part in the body's ability to transport these materials throughout our bodies. Carbohydrates and proteins are used as food which are metabolized and transported by water in the bloodstream. No less important is the ability of water to transport waste material out of our bodies.

Water facts:

DOI: 10.4018/978-1-5225-6111-8.ch019

- Water is also used to fight forest fires; yet, we spray water on coal in a furnace to make it burn better.
- Chemically, water is basically hydrogen oxide.
- At a temperature of 2900°C, substances that contain water cannot be forced to part with it; because others that do not contain water will liberate it when even slightly heated.
- Water is virtually incompressible; as it freezes, it expands by an 11th of its volume.

Treatment of water requires chemical, physical, and sometimes biological processes to remove contaminants. The more common processes used in portable water treatment are the chemical and physical processes. Biological processes for treatment of water. However, the slow sand filteration process is a biological process that has been historically used to remove pathogens from potable water. The biological activated carbon (BAC) process is also a biological process that is used to remove organic contaminants from potable water.

The chemical processes involved in potable water treatment include oxidation, coagulation and disinfection. The physical processes include flocculation, sedimentation, filtration, adsorption, and disinfection using ultraviolet light. The types of processes that are required and the order in which they are used depend on the types and concentrations of contaminants that can be removed. Examples of this include oxidation followed by filtration or sedimentation, followed by filtration. One of the example, is the oxidation process causes the dissolved contaminants to form a precipitate, which can be removed by filtration. Other example is, sedimentation that removes most of the solids by gravity and reduces the solids loading on the down stream filtration process.

The following section describe a brief introduction to each of these basic water treatment processes. Each process will be presented in the order that they are normally used in a treatment train.

OXIDATION

Chemical oxidation is used in water treatment to aid in the removal of inorganic contaminants such as iron (Fe^{2+}), manganese (Mn^{2+}), and arsenic (As^{3+}) to improve removal of particles by coagulation or to destroy taste and odour causing compounds. Oxidation can also be used prior to coagulation, filtration, adsorption, or sedimentation to improve the removal of inorganic, particulates, taste, or odour.

Oxidants

Most commonly used oxidants in small systems include chlorine (Cl_2) and potassium permanganate ($KMnO_4$). To a lesser extent, ozone and chlorine dioxide are also used for this purpose. Chlorine can combine in gas, solid, and liquid forms; and potassium permanganate is usually supplied as a fine granular solid material that is dissolved in water. Ozone is generated onsite using pure oxygen or air. Most desirable oxidants are dependent upon a number of factors, including process requirements, operational cost, chemical safety, and operational complexity.

Mixing

Oxidants are injected as a gas or a liquid. Mixing or diffusion of the gas or liquid into the water stream generally occurs very quickly; and therefore, mixing energy is rarely a significant issue for small systems. As a result, static or mechanical mixers are typically not required, although diffusers or injector assemblies are often used to enhance the diffusion of the oxidant into the water.

Advanced Oxidation Process

Advanced oxidation processes refers to a set of chemical treatment techniques designed to remove organic and inorganic materials in wastewater. One such type of process is called In Situ Chemical oxidation. Contaminants are oxidized by four different reagents: ozone, H_2O_2, O_2, and air. These technique can be combined with UV irradiation and specific catalysts. This will lead to formation of hydroxyl radicals. A commonly known example of AOP is the use of fenton's reagent (Abbas & Zaheer, 2014). The AOP procedure is basically useful for cleaning biologically toxic or non-degradable materials such as aromatics, pesticides, petroleum constituents, and volatile organic matter in wastewater. The contaminants are converted to a large extent into stable inorganic compounds such as water, carbon dioxide and salts. The main mechanism of AOPs is generally the formation of highly reactive free radicals. Hydroxyl radicals are effective in destroying organic chemicals because they are reactive electrophiles that rapidly reacts with electron –rich organic contaminants.

Fenton's Process

It has been demonstrated that Fenton's reagent can destroy toxic compounds in waste waters such as phenols and herbicides. Generation of OH radical occurs by means of addition of H_2O_2 to Fe^{2+} salts. Production of OH radicals by Fenton reagent occurs by means of addition of H_2O_2 to Fe^{2+} salts. (Haber & Weiss)

$$H_2O_2 + Fe^{2+} \rightarrow OH^{\bullet} + OH^{-} + Fe^{3+} \quad (1)$$

This is a very simple way of producing OH radicals neither special reactants nor special apparatus being required. The oxidation using Fenton's reagent has been proven to be a promising and attractive treatment method for the effective decolourization and degradation of dyes (Wang et al., 2005). The Fenton system uses ferrous ions to react with hydrogen peroxide, producing hydroxyl radicals with powerful oxidizing abilities to degrade certain toxic contaminants(Titus et al., 2004). Hydroxyl radicals may react with ferrous ions to form ferric ions or react with organics:

$$OH^{\bullet} + Fe^{2+} \rightarrow OH^{-} + Fe^{3+} \quad (2)$$

$$OH^{\bullet} + \text{organics} \rightarrow \text{products} \quad (3)$$

Hydroxyl radicals can also react with hydrogen peroxide to produce other radicals, and may also combine with each other to produce hydrogen peroxide, which are shown below (4):

$$OH^{\bullet} + H_2O_2 \rightarrow H_2O + HO_2^{\bullet} \quad (4)$$

$$OH^{\bullet} + OH^{\bullet} \rightarrow H_2O_2 \quad (5)$$

Ferrous ions and radicals are produced during the reactions. The reactions are shown in Equations (6)–(9):

$$H_2O_2 + Fe^{3+} \leftrightarrow H^+ + FeOOH^{2+} \quad (6)$$

$$FeOOH^{2+} \rightarrow HO^{2\bullet} + Fe^{2+} \quad (7)$$

$$HO_2^{\bullet} + Fe^{2+} \rightarrow HO^{2-} + Fe^{3+} \quad (8)$$

$$HO_2^{\bullet} + Fe^{3+} \rightarrow O_2 + Fe^{2+} + H^+ \quad (9)$$

Electro-Fenton Process

There is a need to develop effective methods for the degradation of such organic pollutants, either to less harmful compounds or, more desirable, to their complete decolorization. Recently, mainly because of its amenability to automation, high efficiency and environmental compatibility, there is a growing interest in the use of effective direct or indirect electrochemical degradation of organic pollutants in waters (Trasatti, 2013). In the presence of ferrous ions and in acidic aqueous medium the oxidation power will be enhanced due to the production of a very reactive one-electron oxidizing agent hydroxyl radical (•OH) from the Fenton reaction. This electro Fenton process can generate •OH by the simultaneous electrochemical reduction of O_2 in the presence of catalytic amounts of ferrous ions (Wang et al., 2005)

$$O_2 + 2H^+ + 2e- \rightarrow H_2O_2 \quad (10)$$

$$Fe^{2+} + H_2O_2 + H^+ \rightarrow Fe^{3+} + H_2O + OH^{\bullet} \quad (11)$$

$$Fe^{3+} + e- \rightarrow Fe^{2+} \quad (12)$$

$$H_2O \rightarrow 1/2O_2 + 2H^+ + 2e- \quad (13)$$

Photo-Fenton Process

In photo-Fenton process in addition to the above Fenton reactions the formation of hydroxyl radical also occurs by the following reactions (Equations (14) and (15)):

$$H_2O_2 + UV \rightarrow OH^{\bullet} + OH^{\bullet} \quad (14)$$

$$Fe^{3+} + H_2O + UV \rightarrow OH^{\bullet} + Fe^{2+} + H^+ \quad (15)$$

The addition of UV to Fenton's process could be an interesting allied in dye decolorization due to its capacity to influence the direct formation of •OH radicals (Malik & Saha, 2003). The rate of organic pollutant degradation could be increased by irradiation of Fenton with UV light (photo-Fenton process). UV light leads not only to the formation of additional hydroxyl radicals but also to recycling of ferrous catalyst by reduction of Fe^{3+}. In this way, the concentration of Fe^{2+} is increased and the overall reaction is accelerated. Among the AOPs, the oxidation using Fenton's reagent and photo-Fenton's reagent has been found to be a promising and attractive treatment method for the effective decolorization and degradation of dyes. (Malik & Saha, 2003) reported that the removal rate is strongly dependent on the initial concentration of the dye, Fe^{2+} and H_2O_2. Muruganandham and Swaminathan (2004) have carried out studies where similar results were obtained; they suggested a pH of 3 is optimum for fenton and photo-fenton processes.

Advantages of Advanced Oxidation Processes

1. Rapid reaction rates.
2. Small footprint.
3. Potential to reduce toxicity and possibly complete mineralization of organics treated.
4. It does not concentrate waste for further treatment with methods such as membranes.
5. It does not produce materials that require further treatment such as "spent carbon" from activated carbon absorption.
6. It does create sludge as with physical chemical process or biological processes (wasted biological sludge).

Disadvantages of Advanced Oxidation Processes

1. Intensive Capital.
2. Chemistry of complex must be tailored to specific application.
3. For some applications quenching of excess peroxide is require.

Destabilisation is achieved through the addition of chemicals (called coagulants) to the water. Most organic and inorganic material suspended in water and not dissolved will settle out if given enough time.

Coagulation is the process by means of which the colloidal particles in water are destabilised (i.e. the nature of the colloidal particles is changed) so that they form flocs through the process of flocculation that can be readily separated from the water. Destabilisation can be achieved through the addition of chemicals (called coagulants) to the water. Most organic and inorganic material suspended in water and not dissolved will settle out if given enough time.

There are two types of colloidal material:

- **Hydrophobic:** Hydrophobic means water-fearing. Figure 1 Hydrophobic colloidal material is mostly inorganic material that contributes to turbidity and carries a negative electrical surface charge.

Figure 1. Hydrophobic

Figure 2. Hydrophilic

- **Hydrophllic:** Hydrophilic means water-loving. Figure 2 Hydrophilic colloidal material is mostly composed of organic material that is the common source of color in water. Hydrophilic compounds are surrounded by water molecules that tend to make these particles negatively charged as well.

Different chemicals can be used as coagulants. The most common coagulants are:

- Aluminium sulphate, also known as alum $Al_2(SO_4)_3.16H_2O$. The alum is dissolved in water and the aluminium ions, Al^{3+} that form, have a high capacity to neutralise the negative charges which are carried by the colloidal particles and which contribute to their stability. The aluminium ions hydrolise and in the process form aluminium hydroxide, $Al(OH)_3$ which precipitates as a solid.

NOTE Since aluminium may be harmful at high concentrations it must be allowed to precipitate completely as the hydroxide. Complete precipitation is a function of the pH of the water and the pH must therefore be closely controlled between 6,0 and 7,4.

- Ferric chloride, $FeCl_3$ is also commonly used as coagulant. When added to water, the iron precipitates as ferric hydroxide, $Fe(OH)_3$ and the hydroxide flocs enmesh the colloidal particles in the same way as the aluminium hydroxide flocs do. The optimum pH for precipitation of iron is not as critical as with aluminium and pH values of between 5 and 8 give good precipitation.

- Hydrated lime can also be used as coagulant, but its action is different to that of alum and ferric chloride. When lime is added to water the pH increases. This results in the formation of carbonate ions from the natural alkalinity in the water.
- Polymeric coagulants including Dadmacs and polyamines which form white or brown flocs when added to water.
- Polyelectrolytes are mostly used to assist in the flocculation process and are often called flocculation aids. They are polymeric organic compounds consisting of long polymer chains that act to enmesh particles in the water.

Other coagulants are also sometimes used in water treatment. These include:

- Aluminium polymers such as poly-aluminium chloride that provide rapid flocculation, efficient removal of organics, and less sludge than alum under certain conditions, but at a higher cost.
- Activated silica is sometimes used as a flocculant together with alum or hydrated lime as coagulant.
- Bentonite and/or kaolin are sometimes added to water when the water to be flocculated contains too few particles for effective flocculation.

Coagulation is a process of addition of metal salts or polymers to water containing negatively charged contaminants. The positively charged aluminum ions then become attached to the surface of the negatively charged colloid. The overall result is the reduction of the negative surface charges, subsequent formation of agglomerate (floc). This destabilizing factor is the major contribution that coagulation makes to the removal of turbidity, color, and microorganisms.

Flocculation follows coagulation (and is often regarded as part of one process: coagulation-flocculation). The aim of flocculation is to cause the individual destabilised colloidal particles to collide with one another and with the precipitate formed by the coagulant in order to form aggregates which could easily be removed by means of sedimentation or flotation. Flocculation involves the stirring of water to which a coagulant has been added at a slow rate, causing the individual particles to "collide".

Reduction of Microbes by Coagulation-Flocculation

Optimum coagulation to achieve maximum reductions of turbidity and microbes requires careful control of coagulant dose, pH and consideration of the quality of the water being treated, as well as appropriate mixing conditions for optimum flocculation. Under optimum conditions, coagulation-flocculation and sedimentation with alum and iron can achieve microbial reductions of >90 to >99% for all classes of waterborne pathogens (Sproul, 1974; Leong, 1982; Payment & Armon, 1989).

Alum and Iron Coagulation

Coagulation-flocculation treatment with alum, iron and other coagulants requires knowledge, skills to optimize treatment conditions, it is generally considered to be beyond the reach of most consumers. This type of treatment is less likely to be performed reliably at point-of-use for household water treatment. Despite the caveats and limitations, alum coagulation and precipitation to remove turbidity and other

Figure 3. Hydrophobic Particles before and after coagulation

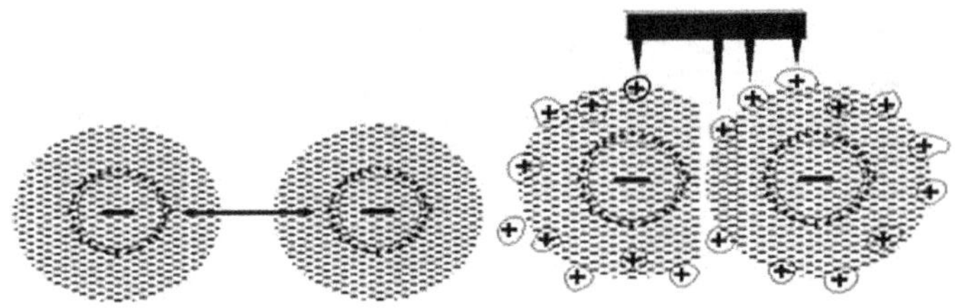

visible contaminants from water at the household level has been traditionally practiced for centuries in many parts of the world (Jahn & Dirar, 1979; Gupta & Chaudhuri, 1992).

Seed Extract Coagulation-Flocculation

Coagulation-flocculation with extracts from natural and renewable vegetation has been widely practiced since recorded time, and appears to be an effective and accepted physical-chemical treatment for household water in some parts of the world. The effectiveness of another traditional seed or nut extract, from the *nirmali* plant or *Strychnos potatorum* (also called the clearing nut) to coagulate-flocculate or precipitate microbes and turbidity in water also has been determined (Tripathi et al., 1976; Able et al., 1984). Despite the potential usefulness of *Moringa oleifera, Strychnos potatorum* and other seed extracts for treatment of turbid water, there has been little effort to characterize the active agents in these seed extracts or evaluate the efficacy as coagulants in reducing microbes from waters having different turbidities.

Filtration

After the flocs are formed, the solution is led to a settling tank, some flocs do not settle. While most of the flocculated material is removed in the settling tank, some floc do not settle. These flocs are removed by the filtration process, which is usually carried out using beds of porous media such as sand or coal. The current trend is to use a mixed -media filter which consists of fine garnet in the bottom layer, silica sand in the middle layer and coarse coal in the top layer which reduces clogging. Most, if not all, water purification systems contain a filter stage. These filter out large things from sewage water, like cotton wool and condoms that should not have been thrown down the toilet in the first place. Filters can also remove smaller particles like silt and suspended solids; dissolved ions and some filters catch bacteria and viruses.

Slow Sand Filters

Slow sand filtration of drinking water has been practiced since the early 19th century and various scales of slow sand filters have been widely used to treat water at the community and sometimes local or household level (Cairncross & Feachem, 1986; Chaudhuri & Sattar, 1990; Droste & McJunken, 1982; Logsdon, 1990). The filters operate with a constant head of overlying water and a flow rate of about 0.1 m/hour. Slow sand filtration is a biological process whereby particulate and microbial removal occurs due the slime layer ("schmutzdecke") that develops within the top few centimeters of sand. . As the filter is used, the schmutzdecke will grow bigger and consequently will reduce the flow rate of the

Figure 4. Slow Sand Filter

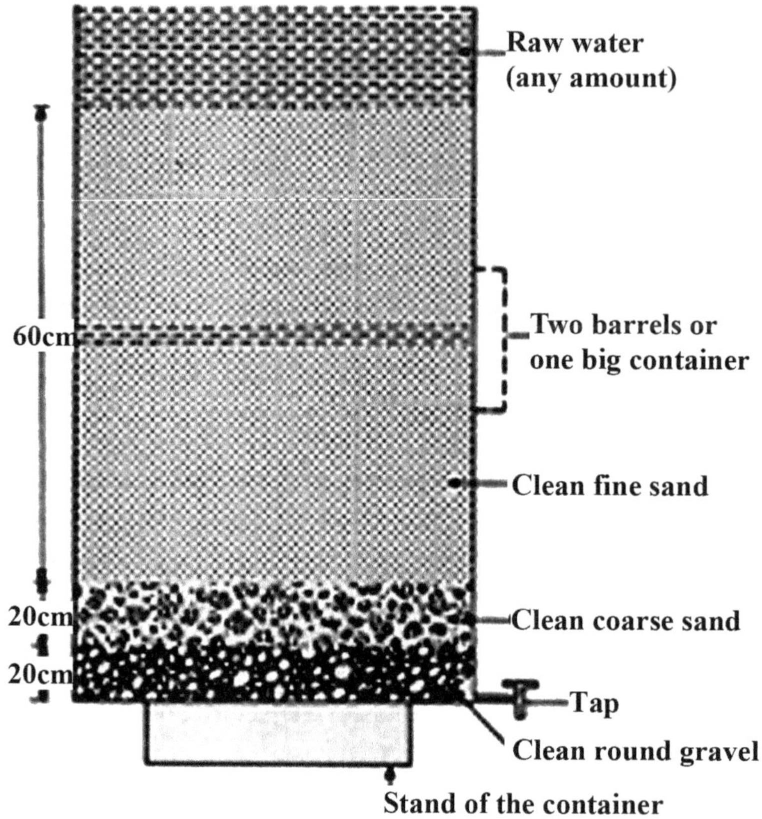

filter. When flow rate becomes too low, the filter has to be cleaned by emptying the filter and scraping off the top layer of sand. Because slow sand filters are slow, the water needs a long time to get through the sand and capacity is limited, although they are relatively easy to manage. Below is a diagram showing a slow sand filter Figure 4 shows the working of a slow sand filter. The biolayer develops above the layers of fine sand at the top. If one were to have a layer of coarse sand on top, to filter bigger particles first, which may seem like a good idea, the biolayer would develop on the finer sand in the middle of the filter, in fact making cleaning much more difficult.

Granular Media, Rapid Rate Filters and Filter Media

Filtration is a physical process of separating suspended and colloidal particles from water by passing the water through a filter media. Filtration involves a number of physical processes. Among these are straining, settling, and adsorption. As particle contaminants pass into the filter, the spaces between the filter grains become clogged, which reduces the openings. Some contaminants are removed merely because they settle onto a media grain. Others are adsorbed onto the surface of individual filter grains. This adsorption process helps to collect the contaminants (floc) and thus re-duces the size of the openings between the media grains.

Figure 5. Granular filteration

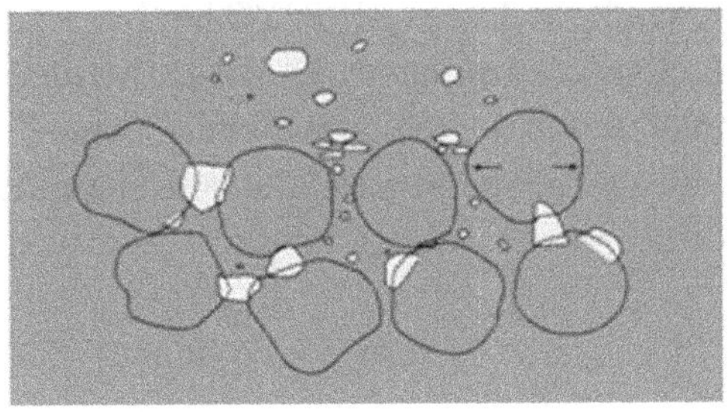

As water and particles (floc) enter the filter, they begin to settle, adsorb, and collect in the upper portion of the filter media. This increases the pressure above the particles, driving them down into the media. Such modified media are positively charged and therefore, more effective for removing and retaining the negatively charged viruses and bacteria by electrostatic adsorption (Chaudhuri & Sattar, 1986). Some improved granular media filter-adsorbers have incorporated bacteriostatic agents, such as silver, in order to prevent the development of undesirable biofilms that release excessive levels of bacteria into the product water (Ahammed & Chaudhuri, 1999). Such media would have to be prepared and distributed to communities and households from specialized facilities.

Gravity filters rely on the depth of water above the filter media to provide the driving force to pass water through the media as it clogs. The amount of available driving force or water depth (head) is limited by the sidewall height of the filter tank above the surface of the filter media. The sidewall height is thus limited by the ceiling height in the building.

Pressure filters are enclosed in pressure vessels and can operate with much higher driving forces. In general, most gravity filters operate with 4 - 6 feet of available head, and pressure filters operate with 10 - 20 feet of head. A major advantage of the pressure filter is that water can be treated under pressure and pumped to a water storage tank at a higher elevation without the need to pump the water after filtration. One disadvantage of the pressure filtration system is the inability to visually observe the condition of the filter media and the backwash process.

Bucket Filters

Bucket filter systems of granular media for household use usually require two or three buckets, one of which has a perforated bottom to serve as the filter vessel. The bucket with the perforated bottom is filled with a layer of sand, layers of both sand and gravel, or other media. Gravel and sand media of specified sizes often can be purchased locally. Buckets are filled with several cm of gravel on the bottom and then a deeper layer of sand (about 40 to 75 cm) on top of the gravel. The granular medium bucket filter is suspended above a similar size empty bucket with a solid bottom to collect the water that drains from the filter as water is poured through it.

Drum or Barrel Filters

Numerous different designs for drum or barrel filters having either up-flow or down-flow of water have been described for use as rapid granular medium filters. These filters are usually 55-gallon (about 200-liter) capacity steel drums and contain sand and gravel media similar to that used for bucket filters (Cairncross & Feachem, 1986; IDRC, 1980; Schiller & Droste, 1982). The filters generally have a cover to prevent the introduction of airborne and other contaminants. Down-flow filters have a perforated pipe at the bottom to collect the water passing through the medium and discharge it from the side of the drum. The outlet pipe for filtered water may discharge the water at the bottom of the drum or it may be configured with an upward bend or loop to discharge the water at the same level as the top of the media in the filter.

Roughing Filters

Simple, low cost, low-maintenance, multi-stage roughing filters for household and community use have been described and characterized (Galvis et al., 2000; Wegelin & Schertenlieb, 1987; Wegelin et al., 1991). Typically, these filters are rectangular, multi-compartment basins constructed of concrete or other materials. Most of the filters are designed to use two different sizes of low cost, coarse granular media in two or three compartments or stages, and such media are generally locally available. Regular backwashing is required to main flow rates and achieve efficient particulate removals, and therefore, some skill and knowledge is required to properly operate and maintain a roughing filter. Removal of indicator bacteria by roughing filters has been reported to be 90-99%.

Filter-Cisterns

Filter-cisterns have been in use since ancient times in areas heavily supplied with rainwater or other water sources but lacking land area for reservoir or basin storage (Baker, 1948). In this filtration system cisterns or large diameter well casings, partially below grade, are surrounded by sand filters, such that water flows through the sand and into the casing or cistern either from the bottom or though side of the casing near the bottom. Such filter-cisterns function as infiltration basins to remove turbidity and other particulates.

Biomass and Fossil Fuel Granular Media Filters

Historically, depth filters composed of filter media derived from vegetable and animal matter have been employed for water treatment. Coal-based and charcoal filter media have been used since ancient times and carbon filter media are widely used today for both point-of-use and community water filtration systems (Argawal & Kimondo, 1981; Baker, 1948; Chaudhuri & Sattar, 1990). Sponge filters imparted objectionable tastes and odors to the water unless they were cleaned regularly, indicating that microbial growths and biofilms probably were a major problem with these filters. Other media also employed in these point-of-use filters included sand, cotton, wool, linen, charcoal and pulverized glass, either individually or in various combinations as successive layers. These media also were used in larger scale filters for community water supply. A large number of other examples of vegetable matter depth filters are those containing burnt rice hulls (as ash) or those consisting of vessels or chambers containing fresh coconut fibers and burnt rice husks in series (Argawal & Kimondo, 1981; Barnes & Mampitiyarachichi, 1983). conditions (Barnes & Mampitiyarachichi, 1983). Technological methods to modify granular media, such as chemical modification to impart positive surface charges, can improve microbial removals by filtration.. Granular media filters are best used at pre-treatment processes to reduce turbidity and provide product water that is more amenable to pathogen reductions by disinfection processes, such as solar radiation or chlorination.

Fiber, Fabric and Membrane Filters

Filters composed of compressed or cast fibers (e.g., cellulose paper), spun threads (cotton) or woven fabrics (cotton, linen and other cloths) have been used to filter water and other beverages (e.g., wine) since ancient times. Simple fiber, fabric, paper and other filters and filter holders for them are available for widespread, practical and affordable household treatment of collected and stored water throughout much of the world. Some waterborne and water-associated pathogens are relatively large, such as the free-swimming larval forms (cercariae) of schistosomes and Faciola species, guinea worm larvae within their intermediate crustacean host (Cyclops), and bacterial pathogens associated with relatively large copepods and other zooplankters in water, such as the bacterium Vibrio cholerae. Paper filters have been recommended for the removal of schistosomes and polyester or monofilament nylon cloth filters are also recommended for the removal of the Cyclops vector of guinea worm (Imtiaz et al., 1990). Such filters have been used successfully at both the household and community levels (Aikhomu et al., 2000). Colwell and colleagues reported various types of sari cloth (fine mesh, woven cotton fabric) and nylon mesh can be used in single or multiple layers to remove from water the zooplankton and phytoplankton harboring V. cholerae, thereby reducing the V. cholerae concentrations by >95 to >99% (Huq et al., 1996). Where waterborne schistosomes, guinea worms, *Faciola* species and zooplanton-associated *V. Cholera* are a problem, use of these simple, point-of-use filter methods are introduced and encouraged, especially if other control measures are not available or difficult to implement. However, typical fabric, paper, monofilament nylon and similar filters are not recommended for general treatment of household water. In these pore sizes are too large to appreciably retain viruses, bacteria and smaller protozoan parasites, especially if such microbes are free and not associated with large particles or organisms.

Porous Ceramic Filters

Porous ceramic filters are made up of clay, carved porous stone and other media have also been used to filter water since ancient times and were cited by Aristotle (322-354 BCE). Modern accounts of ceramic filters for household use date back to at least the 18th century (Baker, 1948). Most modern ceramic filters are made in the form of vessels or hollow cylindrical "candles". Porous ceramic filters can be made in various pore sizes and most modern ceramic filters produced in the developed countries of the world are rated to have micron or sub-micron pore sizes that efficiently remove bacteria as well as parasites. Some of these are estimated to remove at least 99.9999% of bacteria, such as Klebsiella terrigena, 99.99% of viruses, such as polioviruses and rotaviruses, and 99.9% of Giardia cysts and Cryptosporidium oocysts as required for Point-of-Use Microbiological Water Purifiers in the United States (USEPA, 1987). Overall, ceramic filters are recommended for use in water treatment at the household level. The main barriers to the production, distribution and use of fired or unfired ceramic filter-adsorbers are the availability of trained workers, fabrication and distribution facilities and cost. Further efforts are needed to define and implement appropriate manufacturing procedures and product performance characteristics of these filters in order to achieve products of acceptable quality that are capable of adequate microbe reductions from water.

Diatomaceous Earth Filters

Diatomaceous earth (DE) and other fine granular media also can be used to remove particulates and microbial contaminants from water by so-called precoat and body feed filtration. Such filters has achieved great removal efficiencies of a wide range of waterborne microbial contaminants without any chemical pre-treatment of the water (Cleasby, 1990; Logsdon, 1990). A thin layer or cake of fine granular or powdery filter medium is precoated or deposited by filtration onto a permeable material held by a porous, rigid support to comprise a filter element.

Aeration

Aeration of water alone is simple, practical, and affordable, especially done manually in a bottle or other vessel.

Aeration of water has been practiced since ancient times and was believed to improve water quality by "sweetening" and "softening" it (Baker, 1948). However, aeration of water introduces oxygen, which can cause chemical reactions, such as precipitation in anaerobic water containing certain dissolved solutes, and which can contribute indirectly to other process that may lead to microbial reductions.

Disinfection

A large fraction of bacteria and larger micro-organisms are removed during clarification processes, especially by sand filtration. However, many bacteria and viruses still remain in clarified water even at low turbidity levels. It is therefore, essential to disinfect water to prevent the possibility that water-borne diseases are spread by pathogens (disease-causing micro-organisms) in water. Disinfection of water entails the addition of the required amount of a chemical agent (disinfectant) to the water and allowing contact between the water and disinfectant for a pre-determined period of time (under specified conditions of pH and temperature). Physical methods of disinfection of water include irradiation with ultra-violet light and boiling. The most commonly used disinfectant is chlorine gas, Cl_2 that is dissolved in the water at a certain concentration for a certain minimum contact time. Other

disinfectants include ozone, chlorine dioxide and other chlorine compounds such as calcium hypochlorite (HTH), sodium hypochlorite (bleach) and monochloramine. Chlorine is a strong oxidising agent and it reacts and oxidises some of the essential systems of micro-organisms thereby inactivating or destroying them. Chlorine is the most commonly used disinfectant employed for killing bacteria in water. When chlorine is added to water, it rapidly hydrolyzes according to the reaction

$$Cl_2 + H_2O \rightarrow H^+ + Cl^- + HOCl \qquad (1)$$

which has the following equilibrium constant:

$$K = \frac{[H^+][Cl^-][HOCl]}{[Cl_2]} = 4.5 ´ 10\text{-}4 \qquad (2)$$

Hypochlorous acid, HOCl, is a weak acid that dissociates according to the reaction,

$$HOCl \rightarrow H^+ + OCl^- \quad (3)$$

with an ionization constant of 2.7 ´ 10 -8. From the above it can be calculated that the concentration of elemental Cl_2 is negligible at equilibrium above pH 3 when chlorine is added to water at levels below 1.0 g/L. Sometimes, hypochlorite salts are substituted for chlorine gas as a disinfectant. Calcium hypochlorite, $Ca(OCl)_2$, is commonly used. The hypochlorites are safer to handle than gaseous chlorine. The two chemical species formed by chlorine in water, HOCl and OCl^-, are known as free available chlorine. Free available chlorine is very effective in killing bacteria. In the presence of ammonia, monochloramine, dichloramine, and trichloramine are formed:

$$NH_4^+ + HOCl \rightarrow NH_2Cl \text{ (monochloramine)} + H_2O + H^+ \quad (4)$$

$$NH_2Cl + HOCl \rightarrow NHCl_2 \text{ (dichloramine)} + H_2O \quad (5)$$

$$NHCl_2 + HOCl \rightarrow NCl_3 \text{ (trichloramine)} + H_2O \quad (6)$$

The chloramines are called combined available chlorine. Too much ammonia in water is considered undesirable because it exerts excess demand for chlorine. At sufficiently high Cl:N molar ratios in water containing ammonia, some HOCl and OCl^- remain unreacted in solution, and a small quantity of NCl_3 is formed. The ratio at which this occurs is called the breakpoint. Chlorination beyond the breakpoint ensures disinfection. It has the additional advantage of destroying the more common materials that cause odor and taste in water. At moderate levels of NH_3-N (approximately 20 mg/L), when the pH is between 5.0 and 8.0, chlorination with a minimum 8:1 weight ratio of Cl to NH_3-nitrogen produces efficient denitrification:

$$NH_4^+ + HOCl \rightarrow NH_2Cl + H_2O + H^+ \quad (4)$$

$$2NH_2Cl + HOCl \rightarrow N_2(g) + 3H+ + 3Cl^- + H_2O \quad (7)$$

This reaction is used to remove pollutant ammonia from wastewater. However, problems can arise from chlorination of organic wastes. Typical of such by-products is chloroform, produced by the chlorination of humic substances in water. Chlorine is used to treat water other than drinking water. It is employed to disinfect effluent from sewage treatment plants, as an additive to the water in electric power plant cooling towers, and to control microorganisms in food processing.

Chlorine Dioxide

Chlorine dioxide, ClO_2, is an effective water disinfectant that is of particular interest because, in the absence of impurity Cl_2, it does not produce impurity trihalomethanes in water treatment. In acidic and neutral water, respectively, the two half reactions for ClO_2 acting as an oxidant are the following:

$$ClO_2 + 4H+ + 5e- \rightarrow Cl^- + 2H_2O \quad (8)$$

$$ClO_2 + e- \rightarrow ClO_2^- \quad (9)$$

In the neutral pH range, chlorine dioxide in water remains largely as molecular ClO_2 until it contacts a reducing agent with which to react. Chlorine dioxide is a gas that is very reactive with organic matter and explosive when exposed to light. For these reasons, it is not shipped, but is generated on-site by processes such as the reaction of chlorine gas with solid sodium hypochlorite:

$$2NaClO_2(s) + Cl_2(g) \rightarrow 2ClO_2(g) + 2NaCl(s) \quad (10)$$

A high content of elemental chlorine in the product may require its purification to prevent unwanted side-reactions from Cl_2. As a water disinfectant, chlorine dioxide does not chlorinate or oxidize ammonia or other nitrogen-containing compounds. Some concern has been raised over possible health effects of its main degradation byproducts, ClO_2^- and ClO_3^-.

Chlorine can be added to water in different forms.

Chlorine gas, Cl_2 is supplied to the plant in gas cylinders and the chlorine is introduced into the water by means of special dosing devices (chlorinators).

Calcium hypochlorite, $Ca(OCl)_2$ (commonly known as HTH) is available in granular or solid (tablet) form and is therefore a very convenient form in which to apply chlorine, especially for smaller or rural plants. It contains between 65 and 70% of available chlorine, it is relatively stable and can be stored for long periods (months) in a cool dry environment.

Sodium hypochlorite, NaOCl (commonly known as household bleach under different brand names) is available as a solution. Water treatment sodium hypochlorite contains 12 to 13% of hypochlorite, which is equivalent to 10 - 12% available chlorine. Sodium hypochlorite is relatively unstable and deteriorates fairly rapidly, especially when exposed to sunlight. It also forms HOCl and OCl upon dissociation.

Monochloramine (so-called combined available chlorine) is also used for water disinfection. It is formed when HOCl is added to water that contains a small amount of ammonia. The ammonia reacts with HOCl to form monochloramine, NH_2Cl. It is much less effective as a disinfectant than HOCl (the same order of effectiveness as chlorite ion. An important factor that affects disinfection, is the turbidity of the water has to be disinfected. The reason is that when water contains colloidal particles, they may "shield" the micro-organisms from the action of the disinfectant, or alternatively react with the chlorine and in this way prevent effective disinfection.

Disinfection by means of ultra-violet (UV) irradiation is becoming more and more popular because no by-products are formed in the process. UV radiation kills or inactivates micro-organisms provided each organisms gets a minimum amount of irradiation. UV irradiation functions on the principle that each unit of water must be exposed to the irradiation for a minimum amount of time at a minimum dosage intensity (fluence). It is important that the water to be disinfected by UV is properly pre-treated to ensure a low turbidity, preferably lower than 0,5 NTU. If the water contains high turbidity levels the colloids either absorb some of the radiation or shield the microorganisms against radiation which reduces the effectiveness of the process. A further important aspect is that the UV tubes are prone to the formation of layers of scale or other fouling material. This also reduces the effectiveness of radiation.

Stabilisation

Stabilisation of water refers to the chemical stability (specifically with respect to $CaCO_3$) of water. The ponds are usually built in a series of at least three; anaerobic, facultative and aerobic. The first, anaerobic pond is 2-5 m deep and water stays there for 1-7 days only. A combination of processes happen; anaerobic bacteria digest sludge on the bottom and closer to the surface aerobic processes work, receiving oxygen from natural diffusion, algae photosynthesis and wind-mixing. The facultative pond removes up to a further 75% of biological activity. In both these stages, sedimentation occurs and effluent is led to the next pond from above the bottom sludge. The last, aerobic, pond is often called the finishing, maturation or polishing pond, because it finishes the work off. Ponds maturation can be built in series of more ponds for better pathogen removal. Of the three ponds in the stabilization process, the maturation pond is the one that removes actual pathogens. A shallow pond, only 0.5-1.5 m deep so that sunlight can reach to the bottom for photosynthesis. This pond can remove a lot of nitrogen and phosphorous from water if used with algaeal photosynthesis and fish harvesting.

Stabilization ponds need a lot of land area and expert personnel to build and monitor the ponds Water that is not chemically stable may be:

- Corrosive towards metal pipes and fittings causing leaks in distribution systems with substantial cost implications,
- Scale-forming, causing a layer of chemical scale to form in pipes and on heating elements. This also has substantial cost implications because the carrying capacity of pipes is reduced and the

Figure 6. Function of a waste-water stabilization pond with three stages

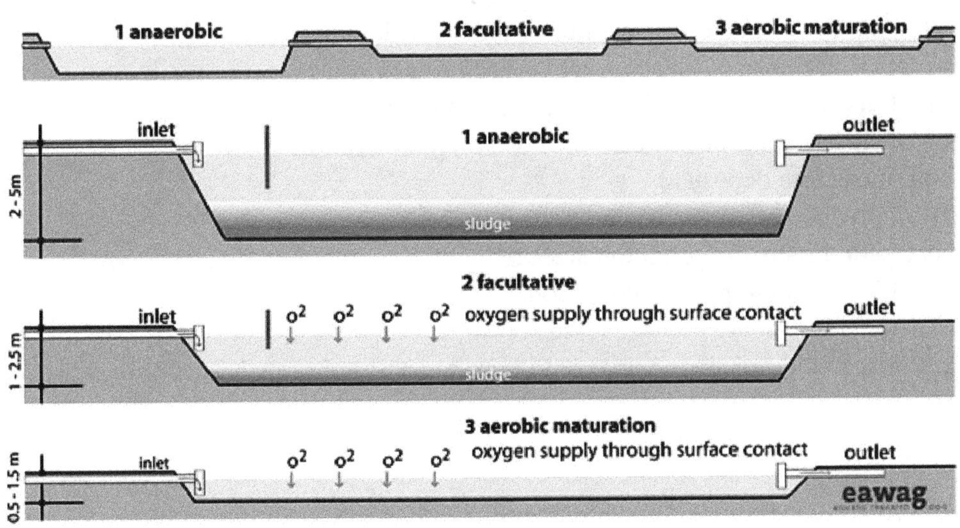

heat transfer in kettles and geysers is impaired. From a cost point of view, it is very important to ensure that water for domestic use is chemically stable. Stabilisation of water involves the addition of chemicals to the water to produce water with a calcium carbonate precipitation potential (CCPP) of about 4 mg/l. This means that the water should be slightly supersaturated with calcium carbonate.

TREATMENT OF WATER FOR INDUSTRIAL USE

Water is widely used in various process applications in industry. Other major industrial uses are boiler feedwater and cooling water. The kind and degree of treatment of water in these applications depends upon the end use. As examples, cooling water may require only minimal treatment, removal of corrosive substances and scale-forming solutes is essential for boiler feed water, and water used in food processing must be free of pathogens and toxic substances. Improper treatment of water for industrial use can cause problems such as corrosion, scale formation, reduced heat transfer in heat exchangers, reduced water flow, and product contamination. These include the following:

- Water requirement
- Quantity and quality of available water sources
- Sequential use of water (successive uses for applications requiring progressively
- lower water quality)
- Water recycle
- Discharge standards

The various specific processes employed to treat water for industrial use are discussed in later sections of this chapter. External treatment, usually applied to the plant's entire water supply, uses processes such as aeration, filtration, and clarification to remove material that may cause problems from water. Such substances include suspended or dissolved solids, hardness, and dissolved gases. Water may also be divided into different streams, some to be used without further treatment, and the rest to be treated for specific applications. Internal treatment is designed to modify the properties of water for specific applications. Examples of internal treatment include the following:

- Reaction of dissolved oxygen with hydrazine or sulfite
- Addition of chelating agents to react with dissolved Ca^{2+} and prevent
- formation of calcium deposits
- Addition of precipitants, such as phosphate used for calcium removal
- Treatment with dispersants to inhibit scale
- Addition of inhibitors to prevent corrosion
- Adjustment of pH
- Disinfection for food processing uses or to prevent bacterial growth in cooling water

REMOVAL OF CALCIUM AND OTHER METALS

Calcium and magnesium salts, which generally are present in water as bicarbonates or sulfates, cause water hardness. Most common manifestations of water hardness is the insoluble "curd" formed by the reaction of soap with calcium or magnesium ions. The formation of these insoluble soap salts. Another problem caused by hard water is the formation of mineral deposits. For example, when water containing calcium and bicarbonate ions is heated, insoluble calcium carbonate is formed:

$$Ca_2^{+} + 2HCO_3^{-} \rightarrow CaCO_3(s) + CO_2(g) + H_2O \quad (1)$$

This product coats the surfaces of hot water systems, clogging pipes and reducing heating efficiency. Dissolved salts such as calcium and magnesium bicarbonates and sulfates can be especially damaging in boiler feed water. Removal of water hardness is essential for many uses of water. Several processes are used for softening water. On a large scale, such as in community water-softening operations, the lime-soda process is used. This process involves the treatment of water with lime, $Ca(OH)_2$, and soda ash, Na_2CO_3. Calcium is precipitated as $CaCO_3$ and magnesium as $Mg(OH)_2$. When the calcium is present primarily as "bicarbonate hardness," it can be removed by the addition of $Ca(OH)_2$ alone:

$$Ca^{2+} + 2HCO_3^{-} + Ca(OH)_2 \rightarrow 2CaCO_3(s) + 2H_2O \quad (2)$$

When bicarbonate ion is not present at substantial levels, a source of CO_3^{2-} must be provided at a high enough pH to prevent conversion of most of the carbonate to bicarbonate. These conditions are obtained by the addition of Na_2CO_3. For example, calcium present as the chloride can be removed from water by the addition of soda ash:

$$Ca^{2+} + 2Cl^{-} + 2Na^{+} + CO_3^{2-} \rightarrow CaCO_3(s) + 2Cl^{-} + 2Na^{+} \quad (3)$$

Note that the removal of bicarbonate hardness results in a net removal of soluble salts from solution, whereas removal of non bicarbonate hardness involves the addition of at least as many equivalents of ionic material as are removed. The precipitation of magnesium as the hydroxide requires a higher pH than the precipitation of calcium as the carbonate:

$$Mg^{2+} + 2OH^{-} \rightarrow Mg(OH)_2(s) \quad (4)$$

The high pH required may be provided by the basic carbonate ion from soda ash:

$$CO_3^{2-} + H_2O \rightarrow HCO_3^{-} + OH^{-} \quad (5)$$

Some large-scale lime-soda softening plants make use of the precipitated calcium carbonate product as a source of additional lime. The calcium carbonate is first heated to at least 825°C to produce quicklime, CaO:

$$CaCO_3 + heat \rightarrow CaO + CO_2(g) \quad (6)$$

The quicklime is then slaked with water to produce calcium hydroxide:

$$CaO + H_2O \rightarrow Ca(OH)_2 \tag{7}$$

The water softened by lime-soda softening plants usually suffers from two defects. First, because of super-saturation effects, some $CaCO_3$ and $Mg(OH)_2$ usually remain in solution. If not removed, these compounds will precipitate at a later time and cause harmful deposits or undesirable cloudiness in water. The second problem results from the use of highly basic sodium carbonate, which gives the product water an excessively high pH, up to pH 11. To overcome these problems, the water is recarbonated by bubbling CO_2 into it. The carbon dioxide converts the slightly soluble calcium carbonate and magnesium hydroxide to their soluble bicarbonate forms:

$$CaCO_3(s) + CO_2 + H_2O \rightarrow Ca^{2+} + 2HCO_3^- \tag{8}$$

$$Mg(OH)_2(s) + 2CO_2 \dashrightarrow Mg^{2+} + 2HCO_3^- \tag{9}$$

The CO_2 also neutralizes excess hydroxide ion:

$$OH^- + CO_2 \rightarrow HCO_3^- \tag{10}$$

The pH generally is brought within the range 7.5-8.5 by recarbonation. The source of CO_2 used in the recarbonation process may be from the combustion of carbonaceous fuel. Scrubbed stack gas from a power plant frequently is utilized. pH, alkalinity, and Ca^{2+} concentration very close to $CaCO_3$ saturation is labelled *chemically stabilized*. Water with Ca^{2+} concentration much below $CaCO_3$ saturation is called an *aggressive* water.

Calcium may be removed from water very efficiently by the addition of orthophosphate:

$$5Ca^{2+} + 3PO_4^{3-} + OH^- \rightarrow Ca_5OH(PO_4)_3(s) \tag{11}$$

It should be pointed out that the chemical formation of a slightly soluble product for the removal of undesired solutes such as hardness ions, phosphate, iron, and manganese must be followed by sedimentation in a suitable apparatus.

Ion Exchange

Water may be purified by ion exchange, the reversible transfer of ions between aquatic solution and a solid material capable of bonding ions. The removal of NaCl from solution by two ion exchange reactions is a good illustration of this process. In first the water is passed over a solid cation exchanger in the hydrogen form, represented by H^+-{Cat(*s*)}:

$$H^+ \text{-}\{Cat(s)\} + Na^+ + Cl^- \rightarrow Na^+ \text{-}\{Cat(s)\} + H^+ + Cl^- \tag{12}$$

Next, the water is passed over an anion exchanger in the hydroxide ion form, represented by

$OH^- + \{An(s\}: OH^- + \{An(s)\} + H^+ + Cl^- \rightarrow Cl^- + \{An(s)\} + H_2O$ (13)

Thus, the cations in solution are replaced by hydrogen ion and the anions are replaced by hydroxide ion, to form water. The softening of water by ion exchange does not require the removal of all ionic solutes, just those cations responsible for water hardness. Among the minerals especially noted for their ion exchange properties are the aluminum silicate minerals, or zeolites. An example of a zeolite which has been used commercially in water softening is glauconite, K2(MgFe)2Al6(Si4O10)3(OH)12. Synthetic zeolites have been prepared by drying and crushing the white gel produced by mixing solutions of sodium silicate and sodium aluminate. It was discovered in the mid-1930s of synthetic ion exchange resins which were composed of organic polymers with attached functional groups marked the beginning of modern ion exchange technology. Structural formulas of typical synthetic ion exchangers are in Figures 7 and 8. The cation exchanger shown in Figure 7 is called a strongly acidic cation exchanger because the parent –SO3$^-$H$^+$ group is a strong acid. When the functional group binding the cation is the –CO2$^-$ group, the exchange resin is called a weakly acidic cation exchanger, because the $-CO_2H$ group is a weak acid. Figure 8 shows a strongly basic anion exchanger in which the functional group is a quaternary ammonium group, N$^+$(CH3)3. In the hydroxide form, –N$^+$(CH3)3OH$^-$, the hydroxide ion is readily released, so the exchanger is classified as strongly basic. The water-softening capability of a cation exchanger is shown in Figure 7, where sodium ion on the exchanger is exchanged for calcium ion in solution. The same reaction occurs with magnesium ion. Sodium chloride contamination results regeneration of a water softener with sodium chloride in order to displace calcium and magnesium ions from the resin and replace these hardness ions with sodium ions:

$Ca^{2+} \text{-}\{Cat(s)\}2 + 2Na+ + 2Cl^- \rightarrow 2Na^+ \text{-}\{Cat(s)\} + Ca^{2+} + 2Cl^-$ (14)

During the regeneration process, a large excess of sodium chloride must be used — several pounds for a home water softener. Appreciable amounts of dissolved sodium chloride can be introduced intosewage by this route. Strongly acidic cation exchangers are used for the removal of water hardness. Alkalinity generally is manifested by bicarbonate ion, a species that is a sufficiently strong base to neutralize the acid of a weak acid cation exchanger:

$2R\text{-}CO_2H + Ca^{2+} + 2HCO_3^- \rightarrow [R\text{-}CO_2^-]2Ca^{2+} + 2H_2O + 2CO_2$ (15)

However, weak bases such as sulfate ion or chloride ion are not strong enough to remove hydrogen ion from the carboxylic acid exchanger. In order to avoid the potential pollution problem caused by the use of excess sodium chloride to regenerate strongly acidic cation changers, these exchangers may be regenerated almost stoichiometrically with dilute strong acids. *Chelation* or *sequestration*, is an effective method of softening water without actually having to remove calcium and magnesium from solution. A complexing agent is added which greatly reduces the concentrations of free hydrated cations, as shown by some of the example calculations. For example, chelating calcium ion with excess EDTA anion (Y4),

$Ca^{2+} + Y^{4-} \rightarrow CaY^{2-}$ (16)

reduces the concentration of hydrated calcium ion, preventing the precipitation of calcium carbonate:

Figure 7. Strongly acidic cation exchanger; sodium exchange for calcium in water is shown.

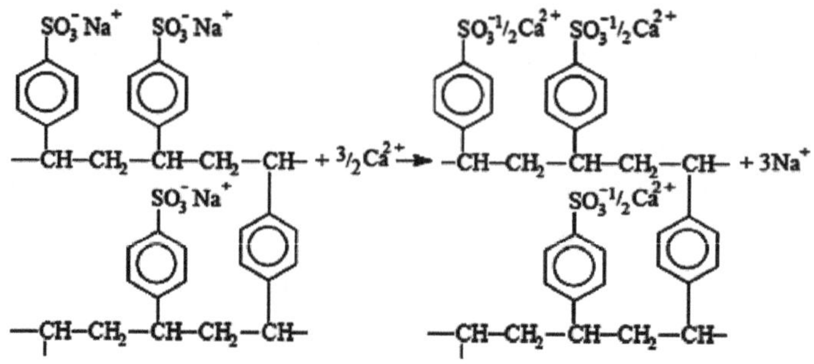

Figure 8. Strongly basic anion exchanger Chloride exchange for hydroxide is shown.

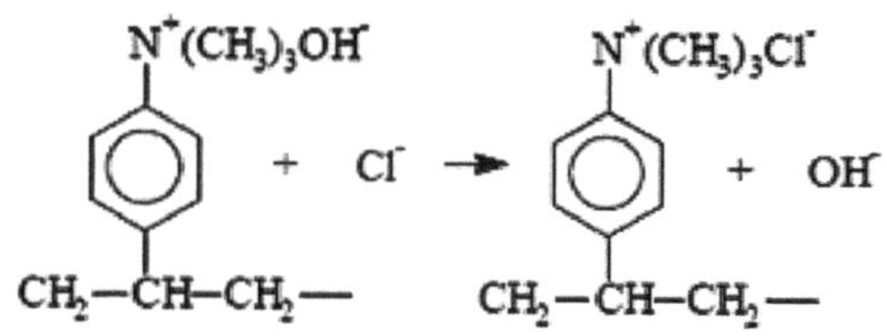

Figure 9. Shows ion exchange demineralization scheme

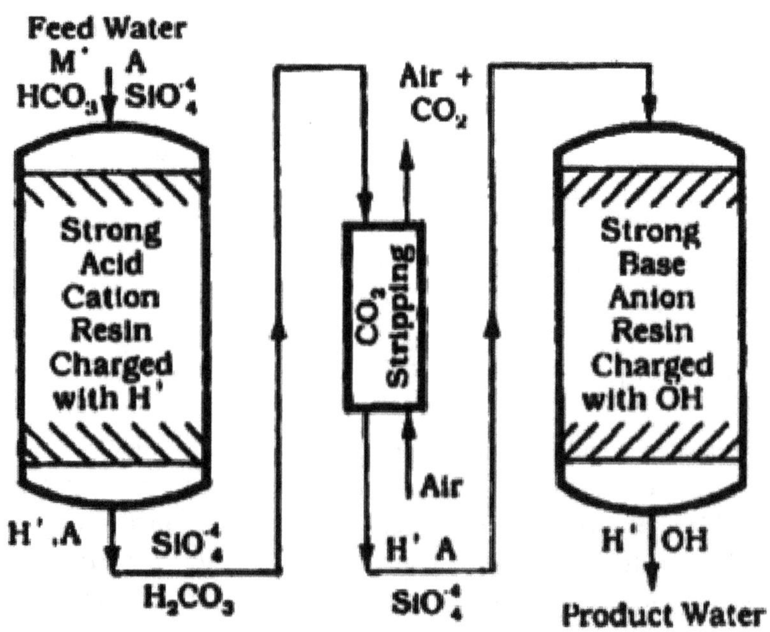

Figure 10. Shows mixed resin demineralization scheme

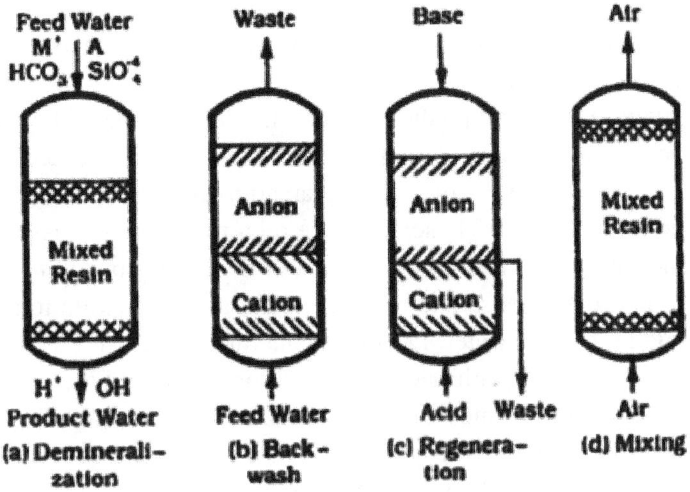

Figure 11. Operational scheme of ion exchange

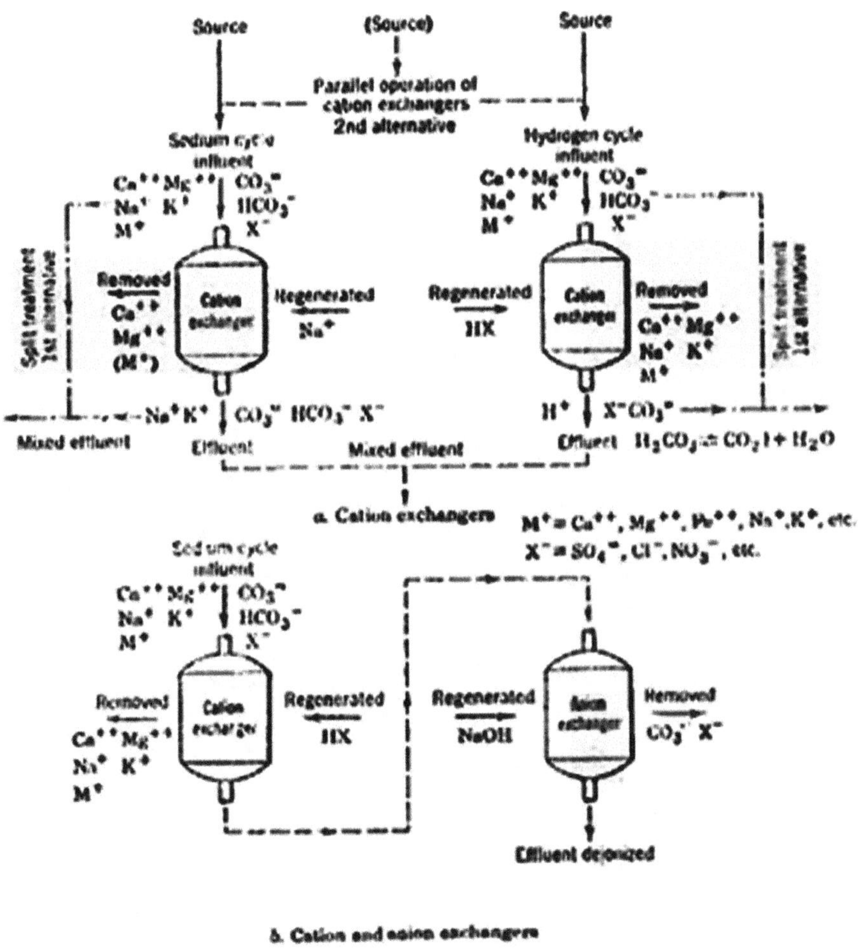

$Ca^{2+} + CO3^{2-} \rightarrow CaCO3(s)$ (17)

Polyphosphate salts, EDTA, and NTA are chelating agents commonly used for water softening. Polysilicates are used to complex iron.

In a typical demineralizer, it is accomplished in the following manner: The influent water is passed through a hydrogen cation-exchange resin which converts the influent salt (e.g., sodium sulfate) to the corresponding acid (e.g., sulfuric acid) by exchanging an equivalent number of hydrogen (H “) ions for the metallic cations (Cae2, Mg’”, Na’). These acids are then removed by passing the effluent through an alkali regenerated anion-exchange resin which replaces the anions in solution (Cl^-, SO,’, NO;) with an equivalent number of hydroxide ions. The hydrogen ions and hydroxide ions neutralize each other to form an equivalent amount of pure water. During regeneration, the reverse reaction takes place. The cation resin is regenerated with either sulfuric or hydrochloric acid and the anion resin is regenerated with sodium hydroxide. Figure 9 illustrates a basic scheme for ion exchange demineralization.

Higher-quality water is obtained from a mixed-bed unit than from a two-bed system. (see Figure 10 for an example). Operation of cation and anion exchanges is shown in Figure 11 (for fundamental processes) and Figure 12 (operation modes for both cation anion exchanges).

Figure 12. Various ion exchange schemes

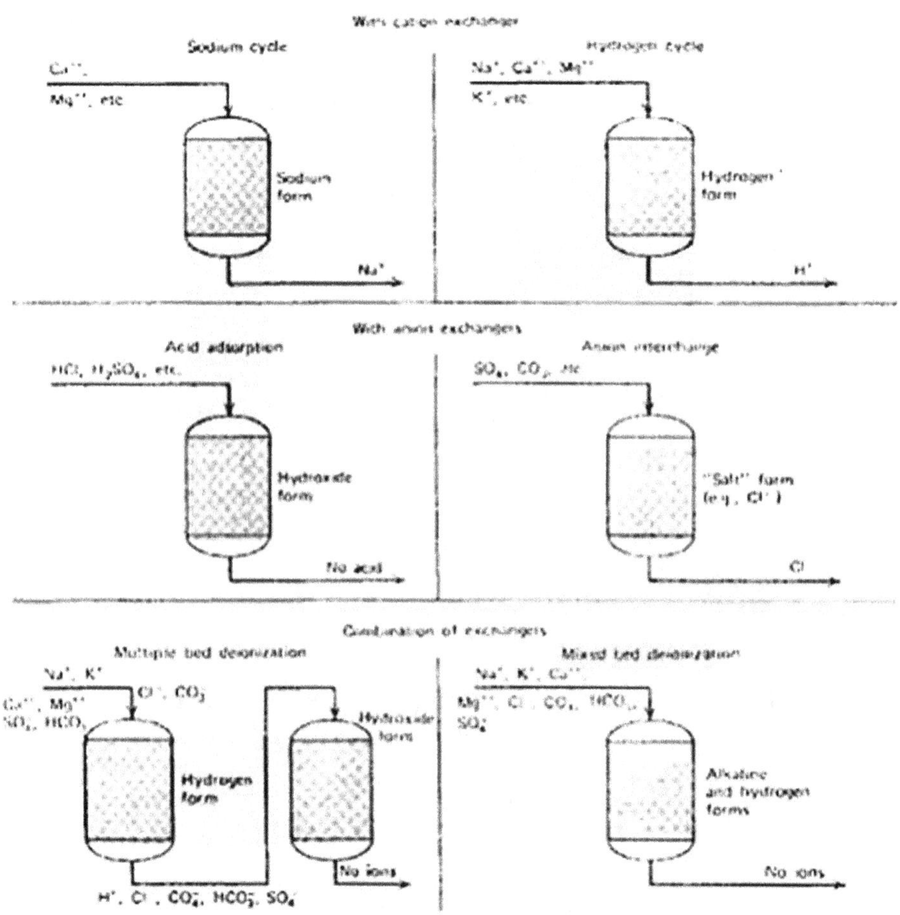

Removal of Iron and Manganese

Soluble iron and manganese are found in many groundwaters because reducing conditions which favors the soluble +2 oxidation state of these metals. Iron is the more commonly encountered of the two metals. In groundwater, the iron seldom exceeds 10 mg/L, and that of manganese is rarely higher than 2 mg/L. The basic method for removing both of these metals depends upon oxidation to higher insoluble oxidation states. The oxidation of soluble Mn(II) to insoluble MnO_2 is a complicated process. It appears to be catalyzed by solid MnO2, which is known to adsorb Mn(II). This adsorbed Mn(II) is slowly oxidized on the MnO_2 surface. Chlorine and potassium permanganate are sometimes employed as oxidizing agents for iron and manganese. There is some evidence that organic chelating agents with reducing properties hold iron(II) in a soluble form in water. In such cases, chlorine is effective because it destroys the organic compounds and enables the oxidation of iron(II). In water with a high level of carbonate, $FeCO_3$ and $MnCO_3$ may be precipitated directly by raising the pH above 8.5 by the addition of sodium carbonate or lime. This approach is less popular than oxidation, however.

Lime treatment, discussed earlier in this section for calcium removal, precipitates heavy metals as insoluble hydroxides, basic salts, or coprecipitated with calcium carbonate or iron(III) hydroxide. This process do not completely remove mercury, cadmium, or lead, so their removal is aided by addition of sulfide (most heavy metals are sulfide-seekers):

$$Cd^{2+} + S^{2-} \rightarrow CdS(s) \qquad (18)$$

Heavy chlorination is frequently necessary to break down metal-solubilizing ligands. Lime precipitation do not normally permit recovery of metals and is sometimes undesirable from the economic view point.

Electrodialysis

Electrodialysis uses ion-selective membranes and an electrical potential difference to separate anions and cations in solution. In the past electrodialysis was most often used for purifying brackish water, but it is now finding a role in hazardous waste treatment. Metal salts from plating rinses are sometimes removed in this way.

Figure 13. Electrodialysis Cell

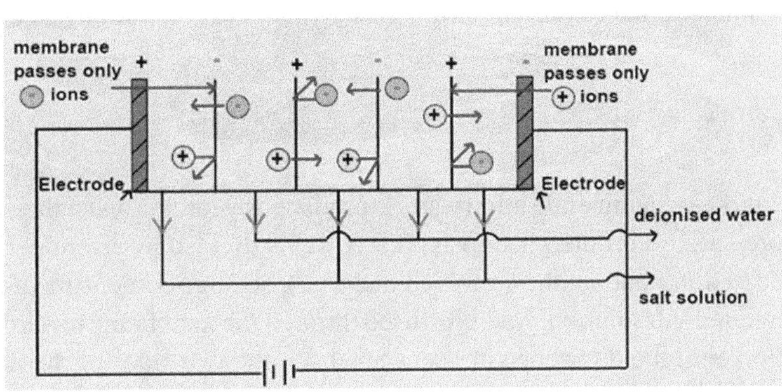

Figure 14. Electrodialysis apparatus for the removal of ionic material from water

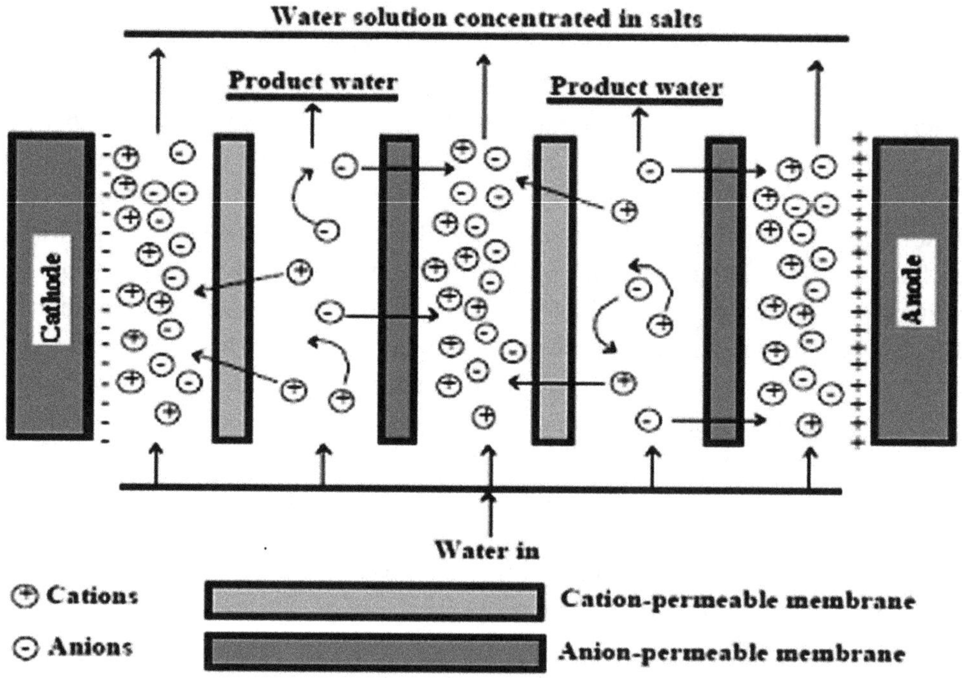

Figure 13 shows a simple dialysis cell in which waste water may be deionised. As shown in the figure two types of membranes (anionic and cationic) are arranged alternatively to form many compartments between the electrodes placed at the two ends. This solution in alternate compartments become more concentrated while that in the remaining becomes more dilute. The electric power requirement is proportional to the number of ions removed from the water. In the electrodialysis process, organic molecules are not removed and they can collect on and clog the membranes. Another disadvantage of this method is that it still leaves concentrated waste water to be disposed of by some appropriate scheme.

In another process of electrodialysis consists of applying a direct current across a body of water separated into vertical layers by membranes alternately permeable to cations and anions. Cations migrate toward the cathode and anions toward the anode. Cations and anions both enter one layer of water, and both leave the adjacent layer. Thus, layers of water enriched in salts alternate with those from which salts have been removed. The water in the brine-enriched layers is recirculated to a certain extent to prevent excessive accumulation of brine. The principles involved in electrodialysis treatment are shown in Figure 14.

Reverse Osmosis

In the reverse osmosis process, demineralisation water is produced by forcing water through semipermeable membranes at high pressure. In ordinary osmosis, if a vessel is divided by a semipermeable membrane (one that is permeable to water but not the dissolved material), and one compartment is filled with water and other with concentrated salt solution, water diffused through the membrane towards the compartment containing salt solution until the difference in water levels on the two sides of the membrane creates a

Figure 15. Osmosis

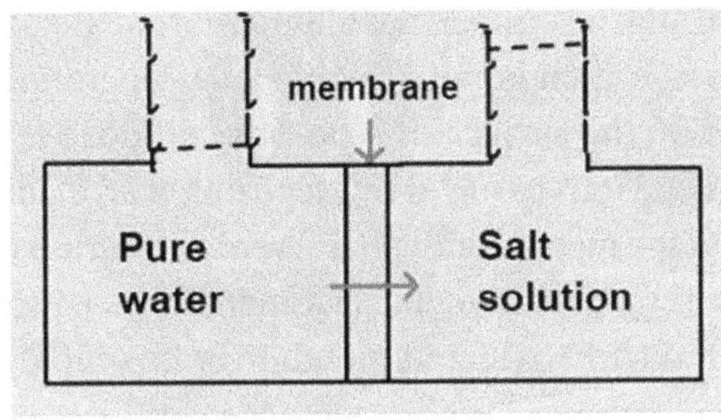

Figure 16. Reversee osmosis

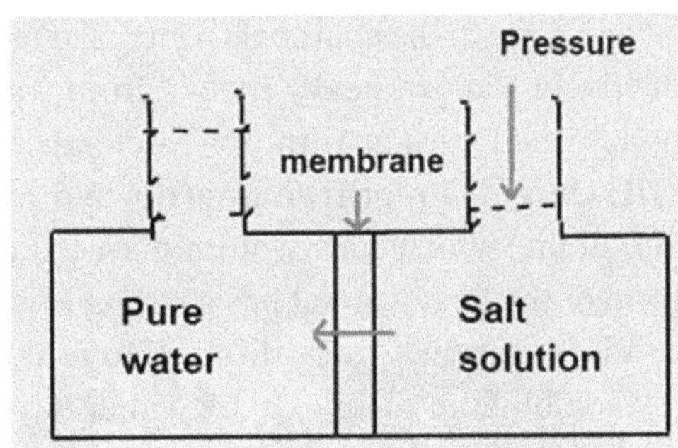

Figure 17. Solute removal from water by reverse osmosis

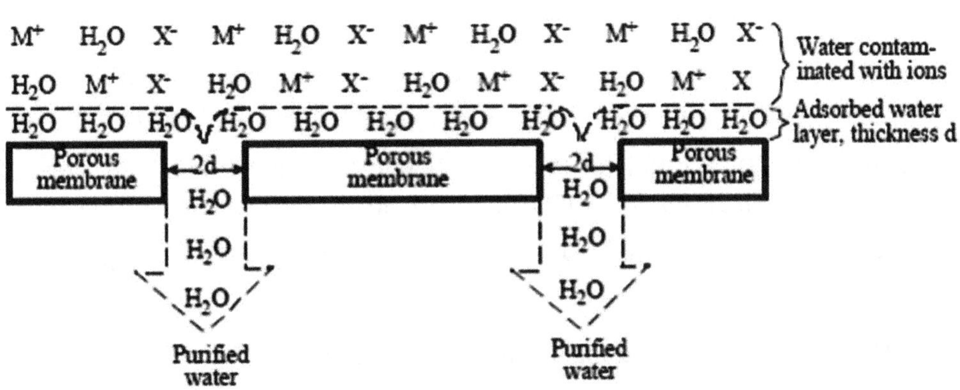

sufficient pressure to counteract the original water flow. The difference in levels represents the osmotic pressure of the solution (Figure 15).

The process can be reversed by applying sufficient pressure to the concentrated solution to overcome the osmotic pressure force the net flow of water through the membrane towards the dilute phase. The solute concentration (impurity) builds up on one side of the membrane while relatively pure water passes through the membrane. In order to obtain adequate solvent (water) flux through the membrane, pressures of the order of 4000 to 7000 kN/m^2 are required. Figure 16 represents the principle of operation of the reverse osmosis unit.

Figure 17, is a very useful and well-developed technique for the purification of water. Basically, it consists of forcing pure water through a semipermeable membrane that allows the passage of water but not of other material. This process, which is not simply sieve separation or ultrafiltration, depends on the preferential sorption of water on the surface of a porous cellulose acetate or polyamide membrane. Electrodeposition (reduction of metal ions to metal by electrons at an electrode), *reverse osmosis*, and *ion exchange* are frequently employed for metal removal. Solvent extraction using organic-soluble chelating substances is also effective in removing many metals. Cementation, a process by which a metal deposits by reaction of its ion with a more readily oxidized metal, may be employed:

$$Cu^{2+} + Fe \text{ (iron scrap)} \rightarrow Fe^{2+} + Cu \quad (19)$$

Activated carbon adsorption effectively removes some metals from water at the part per million level. Sometimes a chelating agent is sorbed to the charcoal to increase metal removal. Various physical-chemical treatment processes effectively remove heavy metals from wastewaters. One such treatment is lime precipitation followed by activated carbon filtration. Activated-carbon filtration may also be preceded by treatment with iron(III) chloride to form an iron(III) hydroxide floc, which is an effective heavy metals scavenger. Similarly, alum, which forms aluminum hydroxide, may be added prior to activated-carbon filtration. The form of the heavy metal has a strong effect upon the efficiency of metal removal. For example, chromium(VI) is normally more difficult to remove than chromium(III). Chelation may prevent metal removal by solubilizing metals. In the ealier days, removal of heavy metals has been largely a fringe benefit of wastewater treatment processes.

REFERENCES

Abbas, I., & Zaheer, S. (2014). Advanced oxidation process for wastewater treatment: A review *American International Journal of Research in Science, Technology, Engineering & Mathematics*, *7*(3), 189–191.

Adin, A., & Asano, T. (1998). The Role of Physical Chemical Treatment in Wastewater Reclamation and Reuse. *Water Science and Technology*, *37*(10), 79–80. doi:10.2166/wst.1998.0381

American Water Works Association. (1998). *Reverse Osmosis and Nanofiltration*. Denver, CO: American Water Works Association.

Ash, M. (1996). *Handbook of Water Treatment Chemicals*. Aldershot, UK: Gower.

Balaban, M. (Ed.). (1991). *Desalination and Water Re-use*. New York: Hemisphere.

Bitton, G. (1999). *Wastewater Microbiology*. New York: Wiley-Liss.

Casey, T. J., & Casey, J. T. (1997). *Unit Treatment Processes in Water and Wastewater Engineering*. New York: John Wiley & Sons.

Connell, G. F. (1996). *The Chlorination/Chloramination Handbook*. Denver, CO: American Water Works Association.

Cowan, R., Ellis, T. G., Higgins, M. J., Alagappan, G., & Park, K. (1996). Treatment Systems. Activated Sludge and Other Aerobic Suspended Culture Processes. *Water Environment Research*, *68*(4), 451–469. doi:10.2175/106143096X135317

Design of Municipal Wastewater Treatment Plants-MOP 8. (1998). 4th ed.). Alexandria, VA: Water Environment Federation.

Droste, R. (1996). *Theory and Practice of Water and Wastewater Treatment*. New York: John Wiley & Sons.

Edell, A., & Morrison, G. M. (1997). Critical Evaluation of Pesticides in the Aquatic Environment and their Removal from Drinking Water. *Sweden Vatten*, *53*, 355-364.

Faust, S. D., & Aly, O. M. (Eds.). (1997). *Chemistry of Water Treatment* (2nd ed.). Denver, CO: American Water Works Association.

Freeman, H. M. (Ed.). (1998). *Standard Handbook of Hazardous Waste Treatment and Disposal* (2nd ed.). New York: McGraw-Hill.

Gallagher, L. M., & Miller, L. A. (1996). *Clean Water Handbook* (2nd ed.). Rockville, MD: Government Institutes.

Gates, D. J. (1998). *The Chlorine Dioxide Handbook*. Denver, CO: American Water Works Association.

Geldreich, E. (1996). *Microbial Quality of Water Supply in Distribution Systems*. Boca Raton, FL: CRC Press/Lewis Publishers.

Haber, F., & Weiss, J. (n.d.). The catalytic decomposition of hydrogen peroxide by iron salts. *Proc.R.Soc. Series A, 147*, 332.

Hahn, H. H., Hoffmann, E., & Hallvard, Ÿ. (1998). *Chemical Water and Wastewater Treatment V: Proceedings of the 7th Gothenburg Symposium 1998*. Springer. Retrieved from https://www.sswm.info/sites/default/files/reference_attachments/SUBRAMANIAN%20Advanced%20Waste%20Water%20Treatment.pdf

Koch, G., & Siegrist, H. (1997). Denitrification with Methanol in Tertiary Filtration. *Water Research*, *31*(12), 3029–3038. doi:10.1016/S0043-1354(97)00174-7

Kuo, J.-F., Stahl, J. F., Chen, C.-L., & Bohlier, P. V. (1998). Dual Role of Activated Carbon Process for Water Reuse. *Water Environment Research*, *70*(2), 161–170. doi:10.2175/106143098X127008

Kurbiel, J. (Ed.). (1991). *Advanced Wastewater Treatment and Reclamation*. London: Pergamon.

Langlais, B., Recknow, D. A., & Brink, D. R. (Eds.). (1991). Ozone in Water Treatment: Application and Engineering. Lewis Publishers.

Malik, P. K., & Saha, S. K. (2003). Oxidation of direct dyes with hydrogen peroxide using ferrous ion catalyst. *Separation and Purification Technology*, *31*(3), 241–250. doi:10.1016/S1383-5866(02)00200-9

Mathie, A. J. (1999). *Chemical Treatment for Cooling Water*. Upper Saddle River, NJ: Prentice Hall.

Mays, L. W. (1999). *Water Distribution Systems Handbook*. New York: McGraw-Hill.

Minear, R. A., & Amy, G. L. (1996). *Disinfection By-Products in Water Treatment: The Chemistry of Their Formation and Control*. Boca Raton, FL: CRC Press/Lewis Publishers.

Montgomery, J. M. (1985). *Consulting Engineers, Water Treatment Principles and Design*. New York: John Wiley & Sons, Inc.

Muruganandham, M., & Swaminathan, M. (2004). Decolourisation of Reactive Orange4 by Fenton and photo-Fenton oxidation technology. *Dyes and Pigments*, *63*(3), 315–321. doi:10.1016/j.dyepig.2004.03.004

Norman, T., & Banuelos, G. (1999). *Phytoremediation of Contaminated Soil and Water*. Boca Raton, FL: CRC Press/Lewis Publishers.

Nyer, E. K. (1993). Groundwater Treatment Technology (2nd ed.). van Nostrand Reinhold.

Patrick, D. R., & White, G. C. (1993). *Handbook of Chlorination and Alternative Disinfectants* (3rd ed.). New York: Nostrand Reinhold.

Polevoy, S. (1996). *Water Science and Engineering*. London: Blackie Academic & Professional.

Rice, R. G. (1999). *Ozone Drinking Water Treatment Handbook*. Boca Raton, FL: CRC Press/Lewis Publishers.

Roques, H. (1996). *Chemical Water Treatment: Principles and Practice*. New York: VCH.

Scholze, R. J. (Ed.). (1988). *Biotechnology for Degradation of Toxic Chemicals in Hazardous Wastes*. Noyes Publications.

Sebastia & Joseph. (1997). Tertiary Treatment of Sewage for Reuse. *Chemical Engineering World*, *32*, 55–57.

Singer, P. C. (1999). *Formation and Control of Disinfection By-Products in Drinking Water*. Denver, CO: American Water Works Association.

Speitel, G. E. (1999). *Advanced Oxidation and Biodegradation Processes for the Destruction of TOC and DBP Precursors*. Denver, CO: AWWA Research Foundation.

Spellman, F. A. (2014). Water and wastewater treatment plant operations. Academic Press.

Spellman, F. R., & Whiting, N. E. (1999). *Water Pollution Control Technology: Concepts and Applications*. Rockville, MD: Government Institutes.

Steiner, V. (1998). UV Irradiation in Drinking Water and Wastewater Treatment for Disinfection. *Wasser Rohrbau*, *49*, 22–31.

Stevenson, D. G. (1997). *Water Treatment Unit Processes*. London: Imperial College Press. doi:10.1142/p063

Tchobanoglous, G., Darby, J., Bourgeous, K., McArdle, J., Genest, P., & Tylla, M. (1998). Ultrafiltration as an Advanced Tertiary Treatment Process for Municipal Wastewater. *Desalination*, *119*(1-3), 315–322. doi:10.1016/S0011-9164(98)00175-1

Thiemann, H., & Weiler, H. (1996). One Year of Operational Experience with the Largest River Water Reverse Osmosis Plant in Germany. *VGB-Kraftwerkstechnik, 76*, 1017–1022.

Titus, M. P., Molina, V. G., Baños, M. A., Giménez, J., & Plugas, S. (2004). Degradation of chlorophenols by means of advanced oxidation processes:a general review, *App.Cat.B. Env., 47*, 219–256.

Tomar, M. (1999). *Laboratory Manual for the Quality Assessment of Water and Wastewater*. Boca Raton, FL: CRC Press.

Trasatti, S. (2013). Electrochemistry and environment: The role of electrocatalysis. *ViaVenezian, 21*, 3.

Türkman, A., & Uslu, O. (Eds.). (1991). *New Developments in Industrial Wastewater Treatment*. Norwell, MA: Kluwer. doi:10.1007/978-94-011-3272-5

Van Der Hoek, J. P., Rijnbende, D. O., Lokin, C. J. A., Bonne, P. A. C., Loonen, M. T., & Hofman, J. A. M. H. (1998). Electrodialysis as an Alternative for Reverse Osmosis in an Integrated Membrane System. *Desalination, 117*(1-3), 159–172. doi:10.1016/S0011-9164(98)00086-1

Wachinski, J. E., & Etzel, A. M. (1997). Environmental Ion Exchange: Principles and Design. CRC Press/Lewis Publishers.

Wang, A., Qu, J., Ru, J., Liu, H., & Ge, J. (2005). Mineralization of anazodye Acid Red by electro-Fenton's reagent using an activated carbon fiber cathode. *Dyes and Pigments, 65*(3), 227–233. doi:10.1016/j.dyepig.2004.07.019

Waste Stabilization Pond. (n.d.). Retrieved from Akvo.org//www.akvo.org/wiki/index.php/Portal:Sanitation retrieved 6/3-2011

White, G. C. (1999). *Handbook of Chlorination and Alternative Disinfectants*. New York: John Wiley & Sons.

KEY TERMS AND DEFINITIONS

Aluminium Sulphate: The alum is dissolved in water and the aluminum ions, Al^{3+} that form, have a high capacity to neutralize the negative charges which are carried by the colloidal particles and which contribute to their stability.

Coagulation: In water treatment, the destabilization and initial aggregation of colloidal and finely divided suspended matter by the addition of a floc-forming chemical.

Flocculation: Flocculation is a physical process of slowly mixing the coagulated water to increase the probability of particle collision.

Hydrophilic: Hydrophilic colloidal material is mostly composed of organic material that is the common source of color in water. Hydrophilic compounds are surrounded by water molecules that tend to make these particles negatively charged as well.

Hydrophobic: Hydrophobic colloidal material is mostly inorganic material that contributes to turbidity and carries a negative electrical surface charge.

Oxidation: The addition of oxygen, removal of hydrogen, or removal of electrons.

Section 5

Biopesticides, Sustainability, Global Warming

Chapter 20
Prospects of Pesticide Contamination and Control Measures in Aquatic Systems:
A Green Approach

Roopa Rani
Manav Rachna University, India

Prem Kishore Patnala
Manav Rachna University, India

Vinayak V. Pathak
Manav Rachna University, India

ABSTRACT

In this chapter, up-to-date knowledge on extenuation strategies to diminish pesticide accumulation in aquatic systems, which has remained a major concern for ground water as well as surface water, adversely affecting aquatic ecosystems and humans through bio-magnification, are included. Several factors affect the toxicity of pesticides like dosage of concentration, relative toxicity, and chemical interactions. The best approach to decrease pesticide pollution in environment is to use safer, non-chemical control methods, and industrial or sewage superfluous should not be dumped into water reservoirs without proper pretreatment. Biological and chemical methods used for the control measures of pesticides pollution in aquatic systems. Thus, a greener approach for remedy of pesticide-contaminated aquatic system could be more cost-effective and sustainable.

INTRODUCTION

Application of chemical pesticide has become a valuable practice to enhance the agricultural productivity and yield. Despite of its beneficial use to control pests in crop fields, they are associated with severe environmental impacts affecting all the dimension of environment i.e. soil, air, water, sediment,

DOI: 10.4018/978-1-5225-6111-8.ch020

sludge, etc (Ali *et al.,* 2014). Resistant nature of these is found to cause environmental degradation of some pesticides. The residues derived from degradation of pesticides have been found to possess significant toxicological effects. Different types of pesticides are used to control pests, insects, weeds, etc. depending upon the type of crop being produced. Among different pesticides organochlorine pesticides such as DDT, hexachlorocyclohexane (HCH), aldrin and dieldrin are widely used in Asian countries to control the undesired agricultural pests or insects and contaminated disease (Abhilash & Singh, 2009). However, most of the developed countries have banned the use of organochlorine pesticides due to their bioaccumulation potential and other biological effects. Recent estimation of pesticide consumption revealed that the consumption of pesticide in India has increased from 55,540 tonnes in 2010-11 to 57,353 tonne in 2014-15 (FICCI, 2015). Pesticides are becoming the largest sub-segment of agrochemicals with 60% market share. Economic survey 2015-16 reported an increase in pesticide residue due to lack of implementation of pesticide use guidelines. Initially synthetic use of DDT was started in 1948-49 for the cure of malaria and locust. Today, Indian pesticide market is dominated by the significant share of insecticide with a projected growth of 2, 29, 884.8 Million INR in financial year 2018 (Ratings, 2017).

In aquatic ecosystem organochlorine pesticide can enter through different ways such as sewerage discharge from industrial sector, runoff from non point sources, wet and dry deposition, etc. A study reports that about 9000 tonnes of pesticides are applied annually in Ganga river basin (Santanu, Chattopadhyay & Vass, 2000). Substantial amount of pesticides are dissipated at the site of application through chemical and biological degradation process, though reasonable fraction of these pesticide residues reaches to the ocean through agricultural runoff, atmospheric transport and sewerage discharge (WHO, 1999). Earlier studies revealed that Ganges water is highly contaminated with hexachlorocyclohexane (HCH), dichlorodiphenyl- trichloroethane (DDT), endosulfan (Samanta, 2013; Kaushik *et al.,* 2010). Simultaneous presence of several pesticides is found to trigger the toxicity of aquatic ecosystem that causes the damage to non target species. Contamination of pesticide is often found as mixture of several pesticides in aquatic ecosystems, the synergistic effect of these became much toxic rather than individual pesticide. Wang *et al.,* found that mixture of 11 pesticides exhibited synergistic effects on zebrafish (*Danio rerio*) (Wang *et al.,* 2018).

The pesticides have different toxicity among different species and thus numerous methods are approached and developed to eradicate these from the aquatic ecosystem. Some of the conventional methods are chemical treatment, thermal degradation, etc but it consumes much energy and thus depletes the resources. In this era, there is a need for rapid, simple, green approaches like microwave radiations (Onuska & Terry, 1993), ultra-sonication (Benitez, Acero & Real, 2002), ozonation (Matouq *et al.,* 2008), adsorption using clay (Manisankar, Selvanathan & Vedhi, 2006), bio-beds (Spliid, Helweg & Heinrichson, 2006) and immobilized on surface of titanium dioxide (Echavia, Matzusawa & Negishi, 2009), carbon slurry (Gupta, Ali & Saini, 2006), biodegradation methods like white rot fungi (Bending, Friloux & Walker, 2002), bacterial (Ogram *et al.,* 1985) and solar-photocatalytic (Konstantinou & Albanis, 2003), etc. This chapter contains a brief discussion of some of the green methods used to eliminate pesticides in aquatic system.

PESTICIDE CLASSIFICATION

The organic compounds which are classified among pesticides (like dichloro diphenyl trichloro ethane, lindane) emerged in the beginning of 19 century, as a result to develop chemical weapons tested against insects. Thereafter, due to urbanization and industrialization it's over usage has been increased affecting human health and environment by means of bio-magnification. Each year 2,50,000- 3,70,000 thousand people die from deliberate ingestion of pesticides. Pesticide exposure poses a potential risk like neuritis, psychiatric problems, hepatorenales disorders and immunological and endocrine problems. Due to the over-usage of these pollutants, WHO proposed classification based on their toxic behavior tested against rats or other laboratory animals administrated orally or dermally estimating the lethal dosage (LD50) (Kim & Smith, 2001). Pesticides are categorized into four main groups based on chemical structure: organochlorines, organophosphates, carbamates and pyrethroids. Organochlorines are synthetic organic compounds having 5 or more chlorine atoms used to control a wide range of insects and they have a long term residual effect in environment as they are resistant to chemical and microbial degradation. Organochlorines, which affect electrophysiological properties and enzymatic neuronal membranes, show problems like accumulation in fatty tissues and are widely used in crops of grapes, lettuce, rice, tomato etc (Cruz-Alcalde, Sans & Esplugas, 2017). Organophosphates, like parathion, malathion, diaznon, etc are not completely persistent in the environment as it can be degraded through various chemicals and biological reactions. Such pesticides, when inhaled, acts on the central nervous system by inhibiting action of the enzyme acetyl cholinesterase and levels of neurotransmitter like acetylcholine. They are also mutagenic, since they act directly on DNA (Kole, Banerjee & Bhattacharyya, 2001). Carbamates are derived from carbamic acid with general formulae with R^1 as alcohol group and R^2 is methyl group and R^3 is hydrogen bonding with C and N and O atoms as shown in figure 1. These are less effective since they cause blocking of the neurotransmitter enzyme through reaction at carbonyl sites of the enzyme. Some of extremely useful carbamates are carbaryl, aminocarb etc. Pyrethins are synthetic analogue to natural occurring organic compounds originated mainly from the plant *chrysanthemum cinerariaefolium*. Pyrethins are acknowledged for their fast knocking down effect against insect, pests, low mammalian toxicity and facile biodegradation like photochemical degradation. Pyrethins act on the central nervous system causing delay in the closure of voltage gated sodium ion channels in the cell causing fluxing of sodium ion in insects and vertebrates. The resistance to itself is bypassed by pairing the insecticide

Figure 1.

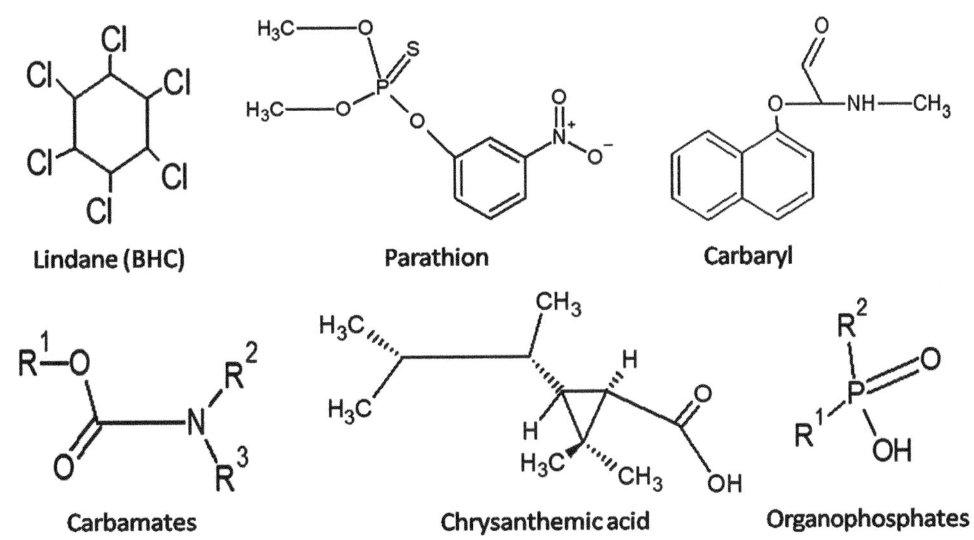

Table 1. Pesticide toxicity classification

Toxicity Class	Toxicity Rating	label
I	Highly toxic	Enormously Poison
II	Moderately toxic	Attention
III	Slightly toxic	Caution
IV	Mostly non toxic	Caution

Table 2. Categories of pesticides based on reactivity

	Pesticide Categories			
Type	Category I	Category II	Category III	Category IV
LD50 (Oral dosage)	Loading up to 50mg/kg	Loading from 50to 500mg/kg	Loading from 500 through 5000mg/kg	Greater than 5000mg/kg
LD 50 (Dermal)	Load up to 200mg/kg	Loading from 200 to 2000mg/kg	Loading from 2000 through 20000mg/ kg	Greater than 20000mg/ kg
Side effects	Corrosive to skin and Corneal opacity	Severe irritation to skin and corneal opacity	Modern irritation and no corneal opacity	Mild or slight irritation for 72hours

Table 3. Comparative toxicity to Natural Products

Pesticide	LD50(mg/kg)	Product Equal to Toxicity
TCDD	0.0002	Ricin, pure (castor bean extract)
Flocoumafen	0.25	Strychmine
Sarin	0.2	Black widow spider venom
Aldicarb	0.9	Nicotine alkaloid
Phorate	1.0	Heroin
Parathion	2.0	Morphine
Carbofuran	8	Codeine
Nicotine sulphate	50	Caffeine
Paraquat	150	Benadryl
Carbaryl	250	Vitamin A
Acephate	833	Salt (KCl, NaCl)
Allethrin	1160	Gasoline
Diazinon	1250	Tobacco
Malathion	5500	Castor oil
Ferbam	16900	Mineral oil
Methoprene	34600	Sugar
Pheromones	103750	Water

with synthetic synergists such as piperonyl butoxide. They prevent detoxification in the insects ensuring insect death. Synergists make pyrethrin more effective allowing lower doses to be effective. They have higher insect nerve sensitivity and are target specific. Some of pyrethroids include permethrin and cypermethrin and deltamerthrin. The relative toxicity ratings of different pesticides are shown in table 1 (Battu, Singh, Kang, & Joia, 2005).

One of the most important properties associated with the toxicity of pesticides is their doses. The pesticide dose (depending on exposure time) is expressed in LD 50 or LC 50. LD50 is meant for rodents and humans, given as LD50*70 (Average kilogram mass of humans). This dosage amount can be categorized into oral, dermal or side effects as shown in Table 2 (Battu, Singh, Kang, & Joia, 2005). A list of comparative toxicity of pesticides to natural products is shown in Table 3 mentioned in terms of $LD_{50.}$

TOXICOLOGICAL ASPECTS OF PESTICIDES

Toxicity is defined as the ability of a pesticide to cause damage to the organs of organisms. It is classified into two types acute and chronic. Acute toxicity defines about subjecting animals to different dosages or concentrations of the active ingredient for short term. These can be assessing the impact through skin or oral inhalation. Where chronic toxicity results by long-term or low exposure level to a toxicant. Major causes of chronic toxicity are Mutagenic toxicants, Oncogenic and Teratogenic toxicants.

Ecological Impact of Pesticides

Pesticides are wide ranged organic micro pollutants having ecological impacts too. They may cause contamination in soil or water which occur either with excessive use of weedicides, herbicides or insecticides that migrates to aquatic system through soil erosion or eutrophication. The major cause of pesticides pollution in water is the running off of contaminated water (Garcia *et al.*, 2012; WHO, 2010). Pesticides also move from one ecosystem to another ecosystem through processes like transformation (degradation) and transfer (mobility). The mobility of pesticides can occur through surface runoff, vapourization to environment, sorption (adsorption/desorption), plant uptake through soil or soil water fluxes. Degradation of pesticides may follow photo-degradation, chemical or biological degradation (Antwi & Reddy, 2015: Lakhani, 2015; Araya, Lherisson & Lomberk, 2014).

In earlier decades, pesticides were considered as “good friends of Human being” as they help us to control numerous insect-borne diseases such as malaria, encephalitis, and bubonic plague. Some of the benefits which are helpful and found to be associated with them are as follows –

1. They are relatively easy to apply to crop,
2. Not much costly,
3. Available easily
4. And in some cases it is the only practical method of control of microorganisms

However, the benefits through the use of pesticides are not devoid of penalty with special reference to aquatic system. The use of pesticides must be with utmost care for the health of humans, animals (aquatic and terrestrial species) and the environment. Toxic effects of pesticides include their toxicity to humans, animals, and useful plants, and the persistence (long life) of some of these chemicals in the environment.

A risk to water bodies by pesticides is dictated by their unique properties. The mechanism of concentration of pesticides in water bodies includes – bioavailability, bioconcentration and biomagnification.

Bioavailability stands for the amount of the pesticide present at the aquatic source freely available to fish and other aquatic wildlife for direct consumption or indirect adsorption. Some of the pesticide residues rapidly breakdown into its non-toxic constituents after application while some remains suspended in the water in their toxic form. However, some are stream bottomed thereby decreasing the risk of availability to organisms whereas some become rapidly diluted in aqueous media and are less available to aquatic life. Bioconcentration involves the accumulation of a chemical/pesticide from the environment to the organism body. For ex – fatty tissue can function as the sink of accumulation of some pesticides. Some of them are metabolized and excreted through excretory system like glyphosate. Pesticides, such as DDT, are "lipophilic", meaning that they are soluble in, and accumulate in fatty tissue such as edible fish tissue and human fatty tissue (Ibrahim, 2016). Some fish may concentrate certain pesticides in their body tissues and organs (especially fats) at levels 10 million times greater than in the water. Biomagnification defines the enhancement of the concentration of a particular chemical as the food energy is transferred to successive trophic level in the food chain. Some sport fish such as bass or trout constantly consume infected animals and bioconcentrate higher levels of pesticides in their body fat. Highest concentrations of chemicals are found among the top predators, including human being (Zacharia, 2011).

The effect of pesticides contamination on environment has severe impact like sudden death of certain species (aquatic or terrestrial), making the soil unfit for cultivation or water non-palatable, tumors or lesions on fish and other aquatic species, reproductive inhibition, weak immune system, hormonal disbalance, DNA damage, egg shell thinning, etc. These ecological impacts of are interrelated and depict the potential risk for human being. The application of pesticides to aquatic organisms is one of the most significant causes to the loss of aquatic biodiversity. Various pesticides have different effect on aquatic life which sometimes makes the aquatic life in stress. The effect may be chronic or acute depending upon the type of pesticides, their dosages, toxicity in terms of LD50 or LC50 and level or persistence.

The excessive use of herbicides or weedicides can also diminish reproductive success of aquatic species. The spray of herbicides near weedy nurseries may diminish the quantity of cover and shelter that smaller fish require in order to protect from predators and to feed. Mostly the smaller fish depends on aquatic phytoplankton as refuge in nursery areas.

Pesticides can be decomposed into its components either through sunlight (photodecomposition) or high air/water temperatures (thermal degradation), moisture conditions, biological action (microbial decay), and soil conditions (pH). Persistent (long-lasting) pesticides break down slowly and may be more available to aquatic animals.

Factors Affecting Toxicity of Pesticides Pollution in Aquatic System

An aquatic organism may have exposure to pesticides through various means like direct contact of pesticide with the habitat or the movement of aquatic species through the contaminated area. The effect of pesticides on human and other aquatic species can be a resultant of numerous factors (Mensah, Palmer & Muller, 2014; Reese *et al.,* 1973; Vassiliou, 2016) as:

1. **Exposure Time:** The toxicity of pesticides among aquatic species depends upon the time of contact with the pesticides content. Once a pesticide is present in an aquatic system, its persistence will affect the species particularly fish. The pesticide content accumulates in the fish through absorp-

tion in gills and can enter in the body of human being through food chain. Long exposure time has much hazards rather than short exposure time. It has been reported in many studies that the pesticides residue level were higher in fish than found in the water.

2. **Exposure Type:** The concentration of pesticide in aquatic system has a direct relation with its toxicity. The effect of its lethal or sub-lethal dose is a major concern for the aquatic species.
3. **Toxicity:** Toxicity of pesticide content depends upon its LD50 value. The lower the LD50, the greater the toxicity. A minute amount of a higher toxic chemical may be more dangerous than a heavier dose of a less toxic chemical.
4. **Persistence:** The persistence of pesticide content in aquatic system has a major impact its toxicity and is determined by living and non-living degradational processes.
5. **Degradates:** The toxicity of pesticide content also varies with the fate of degradation products which may have lesser, equal or greater toxicity than the parent compound.
6. **Mode of Intake:** Pesticide intake in aquatic species or non-aquatic species depends upon the mode of intake and the amount of intake in the body. It may be through skin contact, inhalation or ingestion.

Control Measures for Pesticides Pollution in Aquatic System: Chemical Methods

Pesticides contamination may be controlled by chemical or biological means. The best way of reducing the risk of pesticide contamination is to reduce the use of pesticides or use the other alternate in place of conventional pesticides. Promotion of organic farming is one of the suitable means to reduce the health risk associated with pesticides. The use of pesticides must be certified by the government which is non-toxic (Pimentel *et al.,* 1993). Farmers should be given proper training for the dosage of the pesticides applied to various crop and also the use of proper spraying equipment. The maximum limit of the pesticides must be decided through the consecutive study about the crop. For chemical means of control measures of pesticides pollution, the pesticides may be decided on the basis of the degradation means. For biological means, one of the means include metabolism of pesticides in animals. It is an important mechanism by which organisms defend themselves from the deadly properties of xenobiotics in their food supply. In this process, the chemical is decomposed into a less toxic form and either excreted or stored in the organism (Agrawal, Pandey, & Sharma, 2010). Different organs, especially the liver, may be involved, depending on the chemical. Enzymes play an important role in the metabolic process and the presence of certain enzymes, especially "mixed" function oxygenases (MFOs) in liver, is now used as an indicator that the organism has been exposed to foreign chemicals.

Biological control of pesticide toxicity has been reported by using aquatic plants such as *Lemna minor(L. minor)*, *Elodea canadensis* (*E. canadensis*) and *Cabomba aquatica* (*C. aquatic) (*Aubertot *et al.,* 2007*).* (Bhaskaran *et al.,* 2013) found that free floating macrophytes Pistia is suitable for removing perchlorate (ClO_4^-) from aquatic system. They found that a novel bacterium (*Acinetobacter)* found at the root of Pistia is responsible for the reduction of perchlorate completely in 48 hours. Application of algae for removal of organochlorine pesticide was also investigated by Wu and Kosaric, 1991 (Olette, Couderchet, Biagianti & Eullaffroy, 2008) in alum flocculated algae photobioreactor. Biodegradation was evaluated by using dead as well as living algal cells. The study revealed that relative rate of removal of pesticide was greater with live algal cells than the dead cells. It was concluded that application of continuous upflow photobioreactor system is efficient to remove the organochlorine pesticides.

Numerous methods of applying pesticides to crop are known and most of them govern only 10% to reach the target organism leaving behind the remaining percentage to get deposited on soil, water, sediments and sludge causing serious environmental threats. Evidences are available depicting the side effects of these pesticides to the environment and also the potential risk to human being & other life forms (Bhaskaran, 2013; Wu & Kosaric, 1991; Jeyaratnam, 1985). According to ICMR bulletin report, chronic illness and high mortality rate are being reported per year due toxicity of pesticide globally. Because of the potential hazards to the nature and humans many of the pollutants which are toxic in nature were subsequently banned.

Different studies have been reported describing various methods to reduce the effects of pesticides, their remediation and treatment methods. The conventional technologies includes physical treatments (such as adsorption and percolator filters); chemical treatments (oxidation using hydroxyl ions), heterogeneous photocatalysis with TiO_2 (Igbedioh, 1991), high temperature incineration in special furnaces, etc. However, these methods are associated with several demerits like they are expensive, tedious, time consuming and also has large detrimental impact on environment since their metabolites are persistent and evenly toxic to non-target organisms (Forget, 1993).

Some of the remediation methods are described below

1. **Thermal Desorption:** Thermal desorption is one of the method to remediate pesticides from such pesticides contaminated sites which include soils, sediments or sludge. This method involves the heating of matrix at low temperatures (300-1000°F) that makes the target species volatilized converting them into a gas stream. This gas stream is either condensed or treated by passing through a burner. This process doesn't destruct the compound but remove them from the matrix.
2. **Incineration:** This methodology involves the oxidation of sediments, soils or sludge at high temperature using oxygen. Incineration occurs in two steps. In the first step, the heating is carried out at temperature range of 1000-1800°F that causes partial oxidation and volatilizes the organic pesticides. Secondly, the temperature is further increased in the range 1600-2200°F resulting into complete destruction of target compounds. The ash obtained in the above method is disposed off after concerning the safety regulation with respect to environment.
3. **Phytoremediation:** This methodology of pesticide remediation involves the use of plants that can act as filters to remove pesticides from aqueous ecosystems. Various plant species have been found effective in remediating the pesticides from aqueous system, out of which, few are shown in Table 4.
4. **Bioremediation:** This method is used to treat the polluted sites of solid sediments, sludge, soils and ground water and is economic. This process increases the rate of the natural microbial degradation. Enhancement in degradation process is carried out by supplementing the microorganisms with electron donors and nutrients. Microbes further utilize their intrinsic characteristics to increase the complete rate of degradation of pollutant into H_2O and CO_2 without formation of intermediates. Along with the potentiality of microorganisms to degrade the pollutant, environmental conditions like pH, temperature, humidity and nutrients, also affect the process (Ferrusquía *et al.,* 2008; Singh & Thakur, 2006; Aislabie & Lloyd-Jones, 1995). Most of the pesticides polluted sites are found to be rich in microorganisms that degrade such pollutants. Upon complete biodegradation of the pesticide, the carbon dioxide and water are formed by the oxidation of the parent compound and this process provides the energy to the microbes for their metabolism.

Table 4. List of plants remediating the pesticides from aqueous system

S. No.	Name of the Plant Species	Pesticides Being Removed	Reference
1	*Water Hyacinth (Eichhornia Crassipes)*	Ethion	Xia *et al.*, 2006
2	*Lemna Minor*	Copper sulphate pesticides	Olette *et al.*, 2008
3	*Elodea Canadensis*	Flazasulfuron	Olette *et al.*, 2008
4	*Cabomba Aquatic*	Dimethomorth	Olette *et al.*, 2008
5	*Lemna Minor, Spirodela Polyrhiza, Carya Aquatica, Crocodylus Palustris, Elodea Canadensis*	Dimethomorph and Pyrimethanil	Dosnon-Olette *et al.*, 2009
6	*Pistia Stratiotes L, Dukeweed L. Minor Acorus Calamus*	Chlorpyrifos	Prasertup et al, 2011; Wang *et al.*, 2016
7	*Macrophytes and Algae*	Pyrethroids	Riaz *et al.*, 2017
8	*Saccharaum Officinarum, Candida*	Lindane	Salam *et al.*, 2017
9	*Prairie Grass*	Atrazine	Khrunyk *et al.*, 2017
10	*Festica Arundinacea, Lolium Multiflorum*	Terbuthylazine	Mimmo *et al.*, 2015
11	*Typha Latifolia, Phragmites Australis, Iris Pseudacorus, Juncus Effuses*	Imazalil and Tebuconazole	Lv *et al.*, 2016
12	*A. Thalinana*	Simazine	Azab *et al.*, 2016
13	Solanum lycopersicum, Helianthus annuus, Glycine max, Medicago sativa	Endosulphan	Mitton *et al.*, 2016
14	Plantago Major L	Cyanophos	Romeh, 2014
15	Cytisus Striatus	Hexachlorocyclohexane	Becerra Castro *et al.*, 2013

Organochlorine pesticides are recalcitrant in nature and repelling towards most of the micro-organisms since their breakdown rate is much slow causing long time persistence in environment and also in organisms (Iranzo *et al.*, 2001; Vischetti, Casucci & Perucci, 2002; Díaz, 2004). In many such cases bacteria have proved to be benediction for such polluting sites. These pesticides act mainly on the nervous system of animals causing lethal effects leading to respiratory failure once the nervous system function is disrupted. Organochloride pesticides may cause chronic health effects like cancer, neurological and teratogenic effects (Dua, Singh, Sethunathan & Johri, 2002). *Pseudomonas* sp. and *Spingomonas* sp. Dsp-1 stain found Chloropyrifos, imidacloprid and Profenofos to be an exclusive source of carbon and are isolated from a water waste irrigated agri-soil in India (Chaudhary & Chapalamadugu, 1991; Vaccari, Strom, & Alleman, 2005; Bhagobaty & Malik, 2010; Li, He & Li, 2007). *Xanthomonas* sp. 4R3-M3 and *Pseudomonas* sp. 4H1-M3 uses 3,5,6-trichloro-2-pyridinol and chlorpyriphos as carbon and nitrogen sources (Malghani, Chatterjee, Yu & Luo, 2009). *B. thuringiensis* strain degrades cyhalothrin and pyrethroides from waste water (Pandey, Chauhan, & Jain, 2009). Degradation of acetamiprid was possible using the *Rhodotorula mucilaginosa* strain IM-2 (Rayu, Nielsen, Nazaries & Singh, 2017).

Fungus can also be used to degrade pesticides through various means. *Phanerochaete chrysosporium* fungal species are well known to degrade polycyclic aromatic hydrocarbons and polychlorinated biphenyl. The degradation of pesticides endosulfa and endosulfan sulphage was found to be through white rot fungus *Trametes hirsute* (Chen *et al.,* 2015).

Enzymes like dehydrogenases, cytochrome p450, dioxigenases, ligninases (Dai *et al.,* 2010; Kamei, Takagi & Kondo, 2011; Bourquin, 1977) are also known to degrade certain pesticides. A strain of *Trichoderma harzianum* is known to possess the degradation properties of organochlorines (Castro, Wade, & Belser, 1985).

Algae are the key component of wetland that influences pesticides sorption in a greater extent. Algae participate in a large proportion in primary productivity of wetland. Algae reduce the bioavailability of pesticide by sorption and promote the degradation of pesticide. Degradation of pesticide may occur directly or indirectly by photolysis (Nadeau, Menn, Breen, & Sayler, 1994). Zeng and Arnold (2012) found that photochemically produced reactive intermediates (PPRIs) are the main components in photolysis pathway of pesticide in aquatic environment (Katayama & Matsumura, 1993). It was also reported that dissolve organic material acts as a photo sensitizer in limiting the persistence of pesticides.

Biosorption of pesticide by algae involve metabolic –independent processes such as absorption, adsorption, surface complexation, ion exchange and precipitation. Among these, biosorption is an instant process and also termed as passive process. This passive sorption takes place in the cell wall of both living and dead cells. Algal cell wall is composed of a three dimensional network of macromolecule, which carry negatively charged functional groups such as carboxy, hydroxyl, amine and phosphate. Many studies have performed on the basis of algae-bacteria interactions for degradation of pesticides such as monocrotophos, quinalphos, DDT, atrazine, α-endosulfan and methyl parathion by (Subhash Chandra Bose *et al.,* 2011; Friesen-Pankratz *et al.,* 2003). In algae-bacterial mediated pesticide degradation bacteria provides phytohormones and micronutrients to algae as a result of this supplement algae grow with a higher growth rate (Zeng & Arnold, 2012). Application of algae for removal of organochlorine pesticide was also investigated by Wu and Kosaric, (1991) in alum flocculated algae photobioreactor. Biodegradation was evaluated by using dead as well as living algal cells. The study revealed that relative rate of removal of pesticide was greater with live algal cells than the dead cells. It was concluded that application of continuous upflow photobioreactor system is efficient to remove the organochlorine pesticides. Lindane, carbaryl, chlordane, aldrin, dieldrin, endrin are some of the pesticides which are known to be degraded through algae (Subash Chandra Bose, 2011). *Selenastrum capricornutum,* green algae was found to removed atrazine and lindane from waste water (Ramanan *et al.,* 2016). Microorganisms are used for various purposes like organic pollutants, as the main source of energy for their growth and also detoxification and mineralization of the organic compounds or producing enzymes for enhancing degradation process (Klekner & Kosaric, 1992; Friesen-Pankratz, 2003; Larsson, 1990; Ab Rahman *et al.,* 2017)

APPLICATION OF BIO-PESTICIDES

Thus development of the alternative of this agricultural input has become important now days. In this regard biopesticides and biofertilizers are being developed, which contain non toxic formulations that control the growth of pest by non toxic mechanism. Biopestcides have been used since human civilization.

One of the most widely used biopesticide is derived from microbes such as *Bacillus thuringinesis*. There are various advantages associated with bio-pesticides such as environmental safety, target specificity, efficacy, biodegradability and suitability in integrated pest management (IPM). Microbial bio-control involves microbial antagonisms, plant-microbe interaction, production of antimicrobial compounds, and induced systemic resistance (Shoda, 2000). Antgonistic relationship between microbe and pathogen help in the disease control, in which rizosphere is invaded by microbes, where microbe secrets diffusible antibiotics, volatile organic compounds, toxins and develop extracellular cell wall degrading enzymes such as chitinase, glucanase, pectin methylesterase, etc (Anderson & Kim, 2018).

Among various microbial pesticides, *Pseudomonas chlororaphis* isolates are commonly used pesticide because of its potential for plant protection against the microbial pathogens, insects and nematodes. These isolates are found to produce certain metabolites which limit the pathogen activity through direct antimicrobial activity as well as participate in the iron metabolism (Kramer & Muthukrishnan, 1997). Berini *et al.,* 2018 has developed complete solution for fungicide, insecticide and nematicidal activities by developing chitinases (Berini, 2018). Chitinase have been isolated from tobacco hornworm, Manduca sexta, and several other insect species (Kramer & Muthukrishnan, 1997). Weeks *et al.,* found methionine as a novel bio-pesticide possess low contact and oral toxicity and classified as practically non toxic to bees (Weeks *et al.,* 2018). Methionine serves as essential amino acid required by adult and larval honey bees (*Aphis mellifera*). Emerging biocontrol strategies as an alternative of pesticide, involve the understating of mechanism by which plant itself release broad variety of chemical through its roots and leaves against the pathogenic activity. Holistic study on individual bioactive exudates in the rhizosphere is required for better understanding of attraction of individual microbes within microbiomes.

CONCLUSION AND FUTURE ASPECTS

India being the agrarian society consumed huge amount of pesticide to prevent the crop productivity. However, high biological activity and persistence nature, pesticide consumption causes adverse effect on human health and toxicity to the environment. On the other hand, pesticide use is required to improve the rural health as it is used to control the various vector borne diseases, contaminating insects and other pests. Residues of pesticide are found in food, crops, meat, aquaculture products as well as milk products due to indiscriminate use of pesticide in agriculture and public health. Agrochemical industries should adopt the product innovation strategies to derive the less eco-toxic product. Intervention of information technology should be adopted by these industries to make the farmers more aware towards the pesticide use guidelines and eco-friendly pesticides. Application of biotechnological tools to develop GM (Genetic modified) crops (resistant to the bacterial, fungal and viral infections) should be promoted among the farmers. Use of pesticide with controlled release rate such as micro-encapsulated pesticide or encapsulated pesticide granules should be promoted. In order to track the pesticide residues from agricultural commodities, emphasis should be given to the development of facility for pesticide residue testing near to the agricultural commodities. Promotion of organic farming is considered as potential alternative of pesticide and observed as socially, ecologically and economically sustainable. The key steps in organic farming involves the encourage and enhancement of biological cycle within the farming system, which not only increase the long term fertility of soil but also minimizes all forms of pollution caused by chemical fertilizers and pesticides.

REFERENCES

Ab Rahman, S. F. S., Singh, E., Pieterse, C. M., & Schenk, P. M. (2017). Emerging Microbial Biocontrol Strategies for Plant Pathogens. *Plant Science, 2018*. doi:10.1016/j.plantsci.2017.11.012 PMID:29362088

Abhilash, P. C., & Singh, N. (2009). Pesticide use and application: An Indian scenario. *Journal of Hazardous Materials, 165*(1-3), 1–12. doi:10.1016/j.jhazmat.2008.10.061 PMID:19081675

Agrawal, A., Pandey, R. S., & Sharma, B. (2010). Water pollution with special reference to pesticide contamination in India. *Journal of Water Resource and Protection, 2*(05), 432–448. doi:10.4236/jwarp.2010.25050

Aislabie, J., & Lloyd-Jones, G. (1995). A review of bacterial-degradation of pesticides. *Soil Research (Collingwood, Vic.), 33*(6), 925–942. doi:10.1071/SR9950925

Ali, U., Syed, J. H., Malik, R. N., Katsoyiannis, A., Li, J., Zhang, G., & Jones, K. C. (2014). Organochlorine pesticides (OCPs) in South Asian region: A review. *The Science of the Total Environment, 476*, 705–717. doi:10.1016/j.scitotenv.2013.12.107 PMID:24522007

Anderson, A. J., & Kim, Y. C. (2018). Biopesticides produced by plant-probiotic Pseudomonas chlororaphis isolates. *Crop Protection (Guildford, Surrey), 105*, 62–69. doi:10.1016/j.cropro.2017.11.009

Antwi, F. B., & Reddy, G. V. (2015). Toxicological effects of pyrethroids on non-target aquatic insects. *Environmental Toxicology and Pharmacology, 40*(3), 915–923. doi:10.1016/j.etap.2015.09.023 PMID:26509732

Araya, K., Lherisson, B., & Lomberk, J. (2014). *Pesticides, pollution, and people: an overview of public health and environment in Costa Rica*. Academic Press.

Aubertot, J. N., Barbier, J. M., Carpentier, A., Gril, J. J., Guichard, L., Lucas, P., & Voltz, M. (2007). *Pesticides, agriculture and the environment: reducing the utilization of pesticides and limiting their environmental impacts*. Academic Press.

Azab, E., Hegazy, A. K., El-Sharnouby, M. E., & Abd Elsalam, H. E. (2016). Phytoremediation of the organic Xenobiotic simazine by p450-1a2 transgenic Arabidopsis thaliana plants. *International Journal of Phytoremediation, 18*(7), 738–746. doi:10.1080/15226514.2015.1133559 PMID:26771455

Battu, R. S., Singh, B., Kang, B. K., & Joia, B. S. (2005). Risk assessment through dietary intake of total diet contaminated with pesticide residues in Punjab, India, 1999–2002. *Ecotoxicology and Environmental Safety, 62*(1), 132–139. doi:10.1016/j.ecoenv.2004.12.004 PMID:15978299

Becerra-Castro, C., Kidd, P. S., Rodríguez-Garrido, B., Monterroso, C., Santos-Ucha, P., & Prieto-Fernández, Á. (2013). Phytoremediation of hexachlorocyclohexane (HCH)-contaminated soils using Cytisus striatus and bacterial inoculants in soils with distinct organic matter content. *Environmental Pollution, 178*, 202–210. doi:10.1016/j.envpol.2013.03.027 PMID:23583940

Bending, G. D., Friloux, M., & Walker, A. (2002). Degradation of contrasting pesticides by white rot fungi and its relationship with ligninolytic potential. *FEMS Microbiology Letters, 212*(1), 59–63. doi:10.1111/j.1574-6968.2002.tb11245.x PMID:12076788

Benitez, F. J., Acero, J. L., & Real, F. J. (2002). Degradation of carbofuran by using ozone, UV radiation and advanced oxidation processes. *Journal of Hazardous Materials*, *89*(1), 51–65. doi:10.1016/S0304-3894(01)00300-4 PMID:11734346

Berini, F., Katz, C., Gruzdev, N., Casartelli, M., Tettamanti, G., & Marinelli, F. (2018). Microbial and viral chitinases: Attractive biopesticides for integrated pest management. *Biotechnology Advances*, *36*(3), 818–838. doi:10.1016/j.biotechadv.2018.01.002 PMID:29305895

Bhagobaty, R. K., & Malik, A. (2010). *Utilization of chlorpyrifos as a sole source of carbon by bacteria isolated from wastewater irrigated agricultural soils in an industrial ares of western Uttar Pradesh*. Academic Press.

Bhaskaran, K., Nadaraja, A. V., Tumbath, S., Shah, L. B., & Veetil, P. G. P. (2013). Phytoremediation of perchlorate by free floating macrophytes. *Journal of Hazardous Materials*, *260*, 901–906. doi:10.1016/j.jhazmat.2013.06.008 PMID:23872336

Bourquin, A. W. (1977). Degradation of malathion by salt-marsh microorganisms. *Applied and Environmental Microbiology*, *33*(2), 356–362. PMID:192147

Castro, C. E., Wade, R. S., & Belser, N. O. (1985). Biodehalogenation: Reactions of cytochrome P-450 with polyhalomethanes. *Biochemistry*, *24*(1), 204–210. doi:10.1021/bi00322a029 PMID:4039602

Chaudhry, G. R., & Chapalamadugu, S. (1991). Biodegradation of halogenated organic compounds. *Microbiological Reviews*, *55*(1), 59–79. PMID:2030673

Chen, S., Deng, Y., Chang, C., Lee, J., Cheng, Y., Cui, Z., & Zhang, L. H. (2015). Pathway and kinetics of cyhalothrin biodegradation by Bacillus thuringiensis strain ZS-19. *Scientific Reports*, *5*(1), 8784. doi:10.1038rep08784 PMID:25740758

Cruz-Alcalde, A., Sans, C., & Esplugas, S. (2017). Exploring ozonation as treatment alternative for methiocarb and formed transformation products abatement. *Chemosphere*, *186*, 725–732. doi:10.1016/j.chemosphere.2017.08.040 PMID:28820996

Dai, Y. J., Ji, W. W., Chen, T., Zhang, W. J., Liu, Z. H., Ge, F., & Yuan, S. (2010). Metabolism of the neonicotinoid insecticides acetamiprid and thiacloprid by the yeast Rhodotorula mucilaginosa strain IM-2. *Journal of Agricultural and Food Chemistry*, *58*(4), 2419–2425. doi:10.1021/jf903787s PMID:20112912

Díaz, E. (2004). Bacterial degradation of aromatic pollutants: A paradigm of metabolic versatility. *International Microbiology*, *7*(3), 173–180. PMID:15492931

Dosnon-Olette, R., Couderchet, M., & Eullaffroy, P. (2009). Phytoremediation of fungicides by aquatic macrophytes: Toxicity and removal rate. *Ecotoxicology and Environmental Safety*, *72*(8), 2096–2101. doi:10.1016/j.ecoenv.2009.08.010 PMID:19732953

Dua, M., Singh, A., Sethunathan, N., & Johri, A. (2002). Biotechnology and bioremediation: Successes and limitations. *Applied Microbiology and Biotechnology*, *59*(2-3), 143–152. doi:10.100700253-002-1024-6 PMID:12111139

Echavia, G. R. M., Matzusawa, F., & Negishi, N. (2009). Photocatalytic degradation of organophosphate and phosphonoglycine pesticides using TiO2 immobilized on silica gel. *Chemosphere*, *76*(5), 595–600. doi:10.1016/j.chemosphere.2009.04.055 PMID:19481779

Ferrusquía, C. H., Roa, G., García, M., Amaya, A., & Pavón, T. (2008). Evaluación de la degradación de metilparatión en soluciónusandofotocatálisisheterogénea. Rev. *Latinoamericana de RecursosNaturales*, *4*(2), 285–290.

FICCI. (2015). *Ushering in the 2nd Green Revolution: Role of crop protection chemicals. A report on Indian agrochemical industry*. Author.

Forget, G. (1993). Balancing the need for pesticides with the risk to human health. In Impact of pesticide use on health in developing countries: proceedings of a symposium, Ottawa, Canada.

Friesen-Pankratz, B. B., Doebel, C. C., Farenhorst, A. A., & Gordon Goldsborough, L. (2003). Interactions between algae (Selenastrum capricornutum) and pesticides: Implications for managing constructed wetlands for pesticide removal. *Journal of Environmental Science and Health. Part. B, Pesticides, Food Contaminants, andAgricultural Wastes*, *38*(2), 147–155. doi:10.1081/PFC-120018445 PMID:12617553

Garcia, F. P., Ascencio, S. Y. C., Oyarzun, J. C. G., Hernandez, A. C., & Alavarado, P. V. (2012). Pesticides: Classification, uses and toxicity. Measures of exposure and genotoxic risks. *Int. J. Environ. Sci. Toxic. Res*, *1*, 279–293.

Gupta, V. K., Ali, I., & Saini, V. K. (2006). Adsorption of 2, 4-D and carbofuran pesticides using fertilizer and steel industry wastes. *Journal of Colloid and Interface Science*, *299*(2), 556–563. doi:10.1016/j.jcis.2006.02.017 PMID:16527294

Ibrahim, Y. A. (2016). Health and environmental impacts of pesticides: A responsibility principle and two novel systems for hazard classification and external cost determination. *Journal of Toxicology and Health*, *3*(1), 1. doi:10.7243/2056-3779-3-1

Igbedioh, S. O. (1991). Effects of agricultural pesticides on humans, animals, and higher plants in developing countries. *Archives of Environmental Health: An International Journal*, *46*(4), 218–224. doi:10.1080/00039896.1991.9937452 PMID:2069430

Iranzo, M., Sainz-Pardo, I., Boluda, R., Sanchez, J., & Mormeneo, S. (2001). The use of microorganisms in environmental remediation. *Annals of Microbiology*, *51*(2), 135–144.

Jeyaratnam, J. (1985). Health problems of pesticide usage in the Third World. *British Journal of Industrial Medicine*, *42*(8), 505. PMID:4016001

Kamei, I., Takagi, K., & Kondo, R. (2011). Degradation of endosulfan and endosulfan sulfate by white-rot fungus Trametes hirsuta. *Journal of Wood Science*, *57*(4), 317–322. doi:10.100710086-011-1176-z

Katayama, A., & Matsumura, F. (1993). Degradation of organochlorine pesticides, particularly endosulfan, by Trichoderma harzianum. *Environmental Toxicology and Chemistry*, *12*(6), 1059–1065. doi:10.1002/etc.5620120612

Kaushik, A., Sharma, H. R., Jain, S., Dawra, J., & Kaushik, C. P. (2010). Pesticide pollution of river Ghaggar in Haryana, India. *Environmental Monitoring and Assessment, 160*(1-4), 61–69. doi:10.100710661-008-0657-z PMID:19107563

Khrunyk, Y., Schiewer, S., Carstens, K. L., Hu, D., & Coats, J. R. (2017). Uptake of C14-atrazine by prairie grasses in a phytoremediation setting. *International Journal of Phytoremediation, 19*(2), 104–112. doi:10.1080/15226514.2016.1193465 PMID:27259078

Kim, J. H., & Smith, A. (2001). Distribution of organochlorine pesticides in soils from South Korea. *Chemosphere, 43*(2), 137–140. doi:10.1016/S0045-6535(00)00281-2 PMID:11297392

Klekner, V., & Kosaric, N. (1992). Degradation of phenols by algae. *Environmental Technology, 13*(5), 493–501. doi:10.1080/09593339209385176

Kole, R. K., Banerjee, H., & Bhattacharyya, A. (2001). Monitoring of market fish samples for Endosulfan and Hexachlorocyclohexane residues in and around Calcutta. *Bulletin of Environmental Contamination and Toxicology, 67*(4), 554–559. doi:10.1007001280159 PMID:11779071

Konstantinou, I. K., & Albanis, T. A. (2003). Photocatalytic transformation of pesticides in aqueous titanium dioxide suspensions using artificial and solar light: Intermediates and degradation pathways. *Applied Catalysis B: Environmental, 42*(4), 319–335. doi:10.1016/S0926-3373(02)00266-7

Kramer, K. J., & Muthukrishnan, S. (1997). Insect chitinases: Molecular biology and potential use as biopesticides. *Insect Biochemistry and Molecular Biology, 27*(11), 887–900. doi:10.1016/S0965-1748(97)00078-7 PMID:9501415

Lakhani, L. (2015). How to Reduce Impact of Pesticides in Aquatic Environment. *Social Issues and Environmental Problems, 3*(9), 29–38.

Larsson, T. (1990). *Sidoeffekter av svampbekämpningsmedel*. SLU Info/Redaktionen. Fakta – Mark/Växter, no 6.

Li, X., He, J., & Li, S. (2007). Isolation of a chlorpyrifos-degrading bacterium, Sphingomonas sp. strain Dsp-2, and cloning of the mpd gene. *Research in Microbiology, 158*(2), 143–149. doi:10.1016/j.resmic.2006.11.007 PMID:17306510

Lv, T., Zhang, Y., Casas, M. E., Carvalho, P. N., Arias, C. A., Bester, K., & Brix, H. (2016). Phytoremediation of imazalil and tebuconazole by four emergent wetland plant species in hydroponic medium. *Chemosphere, 148*, 459–466. doi:10.1016/j.chemosphere.2016.01.064 PMID:26841287

Malghani, S., Chatterjee, N., Yu, H. X., & Luo, Z. (2009). Isolation and identification of profenofos degrading bacteria. *Brazilian Journal of Microbiology, 40*(4), 893–900. doi:10.1590/S1517-83822009000400021 PMID:24031438

Manisankar, P., Selvanathan, G., & Vedhi, C. (2006). Determination of pesticides using heteropolyacid montmorillonite clay-modified electrode with surfactant. *Talanta, 68*(3), 686–692. doi:10.1016/j.talanta.2005.05.021 PMID:18970376

Matouq, M. A., Al-Anber, Z. A., Tagawa, T., Aljbour, S., & Al-Shannag, M. (2008). Degradation of dissolved diazinon pesticide in water using the high frequency of ultrasound wave. *Ultrasonics Sonochemistry*, *15*(5), 869–874. doi:10.1016/j.ultsonch.2007.10.012 PMID:18178508

Mensah, P. K., Palmer, C. G., & Muller, W. J. (2014). Lethal and sublethal effects of pesticides on aquatic organisms: the case of a freshwater shrimp exposure to Roundup. In *Pesticides-Toxic Aspects*. InTech.

Mimmo, T., Bartucca, M. L., Del Buono, D., & Cesco, S. (2015). Italian ryegrass for the phytoremediation of solutions polluted with terbuthylazine. *Chemosphere*, *119*, 31–36. doi:10.1016/j.chemosphere.2014.04.114 PMID:24954449

Mitton, F. M., Gonzalez, M., Monserrat, J. M., & Miglioranza, K. S. (2016). Potential use of edible crops in the phytoremediation of endosulfan residues in soil. *Chemosphere*, *148*, 300–306. doi:10.1016/j.chemosphere.2016.01.028 PMID:26814704

Nadeau, L. J., Menn, F. M., Breen, A., & Sayler, G. S. (1994). Aerobic degradation of 1, 1, 1-trichloro-2, 2-bis (4-chlorophenyl) ethane (DDT) by Alcaligenes eutrophus A5. *Applied and Environmental Microbiology*, *60*(1), 51–55. PMID:8117093

Ogram, A. V., Jessup, R. E., Ou, L. T., & Rao, P. S. C. (1985). Effects of sorption on biological degradation rates of (2, 4-dichlorophenoxy) acetic acid in soils. *Applied and Environmental Microbiology*, *49*(3), 582–587. PMID:3994366

Olette, R., Couderchet, M., Biagianti, S., & Eullaffroy, P. (2008). Toxicity and removal of pesticides by selected aquatic plants. *Chemosphere*, *70*(8), 1414–1421. doi:10.1016/j.chemosphere.2007.09.016 PMID:17980900

Onuska, F. I., & Terry, K. A. (1993). Extraction of pesticides from sediments using a microwave technique. *Chromatographia*, *36*(1), 191–194. doi:10.1007/BF02263861

Pandey, J., Chauhan, A., & Jain, R. K. (2009). Integrative approaches for assessing the ecological sustainability of in situ bioremediation. *FEMS Microbiology Reviews*, *33*(2), 324–375. doi:10.1111/j.1574-6976.2008.00133.x PMID:19178567

Pimentel, D., McLaughlin, L., Zepp, A., Lakitan, B., Kraus, T., Kleinman, P., & Selig, G. (1993). *Environmental and economic effects of reducing pesticide use in agriculture*. Agriculture and the Environment.

Prasertsup, P., & Ariyakanon, N. (2011). Removal of chlorpyrifos by water lettuce (Pistia stratiotes L.) and duckweed (Lemna minor L.). *International Journal of Phytoremediation*, *13*(4), 383–395. doi:10.1080/15226514.2010.495145 PMID:21598800

Ramanan, R., Kim, B. H., Cho, D. H., Oh, H. M., & Kim, H. S. (2016). Algae–bacteria interactions: Evolution, ecology and emerging applications. *Biotechnology Advances*, *34*(1), 14–29. doi:10.1016/j.biotechadv.2015.12.003 PMID:26657897

Ratings. (2017). *Outlook of Indian pesticide industry*. Credit analysis and research limited, Mumbai, India.

Rayu, S., Nielsen, U. N., Nazaries, L., & Singh, B. K. (2017). Isolation and molecular characterization of novel chlorpyrifos and 3, 5, 6-trichloro-2-pyridinol-degrading bacteria from sugarcane farm soils. *Frontiers in Microbiology*, *8*, 518. doi:10.3389/fmicb.2017.00518 PMID:28421040

Reese, D. C., Ddoson, I. W., Ulrich, V., Becker, D. L., & Kempter, C. J. (1973). Pesticides in the Aquatic Environment. National Technical Advisory Committee on Pesticides in Water Environments. Environmental Protection Agency, Office of Water Programs.

Riaz, G., Tabinda, A. B., Iqbal, S., Yasar, A., Abbas, M., Khan, A. M., & Baqar, M. (2017). Phytoremediation of organochlorine and pyrethroid pesticides by aquatic macrophytes and algae in freshwater systems. *International Journal of Phytoremediation*, *19*(10), 894–898. doi:10.1080/15226514.2017.1303808 PMID:28448165

Romeh, A. A. (2014). Phytoremediation of cyanophos insecticide by Plantago major L. in water. *Journal of Environmental Health Science & Engineering*, *12*(1), 38. doi:10.1186/2052-336X-12-38 PMID:24447385

Salam, J. A., Hatha, M. A., & Das, N. (2017). Microbial-enhanced lindane removal by sugarcane (Saccharum officinarum) in doped soil-applications in phytoremediation and bioaugmentation. *Journal of Environmental Management*, *193*, 394–399. doi:10.1016/j.jenvman.2017.02.006 PMID:28259469

Samanta, S. (2013). Metal and pesticide pollution scenario in Ganga River system. *Aquatic Ecosystem Health & Management*, *16*(4), 454–464. doi:10.1080/14634988.2013.858587

Sánchez-Bayo, F., van den Brink, P. J., & Mann, R. M. (2012). *Ecological Impacts of Toxic Chemicals*. Francisco Sanchez-Bayo.

Santanu, G., Chattopadhyay, G. N., & Vass, K. K. (2000). Environmental impact assessment of lower Ganga system. *Environment and Ecology*, *18*(1), 126–129.

Shoda, M. (2000). Bacterial control of plant diseases. *Journal of Bioscience and Bioengineering*, *89*(6), 515–521. doi:10.1016/S1389-1723(00)80049-3 PMID:16232790

Singh, P., & Thakur, I. S. (2006). Colour removal of anaerobically treated pulp and paper mill effluent by microorganisms in two steps bioreactor. *Bioresource Technology*, *97*(2), 218–223. doi:10.1016/j.biortech.2005.02.022 PMID:16171678

Spliid, N. H., Helweg, A., & Heinrichson, K. (2006). Leaching and degradation of 21 pesticides in a full-scale model biobed. *Chemosphere*, *65*(11), 2223–2232. doi:10.1016/j.chemosphere.2006.05.049 PMID:16815528

Subashchandrabose, S. R., Ramakrishnan, B., Megharaj, M., Venkateswarlu, K., & Naidu, R. (2011). Consortia of cyanobacteria/microalgae and bacteria: Biotechnological potential. *Biotechnology Advances*, *29*(6), 896–907. doi:10.1016/j.biotechadv.2011.07.009 PMID:21801829

Vaccari, D. A., Strom, P. F., & Alleman, J. E. (2005). *Environmental biology for engineers and scientists*. John Wiley & Sons. doi:10.1002/0471741795

Vassiliou, G. (2016). Factors Affecting Risk Assessment of Pesticides in Water Bodies: A Review. *MOJ Toxicol*, *2*(1), 00030. doi:10.15406/mojt.2016.02.00030

Vischetti, C., Casucci, C., & Perucci, P. (2002). Relationship between changes of soil microbial biomass content and imazamox and benfluralin degradation. *Biology and Fertility of Soils*, *35*(1), 13–17. doi:10.100700374-001-0433-5

Wang, Q., Li, C., Zheng, R., & Que, X. (2016). Phytoremediation of chlorpyrifos in aqueous system by riverine macrophyte, Acorus calamus: Toxicity and removal rate. *Environmental Science and Pollution Research International, 23*(16), 16241–16248. doi:10.100711356-016-6673-6 PMID:27154841

Wang, Y., Wu, S., Chen, J., Zhang, C., Xu, Z., Li, G., & Wang, Q. (2018). Single and joint toxicity assessment of four currently used pesticides to zebrafish (Daniorerio) using traditional and molecular endpoints. *Chemosphere, 192*, 14–23. doi:10.1016/j.chemosphere.2017.10.129 PMID:29091792

Weeks, E. N., Schmehl, D. R., Baniszewski, J., Tomé, H. V., Cuda, J. P., Ellis, J. D., & Stevens, B. R. (2018). Safety of methionine, a novel biopesticide, to adult and larval honey bees (Apis mellifera L.). *Ecotoxicology and Environmental Safety, 149*, 211–216. doi:10.1016/j.ecoenv.2017.11.026 PMID:29175348

World Health Organization. (1999). *Food safety issues associated with products from aquaculture: report of a joint FAO/NACA/WHO study group* (Vol. 883). World Health Organization.

World Health Organization. (2010). *The WHO recommended classification of pesticides by hazard and guidelines to classification 2009*. WHO.

Wu, X. F., & Kosaric, N. (1991). Removal of organochlorine compounds in an upflow flocculated algae photobioreactor. *Water Science and Technology, 24*(5), 221–232. doi:10.2166/wst.1991.0128

Xia, H., & Ma, X. (2006). Phytoremediation of ethion by water hyacinth (Eichhornia crassipes) from water. *Bioresource Technology, 97*(8), 1050–1054. doi:10.1016/j.biortech.2005.04.039 PMID:15982870

Zacharia, J. T. (2011). Ecological Effects of Pesticides. In Pesticides in the Modern World-Risks and Benefits. InTech.

Zeng, T., & Arnold, W. A. (2012). Pesticide photolysis in prairie potholes: Probing photosensitized processes. *Environmental Science & Technology, 47*(13), 6735–6745. doi:10.1021/es3030808 PMID:23116462

KEY TERMS AND DEFINITIONS

Bioremediation: The use of either naturally occurring or deliberately introduced microorganisms to consume and break down environmental pollutants, in order to clean a pesticide-polluted site.

Control Measures: Control measures refer to the steps taken to prevent the pesticide pollution and then measures taken to remove the pesticides from contaminated sites. They may lead to complete removal of pesticides or diluting to a nontoxic level.

Phytoremediation: Phytoremediation is the direct use of living green plants for in situ, or in place, removal, degradation, or containment of contaminants in soils, sludges, sediments, surface water, and groundwater. It is a low-cost, solar-energy-driven cleanup technique.

Chapter 21
Biopesticide Techniques to Remediate Pesticides in Polluted Ecosystems

Rouf Ahmad Bhat
Sher-e-Kashmir University of Agricultural Sciences and Technology, India

Bilal A. Beigh
Barkatullah University, India

Shafat A. Mir
Sher-e-Kashmir University of Agricultural Sciences and Technology, India

Shakeel Ahmad Dar
Sher-e-Kashmir University of Agricultural Sciences and Technology, India

Moonisa Aslam Dervash
Sher-e-Kashmir University of Agricultural Sciences and Technology, India

Asmat Rashid
Sher-e-Kashmir University of Agricultural Sciences and Technology, India

Rafiq Lone
SBBS University, India

ABSTRACT

Indiscriminate and incessant use of synthetic pesticides is becoming an increasing global concern. No doubt, the application of conventional synthetic pesticides has enhanced the quality and quantity of agricultural products. However, accumulation of pesticides in freshwater resources has negative effects on aquatic ecosystem and human health. The persistent and toxic nature of pesticides has led to direct or indirect exposure on the biota in aquatic ecosystems resulting in acute (mortality of organisms) and chronic effects (decreased production and change in community structure), thus posing serious consequences for the ecosystem. Biopesticides provides a cost-effective and innovative approach employing bioremediation techniques for the removal of pesticides in water because of its advantage linked with

DOI: 10.4018/978-1-5225-6111-8.ch021

environmental safety, biodegradability, effectiveness, and target-specificity. Furthermore, biopesticides provide an efficient method for detoxification of pesticides and appropriateness in the integrated pest management (IPM) programs.

INTRODUCTION

The population of the planet earth is showing increasing trend and is projected to grow around 10.12 billion by the end of 21st Century. The time demands an intensive farming approach to fulfil the food requirements of the growing population. The highest yield of crops is based on the improved variety, the appropriate pest and disease management, and recommended fertilization (Birch, 2011; Nawaz, 2016). The role of chemical pesticides cannot be ignored in terms of increase in crop protection and production over the years. However at the same time chemical pesticides are considered as main causative agents for accelerated contamination of environment. Similarly, they have been the main cause of insect resistance as well as adverse impacts on natural enemies and humans (Alzaidi, 2011; Ishtiaq, 2012).

Before the use of pesticides, pests were responsible for enormous losses of agricultural produce and had grave impact on farming and agricultural practices. About 30% of agricultural produce is lost due to pests. Hence, the use of pesticides has become indispensable in agriculture. As agricultural production increased over the past few decades, farmers became more and more dependent on synthetic pesticides. Intensive use of pesticides over the years for increasing the overall production has resulted diverse types of hazards and toxicity and thus has affected the environment and non-target organisms (Tripathi & Tripathi, 2000). The pattern of pesticide usage in India is different from that for the world in general. As can be seen in Figure 1, in India 76% of the pesticide used is insecticide, as against 44% globally (Mathur, 1999). The use of herbicides and fungicides is correspondingly less heavy. The main use of pesticides in India is for cotton crops (45%), followed by paddy and wheat.

IMPACTS OF PESTICIDES ON ENVIRONMENTAL QUALITY

1. Impact on Humans

If the credits of pesticides include enhanced economic potential in terms of increased production of food and fibre, and amelioration of vector-borne diseases, then their debits have resulted in serious health implications to man and his environment (Aktar et al., 2009). There is now overwhelming evidence that some of these chemicals do pose a potential risk to humans and other life forms and unwanted side effects to the environment (Forget, 1993; Igbedioh, 1991; Jeyaratnam, 1981). No segment of the population is completely protected against exposure to pesticides and the potentially serious health effects, though a disproportionate burden is shouldered by the people of developing countries and by high risk groups in each country (WHO, 1990). The high risk groups exposed to pesticides include production workers, formulators, sprayers, mixers, loaders and agricultural farm workers. During manufacture and formulation, the possibility of hazards may be higher because the processes involved are not risk free. In industrial settings, workers are at increased risk since they handle various toxic chemicals including pesticides, raw materials, toxic solvents and inert carriers (Aktar et al., 2009).

Figure 1. Consumption pattern of pesticides

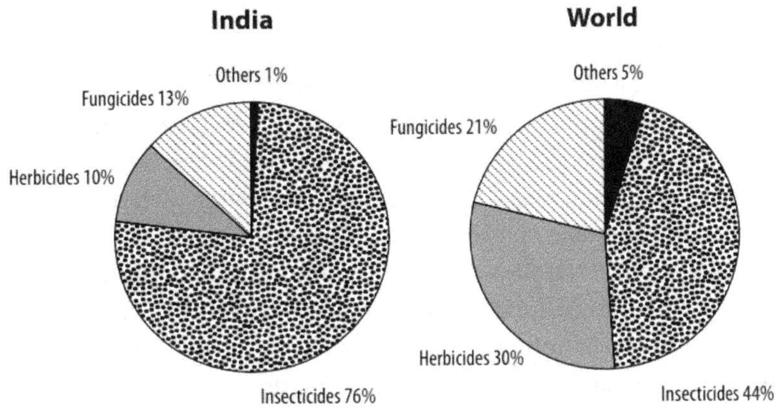

A study on those affected in the Seveso disaster of 1976 in Italy during the production of 2,4,5 T, a herbicide, concluded that chloracne (nearly 200 cases with a definite exposure dependence) was the only effect established with certainty as a result of dioxin formation (Pier et al., 1998). Early health investigations including liver function, immune function, neurologic impairment, and reproductive effects yielded inconclusive results. An excess mortality from cardiovascular and respiratory diseases was uncovered, possibly related to the psychosocial consequences of the accident in addition to the chemical contamination. An excess of diabetes cases was also found. Results of cancer incidence and mortality follow-up showed an increased occurrence of cancer of the gastrointestinal sites and of the lymphatic and haematopoietic tissue. Results cannot be viewed as conclusive, however, because of various limitations: few individual exposure data, short latency period, and small population size for certain cancer types. A similar study in 2001 observed no increase in all-cause and all-cancer mortality. However, the results support the notion that dioxin is carcinogenic to humans and corroborate the hypotheses of its association with cardiovascular- and endocrine-related effects (Pier et al., 2001).

2. Impact on Surface Water Quality

Pesticides can reach surface water through runoff from treated plants and soil. Contamination of water by pesticides is widespread. More than 90 percent of water and fish samples from all streams contained one, or more often, several pesticides (Kole et al.; 2001). The herbicides 2, 4-D, diuron, and prometon, and the insecticides chlorpyrifos and diazinon, all commonly used by urban homeowners and school districts, were among the 21 pesticides detected most often in surface and ground water across the nation (U.S. Geological Survey, 1998). Trifluralin and 2,4-D were found in water samples collected in 19 out of the 20 river basins studied (Bevans et al., 1998; Fenelon et al., 1998; Levings et al., 1998; Wall et al., 1998). The USGS also found that concentrations of insecticides in urban streams commonly exceeded guidelines for protection of aquatic life (U.S. Geological Survey, 1999).

3. Impact on Soils

A large number of transformation products (TPs) from a wide range of pesticides have been documented (Barcelo' an&d Hennion, 1997; Roberts, 1998; Roberts & Hutson, 1999). Not many of all possible pesticide TPs have been monitored in soil, showing that there is a pressing need for more studies in this field. Pesticides and TPs could be grouped into:(a) Hydrophobic, persistent, and bioaccumulable pesticides that are strongly bound to soil. Pesticides that exhibit such behavior include the organochlorine DDT, endosulfan, endrin, heptachlor,lindane and their TPs. Most of them are now banned in agriculture but their residues are still present. (b) Polar pesticides are represented mainly by herbicides but they include also carbamates, fungicides and some organophosphorus insecticide TPs. They can be moved from soil by runoff and leaching, thereby constituting a problem for the supply of drinking water to the population. The pesticides and their TPs are retained by soils to different degrees, depending on the interactions between soil and pesticide properties. The most influential soil characteristic is the organic matter content. The larger the organic matter content, the greater the adsorption of pesticides and TPs. The capacity of the soil to hold positively charged ions in an exchangeable form is important with paraquat and other pesticides that are positively charged. Strong mineral acid is required for extracting these chemicals, without any analytical improvement or study reported in recent years. Soil pH is also of some importance. Adsorption increases with decreasing soil pH for ionizable pesticides (Andreu & Pico', 2004).

Biomagnification, Bioaccumulation of Harmful Pesticides and Their Harmful Effects on Food Chain

Due to recalcitrant and non-degradable nature of pesticides, they show bio-accumulation as well as bio-magnification in the environment, damaging ecosystem stability, triggering pollution and inflicting diseases at alarming levels (Gerhardson, 2002; Arora et. al, 2010, Mishra et al., 2015). Decontaminating pesticide-polluted areas is a very complex task (Gavrilescu, 2005). Organochloride pesticides are cumulative in the organisms and pose chronic health effects, such as cancer and neurological and teratogenic effects (Vaccari et al., 2006). Many xenobiotic compounds are recalcitrant and resistant to biodegradation, especially the organochloride pesticides (Diaz, 2004). In general, these highly toxic and carcinogenic compounds persist in the environment for many years. Organophosphorus pesticides are actually more widely used in the United States. These pesticides affect the nervous system of insects and humans, in addition to influencing the reproductive system (Colosio et al., 2009). These chemical agents block the prolonged inhibition of the cholinesterase enzyme activity of non-target organisms as well. These chemical agents block prolonged inhibition the activity of the enzyme cholinesterase (ChE), responsible for the nervous impulse in organisms (Yair et al., 2008). The excessive use of organophosphorus in agriculture has originated serious problems in the environment (Singh and Walker, 2006). Although, these pesticides degrade quickly in water, there is always the possibility that residues and byproducts will remain, in relatively harmful levels in the organisms (Ragnarsdottir, 2000). The abusive use of pesticides for pest control has been widely used in agriculture. However, the indiscriminate use of pesticides has inflicted serious harm and problems to humans as well as to the biodiversity (Hussain et al., 2009). The problem of environmental contamination by pesticides goes beyond the locality where it is used. The agricultural pesticides that are exhaustively applied to the land surface travel long distances and can move downward until reaching the water table at detectable concentrations, reaching aquatic environments at significantly longer distances. Therefore, the fate of pesticides is often uncertain; they can contaminate other areas that are distant from where they were originally used. Any unwanted substance introduced into the environment is referred to as a 'contaminant'. Harmful effects or damages by the contaminants lead to 'pollution', a process by which a resource (natural or man-made) is rendered unfit for use, more often than not, by humans. Pollutants are present since time immemorial, and life on the earth as we define now has always evolved amongst them. Relative to the pre-industrialization era; industrialization and intensive use of chemical substances, Pesticides and heavy metals are contributing to environmental pollution. Large-scale pollution due to man-made chemical substances is of global concern now. Seepage and run-offs due to the mobile nature, and continuous cycling of volatilization and condensation of many organic chemicals such as pesticides have even led to their presence in rain, fog and snow (Dubus et al., 2000).

Negative impacts of conventional synthetic pesticides led farmers to rethink and use biopesticides which are environmentally safe and effective for integrated pest management Nowadays, a lot of biopesticides have been developed from microorganisms (bacteria, fungi, viruses, etc.), plant, animal derived products (pheromones, hormones, insect-specific toxins, etc.) and genetically modified organisms and used worldwide for insect pest management (Mazhabi, 2011; Islam, 2012). Keeping in view the environmental impacts associated with the use of synthetic pesticides, bio-pesticides can prove to be the best alternative for sustainable agricultural production.

BIO-PESTICIDES

The different types of bio-pesticides are depicted below:

Microbial Pesticides

This group includes microorganisms such as bacteria, fungi, viruses or protozoan communities as an active ingredients against target pests. For instance there are specific fungi and bacteria that target specific weeds and insects (Khanday et al., 2016; Bhatti et al., 2017)

Plant Incorporated Protectants (PIP's)

These are the pesticidal substances that plants are able to produce from genetic material that has been incorporated in it by using genetic engineering tools. For example *Bacillus thuringiensis* (Bt.) gene is incorporated in plant genetic material which gives it strength to cope with the specific pest.

Biochemical Pesticides

Biochemical pesticides are produced naturally and provide defence against pests by non-toxic mechanisms. This category includes insect sex pheromones, which inhibit mating, and also comprise various scented plant extracts to trap insect pests.

Microbial Pesticides and Their Role in Control of Conventional Pesticide Pollution

Microbial pesticide group comprises the largest group and this group has a particular property of killing or inhibiting only target specific pests. There are at least 1500 naturally occurring insect-specific microorganisms, 100 of which are insecticidal (Khachatourians, 2009). The foundation for biopesticide development is a result of earlier studies on symbiotic relationship between Rhizobium and leguminous plants, Plant Growth Promoting rhizobacteria (PGPR) and PGP fungal associations (Hyakumachi & Kubota, 2004; Hynes & Boyetchko, 2006). Various types of microbes with their mode of action which have been used as biopesticides for controlling pest infestation are listed in Table 1 (Koul, 2011).

Table 1. List of some microbial pesticides developed for commercial use

Microorganism	Target Pest	Action
	Bacteria	
A. radiobacter	Crow galls	Antagonist
B. popilliae	Larvae of various beetles	Stomach Poison
B. pumilus	Effective against rust, downy and powdery mildues	Fungicide
B. sphaercus	Mosquitos	Stomach poison
B. subtilis	Effective against root rot caused by *Rhizoctonia, Fusarium, Alternaria, Aspergillus and Pythium.* Also effectiveagainst some foliar diseases	Fungicide and antagonist
B. subtilis FZB24	Effective against R*hizoctonia, Fusarium, Alternaria, Verticiliumand Streptomyces* on vegetables and ornamental plants	Fungicide
B. thuringiensis var. aizawai	Effective against lapidopterans in vegetables and maize	Stomach poison
B. thuringiensis var. galleriae	Effective against American Boll worm on cotton, tobacco caterpillar on chillies, leaf folder of rice and dimond back moth on vegetables	Stomach poison
B. thuringiensis var. israelensis	Effective against masquitos and black fly larvae and midges	Stomach poison
B. thuringiensis var. kurstaki	Effective against most lapidoptiran larvae and some leaf beetles	Stomach poison
B. thuringiensis var. tenebriunis	Effective against coleopteran beetles on vegetables	Stomach poison
Erwiniaamylovora (HrpNharpin protein)	Multispectrum	Insecticide, fungicide and nematicide
P. cepacia	Effective against soil pathogenic fungi	Toxic
P. chlororaphis	Effective against fungal pathogens of barley and oats	Seed treatment control
P. fluorescence	Effective against *P. Tolasi* on mashrooms and Erwinia on fruit crops	Antibacterial
P. solanacearum	Bacterial control in vegetables	Antibacterial
P. syringae	Effective against post- harvest pathogens on apples, pears and citrus	Antagonist
Pseudomonas+Azospirillum	Effective against brown patch and dollar spot pathogens	Antagonist

continued on following page

Table 1. Continued

Microorganism	Target Pest	Action
Fungi		
Alternariadestruens	Cuscuta control	Herbicide
Ampelomycesquisqualis	Powdery mildew control and damping off disease control	Hypeparasitic
B. Bassiana	Effective against variety of insects such as crickets, white grubs, fire ants, flea beetles, white flies, plant bugs, grasshoppers, thrips, aphids, mites, masquito larva and many others	Insect-specific Fungal Disease Inducers
Burkholderiacepacia	Effective against soil fungal pathogens	Insecticide Controls fungi via. seed treatment
Candida oleophila	Effective against postharvest pathogens like *Botrytis* and *Penicillium*	Colonization of diseased tissues
Coniothyriumminitans	Effective against *Sclerotenia species* on Canola, Sunflower, Peanut and Vegetables	Mode of action not clear
Fusariumoxisporum (non- pathogenic)	Effective against pathogenic *Fusarium* on basil, carnation, cyclamen, tomato	Seed treatment and soil incorporation
*Gliocladiumcatenulatum Gliocladium*spp.	Effective against pythium, *Rhizoctonia,Botrytis* and *Didymella* species on greenhouse crops	Mode of action not clear
Gliocladiumvirens	Effective against soil pathogens causing damping off and root rot	Antagonist
Hirsutellathompsonii	Effective against mites	Stimulates premature fungal epizootics
Lagenidiumgiganteum	Effective against mosquito larva and related dipterans	Kill through zoospores
M. anisopilae	Effective against range of pests. Green muscle is specific for locusts and grasshopers	Disease-causing fungus
Myrotheciumvaerrcaria	Effective against many nematodes	Nematicidal
Paecilomycesfumosoroseus	Effective against white flies in green house	Hyparasitic
Paecilomyceslilacinus	Effective against nematodes	Antagonist
Phelbiagigantea	Effective against pine and spruce rust	Biofungicide
Pythiumoligandrum	Management of fungal pathogens in various crops	Seed treatment or soil incorporation
Streptomyces grisuviridis	Effective against wilt, seed rot and stem rot	Antagonist
Streptomyces lydicus	Effective agaist soil born diseases of turf nursery crops	Fungicide
Talaromycesflavus V117b	Effective against fungal pathogens of tomato, cucumber, strawberry and rape oil seeds.	Antagonist
Trichodermaharzianum	Effective against variety of soil pathogens and wound pathogens	Mycoparasitic, antagonistic
T. harzianum+ T. viride	Effective against *Armillaria* and *Botryoshaeria* and others	Mycoparasitic
*Trichoderma*spp.	Supresses root pathogens	Mycoparasitic
Trichodermaviridae	Effective against rot disease	Mycoparasitic
V. lecanii	Effective against other microbes, aphids, white flies and thrips	Antibiotic and Insect eating fungus Insecticide

continued on following page

Table 1. Continued

Microorganism	Target Pest	Action
	Viruses	
Granulosis virus	Effective against leaf roller and codling moth and other lepidopterans	Disease causing virus
NPV for *Anagraphafalcifera*	Effective against lepidopterans	Disease causing virus
NPV for *A.gematalis*	Effective against velvetbean caterpillar and sugarcane borer	Disease causing virus
NPV for *Autographacaloformica*	Effective against Alfafalooper	Disease causing virus
NPV for *H. zea and H. virescens*	Effective against bollworms	Disease causing virus
NPV for *L. dispar*	Effective against Gypsi moth	Disease causing virus
NPV for *Mamestrabrassicae*	Effective against lapidopterans	Disease causing virus
NPV for *Neodiprionsertifer, N. lecontei and N. abietis*	Effective against sawfly larvae	Disease causing virus
NPV for *Orgyiapsudotsugata*	Effective against Douglas-fir tussock Moth	Insecticide
NPV for *SpodopteraExigua*	Effective and lesser army worms, pigweed caterpillar and mottled willow moth	Disease causing virus
NPV for *Syngraphafalcif*	Effective against *Helicoverpa and cydia*spp.	Larva disease causing virus
	Nematodes	
H. bacteriophora	Effective against many lapidopteran larvae, turf and Japanes beetles and soil insects	Entamopathogen
S. glaseri	Effective against root weevils, cut worms, fleas, borers and fungal gnats	Entamopathogen
H. megidis	Effective against black vine weevils soil insects	Entamopathogen
P. hermophrodita	Effective against slugs	Slug eating nematodes
S. carpocapsae	Effective against black vine weevils, Strawberry root weevils, cranberry girdler and termites	Entamopathogen
S. feltiae	Effective against vine weevils, fungal gnats, sciarid flies and soil insects	Entamopathogen
Steinernemariobravis	Effective against citrus weevils	Entamopathogen
S. scapterisci	Effective against mole crickets	Entamopathogen
	Protozoa	
N. Locustae	Effective against grasshoppers and crickets	Insecticide, chronic feeder

Koul, 2011.

MECHANISM OF ACTION OF SOME IMPORTANT MICROBES

Mechanism of Action of *Bacillus thuringiensis*

Bacillus thuringiensis (Bt) releases some crystalline protein precursors (δ-endotoxins) that kill certain target insects such as lepidopteran species (Kumar, 2012). δ-endotoxins also called as cry proteins and exhibit cry gene. When insects ingest these toxin crystals, their digestive tract is denatured and solubilized and thus being susceptible to proteases. As a result of action of proteases found in insect gut, toxins from crystals are released. The cry toxin enters the insect gut and paralyzes the insect by forming a pore. The

Figure 2. Mode of action of Bt. Toxin against lepidopteran insects

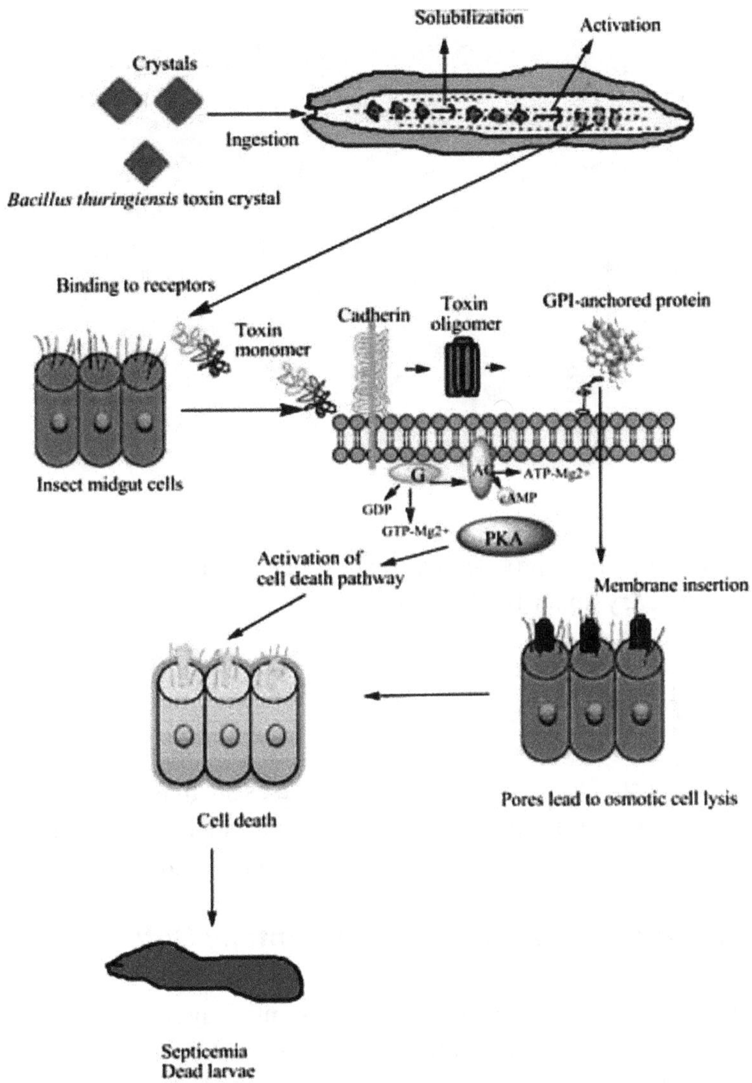

insect stops feeding and dies of starvation and midgut epithelial impairment (Betz *et al.*,2000 ; Zhu *et al.*, 2000 ; Darboux *et al.*, 2001).

Mechanism of Action of Fungi (*Metarhizium anisopliae*)

A number of fungi are in use as microbial pesticides. *Metarhizium anisopliae* need optimum moisture for filamentous growth and production of conidia that are responsible for infection to soil doweling insects. *M. anisopliae* has the potential to be used as a biocontrol agent, particularly for malaria vector species, and is also a suitable candidate for further research and development (Mnyone *et al.*, 2010; Nathan, 2015). There are about 750 estimated species of entomopathogenic fungi belonging to 90 genera (Roberts & Humber, 1981). Some commercially produced fungal species are *Beauveria*, *Metarhizium*,

Figure 3. Trichoderma harzianum
Islam et al., 2010.

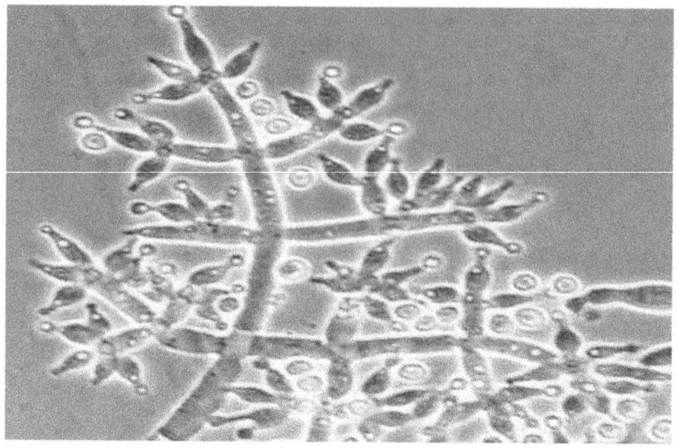

Lecanicillium and *Isaria* that can be produced at mass level. Production of fungi needs high cost and technical knowhow and efficacy under field conditions.

TRICHODERMA AS FUNGICIDE

Trichoderma is used as biopesticide against soil borne diseases such as root rot specifically in dry area crops such as groundnut, black gram, green gram, chickpea which are susceptible to these diseases (Figure 3). Production of *Trichoderma* bio-pesticide is relatively cheaper and easy (Islam *et al.*, 2010).

Use of Microbial Bio-Pesticides as Environmentally Safe Pest Control Agents

Microbial biopesticides are eco-friendly agricultural pest control agents with clean environmental and agricultural application as compared to conventional pesticides which are now considered to be toxic for ecosystems and human health (Khanday et al., 2016; Bhatti et al., 2017).Extensive studies by researchers indicate that microbial biopesticides used in augmentation biological control can give effective pest control with minimal detectable negative impact on the environment (Vestergaard, 2003). A wide array of products are derived from fungi, bacteria etc. which are used as pest control agents. Such as *Blasticidin* was for the first time isolated from *Streptomyces griseochromogenes* which is the soil actinomycete and it is available in market as salt of benzylamino benzene sulfonate Figure 4 (Copping & Men, 2000).

BACULOVIRUS

Baculoviruses are pathogenic in nature and have been used effectively as biocontrol agents against insect pests (Moscardi, 1999). Baculoviruses cause mortality in Lepidopteran larvae (Cory, 2000). Mechanism of action involves ingestion. After ingestion, they enter the insect's body through the midgut and from there they spread throughout the body, although in some insects, infection can be limited to the insect

Figure 4. Structure of Blasticidin

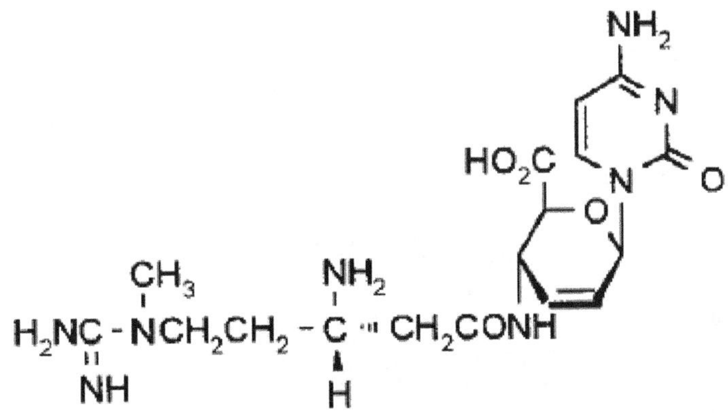

Figure 5. Mode of action of baculoviruses against lepidopteran insects

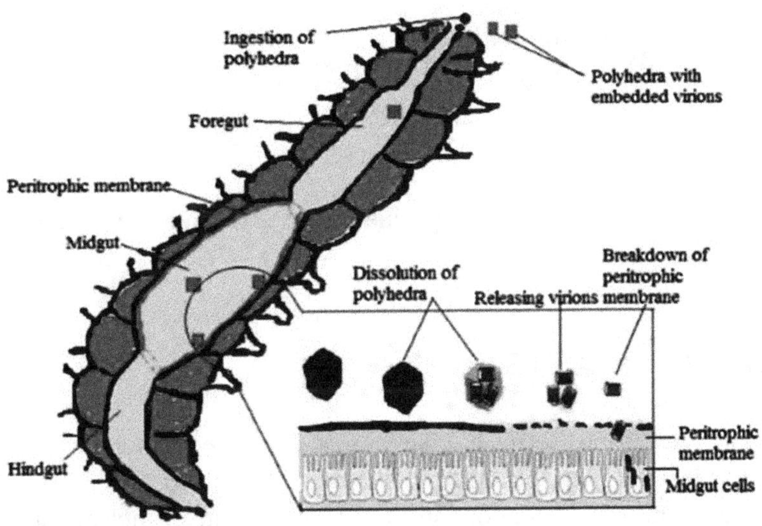

midgut or the fat body (Figure 5). Two groups of baculovirusesare the nucleo polyhedro viruses (NPVs) and granuloviruses (GVs). In NPVs, occlusion bodies comprise numerous virus particles, but in GVs, occlusion bodies ordinarily contain just one virus particle. A common feature of baculoviruses is that they are occluded, i.e., the virus particles are embedded in a protein matrix. The presence of occlusion bodies plays an essential role in baculovirus biology as it allows the virus to survive outside the host (Cory, 2000).

MODE OF ACTION OF NEMATODES

Nematodes are one of the best defence organisms used against pests. In nematodes the parasitic cycle starts at third stage. The non-feeding juveniles infest suitable insect host and enter through insects natural body openings (Grewal *et al.*, 1997). After entry into the host, nematodes cause infection of the hemo-

Figure 6. Mode of action of entamopathogenic nematodes

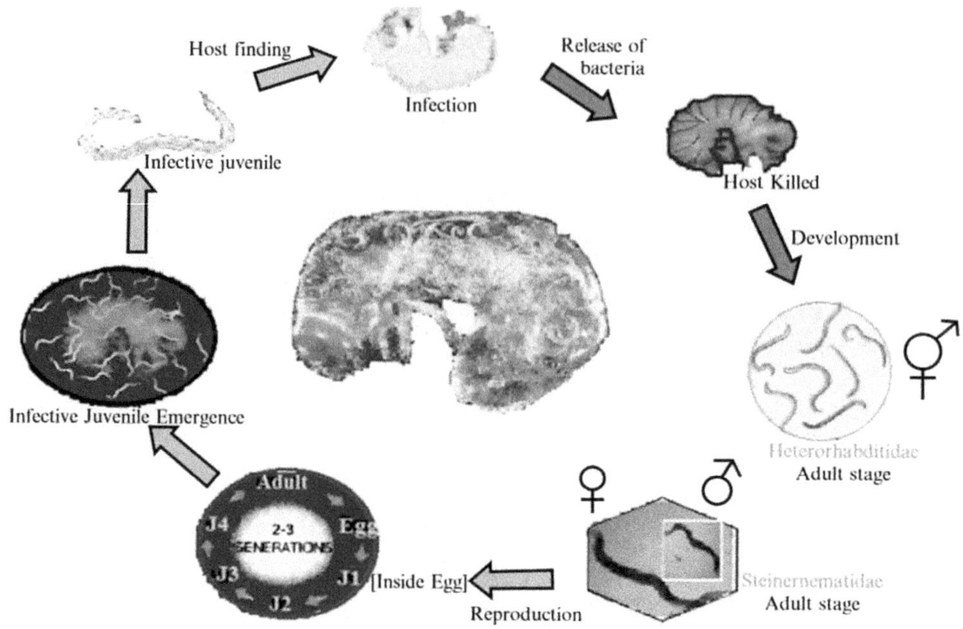

coel and release symbiotic bacteria into the intestine of host. The released bacteria cause Septicemia and cause host to die (Figure 6).

BIOPESTICIDES FROM PLANTS

From ancient times plants have been used by humans as insect repellents. Botanical pesticides are isolated from plant parts such as leaves, roots, barks, fruits, seeds and seed kernels. Several higher plants have the ability to synthesize and produce numerous secondary metabolites which are unwanted for pests. For example *Chrysanthemum species* bud flowers are used as a source of insecticidal compounds. It produces some terpinoid esters which are successfully patented and marketed (Raja & Masresha, 2015.). Some examples of plants and their extracts as biopesticides with mechanism of action are presented in table-2 (Rattan, 2010). The effective plant species have the potential to use toxic substances which they prepare to defend against pests. The first generation phytochemicals particularly include pyrethrins, nicotine and rotenone. The lipophilic compound pyrithrins are 3-Carbon esters bonded with 5-Carbon aromatic alcohols (Figure 7). Pyrithrins are extracted from daisy plant flowers *Tanacetum cinerariefolium* (Schleier & Peterson, 2011). It is used for fogging cocoa and agro-produce against moths (Ephestiacantella) and is a potent fungicide (Enyiukwu, *et al.*, 2016). In addition modifying agricultural crops with genetic capacity to synthesize pyrethrins for improved resistance against a wide array of pests and diseases had been advocated (Enyiukwu *et al.*, 2014).

Till date nine types of azadirachtin have been identified in azadiracta species (neem). It has been revealed that azadirachtin blocks the mitotic cell microtubule polymerization. It also damages DNA and binds to complex proteins such as shock protein (Salehzadeh *et al.*, 2003;Robertson *et al.*, 2007). It is

Figure 7. Structure of pyrethrin
Enyiukwu et al., 2016.

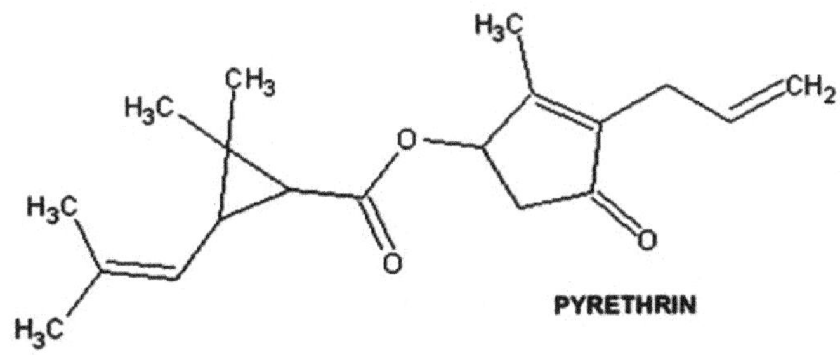

Figure 8. Structure of Azadirachtin
Silva-Aguayo, 2015.

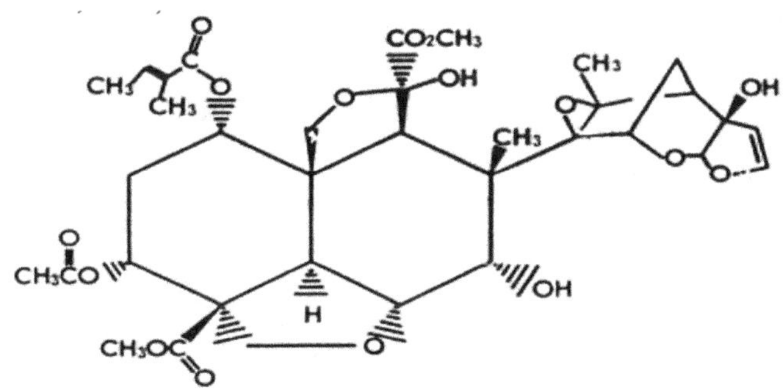

highly reactive and damage many cellular molecules in cytoplasm and nucleus including interference with genes and proteins (Figure 8).

Alkaloids

Alkaloids act as insecticides at low concentrations and are toxic to vertebrates as well. Alkaloids such as nicotine mainly interfere with acetylcholine in receptors in the nervous system of target pest (Figure 9). These are also known to affect membrane sodium channels of nerves. Calaber bean (*Physostigma-venenosum*) is used to extract physostigmine which is used for the production of carbamate insecticide. These chemicals being non-volatile may be used as insect repellents after burning plant materials to produce smoke that can repel insects. Several families such as *Berberidaceae,.Fabaceae, Solanaceae* are known to contain large concentrations of alkaloids traditionally used as insect repellents (Johnson, 1998; Rattan, 2010).

Essential oils such as terpenoids from plant origin have been proved to have insecticidal activity (Figure 10). Essential oils from plant families such as Asteraceae, Myrtaceae, Apiaceae, Rutaceae etc. have insecticidal activity. Essential oils may act as larvicidal and adulticidal, repellent, fumigant against insects of Lepidopteran, Dipteran, Coleopteran, Hemipteran, Isopteran order (Tripathi *et al.*, 2009). Terpenes synergise the effects of other toxins by acting as solvents to facilitate their passage through membranes. An example of such synergism seems to occur in conifer resin (Rattan, 2010).

Figure 9. Structure of nicotine

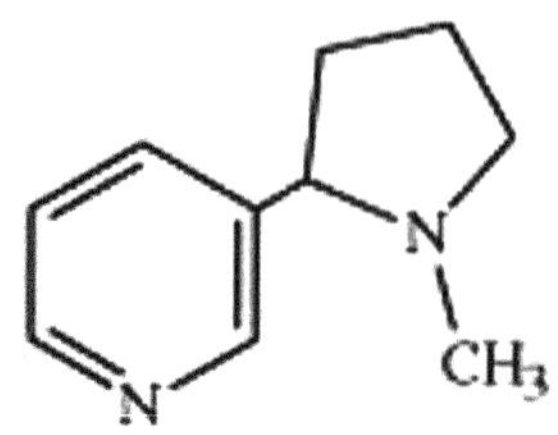

nicotine

Figure 10. Chemical structure of some monoterpenes found in essential oils with insecticidal activities
Mossa, 2016.

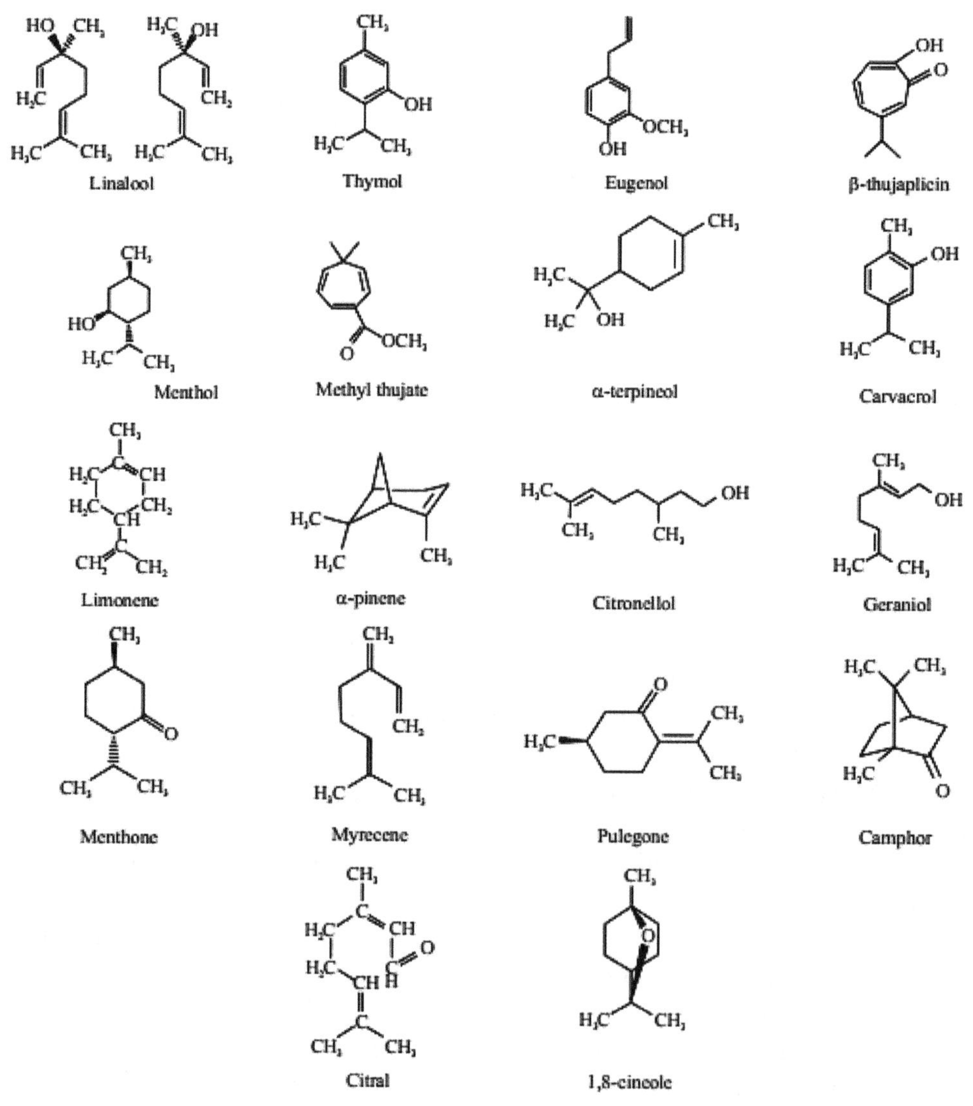

Table 2. Mechanism of action of insecticides of plant origin

S. No.	System	Mechanism of Action	Compound	Plant Source
1.	Cholinergic system	Inhibition of acetylcholineesterase (AChE)	Essential oils	*Azadirachtaindica, Mentha spp., Lavendula spp.*
		Cholinergic acetylcholine nicotinic receptor agonist/antagonist	Nicotine	*Nicotiana spp., Delphinium spp., Haloxylonsalicornicum, Stemonajaponicum*
2.	GABA system	GABA-gated chloride channel	Thymol, Silphinenes	*Thymus vulgaris*
3.	Mitochondrial system	Sodium and potassium ion exchange Disruption	Pyrethrin	*Crysanthemumcinerariaefolium*
		Inhibitor of cellular respiration (mitochondrial complex I electron transport inhibitor or METI)	Inhibitor of cellular respiration (mitochondrial complex I electron transport inhibitor or METI)	*Lonchocarpus spp.*
		Affect calcium channels	Ryanodine	*Ryania spp.*
		Affect nerve cell membrane action	Sabadilla	*Schoenocaulonofficinale*
4.	Octopaminergic System	Octopaminergic receptors,	Essential Oils	*Cedrus spp., Pinus spp., Citronella spp., Eucalyptus spp.*
		Block octopamine receptors by working through tyramine receptors cascade	Thymol	*Thymus vulgaris*
5.	Miscellaneous	Hormonal balance disruption	Azadirachtin	Azadiractaindica

Ratan, 2010.

ROLE OF BIOPESTICIDES IN INTEGRATED PEST MANAGEMENT (IPM)

The United Nations Food and Agricultural Organization defines IPM as "the careful consideration of all available pest control techniques and subsequent integration of appropriate measures that discourage the development of pest populations and keep pesticides and other interventions to levels that are economically justified and reduce or minimize risks to human health and the environment. Its basic components are crop management, nutrient management, pest management and ultimately financial management. IPM strategies combine range of complementary methods to reduce pest populations below economic injury level while minimizing impacts on other components of agro-ecosystems and environmental conditions of the area (Kumar & Singh, 2014). Synthetic pesticides are intensively used for pest management and control pests quickly but at the same time are leaving serious environmental impacts. The alternatives to synthetic chemical pest control include Natural control, Cultural control, Biological control. To overcome the disadvantages of these techniques IPM was used to control pests (Bekele *et al.*, 2014; Kasiotis, 2013). Integrated Pest Management is an effective and environmentally sensitive approach to pest management.IPM is used in agriculture, horticulture and pest control. It plays role in long term prevention of pests, natural pest control mechanism and growth of healthy crops with least disruption to ecosystem (Gautami et al., 2015).

ROLE OF BIOPESTICIDES IN ENVIRONMENTAL SAFETY

Consistent use of conventional pesticides and fertilizers for increased food production and for meeting the demands of growing population has benefited humankind but at the cost of environmental degradation. Majority of the synthetic pesticides are non-biodegradable and thus persist in the environment. Less than 0.1% pesticides reach the target pest and remainder negatively affect humans, livestock and natural biota (Pimentel, 1992). Biopesticides play important role in:

- Decreasing impact of pollution by conventional pesticides.
- Maintain ecological balance and ecosystem health.
- Pesticide residual effect is reduced.
- Natural resources are conserved.
- Biodiversity conservation, conserve pollinators and non-target species.
- Reduces human and animal health toxicity associated with the synthetic pesticides.
- Production and use of bio pesticides generates employment at rural level.

FUTURE PROSPECTS

Biopesticides are gaining importance as future pesticide agents due to their safer and eco-friendly nature. Techniques, such as recombinant DNA technology is being used to increase efficacy of biopesticides. Fusion proteins are being developed to generate next generation biopesticides (Kumar, 2013). A career protein is allowed to combine with a specific toxin (non-toxic to higher animals) which makes them toxic to insect pests when they consume it (Fitches *et al.*, 2004). Further research is needed to investigate the ways to reduce or eliminate effect of biopesticides on non-target organisms and also increase efficiency of biopesticides against resistant pests. Biopesticides must be used in consonance with other pest control agents to reduce impact of synthetic pesticides on environment. Research in production, formulation and delivery may greatly assist in commercialization of biopesticides. At the same time it is also required to encourage public funded programmes, commercial investors and pesticide companies to take up biopesticide enterprise. At the same time strict regulatory approach to maintain quality of biopesticides for safe use is necessary.

CONCLUSION

Biopesticides are much more useful for agroecosystem and is emerging as best alternative to protect crops and at the same time safeguard the environment. Biopesticide are eco-friendly, less toxic and safe to use. They are also less expensive and usually have no adverse effect on plant health and ecological balance. Biopesticides use naturally developed mechanisms to cope with pests and thus do not harm functioning of non-target organisms. Biopesticide production may also help in generates of employment in rural areas. Besides these properties, biopesticides use natural mechanism to fight with pests and their mode of action could be improved by using proper genetic engineering and biotechnological tools.

REFERENCES

Aktar, M. W., Sengupta, D., & Chowdhury, A. (2009). Impact of pesticides use in agriculture: Their benefits and hazards. *Interdisciplinary Toxicology*, *2*(1), 1–12. doi:10.2478/v10102-009-0001-7 PMID:21217838

Al-Zaidi, A. A., Elhag, E. A., Al-Otaibi, S. H., & Baig, M. B. (2011). Negative effects of pesticides on the environment and the farmer's awareness in Saudi Arabia: A case study. *Journal of Animal and Plant Sciences*, *21*(3), 605–611.

Arora, N. K., Khare, E., & Maheshwari, D. K. (2010). Plant Growth Promoting Rhizobacteria: constraints in bioformulation, commercialization and future strategies. In D. K. Maheshwari (Ed.), *Plant growth and health promoting bacteria* (pp. 97–116). Berlin: Springer. doi:10.1007/978-3-642-13612-2_5

Bekele, D., Petros, B., Tekie, H., & Asfaw, Z. (2014). Larvicidal, and Adulticidal, Effects of Extracts from Some Indigenous Plants against the Malaria Vector, Anopheles Arabiensis (Diptera: Culicidae) in Ethiopia. *Journal of Biofertilizers & Biopesticides*, *5*, 144.

Betz, F. S., Hammond, B. G., & Fuchs, R. L. (2000). Safety and advantages of Bacillus thuringiensis-protected plants to control insect pests. *Regulatory Toxicology and Pharmacology*, *32*(2), 156–173. doi:10.1006/rtph.2000.1426 PMID:11067772

Bevans, H. E., Lico, M. S., & Lawrence, S. J. (1998). Water quality in the Las Vegas Valley area and the Carson and Truckee River basins, Nevada and California, 1992–96. Reston, VA: USGS. *Geological Survey Circular*, 1170.

Bhatti, A. A., Haq, S., & Bhat, R. A. (2017). Actinomycetes benefaction role in soil and plant health. *Microbial Pathogenesis*, *111*, 458–467. doi:10.1016/j.micpath.2017.09.036 PMID:28923606

Birch, A. N. E. (2011). How agro-ecological research helps to address food security issues under new IPM and pesticide reduction policies for global crop production systems. *Journal of Experimental Botany*, *62*, 3251–3261. doi:10.1093/jxb/err064 PMID:21669880

Copping, L. G., & Menn, J. J. (2000). Biopesticides: A review of their action, applications and efficacy. *Pest Management Science*, *56*(8), 651–676. doi:10.1002/1526-4998(200008)56:8<651::AID-PS201>3.0.CO;2-U

Cory, J. S. (2000). Assessing the risk of releasing genetically modified virus insecticides: Progress to date. *Crop Protection (Guildford, Surrey)*, *19*(8-10), 779–785. doi:10.1016/S0261-2194(00)00104-6

Darboux, I., Nielsen-LeRoux, C., Charles, J. F., & Pauron, D. (2001). The receptor of Bacillus sphaericus binary toxin in Culexpipiens (Diptera: Culicidae) midgut: molecular cloning and expression. *Insect Biochemistry and Molecular Biology*, *31*(10), 981–990. doi:10.1016/S0965-1748(01)00046-7 PMID:11483434

Fenelon, J. M. (1998). Water quality in the White River Basin, Indiana, 1992–96. Reston, VA: USGS. *Geological Survey Circular*, 1150.

Fitches, E., Edwards, M. G., Mee, C., Grishin, E., Gatehouse, A. M., Edwards, J. P., & Gatehouse, J. A. (2004). Fusion proteins containing insect-specific toxins as pest control agents: Snowdrop lectin delivers fused insecticidal spider venom toxin to insect haemolymph following oral injection. *Journal of Insect Physiology*, *50*(1), 61–71. doi:10.1016/j.jinsphys.2003.09.010 PMID:15037094

Forget, G. (1993). Balancing the need for pesticides with the risk to human health. In *Impact of Pesticide Use on Health in Developing Countries* (Vol. 2). Ottawa: IDRC.

Gautami, J., Sandhyasree, M., & Revathi, B. (2015). Bio-pesticides Approach for Pest Management. *Journal of Agriculture and Allied Sciences*, *4*(1), 1–5.

Gerhardson, B. (2002). Biological substitutes for pesticides. *Trends in Biotechnology*, *20*(8), 338–343. doi:10.1016/S0167-7799(02)02021-8 PMID:12127281

Grewal, P. S., Lewis, E. E., & Gaugler, R. (1997). Response of infective stage parasites (Nematoda: Steinernermatidae) to volatile from infected hosts. *Journal of Chemical Ecology*, *23*(2), 503–515. doi:10.1023/B:JOEC.0000006374.95624.7e

Hyakumachi, M., & Kubota, M. (2004). Fungi as a plant growth promoter and disease suppressor. In D. K. Arora (Ed.), *Fungal Biotechnology in Agricultural, Food, and Environmental Applications* (pp. 101–110). New York: Marcel Dekker.

Hynes, K. R., & Boyetchko, S. M. (2006). Research initiatives in the art and science of biopesticide formulations. *Soil Biology & Biochemistry*, *38*(4), 845–849. doi:10.1016/j.soilbio.2005.07.003

Igbedioh, S. O. (1991). Effects of agricultural pesticides on humans, animals and higher plants in developing countries. *Archives of Environmental & Occupational Health*, *46*(4), 218–224. doi:10.1080/00039896.1991.9937452 PMID:2069430

Ishtiaq, M., Saleem, M. A., & Razaq, M. (2012). Monitoring of resistance in Spodopteraexigua (Lepidoptera: Noctuidae) from four districts of the Southern Punjab, Pakistan to four conventional and six new chemistry insecticides. *Crop Protection (Guildford, Surrey)*, *33*, 13–20. doi:10.1016/j.cropro.2011.11.014

Islam, M. T., Castle, S. J., & Ren, S. (2010). Compatibility of the insect pathogenic fungus Beauveriabassiana with neem against sweet potato whitefly, Bemisiatabaci, on eggplant. *Entomologia Experimentalis et Applicata*, *134*(1), 28–34. doi:10.1111/j.1570-7458.2009.00933.x

Islam, M. T., & Omar, D. B. (2012). Combined effect of Beuveria bassiana with neem on virulence of insect in case of two application approaches. *Journal of Animal and Plant Sciences*, *22*(1), 77–82.

Jeyaratnam, J. (1985). Health problems of pesticide usage in the third world. *The B M J*, *42*, 505. PMID:4016001

Johnson, T. (1998). *CRC Ethnobotany Desk Reference*. Boca Raton, FL: CRC Press.

Kasiotis, K. M. (2013). Biopesticides Analysis: An Editorial. *Journal of Biofertilizers & Biopesticides*, *4*, 115.

Khachatourians, G.G. (2009). Insecticides, microbials. *Applied Microbiology: Agro/Food*, 95-109.

Khanday, M., Bhat, R. A., Haq, S., Dervash, M. A., Bhatti, A. A., Nissa, M., & Mir, M. R. (2016). Arbuscular Mycorrhizal Fungi Boon for Plant Nutrition and Soil Health. In Soil Science: Agricultural and Environmental Prospectives (pp. 317-332). Springer International Publishing.

Kole, R. K., Banerjee, H., & Bhattacharyya, A. (2001). Monitoring of market fish samples for Endosulfan and Hexachlorocyclohexane residues in and around Calcutta. *Bulletin of Environmental Contamination and Toxicology*, *67*(4), 554–559. doi:10.1007001280159 PMID:11779071

Koul, O. (2011). Microbial biopesticides: Opportunities and challenges. *Perspectives in Agriculture, Veterinary Science, Nutrition and Natural Resources*, *6*, 1–26.

Kumar, S. (2012). The role biopesticides in sustainabley feeding the nine billion global populations. *Journal of Biofertilizers & Biopesticides*, *4*, 114.

Kumar, S., & Singh, A. (2014). Biopesticides for integrated crop management, environmental and regulatory aspects. *Journal of Biofertilizers & Biopesticides*, *5*, 121.

Levings, G. W., Healy, D. F., Richey, S. F., & Carter, L. F. (1998). Water quality in the Rio Grande Valley, Colorado, New Mexico, and Texas, 1992–95. *Geological Survey Circular*, 1162.

Mathur, S. C. (1999). Future of Indian pesticides industry in next millennium. *Pesticide Information*, *24*(4), 9–23.

Mazhabi, M., Nemati, H., Rouhani, H., Tehranifar, A., Moghadam, E. M., Kaveh, H., & ... . (2011). The effect of Trichoderma on polianthes qualitative and quantitative properties. *Journal of Animal and Plant Sciences*, *21*(3), 617–621.

Mishra, J., Tiwari, S., Singh, S., & Arora, N. K. (2015). Biopesticides: Where we stand? In N. K. Arora (Ed.), *Plant microbes symbiosis: applied facets* (pp. 37–75). Springer.

Mnyone, L. L., Koenraadt, C. J. M., Lyimo, I. N., Mpingwa, M. W., Takken, W., & Russell, T. L. (2010). Anopheline and culicine mosquitoes are not repelled by surfaces treated with the entomopathogenic fungi Metarhiziumanisopliae and Beauveriabassiana. *Parasites & Vectors*, *3*(1), 80. doi:10.1186/1756-3305-3-80 PMID:20799937

Moscardi, F. (1999). Assessment of the application of baculoviruses for control of Lepidoptera. *Annual Review of Entomology*, *122*, 309–320. PMID:15012374

Mossa, A. T. H. (2016). Green Pesticides: Essential Oils as Biopesticides in Insect-pest Management. *Journal of Environmental Science and Technology*, *9*(5), 354–378. doi:10.3923/jest.2016.354.378

Nathan, S. S. (2015). A Review of Biopesticides and Their Mode of Action Against Insect Pests. In P. Thangavel & G. Sridevi (Eds.), Environmental Sustainability. Academic Press.

Nawaz, M., Mabubu, J. I., & Hua, H. (2016). Article. *Journal of Entomology and Zoology Studies*, *4*(2), 241–246.

Pier, A. B., Bernucci, I., Brambilla, G., Consonni, D., & Pesatori, C. A. (1998). The Seveso Studies on Early and Long-Term Effects of Dioxin Exposure: A Review. *Environmental Health Perspectives*, *106*(2), 5–20. PMID:9539003

Pier, A. B., Consonni, D., Bachetti, S., Rubagotti, M., Baccarelli, A., Zocchetti, C., & Pesatori, A. C. (2001). Health Effects of Dioxin Exposure: A 20-Year Mortality Study. *American Journal of Epidemiology, 153*(11), 1031–1044. doi:10.1093/aje/153.11.1031 PMID:11390319

Pimental, D. (1992). Foreword. In M. Shiyomi, E. Yano, H. Koizumi, D. A. Andow, & N. Hokyo (Eds.), *Ecological Processes in agroecosystems* (pp. 1–2). Tokyo: Yakendo Publishers.

Raja, N., & Masresha, G. (2015). Plant based biopesticides: Safer alternative for Organic Food Production. *Journal of Biofertilizers & Biopesticides, 6*, 128.

Rattan, R. S. (2010). Mechanism of action of insecticidal secondary metabolites of plant origin. *Crop Protection (Guildford, Surrey), 29*(9), 913–920. doi:10.1016/j.cropro.2010.05.008

Robertson, S. L., Ni, W., Dhadialla, T. S., Nisbet, A. J., McCusker, C., & ... . (2007). Identification of a putative azadirachtin-binding complex from Drosophila Kc167 cells. *Archives of Insect Biochemistry and Physiology, 64*, 200–208. doi:10.1002/arch.20171 PMID:17366600

Salehzadeh, A., Akhkha, A., Cushley, W., Adams, R. L., Kusel, J. R., & Strang, R. H. C. (2003). The antimitotic effect of the neem terpenoidazadirachtin on cultured insect cells. *Insect Biochemistry and Molecular Biology, 200*(33), 681–691. doi:10.1016/S0965-1748(03)00057-2 PMID:12826095

Tripathi, Y. C., & Tripathi, J. (2000). Role of biopesticides in environmental safety. *Science and Culture, 66*(5-6), 171–176.

U.S. Geological Survey. (1999). *The quality of our nation's waters – nutrients and pesticides*. Circular 1225. Reston VA: USGS. Retrieved from http: //water.usgs.gov/pubs/circ/ circ1225/

Vestergaard, S., Cherry, A., Keller, S., & Goettel, M. (2003). Safety of hyphomycete fungi as microbial control agents. In *Environmental impacts of microbial insecticides* (pp. 35–62). Dordrecht, The Netherlands: Academic Press.

Wall, G. R., Riva-Murray, K., & Phillips, P. J. (1998). *Water Quality in the Hudson River Basin, New York and adjacent states, 1992–95*. Reston, VA: USGS. *Geological Survey Circular*, 1165.

WHO. (1990). *Public Health Impact of Pesticides Used in Agriculture. (1990)*. Geneva: World Health Organization.

Zhu, Y. C., Kramer, K. J., Oppert, B., & Dowdy, A. K. (2000). cDNAs of aminopeptidase-like protein genes from Plodiainter punctella strains with different susceptibilities to Bacillus thuringiensis toxins. *Insect Biochemistry and Molecular Biology, 30*(3), 215–2. doi:10.1016/S0965-1748(99)00118-6 PMID:10732989

KEY TERMS AND DEFINITIONS

Biomagnification: Biomagnification, also known as bioamplification or biological magnification, is the increasing concentration of a substance, such as a toxic chemical, in the tissues of tolerant organisms at successively higher levels in a food chain.

Bioremediation: Bioremediation is a process used to treat contaminated media, including water, soil, and subsurface material, by altering environmental conditions to stimulate growth of microorganisms and degrade the target pollutants. In many cases, bioremediation is less expensive and more sustainable than other remediation alternatives. Biological treatment is a similar approach used to treat wastes including wastewater, industrial waste, and solid waste.

Detoxification: Detoxification is the physiological or medicinal removal of toxic substances from a living organism, including the human body, which is mainly carried out by the liver.

Chapter 22
Organic Farming: Challenge for Chemical Pollution in Aquatic Ecosystem

Shivom Singh
ITM University, India

Bhupendra Kumar
Perfetti Van Melle India Pvt. Ltd., India

Neha Sharma
Government KRGPG College, India

Kajal S. Rathore
Government KRGPG College, India

ABSTRACT

Agriculture is one of the significant factors contributing to the economic growth of India. In order to reap a better harvest, farmers inoculate the soil with fertilizers. These fertilizers include pesticides, herbicides, insecticides, fungicides, etc., and are broadly used to control pests and pest-induced diseases. Increasingly high inputs of chemical fertilizers have not only left soils degraded, but it has also increased the adverse effect on aquatic life and other environmental hazards. Organic farming methods would crack these issues and make the ecosystem healthier. Bio-fertilizers and bio-pesticides form a link between the biotic and abiotic factors and can be used to supplement the expensive chemical fertilizers. This chapter focuses on agricultural chemicals (fertilizers and pesticides) that impact the aquatic environment. The aim of the chapter is to improve ecological sustainability and to minimize the effects of pesticides on aquatic ecosystems. In addition, the authors attempt to reveal almost all positive aspects of organic farming in special reference to aquatic pollution.

DOI: 10.4018/978-1-5225-6111-8.ch022

INTRODUCTION

Agriculture is the noteworthy factors contributing to the economic growth of developing countries like, India. In order to reap improved agricultural practices, farmers inoculate the soil with pesticides and chemical fertilizers, which is causing perilous effect on aquatic ecosystem. It would be overcome if the constituent activities of agriculture could be natural and free from chemical ingredients. This would be an impressive model for designing incentive schemes to optimize agricultural practice and for minimizing environmental consequences. Agricultural show complex relationship with freshwater systems and effect the aquatic ecosystem in many dimensions. The entire land surface, much of which is agricultural, forms the catchment area for one or other river system and almost anything that happens on the catchment has an effect on the aquatic ecosystem (Moss, 2008).

Agriculture contributes about 70 percent of water abstractions worldwide and plays a chief role in water pollution. Farms discharge large quantities of pesticides, herbicides, insecticides, fungicides, etc. as they are broadly used to control pests and pest-induced diseases and are drainage into water bodies. Increasingly high inputs of chemical fertilizers and pesticides have not only left soils degraded but also increases adverse effect on aquatic life. The resultant water pollution poses demonstrated risks to aquatic ecosystems, human health and productive activities (UNEP, 2016). Agriculture discharge has direct negative impacts not only on aquatic flora and fauna but also on human health. Pesticide and fertilizers accumulation coming from agricultural field, in water bodies and the food chain, led to acute and likely chronic health effects. Water-quality degradation arising from agricultural practices may also have severe and direct impacts on productive, quality and quantity of agriculture itself.

Population growth and changes in consumption patterns, including new dietary preferences require the production of more and diverse food. This, in turn, is driving agricultural expansion and intensification and bringing new environmental externalities, including impacts on water quality (UNDESA, 2017). To overcome these challenges regenerative agriculture is the most excellent way, that leads to healthy soil, high quality and nutrient dense food, concurrently, it improves, rather than degrading land, and ultimately leading to productive farms and healthy communities and economies. An organic farming practice play a significant role in increasing food production, farmers' income and topsoil but also help in controlling and plummeting the chemical pollution in aquatic ecosystem.

This review focuses on agricultural chemicals (fertilizers and pesticides etc) that impact the aquatic environment. To overcome this dangerous issue, alternate source of bio nutrients supply will open new avenues for era of next-generation biofertilizer/biopesticides from nature and will also be useful in organic production of agriculture.

BACKGROUND

The global growth of crop production has been achieved largely through the intensive use of inputs such as pesticides and chemical fertilizers. India's utilize about 76 per cent of pesticides, against the world average of 44 per cent. However, the use in agriculture is less than 350 gm a hectare as against the world average of 500 gm a hectare. In India pesticides production was started in 1952 with the establishment of a plant for the production of BHC near Calcutta, and now India is the second largest manufacturer of pesticides in Asia after China and hold twelfth position globally (Mathur, 1999). The trend has been amplified by the expansion of agricultural land, with irrigation playing a strategic role in improving productivity and rural livelihoods while also transferring agricultural pollution to water bodies.

Chemicals fertilizers and pesticides employed in agricultural activity enter the aquatic environment through atmospheric deposition, surface run-off or leaching (Kreuger, 1999) and frequently accumulate in soft-bottom sediments and aquatic organisms (Miles & Pfeuffer, 1997; Lehotay, Harman-Fetcho & McConnell, 1998; Kreuger, Peterson & Lundgren, 1999). Pesticides and chemicals have been accumulating in the aquatic ecosystem in all parts of the world and agricultural practices are contributing most in this crisis. Different aspects of pesticide toxicity have already been summarized in many review articles and books (Mulla & Mian, 1981; Matsumura, 1985; Ecobichon, 1991; Barron & Woodburn, 1995; DeLorenzo, Scott & Ross 2001). Louis, Weigmann, Hipkins and Stinson, (1996) states that a pesticide's capacity or a fertilizer's capacity (nutrients) to harm fishes and aquatic animals is largely a function of its toxicity exposure time dosage rate and persistence in the environment. Pesticides from farming were responsible for the majority of acute chemical risks to freshwater life and pollute aquatic ecosystem. The impact of chemical pollution on aquatic life was significantly increased close to agricultural land, sewage treatment works and urban areas where there is run-off of pollutants into rivers.

Insecticides, herbicides and fungicides etc are applied intensively in agriculture worldwide (Schreinemachers & Tipraqsa, 2012). Number of pesticides and other chemical fertilizers (such as DDT) are banned in most countries but are still being used illegally and persistently. In current scenario, millions of tones of pesticide ingredients are used in agriculture (FAO, 2016). Due to improper selection and management of these substances, they are polluting water resources with carcinogens and other toxic substances that can affect aquatic ecosystem. These chemical fertilizers and pesticides poisoning causes major health issues worldwide, especially in developing countries like India, where poor farmers often use highly hazardous pesticide formulations.

Agriculture is not only a source of emerging pollutants; it also put in to the spread and reintroduction of these toxic pollutants into aquatic environments. The potential risks to human health posed by exposure to emerging pollutants via contaminated agricultural products needs attention (Thebo, Drechsel, Lambin & Nelson, 2017). In relationship to this subject, the organic movement may have gained a place in the spotlight of the mainstream media now. Organic agriculture is one of the fastest rising agribusiness sectors in the world, with double-digit annual growth in land under organic cultivation, value of organic produce and number of organic farmers (Wood, Sebastian & Scherr 2001; Willer & Yussefi 2005). Organic farming not only increases the fertility and productivity of the soil and crops respectively, but also helps in reducing the aquatic pollution. Although the concept of organic farming is still regarded with some skepticism (Trewavas, 2001; Kirchmann & Ryan 2004; Trewavas, 2004). The aim of the article is to improve ecological sustainability and to minimize the effects of pesticides on aquatic ecosystem. In addition, authors attempt to reveal almost all positive aspects of organic farming in special reference to aquatic pollution.

AGRICULTURAL PESTICIDE TOXICITY

Agriculture was developed to produce crops and livestock for human consumption. As the human population is augmenting, the demand of food products is also increasing. Unfortunately, many organisms are out there that want to consume the crops that are meant for the human population (Abdullah &

Waleema, 2009). It is expected that chief amount of our agricultural crops produced in the world each year are destroyed by pests, which results in big economic loss.

Due to this high loss in food production, pesticides are often used to try to combat the problem. Pesticides are the hazardous chemical that kills or manage the population of pests. Varieties of pesticides are available in the market today, but the most common are herbicides and insecticides, which kill or manage unwanted plants and insects, which is the main hurdle in the crop productivity (Llewellyn, 2002; Aktar, Sengupta & Chowdhury, 2009). The harm caused by the agricultural pests is a global issue, and from last few decades the amount of pesticides used has increased many folds. The benefits of pesticides include disease control and improved food production, increased profits for farmers. It also increases farm profits by serving the farmer save money on labor costs. Using pesticides reduces the amount of time required to manually remove weeds and pests from fields.

Although there are benefits to the use of pesticides, but the debits associated with their use is much higher than its credits. These chemical fertilizers and pesticides do pose a potential risk to humans and other life forms and unwanted side effects to the environment (Igbedioh, 1991; Forget, 1993). The Green Revolution comes in existence during 1960s. Several hectares of land were brought under cultivation. Natural and organic fertilizers were replaced by chemical fertilizers and locally made pesticides were replaced by chemical pesticides just to increase the productivity. When pesticides are used, they do not always stay in the location where they are applied. They are mobile in the environment and often move through water, air and soil. The problem with pesticide mobility is that when they travel, the pesticides come in contact with other organisms and can cause harm (Hurley, Hill & Whiting, 1998; Aktar et al., 2009).

The prototype of pesticide usage in India is highly different from that of the world. In India 76% of the pesticide used is insecticide, as against 44% globally (Mathur, 1999). The use of herbicides and fungicides is correspondingly less than other part of the world. The main use of pesticides in India is for cotton crops (45%), followed by paddy and wheat (Aktar et al., 2009). Pesticides have contaminated almost every part of our environment. Pesticide residues are found in soil and air, and in surface and ground water across the countries, and urban pesticide uses contribute to the problem. Pesticide contamination poses significant risks to the environment and non-target organisms ranging from beneficial soil microorganisms, to insects, plants, fish, and birds. Contrary to common misconceptions, even herbicides can cause harm to the environment.

An enormous amount of pesticides are used in agricultural field for pest management and to increase crop productivity. The toxicity of these chemicals can be frequently detected in aquatic ecosystem as it is totally dependent on the concentration of the chemical in the organism or even the concentration at the target receptor in the organism (Schreinemachers & Tipraqsa, 2012). Pesticides have also been shown to disrupt the balance of an ecosystem as its use is accountable for bioaccumulation and biological magnification. Many synthetic pesticides are not able to be broken down and once they enter the body of an organism, they are permanently stored in the body tissue and cause harmful and serious effects (Akerblom, 2004). Agricultural areas have the potential to pollute the aquatic ecosystem via the popular use of pesticides (chemicals) and fertilizers (nutrients) salts and sediments. More than 90 percent of aquatic ecosystem contained, several pesticides (Kole, Banerjee & Bhattacharyya, 2001). Samples collected from major rivers with mixed agricultural and urban land use influences carry high percentage of pesticides which is responsible for degradation of aquatic life (Bortleson & Davis, 1987-1995).

Aquatic ecosystem is polluted via drainage and irrigation of the field's water overflow of water careless handling of chemicals by laborers spillage runoff erosion (Geissen et al., 2015). Aquatic plants supply as much as 80% of the dissolved oxygen essential for aquatic life in ponds, rivers and lakes. Algae are one of the most primitive group of organisms and the first oxygen producer of the world the major carbon dioxide consumer's major primary consumers increase, and a key component of the aquatic ecosystem (Sahoo, 2003). Spraying of these chemical pesticides in agricultural field, entering the waterways via runoff can result in low oxygen levels and the suffocation of aquatic organisms as well as significantly reducing aquatic flora and fauna productivity (Louis et al., 1996; Abdullah & Waleema, 2009).

ORGANIC FARMING IN PEST MANAGEMENT

Regular use of pesticides, making pests immune to these chemical fertilizers, to overcome this problem farmer needs to use a stronger and expensive pesticide that is doing more damage to the environment. Due to the increased cost of farming, farmers are falling into the trap of money lenders, who are exploiting them to no end, even forcing some to commit suicide. Both consumers and farmers are now gradually shifting back to organic farming in India. Agriculture practices contribute about 70 percent of water abstractions worldwide and play a key role in water pollution. Farms discharge large quantities of organic chemicals, organic matter, drug residues, sediments and saline drainage into water bodies. The resultant water pollution poses demonstrated risks to aquatic ecosystems, human health and productive activities (UNEP, 2016).

On other hand, agricultural pesticides are also found in areas not adjacent to agricultural land (Hoffman, Capel & Larson, 2000). Pest management needs to be measured not exclusively by the effectiveness of a single component but how well a pest or pest complex is controlled by a set of control measures. After considerate the negative and toxic effect of pesticides many alternative methods have been practiced and organic agriculture is one of them. This method not only improves the ecological sustainability but also diminish the effects of pesticides on the environment.

Organic agriculture does not employ any type of chemicals, and therefore, play a vital role in eliminating the significant source of toxic chemicals and pesticides which are responsible for the contamination of our environment (Kristiansen, 2006). Good organic practices work to build the soil and maintain an ecological balance so that chemical fertilizers and synthetic pesticides are proven unnecessary. It is an excellent approach to achieve the farm sustainability by utilizing natural resources on farmland to increase productivity by replacing hazardous components like pesticides, chemical fertilizers, etc (Gurr, Wratten & Altieri, 2004).

In traditional India, the entire agriculture was practiced using organic techniques; therefore, organic farming is not new to India. Even today, also numerous tribal people practice organic farming, for their crop production as they cannot afford agrochemicals. However, the fact that their beliefs are strongly linked to their agricultural practices and the use of Earth's diverse resources might also be one of the reasons for not switching over to agrochemicals. In organic farming is one of them in which natural materials are used. It lays emphasis on soil fertility and plant health.

In this method of farming, hazardous chemicals like pesticides and fertilizers are not used, thus the product obtained through agricultural practices are pesticide-free. Organic farming has attracted considerable attention from those who see it as a boon and want to do natural and healthy agricultural practices. Many farmers in developing countries like India are shifting to organic farming due to the domestic and international demand for organic food. According to the International Fund for Agriculture and Development (IFAD), about 2.5 million hectares of land were being utilized for organic farming in

Figure 1. Significance of organic farming.

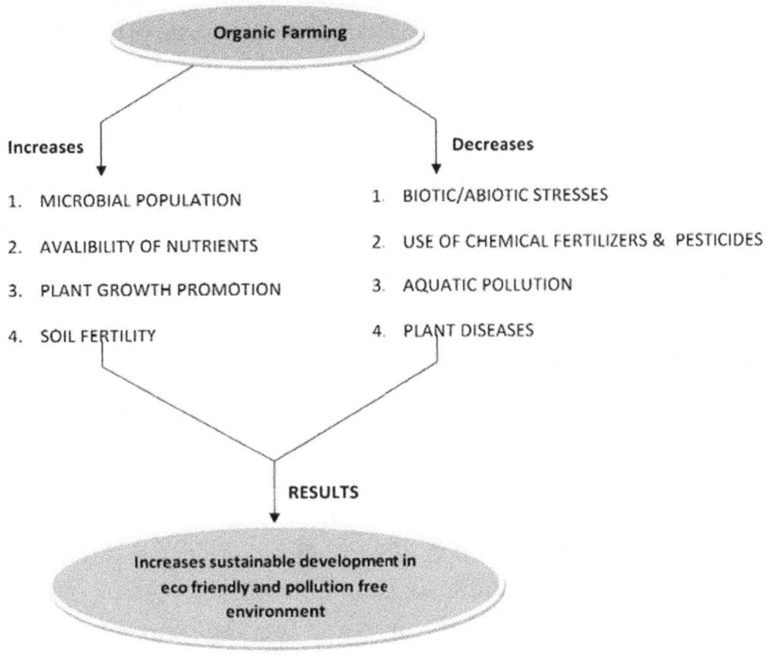

India in 2004. Furthermore, there are over 15,000 certified organic farms in India. Therefore, India is one of the most important suppliers of organic food for developed nations.

Use of chemicals and pesticides has degraded lands and water bodies all over the world (Anurag, Sujata & Goel, 2009; Sundararaman, 2009). Hence, Organic farming not only avoids poisoning our soil and aquatic ecosystem, but, it is possible to turn one's own farmyard waste into value-added products for increasing crop production. Organic farming relies on methods which combine scientific knowledge of ecology and modern technology with traditional farming practices based on naturally occurring biological processes. It is a farming system that sustains the health of soils, ecosystems and people (Folnovic, 2016). The role of organic agriculture, whether in farming, processing, distribution, or consumption, is to sustain and enhance the health of ecosystems and organisms from the smallest in the soil, water to human beings. Hence, such agriculture practices should be managed in an accountable manner to protect the health and well-being of current and future generations and the environment (Figure,1). There is an amazing range of methods of organic farming to overcome the problem of pollution caused by pesticides and chemical fertilizers.

METHODS OF ORGANIC FORMING

Agricultural fertilizers are successful in controlling pathogens, but their applications often also results in the indiscriminate killing of beneficial microorganisms, diminish soil fertility and are accountable for aquatic pollution (Srivastava & Sharma, 2011). It is well-documented that high inputs of chemical fertilizers (especially phosphates and high nitrates) along with certain fungicides (e.g. benomyl) and soil sterilants have negative effects on environment.

Organic farming is an alternative agricultural system which originated early in the 20th century in reaction to rapidly changing farming practices. Organic farming continues to be developed by various organic agriculture organizations today. It relies on fertilizers of organic origin such as compost manure, green manure, and bone meal and places emphasis on techniques such as crop rotation and companion planting. Biological pest control, mixed cropping and the fostering of insect predators are encouraged.

In many nations organic agricultural methods are internationally regulated and legally enforced, as per the standards set by the International Federation of Organic Agriculture Movements (IFOAM), an international umbrella organization for organic farming organizations (Paull, 2010). It can be defined as an integrated farming system that strives for sustainability, the enhancement of soil fertility and biological diversity whilst, with rare exceptions, prohibiting synthetic pesticides, antibiotics, synthetic fertilizers, genetically modified organisms, and growth hormones (Treadwell, Riddle, Barbercheck & Cavanaugh-Grant, 2015; Martin, 2009 ; Mary, 2014).

Organic farming is the most excellent combination of scientific knowledge of ecology and modern technology with traditional farming practices based on naturally occurring biological processes. It is a farming system that maintains the health of soils, ecosystems and human life. Organic farming continues to be developed by various organic agriculture organizations today. The organic agriculture is filled of benefits and opportunities. It provides provision of ecological services and reduced chemical residues in food and the ecosystem (Kristiansen, Taji & Reganold, 2006).

Organic farming makes the use of organic manures and biopesticides with complete escaping of inorganic chemicals and pesticides. In other words it is referred to as uses of natural sources of nutrients instead of using synthetic or inorganic agrochemical. It involves integrated nutrients and pest management in which crops are produced from natural resources having the complete nutritive value and manages to prevent the crop or plants from the pests. To make organic farming well-organized and effectual following techniques are used:

Crop Rotation

It is the vital component of organic farming and will add more crop residues, green manures, other plant debris to the soil and also requires less intensive tillage, which means that soil organic matter does not degrade as quickly. The technique is very beneficial to grow various kinds of crops in the same area, according to the different seasons, in a sequential manner.

Biological Pest Control

It is most effective when used together with other compatible pest control practices in organic farming. There are three main approaches to biological control; conservation, augmentation and importation of natural enemies (Öztemiz, 2008). It is a method of cultivation done in line with nature. For a cultivation which facilitates biodiversity, a cost effective method is to involve Bio-agents. Biological control in organic agricultural systems is very friendly and sustainable to the environment, than the other farming systems.

Compost

Composting is highly rich in nutrients and is a recycled organic matter used as a fertilizer in the agricultural farms. The use of compost in agriculture bring back plant nutrients and organic matter to the soil that otherwise would be lost (Erhart & Hartl, 2010). Organic agriculture is a combination of science, technology and nature which is considered as modernization in agriculture. Regular compost addition enhances soil fauna and soil microbial biomass and stimulates enzyme activity, leading to increased mineralization of organic matter and improved resistance against pests and diseases, both features essential for organic farming.

Crop Diversity

It is fundamental to agricultural growth which is being practicing in organic farming. In this technique in place of one crop (monoculture) variety of crops can be cultivated (polyculture) simultaneously just to meet the increasing demand of crops. They can produce varieties that resist pests and diseases which help in reducing the use of chemical fertilizers and pesticides which are responsible for soil and aquatic pollution.

Soil Management

Organic agriculture is contamination free method for the soil management as it uses natural ways to increase the health of soil. It focuses on the use of bacteria that is present in waste which helps in making the soil nutrients more productive to enhance the soil (Kumar, 2017). After completing one season of cultivation, the soil loses its nutrients value and depletes in quality. To overcome this issue in place of using harmful chemicals to enhance this soil, organic agriculture focuses on implementing natural ways to not only increase the health of soil but also keep the nature and human health unharmed and the environment pollution free.

Weed Management

Weeds are the unnecessary plant that grows in agricultural fields and produce hurdles in the growth of agricultural crops. Organic weed management promotes weed suppression, rather than weed elimination, by enhancing crop competition and phytotoxic effects on weeds (Delate & Hartzler, 2003). Organic farmers integrate cultural, biological, mechanical, physical and chemical tactics to manage weeds without synthetic herbicides. Some naturally sourced chemicals are allowed for herbicidal use. These include certain formulations of acetic acid (concentrated vinegar), corn gluten meal, and essential oils. A few selective bioherbicides based on fungal pathogens have also been developed (Schonbeck, 2010).

Livestock

Organic farming instigates domestic animals use to boostup the sustainability of the farm. There can be no better place for the pet animals to get fresh air, food and a great exercise than the green farm. Since everything is preferred to be in a natural way, just like the animals were used as labor in the earlier times for plowing, organic agriculture encourages the use of domestic animals to increase the sustainability of the farm (Kumar, 2017).

Genetic Modification

Genetic modification is not very well documented with organic farming and is therefore kept away from this kind of agricultural set up. Since organic agriculture encourages the use of natural ways to enhance the farm, crops and soil; there is a discouragement of genetically engineered animals and plants. However, there is an argument keeping in mind that the pollen present in such modified crops are present in the stock of seeds used for organic agriculture, making it impossible to keep this completely out of this agriculture (Kumar, 2017).

Organic farming provides high premium to the farmers as it normally priced about 25% higher than conventional food with low investment. It builds healthy soils by nourishing the living component, microbial inhabitants that contributes to good soil structure and water-holding capacity. In organic farming, unnatural substances such as synthetic chemical, pesticide and fertilizers, GMOs, processing aids, etc., should be limited. An organic farming keeps biodiversity and reduce environmental pollutions such air, water and soil (Nejadkoorki, 2012), it discourages algal bloom and supports water conservation and water health. Water pollution is largely associated with the use and discharge of water in both animal and plant farming and Organic farming helps in lowering down the water pollution level, as there is much reduced eutrophication of chemical inputs (Anh, Kroeze, Bush, & Mol, 2010).

CONCLUSION

Chemical fertilizers and pesticides are applied intensively in agriculture practices now days to increase the productivity. Due to improper management, they are polluting water resources with carcinogens and other toxic substances that can affect humans. Pesticides and chemical fertilizers may also affect biodiversity by killing weeds and insects, with negative impacts up the food chain. Organic farming has emerged as one of the best known alternative farming systems developed in response to the shortcomings of mainstream agriculture. Its principle is to keep out the use of chemical fertilizers and pesticides and not only sustain soil quality and health but is also cooperative in managing the aquatic pollution.

In developing countries like India, the tremendous growth in the use of pesticides and chemicals is due to, dependence on broad spectrum pesticides, weak institutional frameworks; weak rule enforcement, and limited knowledge and awareness among farmers on the use of hazardous chemicals. To overcome these issues and challenges the government must improve the system of controls and enact regulations more efficiently. Secondly, farmers should be given some financial incentives to switch over to organic farming. Such initiatives not only increase the productivity but also help in reducing the toxicity level of soil and aquatic system which are directly related with the adverse effect of chemicals and pesticides. It is one natural method to counter pesticides with biological control. Organic farming can be defined as an approach to agriculture is to create environmentally and economically sustainable agricultural production systems.

Organic farming is done to release nutrients to the crops for increased sustainable production in an eco-friendly and pollution free environment. Research is desirable to appraise policies and instruments for reducing source loads and minimizing pollution along flow paths to the sea. Additional work is also required to quantify the effectiveness of different approaches in reducing the economic impacts of water pollution caused by different agricultural practices. Organic farming ensures an effective approach, thereby helping to optimize the use of resources and reduce pollution.

REFERENCES

Abdullah, A. A., & Waleema, B. B. (2009). Effect of Agricultural Chemicals on Aquatic Ecosystem in Guyana. *Global Journal of Environmental Research, 3*(1), 22–25.

Akerblom, N., Arbjörk, C., Hedlund, M., & Goedkoop, W. (2008). Deltamethrin toxicity to the midge Chironomus riparius Meigen-effects of exposure scenario and sediment quality. *Ecotoxicology and Environmental Safety, 70*(1), 53–60. doi:10.1016/j.ecoenv.2007.09.001 PMID:18255145

Aktar, M. W., Sengupta, D., & Chowdhury, A. (2009). Impact of pesticides use in agriculture: Their benefits and hazards. *Interdisciplinary Toxicology, 2*(1), 1–12. doi:10.2478/v10102-009-0001-7 PMID:21217838

Anh, P. T., Kroeze, C., Bush, S. R., & Mol, A. P. J. (2010). Water pollution by intensive brackish shrimp farming in south-east Vietnam: Causes and options for control. *Agricultural Water Management, 97*(6), 872–882. doi:10.1016/j.agwat.2010.01.018

Anurag, S., & Goel, M. M. (2009). *Enhancing plant defense mechanisms for disease management.* South Asia Conference on 'Outstanding Organic Agriculture Techniques', Bengaluru, India.

Barron, M. G., & Woodburn, K. B. (1995). Ecotoxicology of chlorpyrifos. *Reviews of Environmental Contamination and Toxicology, 144*, 1–93. doi:10.1007/978-1-4612-2550-8_1 PMID:8599033

Bortleson, G. & Davis, D. (1987–1995). *Pesticides in selected small streams in the Puget Sound Basin.* U.S. Geological Survey & Washington State Department of Ecology.

Delate, K. & Hartzler, R. (2003). Weed Management for Organic Farmers. *Iowa State University Extension Bulletin, 1883.*

DeLorenzo, M. E., Scott, G. I., & Ross, P. E. (2001). Toxicity of pesticides to aquatic microorganisms: Areview. *Environmental Toxicology and Chemistry, 20*(1), 84–98. doi:10.1002/etc.5620200108 PMID:11351418

Ecobichon, D. J. (1991). Toxic effects of pesticides. In *Casarett and Doull's Toxicology- The basic science of poisons* (4th ed.; pp. 565–622). New York: McGraw-Hill, Inc.

Erhart, E., & Hartl, W. (2010). Compost Use in Organic Farming. In E. Lichtfouse (Ed.), Genetic Engineering, Biofertilisation, Soil Quality and Organic Farming, Sustainable Agriculture Reviews. Springer. DOI: doi:10.1007/978-90-481-8741-6-2

FAO. (2016). *FAOSTAT Database*. Available at http://faostat3.fao.org/browse/R/RP/E

Folnovic, T. (2016). *Best Practices for Pest Management in Organic Farm Production*. Agrivi.

Forget, G. (1993). Balancing the need for pesticides with the risk to human health. In G. Forget, T. Goodman, & A. de Villiers (Eds.), *Impact of Pesticide Use on Health in Developing Countries* (Vol. 2). Ottawa: IDRC.

Geissen, V., Mol, H., Klumpp, E., Umlauf, G., Nadal, M., van der Ploeg, M., ... Ritsema, C. J. (2015). Emerging pollutants in the environment: A challenge for water resource management. *International Soil and Water Conservation Research, 3*(1), 57–65. doi:10.1016/j.iswcr.2015.03.002

Gurr, G. M., Wratten, S. D., & Altieri, M. A. (2004). Ecological Engineering for Pest Management: Habitat Manipulation for Arthropods. CSIRO Publishing.

Hoffman, R. S., Capel, P. D., & Larson, S. J. (2000). Comparison of pesticides in eight U.S. urban streams. *Environmental Toxicology and Chemistry*, *19*(9), 2249–2258. doi:10.1002/etc.5620190915

Hurley, P. M., Hill, R. N., & Whiting, R. J. (1998). Mode of carcinogenic action of pesticides inducing thyroid follicular cell tumours in rodents. *Environmental Health Perspectives*, *106*(8), 437–445. doi:10.1289/ehp.98106437 PMID:9681970

Igbedioh, S. O. (1991). Effects of agricultural pesticides on humans, animals and higher plants in developing countries. *Archives of Environmental Health*, *46*(4), 218–224. doi:10.1080/00039896.1991.9937452 PMID:2069430

Kirchmann, H., & Ryan, M. H. (2004). Nutrients in organic farming – are there advantages from the exclusive use of organic manures and untreated minerals? In *New directions for a diverse planet. Proceedings of the 4th International Crop Science Congress*. Brisbane, Australia: The Regional Institute.

Kole, R. K., Banerjee, H., & Bhattacharyya, A. (2001). Monitoring of market fish samples for Endosulfan and Hexachlorocyclohexane residues in and around Calcutta. *Bulletin of Environmental Contamination and Toxicology*, *67*(4), 554–559. doi:10.1007001280159 PMID:11779071

Kreuger, J. (1999). *Pesticides in the environment- Atmospheric deposition and transport to surface waters* (Doctoral thesis). Swedish University of Agricultural Sciences, Uppsala, Sweden.

Kreuger, J., Peterson, M., & Lundgren, E. (1999). Agricultural inputs of pesticide residues to stream and pond sediments in a small catchment in southern Sweden. *Bulletin of Environmental Contamination and Toxicology*, *62*(1), 55–62. doi:10.1007001289900841 PMID:9870990

Kristiansen, P., Taji, A., & Reganold, J. (2006). Organic agriculture: opportunities and challenges. In P. Kristiansen, A. Taji, & J. Reganold (Eds.), *Organic Agriculture: A Global Perspective* (pp. 421–441). CSIRO Publishing. doi:10.1079/9781845931698.0421

Kumar, P. (2017). *Integrated Soil Fertility Management: Converting Subsistence Farming to Productive Farming*. Retrieved from http://www.pradan.net/wp-content/uploads/2017/12/Article-No.-3-Integrated-Soil-Fertility-Management-Converting-Subsistence-Farming-to-Productive-Farming-.pdf

Lehotay, S. J., Harman-Fetcho, J. A., & McConnell, L. L. (1998). Agricultural pesticide residues in oysters and water from two Cheaspeake Bay tributaries. *Marine Pollution Bulletin*, *37*(1-2), 32–44. doi:10.1016/S0025-326X(98)00129-5

Llewellyn, R. (2002). The Good Bug Book. IPM Pty Ltd.

Louis, A. H., Weigmann, D. L., Hipkins, P., & Stinson, E. R. (1996). *Pesticides and Aquatic Animals: A Guide to Reducing Impacts on Aquatic Systems*. Retrieved from http://www.ext.vt.edu/pubs/waterquality/420- 013/420-013.html

Martin, H. (2009). *Introduction to Organic Farming*. Ontario Ministry of Agriculture, Food and Rural Affairs.

Mary, G. (2014). *What is organic production?* National Agricultural Library. USDA.

Mathur, S. C. (1999). Future of Indian pesticides industry in next millennium. *Pesticide Information.*, *24*(4), 9–23.

Matsumura, F. (1985). *Toxicology of insecticides* (2nd ed.). New York: Plenum Press. doi:10.1007/978-1-4613-2491-1

Miles, C. J., & Pfeuffer, R. J. (1997). Pesticides in canals of South Florida. *Archives of Environmental Contamination and Toxicology*, *32*(4), 337–345. doi:10.1007002449900194 PMID:9175497

Moss, B. (2008). Water pollution by agriculture. *Philosophical Transactions of the Royal Society of London. Series B, Biological Sciences*, *363*(1491), 659–666. doi:10.1098/rstb.2007.2176 PMID:17666391

Mulla, M. S., & Mian, L. S. (1981). Biological and environmental impacts of the insecticides malathion and parathion on nontarget biota in aquatic ecosystems. *Residue Reviews*, *78*, 100–135. PMID:7017857

Nejadkoorki, F. (2012). Environmental Benefits of Organic Farming. *International Conference on Applied Life Sciences (ICALS2012).*

Öztemiz, C. S. (2008). Biological control in organic farming. *Journal of the AgriculturalFaculty of Gaziosmanpaşa University*, *25*, 19–27.

Paull, J. (2010). From France to the World: The International Federation of Organic Agriculture Movements (IFOAM). *Journal of Social Research & Policy*, *1*(2), 93–102.

Sahoo, D. (2003). *Role of Algae in Environment and Zoology-Proceedings of the National Seminar on Environmental Biotechnology: Envirotect*. Azra Printers.

Schonbeck, M. (2010). An Organic Weed Control Toolbox. Virginia Association for Biological Farming.

Schreinemachers, P., & Tipraqsa, P. (2012). Agricultural pesticides and land use intensification in high, middle and low income countries. *Food Policy*, *37*(6), 616–626. doi:10.1016/j.foodpol.2012.06.003

Srivastava, K., & Sharma, A. K. (2011). Arbuscular Mycorrhizal Fungi in challenging Environment– A Prospective. In S. M. Fulton (Ed.), *Mycorrhizal Fungi: Soil, Agriculture and Environmental Implications* (pp. 1–34). Nova Science Publishers, Inc.

Sundararaman, S. R. (2009). *Crop Production and Plant Protection in Organic Farming*. South Asia Conference on 'Outstanding Organic Agriculture Techniques', Bengaluru, India.

Thebo, A. L., Drechsel, P., Lambin, E. F., & Nelson, K. L. (2017). A global, spatiallyexplicit assessment of irrigated croplands influenced by urban wastewater flows. *Environmental Research Letters*, *12*(7), 074008. doi:10.1088/1748-9326/aa75d1

Treadwell, D., Riddle, J., Barbercheck, M., Cavanaugh-Grant, D., & Zaborski, E. (2015). *Cooperative Extension System, What is organic farming?* Retrieved from http://articles.extension.org/pages/18655/what-is-organic-farming

Trewavas, A. (2001). Urban myths of organic farming. *Nature*, *410*(6827), 409–410. doi:10.1038/35068639 PMID:11260685

Trewavas, A. (2004). A critical assessment of organic farming-and-food assertions with particular respect to the UK and the potential environmental benefi ts of no-till agriculture. *Crop Protection (Guildford, Surrey)*, *23*(9), 757–781. doi:10.1016/j.cropro.2004.01.009

UNDESA. (2017). *World Population Prospects: The 2017 Revision, Key Findings and Advance Tables.* Working Paper No. ESA/P/WP/248. United Nations, Department of Economic and Social Affairs, Population Division.

UNEP. (2016). *A snapshot of the world's water quality: Towards a global assessment.* Nairobi: United Nations Environment Programme (UNEP).

Willer, H., & Yussefi, M. (Eds.). (2005). *The World of Organic Agriculture 2005 - Statistics and Emerging Trends*. Bonn: International Federation of Organic Agriculture Movements.

Wood, S., Sebastian, K., & Scherr, S. J. (2001). *Pilot Analysis of Global Ecosystems: Agroecosystems.* Washington, DC: World Resources Institute.

KEY TERMS AND DEFINITIONS

Biopesticides: Biopesticides are certain types of pesticides derived from such natural materials as animals, plants, bacteria, and certain minerals. For example, canola oil and baking soda have pesticidal applications and are considered biopesticides.

Chemical Fertilizer: A chemical fertilizer is defined as any inorganic material that is partially or completely synthetic in origin that is added to the soil to sustain plant growth. Many artificial fertilizers contain acids, such as sulfuric acid and hydrochloric acid, which tend to increase the acidity of the soil, reduce the soil's beneficial organism population, and interfere with plant growth and contaminate the environment.

Organic Farming: Organic farming is a system which prohibits the use of synthetic fertilizers, pesticides, hormones, etc. and to the maximum extent feasible relies upon crop rotations, crop residues, animal manures, off-farm organic waste, mineral grade rock additives, and biological system of nutrient mobilization and plant protection. The principal goal of organic production is to develop enterprises that are sustainable and harmonious with the environment.

Pesticide: These are the chemical substances that are used to control pests, weeds, insects, etc. These chemical substances are designed to kill or retard the growth of pests that damage or interfere with the growth of agricultural crops and vegetations. Practically all chemical pesticides, however, are poisons and pose long-term danger to the environment and humans through their persistence in nature.

Chapter 23
Global Warming and Pesticides in Water Bodies

Sanjeevi Ramakrishnan
NIMS University, India

Anuradha Jayaraman
NIMS University, India

ABSTRACT

Over the last few decades, significant effort has been undertaken to record the effect, fate, and transport of pesticides into surface water and groundwater. The functional aspect of climate change and pollutant interactions induces vulnerability on species and populations and reveals the onset of adverse events by triggering the nature's threshold. Erstwhile climate change by itself will affect the environmental distribution and induces prominent toxicity of various chemical toxicants like pesticides. Despite of their potential toxicity towards the beneficial organisms and even to human beings, their use is mandatory to improve the productivity and high-quality life standards. In general, climate change alters the efficiency of pesticide use and can also be expected. But their leaching pollutes ground water. Research on the effects of climate change, on the environmental fate and behavior of pesticides and their mechanisms of action between the environmental compartments has been reviewed in this chapter.

INTRODUCTION

In the recent past the rise in global temperature is impacted due to rapid climatic change generated through global warming. Anthropogenic sources, emitting green house gases (GHG) remarkably raised the global temperature (Keer, 2007; Weare, 2009). These raising extreme events which are a real and daunting problem might lead to global average temperature of 2–3ºC by 2050 and by the end of the century reaching 6.5°C (Solomon *et al.*, 2007; Anuradha *et al.*, 2017; Sanjeevi *et al.*, 2017). In general '*Climate change*' is defined as a variation in the statistical properties of climatic systems, on considered it over long period, regardless of its cause (ENSAA, 2011). According to the U.N. Intergovernmental Panel on Climate Change (IPCC) has reckoned its four assessments comprising evidence, impacts, and mitigation of climate change (IPCC, 2007). According to their report unequivocal global warming with

DOI: 10.4018/978-1-5225-6111-8.ch023

the evidence of increases in global mean air and ocean temperature, with consequent effect of widespread melting of snow and ice, and rise in global sea level. The projected raise in temperature of 1.8–4.0 °C by end of this century under a range of probable GHG emission scenarios with greater warming is expected at higher latitudes. These extensive difference pose impact on nature, human health and even the economy, implies that climate change is both spatially and temporally heterogeneous (EEA, 2012).

The functional aspects of climate change and pollutant interactions further induces vulnerability on species and populations and reveals the onset of adverse events by triggering the nature's threshold. Erstwhile climate change by itself will affect the environmental distribution and induces prominent toxicity of various chemical toxicants. Most significantly here we focus on the classes of chemicals with global significance, including persistent organic pollutants (POPs) and other pesticides. As of now, twelve chlorinated organic chemicals (COCs) were listed as POPs as per the U.N. Stockholm Convention, which comprises organochlorine pesticides (e.g. DDT (dichlorodiphenyltrichloroethane) and toxaphene), polychlorinated biphenyls (PCBs), dioxins and furans (UNEP, 2009). Other pesticides, includes aldicarb, atrazine, and chlorpyrifos with special interest due to their wider application in large quantities over a broad spectral area and possessing higher range of toxicological effects.

For several years, pesticides were widely utilized so as to prevent, mitigate and/or to destroy pests and to improve the economy of yield and crop quality. Pesticides provided significant advantage for agricultural crops, than that of weeds and protected the crops from the damage influenced by disease causing agents and pests. Despite of their potential toxicity towards the beneficial organisms and even to the human beings, their use is mandatory to improve the productivity and high quality life standards. In general climate change alters the efficiency of pesticide use and can also be expected. The direction of this action, is however uncertain and has not yet been detailed thoroughly (Noyes *et al.*, 2009). Research on the effects of climate change, on the environmental fate and behavior of pesticides and their mechanisms of action between the environmental compartments has been reviewed in this chapter.

Some of the significant environmental factors posing challenges for sustainability include flooding, drought, temperature extremes, nutrient deficiency, and higher level of pollutants. Presence of lower pesticide residue on crops, induced by climatic effect results in increased pests and diseases attach. This indirectly induces the farmers to spray the pesticides more often in future for better crop yield. Similarly higher pest attach and disease will also enhance the pesticide application frequencies and dosage. As a consequence, the detected residual concentrations will be higher in the products, there by consumer exposure to pesticide residues remains still at the end of food chain. In this respect, the food safety issues related to an increased exposure to pesticide residues, as a consequence of climate change, might occur. On the other hand pesticides and global warming have potentiality to interact, thereby challenges ecological risk assessment (ERA) of pesticides in the warming world (Noyes & Lema, 2015). The toxicity of several pesticides either increases under warming conditions causing "*climate–induce–toxicant–sensitivity*" or the ability to cope up with high temperatures declines efficacy of pesticides causing "*toxicant–induced–climate–change–sensitivity*" (Moe *et al.*, 2013; Noyes & Lema, 2015). The present chapter brings the nut shell on *state–of–the–art* climate change impact on pesticides and their environmental fate. With special reference to surface water quality (from *source to sink*) is that there is a degradation trend of surface and underground water quality leading to an increase of at risk situations with regard to potential health impact, mainly due to climate change.

PESTICIDES

According to U.S. Environmental Protection Agency (EPA) 2006, pesticides are those used to control the pests and are found to contribute to water pollution. Their leaching pollutes ground water. Several factors influencing the leaching process include soil texture, pesticide properties, irrigation and rain fall pattern. Water soluble pesticides peculate into ground water, more in sandy soil. In similar way pesticides does reaches natural water bodies through surface runoff. On reaching natural water bodies the remains of pesticide residues harms the aquatic flora and fauna. Whereas, leaching of persistent organic pollutants (POPs) could cause much more harmful effects in the environment (Pope *et al.*, 2016; Chaudhry & Malik, 2017).

Anthropogenic Organic Chemical Sources of Water Pollution and Its Effects

Anthropogenic and synthetic organic compounds polluting the aquatic systems include compounds such as pesticides, herbicides, numerous industrial chemicals, and compounds derived from energy production and other combustion processes. The most notorious organic contaminant polluting environment is dichlorodiphenyltrichloroethane (DDT). Until the recognition of the adverse environmental impacts of DDT as started in the book Silent Spring by in 1962, DDT is exemplified as class of compounds falling under persistent organic pollutants (POPs). The book Silent Spring evidences the persistent effect of DDT by documenting the reduction of songbird populations in the Midwestern United States and is responsible for complete devastation of the bald eagle populations in southern California Bight on the Channel Islands. Due to this fact DDT has been banned in most countries of the world. Hydrophobic property of DDT favors the binding to suspended particles and molecules in water and gets transported far away from its place of application. Moreover, its hydrophobicity increases the chance of bioaccumulation in organisms and even raises its concentration up in aquatic food chain *via* biomagnifications.

Another class of water pollutant includes polychlorinated biphenyls (PCBs). These are first produced as complex commercial mixtures, primarily used as dielectric fluids in capacitors and transformers. And most prominently, it finds wider application as plasticizers in the plastics industry. The chemical and physical properties of PCBs are prone to long-range atmospheric transport. As of now PCBs are now widely dispersed globally as evidenced by their presence in the fat of polar bears in the Arctic and penguin eggs in the Antarctic region. (Kumar *et al.*, 2002). Some of the toxic effect of POPs and PCBs includes thinning of birds eggshell and causes deformities such as crossed bills in cormorants of the Great Lakes region (Gilbertson *et al.*, 1991; Fox *et al.*, 1991). They do inhibit reproductive system of aquatic species. Most recently, these organic compounds were discovered to have affinity to get bound with estrogen receptor sites and thereby exerting hormonal activity in organisms and interfere with normal biochemical processes (Salama *et al.*, 2003).

Practice of Pesticide Use

In general the term '*Pesticide*' could be defined as the employment of mechanical killing, heat destruction, chemical destruction, biological control or any other means to remove the pest permanently. The practice of pest control through chemicals extends into ancient records of Sumerians (5000–2000 BC) using sulfur compounds to control weeds as early as 4500 years ago. Later, Chinese people also used sulfur, mercury and arsenic compounds to eradicate human lice and household insect pests. But sometimes the remedy was worse than the problem, undoubtedly due to the toxic effect of both mercury and arsenic. Ancient Romans also used sulfur to control insect pests and applied salt for eradicating weeds. Much of pest control was achieved through mechanical means i.e. by employing slaves for pick off insect larvae or to manually remove the weeds.

Usage of chemicals to control insects exploded after World War II with the successful introduction of organochlorines (OCs). Several of the chemicals like dichlorodiphenyltrichloroethane (DDT), lindane, aldrin, dieldrin, endrin, and toxaphene were liberally applied on crop fields, due to fact that their usage was in expensive, and best of all, efficient in controlling insect pests on large variety of crops. Later, the practice of insecticide use emerged to control insects in homes and was found to be effective in controlling mosquitoes.

Starting from 1972 with DDT, the US EPA began to ban organochlorine pesticides (OCPs) including several of OCs and endosulfan. As OCs was being phased out, new pesticides came into the market, namely '*new generation*' pesticides including organophosphates (OPs) and carbamates. Though they were discovered several during 1930s and 1940s but came into existence after banning OCs. In the 1960s, pyrethroids were synthesized from pyrethrum, as a natural insecticide derived from the chrysanthemum flower (*Chrysanthemum cinerariifolium* or *C. coccineum*). Above mentioned chemicals are still in use and are applied in both crop field and for household use. Subsequently, greater development has been achieved in the synthesis and formulation of active ingredients for specific purposes such as herbicides, acaricides, fungicides, or nematicides. Half-lives of these new pesticides are typically measured in days, weeks, or a few months unlike in years or decades.

At present pesticide research continually focus on the production of efficient pesticides characterized with highly effective at lower dosage, species specific and higher degradability. Some control agents with very high species-specificity include synthetic pheromones or sex attractants that lure male insects of a given species to traps where they die before mating. The use of pheromones is an important factor in the control of boll weevils (*Anthonomus grandis*), a serious pest on cotton. Biological organisms such as *Bacillus thurigensis* or Bt, a bacterium that is specific for certain insect larvae such as caterpillars, have been used with success in controlling pests (Sparling, 2016).

Current Use of Pesticides and Its Regulation

The multivariate nature of climate change and nonlinear thresholds in natural processes brings more difficulty in finding links between climate change and pesticide use (Harvell *et al.*, 2002). Here, six factors (fig 1) that directly impact a farmer's use of pesticides include (i) *legislation*; (ii) *economic situation*; (iii) *technological progress*; (iv) *pesticide efficiency*; (v) *crop characteristics* and (vi) *pest occurrence and severity*. Among these first three are not directly influenced by climate, whereas later aspects are directly influenced by climate.

Pesticide use is strongly governed through several regulations, included in legislation, which define the authorized active substance/ crop combinations. The concentration of pesticide residues that may remain on the crops after harvest is regulated by setting Maximum Residue Levels (MRLs). At the national level, legislation is, consequently, an important influencing factor for pesticide use, as it clearly limits the number and scope of pesticides a farmer has at his disposal. Regarding the economic situation, Eid *et al.*, (2007) denoted that high temperatures can constrain agricultural production. In Addition, several other researchers have shown that a shift in precipitation can also slow down the economic development of some nations and particularly affects agricultural production. The economic climate also strongly

Figure 1. Factors influencing the pesticide use.

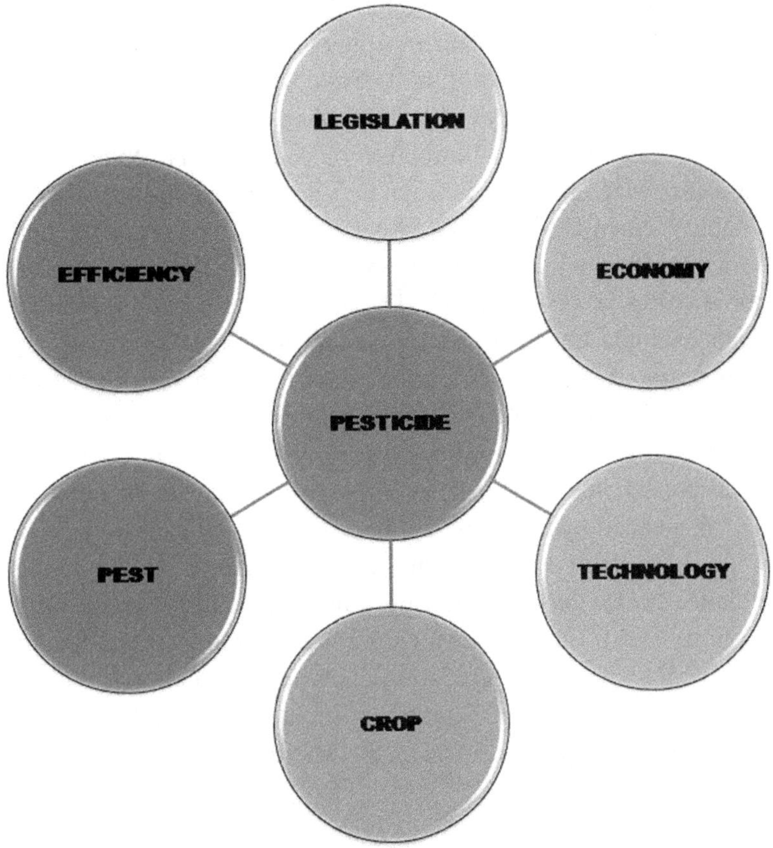

influences on-farm decisions, by limiting farmers' pest control options. Primarily, pesticide manufacturing domains decide on what products, active substances and formulations will be marketed in a specific country. An important issue is the focus on production for major crops, to increase a company's profits. This limits the scope of available products for small-scale crops. Secondly, farmers can use preventive or curative pesticides according to the advice of an officer or seller, but in reality, this choice will largely be influenced by the purchase and application costs. Technological progress also strongly influences plant protection product (PPP) application and residues. For an example, in advanced agricultural technologies, sprayers are equipped with sensors to determine their location and for example, plant numbers, coverage levels, amount of biomass or infection levels. A combination of the recordings of the sensors, with additional information on required pesticide doses, leads to precision farming, which promotes a more efficient pest/disease control (Dworak *et al.*, 2013). Such targeted applications decrease the required PPP volumes, to maintain the same level of crop protection, in comparison with current practices.

Pesticides and Its Fate in the Environment

Persistent organic pollutants will eventually break down in the environment by processes of photolysis, chemical reactions, and biodegradation. But some of their transformation leaves behind the by-products with higher toxicity and are made more water soluble. Such is the case for nitro-poly aromatic hydrocarbons that arises as products of incomplete combustion and from atmospheric transformation of polycyclic aromatic hydrocarbons (PAHs). The atmospheric precipitation of these chemicals contributes to contamination of water bodies. Stockholm (2005) listed the POPs as '*Dirty dozen*' to be eliminated or restricted from production and use includes the compounds of pesticides such as Aldrin, Chlordane, DDT, Dieldrin, Endrin, Heptachlor, Hexachlorobenzene, Mirex and Toxaphene.

Yet another way to categorize chemical pollutants based on their physiochemical properties such as hydrophilic and volatile factors. Pesticide is a broad category that comprises herbicides, fungicides, rodenticides, and others specific to what is intended to be in control/ killed. A better term encompassing to denote all of the above categories is "*biocide*". The daunting effect of biocides is that they are often *non–target–specific*. For example honey bees (*Apis mellifera*) are often a non-target species affected by the exposure to neonicotinoid pesticides found in water bodies nearby agricultural fields using the pesticide. Due to the excessive use of pesticides, surface water runoff can pollute surface water and groundwater, specifically near the agricultural areas.

Fate of plant protection pesticides (PPP) level is determined by the pesticide residue remained on the crop, which in turn influenced by its formulation, dose and application method. Pesticides interact with the plant surface, but are also exposed to environmental factors such as wind, sun radiation and precipitation (Keikotlhaile, 2011). Pesticide transport and degradation occurs in two main routes that affects pesticide availability and efficiency. Pesticide transfer includes volatilization, wash and runoff and leaching processes, while pesticide degradation encompasses photolysis, chemical and microbial breakdown (Figure 2). Next to pesticide dissipation, pesticide eco-toxicity will also determine the efficacy of an applied dose and consequently, influence pesticide use.

Figure 2. Illustration of environmental factors influencing the fate of pesticide after its application Delcour, 2015.

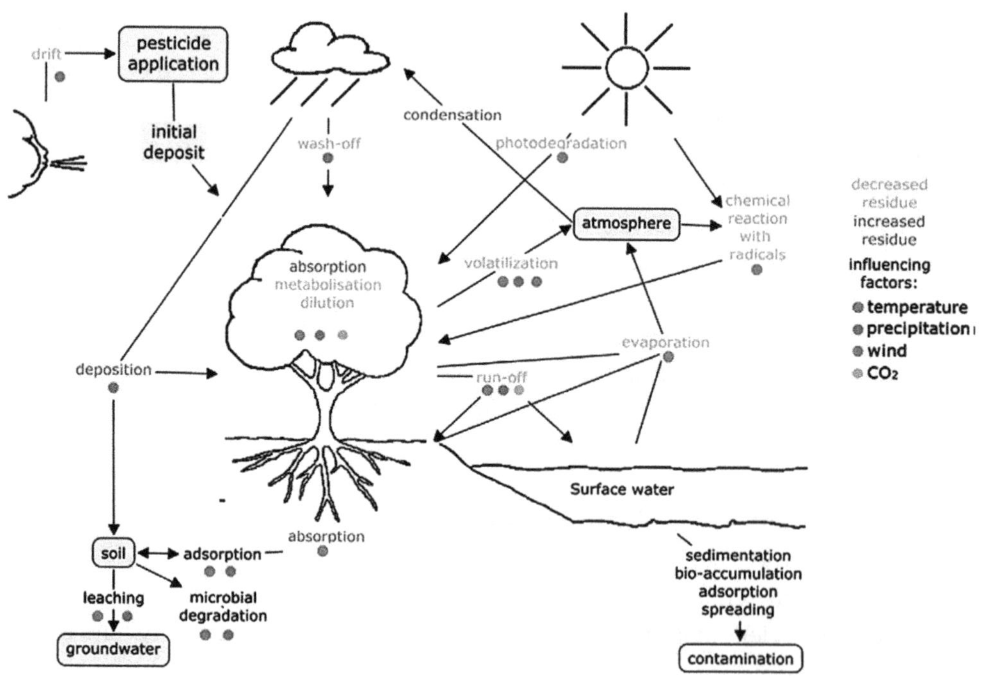

Fate of Pesticide in Terrestrial and Fresh Water Environments

Runoff and drift are the most important transfer pathways of pesticides to other sites or surface waters (Otieno *et al.*, 2013). Pesticide emission and damage by droplet spray drift are defined as the amount of pesticide that is deflected out of the treated area by the action of air currents (De Schampheleire *et al.*, 2007). Farmers can check wind speed and independent of climate change they can apply pesticides at the desired wind speed conditions reducing drift occurrence. Managing run-off under climate change is more difficult. The parcel's slope, soil type, texture and structure combined with crop growth and row directionality strongly influence the runoff rate (Steurbaut, 2009).

Precipitation is the main driving factor for agricultural runoff and soil erosion (Otieno *et al.*, 2013). Runoff of chemicals can vary from 5 to 15% of rainfall (Wauchope *et al.*, 2004) and is generally caused by localized rainfall and surface runoff from nearby fields or canopy through fall at urban sites (Fernandez-Gomez *et al.*, 2013). Several studies indicate that increased precipitation enhances runoff contaminated with pesticides (Oliver *et al.*, 2012). The numbers and concentrations of pesticides have already been proven to rise spectacularly, sometimes resulting in a subsequent release into shallow groundwater (Otieno *et al.*, 2013). In two studies considering herbicides, the effect of precipitation frequency and timing during the planting season was demonstrated (Goel *et al.*, 2005; Paetzold *et al.*, 2007). Paetzold *et al.* (2007) reported a positive correlation with the soil moisture level. A spatial and temporal distribution of pesticide residues in surface waters can thus be expected (Vryzas *et al.*, 2009). Effects of higher temperatures were also shown by Carere *et al.* (2011), who revealed an altered distribution and partitioning of contaminants in water and by Ficklin *et al.* (2010), who demonstrated a decreasing agricultural runoff load. The latter also proved an effect of increasing CO_2 levels on several active ingredients, although found that correlations were contradictive.

Leaching is the downward movement of a chemical through the soil, eventually reaching the groundwater (Keikotlhaile, 2011). Nolan *et al.* (2008) concluded that the transfer of pesticides to depth *via* leaching and to surface water *via* drainage was mostly influenced by interactions between climate and soil–pesticide combinations. Several studies reported an enhancing effect of precipitation volumes, of variable duration, rainfall seasonality, intensity, and timing in relation with pesticide application (Bloomfield *et al.*, 2006; Lewan *et al.*, 2009). Temperature affects soil mineralogy and geochemistry and is consequently a main driver for leaching (Bloomfield *et al.*, 2006). Temperature not only causes a seasonal effect on pesticide transport in leaching (Nolan *et al.*, 2008), but also reduces the influence of winter rainfall (Blenkinsop *et al.*, 2008). This winter rain exhibits an overall strong influence on the more retained and less degraded residues of spring or autumn applications (Bloomfield *et al.*, 2006).

IMPACTS OF CLIMATE CHANGE IN THE FATE AND BEHAVIOR OF PESTICIDES IN WATER BODIES

Over the last two decades significant effort has been dedicated to understanding the fate and transport of pesticides in surface water and groundwater and to use this understanding in the development of environmental policy and regulation. Bloomfield *et al.* (2006) addresses that gap by reviewing how climate change may impact the fate and transport of pesticides in surface and ground waters as a pre-cursor to quantitative studies. Significant research effort has been dedicated to understanding the fate and transport of pesticides in the environment, and the relationships between pesticide fate and transport and specific

Figure 3. Illustration of the source–pathway–receptor scheme
Bloomfield et al., 2006.

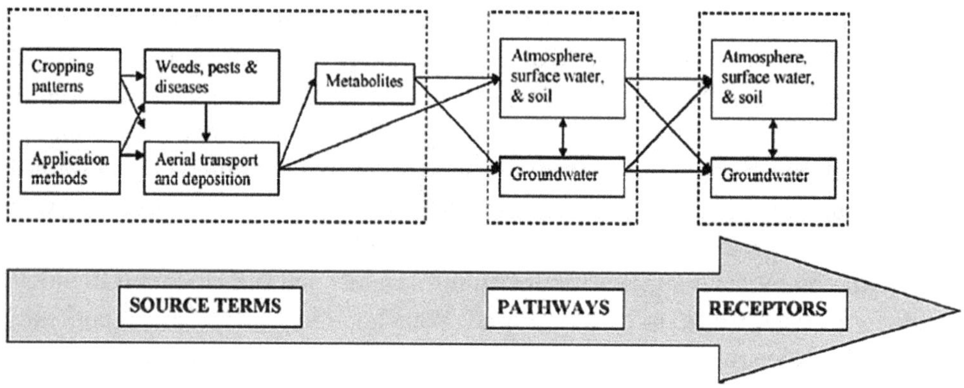

Figure 4. Pesticide source terms, principal environmental pathways and receptors with description of main processes in the soil, and typical pesticide concentrations
Bloomfield et al., 2006.

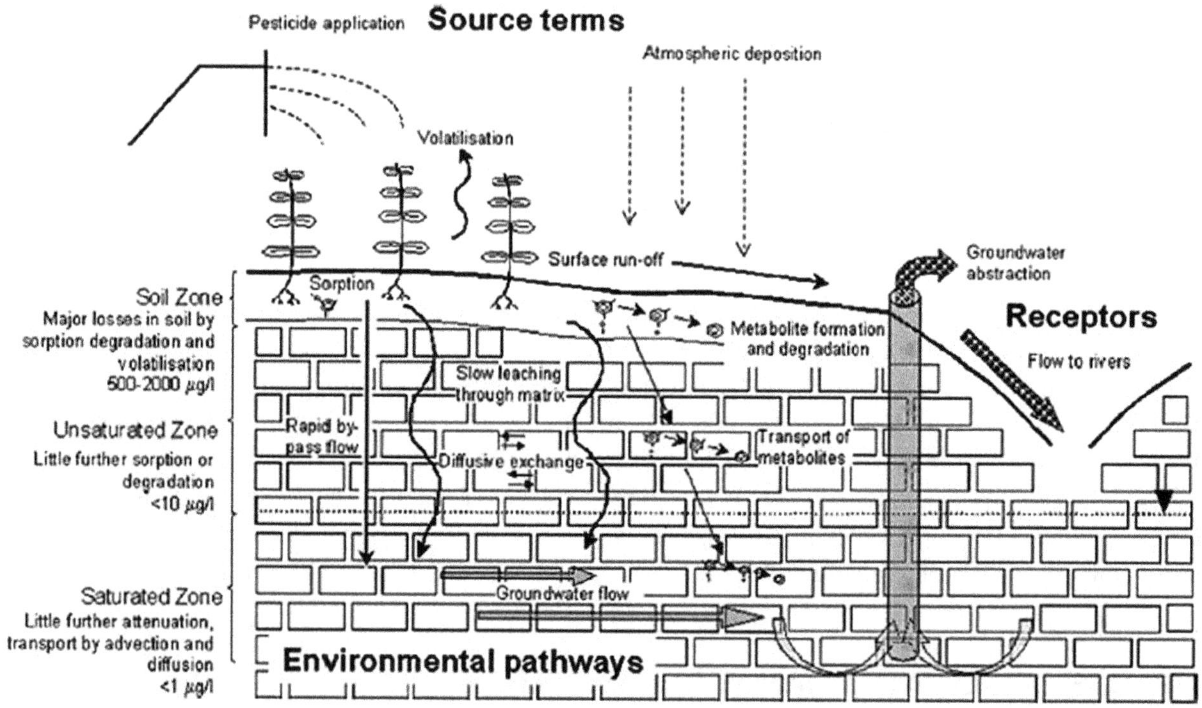

environmental impacts of climate change, particularly investigating the relationships between fate of pesticides in surface and ground water quality influenced by climate change. Authors adopted the use of predicted changes in climate of United Kingdom as the starting point for the source–pathway–receptor analysis.

The Source–Pathway–Receptor Approach

For the study to be tractable it has been necessary to make some simplifying assumptions about pesticide Source terms, pathways and receptors. Figure 3 summarizes some of the main processes associated with the fate and transport of pesticides through the terrestrial aquatic environment.

On the basis of this conceptual model of the fate and transport processes, a simple source–pathway–receptor model, Figure 3, was developed to use as the basis for the sensitivity analysis. In this study, authors identify two principal receptors, groundwater abstraction sources and surface water bodies. This is because it was thought pesticides may follow a variety of transport pathways that could exhibit climate sensitivities but that might have different impacts on each type of receptor. The following analysis is based on the scheme summarized in Figure 4 where components of the analysis are briefly described and the potential climate sensitivities discussed in the context of possible climate changes. In summary the impacts of climate change affects the factors such as pesticide source terms, pathways and surface water and groundwater receptors.

PESTICIDE PATHWAYS

Several factors impacts on climate change and affected pesticide pathways. The following sections provide more detailed descriptions of those possible impacts in the surface water, soil and groundwater compartments.

Surface Water and Soil

The conceptual model (Figure 4) describes the fate of an agriculturally applied pesticide is well established (e.g., Gevao & Jones, 2002): an overview is given here. Pesticides sprayed on fields may be applied to bare soils, intercepted by the crop or lost as spray drift. For that which is deposited on the soil there is a partition between the soil, soil water and air. Pesticide may be lost from the soil/water/air system to the atmosphere through volatilization and this loss is essentially a two-stage process: firstly, the evaporation of pesticide molecules from pesticide residues in the soils or on the crop, and secondly, the dissipation of these molecules into the atmosphere by diffusion and turbulent mixing (Taylor, 1995). Cumulative volatilization losses can range from a few percent to 50% of the applied dose (van den Berg *et al.*, 1999) depending on the properties of the pesticide. Pesticide remaining in the soil will partition between the soil and the soil water; the ratio of the concentration in the soil to the concentration in the water is called the soil sorption coefficient. Sorption is not generally permanent, but is equilibrium between the concentration on the soil and in the soil water. Thus, when rainfall replaces pre-event soil water, presumed to be in equilibrium with the soil, there is the potential for desorption of some of the pesticide from the soil into the event water.

Pesticide is lost from the soil environment through volatilization, uptake by plants, and runoff to rivers, leaching to groundwater, and by microbial and abiotic degradation (Figure 4). Transformation processes change a pesticide into a different compound, often referred to as a degradation product or metabolite, ultimately leading to the complete mineralization of the pesticide. Transformations can occur by a number of processes: oxidation, hydrolysis, photolysis and biodegradation. For the majority of pesticides, biodegradation is the most important transformation process in soil.

In practice the biodegradation process is usually characterized by quoting a half-life (DT_{50}). This is the time it takes for the concentration of a pesticide to decrease by 50%. The half-life of a pesticide has been shown to change with temperature; soil water content and soil organic carbon content (Walker, 1978). Runoff describes the lateral movement of water either over the surface or through the top layers of the soil to a watercourse. In poorly draining soils, artificial drainage systems may be installed to reduce water logging by moving water more rapidly away from the soil surface. The amount of pesticide leaving the soil by this route is greatest during rainfall events and is generally less that1% of the applied dose in each individual rainfall event; although in extreme cases up to 5% loss has been measured (Burgoa & Wauchope, 1995). However, these small amounts can lead to concentrations well in excess of the European Union Drinking Water Directive (98/83/EC) standard of 0.1 µg/l for individual pesticides. Recently the influence of macro-pores on pesticide concentrations has received increased attention. A macro-pore is a planar or tubular pore that traverses the soil and is created by a secondary influence (e.g., earthworms, old root channels). Being of large diameter, they have the capacity to move water rapidly through the soil profile, allowing pesticides to by-pass the soil matrix, where they might have been expected to be adsorbed or degraded. Thus, pesticides moving by this route reach drainage systems in higher concentrations than would be expected by percolation through the soil matrix (Haria *et al.*, 1994; Johnson *et al.*, 1996; Kladivko *et al.*, 2001). Runoff events can also promote erosion of soil particles and the transport of sorbed pesticides. The magnitude of this loss path is generally small compared to that transported in the water phase because of the relative amounts of water moved compared to eroded soil. However, for some highly sorbed pesticides, this erosion becomes the dominant transport path from the field to surface waters (Turnbull *et al.*, 1997; Petersen *et al.*, 2002).

Effect of Pesticides in Aquatic Environment

Aquatic environments receive pesticide inputs by un-intended pollution events where minute amounts enter water bodies by spray drift or run-off. This results in known pesticide effects on larval stages including endocrine disruption by atrazine (Becker *et al.*, 2007) and increased mortality from environmentally relevant glyphosate exposure (Relyea, 2005). Amphibians that migrate to aquatic spawning habitats reveal high population declines (Becker *et al.*, 2007; Denoel, 2012) and pesticides might be a major threat for these species when crossing agricultural areas. Bloomfield (2006) have undertaken a review of climate change impacts on pesticides in surface and ground waters and conclude that changes in temperature, rainfall intensity and seasonality will affect pesticide release and transport. However, long-term land-use change driven by climate change may result in significant changes in pesticide use and release into rivers and lakes. Land-use change and longer growing seasons could increase the use of fertilizers with subsequent leaching to watercourses, rivers and lakes, increasing the risk of eutrophication and loss of biodiversity. Some risks may be countered by the development of pesticides that have fewer side effects and are better targeted.

Pesticide and Water Quality

Groundwater quality changes will be a consequence of changed recharge patterns and land-use. Reduction on soil frost result in more recharge of contaminants and lowers the land flow. This can increases groundwater availability but also increase risk of leaching of contaminants during winter. Warmer climate increase might influence pesticide leaching to groundwater, but the processes are complex and mainly related to land use changes driven by changes in climate (Bloomfield *et al.*, 2006) and increased pest pressures e.g. due to lower winter mortality (Noyes *et al.*, 2009). In cold regions, a milder climate with temperatures around freezing melting point increases the use of salt application for slippery control (Balderacchi *et al.*, 2013). In warmer climate, less recharge can lead to further decline of groundwater levels. Reduced groundwater level shows increase the risk of contamination, mainly from sea water intrusion in coastal aquifers (Werner *et al.*, 2013). Increased flood can lead to river water being more polluted and reduced minimum flow can lead to increase riverside concentration in wastewater effluents as waters are less diluted posing a risk to groundwater in loosing streams with a direct contact to aquifers.

The quality of groundwater may be a limiting factor for some intended uses, such as drinking or irrigation, and for the long- term sustainability of groundwater resources worldwide (Gurdak *et al.*, 2012), and therefore additional research is needed on climate change effects on groundwater quality. Changes in recharge rates and mechanisms may also increase the mobilization of pesticides and other pollutants in the unsaturated zone and reduce groundwater quality (Bloomfield *et al.*, 2006; Sugita & Nakane, 2007). In some semiarid and arid regions, climate change may mobilize naturally occurring salts, such as nitrate and chloride pore water reserves, or enhance denitrification and removal of nitrate from the unsaturated zone prior to recharge (Gurdak *et al.*, 2007;. Stuart *et al.* (2011) noted that nitrate leaching to groundwater as a result of climate change is not sufficiently well understood to make useful predictions without additional monitoring data. Studies on natural soil and agricultural processes in the United Kingdom report a range of nitrate leaching rates from a slight increase to possibly high nitrate concentrations in groundwater by 2100 because of climate change (Stuart *et al.*, 2011). In addition, a possible increase in surface water intrusion and flooding poses a risk to groundwater quality due to the contamination (Silander *et al.*, 2006).

CONCLUSION

The largest impact of climate change on future water demand, however, will potentially be felt by agriculture and horticulture. Several uncertainties underlie these predictions since climate change can impact irrigation water use *via* many different mechanisms, many of which are poorly understood today. The indirect effects of climate-induced changes in demand for water and other natural and agricultural resources and changes in land use may have a greater effect on fate and transport of pesticides in the environment than direct effects. In summary based on the source–pathway–receptor analysis, the main climate drivers for changing pesticide fate and behavior are changing rainfall patterns (changes in seasonality and intensity) and increased temperatures. There is a growing body of evidence that climate change will have broad negative impacts on the distribution and toxicity of environmental contaminants. Climate change will have a powerful effect on the environmental fate and behavior of chemical toxicants by altering physical, chemical, and biological drivers of partitioning between the atmosphere, water, soil/sediment, and biota, including: air-surface exchange, wet/dry deposition, and reaction rates (e.g., photolysis, biodegradation, oxidation in air). Temperature and precipitation, as altered by climate change, are expected to have the largest influence on the partitioning of chemical toxicants.

Global warming will be expected to enhance partitioning of POPs and other pesticides to the atmosphere, though the increase in atmospheric concentrations of these pollutants may be offset by enhanced degradation. Moreover, regions subject to increased storm intensity, frequency, and variability could experience pulses of chemical releases or runoff that might present acute risks to human health and wildlife populations.

Increased temperature and salinity linked to climate change could enhance the toxicity of some POPs and other pesticides in aquatic biota. Altered biotransformation of contaminants to more bioactive metabolites appears to be an important mechanism by which climate change enhances chemical toxicity. Moreover, these climate change and contaminant interactions could compromise homeostasis and physiological responses, potentially impairing species fitness, reproduction, and development.

Improving our understanding of the effects of multiple stressors on natural systems is an important challenge for environmental scientists. It has taken on more urgency as climate change is not only altering the fundamental structure and function of many ecosystems, but is impacting the distribution and toxicity of chemical pollutants. The vulnerability of human and wildlife populations to climate-sensitive chemical exposures, in the context of the many other stressors that are being altered with climate change, is the paramount question that requires more rigorous study.

REFERENCES

Anuradha, Maurya, Haruna, Tripathi, Mishra & Sanjeevi. (2017). Novel Approach Towards Mitigation of Modern Agricultural Scenario: With A Special Reference to Amelioration of GCC Impacts. *International Journal of Engineering Research and Technology*, *5*(12), 53–56.

Balderacchi, M., Benoit, P., Cambier, P., Eklo, O. M., Gargini, A., Gemitzi, A., ... Trevisan, M. (2013). Groundwater pollution and quality monitoring approaches at the European level. *Critical Reviews in Environmental Science and Technology*, *43*(4), 323–408. doi:10.1080/10643389.2011.604259

Becker, C. G., Fonseca, C. R., Haddad, C. F. B., Batista, R. F., & Prado, P. I. (2007). Habitat Split and the Global Decline of Amphibians. *Science*, *318*(5857), 1775–1777. doi:10.1126cience.1149374 PMID:18079402

Blenkinsop, S., Fowler, H. J., Dubus, I. G., Nolan, B. T., & Hollis, J. M. (2008). Developing climatic scenarios for pesticide fate modelling in Europe. *Environmental Pollution*, *154*(2), 219–231. doi:10.1016/j.envpol.2007.10.021 PMID:18045756

Bloomfield, J. P., Williams, R. J., Gooddy, D. C., Cape, J. N., & Guha, P. (2006). Impacts of climate change on the fate and behaviour of pesticides in surface and groundwater —A UK perspective. *The Science of the Total Environment*, *369*(1–3), 163–177. doi:10.1016/j.scitotenv.2006.05.019 PMID:16914182

Burgoa, B., & Wauchope, R. D. (1995). Pesticides in run-off and surface waters. In T. R. Roberts & P. C. Kearney (Eds.), *Environmental Behaviour of Agrochemicals*. Chichester, UK: John Wiley & Sons.

Carere, M., Miniero, R., & Cicero, M. R. (2011). Potential effects of climate change on the chemical quality of aquatic biota. *TrAC Trends in Analytical Chemistry*, *30*(8), 1214–1221. doi:10.1016/j.trac.2011.06.006

Carson, R. (1962). *Silent spring*. Houghton Mifflin Company.

Chaudhry, F. N., & Malik, M. F. (2017). Factors Affecting Water Pollution: A Review. *Journal of Ecosystem Ecography*, *7*, 225.

De Schampheleire, M., Spanoghe, P., Brusselman, E., & Sonck, S. (2007). Risk assessment of pesticide spray drift damage in Belgium. *Crop Protection (Guildford, Surrey)*, *26*(4), 602–611. doi:10.1016/j.cropro.2006.05.013

Delcour, I., Spanoghe, P., & Uyttendaele, M. (2015). Literature review: Impact of climate change on pesticide use. *Food Research International*, *68*, 7–15. doi:10.1016/j.foodres.2014.09.030

Denoel, M. (2012). Newt decline in Western Europe: Highlights from relative distribution changes within guilds. *Biodiversity and Conservation*, *21*(11), 2887–2898. doi:10.100710531-012-0343-x

Donald, W. S. (2016). Current Use Pesticides. In Ecotoxicology Essentials, Environmental Contaminants and their Biological Effects on Animals and Plants (pp. 109–152). Academic Press.

Dworak, V., Selbeck, J., Dammer, K.-H., Hoffmann, M., Akbar Zarezadeh, A., & Bobda, C. (2013). Strategy for the development of a smart NDVI camera system for outdoor plant detection and agricultural embedded systems. *Sensors (Basel)*, *13*(2), 1523–1538. doi:10.3390130201523 PMID:23348037

EEA. (2012). *Climate change. City*. European Environment Agency.

Eid, H. M., El-Marsafawy, S. M., & Ouda, S. A. (2007). Assessing the economic impacts of climate change on agriculture in Egypt: A Ricardian approach. *World Bank Policy Research Working Paper*, 4293.

ENSAA. (2011). *Defining climate change*. Author.

Environmental Databases. Ecotoxicity Database. (2006). Pesticides: Science and Policy. Washington, DC: U.S. Environmental Protection Agency (EPA).

Fernandez-Gomez, C., Antonio Lopez-Lopez, J., Matamoros, V., Diez, S., Garcia-Vargas, M., & Moreno, C. (2013). Atmospheric influence on the distribution of organic pollutants in the Guadalquivir River estuary, SW Spain. *Environmental Monitoring and Assessment*, *185*(4), 3209–3218. doi:10.100710661-012-2784-9 PMID:23010894

Ficklin, D. L., Luo, Y., Luedeling, E., Gatzke, S. E., & Zhang, M. (2010). Sensitivity of agricultural runoff loads to rising levels of CO2 and climate change in the San Joaquin Valley watershed of California. *Environmental Pollution*, *158*(1), 223–234. doi:10.1016/j.envpol.2009.07.016 PMID:19660846

Fox, G. A., Collins, B., Hayakawa, E., Weseloh, D. V., Ludwig, J. P., Kubiak, T. J., & Erdman, T. C. (1991). Reproductive outcomes in colonial fish-eating birds: a biomarker for developmental toxicants in Great Lakes food chains: II. Spatial variation in the occurrence and prevalence of bill defects in young double-crested cormorants in the Great Lakes, 1979-1987. *Journal of Great Lakes Research*, *17*(2), 158–167. doi:10.1016/S0380-1330(91)71353-1

Gevao, B., & Jones, K. C. (2002). Pesticides and persistent organic chemicals. In P. M. Haygarth & S. C. Jarvis (Eds.), *Agriculture, Hydrology and Water Quality*. Wallingford, UK: CABI Publishing. doi:10.1079/9780851995458.0083

Gilbertson, M., Kubiak, T., Ludwig, J., & Fox, G. (1991). Great Lakes embryo mortality, edema and deformities syndrome (GLEMEDS) in clonial fish-eating birds: Similarity to chick-edema disease. *Journal of Toxicology and Environmental Health*, *33*(4), 455–520. doi:10.1080/15287399109531538 PMID:1875430

Goel, A., McConnell, L. L., & Torrents, A. (2005). Wet deposition of current use pesticides at a rural location on the Delmarva Peninsula: Impact of rainfall patterns and agricultural activity. *Journal of Agricultural and Food Chemistry*, *53*(20), 7915–7924. doi:10.1021/jf0507700 PMID:16190650

Gurdak, J. J., Hanson, R. T., McMahon, P. B., Bruce, B. W., McCray, J. E., Thyne, G. D., & Reedy, R. C. (2007). Climate variability controls on unsaturated water and chemical movement, High Plains aquifer, USA. *Vadose Zone Journal*, *6*(3), 533–547. doi:10.2136/vzj2006.0087

Gurdak, J. J., McMahon, P. B., & Bruce, B. W. (2012). Vulnerability of Groundwater Quality to Human Activity and Climate Change and Variability. In H. Treidel, J. J. Martin-Bordes, & J. J. Gurdak (Eds.), *Climate Change Effects on Groundwater Resources: A Global Synthesis of Findings and Recommendations. International Association of Hydrogeologists (IAH) – International Contributions to Hydrogeology* (pp. 145–167). Taylor & Francis Publishing.

Haria, A. H., Johnson, A. C., Bell, J. P., & Batchelor, C. H. (1994). Water Movement and isoproturon behaviour in a drained heavy clay soil:1. Preferential flow processes. *Journal of Hydrology (Amsterdam)*, *163*(3-4), 203–216. doi:10.1016/0022-1694(94)90140-6

Harvell, C. D., Mitchell, C. E., Ward, J. R., Altizer, S., Dobson, A. P., & Ostfeld, R. S. (2002). Climate warming and disease risks for terrestrial and marine biota. *Science*, *296*(5576), 2158–2162. doi:10.1126cience.1063699 PMID:12077394

IPCC (Intergovernmental Panel on Climate Change). (2007). *Climate Change 2007: The Physical Science Basis: Summary for Policymakers. Fourth Assessment Report of the IPCC* (M. L. Parry, O. F. Canziani, J. P. Palutikof, P. J. van der Linden, & C. E. Hanson, Eds.). Cambridge, UK: Cambridge University Press.

Johnson, A. C., Haria, A. H., Bhardwaj, C. L., Williams, R. J., & Walker, A. (1996). Preferential flow pathways and their capacity to transport isoproturon in a structured clay soil. *Pesticide Science*, *48*(3), 225–237. doi:10.1002/(SICI)1096-9063(199611)48:3<225::AID-PS462>3.0.CO;2-D

Keer, R. A. (2007). Global warming is changing the world. *Science*, *316*(5822), 188–190. doi:10.1126cience.316.5822.188 PMID:17431148

Keikotlhaile, B. M. (2011). *Influence of the processing factors on pesticide residues in fruits and vegetables and its application in consumer risk assessment.* Ghent: Ghent University.

Kladivko, E. J., Brown, L. C., & Baker, J. L. (2001). Pesticide transport to subsurface tile drains in humid regions of North America. *Critical Reviews in Environmental Science and Technology*, *31*(1), 1–62. doi:10.1080/20016491089163

Kumar, K. S., Kannan, K., Corsolini, S., Evans, T., Giesy, J. P., Nakanishi, J., & Masunaga, S. (2002). Polychlorinated dibenzop-dioxins, dibenzofurans and polychlorinated biphenyls in polar bear, penguin and south polar skua. *Environment and Pollution*, *119*(2), 151–161. doi:10.1016/S0269-7491(01)00332-3 PMID:12152823

Lewan, E., Kreuger, J., & Jarvis, N. (2009). Implications of precipitation patterns and antecedent soil water content for leaching of pesticides from arable land. *Agricultural Water Management*, *96*(11), 1633–1640. doi:10.1016/j.agwat.2009.06.006

Moe, S. J., De Schamphelaere, K., Clements, W. H., Sorensen, M. T., Van den Brink, P. J., & Liess, M. (2013). Combined and interactive effects of global climate change and toxicants on populations and communities. *Environmental Toxicology and Chemistry*, *32*(1), 49–61. doi:10.1002/etc.2045 PMID:23147390

Nolan, B. T., Dubus, I. G., Surdyk, N., Fowler, H. J., Burton, A., Hollis, J. M., ... Jarvis, N. J. (2008). Identification of key climatic factors regulating the transport of pesticides in leaching and to tile drains. *Pest Management Science*, *64*(9), 933–944. doi:10.1002/ps.1587 PMID:18416432

Noyes, P. D., & Lema, S. C. (2015). Forecasting the impacts of chemical pollution and climate change interactions on the health of wildlife. *Current Zoology*, *61*(4), 669–689. doi:10.1093/czoolo/61.4.669

Noyes, P. D., McElwee, M. K., Miller, H. D., Clark, B. W., Van Tiem, L. A., Walcott, K. C., ... Levin, E. D. (2009). The toxicology fo climate change: Environmental contaminants in a warming world. *Environment International*, *35*(6), 971–986. doi:10.1016/j.envint.2009.02.006 PMID:19375165

Oliver, D. P., Kookana, R. S., Anderson, J. S., Cox, J., Waller, N., & Smith, L. (2012). The off-site transport of pesticide loads from two land uses in relation to hydrological events in the Mt. Lofty Ranges, South Australia. *Agricultural Water Management*, *106*, 70–77. doi:10.1016/j.agwat.2011.07.012

Otieno, P. O., Owuor, P. O., Lalah, J. O., Pfister, G., & Schramm, K. W. (2013). Impacts of climate-induced changes on the distribution of pesticides residues in water and sediment of Lake Naivasha, Kenya. *Environmental Monitoring and Assessment*, *185*(3), 2723–2733. doi:10.100710661-012-2743-5 PMID:22797882

Paetzold, S., Klein, C., & Bruemmer, G. W. (2007). Run-off transport of herbicides during natural and simulated rainfall and its reduction by vegetated filter strips. *Soil Use and Management*, *23*(3), 294–305. doi:10.1111/j.1475-2743.2007.00097.x

Petersen, C. T., Holm, J., Koch, C. B., Jensen, H. E., & Hansen, S. (2002). Movement of pendimethalin, ioxynil and soil particles to field drainage tiles. *Pest Management Science*, *59*(1), 85–96. doi:10.1002/ps.609 PMID:12558103

Pope, C. A. III, Bhatnagar, A., McCracken, J. P., Abplanalp, W. T., Conklin, D. J., & O'Toole, T. (2016). Exposure to fine particulate air pollution is associated with endothelial injury and systemic inflammation. *Circulation Research*, *119*(11), 1204–1214. doi:10.1161/CIRCRESAHA.116.309279 PMID:27780829

Relyea, R. A. (2005). The Lethal Impacts of Roundup and Predatory Stress on Six Species of North American Tadpoles. *Archives of Environmental Contamination and Toxicology*, *48*(3), 351–357. doi:10.100700244-004-0086-0 PMID:15886853

Salama, J., Chakraborty, T. R., Ng, L., & Gore, A. C. (2003). Effects of polychlorinated biphenyls on estrogen receptor-b expression in the anteroventral periventricular nucleus. *Environmental Health Perspectives*, *111*(10), 1278–1282. doi:10.1289/ehp.6126 PMID:12896846

Sanjeevi, Haruna, Tripathi, & Anuradha. (2017). Impacts of Global Carbon Foot Print on the Marine Environment. *International Journal of Engineering Research and Technology*, *5*(12), 57-60.

Silander, J., Vehviläinen, B., Niemi, J., Arosilta, A., Dubrovin, T., Jormola, J., . . . Veijalainen, N. (2006). *Climate Change Adaptation for Hydrology and Water Resources*. FINADAPT Working Paper 6. Finnish Environmental Institute.

Solomon, S.D., Qin, M., Manning, Z., Chen, M., Marquis, K.B., Averyt, M., Tignor, H.L., & Miller. (2007). *Climate Change 2007: The Physical Science Basis. Contribution of Working Group I to the Fourth Assessment Report of the Intergovernmental Panel on Climate Change*. Cambridge University Press.

Sparling, D. W. (2016). Current use of Pesticides. In *Ecotoxicology Essentials, Environmental Contaminants and their Biological Effects on Animals and Plants* (pp. 109–152). Academic Press.

Steurbaut, W. (2009). *Neveneffecten van pesticiden. Fytofarmacie*. Ghent: Ghent University.

Stockholm Convention. (2005). Stockholm Convention on Persistent Organic Pollutants (POPs).

Stuart, M. E., Gooddy, D. C., Bloomfield, J. P., & Williams, A. T. (2011). A review of the impact of climate change on future nitrate concentrations in groundwater of the UK. *The Science of the Total Environment, 409*(15), 2859–2873. doi:10.1016/j.scitotenv.2011.04.016 PMID:21669323

Sugita, F., & Nakane, K. (2007). Combined effects of rainfall patterns and porous media properties on nitrate leaching. *Vadose Zone Journal, 6*(3), 548–553. doi:10.2136/vzj2006.0095

Taylor, A. W. (1995). The Volatilization of Pesticide Residues. In T. R. Roberts & P. C. Kearney (Eds.), *Environmental Behaviour of Agrochemicals*. Chichester, UK: John Wiley & Sons.

Turnbull, A. B., Harrison, R. M., Williams, R. J., Matthiessen, P., Brooke, D. N., Sheahan, D. A., & Mills, M. (1997). Assessment of the Fate of Selected Adsorptive Pesticides at ADAS Rosemaund. *Water and Engironmental Journal, 11*(1), 24–30. doi:10.1111/j.1747-6593.1997.tb00084.x

UNEP. (2009). *Stockholm Convention on Persistent Organic Pollutants (POPs) with Amendments SC 4/10 to SC 4/18*. Secretariat of the Stockholm Convention on Persistent Organic Pollutants.

van den Berg, F., Kubiak, R., Benjey, W. G., Majewski, M. S., Yates, S. R., Reeves, G. L., ... van der Linden, A. M. A. (1999). Emission of Pesticides into the Air. *Water, Air, and Soil Pollution, 115*(1/4), 195–218. doi:10.1023/A:1005234329622

Vryzas, Z., Vassiliou, G., Alexoudis, C., & Papadopoulou-Mourkidou, E. (2009). Spatial and temporal distribution of pesticide residues in surface waters in northeastern Greece. *Water Research, 43*(1), 1–10. doi:10.1016/j.watres.2008.09.021 PMID:18947852

Walker, A. (1978). Simulation of the persistence of eight soil-applied herbicides. *Weed Research, 18*(5), 305–313. doi:10.1111/j.1365-3180.1978.tb01165.x

Wauchope, R. D., Johnson, W. C. III, & Sumner, H. R. (2004). Foliar and soil deposition of pesticide sprays in peanuts and their washoff and runoff under simulated worstcase rainfall conditions. *Journal of Agricultural and Food Chemistry, 52*(23), 7056–7063. doi:10.1021/jf049283v PMID:15537318

Weare, B. C. (2009). How will changes in global climate influence California? *California Agriculture, 63*(2), 59–66. doi:10.3733/ca.v063n02p59

Werner, A. D., Vincent, M. B., Post, E. A., Vandenbohede, A., Lu, C., Ataie-Ashtiani, B., ... Barry, D. A. (2013). Seawater intrusion processes, investigation and management: Recent advances and future challenges. *Advances in Water Resources, 51*, 3–26. doi:10.1016/j.advwatres.2012.03.004

Werner, K. J., & Zedler, J. B. (2002). How sedge meadow soils, microtopography, and vegetation respond to sedimentation. *Wetlands, 22*(3), 451–466. doi:10.1672/0277-5212(2002)022[0451:HSMSMA]2.0.CO;2

Compilation of References

Ab Rahman, S. F. S., Singh, E., Pieterse, C. M., & Schenk, P. M. (2017). Emerging Microbial Biocontrol Strategies for Plant Pathogens. *Plant Science, 2018*. doi:10.1016/j.plantsci.2017.11.012 PMID:29362088

Abbas, I., & Zaheer, S. (2014). Advanced oxidation process for wastewater treatment: A review *American International Journal of Research in Science, Technology, Engineering & Mathematics*, *7*(3), 189–191.

Abdel Rasoul, G. M., Abou Salem, M. E., Mechael, A. A., Hendy, O. M., Rohlman, D. S., & Ismail, A. A. (2008). Effects of occupational pesticide exposure on children applying pesticides. *Neurotoxicology*, *29*(5), 833–838. doi:10.1016/j.neuro.2008.06.009 PMID:18662718

Abdi, H., Lee, J., Ellison, G., Lai, G., & Lam, T. (2017). *Pesticides and primary liver cancer: A systematic review and meta-analysis*. Academic Press.

Abdullah, A. A., & Waleema, B. B. (2009). Effect of Agricultural Chemicals on Aquatic Ecosystem in Guyana. *Global Journal of Environmental Research*, *3*(1), 22–25.

Abhilash, P. C., & Singh, N. (2009). Pesticide use and application: An Indian scenario. *Journal of Hazardous Materials*, *165*(1-3), 1–12. doi:10.1016/j.jhazmat.2008.10.061 PMID:19081675

Abou-Doma, M. B., & Preissig, S. H. (1976). Delayed neurotoxicity from continuous low-dose oral administration of leptophos to hens. *Toxicology and Applied Pharmacology*, *38*(3), 595–608. doi:10.1016/0041-008X(76)90190-3 PMID:65030

Abou-Donia, M. B., & Preissig, S. H. (1976). Delayed neurotoxicity of leptophos; toxic effects on the nervous system of hens. *Toxicology and Applied Pharmacology*, *35*(2), 269–282. doi:10.1016/0041-008X(76)90287-8 PMID:57652

Adachi, A., Ikedia, C., Takagi, S., & Fukao, N. (2001). Efficiency of rice bran for removal of Organochlorine compounds and benzene from industrial waste water. *Journal of Agricultural and Food Chemistry*, *49*(3), 1309–1314. doi:10.1021/jf001147c PMID:11312856

Adamovics, J. A. (Ed.). (2017). *Chromatographic analysis of pharmaceuticals*. Routledge.

Adams, R. M. (1983). *Occupational skin diseases*. New York: Grune and Stratton.

Adin, A., & Asano, T. (1998). The Role of Physical Chemical Treatment in Wastewater Reclamation and Reuse. *Water Science and Technology*, *37*(10), 79–80. doi:10.2166/wst.1998.0381

Afful, S., Anim, A. K., & Serfor-Armah, Y. (2010). Spectrum of organochlorine pesticide residues in fish samples from the Densu Basin. *Res J Environ Earth Sci*, *2*(3), 133–138.

Agarwal, S., Sadeghi, N., Tyagi, I., Gupta, V. K., & Fakhri, A. (2016). Adsorption of toxic carbamate pesticide oxamyl from liquid phase by newly synthesized and characterized graphene quantum dots nanomaterials. *Journal of Colloid and Interface Science*, *478*, 430–438. doi:10.1016/j.jcis.2016.06.029 PMID:27362399

Agatonovic-Kustrin, S. W., Morton, D., & Razic, S. (2014). In silico modelling of pesticide aquatic toxicity. *Combinatorial Chemistry & High Throughput Screening*, *17*(9), 808–818. doi:10.2174/1386207317666141021110738 PMID:25335880

Agency for Toxic Substances and Disease Registry. (2002). *Toxicological profile for DDT, DDE and DDD*. Atlanta, GA: US Department of Health and Human Services.

Agrawal, A., Pandey, R. S., & Sharma, B. (2010). Water pollution with special reference to pesticide contamination in India. *Journal of Water Resource and Protection*, *2*(5), 432–448. doi:10.4236/jwarp.2010.25050

Agriculture Information Network. (2006). Analysis of pesticides demand in China [in Chinese]. *Plant Doctor.*, *19*, 16.

Ahlgren, G., Lundstedt, L., Brett, M., & Forsberg, C. (1990). Lipid com-position and food quality of some freshwater phytoplankton for cladoceran zooplankters. *J Phytoplankton Res*, *12*(4), 809–818. doi:10.1093/plankt/12.4.809

Aislabie, J., & Lloyd-Jones, G. (1995). A review of bacterial-degradation of pesticides. *Soil Research (Collingwood, Vic.)*, *33*(6), 925–942. doi:10.1071/SR9950925

Akcan, R., Hilal, A., Daglioglu, N., Cekin, N., & Gulmen, M. K. (2009). Determination of pesticides in postmortem blood and bone marrow of pesticide-treated rabbits. *Forensic Sci Int.*, *189*(1-3), 82-7.

Akerblom, N., Arbjörk, C., Hedlund, M., & Goedkoop, W. (2008). Deltamethrin toxicity to the midge Chironomus riparius Meigen-effects of exposure scenario and sediment quality. *Ecotoxicology and Environmental Safety*, *70*(1), 53–60. doi:10.1016/j.ecoenv.2007.09.001 PMID:18255145

Akers, K. S., Sinks, G. D., & Schultz, T. W. (1999). Structure–toxicity relationships for selected halogenated aliphatic chemicals. *Environmental Toxicology and Pharmacology*, *7*(1), 33–39. doi:10.1016/S1382-6689(98)00048-9 PMID:21781907

Akhtar, M., Hasany, S. M., Bhanger, M. I., & Iqbal, S. (2007). Low cost sorbents for the removal of methyl parathion pesticide from aqueous solutions. *Chemosphere*, *66*(10), 1829–1838. doi:10.1016/j.chemosphere.2006.09.006 PMID:17109916

Akhtar, M., Iqbal, S., Bhanger, M. I., Zia-Ul-Haq, M., & Moazzam, M. (2009). Sorption of organophosphorous pesticides onto chickpea husk from aqueous solutions. *Colloids and Surfaces. B, Biointerfaces*, *69*(1), 63–70. doi:10.1016/j.colsurfb.2008.10.019 PMID:19091530

Aksu, Z., & Kabasakal, E. (2005). Adsorption characteristics of 2, 4-dichlorophenoxyacetic acid (2, 4-D) from aqueous solution on powdered activated carbon. *Journal of Environmental Science and Health. Part. B, Pesticides, Food Contaminants, and Agricultural Wastes*, *40*(4), 545–570. doi:10.1081/PFC-200061533 PMID:16047879

Aktar, M., Sengupta, D., & Chowdhury, A. (2009). Impact of pesticides use in agriculture: Their benefits and hazards. *Interdisciplinary Toxicology*, *2*(1), 1–12. doi:10.2478/v10102-009-0001-7 PMID:21217838

Alahabadi, A., & Moussavi, G. (2017). Preparation, characterization and atrazine adsorption potential of mesoporous carbonate-induced activated biochar (CAB) from CalligonumComosum biomass: Parametric experiments and kinetics, equilibrium and thermodynamic modeling. *Journal of Molecular Liquids*, *242*, 40–52. doi:10.1016/j.molliq.2017.06.116

Alavanja, M. C. R., & Bonner, M. R. (2012). Occupational pesticide exposures and cancer risk: A review. *Journal of Toxicology and Environmental Health. Part B, Critical Reviews*, *15*(4), 238–263. doi:10.1080/10937404.2012.632358 PMID:22571220

Albanis, T. A., Hela, D. G., Sakellarides, T. M., & Konstantinou, I. K. (1998). Monitoring of pesticide residues and their metabolites in surface and underground waters of Imathia (N. Greece) by means of solid-phase extraction disks and gas chromatography. *Journal of Chromatography. A*, *823*(1–2), 59–71. doi:10.1016/S0021-9673(98)00304-5 PMID:9818393

Alder, L., Greulich, K., Kempe, G., & Vieth, B. (2006). Residue analysis of 500 high priority pesticides: Better by GC–MS or LC–MS/MS? *Mass Spectrometry Reviews*, *25*(6), 838–865. doi:10.1002/mas.20091 PMID:16755599

Alfonso, L. F., Germán, G. V., del Carmen, P. C. M., & Hossein, G. (2017). Adsorption of organophosphorus pesticides in tropical soils: The case of karst landscape of northwestern Yucatan. *Chemosphere*, *166*, 292–299. doi:10.1016/j.chemosphere.2016.09.109 PMID:27700995

Ali, M., & Sreekrishnan, T. R. (2001). Aquatic toxicity from pulp and paper mill effluents: A review. *Advances in Environmental Research*, *5*(2), 175–196. doi:10.1016/S1093-0191(00)00055-1

Ali, U., Syed, J. H., Malik, R. N., Katsoyiannis, A., Li, J., Zhang, G., & Jones, K. C. (2014). Organochlorine pesticides (OCPs) in South Asian region: A review. *The Science of the Total Environment*, *476*, 705–717. doi:10.1016/j.scitotenv.2013.12.107 PMID:24522007

Aliyeva, G., Halsall, C., Alasgarova, K., Avazova, M., Ibrahimov, Y., & Aghayeva, R. (2013). The legacy of persistent organic pollutants in Azerbaijan: An assessment of past use and current contamination. *Environmental Science and Pollution Research International*, *20*(4), 1993–2008. doi:10.100711356-012-1076-9 PMID:22825638

Allinson, G., Zhang, P., Bui, A. D., Allinson, M., Rose, G., Marshall, S., & Pettigrove, V. (2015). Pesticide and trace metal occurrence and aquatic benchmark exceedances in surface waters and sediments of urban wetlands and retention ponds in Melbourne, Australia. *Environmental Science and Pollution Research International*, *22*(13), 10214–10226. doi:10.100711356-015-4206-3 PMID:25697552

Alloway, B., & Ayres, D. C. (1997). *Chemical principles of environmental pollution*. CRC press.

Almazan, L. L., & Robles, D. O. (1956). Exploratory study of the effect of algae upon lowland rice. *Araneta Journal of Agriculture (Philippines)*, *3*(4), 1-27.

Alrahman, K. F., Elbashir, A. A., & El-obid, A. H. (2015). Development and Validation of Spectrophotometric Method for Determination of Oxyfluorfen Herbicide Residues. *Medicinal Chemistry (Shariqah, United Arab Emirates)*, *5*, 383–387.

Al-Saleh, I., Al-Doush, I., Alsabbaheen, A., Mohamed, G. E. D., & Rabbah, A. (2012). Levels of DDT and its metabolites in placenta, maternal and cord blood and their potential influence on neonatal anthropometric measures. *The Science of the Total Environment*, *416*, 62–74. doi:10.1016/j.scitotenv.2011.11.020 PMID:22192892

Altenburger, R., Schmitt, H., & Schüürmann, G. (2005). Algal toxicity of nitrobenzenes: Combined effect analysis as a pharmacological probe for similar modes of interaction. *Environmental Toxicology and Chemistry*, *24*(2), 324–333. doi:10.1897/04-032R.1 PMID:15719992

Al-Zaidi, A. A., Elhag, E. A., Al-Otaibi, S. H., & Baig, M. B. (2011). Negative effects of pesticides on the environment and the farmer's awareness in Saudi Arabia: A case study. *Journal of Animal and Plant Sciences*, *21*(3), 605–611.

Ambridge, E. M. (1988). Pesticide management in relation to user safety. In R. T. Prinsley & P. J. Terry (Eds.), *Crop protection for small-scale farms in East and Central Africa'; A review* (pp. 98–105). Commonwealth Science Council.

American Public Health Association (APHA). (1998). *American Water Works Association (AWWA)* (20th ed.). Washington, DC: Standard Methods for the Contamination of Water and Wastewater.

American Water Works Association. (1998). *Reverse Osmosis and Nanofiltration*. Denver, CO: American Water Works Association.

Ames, R. G., Brown, S. K., Mengle, D. C., Kahn, E., Stratton, J., & Jackson, R. J. (1989). Cholinesterase activity depression among California agricultural pesticide applicators. *The American Journal of Emergency Medicine*, *15*, 143–150. PMID:2729280

Anastassiades, M., Lehotay, S. J., Stajnbaher, D., & Schenck, F. J. (2003). Fast and Easy Multiresidue Method Employing Acetonitrile Extraction/Partitioning and "Dispersive Solid-Phase Extraction" for the Determination of Pesticide Residues in Produce. *Journal of AOAC International, 86*(2), 412–431. PMID:12723926

Andersen, H. R., Schmidt, I. M., Grandjean, P., Jensen, T. K., Budtz-Jørgensen, E., Kjaerstad, M. B., & Main, K. M. (2008). Impaired reproductive development in sons of women occupationally exposed to pesticides during pregnancy. *Environmental Health Perspectives, 116*(4), 566–572. doi:10.1289/ehp.10790 PMID:18414644

Anderson, A. J., & Kim, Y. C. (2018). Biopesticides produced by plant-probiotic Pseudomonas chlororaphis isolates. *Crop Protection (Guildford, Surrey), 105*, 62–69. doi:10.1016/j.cropro.2017.11.009

Andreu, V., & Pico, Y. (2004). Determination of pesticides and their degradation products in soil: Critical review and comparison of methods. *Trends in Analytical Chemistry, 23*(10), 772–789. doi:10.1016/j.trac.2004.07.008

Anh, P. T., Kroeze, C., Bush, S. R., & Mol, A. P. J. (2010). Water pollution by intensive brackish shrimp farming in south-east Vietnam: Causes and options for control. *Agricultural Water Management, 97*(6), 872–882. doi:10.1016/j.agwat.2010.01.018

Anonymous. (2008). *The History of Pesticides, Organic Pesticides*. Retrieved from http://blog.ecosmart.com/index.php/2008/09/19/the-history-of-pesticides/

Anonymous. (2016). *International Code of Conduct on Pesticide Management Guidelines on Highly Hazardous Pesticides*. Food and Agriculture Organization of the United Nations Rome.

Antle, J. M., Pingali, P. L., & Pingali, P. L. (Eds.). (1995). Pesticides, productivity and farmer health: a Philippine case study. Impact of pesticides on farmer health and the rice environment. Bulletin Environmental Contamination Toxicology, 35, 735-738.

Antwi, F. B., & Reddy, G. V. (2015). Toxicological effects of pyrethroids on non-target aquatic insects. *Environmental Toxicology and Pharmacology, 40*(3), 915–923. doi:10.1016/j.etap.2015.09.023 PMID:26509732

Anuradha, Maurya, Haruna, Tripathi, Mishra & Sanjeevi. (2017). Novel Approach Towards Mitigation of Modern Agricultural Scenario: With A Special Reference to Amelioration of GCC Impacts. *International Journal of Engineering Research and Technology, 5*(12), 53–56.

Anurag, S., & Goel, M. M. (2009). *Enhancing plant defense mechanisms for disease management.* South Asia Conference on 'Outstanding Organic Agriculture Techniques', Bengaluru, India.

Araya, K., Lherisson, B., & Lomberk, J. (2014). *Pesticides, pollution, and people: an overview of public health and environment in Costa Rica*. Academic Press.

Arbuckle, T. E., Lin, Z., & Mery, L. S. (2001). An exploratory analysis of the effect of pesticide exposure on the risk of spontaneous abortion in an Ontario farm population. *Environmental Health Perspectives, 109*(8), 851–857. doi:10.1289/ehp.01109851 PMID:11564623

Armaghan, M., & Amini, M. M. (2017). Adsorption of diazinon and fenitrothion on nanocrystalline magnesium oxides. *Arabian Journal of Chemistry, 10*(1), 91–99. doi:10.1016/j.arabjc.2014.01.002

Arnot, J. A., & Gobas, F. A. (2006). A review of bioconcentration factor (BCF) and bioaccumulation factor (BAF) assessments for organic chemicals in aquatic organisms. *Environmental Reviews, 14*(4), 257–297. doi:10.1139/a06-005

Arora, J., & Mehra, N. K. (2003). Species diversity of planktonic and epiphytic rotifers in the backwaters of the Delhi segment of the Yamuna River, with remarks on new records from India. *Zool. Stud. Taipei, 42*, 239–247.

Arora, N. K., Khare, E., & Maheshwari, D. K. (2010). Plant Growth Promoting Rhizobacteria: constraints in bioformulation, commercialization and future strategies. In D. K. Maheshwari (Ed.), *Plant growth and health promoting bacteria* (pp. 97–116). Berlin: Springer. doi:10.1007/978-3-642-13612-2_5

Artemenko, A. G., Muratov, E. N., Kuz'min, V. E., Muratov, N. N., Varlamova, E. V., Kuz'mina, A. V., ... Tropsha, A. (2011). QSAR analysis of the toxicity of nitroaromatics in Tetrahymena pyriformis: Structural factors and possible modes of action. *SAR and QSAR in Environmental Research*, *22*(5-6), 575–601. doi:10.1080/1062936X.2011.569950 PMID:21714735

Arthur, C., & Pawliszyn, L. (1990). Solid-phase microextraction with thermal-desorption using fused-silica optical fibers. *Journal of Analytical Chemistry*, *62*(19), 2145–2148. doi:10.1021/ac00218a019

Artigas, J., Arts, G., Babut, M., Caracciolo, A. B., Charles, S., Chaumot, A., & Friberg, N. (2012). Towards a renewed research agenda in ecotoxicology. *Environmental Pollution*, *160*, 201–206. doi:10.1016/j.envpol.2011.08.011 PMID:22015334

Aruoja, V., Moosus, M., Kahru, A., Sihtmäe, M., & Maran, U. (2014). Measurement of baseline toxicity and QSAR analysis of 50 non-polar and 58 polar narcotic chemicals for the alga Pseudokirchneriella subcapitata. *Chemosphere*, *96*, 23–32. doi:10.1016/j.chemosphere.2013.06.088 PMID:23895738

Asghar, U., Malik, M. F., & Javed, A. (2016). Pesticide exposure and human health: A review. *Journal of Ecosystem and Ecography*, 5-12.

Ashby, J., Tinwell, H., Stevens, J., Pastoor, T., & Breckenridge, C. B. (2002). The effects of atrazine on the sexual maturation of female rats. *Regulatory Toxicology and Pharmacology*, *35*(3), 468–473. doi:10.1006/rtph.2002.1571 PMID:12202059

Ash, M. (1996). *Handbook of Water Treatment Chemicals*. Aldershot, UK: Gower.

Aslam, S., Garnier, P., Rumpel, C., Parent, S. E., & Benoit, P. (2013). Adsorption and desorption behavior of selected pesticides as influenced by decomposition of maize mulch. *Chemosphere*, *91*(11), 1447–1455. doi:10.1016/j.chemosphere.2012.12.005 PMID:23434076

Asmus, C. I. R. F., Camara, V. M., Landrigan, P. J., & Claudio, L. (2016). A Systematic Review of Children's Environmental Health in Brazil. *Annals of Global Health*, *82*(1), 132–148. doi:10.1016/j.aogh.2016.02.007 PMID:27325071

Aspelin, A.L. (1994). Pesticides industry sales and usage, 1992 and 1993 market estimates: U.S. EPA, Office of Pesticides Programs, Biological and Economic Analysis Div. *Economic Analysis Branch Report, 33*, 733-K-94-001.

Assessment, M. E. (2005). *Ecosystems and human well-being: wetlands and water*. Washington, DC: World Resources Institute.

Atreya, K. (2008). Health costs from short-term exposure to pesticides in Nepal. *Social Science & Medicine*, *67*(4), 511–519. doi:10.1016/j.socscimed.2008.04.005 PMID:18514373

ATSDR. (2013). *Draft toxicological profile for endosulfan*. U.S. Department of health and human services, Public Health Service Agency for Toxic Substances and Disease Registry. Retrieved from http://www.atsdr.cdc.gov/toxprofiles/tp41-p.pdf

Aubertot, J. N., Barbier, J. M., Carpentier, A., Gril, J. J., Guichard, L., Lucas, P., & Voltz, M. (2007). *Pesticides, agriculture and the environment: reducing the utilization of pesticides and limiting their environmental impacts*. Academic Press.

Augustijn-Beckers, P.W., Hornsby, A.G., & Wauchope, R.D. (1994). Additional Properties. In *Reviews of Environmental Contamination and Toxicology*. SCS/ARS/CES Pesticide Properties Database for Environmental Decision-making.

Ayranci, E., & Hoda, N. (2005). Adsorption kinetics and isotherms of pesticides onto activated carbon-cloth. *Chemosphere*, *60*(11), 1600–1607. doi:10.1016/j.chemosphere.2005.02.040 PMID:16083766

Aytaç, N., Hilal, A., Yapicioglu, A. B., Daglioglu, N., Gülmen, M. K., & Tanir, F. (2010). Organochlorine Pesticide Level in Breast Milk. *Turkiye Klinikleri J Med Sci, 30*(1).

Azab, E., Hegazy, A. K., El-Sharnouby, M. E., & Abd Elsalam, H. E. (2016). Phytoremediation of the organic Xenobiotic simazine by p450-1a2 transgenic Arabidopsis thaliana plants. *International Journal of Phytoremediation, 18*(7), 738–746. doi:10.1080/15226514.2015.1133559 PMID:26771455

Bai, L., Smuts, J., Walsh, P., Fan, H., Hildenbrand, Z., Wong, D., & Schug, K. A. (2015). Permanent gas analysis using gas chromatography with vacuum ultraviolet detection. *Journal of Chromatography. A, 1388*, 244–250. doi:10.1016/j.chroma.2015.02.007 PMID:25724098

Bailey, H. C., Miller, J. L., Miller, M. J., Wiborg, L. C., Deanovic, L., & Shed, T. (1997). Joint acute toxicity of diazinon and chlorpyrifos to Ceriodaphnia dubia. *Environmental Toxicology and Chemistry, 16*(11), 2304–2308. doi:10.1002/etc.5620161115

Bailey, R., Barrie, L. A., Halsall, C. J., Fellin, P., & Muir, D. C. G. (2000). Atmospheric organochlorine pesticides in the western Canadian Arctic: Evidence of transpacific transport. *Journal of Geophysical Research, D, Atmospheres, 105*(D9), 11805–11811. doi:10.1029/1999JD901180

Baillie-Hamilton, P. F. (2002). Chemical toxins: A hypothesis to explain the global obesity epidemic. *Journal of Alternative and Complementary Medicine (New York, N.Y.), 8*(2), 185–192. doi:10.1089/107555302317371479 PMID:12006126

Baker, B. P., Benbrook, C. M., Groth, E., & Lutz Benbrook, K. (2002). Pesticide residues in conventional, integrated pest management (IPM)-grown and organic foods: Insights from US data sets. *Food Addit Contam A, 19*(5), 427–446. doi:10.1080/02652030110113799 PMID:12028642

Bakour, E. I., & Morillo, H. (2009). usero J and ooassini A, Natural attenuation of pesticide water contamination by using ecological adsorbents, application for chlorinated pesticide included in European water framework directive. *J. of Hydro., 364*, 175–181. doi:10.1016/j.jhydrol.2008.10.012

Balabanic, D., Rupnik, M., & Klemencic, A. K. (2011). Negative impact of endocrine-disrupting compounds on human reproductive health. *Reproduction, Fertility, and Development, 23*(3), 403–416. PMID:21426858

Balaban, M. (Ed.). (1991). *Desalination and Water Re-use*. New York: Hemisphere.

Bala, R., Kumar, M., Bansal, K., Sharma, R. K., & Wangoo, N. (2016). Ultrasensitive aptamer biosensor for malathion detection based on cationic polymer and gold nanoparticles. *Biosensors & Bioelectronics, 85*, 445–449. doi:10.1016/j.bios.2016.05.042 PMID:27208476

Balderacchi, M., Benoit, P., Cambier, P., Eklo, O. M., Gargini, A., Gemitzi, A., ... Trevisan, M. (2013). Groundwater pollution and quality monitoring approaches at the European level. *Critical Reviews in Environmental Science and Technology, 43*(4), 323–408. doi:10.1080/10643389.2011.604259

Ballesteros, E., & Parrado, M. J. (2004). Continuous solid-phase extraction and gas chromatographic determination of organophosphorus pesticides in natural and drinking waters. *Journal of Chromatography. A, 12*(1-2), 267–273. doi:10.1016/j.chroma.2003.12.009 PMID:15032372

Baltazar, M. T., Dinis-Oliveira, R. J., de Lourdes Bastos, M., Tsatsakis, A. M., Duarte, J. A., & Carvalho, F. (2014). Pesticides exposure as etiological factors of Parkinson's disease and other neurodegenerative diseases—a mechanistic approach. *Toxicology Letters, 230*(2), 85–103. doi:10.1016/j.toxlet.2014.01.039 PMID:24503016

Barbash, J. E., & Resek, E. A. (1996). Pesticides in Groundwater, Volume 2 of the series Pesticides in the Hydrologic Environment. Ann Arbor Press, Inc.

Barker, S. A. (2000). Matrix solid-phase dispersion. *Journal of Chromatography. A*, *885*(1-2), 115–127. doi:10.1016/S0021-9673(00)00249-1 PMID:10941670

Barrion, A. T., & Litsinger, J. A. (1982). *The aquatic invertebrate fauna of IRRI ricefields*. Paper presented at the thirteenth National Conference of the Pest Control Council of the Philippines, Baguio City, Philippines.

Barron, M. G. (2003). Bioaccumulation and bioconcentration in aquatic organisms. In Handbook of ecotoxicology (2nd ed.). Lewis Publishers.

Barron, M. G., & Woodburn, K. B. (1995). Ecotoxicology of chlorpyrifos. *Reviews of Environmental Contamination and Toxicology*, *144*, 1–93. doi:10.1007/978-1-4612-2550-8_1 PMID:8599033

Barth, J. A. C., Steidle, D., Kuntz, D., Gocht, T., Mouvet, C., von Tümpling, W., & Grathwohl, P. (2007). Deposition, persistence and turnover of pollutants: First results from the EU project AquaTerra for selected river basins and aquifers. *The Science of the Total Environment*, *376*(1–3), 40–50. doi:10.1016/j.scitotenv.2007.01.065 PMID:17307233

Bartram, J., & Cairncross, S. (2010). Hygiene, sanitation, and water: Forgotten foundations of health. *PLoS Medicine*, *7*(11), e1000367. doi:10.1371/journal.pmed.1000367 PMID:21085694

Bassi, N., & Kumar, M. D. (2012). Addressing the civic challenges: Perspective on institutional change for sustainable urban water management in India. *Environment & Urbanization Asia*, *3*(1), 165–183. doi:10.1177/097542531200300109

Bates, B. C., Kundzewicz, Z. W., Wu, S., & Palutikof, J. P. (Eds.). (2008). Climate Change and Water. Inter governmental Panel on Climate Change, Geneva.

Batiste-Alenton, X. M., Velazquez, N., Creus, A., & Marcos, R. (1986). Mutagenicity of the pyrethroid insecticide cypermethrin in Drisiphila. *Mutageneis*, *1*(5), 343–346. doi:10.1093/mutage/1.5.343

Battu, R. S., Singh, B., Kang, B. K., & Joia, B. S. (2005). Risk assessment through dietary intake of total diet contaminated with pesticide residues in Punjab, India, 1999–2002. *Ecotoxicology and Environmental Safety*, *62*(1), 132–139. doi:10.1016/j.ecoenv.2004.12.004 PMID:15978299

Bayley, P. B., & Li, H. W. (1992). Riverine fishes. In The rivers handbook: hydrological and ecological principles (vol. 1.). Oxford, UK: Blackwell Scientific Publications.

Baylis, A. D. (2000). Why glyphosate is a global herbicide: Strengths, weaknesses and prospects. *Pest Management Science*, *56*(4), 299–308. doi:10.1002/(SICI)1526-4998(200004)56:4<299::AID-PS144>3.0.CO;2-K

Becerra-Castro, C., Kidd, P. S., Rodríguez-Garrido, B., Monterroso, C., Santos-Ucha, P., & Prieto-Fernández, Á. (2013). Phytoremediation of hexachlorocyclohexane (HCH)-contaminated soils using Cytisus striatus and bacterial inoculants in soils with distinct organic matter content. *Environmental Pollution*, *178*, 202–210. doi:10.1016/j.envpol.2013.03.027 PMID:23583940

Becker, C. G., Fonseca, C. R., Haddad, C. F. B., Batista, R. F., & Prado, P. I. (2007). Habitat Split and the Global Decline of Amphibians. *Science*, *318*(5857), 1775–1777. doi:10.1126cience.1149374 PMID:18079402

Bedassa, T., Negussie, M., & Abera, G. (2017). Salting-out Assisted Liquid-Liquid Extraction for the Determination of Multiresidue Pesticides in Alcoholic Beverages by High Performance Liquid Chromatography. *Science Journal of Analytical Chemistry*, *5*(3), 38–45. doi:10.11648/j.sjac.20170503.11

Begon, M., Harper, J. L., & Townsend, C. R. (1990). *Ecology: Individuals, Populations and Communities*. Oxford, UK: Blackwell Scientific Publications.

Behloul, M., Lounici, H., Abdi, N., Drouiche, N., & Mameri, N. (2017). Adsorption study of metribuzin pesticide on fungus Pleurotusmutilus. *International Biodeterioration & Biodegradation*, *119*, 687–695. doi:10.1016/j.ibiod.2016.07.005

Bekele, D., Petros, B., Tekie, H., & Asfaw, Z. (2014). Larvicidal, and Adulticidal, Effects of Extracts from Some Indigenous Plants against the Malaria Vector, Anopheles Arabiensis (Diptera: Culicidae) in Ethiopia. *Journal of Biofertilizers & Biopesticides*, *5*, 144.

Beketov, M. A., Kefford, B. J., Schafer, R. B., & Liess, M. (2013). Pesticides reduce regional biodiversity of stream invertebrates. *Proceedings of the National Academy of Sciences of the United States of America*, *110*(27), 11039–11043. doi:10.1073/pnas.1305618110 PMID:23776226

Beketov, M., & Liess, M. (2012). Ecotoxicology and macroecology - Time for integration. *Environmental Pollution*, *162*, 247–254. doi:10.1016/j.envpol.2011.11.011 PMID:22243871

Belden, J. B., & Lydy, M. J. (2000). Impact of atrazine on organophosphate insecticide toxicity. *Environmental Toxicology and Chemistry*, *19*(9), 2266–2274. doi:10.1002/etc.5620190917

Bell, E. M., Hertz-Picciotto, I., & Beaumont, J. J. (2001). A case-control study of pesticides and fetal death due to congenital anomalies. *Epidemiology (Cambridge, Mass.)*, *12*(2), 148–156. doi:10.1097/00001648-200103000-00005 PMID:11246574

Beltran, J., Peruga, A., Pitarch, E., Lopez, F. J., & Hernandex, F. (2003). Application of solid-phase microextraction for the determination of pyrethroid residues in vegetable samples by GC-MS. *Analytical and Bioanalytical Chemistry*, *376*, 502–511. doi:10.100700216-003-1916-z PMID:12750867

Bempah, C. K., Buah-Kwofie, A., Enimil, E., Blewu, B., & Agyei-Martey, G. (2012). Residues of organochlorine pesticides in vegetables marketed in Greater Accra Region of Ghana. *Food Control*, *25*(2), 537–542. doi:10.1016/j.foodcont.2011.11.035

Bending, G. D., Friloux, M., & Walker, A. (2002). Degradation of contrasting pesticides by white rot fungi and its relationship with ligninolytic potential. *FEMS Microbiology Letters*, *212*(1), 59–63. doi:10.1111/j.1574-6968.2002.tb11245.x PMID:12076788

Benitez, F. J., Acero, J. L., & Real, F. J. (2002). Degradation of carbofuran by using ozone, UV radiation and advanced oxidation processes. *Journal of Hazardous Materials*, *89*(1), 51–65. doi:10.1016/S0304-3894(01)00300-4 PMID:11734346

Benner, J., Helbling, D. E., Kohler, H. P. E., Wittebol, J., Kaiser, E., Prasse, C., & Springael, D. (2013). Is biological treatment a viable alternative for micropollutant removal in drinking water treatment processes? *Water Research*, *47*(16), 5955–5976. doi:10.1016/j.watres.2013.07.015 PMID:24053940

Berini, F., Katz, C., Gruzdev, N., Casartelli, M., Tettamanti, G., & Marinelli, F. (2018). Microbial and viral chitinases: Attractive biopesticides for integrated pest management. *Biotechnology Advances*, *36*(3), 818–838. doi:10.1016/j.biotechadv.2018.01.002 PMID:29305895

Berkowitz, G. S., Wetmur, J. G., Birman-Deych, E., Obel, J., Lapinski, R. H., Godbold, J. H., & Wolff, M. S. (2004). In utero pesticide exposure, maternal paraoxonase activity, and head circumference. *Environmental Health Perspectives*, *112*(3), 388–391. doi:10.1289/ehp.6414 PMID:14998758

Beseler, C., Stallones, L., Hoppin, J., Alavanja, M., Blair, A., Keefe, T., & Kamel, F. (2006). Depression and Pesticide Exposures in Female Spouses of Licensed Pesticide Applicators in the Agricultural Health Study Cohort. *Journal of Occupational and Environmental Medicine*, *48*(10), 1005–1013. doi:10.1097/01.jom.0000235938.70212.dd PMID:17033500

Bester, K., Hunerfuss, H., Brockmann, U., & Rick, H. J. (1995). Biological effects of triazine herbicide contamination on marine phyto-plankton. *Archives of Environmental Contamination and Toxicology*, *29*, 277–283. doi:10.1007/BF00212490

Betz, F. S., Hammond, B. G., & Fuchs, R. L. (2000). Safety and advantages of Bacillus thuringiensis- protected plants to control insect pests. *Regulatory Toxicology and Pharmacology*, *32*(2), 156–173. doi:10.1006/rtph.2000.1426 PMID:11067772

Bevans, H. E., Lico, M. S., & Lawrence, S. J. (1998). Water quality in the Las Vegas Valley area and the Carson and Truckee River basins, Nevada and California, 1992–96. Reston, VA: USGS. *Geological Survey Circular*, 1170.

Bhagobaty, R. K., & Malik, A. (2010). *Utilization of chlorpyrifos as a sole source of carbon by bacteria isolated from wastewater irrigated agricultural soils in an industrial ares of western Uttar Pradesh*. Academic Press.

Bhaskaran, K., Nadaraja, A. V., Tumbath, S., Shah, L. B., & Veetil, P. G. P. (2013). Phytoremediation of perchlorate by free floating macrophytes. *Journal of Hazardous Materials*, *260*, 901–906. doi:10.1016/j.jhazmat.2013.06.008 PMID:23872336

Bhat, A. R., Wani, M. A., Kirmani, A. R., & Raina, T. H. (2010). Pesticides and brain cancer linked in orchard farmers of Kashmir. *Indian Journal of Medical and Paediatric Oncology: Official Journal of Indian Society of Medical & Paediatric Oncology*, *31*(4), 110–120. doi:10.4103/0971-5851.76191 PMID:21584215

Bhattacharyya, A., Barik, S. R., & Ganguly, P. (2009). New pesticide molecules, formulation technology and uses: Present status and future challenges. *Journal of Plant Protection Sciences*, *1*, 9–15.

Bhatti, A. A., Haq, S., & Bhat, R. A. (2017). Actinomycetes benefaction role in soil and plant health. *Microbial Pathogenesis*, *111*, 458–467. doi:10.1016/j.micpath.2017.09.036 PMID:28923606

Bhatt, L. R., Lacoul, P., Lekhak, H. D., & Jha, P. K. (1999). Physico-chemical characteristics and phytoplanktons of Taudaha lake, Kathmandu. *Pollution Research*, *18*, 353–358.

Bhunya, S. P., & Pati, P. C. (1988). Genotoxic effects of a synthetic pyrethroid insecticide cypermethrin in mice *in vivo*. *Toxicology Letters*, *41*(3), 223–230. doi:10.1016/0378-4274(88)90058-6 PMID:3376150

Biggs, J., Von Fumetti, S., & Kelly-Quinn, M. (2017). The importance of small waterbodies for biodiversity and ecosystem services: Implications for policy makers. *Hydrobiologia*, *793*(1), 3–39. doi:10.100710750-016-3007-0

Birch, A. N. E. (2011). How agro-ecological research helps to address food security issues under new IPM and pesticide reduction policies for global crop production systems. *Journal of Experimental Botany*, *62*, 3251–3261. doi:10.1093/jxb/err064 PMID:21669880

Birnbaum, L. (2012). Environmental chemicals: Evaluating low-dose effects. *Environmental Health Perspectives*, *120*(4), 143–144. doi:10.1289/ehp.1205179 PMID:22470049

BIS IS10500. (2012). *Indian Standard Drinking Water–Specification (Second revision)*. Bureau of Indian Standards (BIS).

Bitton, G. (1999). *Wastewater Microbiology*. New York: Wiley-Liss.

Blackburn, S. T. (2007). *Maternal, fetal, & neonatal physiology: A clinical perspective* (3rd ed.). St. Louis, MO: Saunders.

Blanco, M. C., Rodriguez, B. R., Grande, B. C., & Gandara, J. S. (2002). Optimization of solid-phase extraction and solid-phase microextraction for the determination of α- and β- endosulfan in water by gas chromatographyelectron capture detection. *Journal of Chromatography. A*, *976*(1-2), 293–299. doi:10.1016/S0021-9673(02)01232-3 PMID:12462621

Blankespoor, B., Dasgupta, S., & Laplante, B. (2012). *Sea-level rise and coastal wetlands: impacts and costs*. Washington, DC: The World Bank. doi:10.1596/1813-9450-6277

Blankson-Arthur, S., Yeboah, P. O., Golow, A., Tutu, A. O., & Denutsui, D. (2011). Levels of Organochlorine Pesticide Residues in Grasscutter (Thryonomys swinderianus) Tissues. *Research Journal of Environmental and Earth Sciences*, *3*(4), 350–357.

Blenkinsop, S., Fowler, H. J., Dubus, I. G., Nolan, B. T., & Hollis, J. M. (2008). Developing climatic scenarios for pesticide fate modelling in Europe. *Environmental Pollution*, *154*(2), 219–231. doi:10.1016/j.envpol.2007.10.021 PMID:18045756

Bloomfield, J. P., Williams, R. J., Gooddy, D. C., Cape, J. N., & Guha, P. (2006). Impacts of climate change on the fate and behaviour of pesticides in surface and groundwater —A UK perspective. *The Science of the Total Environment*, *369*(1–3), 163–177. doi:10.1016/j.scitotenv.2006.05.019 PMID:16914182

Boccolini, P. M., Boccolini, C. S., Meyer, A., Chrisman, J., & Guimarães, R. M. (2003). Pesticide exposure and low birth weight prevalence in Brazil. *International Journal of Hygiene and Environmental Health*, *216*, 2904. PMID:23021945

Bonner, M. R., & Alavanja, M. C. R. (2017). Pesticides, human health, and food security. *Food and Energy Security*, *6*(3), 89–93. doi:10.1002/fes3.112

Booth, L. H., Bithell, S. L., Wratten, S. D., & Heppelthwaite, V. J. (2003). Vineyard Pesticides and Their Effects on Invertebrate Biomarkers and Bioindicator Species in New Zealand. *Bulletin of Environmental Contamination and Toxicology*, *71*(6), 1131–1138. doi:10.100700128-003-8879-9 PMID:14756280

Bortleson, G. & Davis, D. (1987–1995). *Pesticides in selected small streams in the Puget Sound Basin*. U.S. Geological Survey & Washington State Department of Ecology.

Bourgmann, N., & Morin, A. (1995). Relationship between size structure of invertebrate assemblages and trophy and substrate composition in streams. *Journal of the North American Benthological Society*, *14*(3), 393–403. doi:10.2307/1467205

Bourquin, A. W. (1977). Degradation of malathion by salt-marsh microorganisms. *Applied and Environmental Microbiology*, *33*(2), 356–362. PMID:192147

Bowley, H. J., Gerrard, D. L., Louden, J. D., & Turrell, G. (2012). *Practical Raman spectroscopy*. Springer Science & Business Media.

Bozdogan, A. M. (2014). Assessment of total risk on non-target organisms in fungicide application for agricultural sustainability. *Sustainability*, *6*(2), 1046–1058. doi:10.3390u6021046

Bozdogan, A. M., & Yarpuz-Bozdogan, N. (2008a). Determination of Dermal Bystander Exposure of Malathion For Different Application Techniques. *Fresenius Environmental Bulletin*, *17*, 2103–2210.

Bozdogan, A. M., & Yarpuz-Bozdogan, N. (2008b). Assessment of Buffer Zones to Ditches of Dicofol For Different Applied Doses and Replication Numbers In Pesticide Applications In Adana Province, Turkey. *Fresenius Environmental Bulletin*, *17*, 275–281.

Bozdogan, A. M., & Yarpuz-Bozdogan, N. (2015). Determination of potential risks of pesticide applications in peanut cultivation on human health and environment. *Fresenius Environmental Bulletin*, *24*, 683–689.

Bradbury, S. P. (1994). Predicting modes of toxic action from chemical structure: An overview. *SAR and QSAR in Environmental Research*, *2*(1-2), 89–104. doi:10.1080/10629369408028842 PMID:8790641

Bradbury, S. P. (1995). Quantitative structure-activity relationships and ecological risk assessment: An overview of predictive aquatic toxicology research. *Toxicology Letters*, *79*(1-3), 229–237. doi:10.1016/0378-4274(95)03374-T PMID:7570660

Bradbury, S. P., & Coats, J. R. (1989). Comparative toxicology of the pyrethroid insecticides. *Reviews of Environmental Contamination and Toxicology*, *108*, 133–177. doi:10.1007/978-1-4613-8850-0_4 PMID:2646661

Bradbury, S. P., Russom, C. L., Ankley, G. T., Schultz, T. W., & Walker, J. D. (2003). Overview of data and conceptual approaches for derivation of quantitative structure-activity relationships for ecotoxicological effects of organic chemicals. *Environmental Toxicology and Chemistry*, *22*(8), 1789–1798. doi:10.1897/01-234 PMID:12924578

Bradman, A., & Barr, D. B. (2003). Measurement of pesticides and other toxicants in amniotic fluid as a potential biomarker of prenatal exposure: A validation study. *Environmental Health Perspectives*, 111.

Bradman, A., Barr, D. B., Henn, B. G. C., Drumheller, T., Curry, C., & Eskenazi, B. (2003). Measurement of pesticides and other toxicants in amniotic fluid as a potential biomarker of prenatal exposure: A validation study. *Environmental Health Perspectives*, *111*(14), 1779–1782. doi:10.1289/ehp.6259 PMID:14594631

Bradman, A., Quiros-Alcala, L., Castorina, R., Schall, R. A., Camacho, J., Holland, N. T., ... Eskenazi, B. (2015). Effect of organic diet intervention on pesticide exposures in young children living in low-income urban and agricultural communities. *Environmental Health Perspectives*, *123*(10). doi:10.1289/ehp.1408660 PMID:25861095

Brender, J. D., Felkner, M., Suarez, L., Canfield, M. A., & Henry, J. P. (2010). Maternal pesticide exposure and neural tube defects in Mexican Americans. *Annals of Epidemiology*, *20*(1), 16–22. doi:10.1016/j.annepidem.2009.09.011 PMID:20006272

Brender, J. D., & Weyer, P. J. (2016). Agricultural Compounds in Water and Birth Defects. *Current Environmental Health Reports*, *3*(2), 144–152. doi:10.100740572-016-0085-0 PMID:27007730

Brenner, B., Markowitz, S., Rivera, M., Romero, H., Weeks, M., Sanchez, E., ... Berkowitz, G. (2003). Integrated pest management in an urban community: A successful partnership for prevention. *Environmental Health Perspectives*, *111*(13), 1649–1653. doi:10.1289/ehp.6069 PMID:14527845

Bretveld, R. W., Thomas, C. M., Scheepers, P. T., Zielhuis, G. A., & Roeleveld, N. (2006). Pesticide exposure: The hormonal function of the female reproductive system disrupted? *Reproductive Biology and Endocrinology*, *4*(1), 30. doi:10.1186/1477-7827-4-30 PMID:16737536

Brönmark, C., & Hansson, L.-A. (2002). Environmental issues in lakes and ponds: Current state and perspectives. *Environmental Conservation*, *29*(3), 290–306. doi:10.1017/S0376892902000218

Brouwer, A., Longnecker, M. P., Birnbaum, L. S., Cogliano, J., Kostyniak, P., Moore, J., ... Winneke, G. (1999). Characterization of potential endocrine related health effects at low dose levels of exposure to PCBs. *Environmental Health Perspectives*, *107*(Suppl 4), 639–649. doi:10.1289/ehp.99107s4639 PMID:10421775

Brucker-Davis, F., Wagner-Mahler, K., Delattre, I., Ducot, B., Ferrari, P., Bongain, A., ... Fénichel, P. (2008). Cryptorchidism Study Group from Nice Area. Cryptorchidism at birth in Nice area (France) is associated with higher prenatal exposure to PCBs and DDE, as assessed by colostrums concentrations. *Human Reproduction (Oxford, England)*, *23*(8), 1708–1718. doi:10.1093/humrep/den186 PMID:18503055

Buccini, J. (2003). The Development of a Global Treaty on Persistent Organic Pollutants (POPs). In Persistent Organic Pollutants. The Handbook of Environmental Chemistry (vol. 3, pp. 13-30). Springer.

Burger, J., Mol, F., & Gerowitt, B. (2008). The 'necessary extent' of pesticide use—Thoughts about a key term in German pesticide policy. *Crop Protection (Guildford, Surrey)*, *27*(3-5), 343–351. doi:10.1016/j.cropro.2007.06.006

Burgoa, B., & Wauchope, R. D. (1995). Pesticides in run-off and surface waters. In T. R. Roberts & P. C. Kearney (Eds.), *Environmental Behaviour of Agrochemicals*. Chichester, UK: John Wiley & Sons.

Burton, G. A., Nelson, M. K., & Ingersoll, C. G. (1992). Freshwater benthic toxicity tests. In G. A. Burton Jr., (Ed.), *Sediment toxicity assessment* (pp. 213–240). Chelsea, MA: Lewis Publishers.

Butterfield, P. G., Valanis, B. G., Spencer, P. S., Lindeman, C. A., & Nutt, J. G. (1993). Environmental antecedents of young-onset Parkinson's disease. *Neurology*, *43*(6), 1150–1158. doi:10.1212/WNL.43.6.1150 PMID:8170560

Cairns, J., & Neiderlehner, B. R. (Eds.). (1985). Ecological Toxicity Testing. Lewis Publishers.

California Dept. of Health Services. Environmental Epidemiology and Toxicology Program. (1992, June). *Acute health effects of the Cantara metam sodium spill*. Author.

Callahan, M. A., Slimak, M. W., Gabel, N. W., May, I. P., Fowler, C. F., & Feed, J. R. (1979). Water-related environmental fate of 129 priority pol-lutants, Vol I. EPA 440/4-79-029B. U.S. Environmental Protection Agency.

Calloway, D. (1997). Beer-lambert law. *Journal of Chemical Education*, *74*(7), 744. doi:10.1021/ed074p744.3

Calvert, G. (2001). Surveillance of pesticide-related illness and injury in humans. In R. Krieger (Ed.), *Handbook of Pesticide Toxicology* (2nd ed.; pp. 603–641). New York: Academy Press. doi:10.1016/B978-012426260-7.50030-6

Cameotra, S. S., & Dhanjal, S. (2010). Environmental nanotechnology: nanoparticles for bioremediation of toxic pollutants. In *Bioremediation Technology* (pp. 348–374). Dordrecht: Springer. doi:10.1007/978-90-481-3678-0_13

Cappiello, A., Famiglini, G., Palma, P., & Mangani, F. (2002). Trace level determination of organophosphorus pesticides in water with the new direct-electron ionization LC/MS interface. *Analytical Chemistry*, *74*(14), 3547–3554. doi:10.1021/ac015685f PMID:12139067

Carbone, P., Giordano, F., Nori, F., Mantovani, A., Taruscio, D., Lauria, L., & Figa-Talamanca, I. (2006). Cryptorchidism and hypospadias in the Sicilian district of Ragusa and the use of pesticides. *Reproductive Toxicology (Elmsford, N.Y.)*, *22*(1), 8–12. doi:10.1016/j.reprotox.2006.01.006 PMID:16530380

Carere, M., Miniero, R., & Cicero, M. R. (2011). Potential effects of climate change on the chemical quality of aquatic biota. *TrAC Trends in Analytical Chemistry*, *30*(8), 1214–1221. doi:10.1016/j.trac.2011.06.006

Caron-Beaudoin, É., Denison, M. S., & Sanderson, J. T. (2016). Effects of neonicotinoids on promoter-specific expression and activity of aromatase (CYP19) in human adrenocortical carcinoma (H295R) and primary umbilical vein endothelial (HUVEC) cells. *Toxicological Sciences*, *149*(1), 134–144. doi:10.1093/toxsci/kfv220 PMID:26464060

Carozza, S. E., Li, B., Elgethun, K., & Whitworth, R. (2008). Risk of childhood cancers associated with residence in agriculturally intense areas in the United States. *Environmental Health Perspectives*, *116*(4), 559–565. doi:10.1289/ehp.9967 PMID:18414643

Carson, R. (1962). Silent Spring. Boston: Houghton.

Carson, R. (1962). *Silent spring*. Houghton Mifflin Company.

Carter, A. D. (1999). Leaching Mechanisms. In G. T. Brooks & T. R. Roberts (Eds.), *Pesticide Chemistry and Bioscience, the Food-Environment Challenge* (pp. 291–301). Cambridge, UK: The Royal Society of Chemistry. doi:10.1533/9781845698416.7.291

Carvalho, F. P. (2006). Agriculture, pesticides, food security and food safety. *Environmental Science & Policy*, *9*(7-8), 685–692. doi:10.1016/j.envsci.2006.08.002

Casalegno, M., Sello, G., & Benfenati, E. (2006). Top-priority fragment QSAR approach in predicting pesticide aquatic toxicity. *Chemical Research in Toxicology*, *19*(11), 1533–1539. doi:10.1021/tx0601814 PMID:17112242

Casey, T. J., & Casey, J. T. (1997). *Unit Treatment Processes in Water and Wastewater Engineering*. New York: John Wiley & Sons.

Casida, J. E., & Durkin, K. A. (2013). Neuroactive insecticides: Targets, selectivity, resistance, and secondary effects. *Annual Review of Entomology*, *58*(1), 99–117. doi:10.1146/annurev-ento-120811-153645 PMID:23317040

Casida, J. E., & Quistad, G. B. (1998). Golden age of insecticide research: Past, present, or future? *Annual Review of Entomology*, *43*(1), 1–16. doi:10.1146/annurev.ento.43.1.1 PMID:9444749

Cassardo, C., & Jones, J. A. A. (2011). Managing water in a changing world. *Water (Basel)*, *3*(2), 618–628. doi:10.3390/w3020618

Castillo, L. E., Martinez, E., Ruepert, C., Savage, C., Gilek, M., Pinnock, M., & Solis, E. (2006). Water quality and macroinvertebrate community response following pesticide applications in a banana plantation, Limon, Costa Rica. *The Science of the Total Environment*, *367*(1), 418–432. doi:10.1016/j.scitotenv.2006.02.052 PMID:16643988

Castro, C. E., Wade, R. S., & Belser, N. O. (1985). Biodehalogenation: Reactions of cytochrome P-450 with polyhalomethanes. *Biochemistry*, *24*(1), 204–210. doi:10.1021/bi00322a029 PMID:4039602

Caton, A. R. (2012). Exploring the seasonality of birth defects in the New York State Congenital Malformations Registry. *Birth Defects Research. Part A, Clinical and Molecular Teratology*, *94*(6), 424–437. doi:10.1002/bdra.23006 PMID:22473628

Ccanccapa, A., Masiá, A., Andreu, V., & Picó, Y. (2016). Spatio-temporal patterns of pesticide residues in the Turia and Júcar Rivers (Spain). *The Science of the Total Environment*, *540*, 200–210. doi:10.1016/j.scitotenv.2015.06.063 PMID:26118860

Centers for Disease Control and Prevention. (2010). Racial differences by gestational age in neonatal deaths attributable to congenital heart defects—United States, 2003–2006. *MMWR. Morbidity and Mortality Weekly Report*, *59*, 1208–1211. PMID:20864921

Central Pollution Control Board (CPCB). (2008). *Status of Water Quality in India 2007*. New Delhi: Central Pollution Control Board, Ministry of Environment and Forests, Government of India.

Central Pollution Control Board (CPCB). (2009). *Status of Water Supply, Waste water Generation and Treatment in Class-I Cities and Class-II Towns of India*. New Delhi: Central Pollution Control Board, Ministry of Environment and Forests, Government of India.

Central Pollution Control Board (CPCB). (2010). *Status of Water Quality in India 2009*. New Delhi: Central Pollution Control Board, Ministry of Environment and Forests, Government of India.

Central Water Commission (CWC). (2009). *National Register of Large Dams 2009*. New Delhi: Central Water Commission.

Centre for Science and Environment (CSE). (2012). *Living with changing climate: Impact, Vulnerability and Adaptation Challenges in Indian Sundarbans*. New Delhi: Centre for Science and Environment.

Cervera, M. I., Medina, C., Portolés, T., Pitarch, E., Beltrán, J., Serrahima, E., ... Hernández, F. (2010). Multi-Residue Determination of 130 Multiclass Pesticides in Fruits and Vegetables by Gas Chromatography Coupled to Triple Quadrupole Tandem Mass Spectrometry. *Analytical and Bioanalytical Chemistry*, *397*(7), 2873–2891. doi:10.100700216-010-3597-8 PMID:20237919

Chambers, H. W., Boone, J. S., Carr, R. L., & Chanbers, J. E. (2001). *Handbook of Pesticide Toxicology*. Academic Press. doi:10.1016/B978-012426260-7.50047-1

Chandan, P., Chatterjee, A., & Gautam, P. (2007). Management planning of Himalayan high altitude wetlands. A case study of Tsomoriri and Tsokar Wetlands in Ladakh, India. In *Proceedings of Taal2007: The 12th World Lake Conference* (Vol. 1446, p. 1452). Academic Press.

Chanda, S. M., & Pope, C. N. (1996). Neurochemical and neurobehavioral effects of repeated gestational exposure to chlorpyrifos in maternal and developing rats. *Pharmacology, Biochemistry, and Behavior, 53*(4), 771–776. doi:10.1016/0091-3057(95)02105-1 PMID:8801577

Chandler, D., Bailey, A. S., Tatchell, G. M., Davidson, G., Greaves, J., & Grant, W. P. (2011). The development, regulation and use of biopesticides for integrated pest management. *Philosophical Transactions of the Royal Society of London. Series B, Biological Sciences, 366*(1573), 1987–1998. doi:10.1098/rstb.2010.0390 PMID:21624919

Chandler, D., Davidson, G., Grant, W. P., Greaves, J., & Tatchell, G. M. (2008). Microbial biopesticides for integrated crop management: An assessment of environmental and regulatory sustainability. *Trends in Food Science & Technology, 19*(5), 275–283. doi:10.1016/j.tifs.2007.12.009

Chang, K. L., Lin, J. H., & Chen, S. T. (2011). Adsorption studies on the removal of pesticides (Carbofuran) using activated carbon from rice straw agricultural waste. *World AcadSciEngTechnol, 5*, 4-21.

Chapin, G., & Wasserstrom, R. (1981). Agricultural production and malaria resurgence in Central America and India. *Nature, 292*(5829), 181–185. doi:10.1038/293181a0 PMID:7278974

Chatterjee, S., Das, S. K., Chakravarty, R., Chakrabarti, A., Ghosh, S., & Guha, A. K. (2010). Interaction of malathion, an organophosphorus pesticide with Rhizopusoryzae biomass. *Journal of Hazardous Materials, 174*(1-3), 47–53. doi:10.1016/j.jhazmat.2009.09.014 PMID:19783095

Chaudhry, F. N., & Malik, M. F. (2017). Factors Affecting Water Pollution: A Review. *Journal of Ecosystem Ecography, 7*, 225.

Chaudhry, G. R., & Chapalamadugu, S. (1991). Biodegradation of halogenated organic compounds. *Microbiological Reviews, 55*(1), 59–79. PMID:2030673

Chavan, M., Thacker, N., Tara, J. L., & Gandhi, K. (2013). Monitoring the water quality for organochlorine pesticides in the area of intensive agriculture activity of Nagpur District, India. *Proceedings of the 13th International Conference on Environmental Science and Technology.*

Chávez-Servın, J. L., Castellote, A. I., & López-Sabater, M. C. (2004). Analysis of mono-and disaccharides in milk-based formulae by high-performance liquid chromatography with refractive index detection. *Journal of Chromatography. A, 1043*(2), 211–215. doi:10.1016/j.chroma.2004.06.002 PMID:15330094

Chen, F. (2012). The chronic aquatic toxicity of a microbicide dibromo nitrilopropionamide. *Toxicology and Industrial Health, 28*(2), 181–185. doi:10.1177/0748233711410904 PMID:21937531

Chen, H., Yao, J., Wang, F., Cai, M., & Liu, H. (2013). Toxicity of perfluorooctanoic acid to Pseudomonas putida in the aquatic environment. *Journal of Hazardous Materials, 262*, 726–731. doi:10.1016/j.jhazmat.2013.09.046 PMID:24140521

Chen, J., Dong, D., & Ye, S. (2018). Detection of pesticide residue distribution on fruit surfaces using surface-enhanced Raman spectroscopy imaging. *RSC Advances, 8*(9), 4726–4730. doi:10.1039/C7RA11927E

Chen, M., Chang, C. H., Tao, L., & Lu, C. (2015). Residential exposure to pesticide during childhood and childhood cancers: A meta-analysis. *Pediatrics, 136*(4), 719–729. doi:10.1542/peds.2015-0006 PMID:26371195

Chen, S., Deng, Y., Chang, C., Lee, J., Cheng, Y., Cui, Z., & Zhang, L. H. (2015). Pathway and kinetics of cyhalothrin biodegradation by Bacillus thuringiensis strain ZS-19. *Scientific Reports, 5*(1), 8784. doi:10.1038rep08784 PMID:25740758

Chen, Y., Ren, H. L., Liu, N., Sai, N., Liu, X., Liu, Z., ... Ning, B. (2010). A fluoroimmunoassay based on quantum dot-streptavidin conjugate for the detection of chlorpyrifos. *Journal of Agricultural and Food Chemistry, 58*(16), 8895–8903. doi:10.1021/jf101778t PMID:23654227

Cherkasov, A., Muratov, E. N., Fourches, D., Varnek, A., Baskin, I. I., Cronin, M., & Consonni, V. (2014). QSAR modeling: Where have you been? Where are you going to? *Journal of Medicinal Chemistry*, *57*(12), 4977–5010. doi:10.1021/jm4004285 PMID:24351051

Chevrier, C., Limon, G., Monfort, C., Rouget, F., Garlantézec, R., Petit, C., ... Cordier, S. (2011). Urinary biomarkers of prenatal atrazine exposure and adverse birth outcomes in the PELAGIE birth cohort. *Environmental Health Perspectives*, *119*(7), 1034–1041. doi:10.1289/ehp.1002775 PMID:21367690

Chevrier, C., Warembourg, C., Gaudreau, E., Monfort, C., Le Blanc, A., Guldner, L., & Cordier, S. (2013). Organochlorine pesticides, polychlorinated biphenyls, seafood consumption, and time-to-pregnancy. *Epidemiology (Cambridge, Mass.)*, *24*(2), 251–260. doi:10.1097/EDE.0b013e31827f53ec PMID:23348067

Chicu, S. A., Herrmann, K., & Berking, S. (2000). An approach to calculate the toxicity of simple organic molecules on the basis of QSAR analysis in Hydractinia echinata (Hydrozoa, Cnidaria). *Molecular Informatics*, *19*(3), 227–236.

Chiesa, L. M., Labella, G. F., Panseri, S., Britti, D., Galbiati, F., Villa, R., & Arioli, F. (2017). Accelerated solvent extraction by using an 'in-line' clean-up approach for multiresidue analysis of pesticides in organic honey. *Food Additives & Contaminants. Part A, Chemistry, Analysis, Control, Exposure & Risk Assessment*, *34*(5), 809–818. PMID:28277176

Chitranshi, U. R. (2004). R & D priorities for sustainable development of inland capture fishery sector in India. *Symposium on advances in Fish Physiology: Ecological Considerations, Varanasi*, 12-14.

Choudhury, B. C., & Sathyakumar, S., & Christi, S. (2007). An assessment of the current status of the Indian Peafowl (Pavo cristatus) in India based on questionnaire survey. *Galliformes of India, ENVIS Bulletin. Wildlife and Protected Areas*, *10*, 53–60.

Christos, A. D., & Spyridon, D. K. (2016). Farmers' Exposure to Pesticides: Toxicity Types and Ways of Prevention. *Toxics*, *4*(1), 1. doi:10.3390/toxics4010001 PMID:29051407

CIBRC. (2015). *List of pesticides which are banned, refused registration and restricted in use.* Retrieved March 8, 2018, from http://cibrc.nic.in/ibr2012.doc

Cinelli, G., Avino, P., Notardonato, I., Centola, A., & Russo, M. V. (2014). Study of XAD-2 adsorbent for the enrichment of trace levels of phthalate esters in hydroalcoholic food beverages and analysis by gas chromatography coupled with flame ionization and ion-trap mass spectrometry detectors. *Food Chemistry*, *146*, 181–187. doi:10.1016/j.foodchem.2013.09.064 PMID:24176330

Claeys, W. L., Schmit, J.-F., Bragard, C., Maghuin-Rogister, G., Pussemier, L., & Schiffers, B. (2011). Exposure of several Belgian consumer groups to pesticide residues through fresh fruit and vegetable consumption. *Food Control*, *22*(3-4), 508–516. doi:10.1016/j.foodcont.2010.09.037

Clement, S. L., Grigarick, A. A., & Way, M. O. (1977). The colonization of California rice paddies by chironomid midges. *Journal of Applied Ecology*, *14*(2), 379–389. doi:10.2307/2402551

Clyde, L. O. (2013). *Spray drift of pesticide.* Neb guide. Retrieved from https://en.wikipedia.org/wiki/Pesticide_drift

Coat, S., Bocquené, G., & Godard, E. (2006). Contamination of some aquatic species with the organochlorine pesticide chlordecone in Martinique. *Aquatic Living Resources*, *19*(2), 181–187. doi:10.1051/alr:2006016

Cobas, M., Meijide, J., Sanromán, M. A., & Pazos, M. (2016). Chestnut shells to mitigate pesticide contamination. *Journal of the Taiwan Institute of Chemical Engineers*, *61*, 166–173. doi:10.1016/j.jtice.2015.11.026

Cohen, J. M., & Pinkerton, C. (1966). *Widespread translocation of pesticides by air transport and rain-out.* Academic Press.

Colborn, T. (2006). A case for revisiting the safety of pesticides: A closer look at neurodevelopment. *Environmental Health Perspectives*, *114*(1), 10–17. doi:10.1289/ehp.7940 PMID:16393651

Colborn, T., & Carroll, L. E. (2007). Pesticides, sexual development, reproduction, and fertility: Current perspective and future direction. *Human and Ecological Risk Assessment*, *13*(5), 1078–1110. doi:10.1080/10807030701506405

Colborn, T., Myers, J. P., & Dumanoski, D. (1996). *Our Stolen Future. Penguin Books.*

Colborn, T., VomSaal, F. S., & Soto, A. M. (1993). Developmental effects of endocrine-disrupting chemicals in wildlife and humans. *Environmental Health Perspectives*, *101*(5), 378–384. doi:10.1289/ehp.93101378 PMID:8080506

Cole, D. J. (1985). Mode of action of glyphosate - a literature analysis. In E. Grossbard & D. Atkinson (Eds.), *The herbicide glyphosate* (pp. 48–74). London: Butterworths.

Commission, E. (2016). *Impact Assessment Report on Criteria to identify EDs*. Academic Press.

Connell, G. F. (1996). *The Chlorination/Chloramination Handbook*. Denver, CO: American Water Works Association.

Connors, K. A., Voutchkova-Kostal, A. M., Kostal, J., Anastas, P., Zimmerman, J. B., & Brooks, B. W. (2014). Reducing aquatic hazards of industrial chemicals: Probabilistic assessment of sustainable molecular design guidelines. *Environmental Toxicology and Chemistry*, *33*(8), 1894–1902. doi:10.1002/etc.2614 PMID:24839109

Conrad, A. U., Fleming, R. J., & Crane, M. (1999). Laboratory and field response of Chironomus riparius to a pyrethroid insecticide. *Water Research*, *33*(7), 1603–1610. doi:10.1016/S0043-1354(98)00397-2

Copping, L. G., & Menn, J. J. (2000). Biopesticides: A review of their action, applications and efficacy. *Pest Management Science*, *56*(8), 651–676. doi:10.1002/1526-4998(200008)56:8<651::AID-PS201>3.0.CO;2-U

Corradini, D. (Ed.). (2016). *Handbook of HPLC*. CRC Press.

Cory, J. S. (2000). Assessing the risk of releasing genetically modified virus insecticides: Progress to date. *Crop Protection (Guildford, Surrey)*, *19*(8-10), 779–785. doi:10.1016/S0261-2194(00)00104-6

Council, E. P. A. (2009). Directive 2009/128/EC of the European Parliament and of the Council of 21 October 2009 establishing a framework for Community action to achieve the sustainable use of pesticides. *Official JL*, *309*, 71–86.

Cowan, R., Ellis, T. G., Higgins, M. J., Alagappan, G., & Park, K. (1996). Treatment Systems. Activated Sludge and Other Aerobic Suspended Culture Processes. *Water Environment Research*, *68*(4), 451–469. doi:10.2175/106143096X135317

Cowardin, L. M., Carter, V., Golet, F. C., & LaRoe, E. T. (1979). *Classification of wetlands and deepwater habitats of the United States*. US Department of the Interior, US Fish and Wildlife Service.

Cox, C., & Surgan, M. (2006). Unidentified inert ingredients in pesticides: Implications for human and environmental health. *Environmental Health Perspectives*, *114*(12), 1803–1806. PMID:17185266

CPCB. (2000). *Water quality status of Yamuna river. Assessment & development of river basin series: ADSORBS 32, Central Pollution Control Board*. Delhi, India: CPCB.

Cramer, R. D., Patterson, D. E., & Bunce, J. D. (1988). Comparative molecular field analysis (CoMFA). 1. Effect of shape on binding of steroids to carrier proteins. *Journal of the American Chemical Society*, *110*(18), 5959–5967. doi:10.1021/ja00226a005 PMID:22148765

Crawford, C. G., Wangsness, D. J., & Martin, J. D. (1992). Recovery of benthic invertebrate communities in the White River near Indianapolis, following advance treatment of municipal wastewater. *Archiv für Hydrobiologie*, *126*, 67–84.

Cremonese, C., Freire, C., Camargo, A., Lima, J., Koifman, S., & Meyer, A. (2014). Pesticide consumption, central nervous system and cardiovascular congenital malformations in the South and Southeast region of Brazil. *International Journal of Occupational Medicine and Environmental Health*, *27*(3), 474–486. doi:10.247813382-014-0269-5 PMID:24847732

Criswell, J. T., Shelton, K., & Luper, C. (2014). *Toxicity of pesticides. Pesticide Applicator Certificate Series.* Oklahoma Cooperative Extension Service EPP-7457-8. Available from: http://www.dasnr22.dasnr.okstate.edu/docushare/dsweb/get/.../EPP-7457web.pdf

Crosby, D. G., & Mabury, S. (1992). *Management of pesticide residues in ricefield soil water-a report to the Soil Conservation Service of the US Department of Agriculture*. Department of Environmental Toxicology, University of California-Davis.

Cruz-Alcalde, A., Sans, C., & Esplugas, S. (2017). Exploring ozonation as treatment alternative for methiocarb and formed transformation products abatement. *Chemosphere*, *186*, 725–732. doi:10.1016/j.chemosphere.2017.08.040 PMID:28820996

Culliney, T. W., Pimentel, D., & Pimentel, M. H. (1992). Pesticides and natural toxicants in foods. *Agriculture, Ecosystems & Environment*, *41*(3-4), 297–320. doi:10.1016/0167-8809(92)90117-T

Cummins, K. W. (1974). Structure and function of stream ecosystem. *Bioscience*, *24*(11), 631–641. doi:10.2307/1296676

Cunha, S. C., Fernandes, J. O., Oliveira, B. M., & Oliveira, P.P. (2011). Current Trends in Liquid-Liquid Microextraction for Analysis of Pesticide Residues in Food and Water. *Pesticides Strategies for Pesticides Analysis*.

Cunningham, W. P., & Saigo, B. W. (1998). Environmental Science: A Global Concern (5th ed.). WCB/McGraw-Hill.

Currie, J., Zivin, J. G., Meckel, K., Neidell, M., & Schlenker, W. (2013). Something in the water: Contaminated drinking water and infant health. *The Canadian Journal of Economics. Revue Canadienne d'Economique*, *46*(3), 791–810. doi:10.1111/caje.12039 PMID:27134285

Curtis, K. M., Savitz, D. A., Weinberg, C. R., & Arbuckle, T. E. (1999). The effect of pesticide exposure on time to pregnancy. *Epidemiology (Cambridge, Mass.)*, *10*(2), 112–117. doi:10.1097/00001648-199903000-00005 PMID:10069244

Da Silva, A. C. N., Deda, D. K., da Róz, A. L., Prado, R., Carvalho, C., Viviani, V., & Leite, F. (2013). NanoBiosensors based on chemically modified AFM probes: A useful tool for metsulfuron-methyl detection. *Sensors (Switzerland)*, *13*(2), 1477–1489. doi:10.3390130201477 PMID:23348034

Dabrowski, S., Hanke, W., Polanska, K., Makowiec-Dabrowska, T., & Sobala, W. (2003). Pesticide exposure and birthweight: An epidemiological study in Central Poland. *International Journal of Occupational Medicine and Environmental Health*, *16*(1), 31–39. PMID:12705715

Daglioglu, N., Akcan, R., Gulmen, M. K., Yener, F., & Efeoglu, P. (2011). Pesticide intoxications in Cukurova, Turkey: Three years analysis. *Human and Experimental Toxicology*, *30*(12), 1892–1895. doi:10.1177/0960327111402241 PMID:21393348

Daglioglu, N., Efeoglu, P., Akcan, R., Inandiklioglu, N., Gulmen, M. K., & Demirhan, O. (2013). Polychlorinated biphenyls and organochlorine pesticides in amniotic fluids of pregnant women in South-central Turkey. *Toxicological and Environmental Chemistry*, *95*(6), 954–961. doi:10.1080/02772248.2013.840370

Daglioglu, N., Gulmen, M. K., Akcan, R., Efeoglu, P., Yener, F., & Unal, I. (2010). Determination of organochlorine pesticides residues in human adipose tissue, data from Cukurova, Turkey. *Bulletin of Environmental Contamination and Toxicology*, *85*(1), 97–102. doi:10.100700128-010-0057-2 PMID:20585752

Dagnac, T., Bristeau, S., Jeannot, R., Mouvet, C., & Baran, N. (2005). Determination of chloroacetanilides, triazines and phenylureas and some of their metabolites in soils by pressurised liquid extraction, GC–MS/MS, LC-MS and LC-MS/MS. *Journal of Chromatography. A*, *1067*(1–2), 225–233. doi:10.1016/j.chroma.2004.11.058 PMID:15844528

Dahlawi, S. M., & Siddiqui, S. (2017). Calcium polysulphide, its applications and emerging risk of environmental pollution—a review article. *Environmental Science and Pollution Research International*, *24*(1), 92–102. doi:10.100711356-016-7842-3 PMID:27761854

Dai, Y. J., Ji, W. W., Chen, T., Zhang, W. J., Liu, Z. H., Ge, F., & Yuan, S. (2010). Metabolism of the neonicotinoid insecticides acetamiprid and thiacloprid by the yeast Rhodotorula mucilaginosa strain IM-2. *Journal of Agricultural and Food Chemistry*, *58*(4), 2419–2425. doi:10.1021/jf903787s PMID:20112912

Daly, H., Doyen, J. T., & Purcell, A. H. (1998). *Introduction to insect biology and diversity* (2nd ed.). New York: Oxford University Press.

Damalas, C. A. (2009). Understanding benefits and risks of pesticide use. *Scientific Research and Essays*, *4*, 945–949.

Damalas, C. A. (2018). Current Status and Recent Developments in Biopesticide Use. *Agriculture*, *8*(1), 13. doi:10.3390/agriculture8010013

Damalas, C. A., & Eleftherohorinos, I. G. (2011). Pesticide exposure, safety issues, and risk assessment indicators. *International Journal of Environmental Research and Public Health*, *8*(5), 1402–1419. doi:10.3390/ijerph8051402 PMID:21655127

Damborsky, J., & Schultz, T. W. (1997). Comparison of the QSAR models for toxicity and biodegradability of anilines and phenols. *Chemosphere*, *34*(2), 429–446. doi:10.1016/S0045-6535(96)00361-X PMID:9057301

Dam, K., Seidler, F. J., & Slotkin, T. A. (2000, June). Chlorpyrifos exposure during a critical neonatal period elicits gender-selective deficits in the development of coordination skills and locomotor activity. *Brain Research. Developmental Brain Research*, *121*(2), 179–187. doi:10.1016/S0165-3806(00)00044-4 PMID:10876030

Dam, K., Seidler, F. J., & Slotkin, T. A. (2003). Transcriptional biomarkers distinguish between vulnerable periods for developmental neurotoxicity of chlorpyrifos: Implications for toxicogenomics. *Brain Research Bulletin*, *59*(4), 261–265. doi:10.1016/S0361-9230(02)00874-2 PMID:12464398

Damstra, T. (2002). Potential effects of certain persistent organic pollutants and endocrine disrupting chemicals on the health of children. *Journal of Toxicology. Clinical Toxicology*, *40*(4), 457–465. doi:10.1081/CLT-120006748 PMID:12216998

Daneshvar, N., Aber, S., Khani, A., & Khataee, A. R. (2007). Study of imidaclopride removal from aqueous solution by adsorption onto granular activated carbon using an on-line spectrophotometric analysis system. *Journal of Hazardous Materials*, *144*(1), 47–51. doi:10.1016/j.jhazmat.2006.09.081 PMID:17084525

Darboux, I., Nielsen-LeRoux, C., Charles, J. F., & Pauron, D. (2001). The receptor of Bacillus sphaericus binary toxin in Culexpipiens (Diptera: Culicidae) midgut: molecular cloning and expression. *Insect Biochemistry and Molecular Biology*, *31*(10), 981–990. doi:10.1016/S0965-1748(01)00046-7 PMID:11483434

Darçın, E. S., & Darçın, M. (2017). Health effects of agricultural pesticides. *Biomed Res- India; Special Section: Complex World of Neuroscience,* S13-17.

Das, K. P., & Barone, S. Jr. (1999). Neuronal differentiation in PC12 cells is inhibited by chlorpyrifos and its metabolites: Is acetylcholinesterase inhibition the site of action? *Toxicology and Applied Pharmacology*, *160*(3), 217–230. doi:10.1006/taap.1999.8767 PMID:10544056

David, P. (2002). Encyclopedia of Pest Management (vols. 1-2). CRC Press, Taylor & Francis Group.

Davis, J. R., Brownson, R. C., & Garcia, R. (1992). Family pesticide use in the home, garden, orchard, and yard. *Archives of Environmental Contamination and Toxicology*, *22*(3), 260–266. doi:10.1007/BF00212083 PMID:1616309

Daxenberger, A. (2002). Pollutants with androgen-disrupting potency. *European Journal of Lipid Science and Technology*, *104*(2), 124–130. doi:10.1002/1438-9312(200202)104:2<124::AID-EJLT124>3.0.CO;2-T

Day, K. E. (1989). Acute, chronic and sublethal effects of synthetic pyrethroids on freshwater zooplankton. *Environmental Toxicology and Chemistry*, *8*(5), 411–416. doi:10.1002/etc.5620080507

Day, K. E., Kaushik, N. K., & Solomon, K. R. (1987). Impact of fenvalerate on enclosed freshwater planktonic communities and on in-situ rates of filtration of zooplankton. *Canadian Journal of Fisheries and Aquatic Sciences*, *44*(10), 1714–1728. doi:10.1139/f87-210

de Bruijn, J., & Hermens, J. (1993). Inhibition of acetylcholinesterase and acute toxicity of organophosphorous compounds to fish: A preliminary structure-activity analysis. *Aquatic Toxicology (Amsterdam, Netherlands)*, *24*(3-4), 257–274. doi:10.1016/0166-445X(93)90075-C

De Gerónimo, E., Aparicio, V. C., Bárbaro, S., Portocarrero, R., Jaime, S., & Costa, J. L. (2014). Presence of pesticides in surface water from four sub-basins in Argentina. *Chemosphere*, *107*, 423–431. doi:10.1016/j.chemosphere.2014.01.039 PMID:24548646

De Lorenzo, M. E., Scott, G. I., & Ross, P. E. (2001). Toxicity of pesticides to aquatic microorganisms: A review. *Environmental Toxicology and Chemistry*, *20*(1), 84–98. doi:10.1002/etc.5620200108 PMID:11351418

De Roos, A., Blair, A., Rusiecki, J., Hoppin, J., Svec, M., Dosemeci, M., ... Alavanja, M. (2005). Cancer Incidence among Glyphosate-Exposed Pesticide Applicators in the Agricultural Health Study'. *Environmental Health Perspectives*, *113*(1), 49–54. doi:10.1289/ehp.7340 PMID:15626647

De Schampheleire, M., Spanoghe, P., Brusselman, E., & Sonck, S. (2007). Risk assessment of pesticide spray drift damage in Belgium. *Crop Protection (Guildford, Surrey)*, *26*(4), 602–611. doi:10.1016/j.cropro.2006.05.013

De, A., Bose, R., Kumar, A., & Mozumdar, S. (2014). Worldwide Pesticide Use. In Targeted Delivery of Pesticides Using Biodegradable Polymeric Nanoparticles, Springer Briefs in Molecular Science. Springer. doi:10.1007/978-81-322-1689-6_2

De, A. K. (2003). New Age International. *Environmental Chemistry*, 406.

Dearth, M. A., & Hites, R. A. (1991). Conplete analysis of technical chlordane using negative ionization mass spectrometry. *Environmental Science & Technology*, *25*(2), 245–254. doi:10.1021/es00014a005

Deeb, O., & Goodarzi, M. (2012). In silico quantitative structure toxicity relationship of chemical compounds: Some case studies. *Current Drug Safety*, *7*(4), 289–297. doi:10.2174/157488612804096533 PMID:23062241

Deepa, T. V. G., Lakshmi, P. S., & Lakhsmi, S. K. (2006). *Effects of pesticides*. Retrieved from www. Interhopen.com

Dehghani, M. H., Niasar, Z. S., Mehrnia, M. R., Shayeghi, M., Al-Ghouti, M. A., Heibati, B., ... Yetilmezsoy, K. (2017). Optimizing the removal of organophosphorus pesticide malathion from water using multi-walled carbon nanotubes. *Chemical Engineering Journal*, *310*, 22–32. doi:10.1016/j.cej.2016.10.057

Delaplane, K. S. (2000). *Pesticide usage in the United States: history, benefi ts, risks, and trends. Cooperative Extension Service.* The University of Georgia, College of Agricultural and Environmental Sciences. Bulletin, 1121. Retrieved from http://pubs.caes.uga.edu/ caespubs/pubs/PDF/B1121.pdf

Delate, K. & Hartzler, R. (2003). Weed Management for Organic Farmers. *Iowa State University Extension Bulletin, 1883.*

Delcour, I., Spanoghe, P., & Uyttendaele, M. (2015). Literature review: Impact of climate change on pesticide use. *Food Research International*, *68*, 7–15. doi:10.1016/j.foodres.2014.09.030

DeLorenzo, M. E., Lauth, J., Pennington, P. L., Scott, G. I., & Ross, P. E. (1999b). Atrazine effects on the microbial food web in tidal creek mesocosms. *Aquatic Toxicology (Amsterdam, Netherlands)*, *46*(3-4), 241–251. doi:10.1016/S0166-445X(98)00132-5

DeLorenzo, M. E., Scott, G. I., & Ross, P. E. (1999). Effects of the agricultural pesticides atrazine, deethylatrazine, endosulfan, and chlorpyrifos on an estuarine microbial food web. *Environmental Toxicology and Chemistry*, *18*(12), 2824–2835. doi:10.1002/etc.5620181224

Deneer, J. W. (2000). Toxicity of mixtures of pesticides in aquatic systems. *Pest Management Science*, *56*(6), 516–520. doi:10.1002/(SICI)1526-4998(200006)56:6<516::AID-PS163>3.0.CO;2-0

Denoel, M. (2012). Newt decline in Western Europe: Highlights from relative distribution changes within guilds. *Biodiversity and Conservation*, *21*(11), 2887–2898. doi:10.100710531-012-0343-x

DeNoyelles, F., Kettle, W. D., & Sinn, D. E. (1982). The response of plank-ton communities in experimental ponds to atrazine, the most heavily used pesticide in the United States. *Ecology*, *63*(5), 1285–1293. doi:10.2307/1938856

Derylo-Marczewska, A., Blachnio, M., Marczewski, A. W., Swiatkowski, A., & Buczek, B. (2017). Adsorption of chlorophenoxy pesticides on activated carbon with gradually removed external particle layers. *Chemical Engineering Journal*, *308*, 408–418. doi:10.1016/j.cej.2016.09.082

Deschamps, D., Questel, F., Baud, F. J., Gervais, P., & Dally, S. (1994). Persistent asthma after acute inhalation of organophosphate insecticide. *Lancet*, 344–352. PMID:7996993

Design of Municipal Wastewater Treatment Plants-MOP 8. (1998). 4th ed.). Alexandria, VA: Water Environment Federation.

Dettmer-Wilde, K., & Engewald, W. (2016). *Practical Gas Chromatography*. Springer-Verlag Berlin An.

Dewailly, E., Forde, M., Robertson, L., Kaddar, N., Sidi, E. A. L., Cote, S., ... Ayotte, P. (2014). Evaluation of pyrethroid exposures in pregnant women from 10 Caribbean countries. *Environment International*, *63*, 201–206. doi:10.1016/j.envint.2013.11.014 PMID:24317226

Dewan, P., Jain, V., Gupta, P., & Banerjee, B. D. (2013). Organochlorine pesticide residues in maternal blood, cord blood, placenta, and breast milk and their relation to birth size. *Chemosphere*, *90*(5), 1704–1710. doi:10.1016/j.chemosphere.2012.09.083 PMID:23141556

Diamond, M. L., de Wit, C. A., Molander, S., Scheringer, M., Backhaus, T., Lohmann, R., ... Zetzsch, C. (2015). Exploring the planetary boundary for chemical pollution. *Environment International*, *78*, 8–15. doi:10.1016/j.envint.2015.02.001 PMID:25679962

Díaz, E. (2004). Bacterial degradation of aromatic pollutants: A paradigm of metabolic versatility. *International Microbiology*, *7*(3), 173–180. PMID:15492931

Dich, J., Zahm, S. H., Hanberg, A., Adami, H. O., & Adami, H.-O. (1997). Pesticides and cancer. *Cancer Causes & Control*, *8*(3), 420–443. doi:10.1023/A:1018413522959 PMID:9498903

Dietary Guidelines for Americans. (2005). U.S. Department of Health and Human Services U.S. Department of Agriculture.

Dinh Van, K., Janssens, L., Debecker, S., De Jonge, M., Lambret, P., Nilsson-Örtman, V., ... Stoks, R. (2013). Susceptibility to a metal under global warming is shaped by thermal adaptation along a latitudinal gradient. *Global Change Biology*, *19*(9), 2625–2633. doi:10.1111/gcb.12243 PMID:23640735

Dom, N., Knapen, D., Benoot, D., Nobels, I., & Blust, R. (2010). Aquatic multi-species acute toxicity of (chlorinated) anilines: Experimental versus predicted data. *Chemosphere*, *81*(2), 177–186. doi:10.1016/j.chemosphere.2010.06.059 PMID:20637490

Dom, N., Knapen, D., & Blust, R. (2012). Assessment of aquatic experimental versus predicted and extrapolated chronic toxicity data of four structural analogues. *Chemosphere*, *86*(1), 56–64. doi:10.1016/j.chemosphere.2011.08.050 PMID:21944038

Donald, W. S. (2016). Current Use Pesticides. In Ecotoxicology Essentials, Environmental Contaminants and their Biological Effects on Animals and Plants (pp. 109–152). Academic Press.

dosAnjos, J. P., & de Andrade, J. B. (2015). Simultaneous determination of pesticide multiresidues in white wine and rosé wine by SDME/GC-MS. *Microchemical Journal*, *120*, 69–76. doi:10.1016/j.microc.2015.01.009

Dosnon-Olette, R., Couderchet, M., & Eullaffroy, P. (2009). Phytoremediation of fungicides by aquatic macrophytes: Toxicity and removal rate. *Ecotoxicology and Environmental Safety*, *72*(8), 2096–2101. doi:10.1016/j.ecoenv.2009.08.010 PMID:19732953

Doucet, J., Tague, B., Arnold, D. L., Cooke, G. M., Hayward, S., & Goodyer, C. G. (2009). Persistent organic pollutant residues in human fetal liver and placenta from Greater Montreal, Quebec: A longitudinal study from 1998 through 2006. *Environmental Health Perspectives*, *117*(4), 605–610. doi:10.1289/ehp.0800205 PMID:19440500

Draper, A., Cullinan, P., Campbell, C., Jones, M., & Newman Taylor, A. (2003). Occupational asthma from fungicides fluazinam and chlorothalonil. *Occupational and Environmental Medicine*, *60*(1), 76–77. doi:10.1136/oem.60.1.76 PMID:12499462

Droste, R. (1996). *Theory and Practice of Water and Wastewater Treatment*. New York: John Wiley & Sons.

Dua, M., Singh, A., Sethunathan, N., & Johri, A. (2002). Biotechnology and bioremediation: Successes and limitations. *Applied Microbiology and Biotechnology*, *59*(2-3), 143–152. doi:10.100700253-002-1024-6 PMID:12111139

Du, D., Wang, J., Wang, L., Lu, D., Smith, J. N., Timchalk, C., & Lin, Y. (2011). Magnetic electrochemical sensing platform for biomonitoring of exposure to organophosphorus pesticides and nerve agents based on simultaneous measurement of total enzyme amount and enzyme activity. *Analytical Chemistry*, *83*(10), 3770–3777. doi:10.1021/ac200217d PMID:21462919

Dudek, A. Z., Arodz, T., & Gálvez, J. (2006). Computational methods in developing quantitative structure-activity relationships (QSAR): A review. *Combinatorial Chemistry & High Throughput Screening*, *9*(3), 213–228. doi:10.2174/138620706776055539 PMID:16533155

Dugas, J., Nieuwenhuijsen, M. J., Martinez, D., Iszatt, N., Nelson, P., & Elliott, P. (2010). Use of biocides and insect repellents and risk of hypospadias. *Occupational and Environmental Medicine*, *67*(3), 196–200. doi:10.1136/oem.2009.047373 PMID:19951933

Durhami, W. F., & … . (1965). Organophosphorus insecticides and mental alertness. *Archives of Environmental & Occupational Health*, *10*(1), 55–66. doi:10.1080/00039896.1965.10663953 PMID:14219489

Durovic, R., & Dordevic, T. (2011). Modern Extraction Techniques for Pesticide Residues Determination in Plant and Soil Samples. In M. Stoytcheva (Ed.), *Pesticides in the Modern World Trends in Pesticides Analysis* (pp. 221–246). Rijeka, Croatia: InTech. doi:10.5772/17312

Dworak, V., Selbeck, J., Dammer, K.-H., Hoffmann, M., Akbar Zarezadeh, A., & Bobda, C. (2013). Strategy for the development of a smart NDVI camera system for outdoor plant detection and agricultural embedded systems. *Sensors (Basel), 13*(2), 1523–1538. doi:10.3390130201523 PMID:23348037

Dzudzevic Cancar, H., Soylemez, S., Akpinar, Y., Kesik, M., Göker, S., Gunbas, G., ... Toppare, L. (2016). A Novel Acetylcholinesterase Biosensor: Core-Shell Magnetic Nanoparticles Incorporating a Conjugated Polymer for the Detection of Organophosphorus Pesticides. *ACS Applied Materials & Interfaces, 8*(12), 8058–8067. doi:10.1021/acsami.5b12383 PMID:26956086

Echavia, G. R. M., Matzusawa, F., & Negishi, N. (2009). Photocatalytic degradation of organophosphate and phosphonoglycine pesticides using TiO2 immobilized on silica gel. *Chemosphere, 76*(5), 595–600. doi:10.1016/j.chemosphere.2009.04.055 PMID:19481779

Ecobichon, D. J. (1991). Toxic effects of pesticides. In M. O. Amdur & ... (Eds.), *Casarett and Doull's Toxicology- The basic science of poisons* (4th ed.; pp. 565–622). New York: McGraw-Hill, Inc.

Edell, A., & Morrison, G. M. (1997). Critical Evaluation of Pesticides in the Aquatic Environment and their Removal from Drinking Water. *Sweden Vatten, 53*, 355-364.

Edward, B. (2005). The Bhopal disaster and its aftermath: A review. *Environmental Health, 4*(1), 6. doi:10.1186/1476-069X-4-6 PMID:15882472

Edwards, C. (2013). *Environmental Pollution by Pesticides*. Springer Science & Business Media.

Edwards, C. A. (1973). *Persistent Pesticides in the Environment* (2nd ed.). CRC Press.

Edwards, C. A. (1993). The impact of pesticides on the environment. In *The pesticide question* (pp. 13–46). Boston, MA: Springer. doi:10.1007/978-0-585-36973-0_2

EEA. (2012). *Climate change. City*. European Environment Agency.

Efird, J. T., Holly, E. A., Preston-Martin, S., Mueller, B. A., Lubin, F., Filippini, G., ... Bracci, P. M. (2003). Farm-related exposures and childhood brain tumours in seven countries: Results from the SEARCH International Brain Tumour Study. *Paediatric and Perinatal Epidemiology, 17*(2), 201–211. doi:10.1046/j.1365-3016.2003.00484.x PMID:12675788

Eid, H. M., El-Marsafawy, S. M., & Ouda, S. A. (2007). Assessing the economic impacts of climate change on agriculture in Egypt: A Ricardian approach. *World Bank Policy Research Working Paper*, 4293.

EIFAC. (1968). *Water Quality Criteria for European Freshwater Fish: Report on Extreme pH Values and Inland Fisheries*. Rome: European Inland Fisheries Advisory Commission.

El Bakouri, H., Usero, J., Morillo, J., & Ouassini, A. (2009). Adsorptive features of acid-treated olive stones for drin pesticides: Equilibrium, kinetic and thermodynamic modeling studies. *Bioresource Technology, 100*(18), 4147–4155. doi:10.1016/j.biortech.2009.04.003 PMID:19409776

El Bakouri, H., Usero, J., Morillo, J., Rojas, R., & Ouassini, A. (2009). Drin pesticides removal from aqueous solutions using acid-treated date stones. *Bioresource Technology, 100*(10), 2676–2684. doi:10.1016/j.biortech.2008.12.051 PMID:19186054

El Lakwah, F., Meuser, F., Suckow, P., Abdel Gawaad, A. A., & Abdel Karim, K. (1989). Organochlorine insecticide residues in potatoes used for consumption in Egypt. In *Abstracts of the 3rd World Conference on Environmental and Health Hazards of Pesticides*. Academic Press.

Elliott, S. M., & VanderMeulen, D. D. (2017). A regional assessment of chemicals of concern in surface waters of four Midwestern United States national parks. *The Science of the Total Environment*, *579*, 1726–1735. doi:10.1016/j.scitotenv.2016.11.114 PMID:27932214

ENSAA. (2011). *Defining climate change*. Author.

Environews Forum. Killer environment. (1999). *Environ Health Perspect, 107*, A62-A63. Retrieved from http://www.ncbi.nlm.nih.gov/pmc/articles/PMC1566330/?page=1

Environmental Databases. Ecotoxicity Database. (2006). Pesticides: Science and Policy. Washington, DC: U.S. Environmental Protection Agency (EPA).

EPA. (2009). *Types of Pesticides*. EPA.

EPA. (2009). *Types of Pesticides*. Retrieved from https://en.wikipedia.org/wiki/Pesticide

EPA. (2011). *Pesticides industry sales and usage*. Retrieved from https://en.wikipedia.org/wiki/Environmental_impact_of_pesticides

Erdogrul, O., Covaci, A., Kurtul, N., & Schepens, P. (2004). Levels of organohalogenated persistent pollutants in human milk from Kahramanmaraş region, Turkey. *Environment International*, *30*(5), 659–666. doi:10.1016/j.envint.2003.12.004 PMID:15051242

Erhart, E., & Hartl, W. (2010). Compost Use in Organic Farming. In E. Lichtfouse (Ed.), Genetic Engineering, Biofertilisation, Soil Quality and Organic Farming, Sustainable Agriculture Reviews. Springer. Doi:10.1007/978-90-481-8741-6-2

Eriksson, P. (1996). Developmental neurotoxicology in the neonate-effects of pesticides and polychlorinated organic substances. *Archives of Toxicology. Supplement. Archiv fur Toxikologie. Supplement*, *18*, 81–88.

Eskenazi, B., Bradman, A., & Castorina, R. (1999). Exposures of children to organophosphate pesticides and their potential adverse health effects. *Environmental Health Perspectives*, *1*(Suppl 3), 409–419. doi:10.1289/ehp.99107s3409 PMID:10346990

Eskenazi, B., Harley, K., Bradman, A., Weltzien, E., Jewell, N. P., Barr, D. B., ... Holland, N. T. (2004). Association of in utero organophosphate pesticide exposure and fetal growth and length of gestation in an agricultural population. *Environmental Health Perspectives*, *112*(10), 1116–1124. doi:10.1289/ehp.6789 PMID:15238287

Eskenazi, B., & Maizlish, N. (1988). Effects of occupational exposure to chemicals on neurobehaviourial functioning. In R. E. Tarter (Ed.), *Neuropsychological disorder in neural illness* (pp. 223–264). New York: Plenum Press.

Eskenazi, B., Marks, A. R., Bradman, A., Fenster, L., Johnson, C., Barr, D. B., & Jewell, N. P. (2006). In utero exposure to dichlorodiphenyltrichloroethane (DDT) and dichlo-rodiphenyldichloroethylene (DDE) and neurodevelopment among young Mexican American children. *Pediatrics*, *118*(1), 233–241. doi:10.1542/peds.2005-3117 PMID:16818570

Eskenazi, B., Marks, A. R., Bradman, A., Harley, K., Barr, D. B., Johnson, C., & Jewell, N. P. (2007). Organophosphate pesticide exposure and neurodevelopment in young Mexican-American children. *Environmental Health Perspectives*, *115*(5), 792–798. doi:10.1289/ehp.9828 PMID:17520070

Eurek, A. (2009). *New 'green' pesticides are first to exploit plant defenses in battle of the fungi.* Academic Press.

Evangelou, E., Ntritsos, G., Chondrogiorgi, M., Kavvoura, F. K., Hernández, A. F., Ntzani, E. E., & Tzoulaki, I. (2016). Exposure to pesticides and diabetes: A systematic review and meta-analysis. *Environment International*, *91*, 60–68. doi:10.1016/j.envint.2016.02.013 PMID:26909814

Falcon, M., Oliva, J., Osuna, E., Barba, A., & Luna, A. (2004). HCH and DDT residues in human placentas in Murcia (Spain). *Toxicology*, *195*(2-3), 203–208. doi:10.1016/j.tox.2003.10.009 PMID:14751675

Fan, S., Zhang, F., Deng, K., Yu, C., Liu, S. M., Zhao, P., & Pan, C. (2013). Spinach or Amaranth contains highest residue of metalaxyl, fluazifop-p-butyl, chlorfyrifos, and lamda-cyhalothrin on six leaf vegetables upon open field application. *Journal of Agricultural and Food Chemistry*, *61*(9), 2039–2044. doi:10.1021/jf304710u PMID:23387923

Fan, Y., Lai, K., Rasco, B. A., & Huang, Y. (2015). Determination of carbaryl pesticide in Fuji apples using surface-enhanced Raman spectroscopy coupled with multivariate analysis. *Lebensmittel-Wissenschaft + Technologie*, *60*(1), 352–357. doi:10.1016/j.lwt.2014.08.011

FAO. (2013). *Pesticides as water pollutants in Control of Water Pollution from Agriculture*. Food and Agriculture Organization of the United Nations.

FAO. (2016). *FAOSTAT Database*. Available at http://faostat3.fao.org/browse/R/RP/E

FAO/WHO. (1988). Guidelines for predicting the dietary intake of pesticide residues. *Bulletin of the World Health Organization*, *66*, 429–434. PMID:3262445

FAO/WHO. (2010). *FAO Specifications and Evaluations for Agricultural Pesticides*. Endosulfan FAO/WHO evaluation report 89/2010.

Farhang, L., Weintraub, J. M., Petreas, M., Eskenazi, B., & Bhatia, R. (2005). Association of DDT and DDE with birth weight and length of gestation in the child health and development studies, 1959–1967. *American Journal of Epidemiology*, *162*(8), 717–725. doi:10.1093/aje/kwi276 PMID:16120698

Farr, S. L., Cooper, G. S., Cai, J., Savitz, D. A., & Sandler, D. P. (2004). Pesticide use and menstrual cycle characteristics among premenopausal women in the agricultural health study. *American Journal of Epidemiology*, *160*(12), 1194–1204. doi:10.1093/aje/kwi006 PMID:15583372

Fateixa, S., Raposo, M., Nogueira, H. I. S., & Trindade, T. (2018). A general strategy to prepare SERS active filter membranes for extraction and detection of pesticides in water. *Talanta*, *182*, 558–566. doi:10.1016/j.talanta.2018.02.014 PMID:29501193

Faust, S. D., & Aly, O. M. (Eds.). (1997). *Chemistry of Water Treatment* (2nd ed.). Denver, CO: American Water Works Association.

Fawell, J., & Nieuwenhuijsen, M. J. (2003). Contaminants in drinking waterEnvironmental pollution and health. *British Medical Bulletin*, *68*(1), 199–208. doi:10.1093/bmb/ldg027 PMID:14757718

Fenelon, J. M. (1998). Water quality in the White River Basin, Indiana, 1992–96. Reston, VA: USGS. *Geological Survey Circular*, 1150.

Fenik, J., Tankiewicz, M., & Biziuk, M. (2011). Properties and determination of pesticides in fruits and vegetables. *Trends in Analytical Chemistry*, *30*(6), 814–826. doi:10.1016/j.trac.2011.02.008

Fenske, R. A., & Day, E. W. (2005). Assessment of exposure for pesticide handlers in agricultural, residential and institutional environments. In C. A. Franklin & J. P. Worgan (Eds.), *Occupational and Residential Exposure Assessment for Pesticides. John Wiley & Sons; Chichester* (pp. 13–43). Academic Press. doi:10.1002/0470012218.ch1

Ferguson, A., Penney, R., & Solo-Gabriele, H. (2017). A Review of the Field on Children's Exposure to Environmental Contaminants: A Risk Assessment Approach. *International Journal of Environmental Research and Public Health*, *14*(3), 265–289. doi:10.3390/ijerph14030265 PMID:28273865

Fernández-Alba, A. R., Guil, L. H., López, G. D., & Chisti, Y. (2001). Toxicity of pesticides in wastewater: A comparative assessment of rapid bioassays. *Analytica Chimica Acta*, *426*(2), 289–301. doi:10.1016/S0003-2670(00)00874-6

Fernandez-Casalderrey, A., Ferrando, M. D., & Andreu-Moliner, E. (1992). Filtration and ingestion rates of *Brachionus calyciflorus* after exposure to endosulfan and diazinon. *Comparative Biochemistry and Physiology, C*, 103, 357–361.

Fernandez-Casalderrey, A., Ferrando, M. D., & Andreu-Moliner, E. (1994). Effect of sublethal concentrations of pesticides on the feeding behavior of *Daphnia magna*. *Ecotoxicology and Environmental Safety*, *27*(1), 82–89. doi:10.1006/eesa.1994.1008 PMID:7525207

Fernandez-Gomez, C., Antonio Lopez-Lopez, J., Matamoros, V., Diez, S., Garcia-Vargas, M., & Moreno, C. (2013). Atmospheric influence on the distribution of organic pollutants in the Guadalquivir River estuary, SW Spain. *Environmental Monitoring and Assessment*, *185*(4), 3209–3218. doi:10.100710661-012-2784-9 PMID:23010894

Fernando, C. H., Furtado, J. I., & Lim, R. P. (1980). The ecology of rice fields with special reference to aquatic fauna. In J.I. Furtado (Ed.), *Tropical Ecology and Development: Proceedings of the International Society of Tropical Ecology* (pp. 943-951). Academic Press.

Ferreira, J. D., & Couto, A. C. (2013). Pombo-de-Oliveira MS, Koifman S. Brazilian Collaborative Study Group of Infant Acute Leukemia. In utero pesticide exposure and leukemia in Brazilian children <2 years of age. *Environmental Health Perspectives*, *121*, 269–275. PMID:23092909

Ferrusquía, C. H., Roa, G., García, M., Amaya, A., & Pavón, T. (2008). Evaluación de la degradación de metilparatión en soluciónusandofotocatálisisheterogénea. Rev. *Latinoamericana de RecursosNaturales*, *4*(2), 285–290.

FICCI. (2015). *Ushering in the 2nd Green Revolution: Role of crop protection chemicals. A report on Indian agrochemical industry*. Author.

Ficklin, D. L., Luo, Y., Luedeling, E., Gatzke, S. E., & Zhang, M. (2010). Sensitivity of agricultural runoff loads to rising levels of CO2 and climate change in the San Joaquin Valley watershed of California. *Environmental Pollution*, *158*(1), 223–234. doi:10.1016/j.envpol.2009.07.016 PMID:19660846

Fillâtre, Y., Gray, F. X., & Roy, C. (2017). Pesticides in essential oils: Occurrence and concentration in organic and conventional orange essential oils from eleven geographical origins. *Analytica Chimica Acta*, *992*, 55–66. doi:10.1016/j.aca.2017.08.039 PMID:29054150

Finlayson, C. M., & Davidson, N. C. (1999). Global review of wetland resources and priorities for wetland inventory. *Preface iv Summary Report*, 15.

Fishel, M. (2009). *Pest Management and Pesticides: A Historical Perspective*. UF/IFAS Extension. Retrieved from http://edis.ifas.ufl.edu/

Fishel, F. M. (2005). *43 Document. Pesticide-Organism Interactions, PI: Pesticide Information Office, Florida Cooperative Extension Service*. Institute of Food and Agricultural Sciences, University of Florida. Retrieved from http://edis.ifas.ufl.edu/pdffiles/PI/PI0800.pdf

Fitches, E., Edwards, M. G., Mee, C., Grishin, E., Gatehouse, A. M., Edwards, J. P., & Gatehouse, J. A. (2004). Fusion proteins containing insect-specific toxins as pest control agents: Snowdrop lectin delivers fused insecticidal spider venom toxin to insect haemolymph following oral injection. *Journal of Insect Physiology*, *50*(1), 61–71. doi:10.1016/j.jinsphys.2003.09.010 PMID:15037094

Flessel, P., Quintana, P. J. E., & Hopper, K. (1993). Genetic toxicity of malathion: A review. *Environmental and Molecular Mutagenesis*, *22*(1), 7–17. doi:10.1002/em.2850220104 PMID:8339727

Flury, M., Leuenberger, J., Studer, B., Fluhler, H., Jury, W. A., & Roth, K. (1994). *Pesticide transport through unsaturated field soils:Preferential flow*. Zurich: ETH.

Folnovic, T. (2016). *Best Practices for Pest Management in Organic Farm Production*. Agrivi.

Foo, K.Y., & Hameed, B.H. (2010). *Detoxification ofpesticide waste via activated carbon adsorption process*. Academic Press.

Foo, G. S. (1985). *The pesticide poisoning report: A survey of some Asian countries*. Kuala Lumpur, Malaysia: International Organization of Consumers' Unions.

Forget, G. (1993). Balancing the need for pesticides with the risk to human health. In Impact of pesticide use on health in developing countries: proceedings of a symposium, Ottawa, Canada.

Forget, G. (1993). Balancing the need for pesticides with the risk to human health. In *Impact of Pesticide Use on Health in Developing Countries* (Vol. 2). Ottawa: IDRC.

Fox, G. A., Collins, B., Hayakawa, E., Weseloh, D. V., Ludwig, J. P., Kubiak, T. J., & Erdman, T. C. (1991). Reproductive outcomes in colonial fish-eating birds: a biomarker for developmental toxicants in Great Lakes food chains: II. Spatial variation in the occurrence and prevalence of bill defects in young double-crested cormorants in the Great Lakes, 1979-1987. *Journal of Great Lakes Research*, *17*(2), 158–167. doi:10.1016/S0380-1330(91)71353-1

Francisco, P., Pierre, L., Claude, R., & Jean-P, V. (1988). Modeling the transport and the fate of pesticides in the unsaturated zone considering temperature effects. *Ecological Modelling*, *44*(1-2), 73–88. doi:10.1016/0304-3800(88)90083-X

Frankart, C., Eullaffroy, P., & Vernet, G. (2003). Comparative effects of four herbicides on non-photochemical fluorescence quenching in Lemna minor. *Environmental and Experimental Botany*, *49*(2), 159–168. doi:10.1016/S0098-8472(02)00067-9

Freeman, H. M. (Ed.). (1998). *Standard Handbook of Hazardous Waste Treatment and Disposal* (2nd ed.). New York: McGraw-Hill.

Freitas, L. G., Gotz, C. W., Ruff, M., Singer, H. P., & Muller, S. R. (2004). Quantification of the new triketone herbicides, sulcotrione and mesotrione, and other important herbicides and metabolites, at the ng/l level in surface waters using liquid chromatography-tandem mass spectrometry. *Journal of Chromatography. A*, *1028*(2), 277–286. doi:10.1016/j.chroma.2003.11.094 PMID:14989481

Friberg-Jensen, U., Wendt-Rasch, L., Woin, P., & Christoffersen, K. (2003). Effects of the pyrethroid insecticide, cypermethrin, on a freshwater community studied under field conditions. I. Direct and indirect effects on abundance measures of organisms at different trophic levels. *Aquatic Toxicology (Amsterdam, Netherlands)*, *63*(4), 357–371. doi:10.1016/S0166-445X(02)00201-1 PMID:12758002

Friberg, N., Lindstrom, M., Kronvang, B., & Larsen, S. E. (2003). Macroinvertebrate/sediment relationships along a pesticide gradient in Danish streams. *Hydrobiologia*, *494*(1-3), 103–110. doi:10.1023/A:1025493627272

Friesen-Pankratz, B. B., Doebel, C. C., Farenhorst, A. A., & Gordon Goldsborough, L. (2003). Interactions between algae (Selenastrum capricornutum) and pesticides: Implications for managing constructed wetlands for pesticide removal. *Journal of Environmental Science and Health. Part. B, Pesticides, Food Contaminants, and Agricultural Wastes*, *38*(2), 147–155. doi:10.1081/PFC-120018445 PMID:12617553

Fry, D. M. (1995). Reproductive effects in birds exposed to pesticides and industrial chemicals. *Environmental Health Perspectives*, *103*(Suppl 7), 165–171. doi:10.1289/ehp.95103s7165 PMID:8593865

Fulton, M. H., & Key, P. B. (2001). Acetylcholinesterase inhibition in estuarine fish and invertebrates as an indicator of organophosphorus insecticide exposure and effects. *Environmental Toxicology and Chemistry*, *20*(1), 37–45. doi:10.1002/etc.5620200104 PMID:11351414

Furini, L. N., Constantino, C. J. L., Sanchez-Cortes, S., Otero, J. C., & López-Tocón, I. (2016). Adsorption of carbendazim pesticide on plasmonic nanoparticles studied by surface-enhanced Raman scattering. *Journal of Colloid and Interface Science*, *465*, 183–189. doi:10.1016/j.jcis.2015.11.045 PMID:26674234

Furuhama, A., Hayashi, T. I., & Tatarazako, N. (2016). Acute to chronic estimation of Daphnia magna toxicity within the QSAAR framework. *SAR and QSAR in Environmental Research*, *27*(10), 833–850. doi:10.1080/1062936X.2016.1243151 PMID:27767343

Gaines, T. B. (1969). Acute toxicity of pesticides. *Toxicology and Applied Pharmacology*, *14*(3), 515–534. doi:10.1016/0041-008X(69)90013-1 PMID:5787520

Gallagher, L. M., & Miller, L. A. (1996). *Clean Water Handbook* (2nd ed.). Rockville, MD: Government Institutes.

Galvez, M. P., & Balk, S. J. (2017). Environmental Risks to Children: Prioritizing Health Messages in Pediatric Practice. *Pediatrics*, *38*(6), 263–279. PMID:28572135

Ganesan, M., Natesan, S., & Ranganathan, V. (1979). Spectrophotometric method for the determination of paraquat. *Analyst (London)*, *104*(1236), 258–261. doi:10.1039/an9790400258

Gangesh, S., Anoop, K., Mayank, B., Joseph, P. E., & Ajay, T. (2006). Organochlorine pesticide contamination of ground water in the city of Hyderabad. *Environment International*, *32*(2), 244–247. doi:10.1016/j.envint.2005.08.027 PMID:16183122

Gan, J., Papiernik, S. K., Koskinen, W. C., & Yates, S. R. (1999). Evaluation of Accelerated Solvent Extraction (ASE) for Analysis of Pesticide Residues in Soil. *Environmental Science & Technology*, *33*(18), 3249–3253. doi:10.1021/es990145+

Gao, F., Jia, J., & Wang, X. (2008). Occurrence & ordination of DDT & HCH in agricultural soils from Guangzhou, China. *Archives of Environmental Contamination and Toxicology*, *54*, 155–166. doi:10.100700244-007-9023-3 PMID:17763882

Gao, F., & Lu, X. (2015). Detection of pesticides in foods by enzymatic biosensors. In *Improving and Tailoring Enzymes for Food Quality and Functionality*. Cambridge, UK: Woodhead publishing. doi:10.1016/B978-1-78242-285-3.00007-7

García de Llasera, M. P., & Bernal-González, M. (2001). Presence of carbamate pesticides in environmental waters from the northwest of Mexico: Determination by liquid chromatography. *Water Research*, *35*(8), 1933–1940. doi:10.1016/S0043-1354(00)00478-4 PMID:11337839

Garcia, D. (1991, July 17). Spill threatening Shasta: Response to accident was slow. *San Francisco Chronicle,* pp. A1-4.

Garcia, A. M., Benavides, F. G., Fletcher, T., & Orts, E. (1998). Paternal exposure to pesticides and congenital malformations. *Scandinavian Journal of Work, Environment & Health*, *24*(6), 473–480. doi:10.5271jweh.371 PMID:9988089

Garcia, F. P., Ascencio, S. Y. C., Oyarzun, J. C. G., Hernandez, A. C., & Alavarado, P. V. (2012). Pesticides: Classification, uses and toxicity. Measures of exposure and genotoxic risks. *Int. J. Environ. Sci. Toxic. Res*, *1*, 279–293.

Garg, J. K., Singh, T. S., & Murthy, T. V. R. (1998). *Wetlands of India. SAC.* Ahmadabad: Indian Space Research Organisation.

Garry, V. F. (2004). Pesticides and children. *Toxicology and Applied Pharmacology*, *198*(2), 152–163. doi:10.1016/j.taap.2003.11.027 PMID:15236951

Gasnier, C., Dumont, C., Benachour, N., Clair, E., Chagnon, M. C., & Séralini, G. E. (2009). Glyphosate-based herbicides are toxic and endocrine disruptors in human cell lines. *Toxicology*, *262*(3), 184–191. doi:10.1016/j.tox.2009.06.006 PMID:19539684

Gaspari, L., Paris, F., Jandel, C., Kalfa, N., Orsini, M., Daurès, J. P., & Sultan, C. (2011). Prenatal environmental risk factors for genital malformations in a population of 1442 French male newborns: A nested casecontrol study. *Human Reproduction (Oxford, England)*, *26*(11), 3155–3162. doi:10.1093/humrep/der283 PMID:21868402

Gaspari, L., Sampaio, D. R., Paris, F., Audran, F., Orsini, M., Neto, J. B. S., & Sultan, C. (2012). High prevalence of micropenis in 2710 male newborns from an intensive-use pesticide area of northeastern Brazil. *International Journal of Andrology*, *35*(3), 253–264. doi:10.1111/j.1365-2605.2011.01241.x PMID:22372605

Gates, D. J. (1998). *The Chlorine Dioxide Handbook*. Denver, CO: American Water Works Association.

Gaur, M. S., Sharma, A. K., Sharma, P., Mishra, V., & Tiwari, R. K. (2008). Spectrophotometric assessments of methyl parathion in water samples. *Can J Pure ApplSci*, *2*, 581–587.

Gautami, J., Sandhyasree, M., & Revathi, B. (2015). Bio-pesticides Approach for Pest Management. *Journal of Agriculture and Allied Sciences*, *4*(1), 1–5.

Gavrilescu, M. (2005). Fate of pesticides in the environment and its bioremediation. *Engineering in Life Sciences*, *5*(6), 497–526. doi:10.1002/elsc.200520098

Geissen, V., Mol, H., Klumpp, E., Umlauf, G., Nadal, M., van der Ploeg, M., ... Ritsema, C. J. (2015). Emerging pollutants in the environment: A challenge for water resource management. *International Soil and Water Conservation Research*, *3*(1), 57–65. doi:10.1016/j.iswcr.2015.03.002

Geldreich, E. (1996). *Microbial Quality of Water Supply in Distribution Systems*. Boca Raton, FL: CRC Press/Lewis Publishers.

Gentz, M. C., Murdoch, G., & King, G. F. (2010). Tandem use of selective insecticides and natural enemies for effective, reduced-risk pest management. *Biological Control*, *52*(3), 208–215. doi:10.1016/j.biocontrol.2009.07.012

George, J., & Shukla, Y. (2011). Pesticides and cancer: Insights into toxicoproteomic-based findings. *Journal of Proteomics*, *74*(12), 2713–2722. doi:10.1016/j.jprot.2011.09.024 PMID:21989265

Gerber, F., Krummen, M., Potgeter, H., Roth, A., Siffrin, C., & Spoendlin, C. (2004). Practical aspects of fast reversed-phase high-performance liquid chromatography using 3 μm particle packed columns and monolithic columns in pharmaceutical development and production working under current good manufacturing practice. *Journal of Chromatography. A*, *1036*(2), 127–133. doi:10.1016/j.chroma.2004.02.056 PMID:15146913

Gerhardson, B. (2002). Biological substitutes for pesticides. *Trends in Biotechnology*, *20*(8), 338–343. doi:10.1016/S0167-7799(02)02021-8 PMID:12127281

Gevao, B., & Jones, K. C. (2002). Pesticides and persistent organic chemicals. In P. M. Haygarth & S. C. Jarvis (Eds.), *Agriculture, Hydrology and Water Quality*. Wallingford, UK: CABI Publishing. doi:10.1079/9780851995458.0083

Ghamali, M., Chtita, S., Ousaa, A., Elidrissi, B., Bouachrine, M., & Lakhlifi, T. (2017). QSAR analysis of the toxicity of phenols and thiophenols using MLR and ANN. *Journal of Taibah University for Science*, *11*(1), 1–10. doi:10.1016/j.jtusci.2016.03.002

Ghazy, H. A., Abdel-Razek, M. A. S., El Nahas, A. F., & Mahmoud, S. (2017). Assessment of complex water pollution with heavy metals and Pyrethroid pesticides on transcript levels of metallothionein and immune related genes. *Fish & Shellfish Immunology*, *68*, 318–326. doi:10.1016/j.fsi.2017.07.034 PMID:28734967

Gianessi, L. P., & Puffer, C. (1991). *Herbicide Use in the United States*. Washington, DC: Resources for the Future.

Gilbertson, M., Kubiak, T., Ludwig, J., & Fox, G. (1991). Great Lakes embryo mortality, edema and deformities syndrome (GLEMEDS) in clonial fish-eating birds: Similarity to chick-edema disease. *Journal of Toxicology and Environmental Health*, *33*(4), 455–520. doi:10.1080/15287399109531538 PMID:1875430

Gilden, R. C., Huffling, K., & Sattler, B. (2010). Pesticides and health risks. *Journal of Obstetric, Gynecologic, and Neonatal Nursing*, *39*(1), 103–110. doi:10.1111/j.1552-6909.2009.01092.x PMID:20409108

Gilliom, R. J. (2001). Pesticides in the hydrologic system - What do we know and what's next? *Hydrological Processes*, *15*(16), 3197–3201. doi:10.1002/hyp.501

Gilman, A. P., Peakall, D. B., Hallett, D. J., Fox, G. A., & Norstrom, R. J. (1979). *Herring gulls (Larus argentatus) as monitors of contamination in the Great Lakes. Animals as monitors of environmental pollutants/sponsored by Northeastern Research Center for Wildlife Diseases, University of Connecticut, Registry of Comparative of Lab Animal Resources*. National Academy of Sciences.

Goad, R. T., Goad, J. T., Atieh, B. H., & Gupta, R. C. (2004). Carbofuran-induced endocrine disruption in adult male rats. *Toxicology Mechanisms and Methods*, *14*(4), 233–239. doi:10.1080/15376520490434476 PMID:20021136

Goedkoop, W., & Peterson, M. (2003). The fate, distribution, and toxicity of lindane in tests with Chironomus riparius: Effects of bioturbation and sediment organic matter content. *Environmental Toxicology and Chemistry*, *22*(1), 67–76. doi:10.1002/etc.5620220109 PMID:12503748

Goel, A., McConnell, L. L., & Torrents, A. (2005). Wet deposition of current use pesticides at a rural location on the Delmarva Peninsula: Impact of rainfall patterns and agricultural activity. *Journal of Agricultural and Food Chemistry*, *53*(20), 7915–7924. doi:10.1021/jf0507700 PMID:16190650

Goldman, J. G., Caron, D. A., & Dennett, M. R. (1987). Nutrient cycling in a microflagellate food chain. IV. Phytoplankton-microflagel-late interactions. *Marine Ecology Progress Series*, *38*, 75–87. doi:10.3354/meps038075

Goldman, L. R. (2007). Managing pesticide chronic health risks: U.S. policies. *Journal of Agromedicine*, *12*(1), 67–75. doi:10.1300/J096v12n01_07 PMID:18032337

Golge, O., & Kabak, B. (2018). Pesticide residues in table grapes and exposure assessment. *Journal of Agricultural and Food Chemistry*, *66*(7), 1701–1713. doi:10.1021/acs.jafc.7b05707 PMID:29364655

Gonçalves, C., Esteves da Silva, J. C. G., & Alpendurada, M. F. (2006). Chemometric interpretation of pesticide occurrence in soil samples from an intensive horticulture area in north Portugal. *Analytica Chimica Acta*, *560*(1-2), 164–171. doi:10.1016/j.aca.2005.12.021

González-Martin, M. I., Revilla, I., Vivar-Quintana, A. M., & Salcedo, E. B. (2017). Pesticide residues in propolis from Spain and Chile. An approach using near infrared spectroscopy. *Talanta*, *165*, 533–539. doi:10.1016/j.talanta.2016.12.061 PMID:28153295

Gopal, B., & Sah, M. (1995). Inventory and classification of wetlands in India. In *Classification and Inventory of the World's Wetlands* (pp. 39–48). Springer Netherlands. doi:10.1007/978-94-011-0427-2_5

Gopal, B., Shilpakar, R., & Sharma, E. (2010). *Functions and services of wetlands in the Eastern Himalayas: impacts of climate change. International Centre for Integrated Mountain Development*. ICIMOD.

Gorbach, S., Haaring, R., Knauf, W., & Werner, H. J. (1971). Residue analyses and biotests in ricefields of East Java treated with thiodan. *Bulletin of Environmental Contamination and Toxicology*, *6*(3), 163–199. doi:10.1007/BF01539926

Gorell, J. M., Johnson, C. C., Rybicki, B. A., Peterson, E. L., & Richardson, R. J. (1998). The risk of Parkinson's disease with exposure to pesticides, farming, well water, and rural living. *Neurology*, *50*(5), 1346–1350. doi:10.1212/WNL.50.5.1346 PMID:9595985

Gothwal, A., Beniwal, P., Dhull, V., & Hooda, V. (2014). Preparation of electrochemical biosensor for detection of organophosphorus pesticides. *International Journal of Analytical Chemistry*. PMID:25667593

Gouda, A. A., Amin, A. S., El-Sheikh, R., & Akl, M. A. (2010). Sensitive spectrophotometric methods for determination of some organophosphorus pesticides in vegetable samples. *Chemical Industry & Chemical Engineering Quarterly*, *16*(1), 11–18. doi:10.2298/CICEQ090417003G

Goulson, D., Nicholls, E., Botías, C., & Rotheray, E. L. (2015). Bee declines driven by combined stress from parasites, pesticides, and lack of flowers. *Science*, *347*(6229), 1255957. doi:10.1126cience.1255957 PMID:25721506

Gou, Y. Y., Lin, S., Que, D. E., Tayo, L. L., Lin, D. Y., Chen, K. C., ... Chuang, K. P. (2016). Estrogenic effects in the influents and effluents of the drinking water treatment plants. *Environmental Science and Pollution Research International*, *23*(9), 8518–8528. doi:10.100711356-015-5946-9 PMID:26791027

Graeme, M. (2005). *Pesticide Rotation. Ontario Ministry of Agriculture*. Food and Rural Affairs.

Gramatica, P. (2007). Principles of QSAR models validation: Internal and external. *Molecular Informatics*, *26*(5), 694–701.

Grande, M., Andersen, S., & Berge, D. (1994). Effects of pesticides on fish. *Norwegian Journal of Agricultural Sciences*, *13*, 195–209.

Grant, I. F., Egan, E. A., & Alexander, M. (1983). Pesticides to control ostracods grazing on blue-green algae. *Soil Biology & Biochemistry*, *15*(2), 193–197. doi:10.1016/0038-0717(83)90102-5

Grant, I. F., Roger, P. A., & Watanabe, I. (1986). Ecosystem manipulation for increasing biological nitrogen fixation by blue-green algae (Cyanobacteria) in lowland ricefields. *Biological Agriculture and Horticulture*, *3*(2-3), 299–315. doi:10.1080/01448765.1986.9754477

Grant, I. F., Tirol, A. C., Aziz, T., & Watanabe, I. (1983). Regulation of invertebrate grazers as a means to enhance biomass and nitrogen fixation of cyanophyceae in wetland ricefields. *Soil Science Society of America Journal*, *47*(4), 669–675. doi:10.2136ssaj1983.03615995004700040013x

Greenburg, D., Rusiecki, J., Koutros, S., Dosemeci, M., Patel, R., Hines, C., ... Alavanja, M. (2008). Cancer incidence among pesticide applicators exposed to captan in the Agricultural Health Study. *Cancer Causes & Control*, *19*(10), 1401–1407. doi:10.100710552-008-9187-9 PMID:18587656

Greenop, K. R., Peters, S., Bailey, H. D., Fritschi, L., Attia, J., Scott, R. J., & Milne, E. (2013). Exposure to pesticides and the risk of childhood brain tumors. *Cancer Causes & Control*, *24*(7), 1269–1278. doi:10.100710552-013-0205-1 PMID:23558445

Grewal, P. S., Lewis, E. E., & Gaugler, R. (1997). Response of infective stage parasites (Nematoda: Steinernermatidae) to volatile from infected hosts. *Journal of Chemical Ecology*, *23*(2), 503–515. doi:10.1023/B:JOEC.0000006374.95624.7e

Gribble, G. W. (1994). The natural production of chlorinated compounds. *Environmental Science & Technology*, *28*(7), 310A–319A. doi:10.1021/es00056a712 PMID:22662801

Griffiths, R. W. (1991). Environmental-quality assessment of the St Clair River as refiected by the distribution of benthic macroinvertebrates in 1985. *Hydrobiologia*, *219*(1), 143–164. doi:10.1007/BF00024752

Grigarick, A. A., Webster, R. K., Meyer, R. P., Zalom, F. G., & Smith, K. A. (1990). Effects of pesticide treatments on non-target organisms in California rice paddies. *Hilgardia*, *58*, 1-36. doi:10.3733/hilg.v58n01p036

Grob, D., & Harvey, A. M. (1953). Effects and treatment of nerve gas poisoning. *The American Journal of Medicine*, *14*(1), 52–58. doi:10.1016/0002-9343(53)90358-1 PMID:13016586

Grossmann, K. (2000). Mode of action of auxin herbicides: A new ending to a long, drawn out story. *Trends in Plant Science*, *5*(12), 506–508. doi:10.1016/S1360-1385(00)01791-X PMID:11120460

Gruère, G. P., & Rao, S. R. (2007). *A review of international labeling policies of genetically modified food to evaluate India's proposed rule*. Academic Press.

Guan, H., Brewer, W. E., Garris, S. T., & Morgan, S. L. (2010). Disposable pipette extraction for the analysis of pesticides in fruit and vegetables using gas chromatography/mass spectrometry. *Journal of Chromatography. A*, *1217*(12), 1867–1874. doi:10.1016/j.chroma.2010.01.047 PMID:20144461

Guillette, L. J. Jr, Crain, D. A., Rooney, A. A., & Pickford, D. B. (1995). Organization versus activation: The role of endocrine-disrupting contaminants (EDCs) during embryonic development in wildlife. *Environmental Health Perspectives*, *103*(Suppl 7), 157–164. doi:10.1289/ehp.95103s7157 PMID:8593864

Guimaraes, R. M., & Froes Asmus, C. I. R. (2013). Organochlorine exposure and changes in primary and secondary sexual characters in Cidade dos Meninos. *Br Cad Saúde Colet*, *21*, 475–476.

Guo, Y., Meng, X. Z., Tang, H. L., & Zeng, E. Y. (2008). Tissue distribution of organochlorine pesticides in fish collected from Pearl River Delta, China: Implications for fishery input source & bioaccumulation. *Environmental Pollution*, *155*(1), 150–156. doi:10.1016/j.envpol.2007.10.025 PMID:18063460

Gupta, A. K., & Rajbanshi, V. K. (1991). Toxicity of Copper and Cadmium to Heteropneustes fossils (Bloch). *Acta Hydrochimica et Hydrobiologica*, *19*(3), 331–340. doi:10.1002/aheh.19910190313

Gupta, P. K. (2004). Pesticide exposure: Indian scene. *Toxicology*, *198*(1-3), 83–90. doi:10.1016/j.tox.2004.01.021 PMID:15138033

Gupta, V. K., & Ali, I. (2001). Removal of DDD and DDE from wastewater using bagasse fly ash, a sugar industry waste. *Water Research*, *35*(1), 33–40. doi:10.1016/S0043-1354(00)00232-3 PMID:11257890

Gupta, V. K., Ali, I., & Saini, V. K. (2006). Adsorption of 2, 4-D and carbofuran pesticides using fertilizer and steel industry wastes. *Journal of Colloid and Interface Science*, *299*(2), 556–563. doi:10.1016/j.jcis.2006.02.017 PMID:16527294

Gupta, V. K., Jain, C. K., Ali, I., Chandra, S., & Agarwal, S. (2002). Removal of lindane and malathion from wastewater using bagasse fly ash—a sugar industry waste. *Water Research*, *36*(10), 2483–2490. doi:10.1016/S0043-1354(01)00474-2 PMID:12153014

Gurdak, J. J., Hanson, R. T., McMahon, P. B., Bruce, B. W., McCray, J. E., Thyne, G. D., & Reedy, R. C. (2007). Climate variability controls on unsaturated water and chemical movement, High Plains aquifer, USA. *Vadose Zone Journal*, *6*(3), 533–547. doi:10.2136/vzj2006.0087

Gurdak, J. J., McMahon, P. B., & Bruce, B. W. (2012). Vulnerability of Groundwater Quality to Human Activity and Climate Change and Variability. In H. Treidel, J. J. Martin-Bordes, & J. J. Gurdak (Eds.), *Climate Change Effects on Groundwater Resources: A Global Synthesis of Findings and Recommendations. International Association of Hydrogeologists (IAH) – International Contributions to Hydrogeology* (pp. 145–167). Taylor & Francis Publishing.

Gurr, G. M., Wratten, S. D., & Altieri, M. A. (2004). Ecological Engineering for Pest Management: Habitat Manipulation for Arthropods. CSIRO Publishing.

Gustavsson, M., Kreuger, J., Bundschuh, M., & Backhaus, T. (2017). Pesticide mixtures in the Swedish streams: Environmental risks, contributions of individual compounds and consequences of single-substance oriented risk mitigation. *The Science of the Total Environment*, *598*, 973–983. doi:10.1016/j.scitotenv.2017.04.122 PMID:28468125

Guyton, K. Z., Loomis, D., Grosse, Y., El Ghissassi, F., Benbrahim-Tallaa, L., Guha, N., ... Straif, K. (2015). Carcinogenicity of tetrachlorvinphos, parathion, malathion, diazinon, and glyphosate. *The Lancet Oncology*, *16*(5), 490–491. doi:10.1016/S1470-2045(15)70134-8 PMID:25801782

Guzelian, P. S. (1980). Liver structure and function in patients poisoned with chlordecone. *Gastroenterology*, *78*, 206–213. PMID:6153073

Gwinn, M. R., Whipkey, D. L., Tennant, L. B., & Weston, A. (2005). Differential gene expression in normal human mammary epithelial cells treated with malathion monitored by DNA microarrays. *Environmental Health Perspectives*, *113*(8), 1046–1051. doi:10.1289/ehp.7311 PMID:16079077

Haber, F., & Weiss, J. (n.d.). The catalytic decomposition of hydrogen peroxide by iron salts. *Proc.R.Soc. Series A*, *147*, 332.

Hahn, H. H., Hoffmann, E., & Hallvard, Ÿ. (1998). *Chemical Water and Wastewater Treatment V: Proceedings of the 7th Gothenburg Symposium 1998.* Springer. Retrieved from https://www.sswm.info/sites/default/files/reference_attachments/SUBRAMANIAN%20Advanced%20Waste%20Water%20Treatment.pdf

Hameed, B. H., Salman, J. M., & Ahmad, A. L. (2009). Adsorption isotherm and kinetic modeling of 2, 4-D pesticide on activated carbon derived from date stones. *Journal of Hazardous Materials*, *163*(1), 121–126. doi:10.1016/j.jhazmat.2008.06.069 PMID:18667269

Hamilton, D. D., Ambrus, A., Dieterle, R., Felsot, A., Harris, C., Petersen, B., ... Tanaka, K. (2004). Pesticide residues in food-acute dietary exposure. *Pest Management Science*, *60*(4), 311–339. doi:10.1002/ps.865 PMID:15119595

Hamilton, D. J., Ambrus, A., Dieterle, R. M., Felsot, A. S., Harris, C. A., Holland, P. T., & Wong, S. S. (2003). Regulatory limits for pesticide residues in water (IUPAC Technical Report). *Pure and Applied Chemistry*, *75*(8), 1123–1155. doi:10.1351/pac200375081123

Hanazato, T. (2001). Pesticide effects on freshwater zooplankton: An ecological perspective. *Environmental Pollution*, *112*(1), 1–10. doi:10.1016/S0269-7491(00)00110-X PMID:11202648

Handford, C. E., Elliott, C. T., & Campbell, K. (2015). A review of the global pesticide legislation and the scale of challenge in reaching the global harmonization of food safety standards. *Integrated Environmental Assessment and Management*, *11*(4), 525–536. doi:10.1002/ieam.1635 PMID:25765969

Hanke, W., Romitti, P., Fuortes, L., Sobala, W., & Mikulski, M. (2003). The use of pesticides in a Polish rural population and its effect on birth weight. *International Archives of Occupational and Environmental Health*, *76*(8), 614–620. doi:10.100700420-003-0471-4 PMID:12955525

Hansen, S. H., &Reubsaet, L. (2015). High-Performance Liquid Chromatography (HPLC) and High-Performance Liquid Chromatography–Mass Spectrometry (LC-MS). *Bioanalysis of Pharmaceuticals: Sample Preparation, Separation Techniques, and Mass Spectrometry*, 123-172.

Hapeman, C. J., Dionigi, C. P., Zimba, P. V., & McConnell, L. L. (2002). Agrochemical and nutrient impacts on estuaries and other aquatic systems. *Journal of Agricultural and Food Chemistry*, *50*(15), 4382–4384. doi:10.1021/jf020457n PMID:12105974

Harari, R., Julvez, J., Murata, K., Barr, D., Bellinger, D. C., Debes, F., & Grandjean, P. (2010). Neurobehavioral deficits and increased blood pressure in school-age children prenatally exposed to pesticides. *Environmental Health Perspectives*, *118*(6), 890–896. doi:10.1289/ehp.0901582 PMID:20185383

Hardell, L., Walker, M. J., Walhjalt, B., Friedman, L. S., & Richter, E. D. (2007). Secret ties to industry and conflicting interests in cancer research. *American Journal of Industrial Medicine*, *50*(3), 227–233. doi:10.1002/ajim.20357 PMID:17086516

Harding, G. C. (1986). Organochlorine dynamics between zooplankton and their environment, a reassessment. *Marine Ecology Progress Series*, *33*, 167–191. doi:10.3354/meps033167

Haria, A. H., Johnson, A. C., Bell, J. P., & Batchelor, C. H. (1994). Water Movement and isoproturon behaviour in a drained heavy clay soil:1. Preferential flow processes. *Journal of Hydrology (Amsterdam)*, *163*(3-4), 203–216. doi:10.1016/0022-1694(94)90140-6

Harit, G., & Srivastava, N. (2007). Endosulfan uptake and biochemical profile of the muscle in a freshwater murrel, *Channa punctatus. Bulletin of Pure & Applied Sciences*, *26A*, 23–27.

Harit, G., & Srivastava, N. (2009). Bioaccumulation & elimination of endosulfan in muscle, gills & blood of *Channa punctatus. Journal of Environment & Sociobiology.*, *6*, 13–20.

Harit, G., & Srivastava, N. (2017a). Changes in gill structure induced by uptake of endosulfan in *Channa punctatus* followed by subsequent recovery. *Asian Journal of Applied Science and Technology*, *1*(7), 60–64.

Harit, G., & Srivastava, N. (2017b). Toxicity of endosulfan in *Channa punctatus* (Bloch). *Current Topics in Toxicology*, *13*, 23–30.

Harvell, C. D., Mitchell, C. E., Ward, J. R., Altizer, S., Dobson, A. P., & Ostfeld, R. S. (2002). Climate warming and disease risks for terrestrial and marine biota. *Science*, *296*(5576), 2158–2162. doi:10.1126cience.1063699 PMID:12077394

Hatzios, K. K. (1991). Biotransformation of herbicides in higher plants. Environmental Chemistry of Herbicides, 2, 141-185.

Hayden, K. M., Norton, M. C., Darcey, D., Ostbye, T., Zandi, J. C. S., Breitner, J. C. S., & Welsh-Bohmer, K. A. (2010). Occupational exposure to pesticides increases the risk of incident AD: The cache county study. *Neurology*, *74*(19), 1524–1530. doi:10.1212/WNL.0b013e3181dd4423 PMID:20458069

Hayes, T. B., Case, P., Chui, S., Chung, D., Haeffele, C., Haston, K., & Tsui, M. (2006). Pesticide mixtures, endocrine disruption, and amphibian declines: Are we underestimating the impact? *Environmental Health Perspectives*, *114*(Suppl 1), 40–50. doi:10.1289/ehp.8051 PMID:16818245

Hayo, M. G., & Werf, V. D. (1996). Assessing the impact of pesticides on the environment. *Agriculture, Ecosystems & Environment*, *60*(2-3), 81–96. doi:10.1016/S0167-8809(96)01096-1

Heckman, C. W. (1979). Ricefield ecology in northeastern Thailand: The effect of wet and dry seasons on a cultivated aquatic ecosystem. In J. Illies (Ed.), Monographia Biologicae (Vol. 34). The Hague, The Netherlands: Junk.

Heckman, C. W. (1994). Pesticide chemistry and toxicity to algae. In L. C. Rai, J. P. Gaur, & C. J. Soeder (Eds.), *Algae and Water Pollution* (pp. 205–234). Stuttgart, Germany: E. Schweitzerbart.

Hedlund, M. (2002). *Deltametrins toxiska effekter för Chironomus riparius - en jämförande studie mellan kontaminerat sediment och vatten. Examensarbete, Sveriges Lantbruksuniversitet.* Uppsala, Sweden: SLU. (in Swedish)

Hedman, J. E., Bradshaw, C., Thorsson, M. H., Gilek, M., & Gunnarsson, J. S. (2008). Fate of contaminants in Baltic Sea sediments: Role of bioturbation and settling organic matter. *Marine Ecology Progress Series*, *356*, 25–38. doi:10.3354/meps07218

Hegde, N. (2012). Water scarcity and security in India. *BAIF at the Indian Science Congress.*

Helfrich, L.A. (2009). *Pesticides and Aquatic animals: A guide to reducing impacts on aquatic system.* Virginia, Dept of game and inland fisheries.

Helfrich, L. A. (2009). *Pesticides and Aquatic Animals: A Guide to Reducing Impacts on Aquatic Systems.* Blacksburg, VA: Virginia Department of Game and Inland Fisheries.

Helfrich, L. A., Weigmann, D. L., Hipkins, P., & Stinson, E. R. (1996). *Pesticides and aquatic animals: A guide to reducing impacts on aquatic systems.* Virginia Cooperative Extension.

Hernandez, A. F., Parron, T., & Alarcon, R. (2011). Pesticides and asthma. *Current Opinion in Allergy and Clinical Immunology*, *11*(2), 90–96. doi:10.1097/ACI.0b013e3283445939 PMID:21368619

Hernandez, F., Pitarch, E., Beltran, J., & Lopez, F. J. (2000). Headspace solid-phase microextraction in combination with gas chromatography and tandem mass spectrometry for the determination of organochlorine and organophosphorus pesticides in whole human blood. *Journal of Chromatography. B, Biomedical Sciences and Applications*, *769*, 65–77.

Hernández, L. M., Fernández, M. A., Jiménez, B., González, M. J., & Garcia, J. F. (1994). Organochlorine pollutants in meats and cow's milk from Madrid (Spain). *Bulletin of Environmental Contamination and Toxicology*, *52*(2), 246–253. doi:10.1007/BF00198495 PMID:8123985

Hetmanski, M. T., Fussell, R., Godula, M., & Hübschmann, H. J. (2010, July). Rapid analysis of pesticides in difficult matrices using GC-MS-MS. *LC GC Europe*, 14–15.

Hewitt, M., Ellison, C. M., Enoch, S. J., Madden, J. C., & Cronin, M. T. D. (2010). Integrating (Q) SAR models, expert systems and read-across approaches for the prediction of developmental toxicity. *Reproductive Toxicology (Elmsford, N.Y.)*, *30*(1), 147–160. doi:10.1016/j.reprotox.2009.12.003 PMID:20006701

Hicks, B. (2013). *Agricultural pesticides and human health.* National Association of Geoscience Teachers. Available from http://serc.carleton.edu/NAGTWorkshops/ health / case_studies /pesticides.

Hildebrandt, A., Lacorte, S., & Barceló, D. (2007). Assessment of priority pesticides, degradation products, and pesticide adjuvants in groundwaters and top soils from agricultural areas of the Ebro river basin. *Analytical and Bioanalytical Chemistry*, *87*(4), 1459–1468. doi:10.100700216-006-1015-z PMID:17211597

Hill, M. J., Death, R. G., Mathers, K. L., Ryves, D. B., White, J. C., & Wood, P. J. (2017). Macroinvertebrate community composition and diversity in ephemeral and perennial ponds on unregulated floodplain meadows in the UK. *Hydrobiologia*, *793*(1), 95–108. doi:10.100710750-016-2856-x

Hirang, T., Shiraisshi, K., & Naakano, K. (1955). Studies on the blue-green algae in lowland paddy soil, part 1. On some conditions for growth of blue-green algae in paddy soil and its effects on growth of paddy rice plant. *Shikoku Nogyo Shikenjo Hokoku*, *2*, 121–137.

Hirthe, G., Fisher, T. C., Crane, M., & Callaghan, A. (2001). Short-term exposure to sub-lethal doses of lindane affects developmental parameters in Chironomus riparius Meigen, but has no effect on larval glutathione-S-transferase activity. *Chemosphere*, *44*(4), 583–589. doi:10.1016/S0045-6535(00)00471-9 PMID:11482645

History of pesticide use. (1998). Available from http://www2.mcdaniel.edu/Biology/eh01/pesticides/ historyofpesticidesuse.html

Hodges, R. R., Jr., Hoffman, J. H., Johnson, F. S., & Evans, D. E. (1973). Composition and dynamics of lunar atmosphere. In *Lunar and Planetary Science Conference Proceedings* (*Vol. 4*, p. 2855). Academic Press.

Hodges, G., Roberts, D. W., Marshall, S. J., & Dearden, J. C. (2006). The aquatic toxicity of anionic surfactants to Daphnia magna—a comparative QSAR study of linear alkylbenzene sulphonates and ester sulphonates. *Chemosphere*, *63*(9), 1443–1450. doi:10.1016/j.chemosphere.2005.10.001 PMID:16293293

Hoffman, R. S., Capel, P. D., & Larson, S. J. (2000). Comparison of pesticides in eight U.S. urban streams. *Environmental Toxicology and Chemistry*, *19*(9), 2249–2258. doi:10.1002/etc.5620190915

Hogan, C. M., Patmore, L. L. G., & Seidman, H. (1973). *Computer modeling of pesticide transport in soil for five instrumented watersheds, U.S. Environmental Protection Agency Southeast Water laboratory*. Sunnyvale, CA: Athens, Ga. by ESL Inc.

Holland, N., Furlong, C., Bastaki, M., Richter, R., Bradman, A., Huen, K., ... Eskenazi, B. (2006). Paraoxonase polymorphisms, haplotypes, and enzyme activity in Latino mothers and newborns. *Environmental Health Perspectives*, *114*(7), 985–991. doi:10.1289/ehp.8540 PMID:16835048

Hooda, P. S., Edwards, H. C., Anderson, H. A., & Miller, A. (2000). A review of water quality concerns in livestock farming areas. *The Science of the Total Environment*, *24*(1-3), 143–167. doi:10.1016/S0048-9697(00)00373-9 PMID:10811258

Hooda, P. S., Moynagh, M., Svoboda, I. F., Thurlow, M., Stewart, M., Thomson, M., & Anderson, H. A. (1997a). Streamwater nitrate concentrations in six agricultural catchments in Scotland. *The Science of the Total Environment*, *201*(1), 63–78. doi:10.1016/S0048-9697(97)84053-3 PMID:9232025

Hooda, P. S., Moynagh, M., Svoboda, I. F., Thurlow, M., Stewart, M., Thomson, M., & Anderson, H. A. (1997b). Soil and land use effects on phosphorus in six streams draining small agricultural catchments in Scotland. *Soil Use and Management*, *13*(4), 196–204. doi:10.1111/j.1475-2743.1997.tb00586.x

Hooper, M. J., Ankley, G. T., Cristol, D. A., Maryoung, L. A., Noyes, P. D., & Pinkerton, K. E. (2013). Interactions between chemical and climate stressors: A role for mechanistic toxicology in assessing climate change risks. *Environmental Toxicology and Chemistry*, *32*(1), 32–48. doi:10.1002/etc.2043 PMID:23136056

Hoppin, J., Umbach, D., London, S., Henneberger, P., Kullman, G., Coble, J., ... Sandler, D. (2009). Pesticide use and adult-onset asthma among male farmers in the Agricultural Health Study. *The European Respiratory Journal*, *34*(6), 1296–1303. doi:10.1183/09031936.00005509 PMID:19541724

Horne, A. J., & Goldman, C. R. (1994). Limnology (2nd ed.). McGraw-Hill.

Hornsby, A. G., Wauchope, R. D., & Herner, A. (1995). *Pesticide properties in the environment* (Vol. 1). Springer Science and Business Media.

Horton, M. K., Rundle, A., Camann, D. E., Boyd Barr, D., Rauh, V. A., & Whyatt, R. M. (2011). Impact of prenatal exposure to piperonyl butoxide and permethrin on 36-month neurodevelopment. *Pediatrics*, *127*(3), 699–706. doi:10.1542/peds.2010-0133 PMID:21300677

Hoshi, N., Hirano, T., Omotehara, T., Tokumoto, J., Umemura, Y., Mantani, Y., ... Kitagawa, H. (2014). Insight into the mechanism of reproductive dysfunction caused by neon-icotinoid pesticides. *Biological & Pharmaceutical Bulletin*, *37*(9), 1439–1443. doi:10.1248/bpb.b14-00359 PMID:25177026

Hossain, S. M. Z., Luckham, R. E., McFadden, M. J., & Brennan, J. D. (2009). Reagentless bidirectional lateral flow bioactive paper sensors for detection of pesticides in beverage and food samples. *Analytical Chemistry*, *81*(21), 9055–9064. doi:10.1021/ac901714h PMID:19788278

Hsu, C. C., & Whang, C. W. (2009). Microscale solid phase extraction of glyphosate and aminomethylphosphonic acid in water and guava fruit extract using alumina-coated iron oxide nanoparticles followed by capillary electrophoresis and electrochemiluminescence detection. *Journal of Chromatography. A*, *1216*(49), 8575–8580. doi:10.1016/j.chroma.2009.10.023 PMID:19853856

Huang, B., Chen, F., Shen, Y., Qian, K., Wang, Y., Sun, C., ... Cui, H. (2018). Advances in Targeted Pesticides with Environmentally Responsive Controlled Release by Nanotechnology. *Nanomaterials (Basel, Switzerland)*, *8*(2), 102. doi:10.3390/nano8020102 PMID:29439498

Huang, J., Liang, C., & Zhang, X. (2017). Effects of nano-SiO2 on the adsorption of chiral metalaxyl to agricultural soils. *Environmental Pollution*, *225*, 201–210. doi:10.1016/j.envpol.2017.03.065 PMID:28388518

Huber, W. (1993). Ecotoxicological relevance of atrazine in aquatic systems. *Environmental Toxicology and Chemistry*, *12*(10), 1865–1881. doi:10.1002/etc.5620121014

Huguenot, D., Bois, P., Jézéquel, K., Cornu, J. Y., & Lebeau, T. (2010). Selection of low cost materials for the sorption of copper and herbicides as single or mixed compounds in increasing complexity matrices. *Journal of Hazardous Materials*, *182*(1), 18–26. doi:10.1016/j.jhazmat.2010.05.062 PMID:20594640

Hunt, L., Bonetto, C., Resh, V. H., Buss, D. F., Fanelli, S., Marrochi, N., & Lydy, M. J. (2016). Insecticide concentrations in stream sediments of soy production regions of South America. *The Science of the Total Environment*, *547*, 114–124. doi:10.1016/j.scitotenv.2015.12.140 PMID:26780136

Huo, D., Li, Q., Zhang, Y., Hou, C., & Lei, Y. (2014). A highly efficient organophosphorus pesticides sensor based on CuO nanowires-SWCNTs hybrid nanocomposite. *Sensors and Actuators. B, Chemical*, *199*, 410–417. doi:10.1016/j.snb.2014.04.016

Hu, Q., Noll, R. J., Li, H., Makarov, A., Hardman, M., & Graham Cooks, R. (2005). The Orbitrap: A new mass spectrometer. *Journal of Mass Spectrometry*, *40*(4), 430–443. doi:10.1002/jms.856 PMID:15838939

Hurlbert, Mulla, Kieth, Westlake, & Dusch. (1970). Biological effects and persistence of Dursban in freshwater ponds. *Journal of Economics Entomology, 63*, 43-51.

Hurlbert, S. H., Mulla, M. S., & Wilson, H. R. (1972). Effects of an orthophosphorus insecticide on the phytoplankton, zooplankton and insect populations of freshwater ponds. *Ecological Monographs*, *42*(3), 269–299. doi:10.2307/1942211

Hurley, P. M., Hill, R. N., & Whiting, R. J. (1998). Mode of carcinogenic action of pesticides inducing thyroid follicular cell tumours in rodents. *Environmental Health Perspectives*, *106*(8), 437–445. doi:10.1289/ehp.98106437 PMID:9681970

Hyakumachi, M., & Kubota, M. (2004). Fungi as a plant growth promoter and disease suppressor. In D. K. Arora (Ed.), *Fungal Biotechnology in Agricultural, Food, and Environmental Applications* (pp. 101–110). New York: Marcel Dekker.

Hynes, K. R., & Boyetchko, S. M. (2006). Research initiatives in the art and science of biopesticide formulations. *Soil Biology & Biochemistry*, *38*(4), 845–849. doi:10.1016/j.soilbio.2005.07.003

IARC. (1988). *Overall evaluations of earcinogenicity: An updating of IARC monographs* (Vol. 1-42). International Agency for Research on Cancer.

Ibrahim, Y. A. (2016). Health and environmental impacts of pesticides: A responsibility principle and two novel systems for hazard classification and external cost determination. *Journal of Toxicology and Health*, *3*(1), 1. doi:10.7243/2056-3779-3-1

Igbedioh, S. O. (1991). Effects of Agricultural Pesticides on Humans, Animals and Higher Plants in Developing Countries. *Archives of Environmental Health*, *46*(4), 218–223. doi:10.1080/00039896.1991.9937452 PMID:2069430

Ignatowicz, K. (2011). A mass transfer model for the adsorption of pesticide on coconut shell based activated carbon. *International Journal of Heat and Mass Transfer*, *54*(23-24), 4931–4938. doi:10.1016/j.ijheatmasstransfer.2011.07.005

Indian Chemical Industry. (n.d.). *XI2I*[th] *Five Year Plan: 2012-2017*. Planning Commission of India, Government of India. Retrieved from http://planningcommission.nic.in/aboutus/committee/wrkgrp12/wg_chem0203.pdf

Indian Institute of Technology (IIT). (2011). Trends in Agriculture and Agricultural Practices in Ganga Basin: An Overview. Indian Institute of Technology Kanpur.

Indian National Trust for Art and Cultural Heritage (INTACH). (1998). *Delhi's Dying Water Bodies*. New Delhi: Indian National Trust for Artand Cultural Heritage.

Infante-Rivard, C., & Weichenthal, S. (2007). Pesticides and childhood cancer: An update of Zahm and Ward's 1998 review. *Journal of Toxicology and Environmental Health. Part B*, *10*(1-2), 81–99.

Infante-Rivard, C., & Weichenthal, S. (2007). Pesticides and childhood cancer: An update on Zahm and Ward's 1998 review. *Journal of Toxicology and Environmental Health. Part B, Critical Reviews*, *10*(1-2), 81–99. doi:10.1080/10937400601034589 PMID:18074305

International Rice Research Institute (IRRI). (1986). *Annual reportfor 1985*. Manila, Philippines: IRRI.

Ioannidou, O. A., Zabaniotou, A. A., Stavropoulos, G. G., Islam, M. A., & Albanis, T. A. (2010). Preparation of activated carbons from agricultural residues for pesticide adsorption. *Chemosphere*, *80*(11), 1328–1336. doi:10.1016/j.chemosphere.2010.06.044 PMID:20598734

Ioannis, K., Konstantinou, D. G. H., & Triantafyllos, A. A. (2006). The status of pesticide pollution in surface waters (rivers and lakes) of Greece. Part I. Review on occurrence and levels. *Environmental Pollution*, *141*(3), 555–570. doi:10.1016/j.envpol.2005.07.024 PMID:16226830

IPCC (Intergovernmental Panel on Climate Change). (2007). *Climate Change 2007: The Physical Science Basis: Summary for Policymakers. Fourth Assessment Report of the IPCC* (M. L. Parry, O. F. Canziani, J. P. Palutikof, P. J. van der Linden, & C. E. Hanson, Eds.). Cambridge, UK: Cambridge University Press.

Iranzo, M., Sainz-Pardo, I., Boluda, R., Sanchez, J., & Mormeneo, S. (2001). The use of microorganisms in environmental remediation. *Annals of Microbiology*, *51*(2), 135–144.

Ishibashi, N., & Itoh, S. (1981). Effects of herbicide benthiocarb on fauna in paddy field [in Japanese]. *Proceedings of the Association for Plant Protection of Kyushu*, *27*, 90–93. doi:10.4241/kyubyochu.27.90

Ishihara, A., Sawatsubashi, S., & Yamauchi, K. (2003). Endocrine disrupting chemicals: Interference of thyroid hormone binding to transthyretins and to thyroid hormone receptors. *Molecular and Cellular Endocrinology*, *199*(1), 105–117. doi:10.1016/S0303-7207(02)00302-7 PMID:12581883

Ishtiaq, M., Saleem, M. A., & Razaq, M. (2012). Monitoring of resistance in Spodopteraexigua (Lepidoptera: Noctuidae) from four districts of the Southern Punjab, Pakistan to four conventional and six new chemistry insecticides. *Crop Protection (Guildford, Surrey)*, *33*, 13–20. doi:10.1016/j.cropro.2011.11.014

Islam, M. A., Sakkas, V., & Albanis, T. A. (2009). Application of statistical design of experiment with desirability function for the removal of organophosphorus pesticide from aqueous solution by low-cost material. *Journal of Hazardous Materials*, *170*(1), 230–238. doi:10.1016/j.jhazmat.2009.04.106 PMID:19477587

Islam, M. N., Bint-E-Naser, S. F., & Khan, M. S. (2017). Pesticide Food Laws and Regulations. In *Pesticide Residue in Foods*. Cham, Switzerland: Springer. doi:10.1007/978-3-319-52683-6_3

Islam, M. T., Castle, S. J., & Ren, S. (2010). Compatibility of the insect pathogenic fungus Beauveriabassiana with neem against sweet potato whitefly, Bemisiatabaci, on eggplant. *Entomologia Experimentalis et Applicata, 134*(1), 28–34. doi:10.1111/j.1570-7458.2009.00933.x

Islam, M. T., & Omar, D. B. (2012). Combined effect of Beuveria bassiana with neem on virulence of insect in case of two application approaches. *Journal of Animal and Plant Sciences, 22*(1), 77–82.

Iwata, H., Tanabe, S., Sakai, N., & Tatsukawa, R. (1993). Distribution of persistent organochlorines in the oceanic air and surface seawater and the role of ocean on their global transport and fate. *Environmental Science & Technology, 27*(6), 1080–1098. doi:10.1021/es00043a007

Iyamu, I., Asia, I. O., & Egwakhide, P. A. (2007). Concentrations of residues from organo-chlorine pesticide in water & fish from some rivers in Edo State Nigeria. *International Journal of Physical Sciences, 2*, 237–241.

Jabbar, A., & Mallick, S. (1994). *Pesticides and environment situation in Pakistan* (Working Paper Series No. 19). Sustainable Development Policy Institute (SDPI).

Jacob, J., & Cherian, J. (2013). Review of Environmental and Human Exposure to Persistent Organic Pollutants. *Asian Social Science, 9*(11), 107–120. doi:10.5539/ass.v9n11P107

Jacobs, A. (2015). *Timeline-history of pesticides*. University of Oregon. Available from http://blogs.uoregon.edu/ajacobssu13gateway/timeline/

Jaga, K., & Dharmani, C. (2003). Sources of exposure to and public health implications of organophosphate pesticides. *Pan Am J Public Health, 14*(3), 171–185. doi:10.1590/S1020-49892003000800004 PMID:14653904

Jain, C. K., Singhal, D. C., & Sharma, M. K. (2007a). Estimating nutrient loadings using chemical mass balance approach. *Environmental Monitoring and Assessment, 134*(1-3), 385–396. doi:10.100710661-007-9630-5 PMID:17616829

Jain, S. K., Agarwal, P. K., & Singh, V. P. (2007b). *Hydrology and Water Resources of India*. Springer Netherlands.

Jallow, M. F. A., Awadh, D. G., Albaho, M. S., Devi, V. Y., & Thomas, B. M. (2017). Pesticide risk behaviors and factors influencing pesticide use among farmers in Kuwait. *The Science of the Total Environment, 574*, 490–498. doi:10.1016/j.scitotenv.2016.09.085 PMID:27644027

Jamal, F., Haque, Q. S., Singh, S., & Rastogi, S. (2015). The influence of organophosphate and carbamate on sperm chromatin and reproductive hormones among pesticide sprayers. *Toxicology and Industrial Health*, 1–10. PMID:25647813

Jana, B. B. (1994). Deposition of pollutants and their impacts on fisheries. *Environmental Monitoring, 2*.

Janghel, E. K., Rai, J. K., Rai, M. K., & Gupta, V. K. (2007). A new sensitive spectrophotometric determination of cypermethrin insecticide in environmental and biological samples. *Journal of the Brazilian Chemical Society, 18*(3), 590–594. doi:10.1590/S0103-50532007000300015

Jan, M. R., Shah, J., & Khan, H. (2003). Investigation of new indirect spectrophotometric method for the determination of carbofuran in carbamate pesticides. *Chemosphere, 52*(9), 1623–1626. doi:10.1016/S0045-6535(03)00480-6 PMID:12867195

Jardim, A. N. O., & Caldas, E. D. (2012). Brazilian monitoring programs for pesticide residues in food – results from 2001 to 2010. *Food Control, 25*(2), 607–616. doi:10.1016/j.foodcont.2011.11.001

Jayaraj, R., Megha, P., & Sreedev, P. (2016). Organochlorine pesticides, their toxic effects on living organisms and their fate in the environment. *Interdisciplinary Toxicology, 9*(3-4), 90–100. doi:10.1515/intox-2016-0012 PMID:28652852

Jayashree, R., & Vasudevan, N. (2007). Organochlorine pesticide residues in ground water of Thiruvallur district, India. *Environmental Monitoring and Assessment*, *128*(1-3), 209–215. doi:10.100710661-006-9306-6 PMID:17016752

Jespers, A. B. K., & De Waard, M. A. (1995). Effect of fenpiclonil on phosphorylation of glucose in *Fusarium sulphureum. Pesticide Science*, *44*(2), 167–175. doi:10.1002/ps.2780440210

Jeyaratnam, J. (1985). Health problems of pesticide usage in the Third World. *British Journal of Industrial Medicine*, *42*(8), 505. PMID:4016001

Jeyaratnam, J. (1985). Health Problems of Pesticide Usage in the Third World. *British Medical Journal*, *42*, 505–506. PMID:4016001

Jeyaratnam, J. (1985). Health problems of pesticide usage in the third world. *The B M J*, *42*, 505. PMID:4016001

Jiang, D. G., Fang, C. R., Yang, D. J., & Wang, Z. T. (2005). Determination of 27 organochlorine and pyrethroid pesticide multi-residues in teas by gas chromatography. *Zhongguo Shipin Weisheng Zazhi.*

Jin, L., Trans, D. Q., Ide, C. F., McLachlan, J. A., & Arnold, S. F. (1997). Several synthetic chemicals inhibit progesterone receptor-mediated transactivation in yeast. *Biochemical and Biophysical Research Communications*, *233*(1), 13–146. doi:10.1006/bbrc.1997.6417 PMID:9144411

Jin, Y., Wang, L., Chen, G., Lin, X., Miao, W., & Fu, Z. (2014). Exposure of mice to atrazine and its metabolite diaminochlorotriazine elicits oxidative stress and endocrine disruption. *Environmental Toxicology and Pharmacology*, *37*(2), 782–790. doi:10.1016/j.etap.2014.02.014 PMID:24632104

Johan, N. A., Kutty, S. R. M., Isa, M. H., Muhamad, N. S., & Hashim, H. (2011). Adsorption of copper by using microwave incinerated rice husk ash (MIRHA). *International Journal of Civil and Environmental Engineering*, *3*(4), 211–215.

Johnson, A. C., Haria, A. H., Bhardwaj, C. L., Williams, R. J., & Walker, A. (1996). Preferential flow pathways and their capacity to transport isoproturon in a structured clay soil. *Pesticide Science*, *48*(3), 225–237. doi:10.1002/(SICI)1096-9063(199611)48:3<225::AID-PS462>3.0.CO;2-D

Johnson-Restrepo, B., Kannan, K., Rapaport, D. P., & Rodan, B. D. (2005). Polybrominated diphenyl ethers and polychlorinated biphenyls in human adipose tissue from New York. *Environmental Science & Technology*, *39*(14), 5177–5182. doi:10.1021/es050399x PMID:16082945

Johnson, T. (1998). *CRC Ethnobotany Desk Reference*. Boca Raton, FL: CRC Press.

Jones, K. C., & de Voogt, P. (1999). Persistent organic pollutants (POPs): State of the science. *Environmental Pollution*, *100*(1), 209–221. doi:10.1016/S0269-7491(99)00098-6 PMID:15093119

Jorgenson, J. L. (2001). Aldrin and Dieldrin: A Review of Research on Their Production, Environmental Deposition and Fate, Bioaccumulation, Toxicology, and Epidemiology in the United States. *Environmental Health Perspectives*, *109*(1), 113–139. doi:10.1289/ehp.01109s1113 PMID:11250811

Josefsson, S., Leonardsson, K., Gunnarsson, J. S., & Wiberg, K. (2010). Bioturbation-Driven Release of Buried PCBs and PBDEs from Different Depths in Contaminated Sediments. *Environmental Science & Technology*, *44*(19), 7456–7464. doi:10.1021/es100615g PMID:20831254

Jurewicz, J., & Hanke, W. (2008). Prenatal and childhood exposure to pesticides and neurobehavioural development: Review of epidemiological studies. *International Journal of Occupational Medicine and Environmental Health*, *21*(2), 121–132. doi:10.2478/v10001-008-0014-z PMID:18614459

Jusoh, A., Hartini, W. J. H., & Endut, A. (2011). Study on the removal of pesticide in agricultural run off by granular activated carbon. *Bioresource Technology*, *102*(9), 5312–5318. doi:10.1016/j.biortech.2010.12.074 PMID:21232934

Justino, C., Duarte, A., & Rocha-Santos, T. (2017). Recent Progress in Biosensors for Environmental Monitoring: A Review. *Sensors (Basel)*, *17*(12), 2918. doi:10.339017122918 PMID:29244756

Kalia, A., & Gosal, S. K. (2011). Effect of pesticide application on soil microorganisms. *Archives of Agronomy and Soil Science*, *57*(6), 569–596. doi:10.1080/03650341003787582

Kamble, V. S., & Rao, K. R. (2016). Intentional Study of Bioaccumulative Behaviour of Isomers of Organochlorine Pesticide Thiodan (Endosulfan 35% Ec) In Fresh Water Mollusc, Lamellidens Corrianus During Monsoon Season. *Indian Journal of Applied Research*, *6*(4), 264–266.

Kamei, I., Takagi, K., & Kondo, R. (2011). Degradation of endosulfan and endosulfan sulfate by white-rot fungus Trametes hirsuta. *Journal of Wood Science*, *57*(4), 317–322. doi:10.100710086-011-1176-z

Kamel, F., Engel, L., Gladen, B., Hoppin, J., Alavanja, M., & Sandler, D. (2005). Neurologic symptoms in Licensed Private Pesticide Applicators in the Agricultural Health Study. *Environmental Health Perspectives*, *113*(7), 877–882. doi:10.1289/ehp.7645 PMID:16002376

Kamrin, M. A. (1997). *Pesticide Profiles: toxicity, environmental impact, and fate*. CRC Press. doi:10.1201/9781420049220

Kanagasubbulakshmi, S., Gopinath, A., & Kadirvelu, K. (2017). Enzyme free Thiol capped CdS Quantum dots based sensing method for the detection of Malathion. *Materials Today: Proceedings*, *4*(13), 12448–12456. doi:10.1016/j.matpr.2017.10.015

Kanagasubbulakshmi, S., Kathiresan, R., & Kadirvelu, K. (2018). Structure and physiochemical properties based interaction patterns of organophosphorous pesticides with quantum dots: Experimental and theoretical studies. *Colloids and Surfaces. A, Physicochemical and Engineering Aspects*, *549*, 155–163. doi:10.1016/j.colsurfa.2018.04.007

Kannan, K., Ramu, K., Kajiwara, N., Sinha, R. K., & Tanabe, S. (2005). Organochlorine pesticides, polychlorinated biphenyls, and polybrominateddiphenyl ethers in Irrawaddy dolphins from India. *Archives of Environmental Contamination and Toxicology*, *49*(3), 415–420. doi:10.100700244-005-7078-6 PMID:16170447

Kannan, K., Tanabe, S., Borrell, A., Aguilar, A., Focardi, S., & Tatsukawa, R. (1993). Isomer specifi c analysis and toxic evaluation of polychlorinated biphenyls in food stuff s from India and their implications on human dietary exposure. *Journal of Agricultural and Food Chemistry*, *40*, 158.

Kaprow, M. (1985). Manufacturing danger: Fear and pollution in industrial society. *American Anthropologist*, *87*(2), 342–356. doi:10.1525/aa.1985.87.2.02a00070

Karami-Mohajeri, S., & Abdollahi, M. (2011). Toxic influence of organophosphate, carbamate, and organochlorine pesticides on cellular metabolism of lipids, proteins, and carbohydrates: A systematic review. *Human and Experimental Toxicology*, *30*(9), 1119–1140. doi:10.1177/0960327110388959 PMID:21071550

Karas, P., Metsoviti, A., Zisis, V., Ehaliotis, C., Omirou, M., Papadopoulou, E. S., ... Karpouzas, D. G. (2015). Dissipation, metabolism and sorption of pesticides used in fruit-packaging plants: Towards an optimized depuration of their pesticide-contaminated agro-industrial effluents. *The Science of the Total Environment*, *530–531*, 129–139. doi:10.1016/j.scitotenv.2015.05.086 PMID:26042894

Karunakaran, C. O. (1958). The Kerala food poisoning. *Journal of the Indian Medical Association*, *31*, 204. PMID:13588025

Kasiotis, K. M. (2013). Biopesticides Analysis: An Editorial. *Journal of Biofertilizers & Biopesticides*, *4*, 115.

Kasiotis, K. M., Anagnostopoulos, C., Anastasiadou, P., & Machera, K. (2014). Pesticide residues in honeybees, honey and bee pollen by LC–MS/MS screening: Reported death incidents in honeybees. *The Science of the Total Environment*, *485*, 633–642. doi:10.1016/j.scitotenv.2014.03.042 PMID:24747255

Katayama, A., & Matsumura, F. (1993). Degradation of organochlorine pesticides, particularly endosulfan, by Trichoderma harzianum. *Environmental Toxicology and Chemistry*, *12*(6), 1059–1065. doi:10.1002/etc.5620120612

Katritzky, A. R., & Gordeeva, E. V. (1993). Traditional topological indexes vs electronic, geometrical, and combined molecular descriptors in QSAR/QSPR research. *Journal of Chemical Information and Computer Sciences*, *33*(6), 835–857. doi:10.1021/ci00016a005 PMID:8113335

Kaul, S. (2007). National and International Initiatives in Wetland Management. *Kerala Environment Congress 2007: Thiruvananthapuram, Center for Environment Development.*

Kaushik, A., Sharma, H. R., Jain, S., Dawra, J., & Kaushik, C. P. (2010). Pesticide pollution of river Ghaggar in Haryana, India. *Environmental Monitoring and Assessment*, *160*(1-4), 61–69. doi:10.100710661-008-0657-z PMID:19107563

Kawano, M., Tsuyoshi, I., Wada, T., Hidaka, H., & Tatsukacva, R. (1988). Bioconcentration and residue patterns of chlordane compounds in marine animals: Invertebrates, fish, mammals, and seabirds. *Environmental Science & Technology*, *22*(7), 792–797. doi:10.1021/es00172a008 PMID:22195663

Keer, R. A. (2007). Global warming is changing the world. *Science*, *316*(5822), 188–190. doi:10.1126cience.316.5822.188 PMID:17431148

Keifer, M. C., & Firestone, J. (2007). Neurotoxicity of pesticides. *Journal of Agromedicine*, *12*(1), 17–25. doi:10.1300/J096v12n01_03 PMID:18032333

Keikotlhaile, B. M. (2011). *Influence of the processing factors on pesticide residues in fruits and vegetables and its application in consumer risk assessment*. Ghent: Ghent University.

Keikotlhaile, B. M., Spanoghe, P., & Steurbaut, W. (2010). Effects of food processing on pesticide residues in fruits and vegetables: A meta-analysi approach. *Food and Chemical Toxicology*, *48*(1), 1–6. doi:10.1016/j.fct.2009.10.031 PMID:19879312

Keil, A. P., Daniels, J. L., & Hertz-Picciotto, I. (2014). Autism spectrum disorder, flea and tick medication, and adjustments for exposure misclassification: The CHARGE (Childhood Autism Risks from Genetics and Environment) case –control study. *Environmental Health*, *13*(1), 3. doi:10.1186/1476-069X-13-3 PMID:24456651

Kelloge, R. L., Nehring, R., Grube, A., Goss, D. W., & Plotkin, S. (2000). *Environmental indicators of pesticide leaching and runoff from farm fields*. United States Department of Agriculture Natural Resources Conservation Service.

Kennedy, D., Hurst, V., Konradsdottir, E., & Einarson, A. (2005). Pregnancy outcome following exposure to permethrin and use of teratogen information. *American Journal of Perinatology*, *22*(2), 87–90. doi:10.1055-2005-837736 PMID:15731987

Kettles, M. K., Browning, S. R., Prince, T. S., & Horstman, S. W. (1997). Triazine herbicide expo-sure and breast cancer incidence: An ecologic study of Kentucky counties. *Environmental Health Perspectives*, *105*(11), 1222–1227. doi:10.1289/ehp.971051222 PMID:9370519

Khachatourians, G.G. (2009). Insecticides, microbials. *Applied Microbiology: Agro/Food*, 95-109.

Khan, S. U. (1980). Pesticides in the soil environment. Elsevier Scientific Publishing Company.

Khanday, M., Bhat, R. A., Haq, S., Dervash, M. A., Bhatti, A. A., Nissa, M., & Mir, M. R. (2016). Arbuscular Mycorrhizal Fungi Boon for Plant Nutrition and Soil Health. In Soil Science: Agricultural and Environmental Prospectives (pp. 317-332). Springer International Publishing.

Khandekar, N. (2011). Delhi water bodies go under, almost. *Hindustan Times*.

Khoshnood, M., & Azizian, S. (2012). Adsorption of 2, 4-dichlorophenoxyacetic acid pesticide by graphitic carbon nanostructures prepared from biomasses. *Journal of Industrial and Engineering Chemistry*, *18*(5), 1796–1800. doi:10.1016/j.jiec.2012.04.007

Khrunyk, Y., Schiewer, S., Carstens, K. L., Hu, D., & Coats, J. R. (2017). Uptake of C14-atrazine by prairie grasses in a phytoremediation setting. *International Journal of Phytoremediation*, *19*(2), 104–112. doi:10.1080/15226514.2016.1193465 PMID:27259078

Kim, J. H., & Smith, A. (2001). Distribution of organochlorine pesticides in soils from South Korea. *Chemosphere*, *43*(2), 137–140. doi:10.1016/S0045-6535(00)00281-2 PMID:11297392

Kim, J., Swartz, M. D., Langlois, P. H., Romitti, P. A., Weyer, P., Mitchell, L. E., ... Agopian, A. (2017). The National Birth Defects Prevention Study. Estimated Maternal Pesticide Exposure from Drinking Water and Heart Defects in Offspring. *International Journal of Environmental Research and Public Health*, *14*(8), 889. doi:10.3390/ijerph14080889

Kim, S. D., Cho, J., Kim, I. S., Vanderford, B. J., & Snyder, S. A. (2007). Occurrence and removal of pharmaceuticals and endocrine disruptors in South Korean surface, drinking, and waste waters. *Water Research*, *41*(5), 1013–1021. doi:10.1016/j.watres.2006.06.034 PMID:16934312

Kirchmann, H., & Ryan, M. H. (2004). Nutrients in organic farming – are there advantages from the exclusive use of organic manures and untreated minerals? In *New directions for a diverse planet. Proceedings of the 4th International Crop Science Congress*. Brisbane, Australia: The Regional Institute.

Kirk, K. L. (1991). Persistent Polyhalogenated Compounds: Biochemistry, Toxicology, Medical Applications, and Associated Environmental Issues. Biochemistry of the Elements, 9, 245–254.

Kirkhorn, S. R., & Garry, V. F. (2000). Agricultural Lung Diseases. *Environmental Health Perspectives*, *108*(4), 705–712.

Kishi, M., Hirschhorn, N., Djajadisastra, M., Saterlee, L. N., Strowman, S., & Dilts, R. (1995). Physiological effcts of work stress and pesticide exposure in tree planting by British Columbia silviculture workers. *Ergonomics*, *36*, 951–961.

Kladivko, E. J., Brown, L. C., & Baker, J. L. (2001). Pesticide transport to subsurface tile drains in humid regions of North America. *Critical Reviews in Environmental Science and Technology*, *31*(1), 1–62. doi:10.1080/20016491089163

Klekner, V., & Kosaric, N. (1992). Degradation of phenols by algae. *Environmental Technology*, *13*(5), 493–501. doi:10.1080/09593339209385176

Kmellár, B., Abrankó, L., Fodor, P., & Lehotay, S. J. (2010). Routine approach to qualitatively screening 300 pesticides and quantification of those frequently detected in fruit and vegetables using liquid chromatography tandem mass spectrometry (LC-MS/MS). *Food Additives and Contaminants*, *27*(10), 1415–1430. doi:10.1080/19440049.2010.490791 PMID:20582778

Knauert, S., Dawo, U., Hollender, J., Hommen, U., & Knauer, K. (2009). Effects of photosystem ii inhibitors and their mixture on freshwater phytoplankton succession in outdoor mesocosms. *Environmental Toxicology and Chemistry*, *28*(4), 836–845. doi:10.1897/08-135R.1 PMID:19391680

Koch, G., & Siegrist, H. (1997). Denitrification with Methanol in Tertiary Filtration. *Water Research*, *31*(12), 3029–3038. doi:10.1016/S0043-1354(97)00174-7

Kohler, C. C., & Hubert, W. A. (1993). *Inland fisheries management in North America*. Bethesda, MD: American Fisheries Society.

Köhler, H. R., & Triebskorn, R. (2013). Wildlife ecotoxicology of pesticides: Can we track effects to the population level and beyond? *Science*, *341*(6147), 759–765. doi:10.1126cience.1237591 PMID:23950533

Kole, R. K., Banerjee, H., & Bhattacharyya, A. (2001). Monitoring of market fish samples for Endosulfan and Hexachlorocyclohexane residues in and around Calcutta. *Bulletin of Environmental Contamination and Toxicology*, *67*(4), 554–559. doi:10.1007001280159 PMID:11779071

Kolpin, D. W., Thurman, E. M., & Linhart, S. M. (2001). Occurrence of cyanazine compounds in groundwater: Degradates more prevalent than the parent compound. *Environmental Science & Technology*, *35*(6), 1217–1222. doi:10.1021/es001520x PMID:11347936

Konstantinou, I. K., & Albanis, T. A. (2003). Photocatalytic transformation of pesticides in aqueous titanium dioxide suspensions using artificial and solar light: Intermediates and degradation pathways. *Applied Catalysis B: Environmental*, *42*(4), 319–335. doi:10.1016/S0926-3373(02)00266-7

Koul, O. (2011). Microbial biopesticides: Opportunities and challenges. *Perspectives in Agriculture, Veterinary Science, Nutrition and Natural Resources*, *6*, 1–26.

Kramer, K. J., & Muthukrishnan, S. (1997). Insect chitinases: Molecular biology and potential use as biopesticides. *Insect Biochemistry and Molecular Biology*, *27*(11), 887–900. doi:10.1016/S0965-1748(97)00078-7 PMID:9501415

Kreuger, J. (1999). *Pesticides in the environment- Atmospheric deposition and transport to surface waters* (Doctoral thesis). Swedish University of Agricultural Sciences, Uppsala, Sweden.

Kreuger, J., Peterson, M., & Lundgren, E. (1999). Agricultural inputs of pesticide residues to stream and pond sediments in a small catchment in southern Sweden. *Bulletin of Environmental Contamination and Toxicology*, *62*(1), 55–62. doi:10.1007001289900841 PMID:9870990

Krishnan & Chockalingam. (1989). Toxic and sublethal effects of endosulfan and carbaryl on growth and egg production of Moina micruru Kurz (Cladocera: Moinidae). *Environmental Pollution, 56*, 319-326.

Kristiansen, P., Taji, A., & Reganold, J. (2006). Organic agriculture: opportunities and challenges. In P. Kristiansen, A. Taji, & J. Reganold (Eds.), *Organic Agriculture: A Global Perspective* (pp. 421–441). CSIRO Publishing. doi:10.1079/9781845931698.0421

Kumar, M. D. (2013). Climate in India: key features of the variables. *Climate Change Impacts on Water Resources Systems*, 104.

Kumar, M. D., Panda, R., Niranjan, V., & Bassi, N. (2013). Technology choices and institutions for improving economic and livelihood benefits from multiple uses tanks in western Orissa. Academic Press.

Kumar, P. (2017). *Integrated Soil Fertility Management: Converting Subsistence Farming to Productive Farming*. Retrieved from http://www.pradan.net/wp-content/uploads/2017/12/Article-No.-3-Integrated-Soil-Fertility-Management-Converting-Subsistence-Farming-to-Productive-Farming-.pdf

Kumari, B., Madan, V. K., Kumar, R., & Kathpal, T. S. (2002). Monitoring of seasonal vegetables for pesticide residues. *Environmental Monitoring and Assessment*, *74*(3), 263–270. doi:10.1023/A:1014248827898 PMID:11944799

Kumar, K. S., Kannan, K., Corsolini, S., Evans, T., Giesy, J. P., Nakanishi, J., & Masunaga, S. (2002). Polychlorinated dibenzop-dioxins, dibenzofurans and polychlorinated biphenyls in polar bear, penguin and south polar skua. *Environment and Pollution*, *119*(2), 151–161. doi:10.1016/S0269-7491(01)00332-3 PMID:12152823

Kumar, P., Kim, K.-H., & Deep, A. (2015). Recent advancements in sensing techniques based on functional materials for organophosphate pesticides. *Biosensors & Bioelectronics*, *70*, 469–481. doi:10.1016/j.bios.2015.03.066 PMID:25864041

Kumar, S. (2012). The role biopesticides in sustainabley feeding the nine billion global populations. *Journal of Biofertilizers & Biopesticides*, *4*, 114.

Kumar, S., & Singh, A. (2014). Biopesticides for integrated crop management, environmental and regulatory aspects. *Journal of Biofertilizers & Biopesticides*, *5*, 121.

Kuo, J.-F., Stahl, J. F., Chen, C.-L., & Bohlier, P. V. (1998). Dual Role of Activated Carbon Process for Water Reuse. *Water Environment Research*, *70*(2), 161–170. doi:10.2175/106143098X127008

Kuo, W. S., & Regan, R. W. Sr. (1999). Removal of pesticides from rinsate by adsorption using agricultural residuals as medium. *Journal of Environmental Science and Health. Part. B, Pesticides, Food Contaminants, and Agricultural Wastes*, *34*(3), 431–447. doi:10.1080/03601239909373207

Kurbiel, J. (Ed.). (1991). *Advanced Wastewater Treatment and Reclamation*. London: Pergamon.

Kwiatkowska, M., Huras, B., & Bukowska, B. (2014). The effect of metabolites and impu-rities of glyphosate on human erythrocytes (*in vitro*). *Pesticide Biochemistry and Physiology*, *109*, 34–43. doi:10.1016/j.pestbp.2014.01.003 PMID:24581382

La Farre, M., Pérez, S., Kantiani, L., & Barceló, D. (2008). Fate and toxicity of emerging pollutants, their metabolites and transformation products in the aquatic environment. *TrAC Trends in Analytical Chemistry*, *27*(11), 991–1007. doi:10.1016/j.trac.2008.09.010

Lacasana, M., Vazquez-Grameix, H., Borja-Aburto, V. H., Blanco-Muñoz, J., Romieu, I., Aguilar-Garduño, C., & García, A. M. (2006). Maternal and paternal occupational exposure to agricultural work and the risk of anencephaly. *Occupational and Environmental Medicine*, *63*(10), 649–656. doi:10.1136/oem.2005.023333 PMID:16873458

Lah, K. (2011). Effects of pesticides on human health. *Toxipedia*. Available from http://www.toxipedia.org/display/toxipedia/Effects+of+Pesticides+on+Human+Health

Lakhani, L. (2015). How to Reduce Impact of Pesticides in Aquatic Environment. *Social Issues and Environmental Problems*, *3*(9), 29–38.

Lakshmi, A. (1993). Pesticides in India: Risk assessment to aquatic ecosystem. *The Science of the Total Environment*, *134*(1), 243–253. doi:10.1016/S0048-9697(05)80025-7

Lambropoulou, D. A., & Albanis, T. A. (2007). Liquid-phase micro-extraction techniques in pesticide residue analysis. *Journal of Biochemical and Biophysical Methods*, *70*(2), 195–228. doi:10.1016/j.jbbm.2006.10.004 PMID:17161462

Lampert, W., Fleckner, W., Pott, E., Schober, U., & Storkel, K. U. (1989). Herbicide effects on planktonic systems of different complexity. *Hydrobiologia*, *188/189*(1), 415–424. doi:10.1007/BF00027809

Landrigan, P. J., & Baker, D. B. (2015). The National Children's Studydend or new beginning? *The New England Journal of Medicine*, *372*(16), 16. doi:10.1056/NEJMp1500276 PMID:25875255

Landrum, P. F., Lydy, M. J., & Lee, H. II. (1992). Toxicokinetics in aquatic systems: Model comparisons and use in hazard assessment. *Environmental Toxicology and Chemistry*, *11*(12), 1709–1725. doi:10.1002/etc.5620111205

Langlais, B., Recknow, D. A., & Brink, D. R. (Eds.). (1991). Ozone in Water Treatment: Application and Engineering. Lewis Publishers.

Lari, S. Z., Khan, N. A., Gandhi, K. N., Meshram, T. J., & Thacker, N. P. (2014). Comparison of pesticide residues in surface water and ground water of agriculture intensive areas. *Journal of Environmental Health Science & Engineering, 12*(1), 11. doi:10.1186/2052-336X-12-11 PMID:24398360

Larsson, T. (1990). *Sidoeffekter av svampbekämpningsmedel.* SLU Info/Redaktionen. Fakta – Mark/Växter, no 6.

Leandro, C. C., Bishop, D. A., Fussell, R. J., Smith, F. D., & Keely, B. J. (2006). Semiautomated determination of pesticides in water using solid phase extraction disks and gas chromatography-mass spectrometry. *Journal of Agricultural and Food Chemistry, 54*(3), 645–649. doi:10.1021/jf051874d PMID:16448162

Lee, B. W., London, L., Paulauskis, J., Myers, J., & Christiani, D. C. (2003). Association between human paraoxonase gene polymorphism and chronic symptoms in pesticide-exposed workers. *Journal of Occupational and Environmental Medicine, 45*(2), 118–122. doi:10.1097/01.jom.0000052953.59271.e1 PMID:12625227

Lee, D. H., Lee, I. K., Song, K., Steffes, M., Toscano, W., Baker, B. A., & Jacobs, D. R. Jr. (2006). A strong dose-response relation between serum concentrations of persistent organic pollutants and diabetes: Results from the National Health and Examination Survey 1999-2002. *Diabetes Care, 29*(7), 1638–1644.

Lee, J. J., Seymour, H. H., & Eugene, C. B. (2001). *An Illustrated Guide to the Protozoa* (2nd ed.). Lawrence, KS: Society of Protozoologists.

Lee, M. G., Patil, V., Na, Y. C., Lee, D. S., Lim, S. H., & Yi, G.-R. (2018). Highly stable, rapid colorimetric detection of carbaryl pesticides by azo coupling reaction with chemical pre-treatment. *Sensors and Actuators. B, Chemical, 261*, 489–496. doi:10.1016/j.snb.2018.01.151

Lehmann, E., Turrero, N., Kolia, M., Konaté, Y., & De Alencastro, L. F. (2017). Dietary risk assessment of pesticides from vegetables and drinking water in gardening areas in Burkina Faso. *The Science of the Total Environment, 601*, 1208–1216. doi:10.1016/j.scitotenv.2017.05.285 PMID:28605838

Lehner, B., & Döll, P. (2004). Development and validation of a global database of lakes, reservoirs and wetlands. *Journal of Hydrology (Amsterdam), 296*(1-4), 1–22. doi:10.1016/j.jhydrol.2004.03.028

Lehotay, S. J., Harman-Fetcho, J. A., & McConnell, L. L. (1998). Agricultural pesticide residues in oysters and water from two Cheaspeake Bay tributaries. *Marine Pollution Bulletin, 37*(1-2), 32–44. doi:10.1016/S0025-326X(98)00129-5

Lemaire, G., Terouanne, B., Mauvais, P., Michel, S., & Rahmani, R. (2004). Effect of organochlorine pesticides on human androgen receptor activation in vitro. *Toxicology and Applied Pharmacology, 196*(2), 235–246. doi:10.1016/j.taap.2003.12.011 PMID:15081270

Leonard, R. A. (1990). Movement of pesticides into surface waters. In H. H. Cheng (Ed.), *Pesticides in the Soil Environment: Processes, Impacts and Modelling* (pp. 303–349). Maddison, WI: Soil Science Society of America.

Leroux, P. (1996). Recent developments in the mode of action of fungicides. *Pesticide Science, 47*(2), 191–197. doi:10.1002/(SICI)1096-9063(199606)47:2<191::AID-PS415>3.0.CO;2-I

Leroux, P., Lanen, C., & Fritz, R. (1992). Similarities in the antifungal activities of fenpiclonil, iprodione and tolclofos-methyl against *Botrytis cinerea* and *Fusarium nivale*. *Pesticide Science, 36*(3), 255–261. doi:10.1002/ps.2780360312

Levings, G. W., Healy, D. F., Richey, S. F., & Carter, L. F. (1998). Water quality in the Rio Grande Valley, Colorado, New Mexico, and Texas, 1992–95. *Geological Survey Circular*, 1162.

Lewan, E., Kreuger, J., & Jarvis, N. (2009). Implications of precipitation patterns and antecedent soil water content for leaching of pesticides from arable land. *Agricultural Water Management, 96*(11), 1633–1640. doi:10.1016/j.agwat.2009.06.006

Lewin, R. A. (1974). Biochemical taxonomy. *Blackwell Scientific Publications*.

Lewis, S. E., Brodie, J. E., Bainbridge, Z. T., Rohde, K. W., Davis, A. M., Masters, B. L., ... Schaffelke, B. (2009). Herbicides: A new threat to the Great Barrier Reef. *Environmental Pollution*, *157*(8), 2470–2484. doi:10.1016/j.envpol.2009.03.006 PMID:19349104

Lew, S., Lew, M., Biedunkiewicz, A., & Szarek, J. (2013). Impact of Pesticide Contamination on Aquatic Microorganism Populations in the Littoral Zone. *Archives of Environmental Contamination and Toxicology*, *64*(3), 399–409. doi:10.100700244-012-9852-6 PMID:23229198

Liang, H., Song, D., & Gong, J. (2014). Signal-on electrochemiluminescence of biofunctional CdTe quantum dots for biosensing of organophosphate pesticides. *Biosensors & Bioelectronics*, *53*, 363–369. doi:10.1016/j.bios.2013.10.011 PMID:24184599

Liess, M., Brown, C. D., Dohmen, G. P., Duquesne, S., Hart, A., Heimbach, F., ... Tarazona, J. V. (2005). *Effects of pesticides in the field*. Berlin, Germany: SETAC press.

Liess, M., Brown, C., Dohmen, P., Duquesne, S., Heimbach, F., & Kreuger, J. (2005). *Effects of Pesticides in the Field—EPIF*. Brussels, Belgium: SETAC Press.

Liess, M., & Schulz, R. (1999). Linking insecticide contamination and population response in an agricultural stream. *Environmental Toxicology and Chemistry*, *18*(9), 1948–1955. doi:10.1002/etc.5620180913

Liess, M., & Von Der Ohe, P. C. (2005). Analyzing effects of pesticides on invertebrate communities in streams. *Environmental Toxicology and Chemistry*, *24*(4), 954–965. doi:10.1897/03-652.1 PMID:15839571

Lim, R. P. (1980). Population changes of some aquatic invertebrates in ricefields. In J. I. Furtado (Ed.), *Tropical Ecology and Development* (pp. 971–980). Kuala Lumpur, Malaysia: Proceedings International Society of Tropical Ecology.

Lim, R. P., Abdullah, M. F., & Fernando, C. H. (1984). Ecological studies of cladocera in ricefields of Tanjung Karang, Malaysia, subject to pesticide treatment. *Hydrobiologi'u*, *113*(1), 99–103. doi:10.1007/BF00026596

Lindley, C. E., Stewart, J. T., & Sandstrom, M. W. (1996). Determination of low concentrations of acetochlor in water by automated solid-phase extraction and gas chromatography with mass-selective detection. *Journal of AOAC International*, *79*(4), 962–966.

Lintelmann, J., Katayama, A., Kurihara, N., Shore, L., & Wenzel, A. (2003). Endocrine disruptors in the environment. *Pure and Applied Chemistry*, *75*(5), 631–681. doi:10.1351/pac200375050631

Li, Q., Kobayashi, M., & Kawada, T. (2011). Ziram induces apoptosis and necrosis in human immune cells. *Archives of Toxicology*, *85*(4), 355–361. doi:10.100700204-010-0586-9 PMID:20842346

Liu, G., Li, L., Xu, D., Huang, X., Xu, X., Zheng, S., ... Lin, H. (2017). Metal–organic framework preparation using magnetic graphene oxide–β-cyclodextrin for neonicotinoid pesticide adsorption and removal. *Carbohydrate Polymers*, *175*, 584–591. doi:10.1016/j.carbpol.2017.06.074 PMID:28917904

Liu, H., Cai, X., Wang, Y., & Chen, J. (2011). Adsorption mechanism-based screening of cyclodextrin polymers for adsorption and separation of pesticides from water. *Water Research*, *45*(11), 3499–3511. doi:10.1016/j.watres.2011.04.004 PMID:21529879

Liu, J., & Diamond, J. (2005). China's environment in a globalizing world. *Nature*, *435*(7046), 1179–1186. doi:10.1038/4351179a PMID:15988514

Liu, P., Wu, C., Chang, X., Qi, X., Zheng, M., & Zhou, Z. (2016). Adverse associations of both prenatal and postnatal exposure to organophosphorous pesticides with infant neurodevelopment in an agricultural area of Jiangsu Province. China. *Environmental Health Perspectives*, *124*(10), 1637–1643. doi:10.1289/EHP196 PMID:27153333

Liu, Z., Guo, H., He, H., & Sun, C. (2012). Sorption and cosorption of the nonionic herbicide mefenacet and heavy metals on soil and its components. *Journal of Environmental Sciences (China)*, *24*(3), 427–434. doi:10.1016/S1001-0742(11)60778-X PMID:22655355

Liu, Z., Zhang, H., Tao, M., Yang, S., Wang, L., Liu, Y., ... He, Z. (2010). Organochlorine Pesticides in Consumer Fish and Mollusks of Liaoning Province, China: Distribution and Human Exposure Implications. *Archives of Environmental Contamination and Toxicology*, *59*(3), 444–453. doi:10.100700244-010-9504-7 PMID:20352204

Livingstone, D. J., & Manallack, D. T. (1993). Statistics using neural networks: Chance effects. *Journal of Medicinal Chemistry*, *36*(9), 1295–1297. doi:10.1021/jm00061a023 PMID:8487267

Li, X., He, J., & Li, S. (2007). Isolation of a chlorpyrifos-degrading bacterium, Sphingomonas sp. strain Dsp-2, and cloning of the mpd gene. *Research in Microbiology*, *158*(2), 143–149. doi:10.1016/j.resmic.2006.11.007 PMID:17306510

Lizano-Fallas, V., Masís-Mora, M., Espinoza-Villalobos, D., Lizano-Brenes, M., & Rodríguez-Rodríguez, C. E. (2017). Removal of pesticides and ecotoxicological changes during the simultaneous treatment of triazines and chlorpyrifos in biomixtures. *Chemosphere*, *182*, 106–113. doi:10.1016/j.chemosphere.2017.04.147 PMID:28494353

Llewellyn, R. (2002). The Good Bug Book. IPM Pty Ltd.

Lobe, J. (2006). *WHO urges DDT for malaria control Strategies*. Inter Press Service.

London, L., Beseler, C., Bouchard, M. F., Bellinger, D. C., Colosio, C., Grandjean, P., ... Stallones, L. (2012). Neurobehavioural and mneurodevelopmental effects of pesticide exposures. *Neurotoxicology*, *33*(4), 887–896. doi:10.1016/j.neuro.2012.01.004 PMID:22269431

Long, F., Zhu, A., Shi, H., Sheng, J., & Zhao, Z. (2015). Adsorption kinetics of pesticide in soil assessed by optofluidics-based biosensing platform. *Chemosphere*, *120*, 615–620. doi:10.1016/j.chemosphere.2014.09.072 PMID:25462305

Longnecker, M. P., Klebanoff, M. A., Zhou, H., & Brock, J. W. (2001). Association between maternal serum concentration of the DDT metabolite DDE and preterm and small-for-gestational-age babies at birth. *Lancet*, *358*(9276), 110–114. doi:10.1016/S0140-6736(01)05329-6 PMID:11463412

Loos, R. (2012). *Analytical Methods for the new proposed Priority Substances of the European Water Framework Directive (WFD)*. Revision of the Priority Substance List.

Lopez-Avilla, V., & Young, R. (1994). Microwave-assisted extraction of organic compounds from standards reference soils and sediments. *Analytical Chemistry*, *66*(7), 1, 1097–1106.

Lopez-Carillo, L., & Lopez-Cervantes, M. (1993). Effect of exposure to organophosphate pesticides on serum cholinesterase levels. *Archives of Environmental Health*, *48*(5), 359–363. doi:10.1080/00039896.1993.9936726 PMID:8215602

Lorenz, S., Rasmussen, J. J., Süß, A., Kalettka, T., Golla, B., Horney, P., ... Schäfer, R. B. (2017). Specifics and challenges of assessing exposure and effects of pesticides in small water bodies. *Hydrobiologia*, *793*(1), 213–224. doi:10.100710750-016-2973-6

Louis, A. H., Weigmann, D. L., Hipkins, P., & Stinson, E. R. (1996). *Pesticides and Aquatic Animals: A Guide to Reducing Impacts on Aquatic Systems*. Retrieved from http://www.ext.vt.edu/pubs/waterquality/420- 013/420-013.html

Lozowicka, B., Jankowska, M., & Kaczyński, P. (2012). Pesticide residues in Brassica vegetables and exposure assessment of consumers. *Food Control*, *25*(2), 561–575. doi:10.1016/j.foodcont.2011.11.017

Lu, C., Toepel, K., Irish, R., Fenske, R. A., Barr, D. B., & Bravo, R. (2006). Organic diets significantly lower children's dietary exposure to organophosphorus pesticides. *Environmental Health Perspectives*, *114*(2), 260–263. doi:10.1289/ehp.8418 PMID:16451864

Ludvík, J., & Zuman, P. (2000). Adsorption of 1, 2, 4-triazine pesticides metamitron and metribuzin on lignin. *Microchemical Journal*, *64*(1), 15–20. doi:10.1016/S0026-265X(99)00015-6

Luo, D., Pu, Y., Tian, H., Cheng, J., Zhou, T., Tao, Y., ... Mei, S. (2016). Concentrations of organochlorine pesticides in umbilical cord blood and related lifestyle and dietary intake factors among pregnant women of the Huaihe River Basin in China. *Environment International*, *92*, 276–283. doi:10.1016/j.envint.2016.04.017 PMID:27123771

Luo, Y., Zhang, X., Liu, X., Ficklin, D., & Zhang, M. (2008). Dynamic modeling of organophosphate pesticide load in surface water in the northern San Joaquin Valley watershed of California. *Environmental Pollution*, *156*(3), 1171–1181. doi:10.1016/j.envpol.2008.04.005 PMID:18457909

Lv, T., Zhang, Y., Casas, M. E., Carvalho, P. N., Arias, C. A., Bester, K., & Brix, H. (2016). Phytoremediation of imazalil and tebuconazole by four emergent wetland plant species in hydroponic medium. *Chemosphere*, *148*, 459–466. doi:10.1016/j.chemosphere.2016.01.064 PMID:26841287

Lydia, H. Z. (2015). Stream microbial diversity in response to environmental changes: Review and synthesis of existing research. *Frontiers in Microbiology*, *6*, 454. PMID:26042102

Lyons, G., & Watterson, A. (2010). *A Review of the Role Pesticides Play in some Cancers: Children, Farmers and Pesticide users at Risk?* CHEMTrust, UK. Retrieved from http://www.chemtrust.org.uk/Pesticidesandcancer.php

MacArthur, R. H., & Wilson, E. O. (1967). *The theory of Island Biogeography*. Princeton, NJ: Princeton University Press.

Mackenthun, K. M. (1969). The practice of water pollution biology. U.S. Dept. of Interior, Federal Water Pollution Control Administration.

MacNair, J. E., Patel, K. D., & Jorgenson, J. W. (1999). Ultrahigh-pressure reversed-phase capillary liquid chromatography: Isocratic and gradient elution using columns packed with 1.0-μm particles. *Analytical Chemistry*, *71*(3), 700–708. doi:10.1021/ac9807013 PMID:9989386

Macneale, K. H., Kiffney, P. M., & Scholz, N. L. (2010). Pesticides, aquatic food webs, and the conservation of Pacifi c salmon. *Frontiers in Ecology and the Environment*, *8*(9), 475–482. doi:10.1890/090142

Maheswari, S. T., & Ramesh, A. (2012). Adsorption and Degradation of Anilofos in Different Soils and its Environmental Impact in Soils. *International Journal of Environmental of Research*, *6*, 451–456.

Mahmood, I., Imadi, S. R., Shazadi,K., Gul, A. & Hakeem, K.R. (2016). *Effects of Pesticides on Environment.* Springer International Publishing.

Makelarski, J. A., Romitti, P. A., Rocheleau, C. M., Burns, T. L., Stewart, P. A., Waters, M. A., & Olney, R. S. (2014). Maternal periconceptional occupational pesticide exposure and neural tube defects. *Birth Defects Research. Part A, Clinical and Molecular Teratology*, *100*(11), 877–886. doi:10.1002/bdra.23293 PMID:25124525

Ma, L., He, F., Huang, T., Zhou, Q., Zhang, Y., & Wu, Z. (2016). Nitrogen and phosphorus transformations and balance in a pond-ditch circulation system for rural polluted water treatment. *Ecological Engineering*, *94*, 117–126. doi:10.1016/j.ecoleng.2016.05.051

Malaj, E., von der Ohe, P. C., Grote, M., Kühne, R., Mondy, C. P., Usseglio-Polatera, P. … Schäfer, R. B. (2014). Organic chemicals jeopardize the health of freshwater ecosystems on the continental scale. *Proceedings of the National Academy of Sciences, 111*(26), 9549-9554. Retrieved from http://www.pnas.org/content/111/26/9549.abstract

Malaj, E., Peter, C., Grote, M., Kühne, R., Mondy, C. P., Usseglio-Polatera, P., ... Schäfer, R. B. (2014). Organic chemicals jeopardize the health of freshwater ecosystems on the continental scale. *Proceedings of the National Academy of Sciences of the United States of America, 111*(26), 9549–9554. doi:10.1073/pnas.1321082111 PMID:24979762

Malghani, S., Chatterjee, N., Yu, H. X., & Luo, Z. (2009). Isolation and identification of profenofos degrading bacteria. *Brazilian Journal of Microbiology, 40*(4), 893–900. doi:10.1590/S1517-83822009000400021 PMID:24031438

Malik, P. K., & Saha, S. K. (2003). Oxidation of direct dyes with hydrogen peroxide using ferrous ion catalyst. *Separation and Purification Technology, 31*(3), 241–250. doi:10.1016/S1383-5866(02)00200-9

Malini, T. P., Ramesh, A., Selvi, J. A., Arthanareeswari, M., & Kamaraj, P. (2016). Kinetic Modeling of Photocatalytic Degradation of Alachlor using TiO_2 (Degussa P25) in Aqueous Solution. *Oriental Journal of Chemistry, 32*(6), 3165–3173. doi:10.13005/ojc/320639

Ma, M., Wen, K., Beier, R. C., Eremin, S. A., Li, C., Zhang, S., ... Wang, Z. (2016). Chemiluminescence Resonance Energy Transfer Competitive Immunoassay Employing Hapten-Functionalized Quantum Dots for the Detection of Sulfamethazine. *ACS Applied Materials & Interfaces, 8*(28), 17745–17750. doi:10.1021/acsami.6b04171 PMID:27362827

Mandal, A., & Singh, N. (2017). Optimization of atrazine and imidacloprid removal from water using biochars: Designing single or multi-staged batch adsorption systems. *International Journal of Hygiene and Environmental Health, 220*(3), 637–645. doi:10.1016/j.ijheh.2017.02.010 PMID:28433639

Mandal, A., Singh, N., & Purakayastha, T. J. (2017). Characterization of pesticide sorption behaviour of slow pyrolysis biochars as low cost adsorbent for atrazine and imidacloprid removal. *The Science of the Total Environment, 577*, 376–385. doi:10.1016/j.scitotenv.2016.10.204 PMID:27817922

Manisankar, P., Selvanathan, G., & Vedhi, C. (2006). Determination of pesticides using heteropolyacid montmorillonite clay-modified electrode with surfactant. *Talanta, 68*(3), 686–692. doi:10.1016/j.talanta.2005.05.021 PMID:18970376

Manncsa. (2009). *Pesticides in Agriculture*. Retrieved from http://agropedia.iitk.ac.in/content/pesticides-agriculture

Mansano, A. S., Moreira, R. A., Dornfeld, H. C., Freitas, E. C., Vieira, E. M., Sarmento, H., ... Seleghim, M. H. R. (2017). Effects of diuron and carbofuran and their mixtures on the microalgae Raphidocelis subcapitata. *Ecotoxicology and Environmental Safety, 142*, 312–321. doi:10.1016/j.ecoenv.2017.04.024 PMID:28433596

Manuel, A. E., Eugenio, L. P., Elena, M. C., Jesus, S. G., Juan-Carlos, M., & Luis, G. R. (2008). The mobility and degradation of pesticides in soils and the pollution of groundwater resources. *Agriculture, Ecosystems & Environment, 123*(4), 247–260. doi:10.1016/j.agee.2007.07.011

Margni, M., Rossier, D., Crettaz, P., & Jolliet, O. (2002). Life cycle impact assessment of pesticides on human health and ecosystems. *Agriculture, Ecosystems & Environment, 93*(1-3), 379–392. doi:10.1016/S0167-8809(01)00336-X

Mariyono, J. (2008). Direct and indirect impacts of integrated pest management on pesticide use: A case of rice agriculture in Java, Indonesia. *Pest Management Science, 64*(10), 1069–1073. doi:10.1002/ps.1602 PMID:18493927

Marneffe, Y., Comblin, S., Bussers, J. C., & Thome, J. P. (1996). Biomonitoring of the water quality in the River Warche (Belgium): Impact of tributaries and sewage effluents. *Netherlands Journal of Zoology, 46*(3), 337–350. doi:10.1163/156854295X00294

Maroni, M., Fanetti, A. C., & Metruccio, F. (2006). Risk assessment and management of occupational exposure to pesticides in agriculture. *La Medicina del Lavoro*, *97*, 430–437. PMID:17017381

Martin, H. (2009). *Introduction to Organic Farming*. Ontario Ministry of Agriculture, Food and Rural Affairs.

Martínez-Valenzuela, C., Gómez-Arroyo, S., Villalobos-Pietrini, R., Waliszewski, S., Calderón-Segura, M. E., Félix-Gastélum, R., & Álvarez-Torres, A. (2009). Genotoxic biomonitoring of agricultural workers exposed to pesticides in the north of Sinaloa State, Mexico. *Environment International*, *35*(8), 1155–1159. doi:10.1016/j.envint.2009.07.010 PMID:19665797

Mary, G. (2014). *What is organic production?* National Agricultural Library. USDA.

Mathew, L. L. (2012). *Organochloride Pesticide Toxicity, Drugs, Diseases and Procedures*. Medscape References.

Mathie, A. J. (1999). *Chemical Treatment for Cooling Water*. Upper Saddle River, NJ: Prentice Hall.

Mathur, S. C. (1999). Future of Indian pesticides industry in next millennium. *Pesticide Information*, *24*(4), 9–23.

Mathur, S. C. (1999). Future of Indian pesticides industry in next millennium. *Pesticide Information.*, *24*(4), 9–23.

Matouq, M. A., Al-Anber, Z. A., Tagawa, T., Aljbour, S., & Al-Shannag, M. (2008). Degradation of dissolved diazinon pesticide in water using the high frequency of ultrasound wave. *Ultrasonics Sonochemistry*, *15*(5), 869–874. doi:10.1016/j.ultsonch.2007.10.012 PMID:18178508

Matsumura, F. (1973). Degradation of Pesticide Residues in the Environment. In C. A. Edwards (Ed.), *Environmental Pollution by Pesticides. Environmental Science Research*. doi:10.1007/978-1-4615-8942-6_14

Matsumura, F. (1985). *Toxicology of insecticides* (2nd ed.). New York: Plenum Press. doi:10.1007/978-1-4613-2491-1

Maund, S. J., Peither, A., Taylor, E. J., Jüttner, I., Beyerle-Pfnür, R., Lay, J. P., & Pascoe, D. (1992). Toxicity of lindane to freshwater insect larvae in compartments of an experimental pond. *Ecotoxicology and Environmental Safety*, *23*(1), 76–88. doi:10.1016/0147-6513(92)90023-V PMID:1375150

Maurya, P. K., & Malik, D. S. (2016). Bioaccumulation of xenobiotics compound of pesticides in riverine system and its control technique: A critical review. *Journal of Industrial Pollution Control*, *32*, 580–594.

Mays, L. W. (1999). *Water Distribution Systems Handbook*. New York: McGraw-Hill.

Mazhabi, M., Nemati, H., Rouhani, H., Tehranifar, A., Moghadam, E. M., Kaveh, H., & … . (2011). The effect of Trichoderma on polianthes qualitative and quantitative properties. *Journal of Animal and Plant Sciences*, *21*(3), 617–621.

Mazzatorta, P., Smiesko, M., Lo Piparo, E., & Benfenati, E. (2005). QSAR model for predicting pesticide aquatic toxicity. *Journal of Chemical Information and Modeling*, *45*(6), 1767–1774. doi:10.1021/ci0502471 PMID:16309283

McAllister, D. E., Craig, J. F., Davidson, N., Delany, S., & Seddon, M. (2001). *Biodiversity Impacts of Large Dams*. International Union for Conservation of Nature and United Nations Environmental Programme, Gland and Nairobi.

McArthur, J. V., Leff, L. G., & Smith, M. H. (1992). Genetic diversity of bacteria along a stream continuum. *Journal of the North American Benthological Society*, *11*(3), 269–277. doi:10.2307/1467647

McBride, M. B., Tyler, L. D., & Hovde, D. A. (1981). Cadmium Adsorption by Soils and Uptake by Plants as Affected by Soil Chemical Properties 1. *Soil Science Society of America Journal*, *45*(4), 739–744. doi:10.2136ssaj1981.036159 95004500040013x

McKim, J. M., Bradbury, S. P., & Niemi, G. J. (1987). Fish acute toxicity syndromes and their use in the QSAR approach to hazard assessment. *Environmental Health Perspectives*, *71*, 171–186. doi:10.1289/ehp.8771171 PMID:3297660

McMahon, T. A., Halstead, N. T., Johnson, S., Raffel, T. R., Romansic, J. M., Crumrine, P. W., & Rohr, J. R. (2012). Fungicide-induced declines of freshwater biodiversity modify ecosystem functions and services. *Ecology Letters*, *15*(7), 714–722. doi:10.1111/j.1461-0248.2012.01790.x PMID:22587750

McMichael, P. (2009). Contemporary contradictions of the global development project: Geopolitics, global ecology and the 'development climate'. *Third World Quarterly*, *30*(1), 247–262. doi:10.1080/01436590802622987

Megharaj, M., Venkateswarlu, K., & Rao, A. S. (1988). Tolerance of algal population in rice soil to carbofuran application. *Current Science*, *57*(2), 100–102.

Mehta, P. S., Mehta, A. S., Mehta, S. J., & Makhijani, A. B. (1990). Bhopal Tragedy's Health Effects: A review of Methyl Isocyanate Toxicity. *Journal of the American Medical Association*, *264*(21), 2781–2787. doi:10.1001/jama.1990.03450210081037 PMID:2232065

Memon, G. Z., Bhanger, M. I., Akhtar, M., Talpur, F. N., & Memon, J. R. (2008). Adsorption of methyl parathion pesticide from water using watermelon peels as a low cost adsorbent. *Chemical Engineering Journal*, *138*(1), 616–621. doi:10.1016/j.cej.2007.09.027

Memon, G. Z., Moghal, M., Memon, J. R., Memon, N. N., & Bhanger, M. I. (2014). Adsorption of selected pesticides from aqueous solutions using cost effective walnut shells. *Adsorption*, *4*(10).

Mendez, M. A., Garcia-Etseban, R., Guxens, M., Vrijheid, M., Kogevinas, M., Goñi, F., ... Sunyer, J. (2011). Prenatal organochlorine compound exposure, rapid weight gain, and overweight in infancy. *Environmental Health Perspectives*, *119*(2), 272–278. doi:10.1289/ehp.1002169 PMID:20923745

Meng, X., Wei, J., Ren, X., & … . (2013). Biosensors and Bioelectronics A simple and sensitive fl uorescence biosensor for detection of organophosphorus pesticides using H 2 O 2 -sensitive quantum dots / bi-enzyme. *Biosensors & Bioelectronics*, *47*, 402–407. doi:10.1016/j.bios.2013.03.053 PMID:23612061

Mensah, P. K., Palmer, C. G., & Muller, W. J. (2014). Lethal and sublethal effects of pesticides on aquatic organisms: the case of a freshwater shrimp exposure to Roundup. In *Pesticides-Toxic Aspects*. InTech.

Metcalf, D. H., & Holmes, H. (1969). Toxicology and physiology; EEG, psychological, and neurological alterations in humans with organophosphorus exposure. *Annals of the New York Academy of Sciences*, *160*(1 Biological Ef), 357–365. doi:10.1111/j.1749-6632.1969.tb15857.x PMID:5257403

Metcalfe-Smith, J. L. (1996). Biological water-quality assessment of rivers: use of macroinvertebrate communities. In G. Petts & P. Calow (Eds.), *River Restoration*. Oxford, UK: Blackwell Science.

Michael, H. C. (2010). Water pollution. In Encyclopedia of Earth. Cleveland, OH: National Council on Science and the Environment.

Miles, C. J., & Pfeuffer, R. J. (1997). Pesticides in canals of South Florida. *Archives of Environmental Contamination and Toxicology*, *32*(4), 337–345. doi:10.1007002449900194 PMID:9175497

Millenium Ecosystem Assessment (MEA). (2005). *Ecosystems and Human Well-being: Biodiversity Synthesis*. Washington, DC: World Resources Institute.

Millennium Ecosystem Assessment (MEA). (2005). *Ecosystems and Human Well-being: Wetlands and Water Synthesis*. Washington, DC: World Resources Institute.

Miller, G. T. (2002). Living in the Environment (12th ed.). Belmont: Wadsworth/Thomson Learning.

Miller, S., & Shah, M. A. (1982). Cholinesterase activities of workers exposed to organophosphorus insecticides in Pakistan and Haiti and an evaluation of the tintometric method. *Journal of Environmental Science and Health*, *17*(2), 125–130. doi:10.1080/03601238209372307 PMID:7077053

Milun, V., Grgas, D., & Dragičević, T. L. (2016). Assessment of PCB and chlorinated pesticide accumulation in mussels at Kaštela Bay (Eastern Adriatic). *The Science of the Total Environment*, *562*, 115–127. doi:10.1016/j.scitotenv.2016.03.133 PMID:27096632

Mimmo, T., Bartucca, M. L., Del Buono, D., & Cesco, S. (2015). Italian ryegrass for the phytoremediation of solutions polluted with terbuthylazine. *Chemosphere*, *119*, 31–36. doi:10.1016/j.chemosphere.2014.04.114 PMID:24954449

Minear, R. A., & Amy, G. L. (1996). *Disinfection By-Products in Water Treatment: The Chemistry of Their Formation and Control*. Boca Raton, FL: CRC Press/Lewis Publishers.

Ministry of Environment and Forests (MoEF). (1990). Wetlands of India: a directory. New Delhi: MoEF, Government of India.

Minshall, G. W. (1978). Autotrophy in stream ecosystems. *Bioscience*, *28*(12), 767–771. doi:10.2307/1307250

Mishra, J., Tiwari, S., Singh, S., & Arora, N. K. (2015). Biopesticides: Where we stand? In N. K. Arora (Ed.), *Plant microbes symbiosis: applied facets* (pp. 37–75). Springer.

Mitra, A., Chatterjee, C., & Mandal, F. B. (2011). Synthetic chemical pesticides and their effects on birds. *Res J Environ Toxicol*, *5*(2), 81–96. doi:10.3923/rjet.2011.81.96

Mitton, F. M., Gonzalez, M., Monserrat, J. M., & Miglioranza, K. S. (2016). Potential use of edible crops in the phytoremediation of endosulfan residues in soil. *Chemosphere*, *148*, 300–306. doi:10.1016/j.chemosphere.2016.01.028 PMID:26814704

Mnyone, L. L., Koenraadt, C. J. M., Lyimo, I. N., Mpingwa, M. W., Takken, W., & Russell, T. L. (2010). Anopheline and culicine mosquitoes are not repelled by surfaces treated with the entomopathogenic fungi Metarhiziumanisopliae and Beauveriabassiana. *Parasites & Vectors*, *3*(1), 80. doi:10.1186/1756-3305-3-80 PMID:20799937

Moe, S. J., De Schamphelaere, K., Clements, W. H., Sorensen, M. T., Van den Brink, P. J., & Liess, M. (2013). Combined and interactive effects of global climate change and toxicants on populations and communities. *Environmental Toxicology and Chemistry*, *32*(1), 49–61. doi:10.1002/etc.2045 PMID:23147390

Mohammad, S. G., Ahmed, S. M., & Badawi, A. F. M. (2015). A comparative adsorption study with different agricultural waste adsorbents for removal of oxamyl pesticide. *Desalination and Water Treatment*, *55*(8), 2109–2120. doi:10.1080/19443994.2014.930797

Molur, S., Smith, K. G., Daniel, B. A., & Darwall, W. R. T. (2011). *The status and distribution of freshwater biodiversity in the Western Ghats, India*. Coimbatore, India: IUCN, Cambridge, UK and Glad, Switzerland and Zoo Outreach Organization.

Monirith, I., Uneno, D., Takahashi, S., & Nakata, H. (2003). sudryanto A, Subramanian A, karupiah S, ismail A, Muchtar M, Zhehg J, Richardson B.J., Prudente M, hue N.D., Jana T.s., Tkalin A.V and Tanabe S, Asia –pacific mussel Watch: Monitoring contamination of persistent Organochlorine compounds in coastal water of Asian countries, Mar. pollu. *Bull.*, *46*, 281–300.

Montgomery, M. P., Kamel, F., Saldana, T. M., Alavanja, M. C., & Sandler, D. P. (2008). Incident diabetes and pesticide exposure among licensed pesticide applicators. *Agricultural Health Study, 1993–2003*.

Montgomery, J. M. (1985). *Consulting Engineers, Water Treatment Principles and Design.* New York: John Wiley & Sons, Inc.

Montgomery, M. P., Kamel, F., Saldana, T. M., Alavanja, M. C., & Sandlerm, D. P. (2008). Incident diabetes and pesticide exposure among licensed pesticide applicators: Agricultural Health Study, 1993–2003. *American Journal of Epidemiology, 167*(10), 1235–1246. doi:10.1093/aje/kwn028 PMID:18343878

Montuori, P., Aurino, S., Garzonio, F., Sarnacchiaro, P., Polichetti, S., Nardone, A., & Triassi, M. (2016). Estimates of Tiber River organophosphate pesticide loads to the Tyrrhenian Sea and ecological risk. *The Science of the Total Environment, 559*, 218–231. doi:10.1016/j.scitotenv.2016.03.156 PMID:27065443

Moore, D. R., Breton, R. L., & MacDonald, D. B. (2003). A comparison of model performance for six quantitative structure-activity relationship packages that predict acute toxicity to fish. *Environmental Toxicology and Chemistry, 22*(8), 1799–1809. doi:10.1897/00-361 PMID:12924579

Morillo, E., Undabeytia, T., & Maqueda, C. (1997). Adsorption of glyphosate on the clay mineral montmorillonite: Effect of Cu^{2+} in solution and adsorbed on the mineral. *Environmental Science & Technology, 31*(12), 3588–3592. doi:10.1021/es970341l

Morillo, E., Undabeytia, T., Maqueda, C., & Ramos, A. (2000). Glyphosate adsorption on soils of different characteristics. Influence of copper addition. *Chemosphere, 40*(1), 103–107. doi:10.1016/S0045-6535(99)00255-6 PMID:10665451

Morin, A., & Nadon, D. (1991). Size distribution of epilithic lotic invertebrates and implication for community metabolism. *Journal of the North American Benthological Society, 10*(3), 300–308. doi:10.2307/1467603

Mortensen, G. K., Main, K. M., Andersson, A. M., Leffers, H., & Skakkebæk, N. E. (2005). Determination of phthalate monoesters in human milk, consumer milk, and infant formula by tandem mass spectrometry (LC–MS–MS). *Analytical and Bioanalytical Chemistry, 382*(4), 1084–1092. doi:10.100700216-005-3218-0 PMID:15933851

Moscardi, F. (1999). Assessment of the application of baculoviruses for control of Lepidoptera. *Annual Review of Entomology, 122*, 309–320. PMID:15012374

Mossa, A. T. H. (2016). Green Pesticides: Essential Oils as Biopesticides in Insect-pest Management. *Journal of Environmental Science and Technology, 9*(5), 354–378. doi:10.3923/jest.2016.354.378

Moss, B. (1988). *Ecology of Fresh waters, Man and Medium* (2nd ed.). Blackwell.

Moss, B. (2008). Water pollution by agriculture. *Philosophical Transactions of the Royal Society of London. Series B, Biological Sciences, 363*(1491), 659–666. doi:10.1098/rstb.2007.2176 PMID:17666391

Mostafalou, S., & Abdollahi, M. (2013). Pesticides and human chronic diseases: evidences, mechanisms, and perspectives. *Toxicol. Appl. Pharmacol., 268*(2), 157–77.

Mostafalou, S., & Abdollahi, M. (2012). Concerns of environmental persistence of pesticides and human chronic diseases. *Journal of Clinical and Experimental Pharmacology, 2*(3), 1–2.

Mostafalou, S., & Abdollahi, M. (2013). Pesticides and human chronic diseases: Evidences, mechanisms, and perspectives. *Toxicology and Applied Pharmacology, 268*(2), 157–177. doi:10.1016/j.taap.2013.01.025 PMID:23402800

Mostafalou, S., & Abdollahi, M. (2017). Pesticides: An update of human exposure and toxicity. *Archives of Toxicology, 91*(2), 549–599. doi:10.100700204-016-1849-x PMID:27722929

Moussavi, G., Hosseini, H., & Alahabadi, A. (2013). The investigation of diazinon pesticide removal from contaminated water by adsorption onto NH 4 Cl-induced activated carbon. *Chemical Engineering Journal*, *214*, 172–179. doi:10.1016/j.cej.2012.10.034

Mowbray, D. L. (1988). *Pesticide use In the South Pacific. UNEP Regional Seas Reports and Studies No. 89*. Nairobi: United Nations Environment Programme.

Mugambi, J., Kanja, L., Maitho, T. E., Skaare, J. U., & Lokken, P. (1989). Organochlorine pesticide residues in domestic fowl (Gallus domesticus) eggs from central Kenya. *Journal of the Science of Food and Agriculture*, *48*(2), 165–176. doi:10.1002/jsfa.2740480205

Muirhead, T. R. C. (1971). Pesticides and freshwater fauna. Academic Press.

Mulla, M. S., & Mian, L. S. (1981). Biological and environmental impacts of the insecticides malathion and parathion on nontarget biota in aquatic ecosystems. *Residue Reviews*, *78*, 100–135. PMID:7017857

Munoz-Quezada, M. T., Lucero, B. A., Barr, D. B., Steenland, K., Levy, K., Ryan, P. B., & Vega, C. (2013). Neurodevelopmental effects in children associated with exposure to organophosphate pesticides: A systematic review. *Neurotoxicology*, *39*, 158–168. doi:10.1016/j.neuro.2013.09.003 PMID:24121005

Münze, R., Orlinskiy, P., Gunold, R., Paschke, A., Kaske, O., Beketov, M. A., ... Liess, M. (2015). Pesticide impact on aquatic invertebrates identified with Chemcatcher® passive samplers and the SPEARpesticides index. *The Science of the Total Environment*, *537*, 69–80. doi:10.1016/j.scitotenv.2015.07.012 PMID:26282741

Munz, N. A., Burdon, F. J., De Zwart, D., Junghans, M., Melo, L., Reyes, M., ... Stamm, C. (2017). Pesticides drive risk of micropollutants in wastewater-impacted streams during low flow conditions. *Water Research*, *110*, 366–377. doi:10.1016/j.watres.2016.11.001 PMID:27919541

Muralidharan, S., Ganesan, K., Nambirajan, K., Kirubhanandhini, V., & Dhanajayan, V. (2014). Wetland birds indicators of pesticide contamination- Current and future prospects for research in India. In G. V. Gopi & S. A. Hussain (Eds.), *Waterbirds & Protected Areas* (Vol. 16, pp. 314–323). Dehradun, India: Wildlife Institute of India.

Murphy, G. (2005). *Resistance Management- Pesticide Rotation*. Ontario Ministry of Agriculture, Food and Rural Affairs.

Muruganandham, M., & Swaminathan, M. (2004). Decolourisation of Reactive Orange4 by Fenton and photo-Fenton oxidation technology. *Dyes and Pigments*, *63*(3), 315–321. doi:10.1016/j.dyepig.2004.03.004

Muturi, E. J., Donthu, R. K., Fields, C. J., Moise, I. K., & Kim, C. H. (2017). Effect of pesticides on microbial communities in container aquatic habitats. *Scientific Reports*, *16*(7), 44565. doi:10.1038rep44565 PMID:28300212

Myers, M. J., Ylitalo, G. M., Krahn, M. M., Boyd, D., Calkins, D., Burkanov, V., & Atkinson, S. (2008). Organochlorine contaminants in endangered Steller sea lion pups (Eumetopias jubatus) from western Alaska and the Russian Far East. *The Science of the Total Environment*, *396*(1), 60–69. doi:10.1016/j.scitotenv.2008.02.008 PMID:18384839

Nadeau, L. J., Menn, F. M., Breen, A., & Sayler, G. S. (1994). Aerobic degradation of 1, 1, 1-trichloro-2, 2-bis (4-chlorophenyl) ethane (DDT) by Alcaligenes eutrophus A5. *Applied and Environmental Microbiology*, *60*(1), 51–55. PMID:8117093

Nagaraju, D., & Huang, S. D. (2007). Determination of triazine herbicides in aqueous samples by dispersive liquid-liquid microextraction with gas chromatography-ion trap mass spectrometry. *Journal of Chromatography. A*, *1161*(1–2), 89–97. doi:10.1016/j.chroma.2007.05.065 PMID:17574561

Naidoo, S., London, L., & Burdorf, A. (2010). Pesticide safety training and practices in women working in small-scale agriculture in South Africa. *Occupational and Environmental Medicine*, *67*(12), 823–828. doi:10.1136/oem.2010.055863 PMID:20576924

Naksen, W., Prapamontol, T., Mangklabruks, A., Chantara, S., Thavornyutikarn, P., Robson, M. G., & Panuwet, P. (2016). A single method for detecting 11 organophosphate pesticides in human plasma and breastmilk using GC-FPD. *Journal of Chromatography. B, Analytical Technologies in the Biomedical and Life Sciences*, *1025*, 92–104. doi:10.1016/j.jchromb.2016.04.045 PMID:27232054

Naqvi, S. M., & Vaishnavi, C. (1993). Bioaccumulative potential & toxicity of endosulfan insecticide to non-target animals. *Comparative Biochemistry & Physiology C*, *3*(3), 347–361. doi:10.1016/0742-8413(93)90071-R PMID:7900959

Narahashi, T. (1986). Nerve membrane ionic channels as the target of toxicants. *Archives of Toxicology. Supplement. Archiv fur Toxikologie. Supplement*, *9*, 3–13. doi:10.1007/978-3-642-71248-7_1 PMID:2434060

Narita, K., Matsui, Y., Iwao, K., Kamata, M., Matsushita, T., & Shirasaki, N. (2014). Selecting pesticides for inclusion in drinking water quality guidelines on the basis of detection probability and ranking. *Environment International*, *63*, 114–120. doi:10.1016/j.envint.2013.10.019 PMID:24286868

Nasterlack, M. (2007). Pesticides and childhood cancer: An update. *International Journal of Hygiene and Environmental Health*, *210*(5), 645–657. doi:10.1016/j.ijheh.2007.03.001 PMID:17434797

Nathan, S. S. (2015). A Review of Biopesticides and Their Mode of Action Against Insect Pests. In P. Thangavel & G. Sridevi (Eds.), Environmental Sustainability. Academic Press.

Naushad, M., ALOthman, Z. A., Khan, M. R., ALQahtani, N. J., & ALSohaimi, I. H. (2014). Equilibrium, kinetics and thermodynamic studies for the removal of organophosphorus pesticide using Amberlyst-15 resin: Quantitative analysis by liquid chromatography–mass spectrometry. *Journal of Industrial and Engineering Chemistry*, *20*(6), 4393–4400. doi:10.1016/j.jiec.2014.02.006

Nawaz, M., Mabubu, J. I., & Hua, H. (2016). Article. *Journal of Entomology and Zoology Studies*, *4*(2), 241–246.

Nejadkoorki, F. (2012). Environmental Benefits of Organic Farming. *International Conference on Applied Life Sciences (ICALS2012)*.

Netzeva, T. I., Dearden, J. C., Edwards, R., Worgan, A. D., & Cronin, M. T. (2004). QSAR Analysis of the Toxicity of Aromatic Compounds to Chlorella v ulgaris in a Novel Short-Term Assay. *Journal of Chemical Information and Computer Sciences*, *44*(1), 258–265. doi:10.1021/ci034195g PMID:14741035

Neuwoehner, J., Zilberman, T., Fenner, K., & Escher, B. I. (2010). QSAR-analysis and mixture toxicity as diagnostic tools: Influence of degradation on the toxicity and mode of action of diuron in algae and daphnids. *Aquatic Toxicology (Amsterdam, Netherlands)*, *97*(1), 58–67. doi:10.1016/j.aquatox.2009.12.005 PMID:20044149

Newbold, R. R., Padilla-Banks, E., Snyder, R. J., Phillips, T. M., & Jefferson, W. N. (2007). Developmental exposure to endocrine disruptors and the obesity epidemic. *Reproductive Toxicology (Elmsford, N.Y.)*, *23*(3), 290–296. doi:10.1016/j.reprotox.2006.12.010 PMID:17321108

Nguyen, M.-H., Nguyen, T.-H.-N., Hwang, I.-C., Bui, C.-B., & Park, H.-J. (2016). Effects of the physical state of nanocarriers on their penetration into the root and upward transportation to the stem of soybean plants using confocal laser scanning microscopy. *Crop Protection (Guildford, Surrey)*, *87*, 25–30. doi:10.1016/j.cropro.2016.04.012

Nicolopoulou-Stamati, P., Maipas, S., Kotampasi, C., Stamatis, P., & Hens, L. (2016). Chemical Pesticides and Human Health: The Urgent Need for a New Concept in Agriculture. *Frontiers in Public Health*, *4*, 148. doi:10.3389/fpubh.2016.00148 PMID:27486573

Nicolopoulou-Stamati, P., & Pitsos, M. A. (2001). The impact of endocrine disrupters on the female reproductive system. *Human Reproduction Update*, *7*(3), 323–330. doi:10.1093/humupd/7.3.323 PMID:11392379

Nie, S. M., & Emery, S. R. (1997). Probing single molecules and single nanoparticles by surface-enhanced Raman scattering. *Science, 275*(5303), 1102–1106. doi:10.1126cience.275.5303.1102 PMID:9027306

Niihori, Y., Shima, D., Yoshida, K., Hamada, K., Nair, L. V., Hossain, S., & Negishi, Y. (2018). High-performance liquid chromatography mass spectrometry of gold and alloy clusters protected by hydrophilic thiolates. *Nanoscale, 10*(4), 1641–1649. doi:10.1039/C7NR07840D PMID:29192295

Nikinmaa, M. (2014). *An introduction to aquatic toxicology*. Elsevier. doi:10.1016/B978-0-12-411574-3.00002-5

NIOH (National Institute of Occupational Health). (2002). Final Report of the Investigation of Unusual Illnesses Allegedly Produced by Endosulfan Exposure. In *Padre Village of Kasargod District (N Kerala)*. Ahmedabad: National Institute of Occupational Health, Indian Council for Medical Research.

Niti, S., & Alka, P. (2008). Determination of Persistent Pesticide Residues in Ground Water of Agra Region Using Solid Phase Extraction and Gas Chromatography. *Asian Journal of Water, Environment and Pollution, 5*, 91–94.

Ni, Y., Xiao, W., & Kokot, S. (2009). Application of chemometrics methods for the simultaneous kinetic spectrophotometric determination of aminocarb and carbaryl in vegetable and water samples. *Journal of Hazardous Materials, 168*(2-3), 1239–1245. doi:10.1016/j.jhazmat.2009.03.003 PMID:19349118

Njoku, V. O., Islam, M. A., Asif, M., & Hameed, B. H. (2014). Utilization of sky fruit husk agricultural waste to produce high quality activated carbon for the herbicide bentazon adsorption. *Chemical Engineering Journal, 251*, 183–191. doi:10.1016/j.cej.2014.04.015

Nolan, B. T., Dubus, I. G., Surdyk, N., Fowler, H. J., Burton, A., Hollis, J. M., ... Jarvis, N. J. (2008). Identification of key climatic factors regulating the transport of pesticides in leaching and to tile drains. *Pest Management Science, 64*(9), 933–944. doi:10.1002/ps.1587 PMID:18416432

Norman, T., & Banuelos, G. (1999). *Phytoremediation of Contaminated Soil and Water*. Boca Raton, FL: CRC Press/Lewis Publishers.

Norstrom, R. J., Simon, M., Muir, C. G., & Schweinsburq, R. E. (1988). Organochlorine contaminants in arctic marine food chains: Identification, geographical distribution, and temporal trends in polar bears. *Environmental Science & Technology, 22*(9), 1063–1107. doi:10.1021/es00174a011 PMID:22148661

Norum, U., Friberg, N., Jensen, M. R., Pedersen, J. M., & Bjerregaard, P. (2010). Behavioral changes in three species of freshwater macroinvertebrates exposed to the pyrethroid lambda-cyhalothrin: Laboratory and stream microcosm studies. *Aquatic Toxicology (Amsterdam, Netherlands), 98*(4), 328–335. doi:10.1016/j.aquatox.2010.03.004 PMID:20362345

Nowell, L. H., Moran, P. W., Schmidt, T. S., Norman, J. E., Nakagaki, N., Shoda, M. E., ... Hladik, M. L. (2018). Complex mixtures of dissolved pesticides show potential aquatic toxicity in a synoptic study of Midwestern U.S. streams. *The Science of the Total Environment, 613–614*, 1469–1488. doi:10.1016/j.scitotenv.2017.06.156 PMID:28802893

Noyes, P. D., & Lema, S. C. (2015). Forecasting the impacts of chemical pollution and climate change interactions on the health of wildlife. *Current Zoology, 61*(4), 669–689.

Noyes, P. D., McElwee, M. K., Miller, H. D., Clark, B. W., Van Tiem, L. A., Walcott, K. C., ... Levin, E. D. (2009). The toxicology fo climate change: Environmental contaminants in a warming world. *Environment International, 35*(6), 971–986. doi:10.1016/j.envint.2009.02.006 PMID:19375165

NRA (National Registration Authority). (2000). *Review of Chlorpyrifos*. Commonwealth of Australia. Retrieved from www.nra.gov.au/chemrev/chemrev.shtml

Ntow, W. J. (2005). Pesticide residues in Volta Lake, Ghana. *Lakes and Reservoirs: Research and Management, 10*(4), 243–248. doi:10.1111/j.1440-1770.2005.00278.x

Nwilene, F. E., Nwanze, K. F., & Youdeowei, A. (2008). Impact of integrated pest management on food and horticultural crops in Africa. *Entomologia Experimentalis et Applicata, 128*(3), 355–363. doi:10.1111/j.1570-7458.2008.00744.x

Nyer, E. K. (1993). Groundwater Treatment Technology (2nd ed.). van Nostrand Reinhold.

Nystrom, B., Bjornsater, B., & Blanck, H. (1999). Effects of sulfonylurea herbicides on non-target aquatic micro- organisms: Growth inhibition of micro-algae and short-term inhibition of adenine and thymidine incorporation in periphyton communities. *Aquatic Toxicology (Amsterdam, Netherlands), 47*(1), 9–22. doi:10.1016/S0166-445X(99)00007-7

O'Neil, W., Raucher, R., & Wayzata, M. N. (1998). *Groundwater Policy Education Project; Groundwater public policy leaflet series #4: The costs of groundwater contamination.* Retrieved from http://www.dnr.state.wi.us/org/water/dwg/gw/costofgw.htm

Odum, H. T. (1957). Tropic structure and productivity of Silver Springs, Florida. *Ecological Monographs, 27*, 55-112.

Ogram, A. V., Jessup, R. E., Ou, L. T., & Rao, P. S. C. (1985). Effects of sorption on biological degradation rates of (2, 4-dichlorophenoxy) acetic acid in soils. *Applied and Environmental Microbiology, 49*(3), 582–587. PMID:3994366

Olette, R., Couderchet, M., Biagianti, S., & Eullaffroy, P. (2008). Toxicity and removal of pesticides by selected aquatic plants. *Chemosphere, 70*(8), 1414–1421. doi:10.1016/j.chemosphere.2007.09.016 PMID:17980900

Oliver, D. P., Kookana, R. S., Anderson, J. S., Cox, J., Waller, N., & Smith, L. (2012). The off-site transport of pesticide loads from two land uses in relation to hydrological events in the Mt. Lofty Ranges, South Australia. *Agricultural Water Management, 106*, 70–77. doi:10.1016/j.agwat.2011.07.012

Olu-Owolabi, B. I., Oputu, O. U., Adebowale, K. O., Ogonsolu, O., & Olujimi, O. O. (2012). Biosorption of Cd2+ and Pb2+ ions onto mango stone and cocoa pod waste: Kinetic and equilibrium studies. *Scientific Research and Essays, 7*(15), 1614–1629. doi:10.5897/SRE11.2248

Omri, A., & Benzina, M. (2013). Adsorption characteristics of silver ions onto activated carbon prepared from almond shell. *Desalination and Water Treatment, 51*(10-12), 2317–2326. doi:10.1080/19443994.2012.734585

Ongley, E. D. (1996). *Control of water pollution from Agriculture.* Food and Agriculture Organisation of United Nation Food and Agriculture Org.

Onuska, F. I., & Terry, K. A. (1993). Extraction of pesticides from sediments using a microwave technique. *Chromatographia, 36*(1), 191–194. doi:10.1007/BF02263861

Otero, R., Fernández, J. M., González, M. A., Pavlovic, I., & Ulibarri, M. A. (2013). Pesticides adsorption–desorption on Mg–Al mixed oxides. Kinetic modeling, competing factors and recyclability. *Chemical Engineering Journal, 221*, 214–221. doi:10.1016/j.cej.2013.02.007

Otieno, P. O., Owuor, P. O., Lalah, J. O., Pfister, G., & Schramm, K. W. (2013). Impacts of climate-induced changes on the distribution of pesticides residues in water and sediment of Lake Naivasha, Kenya. *Environmental Monitoring and Assessment, 185*(3), 2723–2733. doi:10.100710661-012-2743-5 PMID:22797882

Öztemiz, C. S. (2008). Biological control in organic farming. *Journal of the AgriculturalFaculty of Gaziosmanpaşa University, 25*, 19–27.

Paetzold, S., Klein, C., & Bruemmer, G. W. (2007). Run-off transport of herbicides during natural and simulated rainfall and its reduction by vegetated filter strips. *Soil Use and Management, 23*(3), 294–305. doi:10.1111/j.1475-2743.2007.00097.x

Palacios Xutuc, C. N. (2010). *Manual to Train Trainers on Safe and Correct Use of Plant Protection Products and Integrated Pest Management (IPM) CropLife Latin America*. Guatemala City, Guatemala: Pesticides and Human Health. 2014. Available from http://www. pesticidereform.org/section.php?id=38

Palis, F. G., Flor, R. J., Warburton, H., & Hossain, M. (2006). Our farmers at risk: Behavior and belief system in pesticide safety. *J. Public Health*, *28*(1), 43–48. doi:10.1093/pubmed/fdi066 PMID:16436451

Pal, O. R., & Vanjara, A. K. (2001). Removal of malathion and butachlor from aqueous solution by clays and organoclays. *Separation and Purification Technology*, *24*(1), 167–172. doi:10.1016/S1383-5866(00)00226-4

Pandey, J., Chauhan, A., & Jain, R. K. (2009). Integrative approaches for assessing the ecological sustainability of in situ bioremediation. *FEMS Microbiology Reviews*, *33*(2), 324–375. doi:10.1111/j.1574-6976.2008.00133.x PMID:19178567

Papa, E., Villa, F., & Gramatica, P. (2005). Statistically Validated QSARs, Based on Theoretical Descriptors, for Modeling Aquatic Toxicity of Organic Chemicals in Pimephales p romelas (Fathead Minnow). *Journal of Chemical Information and Modeling*, *45*(5), 1256–1266. doi:10.1021/ci050212l PMID:16180902

Pape-Lindstrom, P. A., & Lydy, M. J. (1997). Synergistic toxicity of atrazine and organophosphate insecticides contravenes the response addition mixture model. *Environmental Toxicology and Chemistry*, *16*(11), 2415–2420. doi:10.1002/etc.5620161130

Papendick, R. I., Elliott, L. F., & Dahlgren, R. B. (1986). Environmental consequences of modern production agriculture: How can alternative agriculture address these issues and concerns. *American Journal of Alternative Agriculture*, *1*(1), 3–10. doi:10.1017/S0889189300000722

Pareja, L., Cesio, V., Heinzen, H., & Fernández-Alba, A. R. (2011). Evaluation of various QuEChERS based methods for the analysis of herbicides and other commonly used pesticides in polished rice by LC–MS/MS. *Talanta*, *83*(5), 1613–1622. doi:10.1016/j.talanta.2010.11.052 PMID:21238760

Parker, L. W., Rydenvhi, J., Thomas, S., & Whitford, W. G. (1985). Effects of oxamyl and chlordane on the activities of nontarget soil organisms. *Biology and Fertility of Soils*, *1*(3), 141–148. doi:10.1007/BF00301781

Parsons, S. A., Boxall, A. B. A., Sinclair, C. J., & Ramwell, C. (2008). *Pesticide degradates of concern to the drinking water community*. Denver, CO: AWWA Research Foundation.

Pateiro-Moure, M., Perez-Novo, C., Arias-Estevez, M., Lopez-Periago, E., Martinez-Carballo, E., & Simal-Gandara, J. (2007). Influence of copper on the adsorption and desorption of paraquat, diquat, and difenzoquat in vineyard acid soils. *Journal of Agricultural and Food Chemistry*, *55*(15), 6219–6226.

Patel, J. G., Murthy, T. V. R., Singh, T. S., & Panigrahy, S. (2009, December). Analysis of the distribution pattern of wetlands in India in relation to climate change. In *Proceedings of the workshop on impact of climate change on agriculture* (pp. 17-18). Ahmedabad: International Society for Photogrammetry and Remote Sensing.

Pathak, R. K., & Dikshit, A. K. (2012). Atrazine and its Use. *International Journal of Research in Chemistry and Environment*, *2*, 1–6.

Patrick, D. R., & White, G. C. (1993). *Handbook of Chlorination and Alternative Disinfectants* (3rd ed.). New York: Nostrand Reinhold.

Patrick, L. (2009). Thyroid disruption: Mechanism and clinical implications in human health. *Alternative Medicine Review*, *14*, 326–346. PMID:20030460

Paull, J. (2010). From France to the World: The International Federation of Organic Agriculture Movements (IFOAM). *Journal of Social Research & Policy*, *1*(2), 93–102.

Paul, M. B. (2003). *Ground Water in Freshwater-Saltwater Environments of the Atlantic*. Reston, VA: U.S. Geological Survey.

Pavlovic, I., Barriga, C., Hermosín, M. C., Cornejo, J., & Ulibarri, M. A. (2005). Adsorption of acidic pesticides 2, 4-D, Clopyralid and Picloram on calcinedhydrotalcite. *Applied Clay Science*, *30*(2), 125–133. doi:10.1016/j.clay.2005.04.004

Payá, P., Anastassiades, M., Mack, D., Sigalova, I., Tasdelen, B., Oliva, J., & Barba, A. (2007). Analysis of pesticide residues using the Quick Easy Cheap Effective Rugged and Safe (QuEChERS) pesticide multiresidue method in combination with gas and liquid chromatography and tandem mass spectrometric detection. *Analytical and Bioanalytical Chemistry*, *38*(6), 1697–1714. doi:10.100700216-007-1610-7 PMID:17909760

Pearce, P. A., Elliott, J. E., Peakall, D. B., & Norstrom, R. J. (1989). Organochlorine contaminants in eggs of seabirds in the Northwest Atlantic, 1968–1984. *Environmental Pollution*, *56*(3), 217–235. doi:10.1016/0269-7491(89)90039-0 PMID:15092478

Pedersen, T.L. (1997). *Pesticide residues in drinking water.* Retrieved from extoxnet.orst.edu

Pedersen, M. L., & Friberg, N. (2009). Influence of disturbance on habitats and biological communities in lowland streams. *Fundamental and Applied Limnology; Official Journal of the International Association of Theoretical and Applied Limnology*, *174*(1), 27–41. doi:10.1127/1863-9135/2009/0174-0027

Pennington, P. L. (1996). *The toxicity of the herbicides atrazine and alachlor on the estuarine phytoplankter* Pavlova sp. *(*Prymne-siophyceae*) with an emphasis on acute toxicity testing of in-dividual herbicides, herbicide mixtures and multi-generational chronic bioassays* (MS thesis). University of Charleston, Charleston, SC.

Pereira, W. E., & Hostettler, F. D. (1993). Nonpoint source contamina-tion of the Mississippi River and its tributaries by herbicides. *Environmental Science & Technology*, *27*(8), 1542–1552. doi:10.1021/es00045a008

Perera, F. P., Rauh, V., Tsai, W. Y., Kinney, P., Camann, D., Barr, D., ... Whyatt, R. M. (2003). Effects of transplacental exposure to environmental pollutants on birth outcomes in a multiethnic population. *Environmental Health Perspectives*, *111*(2), 201–206. doi:10.1289/ehp.5742 PMID:12573906

Perera, F. P., Rauh, V., Whyatt, R. M., Tang, D., Tsai, W. Y., Bernert, J. T., & Diaz, D. (2005). A summary of recent findings on birth outcomes and developmental effects of prenatal ETS, PAH, and pesticide exposures. *Neurotoxicology*, *26*(4), 573–587. doi:10.1016/j.neuro.2004.07.007 PMID:16112323

Perocco, P., Bolognesi, S., & Alberghini, W. (1983). Toxic activity of seventeen industrial solvents and halogenated compounds on human lymphocytes cultured in vitro. *Toxicology Letters*, *16*(1), 69–75. doi:10.1016/0378-4274(83)90012-7 PMID:6836616

Persistent, P. B. (2000). Environmental Medicine, Part 4: Pesticides–Biologically Persistent and Ubiquitous Toxins. *Alternative Medicine Review*, *5*(5), 432–447. PMID:11056413

Petersen, C. T., Holm, J., Koch, C. B., Jensen, H. E., & Hansen, S. (2002). Movement of pendimethalin, ioxynil and soil particles to field drainage tiles. *Pest Management Science*, *59*(1), 85–96. doi:10.1002/ps.609 PMID:12558103

Peterson, H. G., Boutin, C., Freemark, K. E., & Martin, P. A. (1997). Toxicity of hexazinone and diquat to green algae, diatoms, cyanobacteria and duckweed. *Aquatic Toxicology (Amsterdam, Netherlands)*, *39*(2), 111–134. doi:10.1016/S0166-445X(97)00022-2

Peterson, H. G., Boutin, C., Martin, P. A., Freemark, K. E., Ruecker, N. J., & Moody, M. J. (1994). Aquatic phyto-toxicity of 23 pesticides applied at expected environmental concentrations. *Aquatic Toxicology (Amsterdam, Netherlands)*, *28*(3-4), 275–292. doi:10.1016/0166-445X(94)90038-8

Petrelli, G., Figa-Talamanca, I., Tropeano, R., Tangucci, M., Cini, C., Aquilani, S., & Meli, P. (2000). Reproductive male-mediated risk: Spontaneous abortion among wives of pesticide applicators. *European Journal of Epidemiology, 16*(4), 391–393. doi:10.1023/A:1007630610911 PMID:10959949

Phillips, P. J., Eckhardt, D. A., & Rosenmann, L. (2000). *Pesticides and Their Metabolites in Three Small Public Water-Supply Reservoir Systems, Western New York, 1998-99*. U.S. Department of the Interior, U.S. Geological Survey.

Pier, A. B., Bernucci, I., Brambilla, G., Consonni, D., & Pesatori, C. A. (1998). The Seveso Studies on Early and Long-Term Effects of Dioxin Exposure: A Review. *Environmental Health Perspectives, 106*(2), 5–20. PMID:9539003

Pier, A. B., Consonni, D., Bachetti, S., Rubagotti, M., Baccarelli, A., Zocchetti, C., & Pesatori, A. C. (2001). Health Effects of Dioxin Exposure: A 20-Year Mortality Study. *American Journal of Epidemiology, 153*(11), 1031–1044. doi:10.1093/aje/153.11.1031 PMID:11390319

Pierik, F. H., Burdorf, A., Deddens, J. A., Juttmann, R. E., & Weber, R. F. (2004). Maternal and paternal risk factors for cryptorchidism and hypospadias: A case-control study in newborn boys. *Environmental Health Perspectives, 112*(15), 1570–1576. doi:10.1289/ehp.7243 PMID:15531444

Pihlström, T., Blomkvist, G., Friman, P., Pagard, U., & Österdahl, B. (2007). Analysis of pesticide residues in fruit and vegetables with ethyl acetate extraction using gas and liquid chromatography with tandem mass spectrometric detection. *Analytical and Bioanalytical Chemistry, 389*(6), 1773–1789. doi:10.100700216-007-1425-6 PMID:17609934

Pimental, D. (1992). Foreword. In M. Shiyomi, E. Yano, H. Koizumi, D. A. Andow, & N. Hokyo (Eds.), *Ecological Processes in agroecosystems* (pp. 1–2). Tokyo: Yakendo Publishers.

Pimentel, D. (2005). Environmental and economic costs of the application of pesticides primarily in the United States. *Environment, Development and Sustainability, 7*(2), 229–252. doi:10.100710668-005-7314-2

Pimentel, D., & McLaughlen, L. (1991). Environmental and economic impacts of reducing U.S. agricultural pesticide use. In D. Pimentel & A. A. Hanson (Eds.), *CRC Handbook of Pest Management in Agriculture* (pp. 679–718). Boca Raton, FL: CRC Press.

Pimentel, D., McLaughlin, L., Zepp, A., Lakitan, B., Kraus, T., Kleinman, P., & Selig, G. (1993). *Environmental and economic effects of reducing pesticide use in agriculture*. Agriculture and the Environment.

Pinto, M. D. C. E., Gonçalves, R. G. L., dos Santos, R. M. M., Araújo, E. A., Perotti, G. F., dos Santos Macedo, R., ... Tronto, J. (2016). Mesoporous carbon derived from a biopolymer and a clay: Preparation, characterization and application for an organochlorine pesticide adsorption. *Microporous and Mesoporous Materials, 225*, 342–354. doi:10.1016/j.micromeso.2016.01.012

Pirsaheb, M., Limoee, M., Namdari, F., & Khamutian, R. (2015). Organochlorine pesticides residue in breast milk: A systematic review. *Medical Journal of the Islamic Republic of Iran, 29*, 228. PMID:26478886

Polevoy, S. (1996). *Water Science and Engineering*. London: Blackie Academic & Professional.

Pope, C. A. III, Bhatnagar, A., McCracken, J. P., Abplanalp, W. T., Conklin, D. J., & O'Toole, T. (2016). Exposure to fine particulate air pollution is associated with endothelial injury and systemic inflammation. *Circulation Research, 119*(11), 1204–1214. doi:10.1161/CIRCRESAHA.116.309279 PMID:27780829

Power, E. A., & Chapman, P. M. (1992). Assessing sediment quality. In G. A. Burton Jr., (Ed.), *Sediment toxicity assessment* (pp. 1–18). Chelsea, MA: Lewis Publishers.

Prasad, S. N., Ramachandra, T. V., Ahalya, N., Sengupta, T., Kumar, A., Tiwari, A. K., & Vijayan, L. (2002). Conservation of wetlands of India-a review. *Tropical Ecology, 43*(1), 173–186.

Prasertsup, P., & Ariyakanon, N. (2011). Removal of chlorpyrifos by water lettuce (Pistia stratiotes L.) and duckweed (Lemna minor L.). *International Journal of Phytoremediation*, *13*(4), 383–395. doi:10.1080/15226514.2010.495145 PMID:21598800

Prashanth, M. S., & Neelagund, S. E. (2008). Impact of cypermethrin on enzyme activities in the freshwater fish *Cirrhinus mrigala* (Ham.). *Caspian Journal of Environmental Sciences*, *6*, 91–95.

Prigent, C., Matthews, E., Aires, F., & Rossow, W. B. (2001). Remote sensing of global wetland dynamics with multiple satellite data sets. *Geophysical Research Letters*, *28*(24), 4631–4634. doi:10.1029/2001GL013263

Pruss-Ustun, A., Bos, R., Gore, F., & Bartram, J. (2008). *Safer water, better health: costs, benefits and sustainability of interventions to protect and promote health.* Geneva: World Health Organization.

Qian, S., & Lin, H. (2015). Colorimetric sensor array for detection and identification of organophosphorus and carbamate pesticides. *Analytical Chemistry*, *87*(10), 5395–5400. doi:10.1021/acs.analchem.5b00738 PMID:25913282

Qi, Z., Miller, G. W., & Voit, E. O. (2014). Dopamine metabolism: A computational analysis of pesticide toxicity. *Toxicology*, *6*, 92–101. doi:10.1016/j.tox.2013.11.003 PMID:24269752

Quijano, R. (2002). *Endosulfan poisoning in Kasaragod, Kerala, India.* Report on a fact finding mission, PAN.

Raanan, R., Harley, K. G., Balmes, J. R., Bradman, A., Lipsett, M., & Eskenazi, B. (2015). Early-life exposure to organophosphate pesticides and pediatric respiratory symptoms in the Chamacos cohort. *Environmental Health Perspectives*, *123*(2), 179. PMID:25369257

Radan, B. D., Pennigton, D. W., Eckley, N., & Boethling, A. (1999).. . *Environmental Science & Technology*, 33–3482.

Raghu, K., & MacRae, I. C. (1967). The effect of the gamma isomer of benzene hexachloride upon the microflora of submerged rice soils: II. Effect upon nitrogen mineralization and fixation, and selected bacteria. *Canadian Journal of Microbiology*, *13*(6), 621–627. doi:10.1139/m67-082 PMID:4166540

Raja, N., & Masresha, G. (2015). Plant based biopesticides: Safer alternative for Organic Food Production. *Journal of Biofertilizers & Biopesticides*, *6*, 128.

Raj, P., Singh, A., Kaur, K., Aree, T., Singh, A., & Singh, N. (2016). Fluorescent Chemosensors for Selective and Sensitive Detection of Phosmet/Chlorpyrifos with Octahedral Ni2+ Complexes. *Inorganic Chemistry*, *55*(10), 4874–4883. doi:10.1021/acs.inorgchem.6b00332 PMID:27115348

Raju, J., & Gupta, V. K. (1991). A simple spectrophotometric determination of endosulfan in river water and soil. *Fresenius' Journal of Analytical Chemistry*, *339*(6), 431–433. doi:10.1007/BF00322365

Ramachandra, T. V., & Kumar, U. (2008). Wetlands of greater Bangalore, India: Automatic delineation through pattern classifiers. *Electronic Green Journal*, *1*(26), 1–22.

Ramanan, R., Kim, B. H., Cho, D. H., Oh, H. M., & Kim, H. S. (2016). Algae–bacteria interactions: Evolution, ecology and emerging applications. *Biotechnology Advances*, *34*(1), 14–29. doi:10.1016/j.biotechadv.2015.12.003 PMID:26657897

Ramsar. (2002). *Wetlands: Water, life, and culture.* 8th meeting of the conference of contracting parties to the convention of wetlands, Valencia, Spain.

Randall, C. (2013). *National Pesticide Applicator Certification Core Manual.* Washington, DC: National Association of State Departments of Agriculture Research Foundation.

Ranz, A., Eberl, A., Maier, E., & Lankmayr, E. (2008). Microwave-assisted derivatization of acidic herbicides for gas chromatography-mass spectrometry. *Journal of Chromatography. A*, *1192*(2), 282–288. doi:10.1016/j.chroma.2008.03.076 PMID:18405907

Rao, C. S., Venkateswarlu, V., Surender, T., Eddleston, M., & Buckley, N. A. (2005, June). Pesticide poisoning in south India: Opportunities for prevention and improved medical management. *Tropical Medicine & International Health*, *10*(6), 581–588. doi:10.1111/j.1365-3156.2005.01412.x PMID:15941422

Rao, V. V. S. N., & Lal, R. (1987). Uptake and metabolism of insecticides by blue-green algae *Anabaena* and *Aulosira fertilissima. Mi-crobios Lett*, *36*, 143–147.

Rashid, B., Husnain, T., & Riazuddin, S. (2010). *Herbicides and pesticides as potential pollutants: a global problem. In Plant adaptation phytoremediation* (pp. 427–447). Dordrecht: Springer.

Rathore, M., Bhatnagar, P., Mathur, D., & Saxena, G. N. (2002). Burden of organochlorine pesticide in blood and its effect on thyroid hormones in women. *The Science of the Total Environment*, *295*(1-3), 207–215. doi:10.1016/S0048-9697(02)00094-3 PMID:12186288

Ratings. (2017). *Outlook of Indian pesticide industry*. Credit analysis and research limited, Mumbai, India.

Ratola, N., Botelho, C., & Alves, A. (2003). The use of pine bark as a natural adsorbent for persistent organic pollutants–study of lindane and heptachlor adsorption. *Journal of Chemical Technology and Biotechnology (Oxford, Oxfordshire)*, *78*(2-3), 347–351. doi:10.1002/jctb.784

Rattan, R. S. (2010). Mechanism of action of insecticidal secondary metabolites of plant origin. *Crop Protection (Guildford, Surrey)*, *29*(9), 913–920. doi:10.1016/j.cropro.2010.05.008

Rauh, V. A., Garfinkel, R., Perera, F. P., Andrews, H. F., Hoepner, L., Barr, D. B., & Whyatt, R. W. (2006). Impact of prenatal chlorpyrifos exposure on neurodevelopment in the first 3 years of life among inner-city children. *Pediatrics*, *118*(6), e1845–e1859. doi:10.1542/peds.2006-0338 PMID:17116700

Rauh, V., Arunajadai, S., Horton, M., Perera, F., Hoepner, L., Barr, D. B., & Whyatt, R. (2011). Seven-year neurodevelopmental scores and prenatal exposure to chlorpyrifos, a common agricultural pesticide. *Environmental Health Perspectives*, *119*(8), 1196–1201. doi:10.1289/ehp.1003160 PMID:21507777

Rawat, D. K., Bais, V. S., & Agrawal, N. C. (2002). A correlative study on liver glycogen and endosulfan toxicity in *Heteropneustes fossilis* (Bloch). *Journal of Environmental Biology*, *23*, 205–207. PMID:12602859

Rawtani, D., Khatri, N., Tyagi, S., & Pandey, G. (2018). Nanotechnology-based recent approaches for sensing and remediation of pesticides. *Journal of Environmental Management*, *206*, 749–762. doi:10.1016/j.jenvman.2017.11.037 PMID:29161677

Rayu, S., Nielsen, U. N., Nazaries, L., & Singh, B. K. (2017). Isolation and molecular characterization of novel chlorpyrifos and 3, 5, 6-trichloro-2-pyridinol-degrading bacteria from sugarcane farm soils. *Frontiers in Microbiology*, *8*, 518. doi:10.3389/fmicb.2017.00518 PMID:28421040

Reese, D. C., Ddoson, I. W., Ulrich, V., Becker, D. L., & Kempter, C. J. (1973). Pesticides in the Aquatic Environment. National Technical Advisory Committee on Pesticides in Water Environments. Environmental Protection Agency, Office of Water Programs.

Reffstrup, T. K., Larsen, J. L., & Meyer, O. (2010). Risk assessment of mixtures of pesticides: Current approaches and future strategies. *Regulatory Toxicology and Pharmacology*, *56*(2), 174–192. doi:10.1016/j.yrtph.2009.09.013 PMID:19782118

Reichenberger, S., Bach, M., Skitschak, A., & Frede, H. G. (2007). Mitigation strategies to reduce pesticide inputs into ground-and surface water and their effectiveness; a review. *The Science of the Total Environment*, *384*(1), 1–35. doi:10.1016/j.scitotenv.2007.04.046 PMID:17588646

Reigart, J. R., & Roberts, J. R. (1999). *Recognition and Management of Pesticide Poisonings* (5th ed.). US EPA.

Rekha, K., Thakur, M. S., & Karanth, N. G. (2000). Biosensors for the detection of organophosphorous pesticides. *Critical Reviews in Biotechnology*, *20*(3), 213–235. doi:10.1080/07388550008984170 PMID:11039330

Relyea, R. A. (2005). The impact of insecticides and herbicides on the biodiversity and productivity of aquatic communities. *Ecological Applications*, *15*(2), 618–627. doi:10.1890/03-5342

Relyea, R. A. (2005). The Lethal Impacts of Roundup and Predatory Stress on Six Species of North American Tadpoles. *Archives of Environmental Contamination and Toxicology*, *48*(3), 351–357. doi:10.100700244-004-0086-0 PMID:15886853

Relyea, R. A. (2009). A cocktail of contaminants: How mixtures of pesticides at low concentrations affect aquatic communities. *Oecologia*, *159*(2), 363–376. doi:10.100700442-008-1213-9 PMID:19002502

Relyea, R. A., & Hoverman, J. T. (2008). Interactive effects of predators and a pesticide on aquatic communities. *Oikos*, *117*(11), 1647–1658. doi:10.1111/j.1600-0706.2008.16933.x

Ren, A., Qiu, X., Jin, L., Ma, J., Li, Z., Zhang, L., ... Zhu, T. (2011). Association of selected persistent organic pollutants in the placenta with the risk of neural tube defects. *Proceedings of the National Academy of Sciences of the United States of America*, *108*(31), 12770–12775. doi:10.1073/pnas.1105209108 PMID:21768370

Resh, V. H., Brown, A. V., Covich, A. P., Gurtz, M. E., Li, H. W., Minshall, G. W., ... Wissmar, R. C. (1988). The role of disturbance in stream ecology. *Journal of the North American Benthological Society*, *74*(4), 433–455. doi:10.2307/1467300

Restrepo, M., Munoz, N., Day, N. E., Parra, J. E., De Romero, L., & Nguyen-Dinh, X. (1990). Testicular function in DBCP exposed pesticide workers. *Journal of Occupational Medicine and Toxicology (London, England)*, *21*, 161–166.

Riaz, G., Tabinda, A. B., Iqbal, S., Yasar, A., Abbas, M., Khan, A. M., & Baqar, M. (2017). Phytoremediation of organochlorine and pyrethroid pesticides by aquatic macrophytes and algae in freshwater systems. *International Journal of Phytoremediation*, *19*(10), 894–898. doi:10.1080/15226514.2017.1303808 PMID:28448165

Rice, R. G. (1999). *Ozone Drinking Water Treatment Handbook*. Boca Raton, FL: CRC Press/Lewis Publishers.

Richards, R. P., & Baker, D. B. (1993). Pesticide concentration patterns in agricultural drainage networks in the Lake Erie basin. *Environmental Toxicology and Chemistry*, *12*(1), 13–26. doi:10.1002/etc.5620120104

Rignell-Hydbom, A., Rylander, L., & Hagmar, L. (2007). Exposure to persistent organochlorine pollutants and type 2 diabetes mellitus. *Human and Experimental Toxicology*, *26*(5), 447–452. doi:10.1177/0960327107076886 PMID:17623770

Rippy, M. A., Deletic, A., Black, J., Aryal, R., Lampard, J. L., Tang, J. Y. M., ... Gernjak, W. (2017). Pesticide occurrence and spatio-temporal variability in urban run-off across Australia. *Water Research*, *115*, 245–255. doi:10.1016/j.watres.2017.03.010 PMID:28284091

Ristola, T., Pellinen, J., Ruokolainen, M., Kostamo, A., & Kukkonen, J. V. K. (1999). Effect of sediment type, feeding level, and larval density on growth and development of a midge (Chironomus riparius). *Environmental Toxicology and Chemistry*, *18*(4), 756–764. doi:10.1002/etc.5620180423

Ritter, L., Solomon, K. R., Forget, J., Stemeroff, M., & Leary, C. O. (1995). *An Assessment Report on: DDT, Aldrin, Dieldrin, Endrin, Chlordane, Heptachlor, Hexachlorobenzene, Mirex, Toxaphene, Polychlorinated Biphenyls, Dioxins and Furans. Ritter-Report. International Programme on Chemical Safety*. Genf: IPCS.

Ritter, S. R. (2009). *Pinpointing Trends in Pesticide Use in* E-News. Academic Press.

Roberts, D. A. (2012). Causes and ecological effects of resuspended contaminated sediments (RCS) in marine environments. *Environment International*, *40*, 230–243. doi:10.1016/j.envint.2011.11.013 PMID:22244126

Roberts, J. R., & Karr, C. J. (2012). Pesticide exposure in children. *Pediatrics*, *130*(6), e1765–e1788. doi:10.1542/peds.2012-2758 PMID:23184105

Robertson, S. L., Ni, W., Dhadialla, T. S., Nisbet, A. J., McCusker, C., & … . (2007). Identification of a putative azadirachtin-binding complex from Drosophila Kc167 cells. *Archives of Insect Biochemistry and Physiology*, *64*, 200–208. doi:10.1002/arch.20171 PMID:17366600

Robinson, D. G., Trites, D. G., & Banister, E. W. (1993). Neurotoxic concerns of human pesticide exposures. *American Journal of Industrial Medicine*, *18*, 327–331. PMID:2220839

Rocheleau, C. M., Romitti, P. A., & Dennis, L. K. (2009). Pesticides and hypospadias: A meta-analysis. *Journal of Pediatric Urology*, *5*(1), 17–24. doi:10.1016/j.jpurol.2008.08.006 PMID:18848807

Rodney, S. I., Teed, R. S., & Moore, D. R. J. (2013). Estimating the Toxicity of Pesticide Mixtures to Aquatic Organisms: A Review. *Human and Ecological Risk Assessment*, *19*(6), 1557–1575. doi:10.1080/10807039.2012.723180

Rodrigo, M. A., Oturan, N., & Oturan, M. A. (2014). Electrochemically assisted remediation of pesticides in soils and water: A review. *Chemical Reviews*, *114*(17), 8720–8745. doi:10.1021/cr500077e PMID:24983494

Rodríguez-Cruz, M. S., Andrades, M. S., Parada, A. M., & Sánchez-Martín, M. J. (2008). Effect of different wood pretreatments on the sorption– desorption of linuron and metalaxyl by woods. *Journal of Agricultural and Food Chemistry*, *56*(16), 7339–7346. doi:10.1021/jf800980w PMID:18642844

Rodriguez-Cruz, S., Andrades, M. S., Sanchez-Camazano, M., & Sanchez-Martin, M. J. (2007). Relationship between the adsorption capacity of pesticides by wood residues and the properties of woods and pesticides. *Environmental Science & Technology*, *41*(10), 3613–3619. doi:10.1021/es062616f PMID:17547186

Rodriguez-Lafuente, A., Piri-Moghadam, H., Lord, H. L., Obal, T., & Pawliszyn, J. (2016). Inter-laboratory validation of automated SPME-GC/MS for determination of pesticides in surface and ground water samples: Sensitive and green alternative to liquid–liquid extraction. *Water Quality Research Journal*, *51*(4), 331–343. doi:10.2166/wqrjc.2016.011

Rodríguez-Liébana, J. A., López-Galindo, A., de Cisneros, C. J., Gálvez, A., Rozalén, M., Sánchez-Espejo, R., ... Peña, A. (2016). Adsorption/desorption of fungicides in natural clays from Southeastern Spain. *Applied Clay Science*, *132*, 402–411. doi:10.1016/j.clay.2016.07.006

Roger, P. A., Grant, I. F., & Reddy, P. M. (1985). *Blue-green algae in India: A trip report.* Los Bailos, Philippines: IRRI.

Roger, P. A., Heong, K. L., & Teng, P. S. (1991). Biodiversity and sustainability of wetland rice production: role and potential of microorganisms and invertebrates. In D. L. Hawksworth (Ed.), *The Biodiversity of Microorganisms and Invertebrates: Its Role in Sustainable Agriculture* (pp. 117–136). Wallingford: CABI.

Roger, P. A., Simpson, I., Oficial, R., Ardales, S., & Jimenez, R. (1994). Effects of pesticides on soil and water microflora and mesofauna in wetland ricefields: A summary of current knowledge and extrapolation to temperate environments. *Australian Journal of Experimental Agriculture*, *34*(7), 1057–1068. doi:10.1071/EA9941057

Rojas, R., Morillo, J., Usero, J., Vanderlinden, E., & El Bakouri, H. (2015). Adsorption study of low-cost and locally available organic substances and a soil to remove pesticides from aqueous solutions. *Journal of Hydrology (Amsterdam)*, *520*, 461–472. doi:10.1016/j.jhydrol.2014.10.046

Romeh, A. A. (2014). Phytoremediation of cyanophos insecticide by Plantago major L. in water. *Journal of Environmental Health Science & Engineering*, *12*(1), 38. doi:10.1186/2052-336X-12-38 PMID:24447385

Roques, H. (1996). *Chemical Water Treatment: Principles and Practice*. New York: VCH.

Rosas, L. G., & Eskenazi, B. (2008). Pesticides and child neurodevelopment. *Current Opinion in Pediatrics*, *20*(2), 191–197. doi:10.1097/MOP.0b013e3282f60a7d PMID:18332717

Rosenberg, J., & O'Malley, M. (1997). Pesticides. In Occupational and Environmental Medicine (2nd ed.). Stamford, CT: Appleton Lange.

Rosenberg, D. M. (1975). Food chains concentrations of chlorinated hydrocarbons pesticides in invertebrate communities: A reevaluation. *Quaestiones Entomologicae*, *11*, 97–110.

Rousis, N. I., Bade, R., Bijlsma, L., Zuccato, E., Sancho, J. V., Hernandez, F., & Castiglioni, S. (2017). Monitoring a large number of pesticides and transformation products in water samples from Spain and Italy. *Environmental Research*, *156*, 31–38. doi:10.1016/j.envres.2017.03.013 PMID:28314152

Rousis, N. I., Zuccato, E., & Castiglioni, S. (2017). Wastewater-based epidemiology to assess human exposure to pyrethroid pesticides. *Environment International*, *99*, 213–220. doi:10.1016/j.envint.2016.11.020 PMID:27908456

Rowe, P. H. (2010). *Statistical methods for continuous measured endpoints in in silico toxicology. In Silico Toxicology: Principles and Applications* (pp. 228–251). Cambridge, UK: The Royal Society of Chemistry.

Roy, R. (2015). Risk Assessment in the European Union (EU). In J. Torres & S. Bobst (Eds.), *Toxicological Risk Assessment for Beginners*. Cham: Springer. doi:10.1007/978-3-319-12751-4_9

Rupa, D. S., Reddy, P. P., & Reddi, O. S. (1991). Prevalence of adverse reproduction outcomes in a population occupationally exposed to pesticides in Columbia. *Scandinavian Journal of Work, Environment & Health*, *16*, 232–238.

Russo, R., Becker, J. M., & Liess, M. (2018). Sequential exposure to low levels of pesticides and temperature stress increase toxicological sensitivity of crustaceans. *The Science of the Total Environment*, *610–611*, 563–569. doi:10.1016/j.scitotenv.2017.08.073 PMID:28822923

Rutherford, B. (1996). The synthetic chemical threat - endocrine disruptors. *The Journal of the Pesticide trust. Pesticide News*, *31*, 6–7.

Ruttner, F. (1963). *Fundamentals of Limnology* (3rd ed.). Toronto: University Press Toronto.

Sabater, C., & Carrasco, J. M. (2001a). Effects of pyridaphenthion on growth of five freshwater species of phytoplankton. A laboratory study. *Chemosphere*, *44*(8), 1775–1781. doi:10.1016/S0045-6535(00)00575-0 PMID:11534908

Sabater, C., & Carrasco, J. M. (2001b). Effects of the organophosphorus insecticide fenitrothion on growth in five freshwater species of phytoplankton. *Environmental Toxicology*, *16*(4), 314–320. doi:10.1002/tox.1038 PMID:11501280

Sabbah, S., & Bouguerra, M. L. (1997). Organochlorine pesticides in cow's milk from Tunisia. *Fresenius Environmental Bulletin*, *6*(7), 359–364.

Sadaria, A. M., Sutton, R., Moran, K. D., Teerlink, J., Brown, J. V., & Halden, R. U. (2017). Passage of fiproles and imidacloprid from urban pest control uses through wastewater treatment plants in northern California, USA. *Environmental Toxicology and Chemistry*, *36*(6), 1473–1482. doi:10.1002/etc.3673 PMID:27808432

Sahoo, D. (2003). *Role of Algae in Environment and Zoology-Proceedings of the National Seminar on Environmental Biotechnology: Envirotect*. Azra Printers.

Saikia, S. K. (2011). Review on Periphyton as Mediator of nutrient transfer in aquatic ecosystems. *Ecologia Balkanica, 3*, 65–78.

Saillenfait, A., Ndiaye, D., & Sabate, J. (2015). Pyrethroids: Exposure and health effects—an update. *International Journal of Hygiene and Environmental Health, 218*(3), 281–292. doi:10.1016/j.ijheh.2015.01.002 PMID:25648288

Saiyed, H., Dewan, A., Bhatnagar, V., Shenoy, U., Shenoy, R., Rajmohan, H., ... Lakkad, B. (2003). Effect of endosulfan on male reproductive development. *Environmental Health Perspectives, 111*(16), 1958–1962. doi:10.1289/ehp.6271 PMID:14644673

Sala, M., Ribas-Fitó, N., Cardo, E., de Muga, M. E., Marco, E., Mazón, C., ... Sunyer, J. (2001). Levels of hexachlorobenzene and other organochlorine compounds in cord blood: Exposure across placenta. *Chemosphere, 43*(4-7), 895–901. doi:10.1016/S0045-6535(00)00450-1 PMID:11372882

Salama, J., Chakraborty, T. R., Ng, L., & Gore, A. C. (2003). Effects of polychlorinated biphenyls on estrogen receptor-b expression in the anteroventral periventricular nucleus. *Environmental Health Perspectives, 111*(10), 1278–1282. doi:10.1289/ehp.6126 PMID:12896846

Salameh, P., Waked, M., Baldi, I., Brochard, P., & Saleh, B. A. (2006). Respiratory diseases and pesticide exposure: A case-control study in Lebanon. *Journal of Epidemiology and Community Health, 60*(3), 256–261. doi:10.1136/jech.2005.039677 PMID:16476757

Salameh, R. P., Daldi, I., Brochard, P., & Saleh, B. A. (2004). Pesticides in Lebanon: A knowledge, attitude and practise study. *Environmental Research, 94*(1), 1–6. doi:10.1016/S0013-9351(03)00092-6 PMID:14643280

Salam, J. A., Hatha, M. A., & Das, N. (2017). Microbial-enhanced lindane removal by sugarcane (Saccharum officinarum) in doped soil-applications in phytoremediation and bioaugmentation. *Journal of Environmental Management, 193*, 394–399. doi:10.1016/j.jenvman.2017.02.006 PMID:28259469

Salehzadeh, A., Akhkha, A., Cushley, W., Adams, R. L., Kusel, J. R., & Strang, R. H. C. (2003). The antimitotic effect of the neem terpenoidazadirachtin on cultured insect cells. *Insect Biochemistry and Molecular Biology, 200*(33), 681–691. doi:10.1016/S0965-1748(03)00057-2 PMID:12826095

Salman, J. M., & Hameed, B. H. (2010). Adsorption of 2, 4-dichlorophenoxyacetic acid and carbofuran pesticides onto granular activated carbon. *Desalination, 256*(1), 129–135. doi:10.1016/j.desal.2010.02.002

Salman, J. M., Njoku, V. O., & Hameed, B. H. (2011). Adsorption of pesticides from aqueous solution onto banana stalk activated carbon. *Chemical Engineering Journal, 174*(1), 41–48. doi:10.1016/j.cej.2011.08.026

Salman, J. M., Njoku, V. O., & Hameed, B. H. (2011). Bentazon and carbofuran adsorption onto date seed activated carbon: Kinetics and equilibrium. *Chemical Engineering Journal, 173*(2), 361–368. doi:10.1016/j.cej.2011.07.066

Salvo, L. M., Sinhorini, I. L., Malucelli, B. E., Klemz, C., Sanchez, D. C. O., Nicaretta, L., ... Assis, H. C. S. (2008). Effects of endosulfan sub-lethal concentrations on carp (*Cyprinus carpio, Linnaeus*, 1758): Morphometric, hystologic, ultrastructural analyses & cholinesterase activity evaluation. *Brazilian Journal of Veterinary Research & Animal Science, Sao Paulo, 45*, 87–94. doi:10.11606/issn.1678-4456.bjvras.2008.26704

Samanta, S. (2013). Metal and pesticide pollution scenario in Ganga River system. *Aquatic Ecosystem Health & Management, 16*(4), 454–464. doi:10.1080/14634988.2013.858587

Samsel, A., & Seneff, S. (2015). Glyphosate, pathways to modern diseases III: Manganese, neurological diseases, and associated pathologies. *Surgical Neurology International, 6*(1), 45. doi:10.4103/2152-7806.153876 PMID:25883837

Samsidar, A., Siddiquee, S., & Shaarani, S. M. (2018). A review of extraction, analytical and advanced methods for determination of pesticides in environment and foodstuffs. *Trends in Food Science & Technology*, *71*, 188–201. doi:10.1016/j.tifs.2017.11.011

Sanborn, M., Bassil, K., Vakll, C., Kerr, K., & Ragan, K. (2012). *Systematic Review of Pesticide health Effects.* Ontario College of Family Physicians, Toronto. Retrieved from www.ocfp.on.ca/communications/newpublications/lists/new-publications/ocfp-2012-systematic-review-of-pesticide-health-effects

Sanborn, M., Bassil, K., Vakll, C., Kerr, K., & Ragan, K. (2012). *Systematic Review of Pesticide health Effects*. Toronto: Ontario College of Family Physicians. Retrieved from http://www.ocfp.on.ca/communications/newpublications/lists/new-publications/ocfp-2012-systematic-review-of-pesticide-health-effects

Sanborn, M., Kerr, K. J., Sanin, L. H., Cole, D. C., Bassil, K. L., & Vakil, C. (2007). Non-cancer health effects of pesticides: Systematic review and implications for family doctors. *Canadian Family Physician Medecin de Famille Canadien*, *53*(10), 1712–1720. PMID:17934035

Sánchez-Bayo, F., van den Brink, P. J., & Mann, R. M. (2012). *Ecological Impacts of Toxic Chemicals*. Francisco Sanchez-Bayo.

Sandstrom, M. W. (1995). U.S. Geological Survey Water-Resources Investigations Report: Vol. 95-4105. Filtration of water-sediment samples for the determination of organic compounds. Academic Press.

Sandstrom, M. W., Jelinski, J. C., & Tai, D. Y. (1992). *Apparatus for sampling pesticide residues in run-off with control of sample pump and distributor valve.* U.S. Patent 5, 167, 802.

Sanjeevi, Haruna, Tripathi, & Anuradha. (2017). Impacts of Global Carbon Foot Print on the Marine Environment. *International Journal of Engineering Research and Technology, 5*(12), 57-60.

Sanpera, C., Ruiz, X., Jover, L., Llorente, G., Jabeen, R., Muhammad, A., & Fasola, M. (2003). Persistent organic pollutants in little egret eggs from selected wetlands in Pakistan. *Archives of Environmental Contamination and Toxicology*, *44*(3), 360–368. doi:10.100700244-002-2044-z PMID:12712296

Santanu, G., Chattopadhyay, G. N., & Vass, K. K. (2000). Environmental impact assessment of lower Ganga system. *Environment and Ecology*, *18*(1), 126–129.

Sapkale, G. N., Patil, S. M., Surwase, U. S., & Bhatbhage, P. K. (2010). Supercritical fluid extraction. *International Journal of Chemical Sciences*, *8*(2), 729–743.

Sardar, D., & Kole, R. K. (2005). Metabolism of Chlorpyriphos in relation to its effect on the availability of some plant nutrients in soil. *Chemosphere*, *61*(9), 1273–1280. doi:10.1016/j.chemosphere.2005.03.078 PMID:15894348

Sastry, C. S. P., Vijaya, D., & Rao, K. E. (1986). Sensitive spectrophotometric determinations of carbaryl and propoxur in formulations, water, grains and pulses. *Food Chemistry*, *20*(2), 157–162. doi:10.1016/0308-8146(86)90152-4

Sathiakumar, N., Delzell, E., MacLennan, P., Anne, M., Rosenberg, N., Cheng, H., & Myers, S. (2004). A crosssectional study of triallate exposure and neurological health among workers at a pesticide manufacturing and formulating facility'. *Occupational and Environmental Medicine*, *61*(1), 936–944. doi:10.1136/oem.2004.013128 PMID:15477288

Sathishkumar, M., Binupriya, A. R., Kavitha, D., Selvakumar, R., Jayabalan, R., Choi, J. G., & Yun, S. E. (2009). Adsorption potential of maize cob carbon for 2, 4-dichlorophenol removal from aqueous solutions: Equilibrium, kinetics and thermodynamics modeling. *Chemical Engineering Journal*, *147*(2), 265–271. doi:10.1016/j.cej.2008.07.020

Satpathy, R. (2016). Quantitative structure activity relationship (QSAR) study of some haloalkane compound properties with their aquatic toxicity feature. In Advanced engineering: Current prospective (pp. 111-119). Research India Publication.

Satpathy, R. (2018). Quantitative Structure-Activity Modelling of Toxic Compounds. In *Nanotechnology, Food Security and Water Treatment* (pp. 313–331). Cham: Springer. doi:10.1007/978-3-319-70166-0_10

Savage, E. P., Keefe, T. J., Mounce, L. M., Heaton, R. K., Lewis, J. A., & Bucar, P. J. (1988). Chronic neurological sequelae of acute organophosphate pesticide poisoning. *Archives of Environmental Health*, *43*(1), 38–45. doi:10.1080/00039896.1988.9934372 PMID:3355242

Savonen, C. (1997). Soil microorganisms object of new OSU service. *Good Fruit Grower*. Retrieved from http://www.goodfruit.com/archive/1995/6other.html

Schafer, R. B., Van den Brink, P. J., & Liess, M. (2011). Impacts of pesticides on freshwater ecosystems. *Ecological Impacts of Toxic Chemicals*, 111-137.

Schäfer, R. B., Caquet, T., Siimes, K., Mueller, R., Lagadic, L., & Liess, M. (2007). Effects of pesticides on community structure and ecosystem functions in agricultural streams of three biogeographical regions in Europe. *The Science of the Total Environment*, *382*(2–3), 272–285. doi:10.1016/j.scitotenv.2007.04.040 PMID:17555800

Schäfer, R. B., von der Ohe, P. C., Rasmussen, J., Kefford, B. J., Beketov, M. A., Schulz, R., & Liess, M. (2007). Thresholds for the Effects of Pesticides on Invertebrate Communities and Leaf Breakdown in Stream Ecosystems. *Environmental Science & Technology*, *46*(9), 5134–5142. doi:10.1021/es2039882 PMID:22455566

Schenck, F. J., Lehotay, S. J., & Vega, V. (2002). Comparison of solid-phase extraction sorbents for cleanup in pesticide residue analysis of fresh fruits and vegetables. *Journal of Separation Science*, *25*(14), 883–890. doi:10.1002/1615-9314(20021001)25:14<883::AID-JSSC883>3.0.CO;2-7

Schettler, T., Solomon, G., Valenti, M., & Huddle, A. (1999). *Generations at risk*. Boston: Massachusetts Institute of Technology.

Schmidt, J. R., Shaskus, M., Estenik, J. F., Oesch, C., Khidekel, R., & Boyer, G. (2013). Variations in the microcystin content of different fish species collected from a eutrophic lake. *Toxin*, *5*(5), 992–1009. doi:10.3390/toxins5050992 PMID:23676698

Scholze, R. J. (Ed.). (1988). *Biotechnology for Degradation of Toxic Chemicals in Hazardous Wastes*. Noyes Publications.

Schonbeck, M. (2010). An Organic Weed Control Toolbox. Virginia Association for Biological Farming.

Schreinemachers, D. M. (2003). Birth malformations and other adverse perinatal outcomes in four U.S. wheat-producing areas. *Environmental Health Perspectives*, *111*(9), 1259–1264. doi:10.1289/ehp.5830 PMID:12842783

Schreinemachers, P., & Tipraqsa, P. (2012). Agricultural pesticides and land use intensification in high, middle and low income countries. *Food Policy*, *37*(6), 616–626. doi:10.1016/j.foodpol.2012.06.003

Schulz, R. (2004). Field studies on exposure, effects, and risk mitigation of aquatic nonpoint-source insecticide pollution: A review. *Journal of Environmental Quality*, *33*(2), 419–448. doi:10.2134/jeq2004.4190 PMID:15074794

Schulz, R., & Liess, M. (1995). Chronic effects of low insecticide concentrations on freshwater caddisfly larvae. *Hydrobiologia*, *299*(2), 103–113. doi:10.1007/BF00017562

Schulz, R., & Liess, M. (2000). Toxicity of fenvalerate to caddisfly larvae: Chronic effects of 1-vs 10-h pulse-exposure with constant doses. *Chemosphere*, *41*(10), 1511–1517. doi:10.1016/S0045-6535(00)00107-7 PMID:11057675

Schulz, R., & Liess, M. (2001). Toxicity of aqueous-phase and suspended particle-associated fenvalerate: Chronic effects after pulse-dosed exposure of Limnephilus lunatus (Trichoptera). *Environmental Toxicology and Chemistry*, *20*(1), 185–190. doi:10.1002/etc.5620200121 PMID:11351407

Searles, N. S., McKean-Cowdin, R., Farin, F. M., Holly, E. A., Preston-Martin, S., & Mueller, B. A. (2010). Childhood brain tumors, residential insecticide exposure, and pesticide metabolism genes. *Environmental Health Perspectives, 118*(1), 144–149. PMID:20056567

Sebastia & Joseph. (1997). Tertiary Treatment of Sewage for Reuse. *Chemical Engineering World, 32*, 55–57.

Seiber, J. N., & Woodrow, J. E. (1995). Origin and fate of pesticides in air. In *Eighth International Congress of Pesticide Chemistry–Options 2000*. Washington, DC: American Chemical Society.

Selassie, C., & Verma, R. P. (2015). QSAR of toxicology of substituted phenols. *Journal of Pesticide Science, 40*(1), 1–12. doi:10.1584/jpestics.D14-097

Selevan, S. G., Kimmel, C. A., & Mendola, P. (2000). Identifying critical windows of exposure for children's health. *Environmental Health Perspectives, 108*(Suppl 3), 451–455. doi:10.1289/ehp.00108s3451 PMID:10852844

Selgrade, M. K., Lemanske, R. F. Jr, Gilmour, M. I., Neas, L. M., Ward, M. D., Henneberger, P. K., ... Yeatts, K. B. (2006). Induction of asthma and the environment: What we know and need to know. *Environmental Health Perspectives, 114*(4), 615–619. doi:10.1289/ehp.8376 PMID:16581555

Senthilkumaar, S., Krishna, S. K., Kalaamani, P., Subburamaan, C. V., & Subramaniam, N. G. (2010). Adsorption of organophosphorous pesticide from aqueous solution using "waste" jute fiber carbon. *Modern Applied Science, 4*(6), 67. doi:10.5539/mas.v4n6p67

Seward, J. R., & Schultz, T. W. (1999). QSAR analyses of the toxicity of aliphatic carboxylic acids and salts to Tetrahymena pyriformis. *SAR and QSAR in Environmental Research, 10*(6), 557–567. doi:10.1080/10629369908033224 PMID:10674293

Seyhi, B., Drogui, P., Gortares-Moroyoqui, P., Estrada-Alvarado, M. I., & Alvarez, L. H. (2014). Adsorption of an organochlorine pesticide using activated carbon produced from an agro-waste material. *Journal of Chemical Technology and Biotechnology (Oxford, Oxfordshire), 89*(12), 1811–1816. doi:10.1002/jctb.4256

Shafer, T. J., Meyer, D. A., & Crofton, K. M. (2005). Developmental neurotoxicity of pyrethroid insecticides: Critical review and future research needs. *Environmental Health Perspectives, 113*(2), 123–136. doi:10.1289/ehp.7254 PMID:15687048

Shamsipur, M., Yazdanfar, N., & Ghambarian, M. (2016). Combination of solid-phase extraction with dispersive liquid–liquid microextraction followed by GC–MS for determination of pesticide residues from water, milk, honey and fruit juice. *Food Chemistry, 204*, 289–297. doi:10.1016/j.foodchem.2016.02.090 PMID:26988504

Shao, Y., Jiang, L., Zhou, H., Pan, J., & He, Y. (2016). Identification of pesticide varieties by testing microalgae using Visible/Near Infrared Hyperspectral Imaging technology. *Scientific Reports, 6*(1), 24221. doi:10.1038rep24221 PMID:27071456

Sharma, A. (2018). Science-based mandatory standards and the implementation gap: the case of bottled water regulations in India. *Current Science, 114*(1).

Sharma, D. K., Gupta, A., Kashyap, R., & Kumar, N. (2012). Spectrophotometric method for the determination of Glyphosate in relation to its environmental and toxicological analysis. *Arch. Environ. Sci, 6*, 42–49.

Sharma, D. R., Thapa, R. B., Manandhar, H. K., Shrestha, S. M., & Pradhan, S. B. (2012). Use of pesticides in Nepal and impacts on human health and environment. *J Agric Environ, 13*, 67–72.

Sharpe, R. M., & Irvine, D. S. (2004). How strong is the evidence of a link between environmental chemicals and adverse effects on human reproductive health? *BMJ (Clinical Research Ed.), 328*(7437), 447–451. doi:10.1136/bmj.328.7437.447 PMID:14976101

Shaw, G. M., Swan, S. H., Harris, J. A., & Malcoe, L. H. (1990). Maternal water consumption during pregnancy and congenital cardiac anomalies. *Epidemiology (Cambridge, Mass.)*, *1*(3), 206–211. doi:10.1097/00001648-199005000-00005 PMID:2081254

Shelton, J. F., Geraghty, E. M., & Tancredi, D. J. (2015). Neurodevelopmental disorders and prenatal residential proximity to agricultural pesticides: The CHARGE study. *Environmental Health Perspectives*, *122*, 1103–1109. PMID:24954055

Shelton, J. F., Hertz-Picciotto, I., & Pessah, I. N. (2012). Tipping the balance of autism risk: Potential mechanisms linking pesticides and autism. *Environmental Health Perspectives*, *120*(7), 944–951. doi:10.1289/ehp.1104553 PMID:22534084

Sherma, J. (2016). Thin layer chromatography: Recent advances. *Thin-Layer and Liquid Chromatography and Pesticides of International Importance: Analytical Methods for Pesticides and Plant Growth Regulators*, *7*, 1.

Sherma, J. (2017). Review of thin-layer chromatography in pesticide analysis: 2014–2016. *Journal of Liquid Chromatography & Related Technologies*, *40*(5-6), 226–238. doi:10.1080/10826076.2017.1298024

Shi, J., Tian, F., Lyu, J., & Yang, M. (2015). Nanoparticle based fluorescence resonance energy transfer (FRET) for biosensing applications. *Journal of Materials Chemistry. B, Materials for Biology and Medicine*, *3*(35), 6989–7005. doi:10.1039/C5TB00885A

Shimabukuro, R. H. (1985). Detoxification of herbicides. We Physiology, 2, 215-240.

Shim, Y. K., Mlynarek, S. P., & van Wijngaarden, E. (2009). Parental exposure to pesticides and childhood brain cancer: US Atlantic coast childhood brain cancer study. *Environmental Health Perspectives*, *117*(6), 1002–1006. doi:10.1289/ehp.0800209 PMID:19590697

Shoda, M. (2000). Bacterial control of plant diseases. *Journal of Bioscience and Bioengineering*, *89*(6), 515–521. doi:10.1016/S1389-1723(00)80049-3 PMID:16232790

Shukla, G., Kumar, A., Bhanti, M., Joseph, P.E., & Taneja, A. (2006). Organochlorine pesticide contamination of ground water in the city of Hyderabad. *Environ. Inter.*, *32*, 244-247.

Siddiqui, M., Srivastava, S., Mehrotra, P., Mathur, N., & Tandon, I. (2003). Persistent chlorinated pesticides and intra-uterine foetal growth retardation: A possible association. *International Archives of Occupational and Environmental Health*, *76*(1), 75–80. PMID:12592586

Siddiqui, S., Goddard, R. H., & Bielmyer-Fraser, G. K. (2015). Comparative effects of dissolved copper and copper oxide nanoparticle exposure to the sea anemone, Exaiptasia pallida. *Aquatic Toxicology (Amsterdam, Netherlands)*, *160*, 205–213. doi:10.1016/j.aquatox.2015.01.007 PMID:25661886

Sikorski, J. A., & Gruys, K. J. (1997). Understanding Glyphosate's molecular mode of action with EPSP synthase: Evidence favoring an allosteric inhibitor model. *Accounts of Chemical Research*, *30*(1), 2–8. doi:10.1021/ar950122

Silander, J., Vehviläinen, B., Niemi, J., Arosilta, A., Dubrovin, T., Jormola, J., . . . Veijalainen, N. (2006). *Climate Change Adaptation for Hydrology and Water Resources*. FINADAPT Working Paper 6. Finnish Environmental Institute.

Silva, M. H. (2007). *Endosulfan risk characterization document. Medical toxicology and worker health and safety branches department of pesticide regulation*. California environmental protection. Available from: www.cdpr.ca.gov/docs/emon/pubs/tac/.../endosulfan/endosulfan_doc.pdf

Silva, M. H., & Gammon, D. (2009). An assessment of the developmental, reproductive and neurotoxicity of endosulfan. *Birth Defects Research. Part B, Developmental and Reproductive Toxicology*, *86*(1), 1–28. doi:10.1002/bdrb.20183 PMID:19243027

Silverman, D. T., Devesa, S. S., Moore, L. E., & Rothman, N. (2006). *Cancer epidemiology and prevention*. New York: Oxford University Press.

Simazaki, D., Kubota, R., Suzuki, T., Akiba, M., Nishimura, T., & Kunikane, S. (2015). Occurrence of selected pharmaceuticals at drinking water purification plants in Japan and implications for human health. *Water Research, 76*, 187–200. doi:10.1016/j.watres.2015.02.059 PMID:25835589

Simpson, I. C., Roger, P. A., Oficial, R., & Grant, I. F. (1994). Effects of nitrogen fertilizer and pesticide management on floodwater ecology in a wetland ricefield: I. Experimental design and dynamics of the photosynthetic aquatic biomass. *Biology and Fertility of Soils, 17*(2), 129–137. doi:10.1007/BF00337745

Singer, P. C. (1999). *Formation and Control of Disinfection By-Products in Drinking Water*. Denver, CO: American Water Works Association.

Singer, P. L. (1949). Occupational oligospermia. *Journal of the American Medical Association, 140*, 1249.

Singh, A., Raj, P., & Singh, N. (2016b). *Benzimidazolium-Based Self-Assembled Fluorescent Aggregates for Sensing and Catalytic Degradation of Diethylchlorophosphate*. Academic Press.

Singh, N.N., & Srivastava, A.K. (1995). Sublethal physiological effects on catfish, *Heteropneustes fossilis* (Bloch) during longterm exposure to aldrin and subsequent recovery. *Proceedings of National Academy of Science, India, 65*, 33-38.

Singh, A., Imtiyaz, M., Isaac, R. K., & Denis, D. M. (2012). Comparison of soil and water assessment tool (SWAT) and multilayer perceptron (MLP) artificial neural network for predicting sediment yield in the Nagwa agricultural watershed in Jharkhand, India. *Agricultural Water Management, 104*, 113–120. doi:10.1016/j.agwat.2011.12.005

Singh, A., Raj, P., & Singh, N. (2016a). Benzimidazolium-Based Self-Assembled Fluorescent Aggregates for Sensing and Catalytic Degradation of Diethylchlorophosphate. *ACS Applied Materials & Interfaces, 8*(42), 28641–28651. doi:10.1021/acsami.6b09983 PMID:27731616

Singh, N. K., Banerjee, B. D., Bala, K., Basu, M., & Chhillar, N. (2014). Polymorphism in cytochrome P450 2D6, glutathione S-transferases Pi 1 genes, and organochlorine pesticides in Alzheimer disease: A case–control study in North Indian population. *Journal of Geriatric Psychiatry and Neurology, 27*(2), 119–127. doi:10.1177/0891988714522698 PMID:24584466

Singh, P., & Thakur, I. S. (2006). Colour removal of anaerobically treated pulp and paper mill effluent by microorganisms in two steps bioreactor. *Bioresource Technology, 97*(2), 218–223. doi:10.1016/j.biortech.2005.02.022 PMID:16171678

Sinha, C. P. (2011). Climate change and its impacts on the wetlands of North Bihar, India. *Lakes and Reservoirs: Research and Management, 16*(2), 109–111. doi:10.1111/j.1440-1770.2011.00457.x

Siriwong, W., Thirakhupt, K., Sitticharoenchai, D., Rohitrattana, J., Thongkongowm, P., Borjan, M., & Robson, M. (2009). DDT and Derivatives in Indicator Species of the Aquatic Food Web of Rangsit Agricultural Area, Central Thailand. *Ecological Indicators, 9*(5), 878–882. doi:10.1016/j.ecolind.2008.10.004 PMID:20161116

Slager, R., Poole, J., LeVan, T., Sandler, D., Alavanja, M., & Hoppin, J. (2009). Rhinitis Associated with Pesticide Exposure among Commercial Pesticide Applicators in the Agricultural Health Study. *Ocuppational and Environmental Medication, 66*(11), 718–724. doi:10.1136/oem.2008.041798 PMID:19289390

Smakhtin, V., Revenga, C., & Döll, P. (2004). A pilot global assessment of environmental water requirements and scarcity. *Water International, 29*(3), 307–317. doi:10.1080/02508060408691785

Smith, S. (2014). Global Horticulture (2010 - 2014). *Pink and Healthy, 24*.

Smith, C. (2001). Pesticide exports from US ports, 1997-2000. *International Journal of Occupational and Environmental Health, 7*(4), 266–274. doi:10.1179/oeh.2001.7.4.266 PMID:11783856

Sneyd, A. D., & Hosking, R. J. (1976). Seepage flow through homogeneous soil into a row of drain pipes. *Journal of Hydrology (Amsterdam), 30*(1-2), 127–146. doi:10.1016/0022-1694(76)90094-9

Soares, W. L., & Porto, M. F. D. (2009). Estimating the social cost of pesticide use: An assessment from acute poisoning in Brazil. *Ecological Economics, 68*(10), 2721–2728. doi:10.1016/j.ecolecon.2009.05.008

Solomon, K. R., Harris, S. A., & Stephenson, G. R. (1993). *Applicator and bystander exposure to home garden and landscape pesticides*. Academic Press.

Solomon, S.D., Qin, M., Manning, Z., Chen, M., Marquis, K.B., Averyt, M., Tignor, H.L., & Miller. (2007). *Climate Change 2007: The Physical Science Basis. Contribution of Working Group I to the Fourth Assessment Report of the Intergovernmental Panel on Climate Change*. Cambridge University Press.

Solomon, K. R., Baker, D. B., Richards, R. P., Dixon, K. R., Klaine, S. J., La Point, T. W., ... Williams, W. M. (1996). Ecological risk assessment of atrazine in North American surface waters. *Environmental Toxicology and Chemistry, 15*(1), 31–76. doi:10.1002/etc.5620150105 PMID:23147529

Soloneski, S., Kujawski, M., Scuto, A., & Larramendy, M. L. (2015). Carbamates: A study on genotoxic, cytotoxic, and apoptotic effects induced in Chinese hamster ovary (CHO-K1) cells. *Toxicology In Vitro, 29*(5), 834–844. doi:10.1016/j.tiv.2015.03.011 PMID:25820133

Song, B., Palleroni, N. J., & Häggblom, M. M. (2000). Isolation and characterization of diverse halobenzoate-degrading denitrifying bacteria from soils and sediments. *Applied and Environmental Microbiology, 66*(8), 3446–3453. doi:10.1128/AEM.66.8.3446-3453.2000 PMID:10919805

Song, Y., Chen, M., & Zhou, J. (2017). Effects of three pesticides on superoxide dismutase and glutathione-S-transferase activities and reproduction of Daphnia magna. *Archives of Environmental Protection, 43*(1), 80–86. doi:10.1515/aep-2017-0010

Souza, A. D., Medeiros, A. D. R., Souza, A. C. D., Wink, M., Siqueira, I. R., Ferreira, M. B. C., & Torres, I. L. D. S. (2011). Evaluation of the impact of exposure to pesticides on the health of the rural population: Vale do Taquari, State of Rio Grande do Sul (Brazil). *Ciencia & Saude Coletiva, 16*(8), 3519–3528. doi:10.1590/S1413-81232011000900020 PMID:21860952

Space Applications Centre. (2011). *National Wetland Atlas. SAC*. Ahmedabad: Indian Space Research Organisation.

Sparling, D. W. (2016). Current use of Pesticides. In *Ecotoxicology Essentials, Environmental Contaminants and their Biological Effects on Animals and Plants* (pp. 109–152). Academic Press.

Spear, R. (1991). Recognised and possible exposure to pesticides. In W. J. Hayes & E. R. Laws (Eds.), *Handbook of pesticide toxicology* (pp. 245–274). San Diego, CA: Academic.

Speck-Planche, A., Kleandrova, V. V., & Scotti, M. T. (2012). Fragment-based approach for the in silic discovery of multi-target insecticides. *Chemometrics and Intelligent Laboratory Systems, 111*(1), 39–45. doi:10.1016/j.chemolab.2011.11.010

Speitel, G. E. (1999). *Advanced Oxidation and Biodegradation Processes for the Destruction of TOC and DBP Precursors*. Denver, CO: AWWA Research Foundation.

Spellman, F. A. (2014). Water and wastewater treatment plant operations. Academic Press.

Spellman, F. R., & Whiting, N. E. (1999). *Water Pollution Control Technology: Concepts and Applications*. Rockville, MD: Government Institutes.

Spencer, M. C. (1966). Herbicide dermatitis. *Journal of the American Medical Association*, *198*(12), 1307–1308. doi:10.1001/jama.1966.03110250121037 PMID:5953527

Spliid, N. H., Helweg, A., & Heinrichson, K. (2006). Leaching and degradation of 21 pesticides in a full-scale model biobed. *Chemosphere*, *65*(11), 2223–2232. doi:10.1016/j.chemosphere.2006.05.049 PMID:16815528

Spzir, M. (2006). New thinking on neurodevelopment. *Environmental Health Perspectives*, *114*(2), 100–107. doi:10.1289/ehp.114-a100 PMID:16393665

Squillace, P. J., Thurman, E. M., & Furlong, E. T. (1993). Groundwater as a nonpoint source of atrazine and deethylatrazine in a river during base flow conditions. *Water Resources Research*, *29*(6), 1719–1729. doi:10.1029/93WR00290

Sridhar, R. (2008). *Endosulfan Poisoning and the Struggle of the community in Kasaragod to regain life and the living land*. Retrieved from http://thanal.co.in/uploads/resource/document/endosulfan-poisoning-and-the-struggle-of-the-community-in-kasaragod-to-regain-life-and-the-living-land-a-presentation-at-the-cop3-of-rotterdam-conven-50269185.pdf

Srivastava, N. & Kaushik, N. (2001). Use of fish as bioindicator of aquatic pollution. *Proceeding of ICCE*, 227-229.

Srivastava, N., Agrawal, M., Tyagi, A., & Kaushik, N. (2002). Endosulfan induced changes in the fish *Channa punctatus*. *Proceeding of XI National Symposium of Environment*, 356-358.

Srivastava, A. K., & Srivastava, A. K. (1988). Chlordane induced changes in carbohydrate metabolism of the Indian catfish *Heteropneustes fossilis* (Bloch). *Bulletin of the Institute of Zoology, Academia Sinica*, *27*, 211–215.

Srivastava, K., & Sharma, A. K. (2011). Arbuscular Mycorrhizal Fungi in challenging Environment– A Prospective. In S. M. Fulton (Ed.), *Mycorrhizal Fungi: Soil, Agriculture and Environmental Implications* (pp. 1–34). Nova Science Publishers, Inc.

Stehle, S., Elsaesser, D., Gregoire, C., Imfeld, G., Niehaus, E., Passeport, E., ... Schulz, R. (2011). Pesticide risk mitigation by vegetated treatment systems: A meta-analysis. *Journal of Environmental Quality*, *40*(4), 1068–1080. doi:10.2134/jeq2010.0510 PMID:21712575

Steiner, V. (1998). UV Irradiation in Drinking Water and Wastewater Treatment for Disinfection. *Wasser Rohrbau*, *49*, 22–31.

Stella, K., Charles, F., Lynch, X., & Won, M. A. (2009). Aromatic amine pesticide use and human cancer risk: Results from the U.S. Agricultural Health Study. *International Journal of Cancer*, *124*(5), 1206–1212. doi:10.1002/ijc.24020 PMID:19058219

Stephenson, G. A., & Solomon, K. R. (1993). Pesticides and the environment. Department of Environmental Biology, University of Guelph.

Stephen, W. C. (2011). Determination of organochlorine pesticide residues in fatty foods: A critical review on the analytical methods and their testing capabilities. *Journal of Chromatography. A*, *1218*(33), 5555–5567. doi:10.1016/j.chroma.2011.06.066 PMID:21742333

Steurbaut, W. (2009). *Neveneffecten van pesticiden. Fytofarmacie*. Ghent: Ghent University.

Stevenson, D. G. (1997). *Water Treatment Unit Processes*. London: Imperial College Press. doi:10.1142/p063

Stillerman, K. P., Mattison, D. R., Giudice, L. C., & Woodruff, T. J. (2008). Environmental exposures and adverse pregnancy outcomes: A review of the science. *Reproductive Sciences (Thousand Oaks, Calif.)*, *15*(7), 631–650. doi:10.1177/1933719108322436 PMID:18836129

Stockholm Convention. (2001). *Stockholm Convention on Persistent Organic Pollutants*. Secretariat of the Stockholm Convention Clearing House.

Stockholm Convention. (2005). Stockholm Convention on Persistent Organic Pollutants (POPs).

Stokstad, E., & Grullon, G. (2013). Infographic: Pesticide planet. A global look at the uses, benefits, and drawbacks of pesticides. *Science*, *341*, 730–731. PMID:23950524

Storrs, S. I., & Kiesecker, J. M. (2004). Survivorship patterns of larval amphibians exposed to low concentrations of atrazine. *Environmental Health Perspectives*, *112*(10), 1054–1057. doi:10.1289/ehp.6821 PMID:15238276

Streat, M., & Horner, D. J. (2000). Adsorption of highly soluble herbicides from water using activated carbon and hypercrosslinked polymers. *Process Safety and Environmental Protection*, *78*(5), 363–382. doi:10.1205/095758200530880

Streloke, M., & Kopp, H. (1995). Long-term toxicity tests with Chironomus riparius: development and validation of a new system Heft 315. Biologische Bundesanstalt für Land und Forstwirtschaft (BBA), Blackwell Wissenschafts-Verlag.

Stuart, M. E., Gooddy, D. C., Bloomfield, J. P., & Williams, A. T. (2011). A review of the impact of climate change on future nitrate concentrations in groundwater of the UK. *The Science of the Total Environment*, *409*(15), 2859–2873. doi:10.1016/j.scitotenv.2011.04.016 PMID:21669323

Subashchandrabose, S. R., Ramakrishnan, B., Megharaj, M., Venkateswarlu, K., & Naidu, R. (2011). Consortia of cyanobacteria/microalgae and bacteria: Biotechnological potential. *Biotechnology Advances*, *29*(6), 896–907. doi:10.1016/j.biotechadv.2011.07.009 PMID:21801829

Sugashini, S., & Begum, K. M. S. (2013). Optimization using central composite design (CCD) for the biosorption of Cr (VI) ions by cross linked chitosan carbonized rice husk (CCACR). *Clean Technologies and Environmental Policy*, *15*(2), 293–302. doi:10.100710098-012-0512-3

Sugita, F., & Nakane, K. (2007). Combined effects of rainfall patterns and porous media properties on nitrate leaching. *Vadose Zone Journal*, *6*(3), 548–553. doi:10.2136/vzj2006.0095

Sullivan, T. P., Sullivan, D. S., Lindgren, P. M. F., & Boateng, J. O. (2002). Influence of conventional and chemical thinning on stand structure and diversity of plant and mammal communities in young lodgepole pine forest. *Forest Ecology and Management*, *170*(1), 173–187. doi:10.1016/S0378-1127(01)00775-7

Sultana, R., & Devi, U. (1995). Oxygen consumption in a Cat fish Hystus gulio (Ham) exposed to heavy metals. *Journal of Environmental Biology*, *16*, 207–210.

Su, N. Y., & Scheffrahn, R. H. (1998). A review of subterranean termite control practices and prospects for integrated pest management programmes. *Integrated Pest Management Reviews*, *3*(1), 1–13. doi:10.1023/A:1009684821954

Sundararaman, S. R. (2009). *Crop Production and Plant Protection in Organic Farming*. South Asia Conference on 'Outstanding Organic Agriculture Techniques', Bengaluru, India.

Sun, J., Pan, L., Tsang, D. C., Li, Z., Zhu, L., & Li, X. (2018). Phthalate esters and organochlorine pesticides in agricultural soils and vegetables from fast-growing regions: A case study from eastern China. *Environmental Science and Pollution Research International*, *25*(1), 34–42. doi:10.100711356-016-7725-7 PMID:27738865

Sun, X., Liu, B., & Xia, K. (2011). A sensitive and regenerable biosensor for organophosphate pesticide based on self-assembled multilayer film with CdTe as fluorescence probe. *Luminescence*, *26*(6), 616–621. doi:10.1002/bio.1284 PMID:21567885

Sustain, D. (2013). Soil compaction impact and modelling A review. *Muhammad Farrakh Nawaz & Guilhem Bourrié & Fabienne Trolard Agron*, *33*, 291–309.

Swanson, N. L., Leu, A., Abrahamson, J., & Wallet, B. (2014). Genetically engineered crops, glyphosate and the deterioration of health in the United States of America. *Journal of Organic Systems*, *9*, 6–37.

Swarna, C., Naidu, K. P., Nagendrudu, G., Naidu, N. V. S., & Saraswathi, K. (2011). Spectrophotometric Method for the Determination of Carboxin in its Formulations and Environmental Samples. *Journal of Chemistry*, *8*(4), 1680–1685.

Sweeney, B. W. (1993). Effects of streamside vegetation on microinvertebrate communities of White Clay Creek in Eastern North America. *Proceedings. Academy of Natural Sciences of Philadelphia*, *144*, 291–340.

Swinfield, T., Afriandi, R., Antoni, F., & Harrison, R. D. (2016). Accelerating tropical forest restoration through the selective removal of pioneer species. *Forest Ecology and Management*, *381*, 209–216. doi:10.1016/j.foreco.2016.09.020

Tadeo, J. L. (Ed.). (2008). *Analysis of pesticides in food and environmental samples*. CRC Press. doi:10.1201/9781420007756

Taha, S. M., Amer, M. E., Elmarsafy, A. E., & Elkady, M. Y. (2014). Adsorption of 15 different pesticides on untreated and phosphoric acid treated biochar and charcoal from water. *Journal of Environmental Chemical Engineering*, *2*(4), 2013–2025. doi:10.1016/j.jece.2014.09.001

Takaku, T., Takahashi, M., & Otsuki, A. (1979). Dispersion of an organophosphorus insecticide, fenitrothion, in paddy fields and its effect on microorganisms. *Japanese Journal of Limnology*, *40*(3), 137–144. doi:10.3739/rikusui.40.137

Takamura, K., & Yasuno, M. (1986). Effects of pesticide application on chironomid larvae and ostracods in ricefields. *Applied Entomology and Zoology*, *21*(3), 370–376. doi:10.1303/aez.21.370

Takamura, K., & Yasuno, M. (1986). Effects of pesticide applications on chironomids, ostracods and other benthic organisms in rice fields. *Research Reports of the National Institute Environmental Studies of Japan*, *99*, 81–90.

Tan, G. H., & Chai, M. K. (2011). Sample preparation in the analysis of pesticides residue in food for chromatographic techniques. In M. Stoytcheva (Ed.), *Pesticides—Strategies for Pesticides Analysis* (pp. 28–59). Rijeka, Croatia: InTech.

Tao, S., Xi, X., Xu, F., & Dawson, R. (2002). A QSAR model for predicting toxicity (LC 50) to rainbow trout. *Water Research*, *36*(11), 2926–2930. doi:10.1016/S0043-1354(01)00514-0 PMID:12146883

Tariq, M. I., Afzal, S., Hussain, I., & Sultana, N. (2007). Pesticides exposure in Pakistan: A review. *Environment International*, *33*(8), 1107–1122. doi:10.1016/j.envint.2007.07.012 PMID:17765971

Taylor, A. W. (1995). The Volatilization of Pesticide Residues. In T. R. Roberts & P. C. Kearney (Eds.), *Environmental Behaviour of Agrochemicals*. Chichester, UK: John Wiley & Sons.

Taylor, J. R. (1976). Neurologic disorder induced by Kepone. A preliminary report. *Neurology*, *26*, 358–363.

Tchobanoglous, G., Darby, J., Bourgeous, K., McArdle, J., Genest, P., & Tylla, M. (1998). Ultrafiltration as an Advanced Tertiary Treatment Process for Municipal Wastewater. *Desalination*, *119*(1-3), 315–322. doi:10.1016/S0011-9164(98)00175-1

Tejomyee, S. B., & Pravin, R. P. (2007). Biodegradation of organochlorine pesticide, endosulfan, by a fungal soil isolate, *Aspergillus niger. International Biodeterioration & Biodegradation*, *59*(4), 315–321. doi:10.1016/j.ibiod.2006.09.002

Teklu, B. M., Adriaanse, P. I., Ter Horst, M. M., Deneer, J. W., & Van den Brink, P. J. (2015). Surface water risk assessment of pesticides in Ethiopia. *The Science of the Total Environment*, *508*, 566–574. doi:10.1016/j.scitotenv.2014.11.049 PMID:25481716

Terasaki, M., Abe, R., Makino, M., & Tatarazako, N. (2015). Chronic toxicity of parabens and their chlorinated by-products in Ceriodaphniadubia. *Environmental Toxicology*, *30*(6), 664–673. doi:10.1002/tox.21944 PMID:24376163

Thacker, N. P., Kaur, P., & Rudra, A. (1997). Halogenated compounds in and around Nagpur City water resources, India. *Environmental Education and Information*, *16*, 367.

Thacker, N. P., Vaidya, M. V., Sipani, M., & Kalra, A. (1997). Removal technology for pesticide contaminants in potable water. *Journal of Environmental Science and Health. Part. B, Pesticides, Food Contaminants, and Agricultural Wastes*, *B32*(4), 483–496. doi:10.1080/03601239709373099

Thany, S. H., Reynier, P., & Lenaers, G. (2013). Neurotoxicity of pesticides: Its relationship with neurodegenerative diseases. *Medical Science*, *29*, 273–278. PMID:23544381

Thayer, K. A., Heindel, J. J., Bucher, J. R., & Gallo, M. A. (2012). Role of environmental chemicals in diabetes and obesity: A National Toxicology Programme workshop review. *Environmental Health Perspectives*, *120*(6), 779–789. doi:10.1289/ehp.1104597 PMID:22296744

The Insecticides Act, 1968.

Thebo, A. L., Drechsel, P., Lambin, E. F., & Nelson, K. L. (2017). A global, spatiallyexplicit assessment of irrigated croplands influenced by urban wastewater flows. *Environmental Research Letters*, *12*(7), 074008. doi:10.1088/1748-9326/aa75d1

Thiemann, H., & Weiler, H. (1996). One Year of Operational Experience with the Largest River Water Reverse Osmosis Plant in Germany. *VGB-Kraftwerkstechnik*, *76*, 1017–1022.

Thodal, C. E., Carpenter, J., & Moses, C. W. (2009). *Monitoring for Pesticides in Groundwater and Surface Water in Nevada, 2008 (No. 2009-3093)*. US Geological Survey.

Thompson, T., Fawell, J., Kunikane, S., Jackson, D., Appleyard, S., & Callan, P. (2007). *Chemical safety of drinking water: Assessing priorities for risk management.* WHO.

Thornton, P. K. (2010). Livestock production: Recent trends, future prospects. *Philosophical Transactions of the Royal Society of London. Series B, Biological Sciences*, *365*(1554), 2853–2867. doi:10.1098/rstb.2010.0134 PMID:20713389

Thrupp, L. A. (1991). Sterilization of Workers from Pesticide Exposure: The Causes and Consequences of DBCP-Induced Damage in Costa Rica and Beyond. *International Journal of Health Services*, *21*(4), 731–757. doi:10.2190/25B9-QFMV-Y28P-L736 PMID:1769760

Thurman, E. M., Goolsby, D. A., Meyer, M. T., & Kolpin, D. W. (1992). Her-bicides in surface waters of the midwestern United States: The effect of spring flush. *Environmental Science & Technology*, *25*(10), 1794–1796. doi:10.1021/es00022a018

Thurman, E. M., Goolsby, D. A., Meyer, M. T., Mills, M. S., Pomes, M. L., & Kolpin, D. W. (1992). A reconnaissance study of herbicides and their metabolites in surface water of the midwestern United States using immunoassay and gas chromatography/mass spec-trometry. *Environmental Science & Technology*, *26*(12), 2440–2447. doi:10.1021/es00036a016

Tilman, D., Cassman, K. G., Matson, P. A., Naylor, R., & Polasky, S. (2002). Agricultural sustainability and intensive production practices. *Nature*, *418*(6898), 671–677. doi:10.1038/nature01014 PMID:12167873

Titus, M. P., Molina, V. G., Baños, M. A., Giménez, J., & Plugas, S. (2004). Degradation of chlorophenols by means of advanced oxidation processes:a general review, *App.Cat.B. Env.*, *47*, 219–256.

Tiwari, N., Asthana, A., & Upadhyay, K. (2013). Kinetic-spectrophotometric determination of methyl parathion in water and vegetable samples. *Spectrochimica Acta. Part A: Molecular and Biomolecular Spectroscopy*, *101*, 54–58. doi:10.1016/j.saa.2012.09.084 PMID:23099160

Todeschini, R., & Consonni, V. (2009). Molecular descriptors for chemoinformatics: Vol. 41. *2 volume set*. John Wiley & Sons.

Tomar, M. (1999). *Laboratory Manual for the Quality Assessment of Water and Wastewater*. Boca Raton, FL: CRC Press.

Tomkins, B. A., & Barnard, A. R. (2002). Determination of organochlorine pesticides in ground water using solid phase microextraction followed by dual-column gas chromatography with electron-capture detection. *Journal of Chromatography. A*, *964*(1-2), 21–33. doi:10.1016/S0021-9673(02)00592-7 PMID:12198850

Torstensson, L. (1990). *Natruvaardsverket rapport*, 3536. Solan, Sverige.

Torstensson, L., & Stenström, J. (1990). Persistence of herbicides in forest nursery soils. *Scandinavian Journal of Forest Research*, *5*(1-4), 457–469. doi:10.1080/02827589009382628

Trasatti, S. (2013). Electrochemistry and environment: The role of electrocatalysis. *ViaVenezian*, *21*, 3.

Treadwell, D., Riddle, J., Barbercheck, M., Cavanaugh-Grant, D., & Zaborski, E. (2015). *Cooperative Extension System, What is organic farming?* Retrieved from http://articles.extension.org/pages/18655/what-is-organic-farming

Trewavas, A. (2001). Urban myths of organic farming. *Nature*, *410*(6827), 409–410. doi:10.1038/35068639 PMID:11260685

Trewavas, A. (2004). A critical assessment of organic farming-and-food assertions with particular respect to the UK and the potential environmental benefi ts of no-till agriculture. *Crop Protection (Guildford, Surrey)*, *23*(9), 757–781. doi:10.1016/j.cropro.2004.01.009

Tripathi, Y. C., & Tripathi, J. (2000). Role of biopesticides in environmental safety. *Science and Culture*, *66*(5-6), 171–176.

Tsakiris, I. N., Danis, T. G., Stratis, I. A., Nikitovic, N., Dialyna, I. A., Alegakis, A. K., & Tsatsakis, A. M. (2004). Monitoring of pesticide residues in fresh peaches produced under conventional and integrated crop management cultivation. *Food Addit Contam A*, *21*(7), 670–677. doi:10.1080/02652030410001698715 PMID:15370840

Tsipi, D., Triantafyllou, M., & Hiskia, A. (1999). Determination of organochlorine pesticide residues in honey, applying solid phase extraction with RP-C18 material. *Analyst (London)*, *124*(4), 473–475. doi:10.1039/a809724k PMID:10605876

Turgut, C., Gokbulut, C., & Cutright, T. J. (2009). Contents and sources of DDT impurities in dicofol formulations in Turkey. *Environmental Science and Pollution Research International*, *16*(2), 214–217. doi:10.100711356-008-0083-3 PMID:19052792

Türkman, A., & Uslu, O. (Eds.). (1991). *New Developments in Industrial Wastewater Treatment*. Norwell, MA: Kluwer. doi:10.1007/978-94-011-3272-5

Turnbull, A. B., Harrison, R. M., Williams, R. J., Matthiessen, P., Brooke, D. N., Sheahan, D. A., & Mills, M. (1997). Assessment of the Fate of Selected Adsorptive Pesticides at ADAS Rosemaund. *Water and Engironmental Journal*, *11*(1), 24–30. doi:10.1111/j.1747-6593.1997.tb00084.x

Turner, M. C., Wigle, D. T., & Krewski, D. (2011). Residential pesticides and childhood leukemia: A systematic review and meta-analysis. *Ciencia & Saude Coletiva*, *16*(3), 1915–1931. doi:10.1590/S1413-81232011000300026 PMID:21519680

Turner, R. K., Van Den Bergh, J. C., Söderqvist, T., Barendregt, A., Van Der Straaten, J., Maltby, E., & Van Ierland, E. C. (2000). Ecological-economic analysis of wetlands: Scientific integration for management and policy. *Ecological Economics*, *35*(1), 7–23. doi:10.1016/S0921-8009(00)00164-6

Turusov, V., Rakitsky, V., & Tomatis, L. (2002). Dichlorodiphenyltrichloroethane (DDT): Ubiquity, persistence, and risks. *Environmental Health Perspectives*, *110*(2), 125–128. doi:10.1289/ehp.02110125 PMID:11836138

U.S. Environmental Protection Agency. (1947). *The Federal Insecticide, Fungicide, and Rodenticide Act of 1947*. Author.

U.S. Environmental Protection Agency. (2000). Methods for measuring the toxicity and bioaccumulation of sediment-associated contaminants with freshwater invertebrates (2nd ed.). U.S. Environmental Protection Agency, Office of Research and Development EPA/600/R-99/064.

U.S. Geological Survey. (1999). *The quality of our nation's waters – nutrients and pesticides*. Circular 1225. Reston VA: USGS. Retrieved from http: //water.usgs.gov/pubs/circ/ circ1225/

Ulen, B., Kreuger, J., & Sundin, P. (2002). *Undersökning av bekämpningsmedel i vatten från jordbruk och samhälle år 2001 Rapport 2002:4*. Institutionen för Miljöanalys, Sveriges lantbruksuniversitet. (in Swedish)

Ullah, S., Hasan, Z., & Dhama, K. (2016). Toxic effects of endosulfan on behaviour, protein contents and antioxidant enzyme system in gills, brain, liver and muscle tissues of rohu, *Labeo rohita*. *International Journal of Pharmacology*, *12*(1), 1–10. doi:10.3923/ijp.2016.1.10

UNDESA. (2017). *World Population Prospects: The 2017 Revision, Key Findings and Advance Tables*. Working Paper No. ESA/P/WP/248. United Nations, Department of Economic and Social Affairs, Population Division.

UNECE. (2010). Hemispheric Transport of Air Pollution. Persistent Organic Pollutants, Economic Commission for Europe. UNECE.

Uneke, C. J. (2007). *Integrated pest management for developing countries: a systemic overview*. Nova Publishers.

UNEP. (2005). Ridding the World of POPs. In A guide to the Stockholm Convention on Persistent Organic Pollutants. UNEP.

UNEP. (2009). *Stockholm Convention on Persistent Organic Pollutants (POPs) with Amendments SC 4/10 to SC 4/18*. Secretariat of the Stockholm Convention on Persistent Organic Pollutants.

UNEP. (2009). Stockholm Convention on Persistent Organic Pollutants (POPs). *Secr. Stock. Conv. Persistent Org. Pollut.*, *1*.

UNEP. (2016). *A snapshot of the world's water quality: Towards a global assessment*. Nairobi: United Nations Environment Programme (UNEP).

United Nations Educational Scientific and Cultural Organization (UNESCO). (2007). *Case Studies on Climate Change and World Heritage*. UNESCO World Heritage Centre.

Urban, D. J., & Cook, N. J. (1986). Standard evaluation procedure ecological risk assessment. In *Standard evaluation procedure ecological risk assessment*. EPA.

US Environmental. (2007). *What is a pesticide?* Epa.gov.

US EPA. (2001). Water protection practices bulletin. Washington, DC: Office of Water.

US EPA. (2002). *Reregistration Eligibility Decision for Endosulfan*. EPA, 738-R-02-013. Pollution, Pesticides and Toxic Substances (7508C), United States Environmental Protection Agency. Retrieved from http://www.epa.gov/oppsrrd1/REDs/endosulfan_red.pdf

US EPA. (2006). *About Pesticides*. Retrieved from http://www.epa.gov/pesticides/about/types.htm

US EPA. (2011). Pesticides Industry Sales and Usage: 2006 and 2007 Market Estimates. Washington, DC: Author.

USEPA. (2016). *Pesticides Industry Sales and Usage. 2008-2012 Market estimates*. USEPA.

Vaccari, D. A., Strom, P. F., & Alleman, J. E. (2005). *Environmental biology for engineers and scientists*. John Wiley & Sons. doi:10.1002/0471741795

Vakurov, A., Simpson, C. E., Daly, C. L., Gibson, T. D., & Millner, P. A. (2004). Acetylcholinesterase-based biosensor electrodes for organophosphate pesticide detection: I. Modification of carbon surface for immobilization of acetylcholinesterase. *Biosensors & Bioelectronics*, *20*(6), 1118–1125. PMID:15556357

Vallotton, N., & Price, P. S. (2016). Use of the Maximum Cumulative Ratio As an Approach for Prioritizing Aquatic Coexposure to Plant Protection Products: A Case Study of a Large Surface Water Monitoring Database. *Environmental Science & Technology*, *50*(10), 5286–5293. doi:10.1021/acs.est.5b06267 PMID:27057923

Valvi, D., Mendez, M. A., Martinez, D., Grimalt, J. O., Torrent, M., Sunyer, J., & Vrijheid, M. (2012). Prenatal concentrations of PCBs, DDE, DDT and overweight in children: A prospective birth cohort study. *Environmental Health Perspectives*, *120*(3), 451–457. doi:10.1289/ehp.1103862 PMID:22027556

van den Berg, F., Kubiak, R., Benjey, W. G., Majewski, M. S., Yates, S. R., Reeves, G. L., ... van der Linden, A. M. A. (1999). Emission of Pesticides into the Air. *Water, Air, and Soil Pollution*, *115*(1/4), 195–218. doi:10.1023/A:1005234329622

Van Der Hoek, J. P., Rijnbende, D. O., Lokin, C. J. A., Bonne, P. A. C., Loonen, M. T., & Hofman, J. A. M. H. (1998). Electrodialysis as an Alternative for Reverse Osmosis in an Integrated Membrane System. *Desalination*, *117*(1-3), 159–172. doi:10.1016/S0011-9164(98)00086-1

Van der Linde, D., Konings, E. E., Slager, M. A., Witsenburg, M., Helbing, W. A., Takkenberg, J. J., & Roos-Hesselink, J. W. (2011). Birth prevalence of congenital heart disease worldwide: A systematic review and meta-analysis. *Journal of the American College of Cardiology*, *58*(21), 2241–2247. doi:10.1016/j.jacc.2011.08.025 PMID:22078432

Van Grinsven, H. J. M., Ten Berge, H. F. M., Dalgaard, T., Fraters, B., Durand, P., Hart, A., ... Willems, W. J. (2012). Management, regulation and environmental impacts of nitrogen fertilization in northwestern Europe under the Nitrates Directive; a benchmark study. *Biogeosciences*, *9*(12), 5143–5160. doi:10.5194/bg-9-5143-2012

Van Maele-Fabry, G., Hoet, P., Vilain, F., & Lison, D. (2012). Occupational exposure to pesticides and Parkinson's disease: A systematic review and meta-analysis of cohort studies. *Environment International*, *46*, 30–33. doi:10.1016/j.envint.2012.05.004 PMID:22698719

Van, L. C. J. (2010). Risk Assessment of Chemicals: An Introduction (2nd ed.). Dordrecht, The Netherlands: Springer.

Vandenplas, O., Delwiche, J. P., Auverdin, J., Caroyer, U. M., & Cangh, F. B. (2000). Asthma to tetramethrin. *Allergy*, *55*(4), 417–418. doi:10.1034/j.1398-9995.2000.00599.x PMID:10782541

Vannote, R. L., Minnshall, W. G., Cummins, K. W., Sedell, J. R., & Cushing, C. E. (1980). The river continuum concept. *Can J Aquat Sci*, *37*(1), 130–137. doi:10.1139/f80-017

Vassiliou, G. (2016). Factors Affecting Risk Assessment of Pesticides in Water Bodies: A Review. *MOJ Toxicol*, *2*(1), 00030. doi:10.15406/mojt.2016.02.00030

Veil, J. F. (1990). Public health impact of pesticides used in agriculture. Geneva, Switzerland: WHO (World Health Organization).

Venugopal, P. N. (2008). *Endosulfan victims still suffer in India*. Retrieved from http://southasia.oneworld.net/article/view/157876/1

Verhaar, H. J., Van Leeuwen, C. J., & Hermens, J. L. (1992). Classifying environmental pollutants. *Chemosphere*, *25*(4), 471–491. doi:10.1016/0045-6535(92)90280-5 PMID:10718581

Verhoeven, J. T., Arheimer, B., Yin, C., & Hefting, M. M. (2006). Regional and global concerns over wetlands and water quality. *Trends in Ecology & Evolution*, *21*(2), 96–103. doi:10.1016/j.tree.2005.11.015 PMID:16701481

Verhulst, S. L., Nelen, V., Den Hond, E., Koppen, G., Beunckens, C., Vael, C., ... Desager, K. (2009). Intrauterine exposure to environmental pollutants and body mass index during the first 3 years of life. *Environmental Health Perspectives*, *117*(1), 122–126. doi:10.1289/ehp.0800003 PMID:19165398

Verma, M., Bakshi, N., & Nair, R. P. (2001). *Economic valuation of Bhoj Wetland for sustainable use*. Unpublished project report for World Bank assistance to Government of India, Environmental Management Capacity-Building. Bhopal: Indian Institute of Forest Management.

Verro, R., Finizio, A., Otto, S., & Vighi, M. (2009). Predicting pesticide environmental risk in intensive agricultural areas. II: Screening level risk assessment of complex mixtures in surface waters. *Environmental Science & Technology*, *43*(2), 530–537. doi:10.1021/es801858h PMID:19238990

Vestergaard, S., Cherry, A., Keller, S., & Goettel, M. (2003). Safety of hyphomycete fungi as microbial control agents. In *Environmental impacts of microbial insecticides* (pp. 35–62). Dordrecht, The Netherlands: Academic Press.

Vidal, J. M., Plaza-Bolanos, P., Romero-González, R., & Frenich, A. G. (2009). Determination of pesticide transformation products: A review of extraction and detection methods. *Journal of Chromatography. A*, *1216*(40), 6767–6788. doi:10.1016/j.chroma.2009.08.013 PMID:19720377

Vijayavel, K., Balasubramanian, S. E., & Balasubramanian, M. P. (2006). Effect of urea and ammonium chloride on Aspartate and Alanine transaminases of Oreochromis mossambicus (Peters). *Indian Journal of Fisheries*, *53*(1), 19–22.

Vischetti, C., Casucci, C., & Perucci, P. (2002). Relationship between changes of soil microbial biomass content and imazamox and benfluralin degradation. *Biology and Fertility of Soils*, *35*(1), 13–17. doi:10.100700374-001-0433-5

Vivekanandhan, N., & Duraisamy, A. (2012). Ecological Impact of Pesticides Principally Organochlorine Insecticide Endosulfan: A Review. *Universal Journal of Environmental Research and Technology*, *2*(5), 369–376.

Vizcaino, E., Grimalt, J. O., Lopez-Espinosa, M. J., Llop, S., Rebagliato, M., & Ballester, F. (2010). Maternal origin and other determinants of cord serum organochlorine compound concentrations in infants from the general population. *Environmental Science & Technology*, *44*(16), 6488–6495. doi:10.1021/es101397e PMID:20666461

VoPham, T., Brooks, M. M., Yuan, J.-M., Talbott, E. O., Ruddell, D., Hart, J. E., ... Weissfeld, J. L. (2015). Pesticide exposure and hepatocellular carcinoma risk: A case-control study using a geographic information system (GIS) to link SEER-Medicare and California pesticide data. *Environmental Research*, *143*, 68–82. doi:10.1016/j.envres.2015.09.027 PMID:26451881

Vose, S. C., Holland, N. T., Eskenazi, B., & Casida, J. (2007). Lysophosphatidylcholine hydrolases of human erythrocytes, lymphocytes, and brain: Sensitive targets of conserved specificity for organophosphorus delayed neurotoxicants. *Toxicology and Applied Pharmacology*, *224*(1), 98–104. doi:10.1016/j.taap.2007.06.008 PMID:17663017

Vos, J. G., Dybing, E., Greim, H. A., Ladefoged, O., Lambre, C., Tarazona, J. V., & Vethaak, A. D. (2000). Health effects of endocrine-disrupting chemicals on wildlife, with special reference to the European situation. *Critical Reviews in Toxicology*, *30*(1), 71–133. doi:10.1080/10408440091159176 PMID:10680769

Vryzas, Z., Vassiliou, G., Alexoudis, C., & Papadopoulou-Mourkidou, E. (2009). Spatial and temporal distribution of pesticide residues in surface waters in northeastern Greece. *Water Research, 43*(1), 1–10. doi:10.1016/j.watres.2008.09.021 PMID:18947852

Vukčević, M. M., Kalijadis, A. M., Vasiljević, T. M., Babić, B. M., Laušević, Z. V., & Laušević, M. D. (2015). Production of activated carbon derived from waste hemp (Cannabis sativa) fibers and its performance in pesticide adsorption. *Microporous and Mesoporous Materials, 214*, 156–165. doi:10.1016/j.micromeso.2015.05.012

Wachinski, J. E., & Etzel, A. M. (1997). Environmental Ion Exchange: Principles and Design. CRC Press/Lewis Publishers.

Wafford, K. A., Sattelle, D. B., Gant, D. B., Eldefrawi, A. T., & Eldefrawi, M. E. (1989). Noncompetitive inhibition of GABA receptors in insect and vertebrate CNS by endrin and lindane. *Pesticide Biochemistry and Physiology, 33*(3), 213–219. doi:10.1016/0048-3575(89)90119-3

Waggoner, J. K., Kullman, G. J., Henneberger, P. K., Umbach, D. M., Blair, A., Alavanja, M. C., ... Hoppin, J. A. (2011). Mortality in the Agricultural Health Study. 1993-2007. *American Journal of Epidemiology, 173*(1), 71–83. doi:10.1093/aje/kwq323 PMID:21084556

Walburg, C. H. (1971). Loss of young fish in reservoir discharge and year class survival Lewis and Clark Lake, Missouri River. American Fisheries Society.

Walker, M. K., & Peterson, R. E. (1994). Aquatic toxicity of dioxins and related chemicals. In Dioxins and Health (pp. 347-387). Springer US. doi:10.1007/978-1-4899-1462-0_11

Walker, A. (1978). Simulation of the persistence of eight soil-applied herbicides. *Weed Research, 18*(5), 305–313. doi:10.1111/j.1365-3180.1978.tb01165.x

Wall, G. R., Riva-Murray, K., & Phillips, P. J. (1998). *Water Quality in the Hudson River Basin, New York and adjacent states, 1992–95.* Reston, VA: USGS. *Geological Survey Circular*, 1165.

Wang, X., Yu, S., & Liu, W. (2018). *Molecular Imprinting Based Hybrid Ratiometric Fluorescence Sensor for the Visual Determination of Bovine Hemoglobin.* ACS Sensors acssensors.7b00804.

Wang, A., Padula, A., Sirota, M., & Woodruff, T. J. (2016). Environmental influences on reproductive health: The importance of chemical exposures. *Fertility and Sterility, 106*(4), 905–929. doi:10.1016/j.fertnstert.2016.07.1076 PMID:27513554

Wang, A., Qu, J., Ru, J., Liu, H., & Ge, J. (2005). Mineralization of anazodye Acid Red by electro-Fenton's reagent using an activated carbon fiber cathode. *Dyes and Pigments, 65*(3), 227–233. doi:10.1016/j.dyepig.2004.07.019

Wang, P., Tian, Y., Wang, X. J., Gao, Y., Shi, R., Wang, G. Q., ... Shen, X. M. (2012). Organophosphate pesticide exposure and perinatal outcomes in Shanghai. China. *Environment International, 42*, 100–104. doi:10.1016/j.envint.2011.04.015 PMID:21601922

Wang, Q., Li, C., Zheng, R., & Que, X. (2016). Phytoremediation of chlorpyrifos in aqueous system by riverine macrophyte, Acorus calamus: Toxicity and removal rate. *Environmental Science and Pollution Research International, 23*(16), 16241–16248. doi:10.100711356-016-6673-6 PMID:27154841

Wang, X., Wang, H., Lu, M., Teng, R., & Du, X. (2017). Facile synthesis of phenyl-modified magnetic graphene/mesoporous silica with hierarchical bridge-pore structure for efficient adsorption of pesticides. *Materials Chemistry and Physics, 198*, 393–400. doi:10.1016/j.matchemphys.2016.12.017

Wang, X., Yu, J., Kang, Q., Shen, D., Li, J., & Chen, L. (2016a). Molecular imprinting ratiometric fluorescence sensor for highly selective and sensitive detection of phycocyanin. *Biosensors & Bioelectronics, 77*, 624–630. doi:10.1016/j.bios.2015.10.019 PMID:26485176

Wang, X., Yu, J., Wu, X., Fu, J., Kang, Q., Shen, D., ... Chen, L. (2016b). A molecular imprinting-based turn-on Ratiometric fluorescence sensor for highly selective and sensitive detection of 2,4-dichlorophenoxyacetic acid (2,4-D). *Biosensors & Bioelectronics*, *81*, 438–444. doi:10.1016/j.bios.2016.03.031 PMID:27015146

Wang, Y. J., & Lin, J. K. (1995). Estimation of Selected Phenols in Drinking Water with In Situ Acetylation and Study on the DNA Damaging Properties of Polychlorinated Phenols. *Archives of Environmental Contamination and Toxicology*, *28*(4), 537–542. doi:10.1007/BF00211639 PMID:7755407

Wang, Y., Hu, Q., Tian, T., & Yu, L. (2017a). Simple and sensitive detection of pesticides using the liquid crystal droplet patterns platform. *Sensors and Actuators. B, Chemical*, *238*, 676–682. doi:10.1016/j.snb.2016.07.114

Wang, Y., Wu, S., Chen, J., Zhang, C., Xu, Z., Li, G., & Wang, Q. (2018). Single and joint toxicity assessment of four currently used pesticides to zebrafish (Daniorerio) using traditional and molecular endpoints. *Chemosphere*, *192*, 14–23. doi:10.1016/j.chemosphere.2017.10.129 PMID:29091792

Wang, Y., Zeinhom, M. M. A., Yang, M., Sun, R., Wang, S., Smith, J. N., ... Du, D. (2017b). A 3D-Printed, Portable, Optical-Sensing Platform for Smartphones Capable of Detecting the Herbicide 2,4-Dichlorophenoxyacetic Acid. *Analytical Chemistry*, *89*(17), 9339–9346. doi:10.1021/acs.analchem.7b02139 PMID:28727917

Wang, Y., Zhang, Y., Ji, L., Hu, Y., Zhang, J., Wang, C., ... Tian, Y. (2017). Prenatal and postnatal exposure to organophosphate pesticides and childhood neurodevelopment in Shandong, China. *Environment International*, *108*, 119–126. doi:10.1016/j.envint.2017.08.010 PMID:28843140

Ward, J. (1985). Thermal characteristics of running waters. *Hydrobiol*, *125*(1), 31–46. doi:10.1007/BF00045924

Warngard, L., & Flodstrom, S. (1989). Effects of tetradecanoye phorbol acetate, Pyrethroid and DDT. *Cell Biology and Toxicology*, *5*, 67–75. doi:10.1007/BF00141065 PMID:2920299

Waskom, R. (1994). *Best management practices for private well protection.* Colorado State Univ. Cooperative Extension. Retrieved from http://hermes.ecn.purdue.edu:8001/cgi/convertwq?7488

Waste Stabilization Pond. (n.d.). Retrieved from Akvo.org//www.akvo.org/wiki/index.php/Portal:Sanitation retrieved 6/3-2011

Watanabe, I. (1977). Pentachlorophenol-decomposing and PCP-tolerant bacteria in field soil treated with PCP. *Soil Biology & Biochemistry*, *9*(2), 99–103. doi:10.1016/0038-0717(77)90044-X

Wauchope, R. D. (1978). The Pesticide Content of Surface Water Draining From Agricultural Fields-A Review. *Journal of Environmental Quality*, *7*(4), 459. doi:10.2134/jeq1978.00472425000700040001x

Wauchope, R. D., Johnson, W. C. III, & Sumner, H. R. (2004). Foliar and soil deposition of pesticide sprays in peanuts and their washoff and runoff under simulated worstcase rainfall conditions. *Journal of Agricultural and Food Chemistry*, *52*(23), 7056–7063. doi:10.1021/jf049283v PMID:15537318

Weare, B. C. (2009). How will changes in global climate influence California? *California Agriculture*, *63*(2), 59–66. doi:10.3733/ca.v063n02p59

Weber, J., Halsall, C. J., Muir, D., Teixeira, C., Small, J., Solomon, K., & Bidleman, T. (2010). Endosulfan, a global pesticide: A review of its fate in the environment and occurrence in the Arctic. *The Science of the Total Environment*, *408*(15), 2966–2984. doi:10.1016/j.scitotenv.2009.10.077 PMID:19939436

Weeks, E. N., Schmehl, D. R., Baniszewski, J., Tomé, H. V., Cuda, J. P., Ellis, J. D., & Stevens, B. R. (2018). Safety of methionine, a novel biopesticide, to adult and larval honey bees (Apis mellifera L.). *Ecotoxicology and Environmental Safety*, *149*, 211–216. doi:10.1016/j.ecoenv.2017.11.026 PMID:29175348

Weerathunge, P., Ramanathan, R., Shukla, R., Sharma, T. K., & Bansal, V. (2014). Aptamer-controlled reversible inhibition of gold nanozyme activity for pesticide sensing. *Analytical Chemistry*, *86*(24), 11937–11941. doi:10.1021/ac5028726 PMID:25340286

Weichenthal, S., Moase, C., & Chan, P. (2012). A review of pesticide exposure and cancer incidence in the agricultural health study cohort. *Environmental Health Perspectives*, *118*(8), 1117–1125. doi:10.1289/ehp.0901731 PMID:20444670

Werner, I., and Moran, K. (2008). *Effects of pyrethroid insecticides on aquatic organisms*. Academic Press.

Werner, S. L., Burkhardt, M. R., & DeRusseau, S. N. (1996). Methods of Analysis by the U.S. Geological Survey National Water Quality Laboratory: Determination of Pesticides in a Water by CarboPak-B Solid-Phase Extraction and High-Performance Liquid Chromatography. U.S. Geological Survey.

Werner, A. D., Vincent, M. B., Post, E. A., Vandenbohede, A., Lu, C., Ataie-Ashtiani, B., ... Barry, D. A. (2013). Seawater intrusion processes, investigation and management: Recent advances and future challenges. *Advances in Water Resources*, *51*, 3–26. doi:10.1016/j.advwatres.2012.03.004

Werner, K. J., & Zedler, J. B. (2002). How sedge meadow soils, microtopography, and vegetation respond to sedimentation. *Wetlands*, *22*(3), 451–466. doi:10.1672/0277-5212(2002)022[0451:HSMSMA]2.0.CO;2

Werner, W. (2002). *Fertilizers, Environmental Aspects. In Ullmann's Encyclopedia of Industrial Chemistry*. Weinheim: Wiley-VCH.

Weselak, M., Arbuckle, T. E., Wigle, D. T., & Krewski, D. (2007). In utero pesticide exposure and childhood morbidity. *Environmental Research*, *103*(1), 79–86. doi:10.1016/j.envres.2006.09.001 PMID:17084836

Wesseling, C., Keifer, M., Ahlbom, A., McConnell, R., Moon, J. D., Rosenstock, L., & Hogstedt, C. (2002). Long-term neurobehavioral effects of mild poisonings with organophos-phate and n-methyl carbamate pesticides among banana workers. *International Journal of Occupational and Environmental Health*, *8*(1), 27–34. doi:10.1179/oeh.2002.8.1.27 PMID:11843437

West, I. (1964). Occupational hazards to pest control operators with special reference to pesticides. *Proceedings of the 2nd Vertebrate Pest Control Conference*, 3-6.

WHO. (1967). *Specifications for Pesticides Used in Public Health* (3rd ed.). Geneva: WHO.

WHO. (1979). *Environmental Health Criteria: DDT and its derivatives*. WHO.

WHO. (1990). *Public Health Impact of Pesticides Used in Agriculture. (1990)*. Geneva: World Health Organization.

WHO. (2009). *Health implications from monocrotophos use: A review of the evidence in India*. WHO.

WHO. (2010). *Water for health. WHO guidelines for drinking water quality*. Retrieved from http://www.who.int/water_sanitation_health/WHS_WWD2010_guidelines_2010_6_en.pdf? ua=1

WHO/UNEP. (1990). *Public Health impact of pesticides used in agriculture*. Geneva: WHO.

Widenfalk, A., Svensson, J. M., & Goedkoop, W. (2004). Effects of the pesticides captan, deltamethrin, isoproturon and pirimicarb on the microbial community of a freshwater sediment. *Environmental Toxicology and Chemistry*, *23*(8), 1920–1927. doi:10.1897/03-345 PMID:15352481

Wieczorek, M. V., Bakanov, N., Bilancia, D., Szöcs, E., Stehle, S., Bundschuh, M., & Schulz, R. (2018). Structural and functional effects of a short-term pyrethroid pulse exposure on invertebrates in outdoor stream mesocosms. *The Science of the Total Environment*, *610–611*, 810–819. doi:10.1016/j.scitotenv.2017.08.048 PMID:28826119

Wijngaarden, R. P. A. V., Brock, T. C. M., & Van Den Brink, P. J. (2005). Threshold Levels for Effects of Insecticides in Freshwater Ecosystems: A Review. *Ecotoxicology (London, England), 14*(3), 355–380. doi:10.100710646-004-6371-x PMID:15943110

Willer, H., & Yussefi, M. (Eds.). (2005). *The World of Organic Agriculture 2005 - Statistics and Emerging Trends*. Bonn: International Federation of Organic Agriculture Movements.

Williams, M. K., Barr, D. B., & Camann, D. E. (2006). An intervention to reduce residential insecticide exposure during pregnancy among an inner-city cohort. *Environmental Health Perspectives, 114*, 1684–1689. PMID:17107853

Willson, H. R. (1996). *Pesticides Regulations*. University of Minnesota. Retrieved from https://en.wikipedia.org/wiki/Pesticide

Wilson, C., & Tisdell, C. (2001). Why farmers continue to use pesticides despite environmental, health and sustainability costs. *Ecological Economics, 39*(3), 449–462. doi:10.1016/S0921-8009(01)00238-5

Wilson, J. T., Greene, S., & Alexander, M. (1980). Effects of microcrustaceans on bluegreen algae in flooded soil. *Soil Biology & Biochemistry, 12*(3), 237–240. doi:10.1016/0038-0717(80)90067-X

Wilson, N. K., Chuang, J. C., Lyu, C., Menton, R., & Morgan, M. K. (2003). Aggregate exposures of nine preschool children to persistent organic pollutants at day care and at home. *Journal of Exposure Analysis and Environmental Epidemiology, 13*(3), 187–202. doi:10.1038j.jea.7500270 PMID:12743613

Windham, G., & Fenster, L. (2008). Environmental contaminants and pregnancy outcomes. *Fertility and Sterility, 89*(2), 111–116. doi:10.1016/j.fertnstert.2007.12.041 PMID:18308050

Wohlfahrt-Veje, C., Andersen, H. R., Jensen, T. K., Grandjean, P., Skakkebaek, N. E., & Main, K. M. (2012a). Smaller genitals at school age in boys whose mothers were exposed to non-persistent pesticides in early pregnancy. *International Journal of Andrology, 35*(3), 265–272. doi:10.1111/j.1365-2605.2012.01252.x PMID:22394112

Wohlfahrt-Veje, C., Andersen, H. R., Schmidt, I. M., Aksglaede, L., Sørensen, K., Juul, A., & Main, K. M. (2012b). Early breast development in girls after prenatal exposure to non-persistent pesticides. *International Journal of Andrology, 35*(3), 273–282. doi:10.1111/j.1365-2605.2011.01244.x PMID:22404257

Woin, P. (1998). Short- and long-term effects of the pyrethroid insecticide fenvalerate on an invertebrate pond community. *Ecotoxicology and Environmental Safety, 41*(2), 137–156. doi:10.1006/eesa.1998.1694 PMID:9756702

Wolkers, H., Boily, F., Gremmels, J. F., Bavel, B. V., Hammill, M. O., & Primireerio, R. (2009). Tissue-Specific contaminant accumulation & associated effects on hepatic serum analytes & cytochrome P450 enzyme activities in hooded seals (*Cystophora cristata*) from the gulf of St. Lawrence. *Archives of Environmental Contamination and Toxicology, 56*(2), 360–370. doi:10.100700244-008-9186-6 PMID:18626617

Wong, M. C. (1979). *The ecology of the ostracod, Stenocypris major Baird 1859 with reference to the application of insecticides, carbofuran and endosulfan* (B.S. honour thesis). University of Malaya, Malaysia.

Wong, D. C., Dorn, P. B., & Chai, E. Y. (1997). Acute toxicity and structure-activity relationships of nine alcohol ethoxylate surfactants to fathead minnow and Daphnia magna. *Environmental Toxicology and Chemistry, 16*(9), 1970–1976.

Wood, L. W., Jones, P. A., & Richards, A. (1986). Possible sediment scavenging of chlordane and contamination of the aquatic biota in Belmont Lake, New York. *Bulletin of Environmental Contamination and Toxicology, 36*(1), 159–167. doi:10.1007/BF01623489 PMID:3947750

Woodruff, T., Carlson, A., Schwartz, J., & Giudice, L. (2008). Proceedings of the summit on environmental challenges to reproductive health and fertility: Executive summary. *Fertility and Sterility*, *89*(2), 1–20. doi:10.1016/j.fertnstert.2008.01.065 PMID:18155200

Wood, S., Sebastian, K., & Scherr, S. J. (2001). *Pilot Analysis of Global Ecosystems: Agroecosystems*. Washington, DC: World Resources Institute.

World Bank. (2012). *Turn Down the Heat: Why a 4oC Warmer World Must Be Avoided*. Washington, DC: The World Bank.

World Health Organization. (1986). *The WHO recommended classification of pesticide biohazard. Guideline to classification 1986-1987*. Geneva: World Health Organization.

World Health Organization. (1999). *Food safety issues associated with products from aquaculture: report of a joint FAO/NACA/WHO study group* (Vol. 883). World Health Organization.

World Health Organization. (2010). *The WHO recommended classification of pesticides by hazard and guidelines to classification 2009*. WHO.

World Resources Institute and the International Institute for Environment and Development. (1986). *World Resources*. New York, NY: Basic Books.

Worthing, C. R., & Walker, S. B. (Eds.). (1983). The pesticide manual, a world compendium (7th ed.). The British Crop Protection Council.

Wu, X. F., & Kosaric, N. (1991). Removal of organochlorine compounds in an upflow flocculated algae photobioreactor. *Water Science and Technology*, *24*(5), 221–232. doi:10.2166/wst.1991.0128

Wu, X., Song, Y., Yan, X., Zhu, C., Ma, Y., Du, D., & Lin, Y. (2017). Carbon quantum dots as fluorescence resonance energy transfer sensors for organophosphate pesticides determination. *Biosensors & Bioelectronics*, *94*, 292–297. doi:10.1016/j.bios.2017.03.010 PMID:28315592

Wyman, J. A., Jensen, J. O., Curwen, D., Jones, R. L., & Marquardt, T. E. (1985). Effects of application procedures and irrigation on degradation and movement of aldicarb residues in soil. *Environmental Toxicology and Chemistry*, *4*(5), 641–651. doi:10.1002/etc.5620040508

Xia, H., & Ma, X. (2006). Phytoremediation of ethion by water hyacinth (Eichhornia crassipes) from water. *Bioresource Technology*, *97*(8), 1050–1054. doi:10.1016/j.biortech.2005.04.039 PMID:15982870

Xia, N., Wang, Q., & Liu, L. (2015). Nanomaterials-based optical techniques for the detection of acetylcholinesterase and pesticides. *Sensors (Switzerland)*, *15*(1), 499–514. doi:10.3390150100499 PMID:25558991

Xintaras, C., Burg, J. R., Tanaka, S., Lee, S. T., Johnson, B. L., Cottrill, C. A., & Bender, J. (1978). Occupational exposure to leptophos and other chemicals. Cincinnati, OH: DHEW (NIOSH).

Xu, M. L., Gao, Y., Han, X. X., & Zhao, B. (2017). Detection of Pesticide Residues in Food Using Surface-Enhanced Raman Spectroscopy: A Review. *Journal of Agricultural and Food Chemistry*, *65*(32), 6719–6726. doi:10.1021/acs.jafc.7b02504 PMID:28726388

Xu, W., Wang, X., & Cai, Z. (2013). Analytical chemistry of the persistent organic pollutants identified in the Stockholm Convention: A review. *Analytica Chimica Acta*, *790*, 1–13. doi:10.1016/j.aca.2013.04.026 PMID:23870403

Yadav, I. S., & Devi, N. L. (2017). Pesticides Classification and Its Impact on Human and Environment. *Environment Science and Engineering*, 6.

Yadav, I. C., Devi, N. L., Syed, J. H., Cheng, Z., Li, J., Zhang, G., & Jones, K. C. (2015). Current status of persistent organic pesticides residues in air, water, and soil, and their possible effect on neighboring countries: A comprehensive review of India. *The Science of the Total Environment*, *511*, 123–137. doi:10.1016/j.scitotenv.2014.12.041 PMID:25540847

Yamagishi, A., & Hashizume, A. (1974). Ecology of green algae in paddy fields and their control with chemicals. *Weed Research*.

Yang, D., Howard, A., Bruun, D., Ajua-Alemanj, M., Pickart, C., & Lein, P. J. (2008). Chlorpyrifos and chlorpyrifos-oxon inhibit axonal growth by interfering with the morphogenic activity of acetylcholinesterase. *Toxicology and Applied Pharmacology*, *228*(1), 32–41. doi:10.1016/j.taap.2007.11.005 PMID:18076960

Yang, J. K., Kang, H., Lee, H., Jo, A., Jeong, S., Jeon, S.-J., ... Lee, Y.-S. (2014). Single-step and rapid growth of silver nanoshells as SERS-active nanostructures for label-free detection of pesticides. *ACS Applied Materials & Interfaces*, *6*(15), 12541–12549. doi:10.1021/am502435x PMID:24988366

Yang, T., Zhang, Z., Zhao, B., Hou, R., Kinchla, A., Clark, J. M., & He, L. (2016). Real-time and in situ monitoring of pesticide penetration in edible leaves by surface-enhanced Raman scattering mapping. *Analytical Chemistry*, *88*(10), 5243–5250. doi:10.1021/acs.analchem.6b00320 PMID:27099952

Yang, W., Carmichael, S. L., Roberts, E. M., Kegley, S. E., Padula, A. M., English, P. B., & Shaw, G. M. (2014). Residential agricultural pesticide exposures and risk of neural tube defects and orofacial clefts among offspring in the San Joaquin Valley of California. *American Journal of Epidemiology*, *179*(6), 740–748. doi:10.1093/aje/kwt324 PMID:24553680

Yarpuz-Bozdogan, N., & Bozdogan, A. M. (2016). Pesticide exposure risk on occupational health in herbicide application. *Fresenius Environmental Bulletin, 25*, 3720-3727.

Yarpuz-Bozdogan, N., & Bozdogan, A. M. (2009). Comparison of Field and Model Percentage Drift Using Different Types of Hydraulic Nozzles in Pesticide Applications. *International Journal of Environmental Science and Technology*, *6*, 191–196.

Yarpuz-Bozdogan, N., Bozdogan, A. M., Daglioglu, N., & Erdem, T. (2017). Determination of dermal exposure of operator in greenhouse spraying. *AMA. Agricultural Mechanization in Asia, Africa and Latin America*, *48*, 33–38.

Yi, A. X., Leung, K. M., Lam, M. H., Lee, J. S., & Giesy, J. P. (2012). Review of measured concentrations of triphenyltin compounds in marine ecosystems and meta-analysis of their risks to humans and the environment. *Chemosphere*, *89*(9), 1015–1025. doi:10.1016/j.chemosphere.2012.05.080 PMID:22704212

Yin, S. S., Tang, M. L., Chen, F. F., Li, T. L., & Liu, W. P. (2016). Environmental exposure to polycyclic aromatic hydrocarbons (PAHs): The correlation with and impact on reproductive hormones in umbilical cord serum. *Environment and Pollution*. PMID:27838061

Younis, H. M., Abo-El-Saad, M. M., Abdel-Razik, R. K., & Abo-Seda, S. A. (2002). Resolving the DDT target protein in insects as a subunit of the ATP synthase. *Biotechnology and Applied Biochemistry*, *35*, 9–17. PMID:11834125

Yuan, Y., Chen, C., Zheng, C., Wang, X., Yang, G., Wang, Q., & Zhang, Z. (2014). Residue of chlorpyrifos and cypermethrin in vegetables and probabilistic exposure assessment for consumers in Zhejiang Province, China. *Food Control*, *36*(1), 63–68. doi:10.1016/j.foodcont.2013.08.008

Yu, J., Wang, X., Kang, Q., Li, J., Shen, D., & Chen, L. (2017). One-pot synthesis of a quantum dot-based molecular imprinting nanosensor for highly selective and sensitive fluorescence detection of 4-nitrophenol in environmental waters. *Environmental Science. Nano*, *4*(2), 493–502. doi:10.1039/C6EN00395H

Zacharia, J. T. (2011). Ecological Effects of Pesticides. In Pesticides in the Modern World-Risks and Benefits. InTech.

Zachery, R. S., Valerie, J. H., & Jason, R. R. (2015). A Synthesis of the Effects of Pesticides on Microbial Persistence in Aquatic Ecosystems. *Critical Reviews in Toxicology*, *45*(10), 813–836. doi:10.3109/10408444.2015.1065471 PMID:26565685

Zambonin, C. G., & Palmisano, F. (2000). Determination of triazines in soil leachates by solid-phase microextraction coupled to gas chromatography-mass spectrometry. *Journal of Chromatography A., 874*(2,7), 247-255.

Zare-Shahabadi, V., Abbasitabar, F., Akhond, M., & Shamsipour, M. (2013). Simultaneous determination of chlorpyrifos and carbaryl by spectrophotometry and boosting partial least squares. *Journal of the Brazilian Chemical Society*, *24*(10), 1561–1569.

Zaugg, S. D., Sandstrom, M. W., Smith, S. G., & Fehlberg, K. M. (1995). *Methods of Analysis by the U.S. Geological Survey National Water Quality Laboratory Determination of Pesticides in Water by C-18 Solid-Phase Extraction and Capillary-Column Gas Chromatography/Mass Spectrometry With Selected-Ion Monitoring*. Open-File Rep. U.S. Geol. Surv. 95-181.

Zedler, J. B., & Kercher, S. (2005). Wetland resources: Status, trends, ecosystem services, and restorability. *Annual Review of Environment and Resources*, *30*(1), 39–74. doi:10.1146/annurev.energy.30.050504.144248

Zeng, T., & Arnold, W. A. (2012). Pesticide photolysis in prairie potholes: Probing photosensitized processes. *Environmental Science & Technology*, *47*(13), 6735–6745. doi:10.1021/es3030808 PMID:23116462

Zhang, K., Mei, Q., Guan, G., Liu, B., Wang, S., & Zhang, Z. (2010). Ligand replacement-induced fluorescence switch of quantum dots for ultrasensitive detection of organophosphorothioate pesticides. *Analytical Chemistry*, *82*(22), 9579–9586. doi:10.1021/ac102531z PMID:20973515

Zhang, W., Jiang, F., & Ou, J. (2011). Global pesticide consumption and pollution: With China as a focus. *Proceedings of the International Academy of Ecology and Environmental Sciences*, 1.

Zhang, Z., Zeng, K., & Liu, J. (2017). Immunochemical detection of emerging organic contaminants in environmental waters. TrAC -. *Trends in Analytical Chemistry*, *87*, 49–57. doi:10.1016/j.trac.2016.12.002

Zhao, S., Peng, C., Jiang, H., Tian, D., Lei, X., & Zhou, X. (2006). Land use change in Asia and the ecological consequences. *Ecological Research*, *21*(6), 890–896. doi:10.100711284-006-0048-2

Zheng, S., Chen, B., Qiu, X., Chen, M., Ma, Z., & Yu, X. (2016). Distribution and risk assessment of 82 pesticides in Jiulong River and estuary in South China. *Chemosphere*, *144*, 1177–1192. doi:10.1016/j.chemosphere.2015.09.050 PMID:26461443

Zheng, T., Zahm, S. H., Cantor, K. P., Weisenburger, D. D., Zhang, Y., & Blair, A. (2001). Agricultural exposure to carbamate pesticides and risk of non-Hodgkin lymphoma. *Journal of Occupational and Environmental Medicine*, *43*(7), 641–649. doi:10.1097/00043764-200107000-00012 PMID:11464396

Zheng, W., Guo, M., Chow, T., Bennett, D. N., & Rajagopalan, N. (2010). Sorption properties of greenwastebiochar for two triazine pesticides. *Journal of Hazardous Materials*, *181*(1), 121–126. doi:10.1016/j.jhazmat.2010.04.103 PMID:20510513

Zheng, Z., Li, X., Dai, Z., Liu, S., & Tang, Z. (2011). Detection of mixed organophosphorus pesticides in real samples using quantum dots/bi-enzyme assembly multilayers. *Journal of Materials Chemistry*, *21*(42), 16955. doi:10.1039/c1jm11631b

Zhou, D. M., Wang, Y. J., Cang, L., Hao, X. Z., & Luo, X. S. (2004). Adsorption and cosorption of cadmium and glyphosate on two soils with different characteristics. *Chemosphere*, *57*(10), 1237–1244. doi:10.1016/j.chemosphere.2004.08.043 PMID:15519368

Zhu, H. (2013). From QSAR to QSIIR: searching for enhanced computational toxicology models In Computational Toxicology (vol. 930, pp. 53-65). New York: Humana Press. doi:10.1007/978-1-62703-059-5_3

Zhu, J., Chen, Q., Kutsanedzie, F. Y., Yang, M., Ouyang, Q., & Jiang, H. (2017). Highly sensitive and label-free determination of thiram residue using surface-enhanced Raman spectroscopy (SERS) coupled with paper-based microfluidics. *Analytical Methods*, *9*(43), 6186–6193. doi:10.1039/C7AY01637A

Zhu, Y. C., Kramer, K. J., Oppert, B., & Dowdy, A. K. (2000). cDNAs of aminopeptidase-like protein genes from Plodiainter punctella strains with different susceptibilities to Bacillus thuringiensis toxins. *Insect Biochemistry and Molecular Biology*, *30*(3), 215–2. doi:10.1016/S0965-1748(99)00118-6 PMID:10732989

Zou, Z., Du, D., Wang, J., Smith, J. N., Timchalk, C., Li, Y., & Lin, Y. (2010). Quantum dot-based immunochromatographic fluorescent biosensor for biomonitoring trichloropyridinol, a biomarker of exposure to chlorpyrifos. *Analytical Chemistry*, *82*(12), 5125–5133. doi:10.1021/ac100260m PMID:20507134

Zvinavashe, E., van den Berg, H., Soffers, A. E., Vervoort, J., Freidig, A., Murk, A. J., & Rietjens, I. M. (2008). QSAR models for predicting in vivo aquatic toxicity of chlorinated alkanes to fish. *Chemical Research in Toxicology*, *21*(3), 739–745. doi:10.1021/tx700367c PMID:18254607

About the Contributors

J. Anuradha received her doctoral degree in Pollution control and Environmental Technology from Pondicherry (Central) University. She is now affiliated at Department of Advanced Science and Technology, Nims University Rajasthan. Her technical know-how in the field Environmental management and advanced nanotechnology provides the cutting-edge R&D that strives to carve fresh niches and open new vistas. Her proficiency in field of Environmental Technology has been peer-recognized in the form of several international and national publications. She has thus far published three patents, several papers in peer recognized international and national journals. Twenty-eight of her papers have been taken part in international and nation scientific meet discussions. To add as an exemplary researcher, she has been awarded for eight best paper's in scientific discussions, and does presided as a 'Resource person and Subject specialist' in (ANVESION - West Zone) annual students science meet (2017) in India. She has been 'nominated' for the 'Dr APJ Abdul Kalam Innovation award' in Indian Academy of Biomedical Sciences (2017) and had delivered 'Resource person' lecture on 'Facile synthesis of gold nanoparticles and its application on cancer diagnosis and treatment'.

Bilal A. Beigh is currently pursuing Ph.D., in the department of Environmental Science and Limnology, Barkatullah University, Bhopal. He has vast experience in the field aquatic pollution and has worked on algal bloom diversity index as an indicator of freshwater ecosystem pollution.

Musadiq Hussain Bhat is a Ph.D. student in School of Studies in Botany, Jiwaji University Gwalior, Madhya Pradesh, India. He has obtained his Masters degree from the same University. His areas of interest include plant conservation, plant tissue culture, phytochemistry, phytodiversity, environmental ecology. He has published research papers in various national and international journals.

Rayees Ahmad Bhat did his M.Sc. in Chemistry from Barkatullah University Bhopal India, and pursuing Ph.D. Chemistry Jiwaji University Govt. Science College, Jiwaji University Gwalior India. Gwalior. He has published various research papers in reputed international and national journals.

Rouf Ahmad Bhat has pursued Ph.D. in Environmental Science at Sher-e-Kashmir University of Agricultural Sciences and Technology, Shalimar, Kashmir. Currently working as Lecturer in environmental sciences at Sri Pratap College, Srinagar.

Jyoti Bisht obtained her Ph.D. (Biotechnology) from Kumaun University, Nainital, Uttarakhand, India on bioremediation of pesticide residues in soil using fungi. She is presently working as Assistant

Professor at Department of Biotechnology, Shri Guru Ram Rai (PG) College Dehradun, Uttarakhand. She has 8 years of research experience and published 12 research papers in national and international journals, authored 2 book chapters and presented 7 papers at Seminars and Conferences. Her major fields of research are biodegradation of xenobiotics and to identify the biological pathways', toxicological effect of pesticides and their degradation products, utilization of microbes for bioethanol production, industrial effluents treatment, white biotechnology with the exposure in bio-bleaching of wood pulp and biodeinking by using ligninocellulosic microbes.

Jaya Chaturvedi is Guest Faculty in the Department of School of Studies in Zoology, Jiwaji University, Gwalior, Madhya Pradesh, India. She has obtained her Master's and Ph.D. degree from the same university. Her areas of interest include fish reproduction, Aquatic fauna and effect of various toxic substances on Aquatic life. She has published research papers in various journals with impact factor. She has also presented his work in Various International and National conferences.

Nebile Daglioglu is Assistant Professor in the Department of Forensic Medicine, Faculty of Medicine, Cukurova University, Adana, Turkey. She has completed her M.Sc. and Ph.D. in Forensic Sciences in Cukurova University. She has more than 18 years of teaching and research experience in Department of Forensic Medicine. Her areas of interest include pesticide residue analysis in different biological samples with Gas Chromatography Mass Spectrometry and Liquid Chromatography Tandem Mass Spectrometry devices. Her laboratory placed in The University hospital, she performs forensic and clinical toxicological analysis. She is member of several national and international societies.

Ashaq Ahmad Dar is M.Sc. student in the Department of Environmental Science (IGAEERE), Jiwaji University, Gwalior, Madhya Pradesh, India. He has published research papers in various national journals. He has also presented his work in Various International and National conferences. He has qualified ASRB NET in 2018.

Shakeel Ahmad Dar is currently pursuing Ph.D program in the Division of Environmental Sciences at Sher-e-Kashmir University of Agricultural Sciences and Technology, Shalimar, Kashmir.

Moonisa Aslam Dervash is currently pursuing Ph.D program in the Division of Environmental Sciences at Sher-e-Kashmir University of Agricultural Sciences and Technology, Shalimar, Kashmir.

Sonal Dixit is a Dr. DS Kothari Post Doctoral fellow in Department of Botany, Lucknow University, Lucknow, Uttar Pradesh, India. She has completed her M.Sc. in Environmental Science from CSJM University, Kanpur, India and Ph.D. in Environmental Science from BBA (Central) University, Lucknow, India. Her area of interest include Bioremediation, Phycology, Nanotechnology and waste water management. She has published several research papers in various reputed national and international journals and also an active member of several national and international societies.

Mufida Fayaz is a Ph.D. student in School of Studies in Botany, Jiwaji University Gwalior, Madhya Pradesh, India. She has obtained her M.Sc. and M. Phil. in Botany from the same University. Her areas of interest include Ethnobotany, Nutraceuticals, plant conservation, phytochemistry, phytodiversity, environmental ecology. She has published research papers in various national and international journals.

Azad Gull is M.Sc. Student in the School of Studies in Zoology, Jiwaji University, Gwalior, Madhya Pradesh, India. He is also working in Central Silk Board, Ministry of Textiles – Govt. of India. He has published research papers in various national journals. He has also presented his work in Various International and National conferences.

Garima Harit has completed her research work from Department of Zoology, University of Rajasthan, Jaipur. She has an experience of more than a decade in the field of teaching and research. Her areas of specialization are fish biology, ecotoxicology & fish behavior. She has 10 papers published in renowned journals with impact factor. She has given several oral and poster presentations in various national and international conferences and has also received Best Paper Award in 2016 for her research work from University of Rajasthan. She was also nominated as a Jury Member of Eco film festival in an International Conference. She is an Associate Editor of international journal of fisheries and aquatic studies. Dr. GarimaHarit is presently working as an Assistant Professor in Department of Zoology, The IIS University, Jaipur.

N. S. K. Harsh, a fellow of Indian Phytopathological Society, worked as Scientist-G at Forest Pathology Division and Group coordinator (Research), Forest Research Institute, Dehradun. He has more than 36 years of research experience on fungi with 167 research papers published in national and international journals, authored 6 books and presented 98 papers at seminar and conferences. His major fields of research are Forest Pathology-diseases of nurseries, plantations and forest; wood-decay; control of forestry disease including their biological control; Bioremediation of pollutants; Mushroom cultivation; Taxonomy of higher fungi-wood-decaying fungi, Ascomycetes, Gasteromycetes, Deutromycetes and mushrooms.

Ranjit Hawaldar is currently scientist at centre for materials for electronics technology, Pune, He did his Ph.D. in Physics at department of physics, university of Pune in 2010 and holds masters in physical chemistry He was a Erasmus Mundus Postdoctoral fellow at University of Marie and Perrie Curie Paris He was a FCT postdoctoral fellow at the University of Averio He was also selected for D. S Kothari Postdoctoral fellowship He has 40 publications to his credit and a book chapter He has 3 publications in Prestigious Nature family journal.

Nirmala Kumari Jangid is working as Assistant Professor in the Department of Chemistry, Banasthali Vidyapith, Rajasthan. She has completed her graduation from University of Rajasthan, Jaipur, post graduation and Ph.D. degree from Mohanlal Sukhadia University, Udaipur (Rajasthan). She has published 14 chapters in international books, Encyclopedia of Biomedical Polymers and Polymeric Biomaterials and Encyclopedia of Polymeric applications (Taylor & Francis) and 9 research paper and 2 review article in various national and international journals. Dr. Jangid also presented various research papers in national and international conferences.

K. Kadirvelu has received doctoral degree from Bharathiar University in 1998 and he has completed Post doctoral research in 2000 at France. Currently he is working in DRDO-BU Center for life sciences as Scientist 'F'. He has more than 80 papers which have 6700 citation with 33 h –index for his credits. His current area of research interests includes Heavy metal removal, Environmental remediation, Nanomaterials, Activated carbon, Activated carbon nanofiber, Polmer based nanofibers, Quantum dots,

Bacterail cellulose, Chemical sensing and Biosensing. Currently he is guiding 10 students for Doctoral and Post doctoral degree.

Nesibe Ebru Yaşa Kafkas is a Professor in the Department of Horticulture, Faculty of Agriculture, Cukurova University, Adana, Turkey. She has completed her M.Sc. and Ph.D. at Cukurova University. She has more than 22 years of teaching and research experience in Department of Horticulture and Biotechnology. Her areas of interest fruit quality including flavor, secondary metabolites, sugars, pesticide residue analysis in different biological samples with Gas Chromatography Mass Spectrometry and Liquid Chromatography techniques and manager of the many national and international scientific project in her lab. She is member of several national and international societies such as ISHS etc.

Arun Kalia, Central University of Punjab, Bathinda, is a Ph.D. Research Scholar (UGC-JRF) in the Centre for Environmental Science and Technology Central University of Punjab, Bathinda, presently working on microbial biodegradation and biopolymer technology.

S. Kanagasubbulakshmi has obtained her M.Sc degree in Environmental Science from PSG CAS, Bharathiar University, Coimbatore in 2014. In the same year, she was awarded DRDO-JRF Fellowship, Joined there to pursue her Ph.D. in Environmental Science. Presently she is pursuing her research in semiconductor quantum dots for both chemical and biosensing applications. Her area of research interests are in Quantum dots, Nanoparticles, Green chemistry and Biosensing platforms through Nanotechnology.

Bhupendra Kumar is a food technologist with diversified knowledge and expertise in food R&D, Quality and manufacturing of different food and Nutraceutical products. He worked at India, China, Nigeria and Ghana and has the ability to manufacture different food products from lab to commercial levels of different taste buds. He has developed antimicrobial herbs extracts and commercialized and enriched nutraceutical drink at industrial level. He is known with food business focus, entrepreneur, confident personality, creative thinker, determined, independent, knowledge-seeker, promoter and relationship-builder.

Krishnasamy Lakshmi is one of the author K.Lakshmi, has completed her M.Sc (Organic Chemistry) at Bharathiar University, Coimbatore in 2009. Currently she is a JRF in DRDO BU CLS and pursuing Ph.D.in Bharathiar University. Her area of research interests are Nanofibers, Metal oxide nanoparticles, Environmental remediation, Photocatalysis.

Rafiq Lone is currently working Assistant Professor in the Department of Botany, SBBS University, Khiala, Jalandhar, Punjab, India.

Zubair A. Malik did his Ph. D in Ecology from HNB Garhwal University, Uttarakhand. After Ph. D he worked as an SRF (Senior Research Fellow) in a research project sponsored by Space Application Centre (ISRO) Ahmedabad. He has also worked as a Lecturer (Botany) in different Govt. Degree Colleges of Kashmir. He has published 15 research papers in National and International Journals. He is presently working as a Lecturer in the Department of Botany of Govt. HSS Hardutoroo Anantnag.

Javid Manzoor is a Ph.D. student in the Department of Environmental Science (IGAEERE), Jiwaji University, Gwalior, Madhya Pradesh, India. He has obtained his Master's degree from the same university. He has two years of teaching experience. He is an awardee of University Research Scholarship. He is a permanent member of people for animal. His areas of interest include environmental pollution, occupational health and biomedical waste management. He has published research papers and book chapters in various national and international journals with impact factor.

Shafat A. Mir is a Ph.D Scholar Division of Environmental Sciences at Sher-e-Kashmir University of Agricultural Sciences and Technology, Shalimar, Kashmir.

Zahid Nabi is pursuing his Ph.D. in Environmental Sciences from Central University of Jammu. The author has completed M.Phil. from central University of Punjab and M.Sc. from Jiwaji University Gwalior. The author besides being an environmentalist, contributes write-ups on various issues to multiple newspapers.

Lok Man S. Palni presently working as Vice Chancellor at Graphic Era University. Professor has done his Ph.D. from the University of Wales, Britain and has over 44 years' of distinguished career in research and academics in the field of Biotechnology, in India and abroad. He has been awarded post-doctoral Ph.D. from Australian National University. Prof. Palni is also the Elected Fellow of INSA, NAAS, NASI & International Society of Plant Morphologists, and the member of New York Academy of Sciences.

Vinayak Vandan Pathak is an Assistant Professor in the Department of Chemistry, Faculty of Applied Sciences, Manav Rachna University, Faridabad, India. He has completed in M.Sc. and Ph.D. in Environmental Chemistry from Gurukul Kangri University, Haridwar, India. He has two years of teaching and research experience. His areas of interest include production of biodiesel from waste, Detoxification of heavy metals using natural adsorbents. He has published research papers in various international journals also worked as coauthors in various book chapters.

P. K. Patnala is Assistant Professor in the Department of Chemistry, Faculty of Applied Sciences, Manav Rachna University, Faridabad, India. He has completed in M.Sc. and Ph.D. in Chemistry from Dayalbagh Educational Institute, Agra, India. He has two years of teaching and research experience. His areas of interest include Remediation of industrial effluents, wastewater using ultrasound, Detoxification of heavy metals using natural adsorbents, production of Bio-Hydrogen using natural kitchen waste. He has published research papers in various international journals. He is a member of society of materials chemistry and Indian council of chemists.

Rabia Quadir did her M.Sc. in Environmental Science from HNB Gharwal University Dehradun, Uttarakhand and M. Phil from Jiwaji University Gwalior India. She has published few research articles in various Journals and presented her work in different conferences.

Asmat Rashid has pursued Ph.D program in Environmental Science at Sher-e-Kashmir University of Agricultural Sciences and Technology, Shalimar, Kashmir. Currently she is working as lecturer in Environmental Sciences at Moulana Azad University deputed at Srinagar Region.

Rupali Rastogi is Assistant Professor in Department of Chemistry, ITM University, Gwalior. She has teaching and research experience of 10 years. She has published 7 research papers and participated in about 15 conferences and workshops. She has also worked as Principal Investigator under MPCST funded project. Her specialization is in synthesis, characterization of chalcogen bearing ligands and dithiophosphates and dithiocarbamates. She has also studied the applications of newly synthesized compounds as fertilizers, anticancer, antibacterial and antifungal activity. She is also expertise in various analytical techniques like UV, IR, ESI mass, multinuclear NMR spectroscopy, DTA-TGA, DSC, SEM, TEM and X-ray analysis. At present she is also working on checking the antiwearing properties and purification of water using dithiophosphates and dithiocarbamates.

Varun Rawat was born in Delhi, India. He received both his B.Sc (H) and M.Sc in Chemistry from University of Delhi. He completed his Ph.D. from CSIR-National Chemical Laboratory, Pune. After securing a PBC postdoctoral fellowship from Government of Israel he worked in Professor Arkadi Vigalok's group in Tel Aviv University. He is currently Assistant Professor in School of Chemistry, Amity Institute of Applied Science, Amity University.

Roopa Rani has completed his Masters in Chemistry, M.Phil in Chemistry and Ph.D. in Chemistry from Dayalbagh Educational Institution, Agra. She has numerous publications in national and international journal of repute. She has obtained various awards at national level for her research like Thermax Assett Award through Bhabha Atomic Research Centre, Mumbai. She is now actively working in the field of food authenticity in Delhi-NCR region and bioremediation of pesticides from water systems.

R. Sanjeevi received his doctoral degree in Environmental Science & Engineering from Pondicherry (Central) University and was awarded with Post Doctoral Fellowship form CSIR, New Delhi at selfsame institute. At present he is affiliated at Institute of Advance Science & Technology, Nims University Rajasthan. His technical know-how in the field Water and wastewater quality analysis and treatment; maintenance and operation of anaerobic bioreactor, Biogas collection and analysis using GC; Solid Waste Management, and air quality monitoring provides the momentum for cutting edge R&D that strives to carve fresh niches and open new vistas. The effort has been peer-recognized with Senior Research Fellowship, awarded form CSIR, New Delhi during his doctoral research. His proficiency in field of Bioprocess technology has been peer-recognized in the form of several international and national publications. To add as an exemplary researcher, he has been awarded for three best paper's in scientific discussions, and does presided as a 'Resource person and Subject specialist' in (ANVESION - West Zone) annual students science meet (2017) in India. Further, he is now involved in exploring new avenue of research in Environmental Management for Sustainability.

Raghunath Satpathy received his M.Sc in Botany (with specialization in Biotechnology) from Berhampur University science, India; (Post M.Sc) Advanced P.G Diploma in Bioinformatics from the University of Hyderabad and M. Tech. degree in Biotechnology from VIT University, Vellore India. He has submitted the thesis for the degree of the doctor of philosophy (Ph.D.) in Biotechnology of Sambalpur University, Odisha India. Currently, he is working as an Assistant professor in M.I.T.S Engineering College Odisha, India. He is having more than 9 years of experiences in both teaching and research. He has authored many national and international journal papers and seven book chapters to his credit. In addition to this Mr. Satpathy is associated with many professional bodies like ISRD, APCBEES, SCIEI, SABT,

OBA. His current research interests include computational analysis of protein structure and environmental applications, data mining, biological database and analytical tool development. He is the recipient of many academic and research award such as: prestigious Adarsh Vidya Saraswati Rashtriya Puraskar (A national award of excellence from Global Management council, Ahmedabad, India), Senior Educator and Researcher award (A national award from National foundation for entrepreneurship development (NFED), Coimbatore, India), Best Teacher Award (IRDP group of Journals, Chennai, India), Best Young Teacher Award (Grace India Educational Charitable Trust, New Delhi, India) and recently IRSD- Bright Researcher Award (Biotechnology) (Green ThinkerZ Society, Mohali, India).

Manoj Sharma is an associate professor in pharmacology and toxicology department, school of studies in pharmaceutical science, Jiwaji University Gwalior, India. He has completed his B. Pharma from Barkatulllah University Bhopal, M. Pharma from Rajiv Gandhi University of health sciences, Bangalore, Karnataka and P.hD from Jiwaji University Gwalior. His area of specialization is development of Cytoprotective agent against alkylating anticancer agents and drug discovery and development and nerve agent toxicity. He has published several research papers in national and international journals. He received jewel of India award in 2014. He has authored/ co authored of four books and also applied for one patient.

Neha Sharma got Ph.D. from Jiwaji University, Gwalior (M. P.) in 2009. Insightful academic and research experience of over +10 years. She received "Young Scientist Award in 2018 and junior Scientist of the year 2015. Dr. Sharma published 18 research articles in reputed journals. Her research specifically focuses on Biotechnology and Biochemistry.

Yogesh Kumar Sharma is ex-head and Professor in the Department of Botany, Lucknow University, Lucknow, Uttar Pradesh, India. His areas of interest include Plant Physiology, Plant Nutrition and Molecular Biology. He has completed his M.Sc. and Ph.D.in Botany from *Chaudhary Charan Singh University*, Meerut, Uttar Pradesh, India and Post-Doctoral Research from U.S.A. He has 39 years of Teaching and Research Experience. He has published more than 70 research papers in reputed national and international journals, authored two books and supervised more the 20 Ph.D. students. He is a member of several national and international organizations and also in the editorial board of various journals. He has completed three major research projects between 2003 to 2016 and organized different programmes on Environmental, social, cultural Awareness.

Samreen Siddiqui is a Ph.D. candidate and Graduate Research Assistant at Texas A & M University, Corpus Christi working on a USDA-NIFA funded project towards her dissertation. She has completed her Master's degree in Biology with a thesis focused on ecotoxicology from VSU, GA, USA. Her earlier Master's degree was in Aquatic Ecosystem Management from Edinburgh Napier University, UK. Samreen Siddiqui served as an author as well as reviewer in various peer-reviewed publications and write blogs for American Fisheries Society. She is a Co-Chair in outreach committee within SETAC- NASAC (North America Student Advisory Council). Her research focuses on various environmental related issues.

Meenakshi Singh received her Ph.D. from Banaras Hindu University, Varanasi, India in 1995. She worked as a visiting researcher in the Department of Chemistry, University of Umea (2002–2003). She was appointed as Assistant Professor in 2000 in Department of Chemistry, U.P. College, Varanasi, India. From 2005, she joined Department of Chemistry, MMV, Banaras Hindu University as an Assistant Professor and later on as Professor in 2015. Her research interests include polyelectrolytes, molecular imprinting (MIP) and MIP sensors. She is experienced in designing the novel polymeric formats for crafting molecular recognition elements to device sensors for biomolecules/drugs using molecular imprinting technology. She has developed novel functionalized polymeric frameworks to be incorporated in Analytical technology based on molecularly imprinted polymers for detection of target molecules, peptide and proteins. She has published many papers in peer-reviewed reputed international journals and supervised many Ph.D. and master students in the field of polyelectrolyte chemistry, electrochemical sesnors, piezoelectric sensors and MIP sensors.

Shivom Singh joined Department of Environmental Science, ITM University, Gwalior, India in 2012, where he currently is the Assistant Professor of Environmental Sciences. He also worked as Guest Faculty in Department of Environmental Sciences, G. B. Pant University of Agriculture and Technology, Pantnagar. Following a Postdoctoral Fellowship at the Department of Biological Science, G. B. Pant University of Agriculture and Technology, Pantnagar, where, with Dr. Anil Kumar Sharma he filed Indian patent, Govt. of India. Dr. Shivom Singh received two Ph.D. degrees first in Botany from MJP Rohilkhand University, Bareilly and second in Environmental Sciences from Gurukul Kangri University, Haridwar. His postgraduate degree is from the Gurukul Kangri Univerisity, Haridwar, India in 2003. He is a Fellow of Action for Sustainable, Efficacious Development and Awareness (FASEA) and holds a membership of many national and international professional bodies. Dr. Singh research on Environmental monitoring of different aspects of pollutions and their biological managements and rhizosphere biology. His current work is aimed on biofertilizer and biopesticides. He has written more than thirty five research and review articles in reputed national and international journals.

Madhu Prakash Sivastava is a Research Fellow under the Center of Excellence scheme. He has been assisting B.Sc. and M.Sc. in practical as well as theory class in Department of Botany, Lucknow University, Lucknow, and has shown excellent performance during research work and teaching. He has completed his Ph.D. in Botany at Lucknow University, Lucknow. His area of specialization is Mycology, Plant Pathology and Biological Formulation development. He has 25 research papers published in National /International journals with impact factor and also co-authored several book chapters to his credit. He has been awarded Ruchi Ram Sahni Award (2012) for the best piece of research work in Botany. He is a member of various societies, several national and international organizations and is in the editorial board of various journals.

Irfan Rashid Sofi has done his M.Sc. in Environmental Science from HNB Gharwal University Dehradun, Uttarakhand and M.Phil. from jiwaji University Gwalior and is pursuing Ph.D. from School of Studies in Environmental Science jiwaji university Gwalior M.P. India. He has published research papers and various book chapters in national and international journals.

Anamika Srivastava is Assistant Professor (Chemistry), Department of Chemistry Banasthali Vidyapith, Rajasthan. She obtained her graduation, post-graduation Purvanchal University Jaunpur and Ph D degree from Banasthali Vidyapith, Banasthali. She has published over ten research paper in International Journals, five proceedings in national conference. At present time it is work in field of nanocatalyst, catalyzed organic synthesis. Dr. Srivastava presented research paper in numerous International and national conferences.

Kajal Srivastava, Assistant Professor at KRG PG College, Gwalior. Dr. Srivastava is a Post Doctoral Fellow from G. B. Pant University of Agriculture and Technology, Pantnagar, India. She received her doctoral degree in Botany, from MJP Rohilkhand University, Bareilly, India and had studied post graduation in Industrial Biotechnology from Ch. Charan Singh University, Meerut, India. Dr. Srivastava is the author of over 42 peer-reviewed publications in national and international journals and also holds membership of many professional bodies. She has authored two books as well. Her research specifically focuses on bryomonitoring, plant biochemistry and environmental biotechnology. She has worked on many bilateral research projects viz., indo- German and Indo Swiss Projects sponsored by BDT and DST, Government of India.

Manish Srivastava is Assistant Professor (Chemistry), Department of Chemistry Banasthali Vidyapith, Rajasthan. He obtained his graduation, post graduation and D. Phill (PhD) degree from University of Allahabad, Allahabad. He has published over twenty five research paper in International Journals, ten proceedings in national conference. At present time it is work in field of nanocatalyst, catalyzed organic synthesis and water treatment. At present time six students going to research work under the supervision and two students awarded the PhD degree under the supervision. Dr. Srivastava presented research paper in numerous International and national conferences.

Nazia Tarannum received her B.Sc. and M.Sc. degree from Lucknow University, Lucknow in 2005 and 2007 respectively. She received her Ph.D. from Banaras Hindu University, Varanasi, India in 2012 on the topic 'Synthesis, characterization and analytical applications of certain polysulfobetaines and polycarbobetaines'. She worked as a postdoctoral fellow in Banaras Hindu University in 2012 and as a Research Associate in 2013-2014 in IIT Bombay. She is currently working as Assistant Professor, Department of Chemistry, Chaudhary Charan Singh University, Meerut, India. Her research interests include polyelectrolytes, hydrogels and organogels, molecular imprinting (MIP), chemiresistors and MIP sensors. She is experienced in designing the novel polymeric formats for crafting molecular recognition elements to device sensors for biomolecules/drugs using molecular imprinting technology. She has developed novel functionalized polymeric frameworks to be incorporated in Analytical technology as sensors for detection of small target molecules. She has published many papers and books in peer-reviewed reputed international journals.

Neeta Thacker is a retired Scientist from National Environmental Engineering Research Institute (NEERI). She did her Ph D from Jiwaji University in Synthetic Organic Chemistry. Her research specialization includes Environmental monitoring, Environmental quality evaluation, treatment and management, Analytical techniques development and instrumentation, Quality assurance and quality control with special reference to priority pollutants. Apart from various publications made, she has contributed as national coordinator and team member in the development of National Implementation Plan (NIP) in India as a first step to implement the Stockholm Convention on Persistent Organic Pollutants.

Binota Thokchom is an Institute Post-Doctoral Fellow at Department of Chemical Engineering, Indian Institute of Technology, Guwahati, India. She had graduated her Ph.D. in Environmental Engineering from Korea University, South Korea. She worked at Analytical Instrumentation Division, National Environmental Engineering Research Institute. Her research interests include water treatment using AOPs, nanomaterial synthesis, electrode synthesis, environmental monitoring and assessment etc. Apart from publications, she has achieved few prestigious awards and recognition.

Venkatramanan Varadharajan is currently a graduate student at National Institute of Technology Karnataka. His area of interests includes NanoBiotechnology, Nano materials for environmental and biological applications and Bioengineering. He has published 5 research papers and 2 research communications in peer-reviewed journals.

Nigar Yarpuz-Bozdogan is Professor in Adana Organized Industrial Zone Vocational School of Technical Sciences, Cukurova University, Adana, Turkey. She has 23 years of research experience. She completed her M.Sc and Ph.D.in Agricultural Machinery from Cukurova University, Turkey. She has completed 19 research projects. She has published research papers about pesticide application, plant protecting machinery, residues of pesticides, pesticide exposure risk on occupational health, pesticide drift, environment and human health risk of pesticide applications, assessment of buffer zone in pesticide applications, pesticide applications in greenhouse and biobeds in various national and international journals with impact factor and one chapter of encyclopedia on drift of pesticide.

Mudasir Youssouf is a research scholar at the Central University of Punjab, Centre for Environmental Science and Technology. He got his M.Phil. degree in waste management from the same university and has Masters degree in Environmental Science from Uttrakhand Technical University, Dehradun. He is interested in agricultural waste and food waste management, anaerobic digestion, composting and wastewater treatment.

Mozhgan Zarifikhosroshahi is a Ph.D. student in Department of Biotechnology, Institute of Natural and Applied Science, University of Çukurova, Turkey. She also completed M.Sc. in the same university. She graduated from Plant Pathology, Agriculture Engineering, Urmia University, Iran. Her research focuses on Plant Sciences, Plant Protecting, Horticulture, Fruits Quality including flavor, secondary metabolites, sugars, pesticide residue analysis, and Agricultural Biotechnology. She has published many papers in different national and international journals under supervision of Prof. Dr. Ebru Kafkas.

Index

A

B

C

D

E

F

G

H

I

K

L

M

N

O

P

Q

R

S

T

V

W

X

Z

www.ingramcontent.com/pod-product-compliance
Lightning Source LLC
LaVergne TN
LVHW080619300926
849162LV00005B/1211